A CATALOGUE OF SOUTHERN PECULIAR GALAXIES AND ASSOCIATIONS

VOLUME I

POSITIONS AND DESCRIPTIONS

"The Universe is very strange," said Jeremy.

"Compared to what?" replied the spider.

— with apologies to N. Moss.
Men Who Play God
(London: Golancz)

A CATALOGUE
OF
SOUTHERN PECULIAR GALAXIES
AND
ASSOCIATIONS

VOLUME I
POSITIONS AND DESCRIPTIONS

HALTON C. ARP
Mount Wilson and Las Campanas Observatories
Carnegie Institution of Washington
Pasadena, California

AND

BARRY F. MADORE
David Dunlap Observatory
University of Toronto
Toronto, Canada

The right of the University of Cambridge to print and sell all manner of books was granted by Henry VIII in 1534. The University has printed and published continuously since 1584.

CAMBRIDGE UNIVERSITY PRESS

Cambridge
London New York New Rochelle
Melbourne Sydney

Published by the Press Syndicate of the University of Cambridge
The Pitt Building, Trumpington Street, Cambridge CB2 1RP
32 East 57th Street, New York, NY 10022, USA
10 Stamford Road, Oakleigh, Melbourne 3166, Australia

First published 1987

Printed in the United States of America

Library of Congress Cataloging-in-Publication Data

Arp, Halton C.

A catalogue of southern peculiar galaxies and associations.

Vol. 2 includes photographic reproductions by William E. Roberton.
Contents: v. 1. Positions and descriptions—v. 2. Selected photographs.
1. Galaxies—Catalogs. 2. Stellar associations—Catalogs. I. Madore, B. II. Title.
QB857.A76 1986 523.1′12 86-11700

British Library Cataloging in Publication Data

Arp, Halton C.

A catalogue of southern peculiar galaxies and associations.
1. Galaxies 2. Southern sky (Astronomy)
I. Title II. Madore, Barry F.
523.1′12 QB857

ISBN 0-521-33086-6 volume 1
ISBN 0-521-33087-4 volume 2
ISBN 0-521-12579-X set

PREFACE

When Vincent Reddish was appointed by the United Kingdom Science Research Council (SRC) to plan and build the 48-inch Schmidt telescope in Australia, it was clear that a major opportunity would be created to discover new astronomical objects and phenomena in the relatively little-explored southern hemisphere. One of the first programmes to be undertaken on the telescope would be a systematic survey of the whole southern sky. This project was part of the *ESO/SRC Southern Sky Survey* and was one of the responsibilities of the United Kingdom Schmidt Telescope Unit. The prototype telescope for the UK Schmidt was the 48-inch Schmidt on Palomar Mountain which, in the 1950's, had produced the famous and widely used sky survey of the northern hemisphere. The Schmidt in Australia was to utilize the new Eastman Kodak IIIa-J emulsion, which, together with the fast (f/2.5) optics and the dark skies at Siding Spring, would penetrate to fainter magnitudes than had ever before been systematically observed in either hemisphere.

Early in 1973 Reddish decided that the many new and interesting objects in the southern sky survey should be systematically recorded and published. To that end he invited Arp to Edinburgh to consider the production of a catalogue of the objects then appearing on the first of these survey plates. Arp (1966) had previously published the *Atlas of Peculiar Galaxies*: 333 unusual objects in the northern hemisphere discovered on the *Palomar Sky Survey* by astronomers such as himself, Vorontsov-Velyaminov, A. Wilson, F. Zwicky and others. Arp realized that the deep *U.K. Schmidt Survey* would detect low-surface-brightness features better than any previous survey and would therefore register much larger extensions around the outer edges of galaxies. He felt that these faint outer regions would reveal pecularities of a kind not ordinarily visible and demonstrate associations between objects not normally suspected of being associated. Therefore he enthusiastically began the project on his own in 1973.

After the first session of examining plates it became clear that thoroughly searching all the plates in the southern hemisphere survey was a formidable task for one person. In the fall of 1974, to alleviate the situation Reddish sought participation from some of the other astronomy centres in Britain. As a result of this invitation Donald Lynden-Bell proposed to Barry Madore, then in Cambridge, that he join the project. Madore accepted, and in November 1974 Arp and Madore searched the first set of plates together in Edinburgh. This collaboration was to last more than ten years.

During the first seven years Arp and Madore made visits about once a year to Edinburgh to search each batch of new plates which had been obtained in Australia and shipped to Scotland. At first, each plate took a considerable time to scan; inspecting every galaxy down to a size limit of about 0.2 millimeter (= 10 arcsec) is by no means a trivial task. But gradually, as facility was developed, the plate-scanning speed rose from an average of two per day to about six. Still, even with this improvement the decision was made to limit the Catalogue to zones only as far north as a declination centre of -25 degrees. Therefore the Catalogue is a complete listing of peculiar objects from the south celestial pole to a declination of -22 degrees. There are 537 plates, nominally 6.5 degrees by 6.5 degrees square, comprising the *SRC Sky Survey* coverage of this area in the southern sky.

During all these years the whole staff of the U.K. Schmidt Telescope Unit were enormously helpful. Russell Cannon and Tim Hawarden took the responsibility of setting up plate-scanning and measuring facilities. Russell Eberst and Elizabeth Sim helped to locate the best available plates on which to carry out the survey. When the last plate was scanned in the fall of 1980 the work of collating the tabular descriptions and positions of some seven thousand objects, and producing the photographs of about a sixth of these, was started. It was decided to produce two volumes. The first, containing positions and descriptions for all peculiar objects, would be the quantitative basis for future studies. The second, containing selected photographs, would illustrate the various categories of peculiarity and bring attention to the outstanding members of each category.

Because some of the survey plate centres did not coincide exactly with the nominal centres, some of the early positions of particular objects were inaccurate. It was therefore decided that the positions of all objects should be checked by overlays. In 1980 Petrusia Bojetchko and Ed Anderson spent part of a summer in Toronto correcting positions; and in 1981 Matthew Bates extended the work in a one-month stay at Edinburgh. It later became clear, however, that some of the simple procedures adopted for correcting the co-ordinates were inadequate and so during 1982/83 Madore and Robert Freedman repeated the overlays and off-sets for the entire Catalogue.

This re-analysis was almost as time consuming as

the original search and delayed the publication considerably. But it was essential. The pointing accuracy of most large optical telescopes is now better than a few tenths of an arcmin, and accordingly the TV acquisition finder fields are usually only a minute or two in diameter. If the Catalogue was to serve the practical needs of optical astronomers, the co-ordinates had to be determined to better than the one arc minute accuracy obtained for the first rough reductions. Other catalogues were to be cross-referenced with our Catalogue, thus good positions were essential in eliminating confusion at a later date. Finally, since the names of the objects were to consist of the truncated co-ordinates (the so-called Parkes nomenclature) the derived co-ordinates had to be correct to the limits of the truncation.

The co-ordinate revision also provided unexpected benefits. Because the off-sets were generally made using copies of plates that were not those originally surveyed, ambiguities caused by plate defects were resolved (for instance, "dwarf/defect?" classifications were usually decided by the second plate). Every classification was in this way verified for at least a third time after a considerable period. Very occasionally, a new object was included in the sample if it were of sufficient interest. Although this type of inclusion was not to be desired for reasons of uniformity, it was usually only made for dwarf and compact galaxies which have an intrinsically uncertain discovery rate.

It was decided that in the case of well-separated associations of objects, every effort would be made to assign individual names and positions to the prominent members, while preserving the nature of the association in the description and code. Often it is by no means clear what single co-ordinate should apply to groupings: frequently no single member is dominant, and in the case of pairs, for instance, the midpoint contains neither member. This latter situation was seen as potentially leading to problems in TV acquisition and cross-referencing with other catalogues. Still, because of the numbers involved in such groupings, some practical limit to this process had to be imposed. Thus, for systems less than a few minutes in total extent, or for larger systems in which there is an obvious and dominant centre of symmetry, only one co-ordinate is listed. For larger and for more irregular systems with less than half a dozen dominant members, individual co-ordinates were obtained. We trust that the effort required in making these modifications and extensions will make the Catalogue both a practical and long-lived tool despite the resulting delays.

By the end of 1980, the volume of photographic reproductions was also ready to be prepared. The thousand photographs which were scheduled for publication represented a major undertaking for the ROE Photographic Laboratory. Careful consultation with the head of that laboratory, Brian Hadley, produced a plan for the production of standard sizes and formats for the photographs. Two test runs were made in order to set the optimal contrast and density for the prints. For the final production, Russell Eberst of the Schmidt Unit identified each object on the best available plate or film and marked the area to be photographed. The successful completion of this phase owes much to the good judgement in identification and framing which Eberst made on this long list of objects. Lastly, Bill Roberton of the ROE Photographic Laboratory produced all the final prints. The superb quality of the final prints is a tribute to Roberton's skill and diligence.

During this period, Malcolm Longair assumed the Directorship of ROE. In spite of his many new duties in the vigorous research activities at ROE he found time to see that the grant given by the SRC for the completion of the Catalogue was maintained. He consulted and advised in the publication of the Catalogue, and the whole of the Observatory staff helped with suggestions and criticism of the final edition.

INTRODUCTION AND OVERVIEW

Galaxies do not exist in total isolation. They interact with their environment and they modify their surroundings through nuclear activity, galactic winds, tidal encounters and collisions. While this activity is not necessarily a continuous phenomenon, its effect on the dynamics and long-term evolution of galaxies and galactic systems is undoubtedly significant. Galaxies are the cumulative result of both quiescent and catastrophic evolution. It is the purpose of this Catalogue to highlight the most spectacular phases of galactic evolution, and perhaps formation, by drawing attention to the basic forms in which this activity is manifest.

In addition to cataloguing individual galaxies, the identification and study of associations is important. The relatively recent advent of wide-field, fast Schmidt telescopes has enabled the earlier emphasis on individual galaxies to be shifted to such associations. For example, at the turn of the century Keeler (1900, 1908), Curtis (1918), Pease (1917, 1920) and others published many remarkable large-scale prints of individual galaxies, as photographed by the small-field reflectors of the time. This *individual* approach to galaxies culminated in the production of atlases such as Evans' (1957) *Cape Photographic Atlas of Southern Galaxies*, Sersic's (1968) *Atlas de Galaxies Australes*, Sandage's (1961) *Hubble Atlas of Galaxies* and Sandage and Tammann's (1981) *Revised Shapley-Ames Catalogue* which were basically illustrative of classification systems based on symmetry, and focussed on single galaxies in isolation.

The Schmidt telescopes have therefore provided a new perspective, unavailable to earlier generations of extragalactic astronomers. Pairs, triples and clusters of galaxies became immediately apparent to the casual observer, whereas previously any associations had to be hunted laboriously out of catalogues and listings of uncertain completeness and/or reliability. The wide-field Schmidt also made it possible to survey the whole sky. A complete census of any given kind of object could be made and a statistical analysis could then be performed.

The *Palomar Observatory Sky Survey*, made with the 48-inch Schmidt, resulted in new galaxy and cluster searches being undertaken. For instance Abell (1958) catalogued and classified the relatively conspicuous clusters of galaxies, while Zwicky and his collaborators (Zwicky, Herzog and Wild 1960; Zwicky and Herzog 1963, 1966, 1968; Zwicky, Karpowicz and Kowal 1965; Zwicky and Kowal 1968) identified all galaxies down to $m_{pg} \approx 15.5$ mag, while also noting pairs, triples, etc. and outlining clusters.

Another of the outstanding contributions of the Schmidt telescope has been to locate among myriad galaxies just those that are spectacularly peculiar. The study of these objects has helped us understand some physical processes and at the same time confronts us with many more that we do not understand. Working from *Sky Survey* paper prints Vorontsov-Velyaminov (1959) produced an *Atlas and Catalogue of Interacting Galaxies* which was soon followed by Arp's (1966) *Atlas of Peculiar Galaxies.* While many of these objects were followed up and studied with large-aperture reflectors their relationship to neighbouring galaxies required the wide-field prespective afforded by the Schmidt telescopes.

Photography to slightly fainter surface-brightness levels than usual can often reveal the much larger outer areas of many galaxies. In turn these outer regions, being less dense, can reveal perturbations and distortions more easily. These perturbations can come from internal activity such as explosions or ejections, or from interaction with neighbouring objects. Naturally this means that the regions surrounding galaxies must be carefully studied. It is the combination of fainter-surface-brightness detection and wide-field registration that enabled this new area of astronomical research to be opened up.

The introduction by Kodak of their fine-grained IIIa-J emulsion has made it practical to discover and study these faint features in a systematic way. So in combining the wide-field surveying capabilities of the Schmidt telescope and the deeply penetrating exposures of the IIIa-J emulsion, the *SRC Southern Sky Survey* has provided a unique source of data on peculiar galaxies and associations. We have used this source material to produce what we hope is a complete and detailed catalogue of all the most interesting galaxies and most striking apparent associations of galaxies in the southern sky.

POSITIONS

For each peculiar galaxy an x-y position in millimeters was recorded, as measured from the lower-left (south-east) corner of each plate. These cartesian co-ordinates have been preserved, and are given with the description of each object in Volume I, along with the plate and field to which they pertain. These co-ordinates should be sufficient to locate the object to within a few millimeters. To locate the same objects on the *ESO Quick Blue Survey* films, it is necessary to subtract approximately thirty-five millimeters from both x and y.

Because of uncertainties in the exact centering of the plate at the telescope and in our measuring devices, it was not possible to produce precise right ascension and declination co-ordinates from the x-y co-ordinates directly. Because rotations as well as translations of the plates with respect to the sky were obviously present in some of the earliest plates in the survey, (but not in later plates), a single off-set could not be determined for all of the objects on a given plate. Instead, individual scaled overlays were prepared for every object in the Catalogue after rough positions had been derived from the millimeter positions on the plate. Each object was then re-examined and corrections to the nominal positions were measured to an accuracy of fifteen arc seconds. The final positions are given to the nearest second of time and tenth of a minute of arc. Although these positions still carry a somewhat optimistic appraisal of the errors, they are certainly an order of magnitude better than any derived from the original x-y positions alone.

In some cases the co-ordinates are over-specified. That is, because of the lack of symmetry of some objects, the "centre" is rather subjective. Additionally, because of the extent of some systems or because more than one major object is involved in a system, some visually averaged centre (such as the mid-point between an interacting double) was chosen which does not necessarily contain any member of the system. This type of "centre" was restricted to systems less than a few arc minutes in extent. For larger systems, the co-ordinates for individual members are usually presented. Time constraints precluded this in some instances, and so readers are strongly advised to interpret the listed co-ordinates with care, especially when dealing with multiple systems.

DESCRIPTIONS

A short description accompanies each object listed in the Catalogue to emphasize those aspects that are of most interest and to quantify certain properties, such as the direction of a jet or the number of apparent companions, etc.

Abbreviations Used in the Description

ABS'N.	absorption
APP.	apparent
ASSOC.	association
ASYMM.	asymmetric
BR.	bright
CIRC.	circular
COMP'N.	companion
COMP'NS.	companions
CONDENS.	condensation
CONN.	connected/ing/ion
DBL.	double
DIAM.	diametric
E	elliptical
ELONG.	elongated
ENVEL.	envelope
FILAM.	filament
FNT.	faint
G.D.	grand design
GAL./GALS.	galaxy/galaxies
GRP.	group
HI.S.BR.	high surface brightness
incl.	including
I/A	interacting
IRR.	irregular
LO. S. BR.	low surface brightness
MA. AX.	major axis
MATR'L	material
MDM. S. BR.	medium surface brightness
MI. AX.	minor axis
MULTI.	multiple
PEC.	peculiar
S. BR.	surface brightness
SP.	spiral
V.	very
w.	with
+	and or plus

CODES

A description of the numerical codes shown in the seventh column of the Catalogue is given below. The codes are substantially the same as those listed in Arp and Madore (1977). They provide a fast means of locating objects in a given Category without searching out key words in the description. In many cases the object belongs simultaneously to several Categories. For instance, a galaxy with a jet (Code 7) may have apparent companions (Code 8), one of which may be compact (Code 13) while the dominant galaxy may be part of a wide group (Code 18). In this respect, the numerical code can be supplementary to the description, giving a concise means of noting additional peculiarities.

Table I: The Categories

Numerical Code	Brief General Description	Relative Percentage
1	Galaxies with interacting companion(s)	5.5
2	Interacting doubles (galaxies of comparable size)	12.6
3	Interacting triples	2.0
4	Interacting quartets	0.5
5	Interacting quintets	0.1
6	Ring galaxies (or morphologically similar objects)	3.1
7	Galaxies with (linear) jets	2.4
8	Galaxies with apparent companion(s)	11.5
9	M51-types (companion at end of spiral arm)	2.0
10	Galaxies with peculiar spiral arms	4.1
11	Three-armed spirals and multiple-armed spirals	0.5
12	Peculiar disks (major asymmetry or deformation)	2.8
13	Compact (very high-surface-brightness) galaxies	6.4
14	Galaxies with prominent or unusual dust absorption	1.6
15	Galaxies with tails, loops of material or debris	3.5
16	Irregular or disturbed, (apparantly isolated) galaxies	4.2
17	Chains (four or more galaxies aligned)	4.0
18	Groups (four or more galaxies not aligned)	4.9
19	Clusters (only very conspicuous, rich systems)	1.6
20	Dwarf galaxies (low surface brightness)	6.8
21	Stellar objects with associated nebulosity	0.7
22	Miscellaneous (rare or distinctive objects)	1.4
23	Close pairs (not visibly interacting)	11.4
24	Close triples (not visibly interacting)	5.6
25	Planetary Nebulae	0.9

PREVIOUS IDENTIFICATIONS

A search of previously published catalogues yielded a number of prior identifications of some of our objects. The sources of these alternative designations are listed below. A tremendous amount of time was required to make these inter-comparisons of catalogues, but rather than have every user of the Catalogue perform the same task we have made the effort to be as complete as possible. That is not to say that our identifications are totally complete or absolutely correct in all cases. We tended to err on the generous side when establishing correspondence between two catalogues. This is because in some cases the co-ordinates are not well determined in the original listing or the object may be a specific member of a wider group whose position is different by a few minutes of arc. Readers are therefore cautioned to use the alternative designations with just care. If an identification is crucial to some study, then the original reference and the survey copies themselves should be consulted directly; what we are offering is the first step in that procedure.

A: Objects listed in the *Second Reference Catalogue of Bright Galaxies* [RC2] by de Vaucouleurs, de Vaucouleurs and Corwin (1964), having no previous identification, were listed by those authors as anonymous, but with a designation following the Parkes scheme as adopted in the present Catalogue (e.g., AM 1113-334 = A 1113-33).

Ag: Aguero (1971) searched for peculiar galaxies using print copies of the *Palomar Sky Survey* fields centered on declination -42 degrees. Fifty percent of his interesting galaxies are also contained and cross-referenced in the present Catalogue (e.g., AM 2240-400 = Ag-75).

Be: In his catalogue, Bergvall (1981) lists some four hundred interacting and distorted galaxies found on glass copies of the *ESO Blue Survey.* Approximately sixty percent of the objects thought by Bergvall to be of interest are also included in the present Catalogue (e.g., AM 2227-613 = B 2227-612). When there was insufficient room to put the entire co-ordinate name given by Bergvall we simply placed [Be] in the space available.

DDO: The dwarf galaxies listed by van den Bergh (1959) overlap to some small degree with those discovered in the course of this survey (e.g., AM 0907-224 = DDO 56) but in general the two studies are complementary, covering the northern and southern hemispheres respectively.

DN: Duus and Newall (1977) compiled and classified nearly 900 groups and clusters of galaxies south of declination -27 degrees; 710 of the clusters were newly discovered (e.g., AM 0001-531 = DN 0001-533 = R 3).

ESO: Identifications from Holmberg, Lauberts, Schuster and West (1974a, 1974b, 1975, 1977, 1978a, 1978b, 1980) are designated by the field number, followed by G or IG (ie, Galaxy or Interacting Galaxy, respectively), and their running number (e.g., AM 0240-614 = 115-IG-26).

Fa: The first five lists of galaxies with compact and bright nuclei found by Fairall (1977, 1979, 1980, 1981, 1983, 1984) are also included in the cross-referencing, although the co-ordinates quoted by Fairall, especially in his early lists, are apparently quite rough approximations (e.g., AM 1004-424 = Fa 426).

Ka: A number of dwarf galaxies in the catalogue of Karachentseva (1973) are sufficiently far south so as to have been included in our survey as well (e.g., AM 2318-420 = Ka-57).

Ko: Since the publication of the main reference catalogue on planetary nebulae by Perek and Kohoutek (1967) a number of additional planetary nebulae have been published by Kohoutek (1971, 1977) (e.g., AM 1032-285 = Ko-28).

Lo: After inspecting the UK Schmidt survey plates, Longmore (1977) published a list of new and suspected planetary nebulae which we cross-reference in our Catalogue (e.g., AM 1111-474 = Lo-5).

Lu: In the course of inspecting the Yale-Columbia proper-motion plates for radio source identifications Lu (1971) noted a number of interesting systems in the southern hemisphere. Approximately forty percent of these objects were also classified as peculiar in our survey (e.g., AM 2047-330 = Lu 2047-33).

MC: A search of the IIIa-J plates by Malin and Carter (1983) for elliptical galaxies with shells resulted in the publication of a list of 137 candidates for further study (e.g., AM 0610-623 = MC 0610-625).

MCG: The *Morphological Catalogue of Galaxies* compiled by Vorontsov-Velyaminov and Krasnogorskaya (1962), Vorontsov-Velyaminov and Arkhipova (1963, 1964, 1968, 1974) overlaps in small measure with our survey and cross-references were taken primarily from the work of Lauberts (1982) (e.g., AM 0530-293 = MCG 5-14-2).

NGC/IC: *New General Catalogue* and *Index Catalogue* object positions were taken from Lauberts (1982) (e.g., AM 0241-600 = NGC 1096).

PKS: Possible radio identifications were searched out using the *Parkes Radio Source Catalogue* (Ekers, 1969) and a master-list of the *New Parkes Catalogue* provided by the staff of the ROE; these identifications are the most tentative (e.g., AM 2020-370 = PKS 2020-370).

PK: To guard against the inclusion of planetary nebulae in the compact galaxy class, Perek and Kohoutek (1967) was consulted and coincidences reclassified and extracted from the Catalogue; objects thought to be planetary nebulae, based on their morphology, but without prior identification, have been left in the Catalogue as they may prove in fact to be extragalactic (e.g., AM 1347-505 = PK 312+10-1).

R: Rose (1976) lists 124 probable clusters of galaxies found by him in selected areas around the south galactic pole (e.g., AM 2222-572 = R 75).

Se: Sersic (1974) gives a selected list of peculiar galaxies and groups found on his own survey plates (e.g., AM 2046-474 = Se 143/3).

SS: For a comparison with a complete survey of E and S0 galaxies Sadler and Sharp (1984) compiled a subset of relatively isolated binary pairs consisting of these early type galaxies (e.g., AM 1113-333 = SS 10a).

To: Emission line objects found in the *Tololo/Curtis Schmidt Survey* have been published by Smith, Aquirre and Zemelman (1976). We find that approximately thirty percent of their objects have identifications in our Catalogue (e.g., AM 1348-333 = To 1348-335).

UKS: The published lists of low-suface-brightness galaxies in the general field given by Longmore *et al.* (1978, 1982), and the list of dwarfs in the NGC 5128 Group found by Webster *et al.* (1979), were consulted for comparison with our dwarf galaxies. Both lists are identified by the same prefix (e.g., AM 1346-354 = UKS 1346-358).

vdB: Several planetary nebulae are listed in the paper by van den Bergh *et al.* (1973) and are cross-referenced here for completeness (e.g., AM 1340-603 = vdB-5).

VV: After searching paper copies of the *Palomar Observatory Sky Survey,* Vorontsov-Velyaminov (1959) issued an *Atlas and Catalogue of Interacting Galaxies* some of which are sufficiently far south to overlap with our new survey (e.g., AM 2225-250 = VV 74a/b).

THE CATEGORIES

a. Definition of Categories

The basic approach adopted in the compilation of our Catalogue was that as we scanned survey plates we inspected all galaxies down to slightly less than 0.2 mm (i.e., about ten seconds of arc at the scale of the Schmidt plates). Of course the majority of these galaxies were more or less regular and approximately symmetric. Only when a galaxy departed substantially from the norm was it catalogued as peculiar. In addition, because we initially scanned survey plates at low (5x) magnification we were aware when galaxies occurred in clusters, groups, and loose associations. When we encountered galaxies which were unusually close together compared to the average, we catalogued them as a pair, group, rich cluster, or galaxy with apparent companions. Of course in not every instance will these galaxies be physically associated, but the association categories furnish a homogeneous selection of those most likely to be associated. They can later be tested statistically or with individual observations.

Within the individual categories there is always some indefinite range where the noteworthiness of the peculiarities becomes rather subjective. While objects in this twilight zone of the Catalogue are undoubtedly incomplete, *down to the size limit adopted here we feel that we have produced a catalogue which, by anyone's standards, includes all of the most peculiar galaxies.*

At the same time, distance-dependent biases do exist in our selections and these will vary from Category to Category. For instance, galaxies showing jets and filaments will be lost from the Catalogue at a larger apparent size than the galaxies having interacting companions; galaxies with apparent companions will be selected from a smaller volume of space than apparent doubles, etc. The reader is therefore warned against making sweeping comparisons between relative frequencies of peculiarities unless this very important selection effect is taken into account.

It is important to stress that we did not start this project with a preconceived set of categories. Rather, we developed the categories empirically by simply listing all unusual objects from a preliminary analysis of a set of thirty-six plates. We then noted what categories these objects themselves defined, and thereby established the categories listed in Table I. It is remarkable, looking back now, to realize that such a large number of peculiar galaxies could be fitted into so few categories of peculiarity. In the end, surprisingly few galaxies needed to be put into Category 22 as miscellaneous objects.

One of the important uses of the Catalogue is to enable astronomers, interested in a certain class of objects, to obtain a sufficiently large list of candidates with similar characteristics as a basis for further investigation of specific phenomena.

In the photographic section of the Catalogue the objects are ordered carefully within each Category. It is hoped that this conveys, to some extent, an empirical continuity of certain characteristics which may help us understand the cause of some peculiarities. As shown in the previous table of codes, the few words that describe each Category furnish a definition of the kinds of objects which comprise each Category. In addition, there are amplifications of the definitions and brief comments at the beginning of each Category when it is introduced. More detailed remarks on the possible significance of the Categories and their relation to each other are given below.

b. Statistical Frequency of the Various Categories.

The present Catalogue was designed to provide some additional information which has been previously unavailable. Specifically, we have counted all galaxies in each field down to the same size limit to which we have catalogued the peculiar galaxies. This means that we can calculate how frequent certain types of peculiarities are, down to a given limiting apparent diameter. For example, we can now say with what absolute frequency we find M51-types, or ring forms, or galaxies in chains. Similarly, it is also possible to calculate the relative frequency with which one Category is likely to be found as compared to another; for instance, how frequently does one encounter interacting doubles as compared to interacting triples? As noted before, however, all such intercomparisons should be done with care.

We have inspected and counted all galaxies to an estimated size limit. There is of course some imprecision inherent in this approach when attempted by visual means. Exposure limits and personal judgement vary from plate to plate. Additionally, some of our fields are obscured by galactic absorption. This led to the expected result that we found only a few galaxies in some fields, and undoubtedly missed some connections and interactions among those galaxies we did look at in the obscured fields. Nevertheless, these cases are relatively few in number and probably do not significantly alter our results. Knowing the total counts of galaxies in each of our fields, we can give, with fair accuracy, what percentage of all the galaxies we looked at actually fall into the various Categories.

In the Appendix, we have listed the number of galaxies that we have inspected and counted in each of the 537 fields which comprise the survey area of the Catalogue. It is clear that the counted number of galaxies varies strongly with the galactic latitude of the field. In fact, the numbers can be taken as indicative of the average obscuration for each field. Of course it should be realized that the obscuration can vary dramatically within a single 6.5 degree by 6.5 degree field and that the counts can also be dramatically affected by the presence of galaxy clusters within a given field.

At the same time it should be noted that because of the limited dynamic range of the photographic plates and because we were working with deep exposures, we were especially sensitive to peculiarities in the faint outer reaches of most galaxies; the inner disk often being severely over-exposed. However, in fields obscured by foreground galactic absorption, galaxies can appear "ghost-like", with much lower apparent surface brightnesses. For these objects at low galactic latitudes our basis for judging peculiarities would necessarily shift from the outer to the inner regions. This should be borne in mind by the reader, since this particular surface-brightness effect makes the discovery of "dwarf galaxies" at low galactic latitudes suspect, unless there is some additional property that makes the identification compelling.

The total number of galaxies which are classified into each Category is obtained by counting the number of times each code number appears in the positions and descriptions part of the Catalogue (Volume I). As noted previously, the same galaxy could be classified under more than one Category. For example, one galaxy could have an interacting companion and several apparent companions, and some of these companions in turn could have jets or tails or other peculiarities. By summing the number of times each Category appears, we obtain the total number of galaxies exhibiting that particular peculiarity. These statistics, expressed as percentages, are listed in the final column of Table I. For the 6,445 individual entries there are 8,954 classification codings.

A further quantity of interest is a measure of the frequency with which galaxies in a diameter-limited sample are classified as peculiar. This statistic can be readily obtained by dividing the total number of entries in the Catalogue by the total count of galaxies examined, appropriately scaled for plate overlap. The total number of galaxies found on each survey plate is given in the Appendix. These are the raw counts obtained by us as we scanned each plate. In order to correct these for overlap the numbers listed should be decreased by a factor of 0.83 determined by us empirically from the numbers of peculiar galaxies discovered more than once in the course of the survey. In total we inspected 93,556 galaxies (77,838 when corrected for overlap) out of which we classified 6,445 as peculiar, indicating that eight percent of all galaxies looked at are peculiar.

c. The Atlas of Peculiar Galaxies and other Catalogues

Although the present Catalogue was developed without specific reference to the earlier *Atlas of Peculiar Galaxies* (Arp 1966), looking back now we can see many of the present Categories illustrated in that *Atlas*. Some Categories were given slightly different names in the *Atlas*. For example, the *Atlas* had an important class of objects called "Companions on Arms", which is now Category 9: M51-Types. But many Categories turned out to be essentially the same; some examples from the earlier *Atlas* being "galaxies with associated rings", "galaxies with jets", "disturbed galaxies with interior absorption", and "chains of galaxies". New categories were able to be added to the present Catalogue because of the large number of galaxies included and because we took advantage of the ability to compare surrounding sky areas. The new Categories include the apparent associations (Categories 1, 8, 23, 24), and the galaxies of very high or very low surface brightness (Categories 13 and 20).

In addition to furnishing examples of galaxies which illustrate the standard classification of types of galaxies, the *Hubble Atlas* (Sandage 1961) presents some well-known examples of peculiar galaxies. The *Hubble Atlas* also contains some objects from the southern hemisphere. Modern systematic surveys of southern hemisphere galaxies start with de Vaucouleurs (1956), who gives types and descriptions for the listed galaxies. Classification information is also given in Evans (1957) and Shobbrook (1966), while Sersic (1968) has published an atlas which includes the most peculiar bright galaxies in the southern hemisphere.

The most recent morphological classifications of southern galaxies are given in the *Las Campanas Survey* (Dressler and Sandage 1978; Sandage and Brucato 1979). Of course, the extensive classification of both normal and peculiar galaxies is available from the *ESO/Uppsala Survey* of the *ESO (B) Atlas of the Southern Sky* (for recent references in this series, see Holmberg *et al.* (1980) and the final compilation by Lauberts (1982)).

d. General Introduction to Discussion of Individual Categories of Peculiarity

One of the reasons for the interest in peculiar galaxies arises from the possibility that the perturbation of a galaxy can be treated as a departure from dynamic equilibrium. Physical properties of the galaxy can then be derived by analyzing the event as an experiment performed on the galaxy. This is particularly true if we can assemble sufficient examples of the phenomenon to separate the various parameters which differ between systems, as for example the types of initial impulses and the time since the last disturbance. Perhaps an even more important question is what causes the peculiarity in the first place. Is it a disturbance due to an interaction with a nearby object or is it an internally generated disturbance such as an explosion or ejection?

The Categories which are empirically assembled in the present Catalogue point to the existence of various kinds of events and forces. First, there are the interaction events where two galaxies are found close to each other and deformed, presumably by gravitational or other long-range forces. Second, there are apparently isolated galaxies where the disturbance appears to arise from within the galaxy itself, possibly due to internal dynamical instabilities, nuclear activity, or other explosive events including supernovae or extensive star formation. Third, there are objects which appear to differ from the norm because they reflect differences in the wider environment in which they formed and evolved; for instance, differences noted in two galaxies could be attributed to the effect of a hot gaseous medium, higher mean velocity dispersion or a rapidly fluctuating gravitational potential for a galaxy in a cluster, as compared to a similar object in the field. Peculiarities could then be ascribed to three main causes, immediate neighbourhood, internal activity and general environment.

Of course, the assigning of probable causes for peculiarity can only be tentative when one is dealing with the appearance alone. What appears as a single peculiar galaxy, for example, might be an accidental superposition of two galaxies along the line of sight. In other instances an apparently isolated but disturbed galaxy might be the physical merging of two systems. Alternatively, what appears to be a gravitational interaction might be enhanced internal activity (such as the fueling of an active nucleus) only triggered, but not caused, by the proximity of another galaxy. Or internal activity may be simultaneously occurring in two or more galaxies which are only coincidentally close together. Correlation and cause are rarely easy to disentangle in astronomy. As a starting point, however, the Categories which most suggest certain causes are listed below:

Interaction Candidates: Categories 1-6, 9.
Possible Internal Activity: Categories 6-16.
Initial Conditions or Evolution: Categories 8, 13, 14, 16-20, 23, 24.

Like the *Atlas* which preceded it, the present Catalogue poses intriguing questions as to whether some of the different Categories are physically related to each other, and if so, how. The following detailed discussion of each Category is meant to highlight some of the important peculiarities in galaxies and suggest ways in which possible relationships between the Categories might help us to make progress towards explaining these systems.

Category 1: Galaxies With Interacting Companions

The first examples of this Category in Volume II are spirals with well-defined spiral arms. The reason for starting with these cases is to insure that the effects of the interaction on the arms can be well seen. The experiments which are thus presented to us deal with the phenomenon of spiral structure. The nature and cause of spiral arms has long been debated, with the density wave theories of Lin (Lin and Shu 1964), the magnetic tube concepts of Piddington (1970), the ejection track theories of Ambartsumian (1958), Arp (1963, 1969b, 1978), and Oort (1975), and most recently, the self-propagating star formation theories of Mueller and Arnett (1976), all attempting to explain spiral structure. One way of making progress on this problem is to examine spiral arms under physical stress. Of particular interest in this regard are new cases shown here where the spiral arm appears retarded in rotation by the nearby companion as illustrated by AM 2011-705 and AM 2203-381. In the present Catalogue, there is also one case where the spiral arm seems to flow around the companion (i.e., AM 0244-302).

Later in Category 1, we encounter sub-category 1b, also illustrated Volume II. Here the companions are interacting strongly with the whole body of the main galaxy. They furnish some of the most certain physical interactions and the companions might be considered in the role of test probes. As we go still later into this category we encounter companions relatively more widely separated and central galaxies relatively less disturbed. Among these later sub-categories, we encounter some very unusual shapes in the companions themselves (e.g., AM 0036-432). Some of these suggest that the companions are internally active in their own right, although that internal activity may or may not have been triggered by an interaction.

By the end of the category, the evidence for interaction is more subtle. This arrangement emphasizes the important point that the present category is perfectly continuous with Category 8: Galaxies with Apparent Companions. Whatever the important considerations in Category 8 are, be they initial conditions, evolution, or interaction, some of the same forces may be acting here.

Category 2: Interacting Doubles

The continuity of this Category with the preceding is made empirically clear by considering examples of galaxies with interacting companions; when the companion is larger than approximately half the size of the main galaxy, we classed it as an interacting double. It is by no means certain, however, that all such systems arise from the same causes or initial conditions as those in Category 1. It is interesting to note that some galaxies with small companions can appear to be just as violently disturbed as those interacting with galaxies of much larger size. This raises questions about how the interaction takes place: Is it strictly a gravitational interaction, primarily between stars or do plasma, gas, dust or magnetic fields also play a role? It also raises questions concerning timescales and frequencies: Over what fraction of their lifetimes are interacting galaxies in contact? And what fraction of all galaxies interact at all?

By consulting the statistics in Table I, it is seen that this Category is the most frequently used classification in the whole Catalogue. There are simply a very large number of galaxies encountered where the two galaxies are roughly the same size, a few diameters apart, and apparently disturbed. The forms of these disturbances are endlessly varied and Volume II attempts to give a feeling for these forms.

The E + E sub-category, which leads off the interacting doubles, contains examples of systems where galaxies without much apparent gas and dust are interacting. There the forms of the interaction suggest primarily gravitational perturbations of stars in orbits. Undoubtedly, many of these cases can be treated by the classic precepts of gravitational interaction, as most extensively studied by Toomre and Toomre (1972).

The E + S sub-category, which follows, shows spirals in the gravitational field of a companion E. Presumably here we can see spirals deform in a purely gravitational perturbing field. This sub-category is interesting also from the standpoint that E galaxies have a tendency to occur in clusters and groups with other E or lenticular galaxies. The question is, how do these E + S associations come about? There is also a puzzling suggestion in the literature that, when ellipticals and spirals are in physical pairs, the spirals have systematically higher redshifts (Tifft 1972; Tifft and Tarenghi 1975; Jaakkola 1971, 1973; Arp 1976a; Sulentic 1977).

The final photographic sub-category shows a selection of S + S interactions. These are usually the most spectacular because deviations from equilibrium can be most easily seen in spirals. Also, in many cases, the gas and dust, which have stronger interactive properties than stars, are involved. Study of this sub-category should yield some new information on the physical differences between different types of spirals. Moreover, the relation between this interacting sub-set of spirals and the apparently non-interacting double spirals in Category 23 needs to be clarified. This is especially interesting in light of the strong correlation of Hubble type (Page 1975, Noerdlinger 1979) and colour (Tomov 1978) between pairs of spirals in binary systems. This correlation could be an initial condition of the system, but it is also possible that in these cases the Hubble type and colours are determined by the on-going interaction rather than the internal dynamics of the individual galaxies (Madore 1980).

A whole sub-category of interacting doubles is not included in Category 2. It consists of those interacting doubles which are apparent companions of spirals (Sulentic, Arp and di Tullio 1978). This sub-category is illustrated in Category 8. There are also examples of interacting doubles in Category 6: Ring Galaxies, and in other Categories as well.

In general, there remains an enormous amount of observational work to be done on these interacting doubles. Large numbers of accurate redshifts need to be obtained. Morphologies need to be specified more precisely. Sizes and apparent magnitudes need to be measured. Perhaps new detailed information on this large Category of doubles in the present Catalogue will help to interpret the data which has already been gathered on the Karachentsev doubles in the northern hemisphere (Karachentsev 1974, 1980; Peterson 1979; Tifft 1980) and those of Schweizer (1979) in the south.

Throughout the examples of interacting galaxies runs the question of whether the galaxies are suffering temporary encounters or are merging, fissioning, or ejecting. Considerable study will be needed to decide eventually whether one or more of these processes is working, and if there is more than one, whether any one of them is dominant. One statistic of importance, which is derived in the present Catalogue, is the percentage of all galaxies that are interacting doubles. This must be an important number in constructing theories of how the interactions occur in terms of initial and present conditions. Moreover, in the multiply interacting systems, we frequently encounter linear arrangements. This surprising feature of the Catalogue implies that something more than random encounters of galaxies is taking place. Comparisons with similar samples at higher redshift will be especially interesting.

Categories 3 through 5: Interacting Triples to Quintets

When three galaxies fall within a few diameters of each other on the sky and are disturbed, they have been classed as an interacting triple. When the number of apparently interacting galaxies is greater, they have been classified as interacting quartets and quintets. It is significant that no unequivocal examples of six, or more, interacting galaxies of roughly equal size were encountered in our study. Any candidates for interacting galaxies in greater number than five will have to be sought in the present Catalogue in Categories 17 and 18, Chains and Groups of Galaxies. Sometimes there are smaller galaxies present in interacting systems, which would take the group into a higher order interacting classification if they were considered of dynamical importance in the group. The classification here, however, has to be made on the subjective impression as to what the main components of the group are.

Considering that the interacting double category is by far the largest category in the Catalogue, it is perhaps surprising that there are so few interacting triples. The reduction in numbers as we proceed to interacting quartets and higher orders, is even more drastic. Theories explaining the frequency of interacting doubles should also lead to predictions of the numbers of interacting triples, quadruples and higher-order interactions; the present data should help to select between such theories.

Among the triples, there are two sub-categories distinguished in Volume II: the first, where three galaxies of any type which are roughly the same size seem to be interacting; and the second, where there is rather striking alignment. This tendency towards alignment persists also into the quartets and quintets: AM 1303-371, which consists of a group almost in contact, shows linear elongation. In the northern hemisphere, interacting or disturbed groups have already been shown to have a tendency to be aligned (Arp 1973b). In the well-known Stephan's Quintet, for example, the four high-redshift members are aligned very well, extending even to a fifth high-redshift member to the east. The tendency for the groupings of disturbed galaxies to align is clearly related to Category 17: Chains, because many of the galaxies in chains are also disturbed.

In some groups of disturbed galaxies ring forms appear. Since rings are one of the most striking non-equilibrium forms found, the ring classification generally took precedence over other Categories. Therefore, some apparently interacting groups can also be found in Category 6: Ring Galaxies, which will now be discussed.

Category 6: Ring Galaxies

Any galaxy with a well-defined ring around it, and any galaxies apparently related to this form, have been included in this Category. We have tried to exclude galaxies where spiral arms have become wound up tightly into a circle or near circle (they are in Category 10: Galaxies with Peculiar Spiral Arms) but there is some question as to whether the latter forms are related. There may be some overlap or there may simply be an ambiguity in the classifications.

There have been three major attempts to explain the origin of ring galaxies (Freeman and de Vaucouleurs 1974; Theys and Spiegel 1976; Lynds and Toomre 1976). All of these approaches involve the collision of a cloud or galaxy with another galaxy in a special way, usually through the center of the galaxy which becomes the ring galaxy. The resulting, nearly symmetrical disturbance is proposed to account for the formation of the ring. Naturally, the presence of nearby companions is of great importance to these kinds of theories, and the present examples of ring galaxies with companions furnish many cases where redshifts, magitudes, and spectra can be obtained in order to check these theories. Conversely, it will be very interesting to uncover examples where there are no apparent companions in the vicinity of the ring. Some examples which deserve further study are shown in Volume II under this Category. Perhaps they have an independent origin.

The first sub-category starts with examples of empty rings, that is rings lacking any apparent nucleus. These are particularly interesting because it seems difficult, although by no means impossible (see

Theys and Spiegel 1976), to conceive of a collision which would remove a nucleus entirely. Alternatively one could postulate that the target galaxy had no dominant nucleus to begin with. A further alternative is to postulate that the radial perturbation which formed the ring was caused by internal activity of the nucleus (Saslaw, private communication 1978; Williams, in preparation 1985). Later in this sub-category, we encounter some galaxies with nuclei which are quite off-center, while toward the end of this sub-category the nuclei become fairly well centered. In the examples showing moderately off-center nuclei, we encounter the spectacular rings such as AM 0644-741 and AM 1006-311 and the remarkable "Cartwheel" (previous discussions of some of these objects can be found in Graham (1974), Fosbury and Hawarden (1977), Few, Madore and Arp (1980)). Morphologically, the objects toward the end of these illustrations lead more or less continuously into the second sub-category.

The second sub-category has rings with centrally located nuclei. Some resemble spiral galaxies in which the arms have become wound up by differential rotation, as arms would unless sustained by whatever the spiral arm generating mechanism is. From the pictures available to us, we put objects into the present Category if their rings are more or less unbroken circles with no hint of spiral connections to the central nucleus. The existence of this sub-category, however, raises the interesting question of whether some rings can originate from the winding up of spiral arms or luminous ejected filaments by differential rotation.

Finally, there is a sub-category consisting predominantly of barred spirals which have a high-surface-brightness inner ring and a faint outer ring as well. These "bull's-eye" shapes, to our knowledge, have received no analytical attention to date. Kormendy (1979) discusses some ring-form galaxies which may have some relation to "bull's-eye" forms as well as to rings in spirals in general.

One rather interesting object in this Ring Galaxy Category, is AM 0226-320. It represents the best example of either a prolate galaxy with a fine, well-delineated equatorial ring, or an edge-on galaxy with a ring circling the poles. For further discussion of this class of objects and references to previous work see Schweizer, Whitmore and Rubin (1983).

Category 7: Galaxies with Jets

We consider jets here as straight or nearly straight linear features. Because of the importance and relative rarity of jets, classification into this Category tended to override classification into other Categories. This Category perforce includes many objects which may only possibly have jets. Accidentally superposed companions or stars can fool the eye, so higher resolution or photography to lower surface brightness will be needed to confirm whether many of these cases are actually jets or not. In general, any linear feature that might be a jet was included in this Category, with the expectation that further confirmatory work would be done.

It should be remembered that one of the most famous cases of jets from a galaxy, NGC 1097, was discovered initially as an uncertain possibility on a very deep Schmidt plate, and only confirmed after special efforts (Wolstencroft and Zealey 1975). Such information-stacked plates taken with the CTIO 4m reflector, eventually revealed a total of 4 jets emerging from the nucleus of this galaxy (Arp 1976b; Lorre 1978). The picture of NGC 1097 = AM 0244-302, shown in Volume II, is from a standard IIIa-J survey plate and the print here does not show even the brightest jet.

In general, counterparts to high-surface-brightness jets known in the northern hemisphere, such as in M87 and in 3C273, were not found in our southern hemisphere survey. Because the deep exposure IIIa-J plates we surveyed were saturated over a moderately large fraction of the central galaxy images, there may be some such jets that we have missed in interior regions. But results from the ESO *Quick Blue Survey*, which was made using shorter exposures, turned up no outstanding new cases (Holmberg *et al.* 1980), so we can conclude that such strong jets are very rare and unusual (see also the *Proceedings of Second ESO/ESA Workshop*, Munich, 18-19 Feb. 1981, *ESA Sp-162*, March 1981).

Remarks on some of the individual candidates for jets which we have illustrated, are made in the introduction to the category in Volume II. We should note that Category 15: Galaxies with Tails, Loops of Material or Debris, may be allied closely in some cases with Category 7: Galaxies with Jets. If we consider an initially straight jet deforming under the action of rotation, then some tails or loops could have been jets. Conversely, it could be that some filaments are being viewed in the plane of their orbits, and therefore appear like straight jets.

Category 8: Galaxies with Apparent Companions

The galaxies that are nearest in space to us, and that we therefore know the most about, are almost all grouped together in units of one large, dominant galaxy, and a retinue of smaller companions. Examples are the Local Group (dominated by M31) and the M81 and M101 groups. It is important to find out how general a phenomenon this kind of grouping is over large regions of the universe. The present Catalogue, including all of the sky south of declination -22 degrees, presents extensive lists of the most conspicuous, apparent groupings. These are the cases most likely to turn out to be physical associations, when more detailed observations become available.

The forms of the companions, and their distribution around the central galaxy, present many puzzling problems. Holmberg (1940, 1969) shows that fainter galaxies, many of them presumably physical companions, tend to concentrate in the vicinity of the minor axes of large spiral galaxies. Later, Sulentic, Arp, and de Tullio (1978) showed that disturbed or non-equilibrium galaxies group preferentially around large spirals. A striking local example of this phenomenon is seen in the two closest companions to M81, (i.e., M82 and NGC 3077). Lynds and Sandage (1963), Sandage (1961), Bertola (1981b), and Arp (1980) established the extremely peculiar nature of these companions. Arp (1981a, 1981b, 1982) has shown that companion galaxies, including many of the ones shown in Volume II, tend to have early spectral types and a high incidence of emission. What causes these peculiarities: Is it the interaction of the companions with the main galaxy? Or is there some more fundamental reason for the differences in the companions? Are they formed at a later epoch than the main galaxy? Or are they formed in some basically different process?

Some clues to these puzzles may be obtained by further studies of this Category. It is the one which is illustrated with the largest number of examples in Volume II, and the reason for this is to supply observers with many more candidate groups from which they can select new cases for study. Of particular interest is the tendency for many companions to be interacting doubles, as shown in the third sub-category. What is the connection of interacting double companions to Category 2: Interacting Doubles? Are some of the interacting doubles in Category 2 companions to larger galaxies which are projected some appreciable distance away on the sky?

In the second sub-category, a number of examples are shown where the companions are aligned across the central galaxy. If this is a significant trend, it may be telling us something about the origin or dynamical evolution of the companions and their orbits. Whatever the reasons for the alignment, it is tempting to suppose that there is a common explanation for the alignments observed in the interacting triples, quartets, and quintets, and also, for the alignments observed in Category 17: Chains.

Category 9: M51-Types

In the *Atlas of Peculiar Galaxies* (Arp 1966), the M51-type galaxies were called "Companions on the ends of spiral arms." They made up one of the most important sections of the original *Atlas*, and were ordered from the smallest to the largest companions. The purpose was to show a sequence of forms. One interpretation of this sequence was made by Arp (1969b) in which he postulated that compact bodies were ejected from the nucleus of the parent galaxy. During their passage outward these compact objects formed the spiral arms in the underlying galaxy and then evolved into larger, companion-sized galaxies.

Subsequently Toomre and Toomre (1972) published calculations which showed that certain kinds of encounters between galaxies could gravitationally draw out filaments of stars which would connect them. The Toomres' simulations modelled some of the observed objects so well that there could be little doubt that such a mechanism was acting to some degree. One of the cases most successfully modelled was M51. In recognition of that analysis and its most famous prototype, we renamed the Category, M51-Types.

It is not necessarily true, however, that in every case companions on the ends of spirals arms have to be the result of gravitational encounters (note particularly the effects of small companions; see discussion by Vorontsov-Velyaminov (1957, 1975a, 1975b)). Is a companion originating in the nucleus of a galaxy distinguishable dynamically from a collisional encounter where the companion passes very close to the nucleus? Again, reliable masses and densities of the objects involved are needed to make further analysis. We present many new cases of M51-types, where the properties of a large number of members of this Category can be studied.

Finally, there is the question of the spiral arms themselves. The implication from the Toomre work was that the spiral arms were caused by the encounter (although, in the theoretical calculations, there tended to be appreciable differences between the tidal arm and the counter-tidal arm which are not generally typical of the observations). Nevertheless, if some spiral arms were formed by encounters, there is an obvious temptation to relate all spiral formation to encounters with companions; otherwise there would have to be two or more separate mechanisms for the formation of spiral arms! The all-collision origin for spiral arms is contra-indicated, however, by the existance of spiral galaxies which do not have any companions. It is, of course, difficult to exclude completely the possibility that there might always be companions around a galaxy, postulating that they are of sufficiently high mass-to-light ratio so that they are not conspicuous optically. But we are disinclined to invoke the unobservable too soon.

So we return to the question of the nature of spiral arms touched upon briefly in Category 1: Are spiral arms of such a nature that they are only formed by gravitational encounters? Are they even generally formed by gravitational encounters? Are they formed by perturbations of an internal origin? Or are there a variety of physical effects all of which can lead to the formation of spiral arms? The further study of the nature of spirals will be assisted by examples in the next two Categories. There we illustrate both the variety of arms that can be found, and the conditions under which they exist.

Category 10: Galaxies with Peculiar Spiral Arms

This Category deals with galaxies where the major peculiarity is in the spiral arms and is to be distinguished from Category 12: Peculiar Disks, where any arm peculiarity seems to be just a general consequence of the disturbance of the main body of the galaxy.

The largest sub-category illustrated here is that of one-armed spirals. In one portion of this sub-category, one arm is simply much stronger than the other arm; but, in other cases, there seems to be truly only a single arm. It would seem very difficult, in the density-wave theory, for spiral arms to be formed by an asymmetrical density perturbation which extended on only one side of the nucleus. If these cases, on further examination, turn out to be truly one-armed spirals, they would seem to pose a formidable challenge for any theory of spiral-arm formation. The stability and persistence of the phenomenon in any one example is, of course, an open question.

In Volume II, examples of spirals are given both with very narrow arms and with very broad arms. It is possible that by studying the different conditions in the disks of these galaxies, some clue as to the determining causes of spiral-arm width may emerge.

Another puzzling phenomenon, of which three examples are given in the photographs, is that of barred spirals where the arms are double as they come off one end of the bar. In some cases, they remain closely parallel for a considerable distance.

Finally, there is a sub-category of what are essentially elliptical galaxies with spiral arms around them. Since these spiral arms (rings?) are usually only discovered with deep photography, there may be further examples of this sub-category just below the level of usual detection. (It should be noted that these features are not the plateaus in surface brightness or "ringing" structures originally described in the *Atlas of Peculiar Galaxies* (Arp 1966) and later popularized by Malin and Carter (1980)).

This is a rather amusing Category to classify because the terminology "E galaxy with spiral arms" would seem to be a contradiction in terms. Accordingly, we have placed such objects in Category 10: Galaxies with Peculiar Spiral Arms by using the rationale that if arm regeneration were to stop in a spiral, the arms would grow old, fade, and possibly wind up. We then might be seeing examples of large-bulge systems whose arm formation has been discontinued within the last few hundred million years. The spectrum of the nucleus of AM 0148-472 (Arp 1981a), for example, shows an old stellar population; but the nuclear-bulge population of an early-type spiral could not easily be distinguished from that of an E galaxy. Detailed surface photometry and/or kinematic studies will help to decide between the elliptical versus bulge hypotheses. But these objects possibly represent transitional systems between elliptical and spiral galaxies, and are therefore very interesting in their own right.

Category 11: Three-Armed Spirals and Multiple-Armed Spirals

Three-armed spirals are a surprising Category to encounter because the majority of spiral galaxies appear to have some kind of two-armed, grand design, while the remainder have highly bifurcated or patchy arms. The density-wave theory, as mentioned in the discussion of one-armed spirals, usually gives symmetric density perturbations that are long-lived. Radio ejection is usually in the form of jet and counter jet, or lobe and counter lobe. There are, however, examples of short-lived minor density asymmetries and of triple ejections. In any case, as shown in Volume II we find a number of spiral galaxies with three, fairly equal arms, often symmetrically arranged. We expect they should be important tests for theories of spiral structure and in fact some work in the direction of modal analysis of spiral arms, from the point of density wave theory, has already beens begun (e.g., Haas 1983).

It is to be noted that often the clearest examples of three-armed spirals are the more distant objects where resolution does not reveal details of the arms, but rather emphasizes their general aspects. Some of these galaxies are illustrated in the photographs, as well as cases where more resolution in the arms is possible.

One commonly noticed property of spiral arms is their tendency to bifurcate somewhere along their length. We show some cases here where the bifurcation is so strong that the galaxy has the appearance of being a three-armed spiral. This may have some implications for the interpretation of all three-armed spirals and undoubtedly makes the challange of supplying a general mechanism for spiral-arm formation even more difficult.

We also encounter some multiple-armed spirals and give some examples of these in Volume II. They may or may not form a significant transition to flocculent-arm spirals. The case of AM 0545-431 is shown here as an example of the latter kind of spiral. The subject of grand design and patchy-armed spirals is treated in Gerola and Seiden (1978), while the role of dust in spiral arms is discussed in Elmegreen (1980a), and the subject of spurs on the arms of spiral galaxies is treated in Elmegreen (1980b).

Finally, we should note that spiral arms usually remain fairly distinct, but sometimes they can apparently interact or even cross over each other. In this Catalogue AM 0102-274 is an example where some arms seem to cross each other and seem almost braided or intertwined. Other northern hemisphere examples of galaxies with crossing arms are discussed in Arp (1969b).

Category 12: Peculiar Spirals

This Category is meant to illustrate spirals with peculiarities other than in their arms. The classification is made when the main body of the spiral shows unusual asymmetries or global distortions. There are no obvious sub-categories. The pecularities tend to be quite individual and not readily allied to other kinds of peculiarities. An example is the rectangular nucleus of AM 0902-680, similar to the peculiar northern hemisphere galaxy NGC 128 (Bertola and Cappacioli 1977).

The significance of this Category may lie in illuminating what happens to the arms of a galaxy when the main body of the system is disturbed. If the inner regions of the galaxy become irregular, presumably there are irregular gravitational perturbations which are transmitted outward and should effect the spiral arms if the arms are a result of gravitational density waves. By studying this correlation (or lack of correlation) between the behaviour of the spiral arms and the structure of the interior, some essential properties of spiral galaxies may eventually be derived.

It is also true that strong disturbances are more apparent in spiral galaxies than ellipticals because of the gas, dust and rotation in spirals. Some of the mysterious internal activity which takes place in the centers of galaxies, therefore, can best be studied in the Catalogue examples, given here, of peculiar spirals.

Category 13: Compact Galaxies

Even on a survey taken under the uniform conditions of the *SRC Southern Schmidt Survey*, inevitable differences in registered density of sky background make it difficult to estimate surface brightness of galaxies by inspection. Occasionally, however, the surface brightness of an object becomes so high that it burns out an area of the image. In this case, galaxies can be recognized if they are either a) bigger in diameter than stars, b) appear like bright stellar objects without diffraction spikes, or c) if there is a slight fuzz around the edge of a burned-out image. The above definition of compact objects also fits, rather well, one class of plate defects which can be encountered; therefore, particular effort has been made to check that our candidates are real, by assuring that they are present on more than one plate.

In general, we have not listed faint images which are only suspected of being non-stellar. Such faint objects generally comprise the interesting class of compact objects and groups of compact objects investigated by Zwicky (1971) and Shakbazian (1978). Our objects are more like the *brighter* objects catalogued by Zwicky. Some investigators (Massey 1975; Kormendy 1977; Fairall 1978) claim that compact galaxies are not different from ordinary galaxies. From our experience with the brighter compacts, this seems difficult to accept because, for the same diameter, our compact galaxies have surface brightnesses which burn out or saturate the plate, whereas the vast majority of galaxies do not have this effect. In any case, the present Catalogue lists galaxies that represent the largest, highest-surface-brightness compact galaxies encountered in the southern hemisphere.

An initial investigation of the spectroscopic properties of thirty-one compact galaxies, selected from *U.K. Schmidt Survey* plates, is given in Allen *et al.* (1978) and also Martin, Penfold and Glass (1978). Spectroscopy of southern compact and bright-nucleus galaxies has been reported by Fairall (1977, 1979, 1980, 1981, 1983, 1984). Galaxies with high-surface-brightness "hot spots" in their nuclei have been catalogued by Sersic and Pastoriza (1965), but in general these are not a sub-set of our compact galaxy Category which usually implicates the entire galaxy. Some factors which might be involved in producing the high surface brightness are: strong gaseous emission lines, high mass density, extensive non-thermal emission, or recent, high-mass star formation. Like most extragalactic objects, the present compact objects tend to be associated with other galaxies. Further investigations will have to be made to determine more accurately if these apparent associations are real, and what they might tell us about the cause of the high surface brightness. It would be particularly interesting to answer the inverse question: Are there any compact galaxies which are truly isolated in space?

Category 14: Galaxies with Prominent or Unusual Dust Absorption

Dust absorption is usually most conspicuous in edge-on spirals. In general, this kind of object has not been catalogued here, unless the dust appears to deviate from a flat, planar distribution. When dissipative particles like dust accumulate in a thin rotational plane, as is presumed to be the case where rotating spirals are concerned, the phenomenon is not especially a curiosity. But when we encounter irregular or asymmetric absorption features, we have good reason to believe that something must have perturbed them. Moreover, when we see filaments or tubes of dust, such as in the famous northern hemisphere galaxy NGC 4753, the puzzling question arises as to what mechanism could have formed such features. As in the case of stellar configurations, it is also true that non-equilibrium dust features should be differentially wound up, twisted, or dissipated by the normal motions within a galaxy. Therefore their presence would appear to indicate recent formation.

Another kind of object, included in Category 14, are galaxies where the dust marks an unexpected orientation of the rotation plane. Most elongated galaxies have been assumed to be oblate spheroids, by analogy to the many flattened spirals seen at varying degrees of edge-on projection. A few rare cases of elliptical galaxies are known, however, where the dust indicates a rotational plane that suggests that the object is a prolate spheroid. One famous case is that of the giant elliptical radio galaxy, NGC 5128 = Cen A. Some additional cases like this are noted in the present Catalogue. The possibility of triaxial shapes for galaxies has also received recent theoretical and observational attention; an overview of work on the three-dimensional shapes of elliptical galaxies can be found in Bertola (1981a, 1981b).

Category 15: Galaxies with Tails, Loops of Material or Debris

Galaxies with luminous filaments which are more or less straight have been classified into Category 7: Galaxies with Jets. Luminous features which are appreciably curved have been called tails and when there are any signs of re-entrant behaviour, that is arms returning in the direction from which they originated, they have been called loops. These latter types are put into the present Category 15: Galaxies with Tails, Loops of Material or Debris.

Many doubles, and galaxies with interacting companions, show tails or loops. Presumably this is because of their mutual gravitational interaction, as can be seen in Categories 1 through 5. However, there is a category of galaxies whose members appear single, but also show tails or loops. This is an important Category because it implies that if these galaxies are truly single, then internal activity is a good candidate for producing the phenomenon. In this regard it is useful to consider that an initially straight jet could drift, fall back or be perturbed by other objects in the vicinity, and, with time, evolve into a tail or loop.

Of course there is always the possibility that what looks like a single galaxy may, in fact, be a merged, double galaxy. This could possibly enable the tails and loops to be accounted for with the usual gravitational scenario. In the Catalogue, if there was any chance that the filament could be emanating from a single galaxy, the object was put into the present Category.

Category 16: Irregular or Disturbed Galaxies

In this Category, as in the previous one, we have tried to select galaxies that have a good chance of being only one entity. But in this case we require more than exterior peculiar features, like arms or filaments. Rather we require that the entire galaxy must be asymmetrical, irregular, or of an unusual nature. Again the implication is that single irregular galaxies may have been subjected to irregular conditions of formation or that the more disturbed cases may be reacting to some recent internal perturbation.

Single galaxies are defined here as galaxies which have no comparably-sized galaxy in the immediate neighbourhood which appears to be affecting them. These Category 16 galaxies can, however, be companions to larger galaxies. The prototype for a Category 16: Irregular or Disturbed Galaxies, is the famous northern hemisphere irregular, M82 (see *The Hubble Atlas*, Sandage 1961; *Atlas of Peculiar Galaxies*, No. 337, Arp 1966). An interpretation of M82 is that it is not its proximity to M81, but rather its own internal activity, which accounts for its fractured, asymmetrical appearance. As emphasized in previous Catagories of this Catalogue and elsewhere (Arp 1973b, 1982; Sulentic, Arp and di Tullio 1978), companion galaxies tend often to occur in non-equilibrium forms.

As a future step in the study of the various disturbed categories in this Catalogue, it will be of interest to investigate more quantitatively the relative occurrence of disturbances, as a function of the galaxy's proximity to larger galaxies, as a function of membership in groups, in clusters, etc. Similarly, it will be of importance to know whether any disturbed galaxies are completely isolated in space. While such a study was beyond the scope of this investigation, some feeling for the trends can be had by consulting the relative frequencies of the various Categories as listed in Table 1, since the classification codes themselves (as well as the comments in the main Catalogue) do incorporate some statement as to the local environment defined by other nearby galaxies.

Category 17: Chains

For some investigations this is possibly one of the most significant and least expected categories in the Catalogue. It was noted some time ago that chaining or linear alignments of galaxies appeared to be a frequent occurrence (Arp 1968; Oort 1970). But until we started to list such configurations for the present Catalogue it was not realized just how prevalent these a chains of galaxies were. As Table 1 shows, they are almost as frequent as apparently unaligned groups of four or more galaxies. These groups of unaligned galaxies serve as a control in the sense that we assume four galaxies in a group will be equally recognizable, whether they are aligned or not aligned. Although we feel that the tendency for chaining is very striking visually, we must leave for future statistical analysis the problem of deriving the exact frequency and accuracy of alignment which is in excess of that expected from random arrangements of unaligned groups. Of course, some opinion has already been expressed that chains are accidental projections without significance (Turner and Sargent 1973).

The other striking aspect of this Category is how frequently the galaxies in these chains seem to be disturbed or in non-equilibrium forms. The empirical fact of the disturbance supports the conclusion that the chains are not accidental configurations, because the chance of finding successive random galaxies in such disturbed forms is small. Another result, which bolsters the physical reality of the chains, is the tendency for interacting triples and quadruples to be aligned, as was discussed earlier under those Categories. Finally, the diametric companions of Category 8, many of which are themselves non-equilibrium galaxies, form aligned chains in the sense that the large galaxy is at the center of the chain. In Arp (1968) it was shown that chains of galaxies often originate on either side of radio galaxies, which tended to be the largest and/or the most active galaxies in the chain.

The reason why we feel that the general existence of chains of galaxies, if validated, is so important is because of the following argument: If chains were formed on the order of ten billion years ago, their presently indicated redshift dispersions should have dissipated the chains long ago. So, either the indicated redshift dispersions are wrong, or the groupings are accidental, or the chains are considerably younger than ten billion years.

To have galaxies recently formed is a radical departure from conventional theories of an epoch of galaxy formation in the far past. As for those chains which contain disturbed galaxies, they could have been collisionally disturbed by interactions on plunging orbits through the center. If such a circumstance seems unlikely, we are left with galaxies which are disturbed because they are internally active; this in turn may or may not be further evidence for recent formation.

One possible method to form galaxies in chains is to have the pregalactic material collapse through linear instabilities in the pregalactic medium (Oort 1970). Perhaps magnetic flux tubes could guide such collapses. However we again encounter the age difficulty, in that with the measured redshift dispersions the chains would have to have been formed recently and, as far as we know, we see no evidence for such recent condensations out of the intergalactic medium. An alternative is to eject protogalaxies from central parent galaxies (Arp 1976a). But this also would have to be a recent activity, unless the velocity dispersion along the chain was incorrectly inferred from the measured redshifts. If the physical significance of the candidate chains given in this Catalogue is confirmed, they will perhaps force profound changes on our ideas about the formation, evolution, and nature of galaxies.

Finally, it should be remarked that there is some evidence (Arp 1973b) that non-equilibrium chains like that in Stephan's Quintet, for example, fall close to and appear to be associated with larger nearby, low-redshift spirals. If this line of argument is valid, it would mark the non-equilibrium chains as having large components of non-velocity redshifts.

Category 18: Groups

Objects classified into this Category fit the definition of being four or more galaxies which are not aligned. Additionally they are galaxies which stand out conspicuously by virtue of their density contrast with respect to the surrounding field density of similar sized objects. The distinction between a large group and a small cluster is somewhat arbitrary, but is made along the following lines: a cluster is regarded as a populous, dense grouping of galaxies the membership of which generally shows a gradation in galaxy sizes from large to small; a group tends to be a small aggregate of galaxies each more or less the same size. Of course, some group members may have additional small companions, but they are not a necessary or dominant component.

We have distinguished three broad types of galaxy aggregates in the present Catalogue: galaxies with relatively small companions, galaxies in groups with each member being about the same size, and populous clusters which have a broad range of sizes. We are making a rough statement about the luminosity function of galaxies within these aggregates; saying that there appear to be distinctly different luminosity functions. The challenge now becomes that of systematizing this impression and discovering the reasons why different aggregates of galaxies have different luminosity functions, if such is the case. Our definitions may of course force these correlations, but one must then wonder why the missing categories, such as populous, high-density groupings of low-luminosity galaxies are not found.

Some conspicuous groups of galaxies in the southern hemisphere have been prevously listed by Klemola (1969) and Sersic (1968). Rose (1976), Snow (1970), and Duus and Newall (1977) have each catalogued groups and clusters of galaxies. Some of their poorer clusters appear in our catalogue under Category 8: Galaxies with Apparent Companions; this difference in point of view is interesting to note.

Category 19: Clusters

Over the area of the southern sky which we catalogued, we naturally saw many clusters of galaxies. We have made no systematic attempt to record all of these since a complete catalogue of southern galaxy clusters has been made by Abell and Corwin (*Clusters of Galaxies in the Southern Hemisphere*, in preparation) and we refer the reader to that source. However, in the course of inspecting the southern sky, we did encounter some clusters that were outstanding. In the spirit of including all very unusual objects in the present Catalogue, we have listed all these most conspicuous galaxy clusters. A number of these are clusters dominated by cD galaxies and double, interacting, giant elliptical galaxies. Not surprisingly, many of the identifications of Parkes radio sources with objects in our Catalogue are with these clusters. Some overlap is also found with the previous brief catalogues published by Snow (1970), Rose (1976), and Duus and Newall (1977).

Category 20: Dwarf Galaxies (Low Surface Brightness)

The characteristic of low surface brightness is usually taken as a sufficient condition for low intrinsic luminosity. It is apparently not a necessary condition for low luminosity because there are intrinsically faint galaxies known with high surface brightness (Arp 1965; Sargent and Searle 1970). The dwarfs which are large in angular extent usually reveal a very irregular morphology. It is difficult to identify specific classes of irregularities, however, and although such objects are completely listed in the descriptive part of the Catalogue Volume I, only a few are illustrated in Volume II.

It should be noted at the outset that there is a technical problem in discovering small, low-surface-brightness objects: many times plate defects can look very much like real objects, and especially like dwarf galaxies. We have made every effort to check such images on two independent plates in order to minimize this error.

The outstanding problem in dwarf galaxies seems to be to explain why the grand spiral design becomes more and more disorganized and irregular as the dwarfs become lower in surface brightness, lower in luminosity, and presumably lower in mass. The examples of dwarfs given here, it is hoped, will furnish a varied enough list of objects so that this problem can begin to be investigated.

Some analyses of dwarfs from *U.K. Schmidt Survey* plates have already been made. Webster *et al.* (1979) have searched for dwarfs belonging to NGC 5128. Longmore *et al.* (1982) have presented a list of optical and neutral hydrogen data for 151 low-surface-brightness galaxies. An example of a low-surface-brightness resolved galaxy, which turned out to be a member of the Local Group, is AM 0640-505, given in Cannon, Hawarden, and Tritton (1977) and now known as the "Carina dwarf". Other possibilities have been cited by Longmore *et al.* (1978) and by Madore and Arp (1979). Examples of dwarf galaxies found by ESO astronomers can be found in the general compilation by Lauberts (1982) as well as in selected announcements by Schuster and West (1976), Lausten *et al.* (1977) and Cesarsky *et al.* (1977).

The low-surface-brightness objects of small apparent diameter are particularly interesting. Is there a minimum absolute diameter below which dwarf galaxies do not form (or survive)? To answer this, distances are needed. Since much less morphological detail is visible, it is not clear from the appearance alone whether any particular object is a small, extremely low-luminosity object seen fairly nearby (i.e., loose, inter-galactic star clusters) or whether it is a more typical low-luminosity extragalactic system seen at greater distances (i.e., dwarf galaxies at two to ten megaparsecs, say). How many star clusters are there in the Local Group large enough to be seen as fuzzy patches, but far enough away to be poorly resolved? The statistics are poor, but the searches are far from complete. One such object is AM-1 (Madore and Arp 1979). Since such star clusters are of particular interest in filling out the census of objects belonging to the Milky Way and other Local Group galaxies, they were especially looked for in the course of our survey. However, the very faintest and smallest could easily be missed, even in a careful search (e.g., AM-4, Madore and Arp 1980); so we are probably complete only to the level of AM-1, which is fairly bright. Nevertheless, some of the "dwarfs" listed in our Catalogue may prove to be Local Group globular clusters.

Category 21: Stellar Objects with Associated Nebulosity

Although it became apparent that the vast majority of the members of this Category which we encountered were simply reflection nebulae (of which there are a great many of varied form at low galactic latitudes) we have retained the Category for the few high galactic latitude objects which may be of future interest to others. Some of these objects may yet prove to be more than their morphology suggests. One reflection nebula is illustrated in Category 22 in Volume II.

Category 22: Miscellaneous

The number of objects necessarily classified into the miscellaneous Category are very few. This is a simple demonstration that the other twenty-three Categories encompassed most of the peculiar and unusual aspects of galaxies. It is surprising that so many different peculiar objects can be classified into so few Categories. Perhaps this does mean that the subject is less complicated than it could be, and that only a few fundamental factors are causing the non-equilibrium forms with which we have to deal.

There are two examples of galaxies with "ringing" or stepped, concentric rings illustrated in the miscellaneous Category in Volume II. Objects of this kind were first shown in the *Atlas of Peculiar Galaxies* (Arp 1966); for example, No. 227 through 231. Later, Malin and Carter (1980) printed further high-contrast examples of this type of system. Generally, the rings imply some shocked star formation, but the cause of shock, whether it be internal explosion, merger, or encounter, still needs to be explored. It might be argued that these belong in Category 6: Ring Galaxies or even in Category 15: Galaxies with Loops, Tails of Material or Debris, but they are, we believe, distinct from these Categories, yet sufficiently rare that they do not demand a separate Category to themselves. Thus their present location in the Catalogue.

Category 23 and 24: Close Pairs and Triples

Sometimes large galaxies are encountered which are strikingly close together, relative to comparable sized galaxies in the general vicinity, and the supposition is strong that they are close together in space and probably related. If they are large enough for details to be resolved, and do not show signs of deformation or interaction, this is interesting in itself: Are they only temporarily close with insufficient time to effect gravitational perturbation? Or are all these cases separated much further in the line-of-sight dimension than in their projection on the sky? In any case, it was felt useful to have examples of galaxies which seem to belong together, but which are apparently not perturbed. This Category has also been included to complete the reporting of all galaxies which, from their distribution on the sky, appear to be associated together. For groups of galaxies greater than three in number, one should consult Categories 17, 18, and 19.

FORMATTING OF THE CATALOGUED DATA

Each object in the Catalogue has a two-line entry giving positional and descriptive data as follows:

The first line of the first column contains the Catalogue name of the object. In accordance with the Parkes identification procedure the name is made up from the hours and minutes of the right ascension (1950.0), followed by a minus sign and the truncated three most significant digits of the declination. An asterisk preceeding the Catalogue name indicates that the object is illustrated in Volume II. Occasionally an object is part of a pair, a triple or a larger association of objects which are also listed and described in the Catalogue as separate entries; to find the other members of such associations the second line of the first column gives, in square brackets, the abbreviated right ascension and declination of the member immediately preceeding the current object in order of right ascension. If the current object is the first object in the group, in order of right ascension, then the reference is to the last member of group. Accordingly, all members of a group can be recovered by sequentially following the referenced co-ordinates.

The second column contains the 1950.0 right ascension in hours, minutes and seconds of time, followed in the second line by the declination in degrees, minutes and tenths of minutes of arc.

The third column gives the annual precession in seconds of time (first line) for the right ascension, and in minutes of arc (second line) for the declination. (Note: In the few rare instances where the declination was further south than -85 degrees the precession in right ascension is given in minutes of time per year, rather than seconds).

The fourth column contains the galactic longitude and galactic latitude in the first and second lines, respectively.

The fifth column contains the supergalactic longitude and supergalactic latitude in the first and second lines, respectively.

The sixth column contains, in the first line, a brief description of the object. Abbreviations used in the description are explained on page 2. Below the description, in the second line, are cross-identifications to other catalogues, as listed in the section entitled Previous Identifications. Occasionally, when many cross-identifications were found, additional catalogue references were put at the end of the first line.

The seventh column gives the numerical coding of the description as listed and explained in Table I. Below is an S or D, indicating that the following number (given in arc minutes) is a measure of either the separation of major components, or a characteristic diameter of the system as a whole, respectively.

Finally the eighth column gives the ESO/SERC field number, followed by the number of the actual SERC J plate scanned by the authors. In the second line, are the cartesian co-ordinates of the object, measured in millimeters from the lower left-hand (south-east) corner of the original SERC plate.

ACKNOWLEDGEMENTS

Inevitably in the course of a ten-year programme many individuals and institutions deserve thanks for their help, encouragement and financial support.

In Edinburgh:
Without a doubt this project would not have come into existence had it not been for the foresight and work of Vincent Reddish, who was Astronomer Royal for Scotland and Director of the Royal Observatory, Edinburgh during the early years of this project. Brian Hadley, Callum Barclay and most particularly Bill Roberton at the Photo Lab in Edinburgh, were responsible for all of the photographic work. Their effort in producing the prints used in Volume II was enormous; we hope that our thanks and the appreciation of those who use the Catalogue will be in proportion. Russell Cannon, Russell Eberst, Tim Hawarden, Andy Longmore, Elizabeth Sim, Keith and Sue Tritton, all from the U.K. Schmidt Telescope Unit in Edinburgh and in Siding Spring, Australia are certainly to be thanked. Without their efforts the original plates would not have existed nor would they have been as easily accessible to visiting scientists. Malcolm Longair, who became the next Astronomer Royal for Scotland, continued the support which carried us through to the final plate searching sessions; we thank him for that much-needed continuity and his patience in waiting for the eventual publication.

In Cambridge and Toronto:
Barry Madore especially thanks Donald Lynden-Bell for first suggesting the collaborative effort between Cambridge, Edinburgh and Pasadena. We are indebted to Noel Argue in Cambridge, who produced the overlay programme used to refine the co-ordinates, and to Stuart Button in Toronto, who kept the overlay programme running. Ed Anderson, Matthew Bates, Petrusia Bojetchko and Robert Freedman who worked on the co-ordinate revisions got to see more peculiar galaxies (and overlays) than they perhaps ever wanted to, but their contribution to the success of the final Catalogue was crucial. Esther McLeary cheerfully typed the first draft of the text of the Catalogue. Ed Anderson was invaluable in the final stages of readying the Catalogue for publication and in producing the graphics. BFM also wishes to record his thanks to a life-long friend, Katherine Riordon, who was there when so many things began and who was responsible for starting so much.

The Institute of Astronomy, Cambridge and the Natural Sciences and Engineering Research Council of Canada through the University of Toronto, made travel and subsistence during the course of this project possible. Support from the Henri Chretien Fund awarded by the American Astronomical Society during the course of this work is also very gratefully acknowledged.

Wendy Laurel Freedman deserves more thanks from one of us (BFM) than words or even two volumes of galaxies can say.

In Pasadena:
Halton Arp gratefully acknowledges observing time on the Cerro Tololo 4m and on the Dupont 2.5m reflector in Chile, allocated to him by CTIO and by the Mt. Wilson and Las Campanas Observatories, respectively. All photographs which are not from the *SRC Schmidt Survey* were made with these two telescopes.

Both of the authors thank the Science Research Council in England for financial support administered through the Royal Observatory, Edinburgh. The final publication was made possible through a direct grant from the National Research Council of Canada.

These volumes were compiled under the auspices of the Royal Observatory Edinburgh and the Science and Engineering Research Council of the United Kingdom, in co-operation with the Mount Wilson and Las Campanas Observatories, the Carnegie Institution of Washington, the Institute of Astronomy, University of Cambridge, the David Dunlap Observatory, University of Toronto, the Natural Sciences and Engineering Research Council of Canada and the Cerro Tololo Inter-American Observatory (AURA Inc.).

REFERENCES

Abell, G. O. 1958, *Ap. J. Suppl.*, **3**, 211.

Aguero, E. L. 1971, *P.A.S.P.*, **83**, 310. [Ag]

Allen, D., Longmore, A. J., Hawarden, T. G., Cannon, R. D., and Allen, C. J. 1978, *M.N.R.A.S.*, **184**, 303.

Ambartsumian, V. A. 1958, in *Onzieme Conseil de Physique Solvay*, ed. R. Stoops, (Bruxelles).

Arp, H. C. 1963, *Scientific American*, **208**, 70.

Arp, H. C. 1965, *Ap. J.*, **142**, 402.

Arp, H. C. 1966, *Atlas of Peculiar Galaxies*, (Pasadena: Calif. Inst. of Tech.) = *Ap. J. Suppl.*, **14**, 1.

Arp, H. C. 1968, *P.A.S.P.*, **80**, 129.

Arp, H. C. 1969, *Sky and Telescope*, **38**, 385.

Arp, H. C. 1973, *Ap. J.*, **185**, 797.

Arp, H. C. 1976a, in *L'Evolution des Galaxies et ses Implications Cosmologiques, Colloque C.N.R.S. No. 263, Colloque IAU No. 37*, (Paris: C.N.R.S.), p.391.

Arp, H. C. 1976b, *Ap. J. (Lett.)*, **207**, L147.

Arp, H. C. 1978, in *Problems of Physics and Evolution of the Universe: Ambartsumian Volume*, Armenian Acad. Sci., (Yerevan).

Arp, H. C. 1980, *Ap. J.*, **239**, 469.

Arp, H. C. 1980, *Ninth Texas Symp.*, Ann. N. Y. Acad. Sci., **336**, 94.

Arp, H. C. 1981a, *Ap. J. Suppl.*, **46**, 75.

Arp, H. C. 1981b, *Ap. J.*, **250**, 31.

Arp, H. C. 1982, *Ap. J.*, **256**, 54.

Arp, H. C. 1982, *Ap. J.*, **263**, 54.

Bergvall, N. A. S. 1981, *Uppsala Astr. Obs. Report No. 19*.

Bertola, F. 1981a, *Sky and Telescope*, **61**, 380.

Bertola, F. 1981b, in *The Structure and Evolution of Normal Galaxies*, eds. S.M. Fall and D. Lynden-Bell, (Cambridge: Cambridge Univ. Press), p. 13.

Bertola, F., and Cappacioli, M. 1977, *Ap. J.*, **211**, 697.

Cannon, R. D., Hawarden, T. G., and Tritton, S. B. 1977, *M.N.R.A.S.*, **180**, 81p.

Cesarsky, D. A., Lausten, S., Lequeux, J., Schuster, H. E., and West, R. M. 1977, *Astr. Ap.*, **61**, L31.

Curtis, H. D. 1918, *Lick Obs. Publ.*, vol. 13.

de Vaucouleurs, G. 1956, *Mem. Commonwealth Obs.*, (Mt. Stromlo), No. 13.

de Vaucouleurs, G., de Vaucouleurs, A., and Corwin, H.G. 1976, *Second Reference Catalogue of Bright Galaxies*, (Austin: University of Texas Press). [A] = [RC2]

Duus, A., and Newall, B. 1977, *Ap.J. (Suppl.)*, **35**, 209. [DN]

Dressler, A., and Sandage, A. R., 1978, *P.A.S.P.*, **90**, 5.

Ekers, J. A. 1969, *Austr. J. Phys. (Astr. Suppl.)*, **7**. [PKS]

Elmegreen, D. M. 1980a, *Ap. J. Suppl.*, **43**, 37.

Elmegreen, D. M. 1980b, *Ap. J.*, **242**, 528.

Evans, D. A. 1957, *Cape Photographic Atlas of Southern Galaxies*, (Cape Obs.).

Fairall, A. P. 1977, *M.N.R.A.S.*, **180**, 391.

Fairall, A. P. 1978, *Observatory*, **98**, 1.

Fairall, A. P. 1979, *M.N.R.A.S.*, **188**, 349.

Fairall, A. P. 1980, *M.N.R.A.S.*, **192**, 389.

Fairall, A. P. 1981, *M.N.R.A.S.*, **196**, 417.

Fairall, A. P. 1983, *M.N.R.A.S.*, **203**, 47.

Fairall, A. P. 1984, *M.N.R.A.S.*, **210**, 69.

Few, J. M. A., Madore, B. F., and Arp, H. C. 1982, *M.N.R.A.S.*, s **199**, 633.

Fosbury, R. A. E., and Hawarden, T. G. 1977, *M.N.R.A.S.*, **178**, 473.

Freeman, K. C., and de Vaucouleurs, G. 1974, *Ap. J.*, **194**, 569.

Gerola, H., and Seiden, P. E. 1978, *Ap. J.*, **223**, 129.

Graham, J. A. 1974, *Observatory*, **94**, 290.

Haass, J. 1983, in *Kinematics, Dynamics and Structure of the Milky Way*, ed. W.L.H. Shuter (Dordrecht: Reidel), p.283.

Heidmann, J., Klein, U., and Weilebinski, R. 1982, *Astr. Ap.*, **105**, 188.

Holmberg, E. E. 1940, *Ap. J.*, **92**, 245.

Holmberg, E. E. 1969, *Arkiv. Astr.*, **5**, 20. = *Uppsala Astr. Obs. Medd.*, No. 166.

Holmberg, E. E., Lauberts, A., Schuster, H. E., and West, R. M. 1974a, *Astr. Ap. Suppl.*, **18**, 463.

Holmberg, E. E., Lauberts, A., Schuster, H. E., and West, R. M. 1974b, *Astr. Ap. Suppl.*, **18**, 491.

Holmberg, E. E., Lauberts, A., Schuster, H. E., and West, R. M. 1975, *Astr. Ap. Suppl.*, **22**, 327.

Holmberg, E. E., Lauberts, A., Schuster, H. E., and West, R. M. 1977, *Astr. Ap. Suppl.*, **27**, 295.

Holmberg, E. E., Lauberts, A., Schuster, H. E., and West, R. M. 1978a, *Astr. Ap. Suppl.*, **31**, 15.

Holmberg, E. E., Lauberts, A., Schuster, H. E., and West, R. M. 1978b, *Astr. Ap. Suppl.*, **34**, 28.

Holmberg, E. E., Lauberts, A., Schuster, H. E., and West, R. M. 1980, *Astr. Ap. Suppl.*, **39**, 173.

Jaakola, T. 1971, *Nature*, **234**, 534.

Jaakola, T. 1973, *Astr. Ap.*, **27**, 449.

Karachentseva, V. E. 1973, *Comm. Spec. Ap. Obs.*, USSR, **5**, 10. [Ka]

Karachentsev, I. D. 1972, *Comm. Spec. Ap. Obs.*, USSR, **7**, 1.

Karachentsev, I. D. 1974, *Observatory*, **94**, 290.

Karachentsev, I. D. 1980, *Ap. J. Suppl.*, **44**, 137.
Keeler, J. E. 1900, *Ap. J.*, **11**, 325.
Keeler, J. E. 1908, *Lick Obs. Publ.*, **8**, 1.
Klemola, A. R. 1969, *A. J.*, **74**, 804. [KLEM]
Kohoutek, L. 1971, *Astron. Ap.*, **13**, 493. [Ko]
Kormendy, J. 1977, *Ap. J.*, **218**, 333.
Kormendy, J. 1979, *Ap. J.*, **227**, 714.
Lauberts, A. 1982, *The ESO/Uppsala Survey of the ESO(B) Atlas*, (Munich: European Southern Observatory).
Lausten, S., Richter, W., van der Lans, J., West, R. M., and Wilson, R. N. 1977, *Astr. Ap.*, **54**, 639.
Lin, C. C., and Shu, F. H. 1964, *Ap. J.*, **140**, 646.
Longmore, A. J. 1977, *M.N.R.A.S.*, **178**, 251. [Lo]
Longmore, A. J., Hawarden, T. G., Webster, B. L., Goss, W. M., and Mebold, U.n 1978, *M.N.R.A.S.*, **183**, 97p. [UKS]
Longmore, A. J., Hawarden, T. G., Goss, W. M., Mebold, U., and Webster, B. L. 1982, *M.N.R.A.S.*, **200**, 325. [UKS]
Longmore, A. J., and Tritton, S. B. 1980, *M.N.R.A.S.*, **193**, 521. [LT]
Lorre, J. 1978, *Ap. J. (Lett.)*, **222**, L99.
Lu, P. K. 1971, *A. J.*, **76**, 775. [Lu]
Lynds, C. R., and Sandage, A. R. 1963, *Ap. J.*, **137**, 1005.
Lynds, C. R., and Toomre, A. 1976, *Ap. J.*, **209**, 382.
Madore, B. F., and Arp, H. C. 1979, *Ap. J. (Lett.)*, **227**, L103. [AM]
Madore, B. F., and Arp, H. C. 1982, *P.A.S.P.*, **94**, 40. [AM]
Madore, B. F. 1980, *A. J.*, **85**, 507.
Malin, D. F., and Carter, D. 1980, *Nature*, **285**, 643.
Malin, D. F., and Carter, D. 1983, *Ap. J.*, **274**, 534. [MC]
Martin, W. L., Penfold, J. E., and Glass, I. S. 1978, *M.N.R.A.S.*, **184**, 15p.
Massey, P. L. 1975, *A. J.*, **80**, 778.
Mueller, M. W., and Arnett, W. D., 1976, *Ap. J.*, **210**, 670.
Noerdlinger, P. D 1979, *Ap. J.*, **229**, 877.
Oort, J. H. 1970, *Science*, **170**, 1365.
Oort, J. H. 1975, *Colloque C.N.R.S.*, No. 241, 517.
Page, T. 1975, in *Stars and Stellar Systems*, Vol. 9, *Galaxies and the Universe*, ed. A. Sandage, M. Sandage and J. Kristian (Chicago: Univ. Chicago Press), p.514.
Pease, F. G. 1917, *Ap. J.*, **46**, 24.
Pease, F. G. 1920, *Ap. J.*, **51**, 276.
Perek, L., and Kohoutek, L. 1967, *Catalogue of Galactic Planetary Nebulae*, (Prague). [PK]
Peterson, S. D. 1979, *Ap. J.*, **206**, 38.
Piddington, J. H. 1970, *Austr. J. Phys.*, **23**, 731.
Rose, J. A. 1976, *Astr. Ap. Suppl.*, **23**, 109. [R]
Sadler, E. M., and Sharp, N. A. 1984, *Astr. Ap.*, **133**, 216. [SS]
Sandage, A. R. 1961, *The Hubble Atlas*, (Washington: Carnegie Inst. Washington).
Sandage A. R., and Brucato, R. 1979, *A. J.*, **84**, 472.
Sandage, A. R., and Tammann, G. A. 1981,*A Revised Shapley-Ames Catalog of Bright Galaxies*, (Washington: Carnegie Inst. Washington).
Sargent, W. L. W., and Searle, L. 1970, *Ap. J. (Lett.)*, **162**, L155.
Schweizer, F. 1980, *Ap. J.*, **237**, 303.
Schweizer, F., Whitmore, B.C. and Rubin, V. 1983, *A. J.*, **88**, 909.
Schweizer, L. 1979, unpublished doctoral thesis, U.C. Berkeley.
Schuster, H. E., and West, R. M. 1976, *Astr. Ap.*, **49**, 129.
Sersic, J. L. 1968, *Atlas de Galaxias Australes*, (Cordoba: Observatorio de Universidad de Cordoba).
Sersic, J. L. 1974, *Ap. and Space Sci.*, **28**, 365. [Se]
Sersic, J. L., and Pastoriza, M. 1965, *P.A.S.P.*, **77**, 287.
Shakbazyan, R. K. 1978, *Astrofisika*, **14**, 273.
Shobbrook, R. R. 1966, *M.N.R.A.S.*, **131**, 351.
Smith, M. G., Aguirre, C., and Zemelman, M. 1976, *Ap. J. Suppl.*, **32**, 217. [To]
Snow, T. P. 1970, *A. J.*, **75**, 237.
Sulentic, J. W. 1977, *Ap. J. (Lett.)*, **211**, L59.
Sulentic, J. W., Arp, H. C., and di Tullio, G. 1978, *Ap. J.*, 3 **220**, 47.
Theys, J. C., and Spiegel, E. A. 1976, *Ap. J.*, **208**, 650.
Tifft, W. G. 1972, *Ap. J.*, **175**, 613.
Tifft, W. G., and Tarenghi, M. 1975, *Ap. J.*, **199**, 10.
Tifft, W. G. 1980, *Ap. J.*, **236**, 70.8
Tomov, A. N. 1978, *Sov. A. J.*, **22**, 540.
Toomre, A., and Toomre, J. 1972, *Ap. J.*, **178**, 623.
Turner, E. L., and Sargent, W. L. W. 1973, *P.A.S.P.*, **85**, 538.
van den Bergh, S., *Publ. David Dunlap Obs.*, **5**, 147.
van den Bergh, S., Marscher, A. P. and Terzian, Y. 1973, *Ap. J. Suppl.*, **26**, 19. [BMT]
van den Bergh, S., Racine, R, van Agt, S., Barnes, T., Coutts, C., Madore, B. F., and Skill, A. 1973, *Ap. J.*, **179**, 863. [vdB]
Vorontsov-Velyaminov, B. A. 1957, *Astr. Z.*, **34**, 8 = 1958, *Sov. Astr.*, **1**, 9.
Vorontsov-Velyaminov, B. A. 1959,*Atlas and Catalogue of Interacting Galaxies* Part I., (Moscow: Moscow University). [VV]
Vorontsov-Velyaminov, B. A. 1975a, *Astr. Z.*, **52**, 692 = 1976, *Sov. Astr.*, **19**, 422.

Vorontsov-Velyaminov, B. A. 1975b, *Proc. Third Europ. Astr. Meeting,* Tibilisi, ed. Kharadze, p.181.

Vorontsov-Velyaminov, B. A., and Krasnogorskaya, A. 1961, *Morphological Catalogue of Galaxies,* I, (Moscow: Moscow State University). [MCG]

Vorontsov-Velyaminov, B. A., and Arhipova, V. P. 1963, *Morphological Catalogue of Galaxies,* III, (Moscow: Moscow State University). [MCG]

Vorontsov-Velyaminov, B. A., and Arhipova, V. P. 1964, *Morphological Catalogue of Galaxies,* II, (Moscow: Moscow State University). [MCG]

Vorontsov-Velyaminov, B. A., and Arhipova, V. P. 1968, *Morphological Catalogue of Galaxies,* IV, (Moscow: Moscow State University). [MCG]

Vorontsov-Velyaminov, B. A., and Arhipova, V. P. 1974, *Morphological Catalogue of Galaxies,* V, (Moscow: Moscow State University). [MCG]

Webster, B. L., Goss, W. M., Hawarden, T. G., Longmore, A. J., and Mebold, U. 1979, *M.N.R.A.S.*, **186**, 31.

Wolstencroft, R. D., and Zealey, W. J. 1975, *M.N.R.A.S.*, **173**, 51p.

Zwicky, F. 1971, *Catalogue of Selected Compact and Post Eruptive Galaxies,* (Guemlingen).

Zwicky, F., and Herzog, E. 1963 *A Catalogue of Galaxies and of Clusters of Galaxies,* II, (Pasadena: Calif. Inst. of Tech.).

Zwicky, F., and Herzog, E. 1966 *A Catalogue of Galaxies and of Clusters of Galaxies,* III, (Pasadena: Calif. Inst. of Tech.).

Zwicky, F., and Herzog, E. 1968 *A Catalogue of Galaxies and of Clusters of Galaxies,* IV, (Pasadena: Calif. Inst. of Tech.).

Zwicky, F., Herzog, E., and Wild, P. 1961 *A Catalogue of Galaxies and of Clusters of Galaxies,* I, (Pasadena: Calif. Inst. of Tech.).

Zwicky, F., Karpowicz, M., and Kowal, C. 1965 *A Catalogue of Galaxies and of Clusters off Galaxies,* V, (Pasadena: Calif. Inst. of Tech.).

Zwicky, F., and Kowal, C. 1968 *A Catalogue of Galaxies and of Clusters of Galaxies,* VI, (Pasadena: Calif. Inst. of Tech.).

AUTHORS' NOTES

At its worst, cataloguing is a tedious and thankless pursuit; at its best it is simply tedious. However, for the two of us in the process of producing the seemingly endless listings of positions, types, sizes and descriptions that are contained in this Catalogue there was the pleasure of discovery. Being the first to see a newly found galaxy, to be able to wonder, however briefly, at its place in wider context and then to post notice of its existence, is an opportunity that few are privileged to share.

As we worked together on this project, we have always known it to be a joint effort, first at the level of scanning the plates in turn, discussing each new object, proudly showing the collection to anyone who innocently walked into the back corner of the plateroom at the Royal Observatory, Edinburgh, where the actual scanning was done, and finally compiling the list of objects which will be looked to by others seeking the next levels of discovery.

Our individual reasons for investing considerable effort and a great deal of time in this Catalogue were different, and not totally self-sacrificing. For one of us (HCA), there was a desire to find a large sample of companion galaxies and active galaxies in order to investigate the nature of their redshifts. For the other of us (BFM), this was an opportunity to survey systematically the rich variety of forms found among galaxies, from which some new insights and, at the very least, a close familiarity would arise.

The ten years of work that went into the Catalogue provided us both with yet another unique opportunity: an opportunity to become friends, each of us pleased with a mutual effort that we hope will speak for itself.

THE CATALOGUE

Identification	R.A. (1950.0) Declination	Precession R.A./Dec.	Gal. Long. Gal. Lat.	S.G.L. S.G.B.	Description Cross-Identification	Coding Size (Min.)	Field/Plate (x, y)
AM 0000-551	00 00 12 −55 17.3	3.071 0.334	317.25 −60.73	240.6 −5.6	LUMPY DWARF/GALAXY w. KNOTS	20 D 0.6	F192/J1839 (17,158)
AM 0000-803	00 00 14 −80 37.5	3.064 0.334	305.45 −36.53	216.1 −13.3	LO.S.BR. DWARF + COMP'NS. 12-G-14	8,20 D 1.6	F012/J2486 (179,146)
AM 0000-594	00 00 35 −59 47.9	3.067 0.334	314.03 −56.54	236.3 −7.1	CHAIN of 3 111-IG-11	24 D 1.3	F111/J0921 (172,188)
AM 0000-545	00 00 44 −54 50.1	3.067 0.334	317.46 −61.18	241.0 −5.5	SPIRAL + 4 APP. COMP'NS.	8 S 3.6	F192/J1839 (11,182)
AM 0000-500	00 00 49 −50 04.3	3.067 0.334	321.85 −65.52	245.5 −3.9	S0 w. DIAMETRIC COMP'NS. 193-G-11 = Fa 361	8 S 0.5	F193/J3714 (172,175)
AM 0001-435	00 01 15 −43 53.8	3.066 0.334	330.09 −70.90	251.4 −1.9	SP. w. PEC. ARM + COMPACT COMP'N. Fa 628 ?	1,10,13 S 0.7	F241/J3559 (167,239)
AM 0001-531	00 01 33 −53 11.7	3.061 0.334	318.59 −62.74	242.6 −5.1	E + 4 APP. COMP'NS. PKS 0001-531 = R3 = DN 0001-533	8 S 2.5	F149/J1739 (167,276)
AM 0001-505 [0001-5053]	00 01 39 −50 55.5	3.061 0.334	320.68 −64.81	244.8 −4.3	PAIR (#1: SPIRAL) 193-G-14	23 D 0.6	F193/J3714 (165,129)
AM 0001-505 [0001-5055]	00 01 53 −50 53.8	3.059 0.334	320.63 −64.85	244.8 −4.3	PAIR (#2: DISTURBED SPIRAL) 193-G-15	12,23 D 0.6	F193/J3714 (163,131)
AM 0001-323	00 01 59 −32 31.7	3.066 0.334	3.35 −78.60	262.1 1.9	COMPACT	13 D 0.3	F409/J2693 (157,43)
AM 0001-454	00 01 59 −45 43.6	3.061 0.334	326.87 −69.43	249.7 −2.6	GROUP of 5	18 D 4.5	F292/J3733 (48,140)
AM 0002-280	00 02 30 −28 01.2	3.065 0.334	26.72 −79.69	266.4 3.3	COMPACT (SOME STRUCTURE)	13 D 0.3	F472/J3428 (150,16)
AM 0002-510	00 02 33 −51 06.4	3.055 0.334	320.20 −64.70	244.7 −4.5	GROUP of 4 193-G-17	18 D 4.3	F193/J3714 (158,120)
AM 0002-405	00 02 42 −40 53.0	3.059 0.334	335.24 −73.48	254.3 −1.1	3 DISTURBED GALAXIES in a GROUP	24 D 1.8	F293/J3578 (152,132)
AM 0002-305	00 02 50 −30 50.6	3.063 0.334	11.33 −79.35	263.8 2.3	CHAIN of 8 (I/A DBL. at END)	2,17 D 39.4	F409/J2693 (147,134)
AM 0002-503	00 02 54 −50 32.8	3.052 0.334	320.64 −65.24	245.2 −4.4	SPIRAL w. APP. COMP'NS. 193-G-19 = B 0002-505 = Fa 631	8 S 2.8	F193/J3714 (154,150)
AM 0003-834	00 03 13 −83 41.3	2.903 0.334	304.50 −33.57	213.1 −14.2	GALAXY w. MATERIAL to SIDE 2-G-3	15 D 0.6	F002/J2563 (175,250)
AM 0003-473	00 03 21 −47 36.2	3.052 0.334	323.77 −67.91	248.0 −3.5	SMALL I/A DBL. 193-IG-21	2 S 0.3	F193/J3714 (149,308)
*AM 0003-362	00 03 26 −36 24.9	3.058 0.334	347.06 −76.79	258.5 0.3	I/A QUINTET 349-IG-26/27 [A82]	5 D 4.0	F349/J0738 (139,102)
AM 0003-402	00 03 41 −40 29.8	3.055 0.334	335.59 −73.90	254.7 −1.1	I/A DBL. w. TAILS in GROUP of 5 293-IG-33	2,15,18 S 0.2	F348/J3586 (39,151)
AM 0003-414	00 03 48 −41 46.5	3.053 0.334	332.80 −72.90	253.5 −1.6	IRR.(WARPED PLANE) w. I/A COMP'N 293-IG-34 = Ag-1 = B0003-417 [RC2]	1,16 S 1.0	F293/J3578 (141,84)
AM 0003-540	00 03 58 −54 06.7	3.041 0.334	317.12 −62.03	241.9 −5.7	I/A DBL. (R4 = DN 0004-681) 149-IG-17 = Se 164/1?	2 S 0.4	F149/J1739 (148,227)
AM 0004-675	00 04 00 −67 54.3	3.015 0.334	309.41 −48.92	228.6 −10.0	I/A DBL. (in CLUSTER)	2,18 S 0.1	F050/J0955 (156,289)
AM 0004-420	00 04 21 −42 07.2	3.050 0.334	331.84 −72.68	253.2 −1.8	DWARF 293-G-35	20 D 0.5	F348/J3586 (35,64)
AM 0004-500	00 04 31 −50 09.1	3.041 0.334	320.48 −65.70	245.7 −4.5	SMALL I/A PAIR of SPIRALS Fa 633	2 S 0.3	F193/J3714 (140,171)
AM 0004-665	00 04 32 −66 50.0	3.011 0.334	309.77 −49.97	229.7 −9.7	LINE of 3 (E w. FAINT JET)	7,24 D 3.1	F078/J0928 (152,79)
*AM 0004-413	00 04 34 −41 38.1	3.049 0.334	332.68 −73.10	253.7 −1.7	DISTURBED GALAXY w. LOOPS Ag-2 = 293-IG-37	15 D 0.8	F348/J3586 (32,90)
*AM 0004-413	00 04 36 −41 38.2	3.049 0.334	332.67 −73.10	253.7 −1.7	I/A COMP'N. to DISTURBED SPIRAL 293-IG-37	1 S 1.5	F293/J3578 (133,91)
*AM 0004-220	00 04 37 −22 03.8	3.062 0.334	58.44 −78.61	272.2 4.9	GALAXY w. ONE BRIGHT ARM MCG-4-1-16 = 538-G-22	1,10 D 1.6	F472/J3428 (121,337)
AM 0004-443	00 04 37 −44 38.9	3.046 0.334	327.32 −70.59	250.9 −2.7	E + V.CLOSE COMP'NS.	8 S 0.4	F292/J3733 (21,197)
AM 0004-525	00 04 44 −52 53.8	3.037 0.334	317.88 −63.20	243.1 −5.4	DWARF (SOME RESOLUTION) 149-G-18	20 D 1.1	F193/J3714 (141,23)
AM 0004-374	00 04 53 −37 43.8	3.051 0.334	342.08 −76.12	257.4 −0.4	I/A DBL. (incl. DWARF) 293-G-38 incl. 293-G-40	2,20 S 1.0	F293/J3578 (128,301)
AM 0005-393	00 05 09 −39 34.3	3.048 0.334	336.97 −74.80	255.7 −1.1	GROUP of 5 DISTURBED GALAXIES	18 D 3.9	F293/J3578 (126,202)
AM 0005-722 [0005-7228]	00 05 10 −72 27.2	2.978 0.334	307.61 −44.54	224.2 −11.4	I/A DBL.(#1: HI.S.BR. SPIRAL)	2 S 1.6	F050/J0955 (156,45)
AM 0005-722 [0005-7227]	00 05 10 −72 28.7	2.977 0.334	307.60 −44.52	224.2 −11.4	I/A DBL.(#2: EDGE-ON SPIRAL) 50-G-5	2 S 1.6	F050/J0955 (156,44)
AM 0005-705	00 05 34 −70 55.1	2.979 0.334	308.09 −46.04	225.7 −11.0	GALAXY w. APP. COMP'N.	8 S 0.4	F050/J0955 (152,127)

Identification	R.A. (1950.0) Declination	Precession R.A./Dec.	Gal. Long. Gal. Lat.	S.G.L. S.G.B.	Description Cross-Identification	Coding Size (Min.)	Field/Plate (x, y)
*AM 0005-301	00 05 49 −30 11.6	3.053 0.334	13.93 −80.14	264.6 1.9	IRR. w. SILOUETTE ABS'N. NGC 7 = MCG-5-1-37 = 409-G-22	14,16 D 2.5	F409/J2693 (112,168)
*AM 0006-340	00 06 02 −34 08.3	3.049 0.334	354.15 −78.58	260.9 0.6	SPIRAL w. PECULIAR ARM NGC 10 = 349-G-32	10 D 2.5	F349/J0738 (109,223)
AM 0006-311	00 06 12 −31 19.3	3.051 0.334	7.55 −79.89	263.5 1.5	LONG-ARMED SPIRAL in CHAIN of 4	10,17 D 2.2	F409/J2693 (108,108)
AM 0006-373 [0006-3738]	00 06 26 −37 36.6	3.044 0.334	341.52 −76.42	257.6 −0.7	PAIR (#1: SPIRAL) 293-G-42	23 S 2.1	F293/J3578 (112,307)
AM 0006-373 [0006-3736]	00 06 35 −37 38.2	3.043 0.334	341.35 −76.42	257.6 −0.7	PAIR (#2: E) 293-G-43	23 S 2.1	F293/J3578 (110,306)
AM 0006-602	00 06 41 −60 23.2	3.004 0.334	312.39 −56.22	236.0 −8.0	DWARF (RESOLUTION) 111-G-15	20 D 1.1	F111/J0921 (132,156)
AM 0006-375	00 06 57 −37 52.8	3.041 0.334	340.42 −76.29	257.4 −0.8	GROUP of 4 (DISTURBED)	18 D 5.8	F293/J3578 (106,293)
AM 0007-251	00 07 24 −25 14.8	3.053 0.334	43.59 −80.44	269.4 3.2	SPIRAL + APP. COMP'NS. off MI.AX. NGC 24 = 472-G-16	8 S 9.0	F472/J3428 (90,165)
AM 0007-355	00 07 24 −35 58.0	3.042 0.334	346.25 −77.69	259.2 −0.3	COMPACT CLUSTER of GALAXIES	19 D 0.9	F349/J0738 (96,125)
*AM 0007-571	00 07 28 −57 18.8	3.005 0.334	313.95 −59.18	239.0 −7.2	GROUP (LARGE E) PKS 0007-572 [DN] NGC 25/28/31/37 = Se 164/4 = Fa 1/2	2,18 D 9.0	F111/J0921 (122,321)
AM 0007-464 [0007-4646]	00 07 51 −46 41.7	3.024 0.334	323.04 −69.06	249.1 −3.9	PAIR (#1: EDGE-ON + COMP'NS.) 241-G-21	8,23 D 1.8	F241/J3559 (107,88)
AM 0007-464 [0007-4641]	00 07 55 −46 46.0	3.024 0.334	322.93 −69.00	249.1 −3.9	PAIR (#2: Sc) 241-G-22	23 D 1.0	F241/J3559 (106,84)
AM 0008-685	00 08 14 −68 56.9	2.948 0.334	308.45 −48.01	227.7 −10.7	COMPACT	13 D 0.2	F050/J0955 (137,233)
AM 0008-581	00 08 26 −58 10.4	2.994 0.334	313.21 −58.41	238.2 −7.6	CLOSE TRIPLE	24 D 1.5	F111/J0921 (116,274)
AM 0008-324	00 08 55 −32 41.6	3.040 0.334	359.10 −79.84	262.4 0.5	I/A TRIPLE	3 S 1.8	F409/J2693 (79,33)
*AM 0009-371	00 09 05 −37 13.0	3.033 0.334	341.06 −77.07	258.2 −1.0	SPIRAL w. I/A COMP'N. 349-G-39	1 S 0.6	F293/J3578 (83,328)
AM 0009-221	00 09 15 −22 11.0	3.051 0.334	60.82 −79.59	272.4 3.8	CONTACT DBL.	23 S 0.1	F472/J3428 (64,329)
AM 0009-273	00 09 17 −27 38.9	3.045 0.334	29.01 −81.19	267.2 2.0	I/A SPIRALS in GROUP of 4 409-IG-27	2,18 D 1.8	F409/J2693 (69,304)
AM 0009-272	00 09 42 −27 28.4	3.044 0.334	30.17 −81.28	267.4 2.0	SPIRAL w. I/A COMP'N. MCG-5-1-42 = 472-G-18	1 S 2.3	F409/J2693 (63,314)
AM 0009-621	00 09 42 −62 17.9	2.965 0.334	310.85 −54.49	234.3 −8.9	CLOSE I/A DBL. (JET?) 111-IG-18	2,7 S 0.1	F111/J0921 (116,52)
AM 0009-372	00 09 47 −37 21.6	3.029 0.334	340.16 −77.06	258.1 −1.2	I/A DBL. (ONE COMPACT)	2,13 S 0.3	F293/J3578 (76,320)
AM 0009-840	00 09 47 −84 09.3	2.516 0.334	304.19 −33.14	212.7 −14.5	I/A DBL. 2-G-4? = PKS 0010-840?	2 S 1.4	F002/J2563 (166,225)
AM 0009-265	00 09 50 −26 52.3	3.044 0.334	214.15 −81.28	272.1 −14.8	COMPACT	13 D 0.2	F410/J0778 (334,343)
AM 0009-620	00 09 55 −62 03.5	2.964 0.334	310.91 −54.73	234.5 −8.9	I/A DBL.	2 S 0.2	F078/J0928 (113,334)
*AM 0010-242	00 10 31 −24 29.5	3.045 0.334	229.36 −80.88	269.7 −15.4	SPIRAL w. PECULIAR ARMS MCG-4-1-20 = 472-G-20	11 D 1.2	F473/J3523 (332,205)
AM 0010-493	00 10 47 −49 38.2	2.999 0.334	318.69 −66.57	246.5 −5.3	CLOSE PAIR 193-G-36/37 = Fa 637/638	23 S 1.4	F193/J3714 (86,197)
*AM 0010-265	00 10 48 −26 52.3	3.041 0.334	214.28 −81.49	272.0 −14.6	GALAXY w. JET + CONDENSATION 472-IG-21	7 D 1.0	F473/J3523 (325,77)
AM 0011-415 [0011-4152]	00 11 07 −41 53.0	3.015 0.334	328.66 −73.57	253.9 −2.9	I/A DBL. (#1)	2 S 1.7	F294/J3551 (327,74)
AM 0011-415 [0011-4153]	00 11 13 −41 52.1	3.014 0.334	328.63 −73.59	253.9 −2.9	I/A DBL. (#2) 293-G-46 = B 0011-418	2 S 1.7	F294/J3551 (327,75)
AM 0011-335	00 11 17 −33 51.9	3.029 0.334	352.06 −79.63	261.5 −0.4	DBL. cD in MEDIUM-DISTANT CLUSTER	2,19 D 3.1	F350/J2633 (321,238)
AM 0011-431	00 11 29 −43 10.8	3.010 0.334	326.24 −72.48	252.7 −3.4	I/A DBL. (SMALL SPIRAL) R 11 = DN 0011-433	2 S 0.2	F242/J1048 (338,270)
*AM 0011-232	00 11 33 −23 27.4	3.044 0.334	55.87 −80.68	271.4 2.9	RESOLVED SPIRAL + APP. COMP'NS. NGC 45 = 473-G-1	8 S 4.3	F472/J3428 (37,260)
AM 0011-701	00 11 34 −70 17.9	2.885 0.334	307.59 −46.75	226.5 −11.3	ISOLATED G.D. SPIRAL 50-G-6	10 D 1.9	F050/J0955 (124,160)
*AM 0011-303	00 11 42 −30 39.5	3.033 0.334	8.92 −81.22	264.6 0.6	V.RICH, V.V.DISTANT CLUSTER	19 D 9.0	F409/J2693 (44,142)
AM 0011-620	00 11 46 −62 00.4	2.944 0.334	310.57 −54.83	234.6 −9.1	DWARF (BARRED) 111-G-19	20 D 0.7	F078/J0928 (102,336)
AM 0011-382	00 11 56 −38 29.2	3.018 0.334	335.58 −76.47	257.1 −2.0	LINE of 3 (GROUP of 5 (COMPACTS)) 293-IG-47 = DN 0011-380	13,17,18 D 3.1	F294/J3551 (328,257)

Identification	R.A. (1950.0) Declination	Precession R.A./Dec.	Gal. Long. Gal. Lat.	S.G.L. S.G.B.	Description Cross-Identification	Coding Size (Min.)	Field/Plate (x, y)
*AM 0012-603	00 12 16 −60 36.6	2.946 0.334	311.09 −56.20	236.0 −8.7	SB (DISTURBANCE at END of BAR) NGC 53 = 111-G-2 = Fa 3 [A82]	12 D 2.2	F111/J0921 (95,142)
AM 0012-242	00 12 24 −24 22.1	3.040 0.334	51.11 −81.24	270.6 2.4	GROUP of 8 + PECULIAR SPIRAL MCG-4-1-22 = 473-G-2	12,18 D 4.0	F472/J3428 (28,211)
AM 0012-384	00 12 32 −38 45.8	3.014 0.334	334.48 −76.32	256.9 −2.2	IRR. KNOTTY GALAXY	16 D 0.7	F293/J3578 (49,244)
*AM 0012-235	00 12 38 −23 59.6	3.040 0.333	233.48 −81.14	269.0 −15.1	E + INTEGRAL SPIRAL in CLUSTER	1,19 S 1.0	F473/J3523 (306,232)
AM 0012-573	00 12 48 −57 31.5	2.956 0.333	312.49 −59.19	239.0 −7.9	SPIRAL w. BRIGHT INNER RING + 111-G-22 = Fa 4 FAINT OUTER RING	6,10 D 1.6	F150/J1804 (324,38)
AM 0012-574	00 12 52 −57 44.0	2.954 0.333	312.36 −58.99	238.8 −8.0	SPIRAL w. I/A COMP'N.	1 S 0.2	F111/J0921 (84,296)
AM 0012-422	00 12 56 −42 25.5	3.004 0.333	326.69 −73.27	253.5 −3.4	I/A DBL.	2 S 0.2	F242/J1048 (326,311)
AM 0012-392	00 12 59 −39 24.1	3.011 0.333	332.64 −75.85	256.3 −2.5	ARC of 5 DISTURBED GALAXIES DN 0012-390	17 D 4.0	F293/J3578 (45,210)
*AM 0013-322	00 13 00 −32 27.6	3.025 0.333	357.77 −80.70	262.9 −0.3	DWARF w. 6 STARS to SIDE 410-G-5	20 D 1.8	F409/J2693 (32,45)
AM 0013-424	00 13 00 −42 40.0	3.003 0.333	326.25 −73.06	253.2 −3.5	SMALL CONTACT DBL. w. PLUMES	2,15 S 0.1	F294/J3551 (307,33)
AM 0013-675	00 13 03 −67 50.6	2.886 0.333	308.17 −49.18	228.9 −10.8	GAL. w. I/A COMP'N. (NARROW LOOP)	1 D 0.4	F050/J0955 (110,291)
AM 0013-691	00 13 21 −69 16.3	2.867 0.333	307.68 −47.79	227.5 −11.2	CHAIN of FOUR	17 D 1.0	F050/J0955 (113,214)
AM 0013-620	00 13 22 −62 00.4	2.926 0.333	310.26 −54.88	234.7 −9.3	DWARF?	20 D 0.9	F111/J0921 (92,67)
AM 0013-394	00 13 31 −39 43.2	3.008 0.333	331.56 −75.64	256.1 −2.6	GROUP of 4 PECULIARS	18 D 3.6	F294/J3551 (308,191)
AM 0013-550	00 13 33 −55 04.0	2.960 0.333	313.68 −61.58	241.4 −7.3	I/A TRIPLE (SMALL ONE-ARMED 149-IG-26 SPIRAL)	3,10 D 0.9	F149/J1739 (75,173)
*AM 0013-562	00 13 33 −56 21.5	2.954 0.333	312.93 −60.34	240.1 −7.7	M51-TYPE in CHAIN of 4 149-IG-25 = B 0013-563	2,9,17 D 2.7	F149/J1739 (78,104)
AM 0013-272	00 13 52 −27 24.3	3.031 0.333	30.85 −82.21	267.8 1.2	GALAXY w. I/A COMP'N.	1 S 0.5	F410/J0778 (285,316)
AM 0014-270	00 14 09 −27 07.7	3.031 0.333	32.91 −82.25	268.1 1.2	V. CLOSE PAIR (CONTACT)	23 S 0.4	F473/J3523 (285,64)
AM 0015-415	00 15 00 −41 59.8	2.994 0.333	326.22 −73.83	254.0 −3.6	CLOSE PAIR	23 S 0.2	F293/J3578 (29,70)
*AM 0015-315	00 15 01 −31 53.9	3.019 0.333	359.53 −81.35	263.6 −0.5	GALAXY w. 2 TAILS 410-IG-6	15 D 0.5	F410/J0778 (268,75)
*AM 0015-324	00 15 02 −32 47.7	3.017 0.333	354.56 −80.89	262.8 −0.8	CHAIN of 5 (PEC. SP. + COMPACT)	12,13,17 D 10.3	F410/J0778 (267,26)
*AM 0015-352	00 15 06 −35 28.9	3.010 0.333	342.64 −79.16	260.2 −1.6	CONSPICUOUS CLUSTER	19 D 4.0	F350/J2633 (276,152)
AM 0015-615	00 15 12 −61 59.9	2.906 0.333	309.89 −54.94	234.7 −9.5	CLOSE DBL. B 0015-620 in DN 0015-620	23 S 1.3	F111/J0921 (81,67)
AM 0015-372	00 15 25 −37 26.8	3.004 0.333	335.98 −77.71	258.4 −2.3	COMPACT + PLUMES	13,15 D 0.2	F294/J3551 (293,314)
AM 0015-340	00 15 34 −34 09.2	3.011 0.333	347.62 −80.14	261.5 −1.3	RINGED SPIRAL + COMP'NS. 350-G-7	8 S 8.1	F350/J2633 (273,224)
AM 0015-550	00 15 38 −55 06.1	2.942 0.333	313.06 −61.63	241.4 −7.6	FAN-SHAPED ARMS 149-IG-28	10 D 0.6	F150/J1804 (312,169)
AM 0015-434	00 15 43 −43 42.6	2.985 0.333	323.18 −72.35	252.4 −4.3	M51-TYPE 242-IG-1	9 S 0.2	F242/J1048 (295,243)
AM 0015-450	00 15 49 −45 08.5	2.980 0.333	321.26 −71.05	251.0 −4.7	3-ARMED SPIRAL 242-IG-2 = B 0015-451	11 S 1.1	F242/J1048 (292,167)
AM 0015-571	00 15 59 −57 13.9	2.928 0.333	311.83 −59.59	239.4 −8.3	CLOSE GROUP of 4	17,18 D 4.0	F150/J1804 (302,55)
AM 0016-674	00 16 02 −67 40.1	2.846 0.333	307.80 −49.41	229.2 −11.0	I/A TRIPLET	3 D 1.0	F050/J0955 (95,300)
AM 0016-475	00 16 04 −47 56.0	2.969 0.333	318.15 −68.48	248.4 −5.6	DWARF/SPIRAL? 194-G-2	20 D 0.6	F242/J1048 (284,17)
AM 0016-361	00 16 07 −36 13.2	3.004 0.333	339.24 −78.75	259.6 −2.0	I/A DBL. + I/A COMP'N. 350-IG-8	1,2 D 1.3	F350/J2633 (264,113)
AM 0016-670	00 16 09 −67 03.5	2.851 0.333	307.97 −50.01	229.8 −10.9	SPIRAL w. PECULIAR ARM	10 D 0.4	F050/J0955 (92,332)
AM 0016-310	00 16 12 −31 00.6	3.016 0.333	4.19 −81.98	264.5 −0.5	PAIR of DBL's.	2,23 S 3.4	F410/J0778 (255,122)
AM 0016-231	00 16 33 −23 12.8	3.032 0.333	240.87 −81.57	268.0 −14.5	CHAIN of 6 (incl. PEC. SPIRAL) NGC 66 = MCG-4-2-2 = 473-G-10	12,17 D 17.9	F473/J3523 (259,275)
AM 0016-395	00 16 36 −39 56.6	2.992 0.333	329.05 −75.77	256.0 −3.3	E + SPIRAL	2 S 0.3	F294/J3551 (276,181)

Identification	R.A. (1950.0) Declination	Precession R.A./Dec.	Gal. Long. Gal. Lat.	S.G.L. S.G.B.	Description Cross-Identification	Coding Size (Min.)	Field/Plate (x, y)
AM 0016-270	00 16 39 −27 01.4	3.024 0.333	214.05 −82.80	271.7 −13.3	I/A TRIPLE in a LINE	3 D 3.8	F410/J0778 (252,337)
AM 0016-614	00 16 40 −61 40.5	2.893 0.333	309.73 −55.29	235.1 −9.5	I/A DBL. Fa 6	2 S 0.2	F111/J0921 (70,84)
AM 0016-365	00 16 42 −36 52.6	3.000 0.333	336.69 −78.32	259.0 −2.4	CONTACT DBL.	23 S 0.2	F350/J2633 (257,78)
AM 0016-575	00 16 43 −57 55.4	2.917 0.333	311.31 −58.94	238.7 −8.5	'ELLIPTICAL' w. VERY FAINT OUTER SPIRAL STRUCTURE + COMP'N.	6,8,12 D 1.7	F150/J1804 (295,19)
AM 0016-265	00 16 48 −26 59.2	3.023 0.333	214.36 −82.83	271.7 −13.3	CLOSE PAIR in GROUP of 3	23,24 S 1.2	F473/J3523 (253,72)
AM 0016-694	00 16 54 −69 43.9	2.806 0.333	307.10 −47.40	227.2 −11.6	I/A DBL. (in a GROUP) THIN ARC 50-IG-8	2,15,18 S 0.2	F050/J0955 (98,189)
AM 0016-373	00 16 57 −37 31.2	2.997 0.333	334.63 −77.83	258.4 −2.6	COMPACT	13 D 0.1	F350/J2633 (254,43)
AM 0017-615	00 17 10 −61 58.6	2.885 0.333	309.51 −55.01	234.8 −9.7	CLOSE DBL.	23 S 1.6	F111/J0921 (68,67)
AM 0017-513	00 17 12 −51 32.6	2.947 0.333	314.79 −65.11	244.9 −6.8	GROUP of 5 194-G-4	18 D 6.7	F193/J3714 (36,92)
*AM 0017-454	00 17 23 −45 44.2	2.969 0.333	319.80 −70.61	250.5 −5.2	M51-TYPE (2 COMP'NS. on ARM ?)	9 S 0.7	F242/J1048 (276,135)
AM 0017-772	00 17 54 −77 22.3	2.608 0.333	305.21 −39.88	219.6 −13.4	LO.S.BR. SBd + 6 COMP'NS 28-G-14 = UKS 0017-773	20 D 1.1	F012/J2486 (128,319)
AM 0018-370	00 18 02 −37 00.6	2.994 0.333	335.24 −78.37	258.9 −2.7	GALAXY w. 4 COMP'NS.	18 S 0.1	F294/J3551 (266,339)
*AM 0018-313	00 18 07 −31 39.1	3.008 0.333	358.73 −82.05	264.0 −1.0	CLOSE TRIPLE(RING + MULTI-NUCLEUS) 410-IG-8	24 D 5.4	F410/J0778 (232,88)
AM 0018-344	00 18 07 −34 44.9	3.000 0.333	343.00 −80.11	261.1 −2.0	I/A DBL. 350-IG-10	2 S 0.4	F350/J2633 (244,192)
AM 0018-255	00 18 12 −25 59.5	3.021 0.333	222.72 −82.98	270.6 −13.3	TRIPLE cD (V. RICH CLUSTER)	19 D 9.9	F473/J3523 (237,126)
AM 0018-640	00 18 35 −64 08.1	2.850 0.333	308.49 −52.93	232.7 −10.4	PEC. GALAXY w. CONDENS.? in JET 78-G-22	7,16 D 2.0	F078/J0928 (67,219)
AM 0018-412	00 18 50 −41 23.6	2.976 0.333	324.89 −74.69	254.8 −4.1	ONE-ARMED SPIRAL w. COMP'N.?	8,10 D 0.5	F294/J3551 (251,103)
AM 0018-391	00 18 51 −39 16.3	2.983 0.333	328.91 −76.56	256.8 −3.5	I/A DBL. (FILAMENT) 294-IG-3	2 S 0.4	F294/J3551 (254,217)
*AM 0018-485	00 18 55 −48 55.2	2.947 0.333	316.10 −67.70	247.6 −6.3	ROBERT'S I/A QUARTET DN 0019-489 NGC 87/8/9/92 = 194-G-8/10/11/12	4 D 3.6	F194/J1861 (277,235)
AM 0019-570	00 19 46 −57 05.3	2.895 0.333	310.91 −59.85	239.7 −8.7	I/A CONTACT DBL. 150-IG-4	2 S 0.2	F150/J1804 (275,65)
AM 0019-812	00 19 56 −81 23.9	2.306 0.333	304.34 −35.91	215.5 −14.2	I/A DBL. 12-IG-16 = B 0019-813	2 S 0.3	F013/J2551 (317,83)
AM 0020-535	00 20 01 −53 55.2	2.913 0.333	312.40 −62.94	242.8 −7.9	Sd/DWARF + COMPACT 150-G-5 = UKS 0020-539	13,20 D 2.7	F149/J1739 (21,232)
AM 0020-242	00 20 08 −24 24.5	3.020 0.333	236.05 −82.86	268.9 −13.4	GALAXY w. I/A COMP'N. (JET ?) MCG-4-2-8/9 = 473-G-18	1,7 S 0.3	F473/J3523 (214,211)
AM 0020-605	00 20 16 −60 53.1	2.861 0.333	309.27 −56.15	236.0 −9.8	I/A DBL. (SPIRAL + LENTICULAR) B 0020-608?	2 S 0.8	F111/J0921 (44,124)
AM 0020-395	00 20 31 −39 58.8	2.973 0.333	326.28 −76.10	256.2 −4.0	I/A DBL. (SMALL)	2 S 0.3	F294/J3551 (236,180)
AM 0020-425	00 20 32 −42 58.2	2.962 0.333	321.48 −73.38	253.4 −4.9	SMALL CONTACT DBL. 242-IG-6	20 S 0.3	F242/J1048 (250,285)
AM 0020-695	00 20 32 −69 58.4	2.745 0.333	306.58 −47.22	227.0 −12.0	M51-TYPE	9 S 0.2	F050/J0955 (82,174)
AM 0020-571	00 20 43 −57 17.3	2.885 0.333	310.57 −59.68	239.5 −8.9	E w. 6 APP. COMP'NS. (in LINE) DN 0020-573	8 S 1.6	F150/J1804 (268,54)
AM 0021-343	00 21 02 −34 31.4	2.989 0.333	341.28 −80.69	261.5 −2.5	CLOSE GROUP of 4	18 D 2.0	F350/J2633 (212,205)
AM 0021-624	00 21 05 −62 46.3	2.834 0.333	308.47 −54.32	234.1 −10.3	I/A DBL. (COMPACT)	2,13 S 0.1	F111/J0921 (47,23)
*AM 0021-571	00 21 29 −57 10.0	2.879 0.333	310.42 −59.82	239.7 −9.0	I/A SPIRALS in LINE of 3 150-G-6 (I/A COMP'N.)	1,2,9,18 S 1.3	F150/J1804 (263,61)
*AM 0021-623	00 21 33 −62 33.0	2.831 0.333	308.45 −54.55	234.4 −10.3	M51-TYPE w. COMPACT 78-IG-23 [A82]	9,13 S 0.8	F111/J0921 (43,34)
AM 0021-395 [0022-4000]	00 21 50 −39 58.2	2.966 0.332	325.35 −76.22	256.3 −4.2	DISTORTED PAIR in TRIPLE (#1)	23 S 1.6	F294/J3551 (222,181)
AM 0022-402	00 22 04 −40 24.2	2.964 0.332	324.41 −75.85	255.9 −4.4	I/A DBL. in CLUSTER	2,19 S 0.1	F294/J3551 (219,157)
AM 0022-395 [0021-3958]	00 22 14 −39 59.1	2.964 0.332	325.03 −76.24	256.3 −4.3	DISTORTED PAIR in TRIPLE (#2)	2,23 S 1.6	F294/J3551 (218,180)
AM 0022-400 [0022-3959]	00 22 19 −40 00.7	2.964 0.332	324.92 −76.22	256.3 −4.3	DISTORTED PAIR in TRIPLE (#3)	2,23 S 1.6	F294/J3551 (217,178)

Identification	R.A. (1950.0) Declination	Precession R.A./Dec.	Gal. Long. Gal. Lat.	S.G.L. S.G.B.	Description Cross-Identification	Coding Size (Min.)	Field/Plate (x, y)
AM 0022-572	00 22 20 −57 29.7	2.869 0.332	310.06 −59.53	239.4 −9.1	CHAIN of 7 (SPIRAL to I/A TRIPLE) R 25 = DN 0022-571	3,17 D 9.8	F111/J0921 (14,305)
AM 0022-422	00 22 32 −42 29.2	2.953 0.332	320.92 −73.97	253.9 −5.1	GALAXY w. I/A COMP'N.	1 S 0.7	F242/J1048 (230,311)
AM 0022-694	00 22 38 −69 42.3	2.717 0.332	306.38 −47.51	227.3 −12.1	DWARF (RESOLVED) LOOP?	15,20 D 0.6	F050/J0955 (71,188)
AM 0023-690	00 23 01 −69 04.8	2.723 0.332	306.47 −48.13	227.9 −12.0	I/A TRIPLE + COMPACT COMP'NS 50-IG-11	3,8,13,17 D 0.4	F051/J1877 (318,220)
AM 0023-331	00 23 02 −33 19.4	2.985 0.332	344.73 −81.84	262.8 −2.5	cD + 14 GALAXIES 350-G-15 = PKS 0023-33	19 D 11.2	F350/J2633 (190,269)
AM 0023-332	00 23 04 −33 24.9	2.984 0.332	344.25 −81.78	262.7 −2.6	BRIGHT-RIMMED SPIRAL 350-G-16	11 D 0.7	F350/J2633 (190,265)
AM 0023-451	00 23 51 −45 18.6	2.933 0.332	316.96 −71.39	251.3 −6.1	CLOSE PAIR	23 S 0.1	F242/J1048 (215,160)
AM 0023-631	00 23 54 −63 12.8	2.797 0.332	307.80 −53.94	233.8 −10.7	GROUP of 5	18 D 5.4	F079/J1793 (299,267)
AM 0024-420	00 24 05 −42 8.2	2.946 0.332	320.40 −74.40	254.4 −5.3	DWARF 294-G-10	20 D 0.9	F294/J3551 (198,65)
AM 0024-602	00 24 08 −60 20.0	2.826 0.332	308.61 −56.78	236.6 −10.1	PECULIAR SPIRAL (BRIGHT KNOTS)	12 D 0.7	F112/J1731 (268,157)
AM 0024-684	00 24 44 −68 42.2	2.704 0.332	306.32 −48.53	228.3 −12.0	COMPACT (FUZZ)	13 D 0.1	F050/J0955 (56,241)
AM 0024-473	00 24 54 −47 37.0	2.914 0.332	314.51 −69.23	249.1 −6.9	DISTURBED SINGLE Fa 653	16 D 0.4	F194/J1861 (226,307)
AM 0025-472	00 25 07 −47 23.7	2.914 0.332	314.57 −69.45	249.3 −6.9	DWARF/DEFECT?	20 D 0.2	F194/J1861 (224,319)
AM 0025-385	00 25 20 −38 57.9	2.954 0.332	324.58 −77.42	257.5 −4.6	VERY SMALL I/A KNOTTY DWARFS 294-IG-13	2,20 S 0.3	F294/J3551 (187,235)
AM 0025-443	00 25 25 −44 32.3	2.927 0.332	316.86 −72.21	252.1 −6.2	CLOSE I/A TRIPLE	3 D 0.3	F242/J1048 (201,201)
AM 0025-505	00 25 28 −50 51.7	2.891 0.332	312.19 −66.11	246.0 −7.9	I/A DBL. w. MATERIAL 194-IG-18	2 D 0.4	F194/J1861 (217,133)
AM 0025-472	00 25 47 −47 20.0	2.910 0.332	314.30 −69.54	249.4 −7.0	DWARF	20 D 0.4	F194/J1861 (218,322)
AM 0026-274	00 26 03 −27 47.3	2.993 0.332	27.16 −84.90	268.2 −1.5	I/A TRIPLE in a LINE MGC-5-2-9 = 410-G-13	3 D 1.1	F410/J0778 (140,296)
AM 0026-235	00 26 06 −23 55.4	3.006 0.332	245.61 −83.79	268.0 −12.2	VERY RICH, FAINT CLUSTER	19 D 9.0	F473/J3523 (141,237)
AM 0026-411	00 26 12 −41 11.2	2.940 0.332	320.28 −75.43	255.4 −5.4	I/A TRIPLE 294-IG-14	3 D 0.7	F294/J3551 (177,116)
AM 0026-324	00 26 17 −32 41.0	2.975 0.332	344.48 −82.78	263.6 −3.0	G.D. SPIRAL + 2 DIAMETRICS 350-G-20	8 S 5.8	F350/J2633 (154,304)
AM 0026-272	00 26 32 −27 26.1	2.993 0.332	31.16 −85.02	268.6 −1.5	GAL. w. I/A COMP'N. (LINE of 3)	1,24 S 0.2	F410/J0778 (134,315)
AM 0026-310	00 26 42 −31 07.8	2.979 0.332	354.08 −83.87	265.1 −2.6	I/A DBL.	2 S 0.7	F410/J0778 (134,116)
AM 0027-370	00 27 40 −37 09.4	2.951 0.332	326.38 −79.25	259.3 −4.5	SPIRAL w. 3 COMP'NS. 350-G-22	18 S 2.3	F294/J3551 (162,332)
AM 0027-333	00 27 53 −33 31.3	2.966 0.332	338.26 −82.37	262.9 −3.6	SP. w. MATR'L + DIAMETRIC COMP'NS NGC 134 = 350-G-23	8 S 17.9	F350/J2633 (136,259)
AM 0027-680	00 27 53 −68 01.2	2.671 0.332	306.02 −49.24	229.1 −12.2	I/A DBL.	2 S 0.3	F050/J0955 (37,275)
AM 0028-410	00 28 00 −41 09.6	2.931 0.332	319.04 −75.57	255.5 −5.7	I/A DBL. in WIDE GROUP of 8 294-IG-18 = B 0028-411	2,18 S 0.4	F294/J3551 (159,117)
AM 0028-372	00 28 02 −37 24.7	2.948 0.332	325.45 −79.05	259.1 −4.7	I/A DBL. in GROUP of 3	2,24 S 1.0	F294/J3551 (158,318)
AM 0028-533	00 28 02 −53 39.1	2.851 0.332	309.95 −63.45	243.3 −9.0	FAINT LINEAR CLUSTER DN 0028-536	19 D 13.4	F150/J1804 (219,251)
AM 0028-370 [0028-3710]	00 28 19 −37 07.5	2.948 0.331	325.82 −79.33	259.4 −4.7	TRIPLE (#1: E)	24 S 5.4	F350/J2633 (133,65)
AM 0028-424	00 28 28 −42 45.6	2.920 0.331	316.84 −74.07	254.0 −6.2	M51-TYPE (SMALL)	9 S 0.3	F242/J1048 (172,297)
AM 0028-443	00 28 33 −44 33.1	2.910 0.331	315.09 −72.35	252.2 −6.7	COMPACT w. I/A COMP'N.	1,13 S 0.1	F242/J1048 (171,201)
AM 0028-333	00 28 35 −33 32.7	2.963 0.331	337.28 −82.44	262.9 −3.7	NUCLEATED DWARF 350-G-26	20 D 1.0	F350/J2633 (128,257)
AM 0028-371 [0028-3707]	00 28 35 −37 10.0	2.947 0.331	325.48 −79.32	259.4 −4.7	TRIPLE (#2: S0) 350-G-27	24 S 5.4	F350/J2633 (130,63)
AM 0028-225 [0028-2255]	00 28 37 −22 53.9	3.003 0.331	256.14 −83.57	266.8 −11.9	TRIPLE (#1: SPIRAL w. STRONG ARM) NGC 142 = 473-G-21	10,24 D 5.6	F473/J3523 (110,292)
AM 0028-225 [0028-2253]	00 28 46 −22 50.4	3.003 0.331	256.71 −83.55	266.8 −11.9	TRIPLE (#2: EDGE-ON SPIRAL) NGC 143 = 473-G-22	24 D 5.6	F473/J3523 (108,295)

Identification	R.A. (1950.0) Declination	Precession R.A./Dec.	Gal. Long. Gal. Lat.	S.G.L. S.G.B.	Description Cross-Identification	Coding Size (Min.)	Field/Plate (x, y)
AM 0028-564	00 28 46 −56 43.6	2.818 0.331	308.63 −60.44	240.3 −9.8	GANG of 4 incl. 150-G-10	18 D 4.0	F150/J1804 (210,86)
AM 0028-370 [0028-3710]	00 28 48 −37 07.7	2.946 0.331	325.35 −79.37	259.4 −4.7	TRIPLE (#3: S0/a) 350-G-28	24 S 5.4	F350/J2633 (128,65)
AM 0028-225 [0028-2250]	00 28 51 −22 55.5	3.002 0.331	256.30 −83.62	266.9 −11.9	TRIPLE (#3: SPIRAL) NGC 144 = 473-G-23	24 D 5.6	F473/J3523 (107,290)
AM 0028-230	00 28 53 −23 02.4	3.002 0.331	255.60 −83.71	267.0 −11.9	DWARF 473-G-24	20 D 1.1	F473/J3523 (107,284)
AM 0029-333	00 29 27 −33 32.5	2.959 0.331	336.18 −82.55	262.9 −3.9	DWARF 350-G-31	20 D 0.4	F350/J2633 (118,258)
AM 0029-442	00 29 30 −44 26.2	2.905 0.331	314.64 −72.51	252.4 −6.8	GALAXY w. I/A COMP'N. Ag-4 = 242-IG-11	1 S 0.3	F242/J1048 (162,207)
AM 0029-280	00 29 38 −28 00.2	2.981 0.331	23.70 −85.67	268.3 −2.4	SMALL I/A DBL. w. TAIL	2,15 S 1.1	F410/J0778 (98,284)
AM 0029-403	00 29 44 −40 32.4	2.925 0.331	318.58 −76.26	256.2 −5.9	Sd-DWARF (RESOLVED) 294-G-20	20 D 1.3	F294/J3551 (141,150)
AM 0029-643	00 29 45 −64 35.5	2.709 0.331	306.40 −52.67	232.6 −11.7	CLOSE PAIR 79-G-2/3	23 S 10.3	F079/J1793 (259,196)
*AM 0029-664	00 29 45 −66 40.8	2.672 0.331	306.00 −50.60	230.5 −12.1	COMPACT CHAIN of COMPACTS DN 0030-666	13,17 D 2.9	F078/J0928 (19,77)
AM 0029-643	00 29 46 −64 31.8	2.710 0.331	306.41 −52.73	232.6 −11.7	GALAXY + PECULIAR COMP'N. 79-G-3 (JET w. CONDENSATION)	1,7,8 S 9.0	F078/J0928 (5,192)
AM 0030-453	00 30 13 −45 30.5	2.894 0.331	313.41 −71.50	251.4 −7.2	GALAXY w. 2 I/A COMP'NS. 242-G-12	1 S 0.6	F242/J1048 (155,149)
AM 0030-394	00 30 19 −39 45.4	2.926 0.331	319.20 −77.04	257.0 −5.7	PAIR of E's	23 S 1.2	F294/J3551 (135,192)
AM 0030-681	00 30 30 −68 14.8	2.629 0.331	305.61 −49.05	228.9 −12.5	CLOSE PAIR (DWARF) 51-IG-1	20,23 S 0.7	F050/J0955 (25,262)
AM 0030-323	00 30 41 −32 32.4	2.959 0.331	339.46 −83.50	264.0 −3.8	GALAXY w. PECULIAR ABS'N. IC 1554 = MGC-5-2-15 = 350-G-33?	14 D 1.1	F410/J0778 (90,40)
AM 0030-565	00 30 42 −56 57.8	2.799 0.331	308.03 −60.25	240.2 −10.1	SPIRAL w. PECULIAR ARMS (SMALL)	10 D 0.4	F112/J1731 (229,340)
AM 0030-260	00 30 46 −26 06.5	2.985 0.331	229.41 −85.73	269.8 −10.6	COMPACT	13 D 0.2	F474/J1915 (336,116)
AM 0030-223	00 30 49 −22 38.2	2.998 0.331	261.10 −83.71	266.4 −11.5	GALAXY w. TAIL 473-G-27	15 D 1.3	F473/J3523 (82,306)
AM 0030-565 [0031-5701]	00 30 53 −56 57.6	2.797 0.331	307.98 −60.25	240.2 −10.1	GROUP of 3 (#1)	24 S 0.6	F150/J1804 (194,74)
AM 0030-570 [0030-5657]	00 30 53 −57 07.8	2.795 0.331	307.93 −60.08	240.0 −10.2	GROUP of 3 (#2)	24 S 0.6	F150/J1804 (194,65)
AM 0030-782	00 30 57 −78 29.4	2.190 0.331	304.16 −38.86	218.5 −14.2	I/A TRIPLE in GROUP of 6 12-IG-19	3,18 D 0.9	F012/J2486 (97,255)
AM 0031-420	00 31 06 −42 06.3	2.910 0.331	315.78 −74.84	254.7 −6.5	HI.S.BR. LINEAR OBJECT	22 D 0.6	F242/J1048 (145,332)
*AM 0031-280	00 31 09 −28 04.4	2.976 0.331	22.11 −85.99	268.3 −2.7	I/A DBL. 410-G-17	2 S 1.9	F410/J0778 (80,280)
*AM 0031-570 [0030-5707]	00 31 10 −57 01.2	2.794 0.331	307.88 −60.20	240.1 −10.2	GROUP of 3 (#3: DISTURBED GAL. w. CONDENSATION in TAIL [S])[A82]	15,24 S 0.6	F150/J1804 (192,71)
AM 0031-530	00 31 15 −53 04.5	2.831 0.331	309.10 −64.10	244.0 −9.3	PECULIAR SPIRAL w. I/A COMP'N.	1,12 S 1.3	F150/J1804 (194,283)
AM 0031-343	00 31 33 −34 33.1	2.947 0.331	329.45 −81.92	262.1 −4.6	I/A DBL. (MIDDLE of CLUSTER) 350-IG-34 = PKS 0031-345	2,19 S 0.2	F350/J2633 (96,203)
AM 0031-381 [0031-3812]	00 31 41 −38 10.5	2.928 0.331	320.49 −78.62	258.6 −5.6	GROUP of 3 (#1)	24 D 3.3	F294/J3551 (120,277)
AM 0031-310	00 31 43 −31 03.0	2.962 0.331	347.80 −84.74	265.5 −3.6	DWARF (CLUMPS) UKS 0031-310 MCG-5-2-16 = 410-G-18	20 D 1.8	F411/J0768 (341,118)
*AM 0031-280	00 31 47 −28 05.0	2.974 0.331	21.64 −86.13	268.3 −2.8	HI.S.BR. 'MERGER' + 5 COMP'NS. NGC 150 = 410-G-19 [A82]	8,10 D 4.0	F473/J3523 (75,13)
AM 0031-512	00 31 47 −51 24.2	2.842 0.331	309.55 −65.77	245.7 −9.0	PECULIAR SPIRAL	10 D 0.5	F194/J1861 (164,104)
AM 0031-320	00 31 48 −32 03.8	2.957 0.331	340.56 −84.02	264.5 −3.9	DOMINANT LENS NGC 148 = 410-G-20	13,22 D 1.9	F350/J2633 (91,337)
AM 0031-380 [0031-3810]	00 31 50 −38 09.7	2.928 0.331	320.38 −78.64	258.6 −5.6	GROUP of 3 (#2)	24 D 3.3	F294/J3551 (118,278)
AM 0031-381 [0031-3809]	00 31 54 −38 12.2	2.927 0.331	320.24 −78.61	258.6 −5.6	GROUP of 3 (#3)	24 D 3.3	F294/J3551 (118,276)
*AM 0031-782	00 31 58 −78 24.1	2.168 0.331	304.10 −38.95	218.6 −14.3	2 I/A TRIPLES in CHAIN of 13	3,17 D 0.6	F013/J2551 (335,245)
*AM 0031-782	00 31 59 −78 27.9	2.162 0.331	304.09 −38.89	218.6 −14.3	I/A TRIPLE (in AM 0031-7824)	3,17 D 0.5	F013/J2551 (334,242)
AM 0032-255	00 32 03 −25 50.9	2.983 0.331	234.16 −85.91	269.5 −10.4	CLOSE I/A DBL. (2 E's)	2 S 0.3	F473/J3523 (70,133)

Identification	R.A. (1950.0) Declination	Precession R.A./Dec.	Gal. Long. Gal. Lat.	S.G.L. S.G.B.	Description Cross-Identification	Coding Size (Min.)	Field/Plate (x, y)
*AM 0032-301	00 32 08 −30 17.7	2.964 0.331	353.78 −85.30	266.2 −3.5	SPIRAL w. DIAMETRIC COMP'NS. IC 1555 = 410-G-21	8 S 4.5	F410/J0778 (71,160)
AM 0032-224	00 32 17 −22 46.6	2.994 0.331	262.73 −84.04	266.5 −11.2	CLOSE GROUP of 6	18 D 2.7	F474/J1915 (321,295)
AM 0032-675	00 32 31 −67 55.7	2.607 0.331	305.38 −49.38	229.3 −12.6	CHAIN of 8 DN 0033-679	17 D 6.7	F051/J1877 (278,286)
AM 0032-424	00 32 39 −42 41.9	2.898 0.331	314.18 −74.33	254.2 −6.9	COMPACT? CLOSE TRIPLE?	13,24 D 0.1	F242/J1048 (130,300)
AM 0032-442	00 32 55 −44 21.4	2.886 0.331	312.71 −72.73	252.6 −7.4	M51-TYPE (STAR?) 242-G-17	9 D 0.7	F242/J1048 (129,211)
AM 0033-441	00 33 14 −44 18.7	2.884 0.330	312.56 −72.78	252.7 −7.5	GALAXY w. I/A COMP'N.	1 S 0.2	F242/J1048 (126,213)
AM 0033-253	00 33 19 −25 39.1	2.980 0.330	238.45 −86.09	269.2 −10.2	DWARF SPIRAL UKS 0033-256	20 D 4.0	F473/J3523 (54,143)
AM 0033-372	00 33 24 −37 24.8	2.925 0.330	320.15 −79.45	259.4 −5.7	CONTACT I/A DBL.	2 S 0.1	F294/J3551 (101,318)
AM 0033-455	00 33 26 −45 56.8	2.872 0.330	311.41 −71.19	251.1 −7.9	I/A DBL.	2 S 0.2	F242/J1048 (126,125)
AM 0033-312	00 33 34 −31 28.6	2.954 0.330	341.47 −84.71	265.2 −4.1	COMPACT	13 D 0.2	F411/J0768 (319,96)
AM 0033-262	00 33 37 −26 23.8	2.976 0.330	228.23 −86.43	269.9 −9.9	E + SPIRAL in GROUP of 13	2,18,23 D 6.7	F474/J1915 (301,101)
*AM 0033-324 [0033-3252]	00 33 37 −32 49.2	2.947 0.330	333.53 −83.63	263.9 −4.5	TRIPLE (#1: E)	24 S 4.0	F350/J2633 (71,296)
AM 0033-315	00 33 38 −31 54.7	2.951 0.330	338.47 −84.38	264.7 −4.3	CLOSE TRIPLE	24 D 2.0	F410/J0778 (55,73)
AM 0033-262	00 33 40 −26 21.2	2.976 0.330	228.97 −86.43	269.9 −9.9	I/A TRIPLE	3 D 0.4	F474/J1915 (301,103)
AM 0033-325 [0033-3251]	00 33 41 −32 52.5	2.947 0.330	333.17 −83.59	263.8 −4.5	TRIPLE (#2: SPIRAL) 350-G-37	24 S 4.0	F350/J2633 (70,293)
AM 0033-325 [0033-3249]	00 33 44 −32 51.1	2.946 0.330	333.19 −83.62	263.8 −4.5	TRIPLE (#3: SPIRAL)	24 S 4.0	F350/J2633 (70,294)
AM 0033-325	00 33 56 −32 50.7	2.946 0.330	332.89 −83.65	263.9 −4.6	CLOSE TRIPLE	24 D 4.0	F410/J0778 (53,23)
AM 0033-280 [0034-2805]	00 33 59 −28 03.7	2.968 0.330	20.66 −86.61	268.5 −3.3	TRIPLE (#1: SPIRAL) MCG-5-2-23 = 410-G-24	24 D 3.9	F473/J3523 (49,14)
AM 0034-592	00 34 03 −59 28.0	2.738 0.330	306.56 −57.82	237.8 −11.1	GALAXY w. I/A COMP'N. B 0034-594	1 S 0.1	F112/J1731 (203,206)
AM 0034-225 [0035-2249]	00 34 09 −22 52.2	2.989 0.330	265.55 −84.37	266.5 −10.7	GROUP of 3 EDGE-ONS (#1) NGC 168 = 474-G-4	24 D 13.4	F473/J3523 (41,293)
AM 0034-260	00 34 10 −26 02.9	2.976 0.330	234.19 −86.43	269.6 −9.9	COMPACT	13 D 0.2	F474/J1915 (295,120)
*AM 0034-280 [0033-2803]	00 34 10 −28 04.0	2.967 0.330	20.45 −86.65	268.5 −3.3	TRIPLE (#2: SP. w. LO.S.BR. ARMS) MCG-5-2-24 = 410-G-25	24 D 3.9	F473/J3523 (46,13)
*AM 0034-280 [0034-2804]	00 34 13 −28 05.6	2.967 0.330	19.95 −86.66	268.5 −3.4	TRIPLE (#3: EDGE-ON SPIRAL) MCG-5-2-26 = 410-G-26	24 D 3.9	F473/J3523 (46,12)
*AM 0034-283	00 34 22 −28 38.7	2.964 0.330	10.86 −86.54	267.9 −3.5	SPIRAL w. PEC. ARMS + APP.COMP'NS MGC-5-2-27 = 410-G-27	8,10 S 4.5	F410/J0778 (42,248)
AM 0034-334	00 34 24 −33 49.9	2.939 0.330	328.06 −82.83	262.9 −4.9	COMPACT I/A DBL. 350-IG-38	2,13 S 0.3	F351/J0739 (327,239)
AM 0034-294	00 34 32 −29 45.4	2.958 0.330	355.42 −86.04	266.9 −3.9	SPIRAL + COMPACTS NGC 174 = 411-G-1	8,13 S 13.0	F411/J0768 (310,188)
AM 0034-324	00 34 35 −32 48.4	2.943 0.330	332.01 −83.75	263.9 −4.7	DISTURBED SINGLE w. JETS 350-G-39	7,16 D 0.4	F411/J0768 (305,25)
AM 0034-351	00 34 38 −35 12.9	2.931 0.330	323.48 −81.59	261.6 −5.4	COMPACT w. HI.S.BR. RING(?)	6,8,13 D 0.6	F350/J2633 (63,167)
AM 0034-225 [0034-2252]	00 34 44 −22 51.9	2.988 0.330	266.72 −84.45	266.4 −10.6	GROUP of 3 EDGE-ONS (#2) NGC 172 = 474-G-5	24 D 13.4	F473/J3523 (34,293)
AM 0034-465 [0035-4647]	00 34 44 −46 55.2	2.857 0.330	310.22 −70.27	250.2 −8.4	CLOSE PAIR of SPIRALS (#1) 242-G-18	23 D 4.0	F242/J1048 (114,73)
AM 0034-434	00 34 45 −43 48.9	2.879 0.330	311.99 −73.32	253.2 −7.6	I/A DBL. w. MATERIAL 242-IG-19	2 S 0.2	F242/J1048 (111,239)
AM 0034-533	00 34 57 −53 32.1	2.798 0.330	307.71 −63.72	243.7 −9.9	MULTI-ARMED ("PARALLEL") SPIRAL 150-G-15 + PECULIAR COMP'N.	8,10 D 0.9	F150/J1804 (164,258)
AM 0035-284	00 35 04 −28 48.2	2.961 0.330	7.45 −86.62	267.8 −3.7	RICH CLUSTER	19 D 9.0	F410/J0778 (34,239)
AM 0035-224 [0034-2251]	00 35 05 −22 49.4	2.987 0.330	267.63 −84.46	266.4 −10.5	GROUP of 3 EDGE-ONS (#3) NGC 177 = 474-G-6	24 D 13.4	F473/J3523 (30,295)
AM 0035-515	00 35 08 −51 57.3	2.812 0.330	308.11 −65.29	245.3 −9.6	SPIRAL w. I/A COMP'N.	1 S 0.1	F150/J1804 (162,343)
*AM 0035-335	00 35 14 −33 59.4	2.935 0.330	326.26 −82.77	262.8 −5.2	'CARTWHEEL' RING + I/A COMP'NS. 350-G-40= PKS0035-340= Lu 0035-34	1,6 D 1.8	F350/J2633 (54,232)

Identification	R.A. (1950.0) Declination	Precession R.A./Dec.	Gal. Long. Gal. Lat.	S.G.L. S.G.B.	Description Cross-Identification	Coding Size (Min.)	Field/Plate (x, y)
AM 0035-434	00 35 18 −43 46.8	2.877 0.330	311.67 −73.38	253.3 −7.7	DWARF	20 D 0.4	F242/J1048 (106,241)
AM 0035-335	00 35 19 −33 56.5	2.935 0.330	326.32 −82.82	262.9 −5.2	OVAL COMPACT	13 D 0.3	F350/J2633 (53,235)
AM 0035-252	00 35 31 −25 21.4	2.975 0.330	246.19 −86.36	268.8 −9.8	GROUP of 10 E's PKS 0035-252?	18 D 2.7	F474/J1915 (279,157)
AM 0035-340	00 35 40 −34 00.1	2.933 0.330	325.58 −82.79	262.8 −5.2	COMPACT	13 D 0.1	F350/J2633 (49,232)
AM 0035-375	00 35 41 −37 57.3	2.911 0.330	316.96 −79.07	259.0 −6.3	COMPACT	13 D 0.3	F350/J2633 (55,19)
AM 0035-464 [0034-4655]	00 35 41 −46 47.6	2.852 0.330	309.80 −70.42	250.4 −8.5	CLOSE PAIR of SPIRALS (#2) 242-G-20	23 D 4.0	F242/J1048 (106,79)
AM 0035-431	00 35 42 −43 11.8	2.878 0.330	311.79 −73.96	253.9 −7.6	I/A DBL.	2 S 0.2	F242/J1048 (101,272)
AM 0035-555	00 35 50 −55 59.5	2.765 0.330	306.83 −61.30	241.3 −10.6	SMALL I/A TRIPLE	3 D 0.6	F150/J1804 (158,126)
*AM 0036-243 [0036-2433]	00 36 05 −24 37.0	2.977 0.330	255.97 −85.99	268.1 −9.8	CLOSE PAIR of SPIRALS (#1) IC 1561 = 474-G-8	23 S 4.2	F473/J3523 (19,198)
*AM 0036-243 [0036-2437]	00 36 07 −24 33.1	2.977 0.330	256.69 −85.95	268.0 −9.8	CLOSE PAIR of SPIRALS (#2) IC 1562 = 474-G-9 [A82]	23 S 4.2	F473/J3523 (19,202)
AM 0036-660	00 36 13 −66 04.1	2.599 0.330	305.07 −51.27	231.2 −12.6	I/A DBL. (incl. PECULIAR SPIRAL) 79-IG-4 = B 0036-660	2 S 0.9	F079/J1793 (220,118)
*AM 0036-391 [0036-3907]	00 36 18 −39 14.4	2.901 0.330	314.75 −77.86	257.8 −6.7	CHAIN of 4 (#1: DISTURBED SPIRAL) DN 0036-393	6,17 D 8.1	F295/J2542 (339,215)
*AM 0036-391 [0036-3914]	00 36 19 −39 13.6	2.901 0.330	314.76 −77.87	257.8 −6.7	CHAIN of 4 (#2: SPIRAL)	17 D 8.1	F295/J2542 (339,216)
AM 0036-390 [0036-3913]	00 36 28 −39 09.7	2.901 0.330	314.70 −77.94	257.9 −6.7	CHAIN of 4 (#3: EDGE-ON SPIRAL)	17 D 8.1	F295/J2542 (338,219)
AM 0036-390 [0036-3909]	00 36 39 −39 07.9	2.900 0.330	314.57 −77.98	257.9 −6.8	CHAIN of 4 (#4: E)	17 D 8.1	F295/J2542 (336,221)
*AM 0036-432	00 36 52 −43 21.1	2.871 0.330	310.95 −73.85	253.8 −7.9	DISTURBED w. I/A DBL. COMP'N. 242-G-23 [A82]	1,16 S 1.8	F242/J1048 (90,264)
AM 0037-375 [0037-3749]	00 37 05 −37 50.1	2.906 0.330	315.71 −79.26	259.2 −6.5	I/A DBL. (#1: E)	2 S 1.0	F294/J3551 (62,295)
AM 0037-374 [0037-3750]	00 37 09 −37 49.3	2.906 0.330	315.67 −79.27	259.2 −6.5	I/A DBL. (#2: SPIRAL)	2 S 1.0	F294/J3551 (62,295)
AM 0037-245	00 37 19 −24 52.7	2.973 0.330	256.10 −86.37	268.2 −9.5	I/A DBL.	2 S 0.5	F474/J1915 (258,183)
*AM 0037-305	00 37 38 −30 52.3	2.942 0.330	337.76 −85.72	266.0 −4.8	THIN CONN. to COMP'N. (PEC. ARMS) MCG-5-2-34 = 411-G-6	9 S 1.6	F411/J0768 (273,129)
AM 0038-721	00 38 02 −72 15.5	2.383 0.329	304.18 −45.10	225.0 −13.8	CLOSE PAIR	23 S 1.2	F029/J1948 (294,320)
AM 0038-372	00 38 14 −37 28.4	2.903 0.329	314.98 −79.66	259.6 −6.6	COMPACT in GROUP of 4	13,18 D 0.2	F295/J2542 (323,311)
AM 0038-562	00 38 18 −56 27.3	2.738 0.329	306.04 −60.87	240.9 −11.0	CLUSTER of BRIGHT GALAXIES incl. NGC 212/215 = PKS 0038-564	19 D 5.4	F150/J1804 (140,101)
AM 0038-383 [0039-3838]	00 38 20 −38 37.1	2.895 0.329	313.54 −78.55	258.5 −7.0	TRIPLE (#1: S0)	24 D 11.4	F295/J2542 (319,249)
AM 0038 375 [0038-3747]	00 38 27 −37 51.3	2.900 0.329	314.28 −79.30	259.2 −6.8	CHAIN of 3 (#1)	24 S 1.0	F294/J3551 (48,293)
AM 0038-634	00 38 28 −63 42.9	2.621 0.329	304.97 −53.63	233.6 −12.4	LO.S.BR. SPIRAL + LO.S.BR. COMP'N 79-G-5	8 S 1.8	F079/J1793 (210,245)
AM 0038-374 [0038-3810]	00 38 30 −37 48.3	2.900 0.329	314.29 −79.35	259.3 −6.8	I/A DBL. w. FAINT MATERIAL (CHAIN of 3: (#2))	2,24 S 1.0	F294/J3551 (47,296)
AM 0038-374 [0038-3748]	00 38 33 −37 47.4	2.900 0.329	314.25 −79.37	259.3 −6.8	I/A DBL. w. FAINT MATERIAL (CHAIN of 3: (#3))	2,24 S 1.0	F294/J3551 (47,297)
AM 0038-384 [0038-3837]	00 38 50 −38 40.5	2.893 0.329	313.01 −78.51	258.5 −7.1	TRIPLE (#2: S0/E)	24 D 11.4	F295/J2542 (314,246)
AM 0039-380 [0039-3809]	00 39 00 −38 07.9	2.895 0.329	313.40 −79.05	259.0 −7.0	CLOSE PAIR (#1: SPIRAL)	23 S 2.7	F295/J2542 (313,276)
AM 0039-380 [0039-3807]	00 39 07 −38 09.6	2.895 0.329	313.25 −79.03	259.0 −7.0	CLOSE PAIR (#2: E/S0)	23 S 2.7	F295/J2542 (312,274)
AM 0039-335	00 39 14 −33 53.7	2.920 0.329	320.26 −83.16	263.1 −5.9	COMPACT I/A DBL.	2,13 S 0.2	F351/J0739 (273,236)
AM 0039-383 [0038-3840]	00 39 17 −38 38.7	2.891 0.329	312.60 −78.56	258.5 −7.1	TRIPLE (#3: S/S0)	24 D 11.4	F295/J2542 (309,248)
AM 0039-815	00 39 34 −81 50.8	1.471 0.329	303.41 −35.53	215.1 −15.0	LONG, ONE-ARMED SPIRAL 12-6-22	10 D 0.8	F013/J2551 (274,70)
*AM 0040-234	00 40 20 −23 49.7	2.970 0.329	273.78 −85.94	267.0 −9.1	I/A TRIPLE [A82] NGC 230/32/35a = 474-G-15/16/17	3 D 2.2	F474/J1915 (221,240)
AM 0040-223	00 40 35 −22 31.8	2.975 0.329	281.14 −84.78	265.7 −9.4	DWARF (RESOLUTION) + PEC. COMP'N. IC 1574 = 474-G-18 = UKS 0040-225	8,13,20 D 1.8	F474/J1915 (218,310)

Identification	R.A. (1950.0) Declination	Precession R.A./Dec.	Gal. Long. Gal. Lat.	S.G.L. S.G.B.	Description Cross-Identification	Coding Size (Min.)	Field/Plate (x, y)
AM 0040-274	00 40 47 −27 46.0	2.948 0.329	20.76 −88.14	269.2 −4.7	M51-TYPE MCG-5-2-36 = 411-G-8	9 S 0.4	F411/J0768 (238,297)
*AM 0041-390	00 41 01 −39 01.8	2.880 0.329	310.62 −78.24	258.2 −7.6	ISOLATED SPIRAL w. TAIL/RING? 295-G-2 = B 0041-388	6,10 D 1.3	F295/J2542 (291,228)
AM 0041-364	00 41 04 −36 44.0	2.895 0.329	312.67 −80.50	260.5 −7.0	COMPACT	13 D 0.2	F351/J0739 (251,84)
*AM 0041-502	00 41 05 −50 27.5	2.784 0.329	306.21 −66.88	247.0 −10.2	SB (BROKEN BAR) + I/A COMP'NS. NGC 238 = 194-G-31 = Fa 663 [A82]	1,12 D 1.8	F195/J0937 (338,148)
*AM 0041-243	00 41 45 −24 37.4	2.962 0.328	272.11 −86.79	267.7 −8.6	I/A DBL.	2 S 0.4	F474/J1915 (204,197)
AM 0041-393	00 41 45 −39 37.8	2.873 0.328	309.54 −77.67	257.7 −7.8	SMALL I/A DBL. + COMP'NS.	2,8 S 0.3	F295/J2542 (282,196)
AM 0041-383	00 41 46 −38 34.1	2.880 0.328	310.24 −78.72	258.7 −7.6	COMPACT	13 D 0.2	F294/J3551 (14,254)
AM 0042-563	00 42 46 −56 39.2	2.696 0.328	304.75 −60.71	240.8 −11.7	GALAXY w. RINGS + I/A COMP'N. 150-G-21	1,6 S 0.3	F150/J1804 (107,90)
*AM 0043-373	00 43 02 −37 38.5	2.881 0.328	309.58 −79.67	259.7 −7.6	I/A TRIPLE in GROUP of 4	3,18 D 0.4	F295/J2542 (271,303)
AM 0043-232	00 43 18 −23 29.2	2.964 0.328	284.47 −85.89	266.5 −8.5	CONSPICUOUS CLUSTER	19 D 4.5	F474/J1915 (185,258)
AM 0043-312	00 43 25 −31 28.5	2.919 0.328	319.20 −85.74	265.7 −6.2	SPIRAL w. PECULIAR ARM MCG-5-3-3 = 411-G-10	10 D 1.3	F411/J0768 (206,98)
AM 0043-635	00 43 47 −63 50.1	2.557 0.328	303.97 −53.54	233.6 −13.0	RICH CLUSTER PKS 0043-63 = DN 0044-638	19 D 11.2	F079/J1793 (179,239)
AM 0043-385	00 43 57 −38 57.5	2.867 0.328	307.87 −78.38	258.4 −8.1	I/A DBL. 295-IG-03 = B 0044-388	2 S 1.1	F295/J2542 (260,233)
AM 0043-474	00 43 59 −47 42.1	2.793 0.328	305.43 −69.66	249.8 −10.1	I/A DBL. in GROUP of 4 194-IG-36	2,18 S 0.4	F195/J0937 (321,297)
*AM 0044-511	00 44 24 −51 16.8	2.752 0.328	304.77 −66.09	246.3 −10.9	SPIRAL w. I/A COMP'NS. B 0044-512 = 194-IG-37 = Fa 665	1 D 1.1	F195/J0937 (308,105)
AM 0044-353	00 44 37 −35 32.8	2.888 0.328	309.25 −81.79	261.8 −7.4	COMPACT	13 D 0.1	F351/J0739 (213,148)
AM 0044-225	00 44 39 −22 52.0	2.964 0.328	290.50 −85.37	265.8 −8.4	DWARF MCG-4-3-6 = 474-G-25	20 D 1.3	F474/J1915 (168,292)
*AM 0044-243	00 44 40 −24 38.7	2.954 0.328	283.25 −87.09	267.6 −7.9	GALAXY w. LOOPS PAIRED w. RING 474-G-26	6,15,23 D 0.9	F474/J1915 (168,196)
*AM 0044-521	00 44 48 −52 19.5	2.737 0.328	304.52 −65.05	245.2 −11.1	ONE-ARMED I/A SP.(COMPACT at END) 194-IG-39 = Fa 293 = B 0044-523	2,9,10,13 S 0.5	F150/J1804 (83,321)
AM 0044-554	00 44 55 −55 44.1	2.691 0.328	304.21 −61.64	241.8 −11.8	ONE-ARMED SPIRAL + APP. COMP'NS.	8,10 D 0.8	F151/J1693 (336,132)
AM 0045-314	00 45 02 −31 41.9	2.912 0.328	314.03 −85.60	265.6 −6.6	E w. RING + 6 APP.COMP'NS (2 I/A) NGC 254 = 411-G-15	2,6,8 D 1.8	F411/J0768 (187,86)
AM 0045-230	00 45 07 −23 4.5	2.962 0.328	291.29 −85.60	266.0 −8.2	DISTURBED SINGLE + APP. COMP'N. MCG-4-3-8 = 474-IG-28	8,16 D 1.1	F474/J1915 (162,280)
AM 0045-483	00 45 09 −48 30.5	2.777 0.328	304.77 −68.87	249.0 −10.4	E w. 6 COMP'NS. 195-G-1 = DN 0045-484	8 S 2.7	F195/J0937 (309,254)
AM 0045-521	00 45 30 −52 12.0	2.733 0.327	304.28 −65.18	245.4 −11.2	GALAXY w. COMPACT COMP'N. in TAIL	1,13 S 0.2	F150/J1804 (77,328)
AM 0045-400	00 45 51 −40 00.9	2.850 0.327	305.75 −77.36	257.5 −8.7	COMPACT	13 D 0.4	F295/J2542 (239,176)
*AM 0045-421	00 45 52 −42 17.8	2.831 0.327	305.25 −75.08	255.2 −9.2	cD in NEST of 7 GALAXIES 295-G-5	18 D 0.9	F295/J2542 (238,54)
AM 0046-572	00 46 02 −57 23.6	2.656 0.327	303.80 −59.99	240.2 −12.2	ELONGATED COMPACT	13 D 0.4	F151/J1693 (322,44)
AM 0046-413	00 46 24 −41 33.9	2.835 0.327	304.97 −75.81	256.0 −9.1	COMPACT	13 D 0.3	F295/J2542 (233,93)
AM 0046-473	00 46 30 −47 34.2	2.778 0.327	304.22 −69.81	250.0 −10.4	CLOSE TRIPLET Fa 369	24 D 1.2	F195/J0937 (299,305)
AM 0046-372	00 46 31 −37 29.6	2.866 0.327	305.80 −79.88	260.0 −8.2	IRR. COMPACT	13,16 D 0.4	F351/J0739 (192,44)
AM 0046-394 [0047-3954]	00 46 48 −39 48.5	2.847 0.327	304.96 −77.57	257.7 −8.8	WIDE PAIR (#1: SPIRAL)	23 D 1.6	F295/J2542 (230,187)
*AM 0047-494	00 47 00 −49 45.1	2.751 0.327	303.84 −67.63	247.9 −11.0	SPIRAL w. 2 I/A COMP'NS. 195-G-3 = Fa 666	1 S 1.4	F195/J0937 (289,188)
AM 0047-553	00 47 07 −55 38.5	2.674 0.327	303.56 −61.74	242.0 −12.1	I/A GROUP of 5 incl. 150-G-23	5 D 2.7	F150/J1804 (72,143)
AM 0047-513	00 47 25 −51 30.2	2.728 0.327	303.60 −65.88	246.1 −11.4	SPIRAL w. APP. COMP'N.	8 S 1.5	F194/J1861 (33,94)
AM 0047-395 [0046-3948]	00 47 33 −39 54.7	2.843 0.327	304.28 −77.47	257.6 −9.0	WIDE PAIR (#2: GALAXY w. COMP'N.) 295-G-10	8,23 D 1.6	F295/J2542 (222,182)
AM 0047-473	00 47 39 −47 39.7	2.770 0.327	303.65 −69.72	250.0 −10.7	I/A DBL. w. 2 COMP'NS + DWARF 195-IG-4 = Fa 667	2,8,20 S 0.6	F195/J0937 (288,301)

Identification	R.A. (1950.0) Declination	Precession R.A./Dec.	Gal. Long. Gal. Lat.	S.G.L. S.G.B.	Description Cross-Identification	Coding Size (Min.)	Field/Plate (x, y)
AM 0047-304	00 47 53 −30 45.6	2.908 0.327	307.02 −86.62	266.6 −6.9	SPIRAL w. PECULIAR ARM	10 D 0.4	F411/J0768 (154,136)
AM 0048-230	00 48 15 −23 09.5	2.954 0.327	300.66 −85.77	265.9 −7.5	COMPACT	13 D 0.3	F474/J1915 (124,276)
AM 0048-270	00 48 16 −27 04.2	2.930 0.327	275.75 −89.64	269.7 −6.6	COMPACT + SPIRAL	13,23 S 1.7	F474/J1915 (125,66)
AM 0048-313	00 48 17 −31 39.4	2.901 0.327	305.01 −85.73	265.8 −7.2	SPIRAL w. I/A COMP'N. MCG-5-3-8 = 411-IG-21/22	1 S 3.1	F411/J0768 (150,88)
AM 0048-390	00 48 39 −39 01.9	2.845 0.327	303.33 −78.35	258.6 −9.0	LARGE COMPACT 295-G-13	13 D 0.9	F295/J2542 (211,229)
AM 0048-502	00 48 50 −50 29.2	2.730 0.326	303.06 −66.90	247.2 −11.4	I/A DBL. 195-IG-8	2 S 0.4	F195/J0937 (272,149)
*AM 0048-381	00 48 54 −38 11.3	2.850 0.326	303.09 −79.20	259.4 −8.9	KNOTTY GALAXY in I/A DBL. 295-G-14	2 S 1.1	F295/J2542 (209,275)
AM 0048-284	00 48 57 −28 48.0	2.917 0.326	303.41 −88.59	268.6 −6.7	CONSPICUOUS CLUSTER	19 D 6.7	F411/J0768 (141,241)
AM 0049-225	00 49 39 −22 57.0	2.951 0.326	304.98 −85.56	265.6 −7.3	SPIRAL w. LONG ARMS + PEC. COMP'N NGC 276 = IC 1591 = 474-G-34	8,10 D 1.0	F474/J1915 (106,287)
AM 0049-482	00 49 44 −48 29.0	2.748 0.326	302.66 −68.90	249.2 −11.2	CLOSE PAIR 195-IG-10	24 D 1.1	F195/J0937 (268,257)
*AM 0049-273	00 49 50 −27 37.3	2.922 0.326	265.25 −89.70	269.8 −6.6	I/A TRIPLE 411-G-24 [A82]	3 D 1.3	F474/J1915 (107,36)
AM 0049-351 [0049-3516]	00 49 56 −35 18.7	2.868 0.326	301.61 −82.07	262.3 −8.4	PAIR (#1: SPIRAL w. I/A COMP'N.)	1,9,23 S 0.6	F351/J0739 (155,161)
*AM 0049-412 [0049-3518]	00 49 57 −41 20.1	2.819 0.326	302.25 −76.05	256.3 −9.7	DISTURBED GALAXY w. JET 295-IG-17	7,16 D 0.7	F295/J2542 (197,106)
AM 0049-351	00 49 59 −35 16.4	2.868 0.326	301.52 −82.11	262.3 −8.4	PAIR (#2: EDGE-ON SPIRAL (DUST)) 351-G-16	23 D 0.7	F351/J0739 (154,163)
AM 0050-652	00 50 16 −65 29.7	2.435 0.326	302.79 −51.89	232.0 −14.0	GALAXY w. MATERIAL + APP. COMP'N. 79-GA-7 = MC 0050-654	8,15 S 1.3	F079/J1793 (142,150)
*AM 0050-312	00 50 18 −31 28.6	2.895 0.326	299.09 −85.90	266.1 −7.6	SP. w. I/A COMP'N.(DISTURBED ARM) NGC 289 = 411-G-25 [A82]	1,10 D 9.0	F411/J0768 (127,97)
*AM 0050-370	00 50 26 −37 02.9	2.853 0.326	301.28 −80.34	260.6 −8.9	BROKEN RING + COMPACT 351-IG-17	9,13 D 1.6	F295/J2542 (192,336)
AM 0050-222	00 50 27 −22 23.0	2.953 0.326	306.86 −84.98	265.0 −7.2	GALAXY w. I/A COMP'N.	1 S 0.2	F474/J1915 (96,317)
AM 0050-365	00 50 50 −36 58.9	2.852 0.326	300.80 −80.40	260.7 −9.0	VERY COMPACT	13 D 0.2	F295/J2542 (188,339)
*AM 0051-473	00 51 11 −47 38.8	2.748 0.326	301.93 −69.74	250.1 −11.2	DBL. w. I/A COMP'N. (I/A TRIPLE?) 195-G-11	1,3,23 S 0.6	F243/J3492 (222,37)
AM 0051-271 [0051-2725]	00 51 15 −27 19.7	2.920 0.326	26.40 −89.49	269.8 −5.8	PAIR (#1: EDGE-ON) MCG-5-3-11 = 474-G-39	23 D 1.4	F411/J0768 (113,320)
AM 0051-475	00 51 30 −47 55.1	2.743 0.326	301.81 −69.46	249.8 −11.3	SPIRAL + COMP'NS. IC 1594 = 195-G-12	8 D 1.0	F195/J0937 (253,288)
*AM 0051-272 [0051-2719]	00 51 32 −27 25.0	2.918 0.326	216.10 −89.43	270.1 −6.9	PAIR (#2: RING) 474-G-40	6,23 D 0.9	F411/J0768 (110,315)
AM 0051-234	00 51 54 −23 49.2	2.941 0.325	313.54 −86.37	266.4 −6.5	I/A QUARTET [A82]	4 D 2.5	F474/J1915 (79,240)
AM 0052-431	00 52 00 −43 11.7	2.791 0.325	300.99 −74.18	254.6 −10.5	GALAXY w. I/A COMP'N.	1 S 0.4	F243/J3492 (218,276)
AM 0052-660	00 52 19 −66 07.6	2.390 0.325	302.46 −51.26	231.4 −14.2	I/A DBL. 79-IG-9	2 S 0.8	F079/J1793 (132,115)
AM 0052-603	00 52 26 −60 33.4	2.536 0.325	302.23 −56.83	237.1 −13.5	E w. I/A COMPACT (CONNECTION?)	1,13 S 0.7	F113/J1870 (335,142)
AM 0052-575	00 52 27 −57 59.8	2.588 0.325	302.10 −59.38	239.7 −13.2	COMPACT	13 D 0.2	F113/J1870 (347,279)
AM 0052-840 [0055-8410]	00 52 29 −84 07.5	0.125 0.325	302.89 −33.26	212.8 −15.6	CLOSE TRIPLE (#1: LARGE S0) 2-G-10	24 S 6.3	F003/J2565 (302,202)
*AM 0052-321	00 52 32 −32 17.8	2.881 0.325	294.32 −85.03	265.4 −8.2	I/A DBL. (PECULIAR ABS'N.) MCG-5-3-13/14 = 411-G-29/30	2,14 S 0.8	F411/J0768 (102,53)
*AM 0052-375	00 52 38 −37 56.9	2.836 0.325	299.09 −79.41	259.8 −9.5	SUPER-ASSOC. on ENDS of ARMS NGC 300 = 295-G-20	10 D 9.9	F351/J0739 (127,19)
AM 0053-242	00 53 05 −24 25.8	2.934 0.325	320.53 −86.90	266.9 −6.1	I/A DBL. IC 1601 = 474-G-44	2 S 1.0	F474/J1915 (65,207)
AM 0053-243	00 53 18 −24 30.9	2.933 0.325	321.87 −86.97	267.0 −6.1	I/A DBL. (HI.S.BR.)	2 S 1.1	F475/J1916 (332,202)
AM 0053-304	00 53 18 −30 48.5	2.889 0.325	287.89 −86.45	266.9 −8.1	DWARF w. CONDENSATIONS 411-G-31	20 D 0.5	F411/J0768 (92,133)
*AM 0053-374	00 53 26 −37 40.2	2.835 0.325	298.09 −79.67	260.1 −9.6	I/A DBL. off MI.AX. of NGC 300 295-IG-22	2 S 1.5	F295/J2542 (161,302)
*AM 0053-504	00 53 31 −50 44.4	2.695 0.325	301.20 −66.63	247.1 −12.2	I/A DBL. 195-G-17	2 S 1.6	F195/J0937 (232,137)

Identification	R.A. (1950.0) Declination	Precession R.A./Dec.	Gal. Long. Gal. Lat.	S.G.L. S.G.B.	Description Cross-Identification	Coding Size (Min.)	Field/Plate (x, y)
AM 0053-530 [0055-5302]	00 53 56 −53 06.2	2.658 0.325	301.29 −64.27	244.7 −12.6	PAIR in GROUP of 4 (#1: SPIRAL) 151-G-5	18,23 D 0.6	F195/J0937 (226,10)
AM 0053-532	00 53 56 −53 27.4	2.653 0.325	301.33 −63.92	244.3 −12.7	EDGE-ON w. "BOX" NUCLEUS 151-G-4	22 D 1.1	F150/J1804 (12,257)
AM 0053-353	00 53 57 −35 31.3	2.851 0.325	295.91 −81.80	262.3 −9.3	RING GALAXY (MEMBER of PAIR) 351-G-23	6,23 D 0.6	F351/J0739 (111,149)
AM 0053-472 [0055-4714]	00 53 58 −47 20.0	2.735 0.325	300.53 −70.03	250.5 −11.6	CLOSE TRIPLE (#1: Sa) 243-G-12 = Fa 669	24 D 0.7	F195/J0937 (231,320)
AM 0054-530 [0053-5306]	00 54 03 −53 03.4	2.658 0.325	301.25 −64.31	244.7 −12.6	PAIR in GROUP of 4 (#2: E) NGC 312 = 151-G-6	18,23 D 0.9	F195/J0937 (225,12)
AM 0054-744	00 54 23 −74 46.9	1.918 0.325	302.52 −42.60	222.5 −15.2	GALAXY w. APP. COMP'N.SC,103	8 S 0.8	F029/J1948 (220,190)
AM 0054-533	00 54 27 −53 32.7	2.647 0.325	301.17 −63.82	244.3 −12.8	KNOTTY IRR. 151-IG-7 ?	16 D 0.4	F150/J1804 (8,251)
*AM 0054-531	00 54 28 −53 14.6	2.652 0.325	301.12 −64.12	244.6 −12.7	CHAIN of 9 (incl. I/A's) NGC 323 = 151-G-9 (G-13 Group)	17 D 22.4	F151/J1693 (269,270)
AM 0054-310	00 54 30 −31 08.5	2.883 0.325	285.60 −86.06	266.6 −8.4	CHAIN of 8 (DISTURBED + TRIPLE) incl. MCG-5-3-19 = 411-G-33	3,16,17 D 10.8	F411/J0768 (79,115)
AM 0054-471 [0053-4720]	00 54 40 −47 14.5	2.732 0.325	300.16 −70.11	250.6 −11.7	CLOSE TRIPLE (#2: E)	24 D 0.4	F195/J0937 (225,325)
AM 0054-634	00 54 40 −63 44.8	2.433 0.325	301.94 −53.63	233.9 −14.2	I/A DBL. (KNOTTY RING + GALAXY) 79-IG-13 = B 0054-637	2,6 S 0.9	F079/J1793 (114,243)
*AM 0054-240	00 54 42 −24 05.5	2.932 0.325	324.60 −86.46	266.5 −5.9	COMPACT	13 D 0.1	F475/J1916 (315,225)
AM 0054-241	00 54 45 −24 12.8	2.931 0.325	325.55 −86.57	266.6 −5.8	SPIRAL w. JET(?) to COMP'N.	7 D 0.6	F475/J1916 (314,218)
*AM 0054-401	00 54 48 −40 15.0	2.805 0.324	298.04 −77.08	257.6 −10.4	GALAXY w. JET + VERY THIN ARM	7,10 D 1.6	F295/J2542 (148,164)
*AM 0054-531 [0054-5303]	00 54 48 −53 11.8	2.650 0.324	301.01 −64.17	244.6 −12.8	GROUP of 4 (#3: SPIRAL) NGC 328 = 151-G-13	18 D 1.8	F195/J0937 (219,5)
AM 0054-435	00 54 53 −43 59.9	2.767 0.324	299.31 −73.35	253.9 −11.2	GALAXY w. I/A COMP'N. NGC 322 = 243-G-15 = 0054-439	1 S 0.2	F243/J3492 (190,233)
*AM 0054-411	00 54 55 −41 13.9	2.795 0.324	298.36 −76.10	256.6 −10.6	S0 w. APP. COMP'NS. NGC 324 = 295-G-25	8 S 1.8	F295/J2542 (147,111)
AM 0055-530 [0054-5311]	00 55 00 −53 02.6	2.651 0.324	300.92 −64.32	244.8 −12.8	GROUP of 4 (#4: E)	18 D 0.3	F195/J0937 (217,13)
AM 0055-494	00 55 05 −49 46.9	2.697 0.324	300.42 −67.57	248.1 −12.3	I/A TRIPLE	3 D 1.4	F195/J0937 (219,188)
AM 0055-274	00 55 22 −27 46.4	2.905 0.324	228.65 −88.54	270.0 −7.8	SPIRAL w. RE-ENTRANT ARMS MCG-5-3-20 = 411-G-34	10 D 2.5	F475/J1916 (303,27)
*AM 0055-464	00 55 23 −46 43.1	2.733 0.324	299.70 −70.63	251.2 −11.8	SPIRAL w. PECULIAR ARM 243-G-17 = Fa 670	10 D 1.1	F243/J3492 (185,87)
*AM 0055-491	00 55 23 −49 10.5	2.703 0.324	300.19 −68.18	248.7 −12.2	SB w. I/A DBL's. ([N] and [S]) IC 1605 = 195-G-19 = Fa 671 [A82]	2,8 D 1.3	F195/J0937 (217,221)
AM 0055-471 [0054-4714]	00 55 24 −47 14.7	2.727 0.324	299.81 −70.10	250.6 −11.9	CLOSE TRIPLE (#3: Sc)	24 D 0.6	F195/J0937 (218,325)
AM 0055-840 [0052-8407]	00 55 33 −84 09.6	−0.063 0.324	302.80 −33.22	212.8 −15.6	CLOSE TRIPLE (#2: S0)	24 S 6.3	F003/J2565 (297,201)
AM 0055-841 [0055-8409]	00 55 46 −84 10.6	−0.083 0.324	302.79 −33.21	212.8 −15.6	CLOSE TRIPLE (#3: Sd)	24 S 6.3	F003/J2565 (296,201)
AM 0055-232	00 55 56 −23 29.7	2.933 0.324	325.32 −85.80	265.8 −5.7	I/A DBL.	2 S 0.4	F475/J1916 (300,257)
AM 0056-535	00 56 17 −53 53.6	2.628 0.324	300.60 −63.46	243.9 −13.1	COMPACT	13 D 0.1	F151/J1693 (253,236)
AM 0056-670	00 56 17 −67 04.0	2.305 0.324	301.89 −50.30	230.5 −14.7	RICH CLUSTER (13 E's + SPIRAL) DN 0056-670 (incl. Fa 216)	19 D 9.0	F079/J1793 (113,64)
AM 0056-535	00 56 18 −53 53.4	2.628 0.324	300.59 −63.46	243.9 −13.1	3-ARMED SPIRAL + I/A COMP'N.	1,11,13 D 0.3	F151/J1693 (253,236)
AM 0056-283	00 56 30 −28 35.0	2.895 0.324	249.26 −87.96	269.2 −8.2	CLOSE PAIR	23 S 1.6	F412/J3516 (327,253)
AM 0056-535	00 56 56 −53 59.7	2.621 0.324	300.40 −63.35	243.9 −13.2	DISTURBED SINGLE GALAXY	16 D 0.4	F151/J1693 (248,230)
AM 0057-425	00 57 09 −42 55.6	2.766 0.324	297.45 −74.37	255.0 −11.4	GALAXY w. I/A COMP'N.	1 S 0.9	F295/J2542 (126,20)
AM 0057-541	00 57 21 −54 17.6	2.613 0.324	300.31 −63.05	243.6 −13.3	COMPACT (ELONGATED)	13 D 0.3	F151/J1693 (244,214)
AM 0057-222	00 57 23 −22 25.8	2.936 0.324	324.43 −84.69	264.7 −5.6	CLOSE PAIR (incl. M51-TYPE)	9,23 S 1.1	F475/J1916 (283,315)
AM 0057-234	00 57 34 −23 48.1	2.926 0.324	331.82 −85.93	266.0 −5.3	V.RICH CLUSTER of FAINT GALAXIES	19 D 5.4	F475/J1916 (280,241)
AM 0057-304	00 57 35 −30 45.4	2.875 0.324	274.38 −86.14	267.1 −8.9	COMPACT	13 D 0.2	F412/J3516 (311,137)

Identification	R.A. (1950.0) Declination	Precession R.A./Dec.	Gal. Long. Gal. Lat.	S.G.L. S.G.B.	Description Cross-Identification	Coding Size (Min.)	Field/Plate (x, y)
AM 0058-682	00 58 07 −68 24.2	2.226 0.323	301.72 −48.96	229.1 −15.0	ONE-ARMED SPIRAL 51-G-11 = Fa 217?	10 D 1.3	F080/J1004 (332,6)
*AM 0058-311	00 58 20 −31 12.0	2.869 0.323	275.46 −85.68	266.7 −9.2	BROKEN RING w. I/A COMP'N. 412-G-2 [A82]	1,6 S 0.8	F412/J3516 (301,113)
AM 0058-430 [0058-4301]	00 58 20 −43 03.5	2.759 0.323	296.72 −74.22	254.9 −11.6	CLOSE PAIR (#1: EDGE-ON SPIRAL)	23 S 1.0	F295/J2542 (115,12)
AM 0058-220	00 58 21 −22 01.3	2.937 0.323	325.09 −84.23	264.3 −5.5	RING + COMP'N. 541-IG-9	6,8 S 0.3	F475/J1916 (272,337)
*AM 0058-480	00 58 21 −48 08.3	2.697 0.323	298.60 −69.17	249.8 −12.5	I/A DBL. w. MATERIAL 195-IG-21 = B 0058-481	2 S 0.4	F195/J0937 (191,277)
AM 0058-665	00 58 21 −66 53.3	2.284 0.323	301.56 −50.47	230.7 −14.9	RING (NUCLEUS in RIM) GROUP of 4	6,18 D 0.5	F051/J1877 (146,346)
AM 0058-430 [0058-4303]	00 58 23 −43 01.1	2.759 0.323	296.66 −74.25	255.0 −11.6	CLOSE PAIR (#2: SPIRAL)	23 S 0.8	F295/J2542 (114,14)
*AM 0058-402	00 58 26 −40 22.4	2.786 0.323	295.06 −76.87	257.6 −11.1	M51-TYPE + COMP'NS. [A82]	8,9 S 0.5	F295/J2542 (111,157)
*AM 0058-580	00 58 27 −58 00.8	2.533 0.323	300.54 −59.33	239.8 −14.0	COMPACT GALAXY w. DIAMETRIC JETS 113-IG-4	7,13 D 0.3	F113/J1870 (304,282)
AM 0058-504	00 58 33 −50 48.9	2.659 0.323	299.21 −66.50	247.1 −13.0	CLUSTER of FAINT GALS. (DBL. cD) 195-IG-22 = DN 0059-508	19 D 8.1	F195/J0937 (189,133)
AM 0058-391	00 58 41 −39 16.3	2.796 0.323	293.99 −77.95	258.7 −11.0	SMALL I/A DBL.	2 S 0.4	F295/J2542 (107,216)
AM 0059-374 [0100-3746]	00 59 18 −37 44.7	2.808 0.323	291.83 −79.42	260.3 −10.8	ARC of 4 (#1: S0)	17 S 9.2	F295/J2542 (98,297)
AM 0059-374 [0059-3744]	00 59 24 −37 45.8	2.808 0.323	291.76 −79.40	260.3 −10.8	ARC of 4 (#2: SPIRAL)	17 S 9.2	F295/J2542 (97,296)
AM 0059-374 [0059-3745]	00 59 33 −37 46.7	2.807 0.323	291.61 −79.37	260.3 −10.8	ARC of 4 (#3: E/S0)	17 S 9.2	F295/J2542 (96,296)
AM 0059-673	00 59 33 −67 34.1	2.241 0.323	301.44 −49.79	230.0 −15.1	GALAXY w. TAIL to COMP'N. Fa 219?	15 S 0.4	F079/J1793 (97,36)
AM 0059-384	00 59 41 −38 47.6	2.796 0.323	292.61 −78.38	259.3 −11.1	COMPACT	13 D 0.4	F295/J2542 (96,241)
AM 0059-254	00 59 46 −25 47.9	2.906 0.323	0.21 −87.11	267.9 −4.4	3 in a LINE (COMPACT CENTRAL)	13,24 D 1.8	F475/J1916 (252,134)
AM 0059-385	00 59 51 −38 55.8	2.794 0.323	292.59 −78.24	259.1 −11.1	COMPACT	13 D 0.4	F295/J2542 (94,234)
AM 0100-321	01 00 00 −32 12.2	2.855 0.323	277.29 −84.62	265.8 −9.8	CHAIN of 8 (incl. I/A DBL.) PKS 0100-322	2,17 D 4.0	F351/J0739 (39,325)
AM 0100-374 [0059-3747]	01 00 00 −37 46.7	2.805 0.323	291.18 −79.36	260.3 −10.9	ARC of 4 (#4: E:)	17 S 9.2	F295/J2542 (91,295)
AM 0100-543	01 00 10 −54 35.5	2.585 0.323	299.47 −62.72	243.3 −13.8	Sd-DWARF 151-G-19 = UKS 0100-545	20 D 1.6	F151/J1693 (222,199)
AM 0100-472	01 00 35 −47 27.1	2.693 0.322	297.31 −69.81	250.6 −12.8	DISTURBED (NGC 2685-TYPE) 243-IG-19	16 D 0.5	F195/J0937 (171,314)
AM 0100-614	01 00 37 −61 41.5	2.425 0.322	300.56 −55.64	236.1 −14.7	LUMPY COMPACT	13 D 0.3	F113/J1870 (277,85)
AM 0100-820	01 00 54 −82 04.8	0.551 0.322	302.50 −35.30	214.9 −15.8	LO.S.BR. SPIRAL 13-G-8	20 D 2.7	F012/J2486 (67,54)
AM 0101-401	01 01 20 −40 10.8	2.775 0.322	292.50 −76.96	257.9 −11.6	CLUSTER (7 BR. GAL. + FAINTER) PKS 0101-401	19 D 3.6	F295/J2542 (81,166)
AM 0101-333	01 01 21 −33 32.3	2.839 0.322	280.35 −83.30	264.6 −10.3	COMPACT	13 D 0.4	F352/J0749 (296,254)
AM 0101-280	01 01 22 −28 01.1	2.885 0.322	226.70 −87.19	270.0 −9.1	SMALL IRR. w. TAIL toward COMP'N. 412-IG-3	1,12,13 S 0.9	F412/J3516 (270,285)
AM 0101-234	01 01 28 −23 45.3	2.917 0.322	341.44 −85.41	265.8 −4.4	SPIRAL w. I/A COMP'N.	1 S 0.4	F475/J1916 (232,244)
AM 0101-591 [0101-5915]	01 01 34 −59 11.8	2.478 0.322	299.95 −58.12	238.6 −14.5	CLOSE PAIR [N.S.] (#1)	23 D 0.4	F113/J1870 (279,220)
AM 0101-591 [0101-5911]	01 01 34 −59 15.7	2.477 0.322	299.96 −58.05	238.6 −14.5	CLOSE PAIR [N.S.] (#2)	23 D 0.3	F113/J1870 (278,216)
*AM 0102-402	01 02 01 −40 25.0	2.769 0.322	292.17 −76.70	257.7 −11.8	3-ARMED SPIRAL + COMP'NS. 295-G-31	8,11 D 1.3	F295/J2542 (74,153)
AM 0102-241	01 02 07 −24 15.3	2.912 0.322	347.04 −85.70	266.3 −4.2	V.RICH CLUSTER of FAINT GALAXIES	19 D 11.2	F475/J1916 (224,217)
AM 0102-335	01 02 08 −33 55.2	2.832 0.322	280.43 −82.88	264.2 −10.6	DIFFRACTION SPIKE on NUCLEUS 352-G-2	22 D 0.7	F352/J0749 (287,234)
AM 0102-511	01 02 08 −51 18.2	2.626 0.322	297.96 −65.96	246.7 −13.6	SPIRAL w. 8 COMPN'S + I/A DBL. IC 1617 = 195-G-28	2,8 D 1.0	F195/J0937 (159,107)
AM 0102-464	01 02 10 −46 46.1	2.692 0.322	296.25 −70.45	251.3 −12.9	GAL. w. LONG TAIL and I/A COMP'N.	1,15 S 1.5	F243/J3492 (122,84)
*AM 0102-274	01 02 33 −27 41.6	2.884 0.322	219.60 −86.98	270.4 −9.3	SPIRAL w. PEC. ARMS (INTERLACED) IC 1616 = 412-G-4	10 D 2.0	F475/J1916 (218,33)

Identification	R.A. (1950.0) Declination	Precession R.A./Dec.	Gal. Long. Gal. Lat.	S.G.L. S.G.B.	Description Cross-Identification	Coding Size (Min.)	Field/Plate (x, y)
AM 0102-642	01 02 36 −64 23.5	2.321 0.322	300.56 −52.93	233.3 −15.1	GALAXY w. TAIL 79-G-16 = B 0102-643	15 D 1.1	F080/J1004 (333,223)
*AM 0102-391	01 02 44 −39 18.2	2.777 0.322	290.38 −77.75	258.9 −11.7	DISTANT CLUSTER (STARS?)	22 D 0.7	F295/J2542 (65,213)
AM 0102-573	01 02 45 −57 38.5	2.503 0.322	299.36 −59.65	240.2 −14.5	2 DISRUPTED GALAXY + 2 COMP'NS. 113-IG-9	8,16 D 0.7	F113/J1870 (275,303)
*AM 0102-391	01 02 54 −39 14.1	2.777 0.321	290.15 −77.80	258.9 −11.8	CLOSE GROUP of 4 295-G-34	18 D 3.1	F295/J2542 (63,216)
AM 0102-610	01 02 56 −61 04.8	2.417 0.321	299.97 −56.23	236.7 −14.9	FAINT RING w. NUCLEUS in RIM	2,6 S 0.4	F113/J1870 (264,119)
*AM 0102-231	01 02 59 −23 17.5	2.917 0.321	341.42 −84.83	265.3 −4.2	SP. w. DIAM. COMP'NS. (in CHAIN) MCG-4-3-50 = 475-G-9	8,12,17 D 1.1	F475/J1916 (214,269)
AM 0102-463	01 02 59 −46 38.1	2.689 0.321	295.77 −70.56	251.5 −13.0	CLOSE TRIPLE incl. 243-G-24	24 D 2.0	F243/J3492 (115,91)
AM 0103-224	01 03 00 −22 44.1	2.921 0.321	338.05 −84.37	264.7 −4.3	E w. BRIDGE in GROUP of 3	2,24 D 0.7	F475/J1916 (214,299)
AM 0103-245	01 03 12 −24 56.5	2.904 0.321	356.30 −85.98	266.9 −3.8	RICH LONG CLUSTER of FAINT GALS.	19 D 11.2	F475/J1916 (211,180)
AM 0103-572	01 03 17 −57 29.8	2.501 0.321	299.18 −59.79	240.4 −14.6	DISRUPTED GALAXIES Fa 676	16,23 D 0.7	F113/J1870 (271,311)
AM 0103-415	01 03 19 −41 55.6	2.746 0.321	292.54 −75.17	256.2 −12.3	NEST of 4/5 E's (COMMON ENVELOPE) 295-G-35 = PKS 0103-417?	4,18 D 0.4	F296/J3524 (326,73)
AM 0103-234	01 03 40 −23 40.6	2.912 0.321	345.55 −85.03	265.6 −4.0	E w. HI.S.BR. JET (GAL.?)	7 D 0.4	F475/J1916 (205,248)
*AM 0103-520	01 03 42 −52 01.3	2.603 0.321	297.60 −65.21	246.0 −13.9	I/A TRIPLET 195-IG-30	3 D 0.9	F195/J0937 (146,68)
AM 0103-303 [0103-3044]	01 03 43 −30 39.9	2.856 0.321	259.41 −85.41	267.5 −10.2	PAIR (#1: HI.S.BR. GAL. w. KNOTS)	13,14,23 S 4.9	F412/J3516 (240,143)
AM 0103-620	01 03 44 −62 03.7	2.381 0.321	299.97 −55.24	235.7 −15.1	SMALL I/A TRIPLE (JET?)	3 D 0.8	F113/J1870 (256,67)
AM 0103-304 [0103-3039]	01 03 47 −30 44.5	2.855 0.321	259.95 −85.34	267.5 −10.2	PAIR (#2: RING SPIRAL)	6,23 S 4.9	F412/J3516 (239,139)
*AM 0103-302	01 03 49 −30 26.8	2.857 0.321	257.20 −85.54	267.7 −10.2	SPIRAL w. HI.S.BR. COMP'N. NGC 378 = 412-G-5	8 S 1.4	F412/J3516 (239,155)
AM 0104-642	01 04 05 −64 27.0	2.302 0.321	300.31 −52.86	233.2 −15.3	CLOSE PAIR	23 S 0.3	F080/J1004 (325,221)
AM 0104-565	01 04 10 −56 54.3	2.506 0.321	298.81 −60.36	241.0 −14.6	I/A DBL.	2 S 0.4	F151/J1693 (190,75)
AM 0104-613	01 04 47 −61 33.2	2.385 0.321	299.66 −55.73	236.2 −15.1	LINE of 3 [E.W.]	24 D 8.1	F113/J1870 (251,94)
*AM 0104-421 [0105-4207]	01 04 50 −42 16.7	2.734 0.321	291.79 −74.76	255.9 −12.6	CHAIN of 4 (#1: G.D. SPIRAL)	17 D 9.9	F295/J2542 (49,52)
AM 0104-811	01 04 52 −81 14.3	0.652 0.321	302.25 −36.13	215.8 −15.9	E + SPIRAL + COMP'NS.	8,23 S 3.4	F013/J2551 (230,110)
AM 0104-420 [0104-4216]	01 04 54 −42 08.6	2.735 0.321	291.62 −74.89	256.1 −12.6	CHAIN of 4 (#2: S0) 295-G-38?	17 D 9.9	F295/J2542 (48,59)
AM 0105-420 [0104-4208]	01 05 00 −42 09.5	2.734 0.321	291.56 −74.87	256.1 −12.7	CHAIN of 4 (#3: SB0/a)	17 D 9.9	F295/J2542 (47,59)
*AM 0105-370	01 05 02 −37 01.4	2.791 0.321	284.64 −79.79	261.2 −11.7	GALAXIES w. I/A COMP'N.	1 S 1.3	F295/J2542 (36,334)
AM 0105-420 [0105-4209]	01 05 20 −42 07.4	2.733 0.321	291.30 −74.89	256.1 −12.7	CHAIN of 4 (#4: KNOTTY SB)	17 D 9.9	F295/J2542 (44,60)
AM 0105-471	01 05 29 −47 10.5	2.667 0.320	294.80 −69.95	251.0 −13.5	E + I/A COMP'N. in GROUP of 5 IC 1625 = 243-G-33	1,18 D 0.9	F195/J0937 (126,328)
AM 0105-270	01 05 46 −27 08.0	2.879 0.320	30.07 −86.27	268.9 −2.8	GALAXY w. CONDENSATION in TAIL	15 D 1.1	F475/J1916 (180,63)
*AM 0105-370	01 05 49 −37 00.8	2.788 0.320	283.80 −79.74	261.3 −11.9	RICH CLUSTER	19 D 7.2	F295/J2542 (27,335)
*AM 0105-472	01 05 57 −47 23.9	2.661 0.320	294.70 −69.72	250.8 −13.7	GROUP of 4 + THIN ARM SPIRAL 243-G-35	10,12,18 D 3.6	F195/J0937 (122,316)
AM 0106-470	01 06 02 −47 01.3	2.666 0.320	294.45 −70.08	251.2 −13.6	IRR. IC 1630 = 243-G-36 = B 0106-470	16 D 1.3	F195/J0937 (121,336)
AM 0106-382	01 06 04 −38 28.7	2.771 0.320	286.23 −78.35	259.8 −12.2	KNOTTY IRR.	16 D 0.5	F295/J2542 (28,256)
AM 0106-472	01 06 10 −47 21.9	2.660 0.320	294.57 −69.74	250.8 −13.7	GROUP of 5 (BARS and RINGS) 243-G-38	6,18 D 4.0	F244/J1761 (341,47)
AM 0106-620	01 06 10 −62 03.4	2.356 0.320	299.47 −55.21	235.7 −15.3	GALAXY w. APP. COMP'NS. 113-G-16 = B 0106-620	8 S 0.6	F079/J1793 (38,329)
*AM 0106-803	01 06 21 −80 34.4	0.775 0.320	302.11 −36.78	216.5 −16.0	S0/a? (NARROW DUST) in GROUP 13-G-12	8,13,14,24 D 2.8	F012/J2486 (35,129)
AM 0106-403	01 06 23 −40 36.3	2.746 0.320	288.94 −76.31	257.7 −12.6	RICH CLUSTER 296-G-1	19 D 4.9	F295/J2542 (30,141)

Identification	R.A. (1950.0) Declination	Precession R.A./Dec.	Gal. Long. Gal. Lat.	S.G.L. S.G.B.	Description Cross-Identification	Coding Size (Min.)	Field/Plate (x, y)
AM 0106-285	01 06 24	2.863	234.87	269.5	PARTIAL ARC around COMP'N.[E]	8	F412/J3516
	−28 51.0	0.320	−85.89	−10.4	IC 1628 = 412-G-7	S 1.3	(210,241)
AM 0106-464	01 06 32	2.667	294.04	251.5	I/A DBL. w. TAILS (PEC. SPIRAL?)	2,15	F243/J3492
	−46 44.4	0.320	−70.34	−13.7	IC 1631 = 243-G-40 = B 0106-467	S 0.2	(82,85)
AM 0106-453	01 06 41	2.682	293.22	252.7	CLOSE PAIR of SPIRALS (#1)	23	F244/J1761
[0106-4536]	−45 34.3	0.320	−71.48	−13.5	243-G-43	D 0.4	(342,144)
AM 0106-453	01 06 45	2.681	293.20	252.6	CLOSE PAIR of SPIRALS (#2)	23	F244/J1761
[0106-4534]	−45 36.5	0.320	−71.44	−13.5	243-G-44	D 0.4	(341,142)
AM 0107-390	01 07 04	2.760	286.22	259.3	GALAXY w. I/A COMP'N.	1	F295/J2542
	−39 02.1	0.320	−77.76	−12.5	296-G-3 = B 0107-390	S 0.8	(19,225)
AM 0107-480	01 07 19	2.642	294.42	250.1	DWARF w. LOOP	15,20	F195/J0937
	−48 03.7	0.320	−69.02	−14.0	195-G-32	D 1.1	(111,280)
*AM 0107-375	01 07 23	2.771	284.13	260.4	I/A TRIPLE	3	F296/J3524
	−37 58.6	0.320	−78.73	−12.4	B 0107-379	D 1.3	(291,287)
AM 0107-231	01 07 26	2.906	349.24	265.0	CLOSE GROUP of 3	24	F475/J1916
	−23 15.3	0.320	−84.13	−3.2		D 0.9	(159,271)
AM 0107-744	01 07 41	1.644	301.34	222.5	SP.w. APP.COMP'NS (incl.I/A DBL.)	2,8	F029/J1948
	−74 46.9	0.320	−42.54	−16.1	29-G-34	S 6.9	(173,191)
AM 0107-461	01 07 43	2.667	293.10	252.1	E at CENTRE of RICH CLUSTER	19	F243/J3492
	−46 11.7	0.320	−70.83	−13.8	IC 1633 = 243-G-46 = MC 0107-461	D 26.9	(70,114)
AM 0108-302	01 08 14	2.842	250.51	267.9	PAIR of SPIRALS (#1)	23	F412/J3516
[0108-3042]	−30 29.3	0.319	−84.77	−11.1	NGC 418 = 412-G-9	D 2.2	(188,153)
AM 0108-462	01 08 15	2.661	292.99	251.8	ONE-ARMED SPIRAL	10	F244/J1761
	−46 26.4	0.319	−70.57	−13.9		D 0.5	(325,98)
AM 0108-383	01 08 20	2.760	284.29	259.8	GROUP of 5 (incl. M51-TYPE)	9,18	F296/J3524
	−38 32.1	0.319	−78.14	−12.7		D 1.8	(281,257)
AM 0108-423	01 08 33	2.710	289.66	255.7	DWARF	20	F295/J2542
	−42 38.5	0.319	−74.24	−13.4	243-G-50	D 0.7	(13,31)
AM 0108-485	01 08 34	2.620	294.38	249.2	SMALL GROUP of 6 GALAXIES	18	F195/J0937
	−48 59.3	0.319	−68.07	−14.3	195-IG-33	D 1.8	(101,230)
AM 0108-304	01 08 39	2.839	251.85	267.7	PAIR of SPIRALS (#2)	23	F412/J3516
[0108-3029]	−30 42.3	0.319	−84.57	−11.3	IC 1637 = 412-G-10	D 1.6	(183,141)
AM 0109-382	01 09 09	2.759	283.22	260.1	SPIRAL w. CHAIN of 3 COMP'NS.	8,17	F296/J3524
	−38 20.9	0.319	−78.26	−12.8	NGC 424 = 296-G-4	S 10.8	(272,267)
AM 0109-825	01 09 21	−0.102	302.23	214.1	CLOSE I/A PAIR + COMP'NS.	2,8	F002/J2563
	−82 51.1	0.319	−34.51	−16.0		S 1.3	(66,278)
AM 0109-545	01 09 40	2.503	296.64	243.1	DISTURBED SINGLE ?	16	F151/J1693
	−54 57.5	0.319	−62.17	−15.2		D 0.4	(149,179)
AM 0109-364	01 09 45	2.774	279.18	261.7	COMPACT	13	F352/J0749
	−36 45.0	0.319	−79.66	−12.6		D 0.3	(201,83)
*AM 0110-503	01 10 09	2.582	294.61	247.5	SPIRAL w. CONDENSATION in ARM	10	F195/J0937
	−50 39.9	0.318	−66.38	−14.8	IC 1650 = 195-G-34	S 0.4	(90,140)
AM 0110-583	01 10 15	2.415	297.66	239.4	I/A TRIPLE (#1) [A82]	3	F113/J1870
[0110-5832]	−58 30.6	0.318	−58.66	−15.6	NGC 434 = 113-G-23 = B 0110-585	D 1.0	(220,259)
AM 0110-312	01 10 23	2.827	254.83	267.1	LONG [N.S.] CHAIN	17	F412/J3516
	−31 22.3	0.318	−83.87	−11.7		D 35.8	(163,106)
*AM 0110-240	01 10 28	2.892	359.88	265.7	PAIR of SPIRALS each w. COMP'NS.	8,23	F475/J1916
	−24 05.2	0.318	−84.15	−2.4		S 2.5	(122,226)
*AM 0110-582	01 10 33	2.413	297.58	239.5	I/A TRIPLE (#2)	3	F113/J1870
[0110-5830]	−58 28.4	0.318	−58.69	−15.6	113-G-24	D 0.7	(218,261)
AM 0110-583	01 10 48	2.409	297.54	239.4	I/A TRIPLE (#3)	3	F113/J1870
[0110-5828]	−58 32.9	0.318	−58.61	−15.6	NGC 440 = 113-G-25	D 0.7	(216,257)
AM 0110-242	01 10 55	2.889	2.57	265.9	CLOSE PAIR (#1)	23	F475/J1916
[0111-2419]	−24 20.8	0.318	−84.21	−2.2		D 0.3	(116,212)
AM 0111-302	01 11 07	2.833	246.99	268.0	PAIR (#1: S0)	23	F412/J3516
[0111-3030]	−30 29.5	0.318	−84.25	−11.7		S 2.7	(155,153)
AM 0111-241	01 11 08	2.889	2.62	265.9	CLOSE PAIR (#2)	23	F475/J1916
[0110-2420]	−24 19.3	0.318	−84.15	−2.2		D 0.4	(114,213)
*AM 0111-320	01 11 17	2.817	258.11	266.5	15 GAL. + FNT. COMP'NS. [KLEM. 1]	8,18	F352/J0749
	−32 01.4	0.318	−83.30	−12.1	NGC 439 = 412-G-18 in DN 0112-321	D 11.2	(185,337)
AM 0111-303	01 11 18	2.832	246.86	268.0	PAIR (#2: E)	23	F412/J3516
[0111-3029]	−30 30.0	0.318	−84.21	−11.8		S 2.7	(152,152)
AM 0111-523	01 11 25	2.538	295.07	245.6	OPEN SBc w. OVAL NUCLEUS	10,12	F196/J1851
	−52 32.9	0.318	−64.49	−15.2	151-G-34?	D 0.9	(330,37)
AM 0111-454	01 11 37	2.652	290.68	252.7	I/A E's in WIDE GROUP	2,18	F244/J1761
	−45 40.4	0.318	−71.17	−14.4		S 1.2	(295,141)
*AM 0111-325	01 11 47	2.806	262.69	265.6	SPIRAL + DIAMETRIC COMP'NS.	8	F412/J3516
	−32 55.0	0.318	−82.60	−12.3	IC 1657 = 352-G-24	S 2.1	(147,23)
AM 0111-323	01 11 56	2.810	260.32	266.0	CHAIN of 5 (PART of GROUP)	17,18	F352/J0749
[0112-3232]	−32 31.2	0.318	−82.86	−12.3	MCG-5-4-19 = 352-G-26	D 4.5	(177,310)
AM 0112-581	01 12 02	2.407	297.12	239.8	SPIRAL w. HI.and LO.S.BR. COMP'NS	8	F113/J1870
[0111-3231]	−58 12.0	0.318	−58.92	−15.8	113-G-27	S 3.1	(208,276)

Identification	R.A. (1950.0) Declination	Precession R.A./Dec.	Gal. Long. Gal. Lat.	S.G.L. S.G.B.	Description Cross-Identification	Coding Size (Min.)	Field/Plate (x, y)
AM 0112-312	01 12 10 −31 26.9	2.820 0.318	253.21 −83.52	267.1 −12.1	GALAXY w. I/A COMP'N. MCG-5-4-21 = 412-G-21	1 S 0.9	F412/J3516 (143,102)
*AM 0112-323	01 12 11 −32 32.0	2.809 0.318	260.11 −82.81	266.0 −12.3	CHAIN of 5 (4 SPIRALS) MCG-5-4-20 = 352-G-27 = PKS 0112-325	17 D 16.1	F412/J3516 (143,43)
AM 0112-475	01 12 15 −47 51.7	2.615 0.318	292.05 −69.03	250.5 −14.8	CLUSTER of GALAXIES DN 0112-474	19 D 6.7	F195/J0937 (66,290)
*AM 0112-554	01 12 19 −55 40.1	2.466 0.318	296.13 −61.41	242.4 −15.6	I/A DBL. + RESOLVED KNOTS [S.] NGC 454 = 151-IG-36 = B 0112-556	2 S 0.8	F151/J1693 (129,141)
AM 0112-464 [0113-4646]	01 12 48 −46 47.5	2.629 0.317	291.00 −70.04	251.6 −14.7	CLOSE PAIR (#1) 244-G-5	23 D 0.4	F244/J1761 (282,81)
AM 0112-422	01 12 53 −42 26.6	2.691 0.317	286.64 −74.18	256.0 −14.1	SMALL CONTACT DBL.	2 S 0.1	F243/J3492 (12,313)
AM 0113-464 [0112-4647]	01 13 06 −46 46.6	2.627 0.317	290.84 −70.04	251.6 −14.8	CLOSE PAIR (#2) 244-G-8	23 D 0.4	F244/J1761 (279,82)
AM 0113-514	01 13 10 −51 45.2	2.541 0.317	294.05 −65.21	246.5 −15.4	DWARF	20 D 0.2	F196/J1851 (319,81)
AM 0113-264	01 13 11 −26 42.6	2.862 0.317	27.21 −84.57	268.1 −1.2	GAL. w. LONG TAIL + 6 PEC.COMP'NS MCG-5-4-24 = 475-G-14	8,15 D 3.4	F475/J1916 (90,85)
AM 0113-233	01 13 19 −23 32.2	2.890 0.317	359.30 −83.30	265.0 −1.8	SMALL GALAXY w. JET	7 D 0.7	F475/J1916 (86,255)
*AM 0113-324	01 13 34 −32 44.6	2.802 0.317	259.72 −82.46	265.9 −12.7	SPIRAL + RING COMP'N. MCG-5-4-27 = 352-G-30	6,8 S 2.9	F412/J3516 (127,32)
AM 0113-613 [0113-6134]	01 13 39 −61 31.2	2.295 0.317	297.80 −55.62	236.4 −16.2	I/A DBL. (#1) 113-G-30	2 D 0.5	F113/J1870 (194,98)
AM 0113-613 [0113-6131]	01 13 47 −61 34.4	2.292 0.317	297.79 −55.56	236.3 −16.2	I/A DBL. (#2) 113-G-31	2 D 0.7	F113/J1870 (193,95)
AM 0114-331	01 14 04 −33 12.6	2.795 0.317	261.63 −82.05	265.4 −12.9	COMPACT 352-G-31	13 D 0.2	F352/J0749 (154,273)
AM 0114-245	01 14 09 −24 53.4	2.876 0.317	10.55 −83.83	266.3 −1.4	GALAXY w. DIAMETRIC COMP'NS.	8 S 1.8	F475/J1916 (77,182)
AM 0114-284	01 14 24 −28 49.6	2.838 0.317	228.89 −84.22	269.8 −12.1	COMPACT + ONE-ARMED SP. in I/A 3	3,10,13 D 1.1	F412/J3516 (116,242)
AM 0114-500	01 14 42 −50 01.4	2.563 0.316	292.46 −66.84	248.3 −15.4	DWARF	20 D 0.2	F196/J1851 (311,175)
AM 0114-614	01 14 48 −61 48.0	2.274 0.316	297.65 −55.32	236.1 −16.3	SPIRAL + I/A COMP'N. in LINE of 3 113-G-32/33	24 D 11.2	F113/J1870 (187,83)
AM 0114-371	01 14 49 −37 17.4	2.747 0.316	275.83 −78.70	261.3 −13.7	DBL. M51-TYPE	9 D 0.8	F296/J3524 (213,325)
*AM 0115-340	01 15 02 −34 06.2	2.782 0.316	264.83 −81.26	264.6 −13.2	SPIRAL w. 9 CLOSE COMP'NS. NGC 461 = 352-G-23	8 D 1.3	F352/J0749 (143,225)
AM 0115-591	01 15 13 −59 10.6	2.351 0.316	296.68 −57.89	238.8 −16.3	GALAXY w. 7 APP. COMP'NS. NGC 466 = 113-G-34	8 S 2.9	F113/J1870 (185,224)
AM 0115-260	01 15 22 −26 05.8	2.862 0.316	22.13 −83.97	267.4 −0.9	CLOSE GROUP of 4 PKS 0115-261	18 D 1.0	F475/J1916 (64,117)
AM 0115-365	01 15 46 −36 52.4	2.748 0.316	273.87 −78.96	261.8 −13.8	CHAIN of 5 (SPIRAL in CENTRE) PKS 0115-368	17 D 5.4	F352/J0749 (136,76)
AM 0115-310	01 15 47 −31 02.2	2.812 0.316	246.59 −83.11	267.7 −12.8	I/A PAIR	2 S 3.8	F412/J3516 (101,123)
AM 0115-271	01 15 48 −27 15.9	2.849 0.316	213.36 −84.05	271.5 −12.1	RICH CLUSTER of V. FAINT GALAXIES	19 D 4.5	F475/J1916 (60,54)
AM 0115-420 [0116-4205]	01 15 54 −42 07.0	2.680 0.316	284.29 −74.28	256.4 −14.7	PAIR + COMP'NS. (#1: LO.S.BR.) 296-G-8	8,20,23 D 0.8	F296/J3524 (200,66)
AM 0115-444	01 15 58 −44 43.3	2.642 0.316	287.47 −71.84	253.8 −15.0	DISTURBED (MERGER?) Fa 690 = Ag-5 244-G-12 = VV 827 = B 0015-447	16 D 1.3	F244/J1761 (256,193)
AM 0115-483	01 15 58 −48 31.2	2.581 0.316	290.89 −68.23	249.9 −15.5	DWARF	20 D 0.3	F196/J1851 (304,256)
AM 0115-463	01 15 59 −46 30.2	2.615 0.316	289.22 −70.16	251.9 −15.2	ONE-ARMED SP. + COMP'N. (RING?) 244-IG-13	6,8,10,15 S 0.4	F244/J1761 (253,98)
AM 0116-372	01 16 01 −37 21.9	2.741 0.316	274.99 −78.51	261.3 −14.0	GAL.w. PLUMES, COMP'NS + COMPACT 352-G-38	8,13,15 S 2.1	F296/J3524 (200,321)
AM 0116-420 [0115-4207]	01 16 11 −42 05.5	2.679 0.316	284.07 −74.28	256.5 −14.7	PAIR + COMP'NS. (#2: LO.S.BR.) 296-G-9	8,20,23 D 0.7	F296/J3524 (198,67)
AM 0116-241	01 16 22 −24 12.3	2.877 0.316	7.20 −83.07	265.5 −1.0	SP. w. PEC. ARM + 4 APP. COMP'NS MCG-4-4-3 = 475-G-16	8,10 D 1.1	F475/J1916 (50,219)
AM 0116-402	01 16 35 −40 28.4	2.699 0.316	281.25 −75.73	258.1 −14.5	SMALL I/A SPIRALS 295-G-31	2 S 0.2	F296/J3524 (194,154)
AM 0116-245	01 16 46 −24 54.5	2.869 0.315	12.87 −83.30	266.2 −0.8	CLOSE TRIPLE	24 D 1.6	F475/J1916 (46,181)
AM 0116-370	01 16 52 −37 04.2	2.741 0.315	273.47 −78.67	261.6 −14.1	ARC of 4 COMPACTS (CONN. MATR'L) 352-IG-42 = PKS 0116-373?	4,13,17 D 1.8	F296/J3524 (191,337)
AM 0117-674	01 17 14 −67 47.5	1.991 0.315	298.91 −49.37	229.8 −16.8	PECULIAR SPIRAL	12 D 0.4	F052/J0930 (311,289)

Identification	R.A. (1950.0) Declination	Precession R.A./Dec.	Gal. Long. Gal. Lat.	S.G.L. S.G.B.	Description Cross-Identification	Coding Size (Min.)	Field/Plate (x, y)
AM 0117-681	01 17 21	1.961	299.02	229.3	DWARF (RESOLUTION)	20	F052/J0930
	−68 18.6	0.315	−48.86	−16.8	Se 10/1	D 0.9	(308,261)
*AM 0117-520	01 17 23	2.506	292.73	246.2	ONE-ARMED SPIRAL / RING ?	6,10	F196/J1851
	−52 02.5	0.315	−64.78	−16.0	196-G-3	D 0.8	(283,67)
AM 0117-681	01 17 23	1.960	299.02	229.3	IRREGULAR COMPACT COMP'N.	13,16	F051/J1877
	−68 18.7	0.315	−48.86	−16.8		D 0.7	(54,264)
AM 0117-434	01 17 32	2.648	285.53	254.8	I/A DBL. + COMP'NS. (#1)	2,8	F244/J1761
[0117-4349]	−43 47.6	0.315	−72.61	−15.2	244-G-15	D 0.4	(242,243)
AM 0117-434	01 17 33	2.647	285.57	254.7	I/A DBL. + COMP'NS (#2)	2,8	F244/J1761
[0117-4347]	−43 49.9	0.315	−72.58	−15.2	244-G-16	D 0.4	(242,241)
AM 0117-343	01 17 37	2.768	264.09	264.2	I/A DBL.	2	F352/J0749
	−34 31.3	0.315	−80.60	−13.8		S 0.9	(115,202)
AM 0117-223	01 17 44	2.888	358.48	263.9	ONE-ARMED SPIRAL w. COMP'NS.	8,12	F475/J1916
	−22 38.5	0.315	−81.94	−1.0	NGC 478 = 476-IG-3	D 0.6	(31,302)
AM 0117-303	01 17 45	2.811	241.13	268.3	PAIR (#1: HI.S.BR. SPIRAL)	23	F412/J3516
[0117-3026]	−30 30.8	0.315	−82.98	−13.1		S 4.7	(78,151)
*AM 0117-412	01 17 45	2.680	282.18	257.1	"RIGHT-ANGLE" GALAXY = B 0117-414	2	F296/J3524
	−41 29.9	0.315	−74.71	−14.9	Ag-6 = MCG-7-3-16 = 296-IG-11	S 0.2	(182,99)
*AM 0117-340	01 17 48	2.771	262.39	264.6	SPIRAL w. BR. STARS in CORNER	12	F352/J0749
	−34 09.8	0.315	−80.83	−13.8	NGC 491a = 352-G-46	D 2.0	(112,222)
AM 0117-302	01 17 55	2.811	240.50	268.4	PAIR (#2: Sa:)	23	F412/J3516
[0117-3030]	−30 26.7	0.315	−82.97	−13.2		S 4.7	(76,155)
AM 0117-442	01 17 59	2.636	286.01	254.1	SB + 4 COMP'NS.	8	F244/J1761
	−44 23.6	0.315	−72.03	−15.3	244-G-17	S 2.7	(237,211)
AM 0118-623	01 18 00	2.216	297.23	235.4	COMPACT	13	F113/J1870
	−62 30.1	0.315	−54.56	−16.7		D 0.4	(167,45)
AM 0118-500	01 18 05	2.539	291.19	248.3	DWARF (PAIR) ?	20,23	F196/J1851
	−50 05.5	0.315	−66.62	−16.0	Fa 692	S 1.0	(282,172)
AM 0118-411	01 18 07	2.682	281.51	257.4	EDGE-ON + 3 COMP'NS.	8	F296/J3524
	−41 13.8	0.315	−74.92	−14.9	NGC 482 = 296-G-13	S 6.7	(178,114)
AM 0118-600	01 18 09	2.295	296.36	237.8	CLOSE PAIR	23	F113/J1870
	−60 08.2	0.315	−56.87	−16.7	113-G-37	S 1.8	(165,172)
AM 0118-221	01 18 12	2.890	356.77	263.5	SMALL I/A DBL.	2	F475/J1916
	−22 14.7	0.315	−81.61	−1.0		S 0.3	(25,324)
AM 0118-270	01 18 37	2.843	31.94	268.2	GALAXY w. PECULIAR ABSORPTION	8,14	F475/J1916
	−27 04.2	0.315	−83.41	0.0		D 0.9	(26,64)
AM 0118-265	01 18 45	2.844	31.26	268.2	S0 w. DISTURBED COMP'N.	8	F476/J2653
	−26 59.3	0.314	−83.37	0.1	MCG-5-4-31 = 476-G-4	S 5.8	(290,72)
AM 0118-230	01 18 53	2.881	2.03	264.3	SPIRAL w. COMPACT off MI.AX.	8,13	F476/J2653
	−23 03.7	0.314	−81.98	−0.7	MCG-4-4-6 = 476-G-5	D 2.2	(292,283)
AM 0118-364	01 18 57	2.736	270.75	262.0	E w. DBL. NUCLEUS	8,22	F352/J0749
	−36 44.1	0.314	−78.71	−14.4	352-G-51 (DIAMETRIC SPIRALS)	D 0.6	(102,83)
AM 0119-340	01 19 00	2.767	261.02	264.7	COMPACT w. FAINT RING	6,13	F352/J0749
	−34 06.1	0.314	−80.70	−14.0		D 0.3	(99,225)
AM 0119-444	01 19 10	2.625	285.73	253.9	CLOSE PAIR (E + SPIRAL)	23	F244/J1761
	−44 41.4	0.314	−71.67	−15.6		S 1.7	(225,196)
AM 0119-424	01 19 36	2.654	282.85	255.9	SPIRAL w. HI. and LO.S.BR.ARMS	8,10	F244/J1761
	−42 40.7	0.314	−73.49	−15.4	244-G-20 + APP. COMP'N.	D 0.8	(223,304)
AM 0119-613	01 19 38	2.231	296.59	236.3	E + SPIRAL w. COMP'NS.	2,8	F113/J1870
	−61 37.1	0.314	−55.39	−16.9	113-G-41/42 = B 0119-616	S 1.3	(156,92)
AM 0119-555	01 19 40	2.398	294.18	242.2	DISTANT CLUSTER	19	F151/J1693
	−55 59.4	0.314	−60.87	−16.7		D 4.0	(74,121)
AM 0120-340	01 20 27	2.762	259.89	264.7	COMPACT	13	F352/J0749
	−34 07.7	0.314	−80.46	−14.3		D 0.2	(83,223)
AM 0120-325	01 20 32	2.774	254.37	265.9	M51-TYPE (SMALL)	9	F413/J0859
	−32 59.7	0.314	−81.18	−14.2		S 0.4	(306,16)
AM 0120-540	01 20 34	2.440	292.84	244.2	COMPACT	13	F152/J1862
	−54 00.2	0.314	−62.76	−16.7		D 0.2	(325,228)
AM 0120-305	01 20 41	2.798	241.67	268.0	CHAIN of 5 (IRR. at CENTRE)	16,17	F412/J3516
	−30 52.7	0.314	−82.25	−13.8		D 11.0	(44,131)
AM 0120-424	01 20 43	2.646	282.41	255.8	E + I/A COMP'N.	1	F244/J1761
	−42 49.7	0.313	−73.27	−15.6	244-G-22	S 0.2	(212,296)
AM 0120-341	01 20 45	2.758	260.37	264.6	COMPACT	13	F352/J0749
	−34 17.9	0.313	−80.30	−14.4		D 0.2	(80,214)
AM 0120-432	01 20 50	2.637	283.15	255.2	EDGE-ON + I/A COMP'N.	1	F244/J1761
	−43 23.2	0.313	−72.76	−15.7	244-G-23	S 0.2	(210,266)
AM 0121-272	01 21 15	2.833	214.42	271.6	CLUSTER of FAINT GALAXIES	19	F413/J0859
	−27 20.0	0.313	−82.84	−13.3		D 6.7	(304,320)
AM 0121-340	01 21 39	2.758	258.61	264.8	COMPACT (LUMPY)	13,16	F352/J0749
	−34 03.9	0.313	−80.32	−14.6	352-G-67	D 0.3	(70,226)
*AM 0121-351	01 21 40	2.743	263.77	263.5	I/A DBL. in SUB-GROUP of 4	2,18	F353/J3596
	−35 19.7	0.313	−79.44	−14.8	NGC 526 = 352-IG-66	S 0.3	(336,159)

Identification	R.A. (1950.0) Declination	Precession R.A./Dec.	Gal. Long. Gal. Lat.	S.G.L. S.G.B.	Description Cross-Identification	Coding Size (Min.)	Field/Plate (x, y)
AM 0121-590	01 21 51 −59 04.2	2.293 0.313	295.08 −57.81	239.0 −17.1	GAL. w. HI.S.BR. NUCLEUS + COMP'N 113-IG-45 = Fa 9	8 S 0.4	F113/J1870 (139,229)
AM 0121-452	01 21 58 −45 21.7	2.599 0.313	285.11 −70.87	253.2 −16.1	DISTURBED GALAXY	16 D 0.4	F244/J1761 (198,160)
*AM 0122-332	01 22 16 −33 26.0	2.763 0.313	255.16 −80.62	265.5 −14.6	SPIRAL w. APP. COMP'N. 352-G-71	8 S 3.1	F353/J3596 (333,261)
*AM 0122-320	01 22 31 −32 01.1	2.779 0.313	247.49 −81.38	266.9 −14.4	SPIRAL w. LINEAR CONDENS. in ARM MCG-5-4-36 = 413-G-4	11 D 1.3	F413/J0859 (284,69)
AM 0122-360	01 22 57 −36 04.3	2.728 0.312	265.42 −78.72	262.8 −15.1	CLOSE PAIR 352-G-72	23 S 0.4	F353/J3596 (321,119)
AM 0122-381	01 22 58 −38 19.7	2.699 0.312	272.12 −76.95	260.5 −15.5	I/A DBL.: E + SP. (in GROUP of 6) NGC 544/6 = 296-G-24/5 = B 0122-383	2,18 S 1.2	F296/J3524 (127,269)
*AM 0123-685	01 23 19 −68 52.7	1.843 0.312	298.37 −48.20	228.7 −17.3	CLOSE GROUP of 7 52-IG-1 = Se 10/2 = DN 0124-689	18 D 2.2	F052/J0930 (276,233)
*AM 0123-373	01 23 35 −37 35.6	2.706 0.312	269.69 −77.47	261.3 −15.5	E w. CLOSE DIAMETRIC COMP'NS. 296-G-29	8,13 D 0.5	F353/J3596 (311,38)
AM 0123-403	01 23 51 −40 39.5	2.662 0.312	276.80 −74.91	258.1 −15.9	GROUP of 4	18 D 2.5	F296/J3524 (120,144)
*AM 0124-385	01 24 00 −38 52.8	2.687 0.312	272.78 −76.39	260.0 −15.7	LINE of 3 MC 1024-388	24 S 7.8	F296/J3524 (116,239)
AM 0124-232	01 24 10 −23 29.2	2.865 0.312	8.79 −81.15	264.5 0.6	SPIRAL w. 3 COMP'NS.(1 DISTURBED) MCG-4-4-9 = 476-G-8	8 S 4.9	F476/J2653 (226,260)
AM 0124-282	01 24 27 −28 26.0	2.812 0.312	222.67 −82.10	270.7 −14.2	I/A DBL.	2 S 0.6	F413/J0859 (265,262)
AM 0124-384	01 24 39 −38 43.8	2.686 0.311	271.98 −76.44	260.1 −15.8	DWARF	20 D 0.7	F296/J3524 (109,247)
*AM 0124-225	01 24 46 −22 58.8	2.868 0.311	6.39 −80.79	264.0 0.7	I/A 4 (CONNECTED CHAIN of 6) NGC 554 = 476-IG-11	4,17 D 4.5	F476/J2653 (219,288)
AM 0124-355	01 24 49 −35 52.3	2.724 0.311	263.32 −78.60	263.1 −15.5	GALAXY + DUST + GROUP of COMP'NS	8,14 D 4.5	F352/J0749 (37,128)
*AM 0125-583	01 25 13 −58 36.8	2.277 0.311	294.06 −58.14	239.4 −17.5	SPIRAL w. PEC. ARMS + PEC.COMP'N. 113-G-48 [A82]	8,10 S 2.7	F113/J1870 (115,253)
AM 0125-234	01 25 31 −23 48.0	2.858 0.311	11.48 −81.02	264.8 1.0	GALAXY w. TAILS	15 D 0.4	F476/J2653 (210,244)
AM 0125-365	01 25 32 −36 56.5	2.707 0.311	266.36 −77.72	262.0 −15.8	COMPACT w. BULGE + STAR?	13 D 0.3	F352/J0749 (32,70)
AM 0125-850	01 25 32 −85 00.0	−0.042 0.311	302.06 −32.32	211.9 −16.3	SPIRAL w. I/A COMP'N. (CONDENSATION at END)	1,9 S 0.6	F002/J2563 (82,161)
AM 0126-525	01 26 03 −52 53.1	2.426 0.311	290.38 −63.59	245.5 −17.4	SPIRAL w. DBL. NUCLEUS (STAR ?) 152-IG-3 = Fa 687/8	12 D 0.8	F196/J1851 (211,24)
AM 0126-434	01 26 13 −43 47.3	2.602 0.311	280.85 −71.97	254.9 −16.7	I/A SPIRALS B 0126-437 MCG-7-4-8 = 244-IG-29 = Ag-8	2 S 0.4	F244/J1761 (158,244)
AM 0126-323	01 26 17 −32 33.2	2.759 0.311	247.88 −80.42	266.5 −15.3	GROUP of 3	24 D 1.0	F353/J3596 (289,309)
*AM 0126-680	01 26 21 −68 06.5	1.850 0.311	297.72 −48.89	229.5 −17.6	BULL'S-EYE w. ELLIPTICAL RING	6 D 1.1	F080/J1004 (193,32)
*AM 0126-525	01 26 27 −52 53.5	2.423 0.311	290.26 −63.56	245.5 −17.5	RING or GALAXY w. LOOP 152-IG-3 = PKS 0126-53?	6,15 D 1.0	F151/J1693 (11,284)
*AM 0126-614	01 26 27 −61 41.7	2.159 0.311	295.24 −55.13	236.2 −17.7	HI.S.BR. GALAXY w. TAILS 113-G-49 in 113-G-50 Group	15 D 0.4	F113/J1870 (113,87)
AM 0126-514	01 26 37 −51 40.0	2.449 0.310	289.28 −64.71	246.7 −17.4	GALAXY w. PECULIAR DUST and PLUME 196-G-6 (in 196-G-11 Group)	14,15 D 0.5	F196/J1851 (207,90)
AM 0126-653	01 26 42 −65 32.4	1.988 0.310	296.76 −51.39	232.2 −17.7	DISTURBED E w. JET	7 D 0.7	F080/J1004 (193,170)
AM 0126-515	01 26 57 −51 51.2	2.443 0.310	289.31 −64.52	246.5 −17.5	DISTURBED SPIRAL NGC 576 = 196-G-7 = Fa 295	12 D 0.4	F196/J1851 (204,80)
AM 0127-441	01 27 05 −44 17.2	2.590 0.310	281.14 −71.45	254.4 −16.9	COMPACT	13 D 0.1	F244/J1761 (150,218)
AM 0127-223	01 27 17 −22 34.9	2.866 0.310	6.11 −80.09	263.5 1.2	I/A DBL. MCG-4-4-18 = 476-G-14	2 S 2.0	F476/J2653 (188,309)
AM 0127-524	01 27 25 −52 48.8	2.417 0.310	289.89 −63.59	245.5 −17.6	CLOSE I/A DBL.?	2 S 0.2	F152/J1862 (273,295)
AM 0127-423	01 27 41 −42 34.7	2.615 0.310	278.13 −72.90	256.2 −16.9	DISTURBED SINGLE GALAXY 244-IG-30 = B 0127-425 + COMP'NS	8,16 D 0.9	F244/J1761 (143,309)
AM 0127-614	01 27 44 −61 43.4	2.145 0.310	295.00 −55.06	236.2 −17.9	E + I/A COMP'N. (JET?) 113-G-50 Group	1,7 S 1.3	F113/J1870 (104,85)
AM 0127-514	01 27 52 −51 40.7	2.441 0.310	288.87 −64.63	246.7 −17.6	I/A DBL. 196-IG-10 = B 0128-516	2,15 S 0.2	F196/J1851 (197,89)
AM 0127-331	01 27 52 −33 10.1	2.746 0.310	249.94 −79.82	266.0 −15.7	CONNECTED E's	2 S 1.1	F413/J0859 (223,8)
AM 0128-270 [0128-2707]	01 28 04 −27 02.3	2.817 0.310	32.94 −81.31	267.8 2.1	PAIR (#1: Sc) MCG-5-4-41 = 476-G-16	23 D 1.8	F413/J0859 (222,337)

Identification	R.A. (1950.0) Declination	Precession R.A./Dec.	Gal. Long. Gal. Lat.	S.G.L. S.G.B.	Description Cross-Identification	Coding Size (Min.)	Field/Plate (x, y)
*AM 0128-225	01 28 05 −22 55.6	2.861 0.310	8.38 −80.10	263.8 1.4	SPIRAL w. 6 COMP'NS. NGC 578 = 476-G-15	8 D 4.9	F476/J2653 (178,291)
AM 0128-741	01 28 14 −74 11.1	1.302 0.310	299.36 −42.93	223.1 −17.5	GALAXY w. APP. COMP'N.	8 S 0.8	F029/J1948 (98,220)
AM 0128-413	01 28 15 −41 33.3	2.628 0.310	275.95 −73.71	257.3 −16.9	GALAXY w. APP. COMP'N. 296-G-34	8 S 3.3	F297/J3693 (337,92)
AM 0128-421	01 28 15 −42 15.3	2.617 0.310	277.25 −73.12	256.6 −16.9	DWARF	20 D 0.4	F297/J3693 (335,54)
AM 0128-375	01 28 27 −37 58.7	2.680 0.309	267.48 −76.56	261.0 −16.5	SP. w. 3 PEC. ARMS + APP. COMP'N. 296-G-35	8,10,11 S 0.9	F296/J3524 (68,286)
AM 0128-270 [0128-2702]	01 28 53 −27 07.2	2.814 0.309	213.57 −81.13	272.2 −14.9	PAIR (#2: E) MCG-5-4-42 = 476-G-18	8,23 D 1.2	F413/J0859 (212,333)
AM 0128-235	01 28 55 −23 52.1	2.849 0.309	13.86 −80.34	264.7 1.7	LINE of 3 SPIRALS	24 D 11.6	F476/J2653 (168,240)
AM 0129-514	01 29 28 −51 47.0	2.427 0.309	288.41 −64.45	246.6 −17.9	I/A DBL.	2 S 0.2	F196/J1851 (183,84)
AM 0129-332	01 29 34 −33 22.5	2.738 0.309	249.88 −79.41	265.8 −16.1	SPIRAL w. 4 COMP'NS. 353-G-9	8 S 1.8	F353/J3596 (252,266)
AM 0129-415	01 29 38 −41 56.0	2.615 0.309	275.90 −73.25	256.9 −17.2	CHAIN of 4 in GROUP of 6	17,18 D 5.8	F296/J3524 (63,74)
AM 0129-261	01 29 54 −26 15.7	2.821 0.309	28.30 −80.81	267.0 2.4	CLUSTER of FAINT GAL. (TRIPLE E)	3,19 D 5.4	F476/J2653 (156,112)
AM 0129-421	01 29 59 −42 18.3	2.608 0.309	276.41 −72.91	256.6 −17.3	DWARF	20 D 0.4	F297/J3693 (318,52)
AM 0130-312	01 30 03 −31 20.9	2.761 0.309	239.35 −80.22	267.9 −15.9	SMALL CLUSTER (MIXED TYPES)	19 D 4.5	F413/J0859 (199,106)
AM 0130-540	01 30 04 −54 07.1	2.366 0.309	290.04 −62.23	244.2 −18.1	SPIRAL w. LINEAR ARM + 2 COMP'NS 152-G-7	8,10 S 1.4	F152/J1862 (250,226)
AM 0130-385	01 30 14 −38 56.2	2.659 0.308	268.95 −75.60	260.1 −16.9	CHAOTIC SPIRAL w. APP. COMP'N. 296-G-38	8,16 S 2.0	F297/J3693 (323,233)
AM 0130-293	01 30 29 −29 34.1	2.781 0.308	228.90 −80.63	269.7 −15.7	I/A DBL. (SPIRALS)	2 S 0.4	F413/J0859 (193,201)
*AM 0131-364	01 31 05 −36 42.0	2.688 0.308	261.93 −77.12	262.4 −16.8	I/A TRIPLE PKS 0131-36	3 D 0.5	F353/J3596 (232,88)
AM 0131-434 [0131-4345]	01 31 11 −43 48.0	2.577 0.308	278.37 −71.52	255.0 −17.6	TRIPLE (#1)	24 D 0.2	F244/J1761 (110,243)
AM 0131-715	01 31 26 −71 55.6	1.483 0.308	298.36 −45.08	225.5 −17.9	LINE of 3	24 D 4.4	F029/J1948 (74,339)
AM 0131-434 [0131-4348]	01 31 31 −43 49.7	2.574 0.308	278.25 −71.46	255.0 −17.7	TRIPLE (#2 RING + I/A COMP'N.)	1,6,24 D 0.2	F244/J1761 (107,242)
AM 0131-434 [0131-4349]	01 31 33 −43 45.3	2.575 0.308	278.12 −71.52	255.1 −17.7	TRIPLE (#3) 244-G-33	24 D 0.4	F244/J1761 (106,245)
*AM 0131-364 [0132-3644]	01 31 42 −36 44.9	2.684 0.308	261.70 −77.00	262.4 −17.0	3 in GROUP of 8 (#1: THIN DUST) NGC 612 = 353-G-15	8,14,18,24 S 2.5	F353/J3596 (225,85)
AM 0131-343 [0131-3442]	01 31 44 −34 38.7	2.713 0.308	254.06 −78.33	264.6 −16.7	CLOSE PAIR (#1: Sd) 353-G-14	20,23 S 2.9	F353/J3596 (226,198)
AM 0131-330 [0132-3305]	01 31 46 −33 03.7	2.734 0.308	247.16 −79.17	266.2 −16.5	TRIPLE (#1: Sbc) 353-G-16	24 D 0.8	F413/J0859 (179,13)
AM 0131-392	01 31 46 −39 21.1	2.646 0.308	269.09 −75.10	259.7 −17.3	COMPACT GAL. w. CONDENS. in TAIL 297-IG-1	13,15 D 3.4	F297/J3693 (306,212)
AM 0131-344 [0131-3438]	01 31 48 −34 42.1	2.712 0.308	254.25 −78.28	264.5 −16.7	CLOSE PAIR (#2: HI.S.BR. KNOTTY) 353-G-17	16,23 S 2.9	F353/J3596 (225,195)
*AM 0132-294	01 32 00 −29 40.7	2.775 0.307	229.16 −80.28	269.7 −16.0	MULTI-ARMED SPIRAL NGC 613 = 413-G-11 = PKS 0131-296	10,11 D 5.0	F413/J0859 (176,195)
AM 0132-325 [0131-3303]	01 32 00 −32 59.9	2.734 0.307	246.75 −79.15	266.3 −16.5	TRIPLE (#2: Sc) 353-G-18	24 D 0.6	F413/J0859 (177,17)
AM 0132-385	01 32 01 −38 52.3	2.652 0.307	267.73 −75.43	260.2 −17.3	S0 + 6 COMP'NS. (3 I/A) 297-G-3	3,8 S 6.3	F297/J3693 (305,237)
AM 0132-330 [0132-3259]	01 32 04 −33 05.4	2.733 0.307	247.14 −79.10	266.2 −16.6	TRIPLE (#3: E)	24 D 0.3	F413/J0859 (176,12)
AM 0132-315	01 32 07 −31 51.7	2.748 0.307	241.12 −79.62	267.4 −16.4	SMALL CLUSTER (MIXED TYPES)	19 D 7.2	F413/J0859 (175,78)
AM 0132-600	01 32 31 −60 09.6	2.158 0.307	293.23 −56.40	237.8 −18.5	GROUP of 5	18 D 1.8	F114/J1813 (322,164)
AM 0132-244	01 32 36 −24 40.6	2.832 0.307	19.97 −79.84	265.3 2.7	GALAXY w. MATERIAL (ABS'N.?) 476-G-26 = MC 0132-246	14,16 D 0.7	F476/J2653 (123,197)
AM 0132-364 [0131-3644]	01 32 37 −36 44.7	2.681 0.307	261.15 −76.87	262.4 −17.2	3 in WIDE GROUP of 8 (#2: S(r)) NGC 619 = 353-G-21	18,24 S 2.5	F353/J3596 (215,85)
AM 0132-364 [0132-3644]	01 32 52 −36 44.7	2.680 0.307	261.00 −76.83	262.4 −17.2	3 in WIDE GROUP of 8 (#3: S0) 353-G-23	18,24 S 2.5	F353/J3596 (213,85)
*AM 0132-414	01 32 56 −41 41.7	2.604 0.307	273.69 −73.10	257.3 −17.7	S0?/IRR. + DISTURBED COMP'N. NGC 625 = 297-G-5	8 S 11.2	F297/J3693 (290,86)

Identification	R.A. (1950.0) Declination	Precession R.A./Dec.	Gal. Long. Gal. Lat.	S.G.L. S.G.B.	Description Cross-Identification	Coding Size (Min.)	Field/Plate (x, y)
AM 0133-392 [0133-3936]	01 33 01 −39 24.0	2.640 0.307	268.50 −74.90	259.6 −17.5	WIDE TRIPLE (#1: SPIRAL) 297-G-6	24 S 1.6	F296/J3524 (23,208)
AM 0133-380	01 33 06 −38 05.0	2.659 0.307	264.94 −75.87	261.0 −17.4	CLOSE PAIR (STAR BETWEEN) 297-G-7	23 S 1.8	F297/J3693 (295,280)
AM 0133-393 [0133-3924]	01 33 18 −39 38.4	2.635 0.307	268.93 −74.68	259.4 −17.6	WIDE TRIPLE (#2: S(r)) 297-G-8 = SS-1b	24 S 1.6	F296/J3524 (21,195)
AM 0133-393 [0133-3938]	01 33 25 −39 36.8	2.635 0.307	268.81 −74.69	259.4 −17.6	WIDE TRIPLE (#3: E) NGC 630 = 297-G-9 = SS-1a	24 S 1.6	F296/J3524 (20,196)
AM 0133-293	01 33 34 −29 38.6	2.771 0.307	228.60 −79.96	269.8 −16.4	COMPACT	13 D 0.1	F413/J0859 (158,197)
AM 0133-474	01 33 41 −47 41.3	2.490 0.306	282.71 −67.90	251.0 −18.3	LUMPY COMPACT	13 D 0.3	F196/J1851 (146,304)
*AM 0134-373	01 34 11 −37 35.0	2.662 0.306	262.85 −76.07	261.6 −17.6	SPIRAL MERGER? MCG-6-4-56/57 NGC 633 = 297-G-11 = B 0134-375	1 S 1.2	F297/J3693 (284,307)
AM 0134-825	01 34 11 −82 50.4	−1.176 0.306	301.30 −34.40	214.1 −16.8	CLOSE PAIR	23 S 0.8	F003/J2565 (262,286)
AM 0134-425	01 34 16 −42 50.6	2.578 0.306	275.21 −72.02	256.1 −18.1	SPIRAL + 2 APP. COMP'NS. 244-G-36	8,20 S 1.6	F244/J1761 (78,294)
AM 0134-762	01 34 23 −76 26.1	0.856 0.306	299.51 −40.64	220.7 −17.6	I/A SPIRALS	2 S 0.3	F029/J1948 (90,97)
AM 0134-603	01 34 37 −60 37.8	2.120 0.306	293.04 −55.87	237.3 −18.7	IRR. w. FILAMENT around STAR 113-IG-54	15,16 D 0.8	F114/J1813 (306,140)
AM 0134-803 [0136-8036]	01 34 41 −80 34.2	−0.157 0.306	300.68 −36.61	216.4 −17.1	PAIR (#1: LO.S.BR. GALAXY) 13-G-17	20,23 S 3.6	F013/J2551 (169,149)
AM 0134-464	01 34 43 −46 44.3	2.503 0.306	281.08 −68.66	252.0 −18.5	PAIR	23 S 0.5	F245/J3493 (338,81)
AM 0135-280	01 35 16 −28 09.0	2.784 0.306	219.95 −79.74	271.4 −16.5	COMPACT (LUMPY)	13 D 0.2	F413/J0859 (137,277)
*AM 0135-624	01 35 24 −62 47.5	2.022 0.305	294.05 −53.78	235.0 −18.7	E in PAIR + DIAMETRIC COMP'NS. [A82]	8,23 D 2.4	F114/J1813 (293,25)
AM 0135-330 [0135-3306]	01 35 34 −33 09.8	2.719 0.305	245.73 −78.42	266.2 −17.3	PAIR of SPIRALS (PEC. ARMS) (#1)	10,23 S 4.3	F353/J3596 (184,278)
AM 0135-650	01 35 49 −65 09.0	1.902 0.305	295.12 −51.52	232.6 −18.7	M51-TYPE NGC 646 = 80-G-2 = B 0135-651	9 S 1.1	F080/J1004 (142,191)
AM 0135-401 [0136-4015]	01 35 50 −40 19.5	2.612 0.305	269.20 −73.85	258.7 −18.2	PAIR (#1: EDGE-ON) 297-G-16	23 S 9.0	F297/J3693 (263,161)
AM 0135-330 [0135-3309]	01 35 53 −33 06.5	2.719 0.305	245.33 −78.38	266.3 −17.4	PAIR of SPIRALS (PEC. ARMS) (#2) 353-G-28	10,23 S 4.3	F353/J3596 (181,281)
*AM 0136-401 [0135-4019]	01 36 26 −40 15.9	2.611 0.305	268.76 −73.82	258.8 −18.3	PAIR (#2: Sb + 6 COMP'NS.) 297-G-18	8,23 S 3.1	F297/J3693 (257,164)
AM 0136-424 [0136-4250]	01 36 33 −42 46.9	2.567 0.305	274.00 −71.83	256.2 −18.5	PAIR (#1: E) SS-2a NGC 641 = 244-G-42 = PKS 0136-427	23 D 0.5	F244/J1761 (56,296)
AM 0136-465 [0136-4649]	01 36 35 −46 51.5	2.490 0.305	280.48 −68.40	251.9 −18.8	CLOSE PAIR (#1)	23 D 0.3	F244/J1761 (63,78)
AM 0136-803 [0134-8034]	01 36 40 −80 36.3	−0.232 0.305	300.59 −36.56	216.4 −17.2	PAIR (#2: EDGE-ON) 13-G-18	23 S 3.6	F013/J2551 (164,147)
AM 0136-425 [0136-4246]	01 36 45 −42 50.3	2.565 0.305	274.00 −71.76	256.1 −18.5	PAIR (#2: SPIRAL) Fa 706 NGC 644 = 244-G-43 = SS-2b	23 D 0.7	F244/J1761 (54,293)
AM 0136-301	01 36 49 −30 10.1	2.755 0.305	230.70 −79.16	269.3 −17.1	SPIRAL w. I/A HI.S.BR. COMP'N. NGC 642 = 413-G-14	1 S 1.6	F413/J0859 (120,168)
AM 0136-470	01 36 50 −47 05.0	2.484 0.305	280.68 −68.18	251.6 −18.8	SPIRAL in LONG CHAIN of 6 244-G-44 + APP. COMP'NS.	8,17 D 26.9	F244/J1761 (62,65)
AM 0136-464 [0136-4651]	01 36 53 −46 49.6	2.489 0.305	280.32 −68.40	251.9 −18.8	CLOSE PAIR (#2 + I/A COMP'N) 244-G-45 = MC 0136-468	23 D 0.5	F244/J1761 (61,79)
AM 0136-500	01 36 53 −50 02.7	2.419 0.305	284.22 −65.56	248.5 −19.0	I/A DBL. 196-IG-14	2 S 1.2	F196/J1851 (120,177)
AM 0136-433	01 36 57 −43 36.8	2.550 0.305	275.32 −71.11	255.3 −18.6	I/A SPIRALS 244-G-46/47 = B 0136-436	2,15 S 0.8	F244/J1761 (54,252)
AM 0137-225	01 37 01 −22 58.6	2.840 0.305	193.55 −78.28	276.8 −16.0	GALAXIES CONNECTED w. FILAMENT	2 S 1.0	F477/J3785 (340,286)
AM 0137-281 [0137-2817]	01 37 38 −28 12.6	2.777 0.304	220.20 −79.21	271.4 −17.0	PAIR (#1: Sc) MCG-5-5-5 = 413-G-15	23 D 1.0	F413/J0859 (109,273)
AM 0137-281 [0137-2812]	01 37 42 −28 17.2	2.776 0.304	220.61 −79.20	271.3 −17.0	PAIR (#2: Sa pec) MCG-5-5-6 = 413-G-18	23 D 2.2	F413/J0859 (108,269)
AM 0137-434	01 37 50 −43 40.2	2.545 0.304	275.02 −70.97	255.3 −18.8	I/A QUARTET incl. 244-G-50 = B 0137-436	4 D 1.6	F245/J3493 (317,247)
AM 0137-294	01 37 54 −29 42.5	2.757 0.304	228.08 −79.02	269.9 −17.3	I/A TRIPLET	3 S 0.7	F413/J0859 (107,193)
AM 0138-751	01 38 22 −75 15.9	0.959 0.304	298.82 −41.72	221.9 −18.0	GALAXY w. TAIL NGC 6438 = 29-G-53	15 D 1.3	F030/J1949 (294,157)
*AM 0138-261	01 38 45 −26 16.6	2.797 0.303	30.12 −78.85	266.7 4.3	I/A DBL. w. LOTS of DEBRIS MCG-5-5-5? = 476-IG-31	2 S 0.7	F476/J2653 (50,110)

Identification	R.A. (1950.0) Declination	Precession R.A./Dec.	Gal. Long. Gal. Lat.	S.G.L. S.G.B.	Description Cross-Identification	Coding Size (Min.)	Field/Plate (x, y)
AM 0139-274	01 39 06 −27 40.6	2.779 0.303	217.39 −78.89	272.0 −17.3	GALAXY w. APP. COMP'N. 413-G-22	8 S 1.3	F477/J3785 (309,34)
AM 0139-305	01 39 24 −30 50.2	2.738 0.303	233.40 −78.47	268.7 −17.8	SMALL GROUP of GALAXIES	18 D 1.8	F413/J0859 (91,132)
AM 0139-655	01 39 50 −65 52.1	1.815 0.303	294.84 −50.71	231.8 −19.1	I/A DBL.	2 S 1.9	F080/J1004 (121,151)
AM 0140-402 [0140-4024]	01 40 13 −40 22.0	2.592 0.303	267.10 −73.26	258.8 −19.0	CLOSE PAIR (#1: SPIRAL)	23 S 2.8	F297/J3693 (218,159)
AM 0140-402 [0140-4022]	01 40 13 −40 24.7	2.591 0.303	267.21 −73.23	258.7 −19.0	CLOSE PAIR (#2: SPIRAL)	23 S 2.8	F297/J3693 (218,157)
AM 0140-280	01 40 31 −28 05.1	2.770 0.302	219.50 −78.58	271.6 −17.6	DISTURBED SPIRAL	16 D 0.6	F413/J0859 (75,279)
*AM 0140-300	01 40 54 −30 09.1	2.742 0.302	229.74 −78.30	269.5 −18.0	SP. w. BIFURCATED ARM + COMP'NS 413-G-26	8,10 D 0.9	F414/J0847 (340,168)
AM 0141-342	01 41 10 −34 27.8	2.681 0.302	248.55 −76.78	265.0 −18.6	CHAIN of 5 PKS 0141-344	17 D 5.8	F353/J3596 (122,208)
AM 0141-462	01 41 20 −46 28.5	2.471 0.302	278.06 −68.30	252.3 −19.6	RICH, DISTANT CLUSTER of GALAXIES	19 D 3.6	F245/J3493 (278,98)
AM 0141-624	01 41 29 −62 43.2	1.963 0.302	292.89 −53.63	235.1 −19.4	DISRUPTED SPIRAL 80-IG-04	12 D 0.9	F080/J1004 (104,320)
AM 0141-451	01 41 30 −45 13.9	2.496 0.302	276.07 −69.32	253.6 −19.6	DISTURBED GAL. + COMPACT COMP'N.	13,16 S 0.4	F244/J1761 (14,163)
*AM 0141-362	01 41 31 −36 20.4	2.652 0.302	255.09 −75.72	263.0 −18.9	PEC. DIFFUSE RING w. 4 COMP'NS. 353-G-40	6,8 S 0.8	F353/J3596 (119,107)
AM 0141-635 [0141-6349]	01 41 39 −63 50.2	1.906 0.302	293.49 −52.57	233.9 −19.4	CLOSE TRIPLE (#1)	24 D 1.0	F080/J1004 (106,260)
AM 0141-302	01 41 41 −30 24.2	2.736 0.302	230.79 −78.09	269.3 −18.2	I/A TRIPLET	3 S 1.7	F413/J0859 (64,154)
AM 0141-634 [0141-6350]	01 41 42 −63 49.7	1.906 0.302	293.48 −52.58	233.9 −19.4	CLOSE TRIPLE (#2)	24 D 1.0	F080/J1004 (105,260)
AM 0141-682	01 41 44 −68 28.7	1.618 0.302	295.79 −48.17	229.0 −19.0	I/A DBL. (CLOSE)	2 S 0.1	F052/J0930 (187,259)
AM 0141-634 [0141-6349]	01 41 45 −63 49.7	1.905 0.302	293.47 −52.57	233.9 −19.4	CLOSE TRIPLE (#3)	24 D 1.0	F080/J1004 (105,260)
*AM 0141-353	01 41 59 −35 34.0	2.662 0.301	252.24 −76.07	263.9 −18.9	CHAIN of 5 E's	17 D 3.8	F353/J3596 (114,149)
*AM 0142-435	01 42 59 −43 50.8	2.516 0.301	273.08 −70.28	255.1 −19.7	MAGELLANIC DWARF 245-G-5	20 D 3.6	F245/J3493 (267,240)
AM 0143-731	01 43 13 −73 10.2	1.150 0.301	297.62 −43.65	224.1 −18.6	CHAIN of 5	17 D 2.7	F029/J1948 (35,267)
AM 0143-463	01 43 26 −46 30.1	2.459 0.301	277.29 −68.08	252.3 −19.9	2 ENVELOPED E's + FILAMENT	2 S 0.1	F245/J3493 (258,97)
AM 0143-781	01 43 40 −78 12.5	0.276 0.300	299.45 −38.80	218.8 −17.9	LO.S.BR. Sm in PAIR 13-G-20	20,23 S 4.5	F013/J2551 (142,275)
AM 0143-482	01 43 47 −48 21.9	2.415 0.300	279.78 −66.48	250.3 −20.1	GALAXY w. TAIL 196-G-20	15 D 0.4	F197/J0938 (321,261)
AM 0143-484	01 43 51 −48 44.0	2.406 0.300	280.24 −66.16	249.9 −20.1	I/A TRIPLET	3 D 0.2	F197/J0938 (319,241)
AM 0143-553	01 43 51 −55 39.1	2.217 0.300	287.46 −60.05	242.6 −20.0	CHAIN of 5 PKS 0143-559?	17 D 2.7	F152/J1862 (143,144)
AM 0143-731	01 43 54 −73 16.9	1.125 0.300	297.60 −43.53	223.9 −18.6	GROUP of 7 30-IG-2 = DN 0143-732	18 D 4.9	F029/J1948 (34,260)
AM 0143-330	01 43 58 −33 00.8	2.693 0.300	241.70 −76.87	266.6 −19.0	COMPACT + KNOTTY MATERIAL 353-IG-43	13,15 D 0.4	F413/J0859 (42,13)
AM 0143-561	01 43 58 −56 19.8	2.194 0.300	287.99 −59.43	241.9 −20.0	NEST of 7 152-IG-23	18 D 0.7	F152/J1862 (142,108)
AM 0144-474	01 44 23 −47 42.7	2.427 0.300	278.68 −66.98	251.0 −20.1	GALAXY w. CONDENSATION in DISK 196-G-21	12 D 0.7	F197/J0938 (317,297)
AM 0144-585	01 44 44 −58 55.2	2.095 0.300	289.80 −57.02	239.1 −20.0	KNOTTY DWARF Se 19/6 = 114-G-7	20 D 2.0	F114/J1813 (244,236)
AM 0144-295	01 44 53 −29 51.5	2.734 0.300	227.69 −77.50	269.9 −18.8	CLOSE GROUP of 6	18 D 2.0	F413/J0859 (26,183)
AM 0144-583	01 44 54 −58 30.6	2.109 0.300	289.46 −57.39	239.5 −20.1	COMPACT + VERY FAINT RING	6,13,23 S 1.0	F114/J1813 (243,258)
AM 0144-483	01 44 57 −48 39.5	2.402 0.300	279.76 −66.13	250.0 −20.3	GALAXY w. JET	7 D 0.4	F197/J0938 (310,246)
AM 0145-405	01 45 06 −40 54.7	2.560 0.299	266.13 −72.23	258.3 −20.0	SPIRAL w. ONE STRONG ARM 297-G-29	10 D 0.9	F297/J3693 (168,130)
AM 0145-270	01 45 37 −27 08.4	2.768 0.299	215.14 −77.42	272.8 −18.6	E w. 3 APP. COMP'NS. (DBL.I/A) IC 1729 = 477-G-4	2,8,17,18 S 3.1	F477/J3785 (232,64)
AM 0145-681 [0145-6817]	01 45 37 −68 18.5	1.579 0.299	295.20 −48.21	229.1 −19.4	CLOSE PAIR (#1)	23 S 1.1	F080/J1004 (98,19)

Identification	R.A. (1950.0) Declination	Precession R.A./Dec.	Gal. Long. Gal. Lat.	S.G.L. S.G.B.	Description Cross-Identification	Coding Size (Min.)	Field/Plate (x, y)
AM 0145-612	01 45 38 −61 26.7	1.981 0.299	291.34 −54.64	236.4 −20.0	SPIRAL w. APP. COMP'NS.	8 S 1.3	F114/J1813 (232,101)
AM 0145-681 [0145-6818]	01 45 43 −68 17.8	1.579 0.299	295.18 −48.22	229.2 −19.4	CLOSE PAIR (#2)	23 S 1.1	F080/J1004 (97,19)
*AM 0145-530	01 45 49 −53 00.6	2.283 0.299	284.48 −62.29	245.4 −20.4	RESOLVED GALAXY + COMPACT [A82] NGC 685 = 152-G-24 = PKS 0145-532	8,13 S 4.3	F152/J1862 (124,286)
*AM 0145-530	01 45 49 −53 00.4	2.283 0.299	284.48 −62.29	245.4 −20.4	COMPACT (LUMPY)	13 D 0.2	F197/J0938 (291,13)
AM 0145-320	01 45 55 −32 09.6	2.698 0.299	237.57 −76.77	267.5 −19.3	RICH CLUSTER	19 D 3.6	F354/J0832 (337,327)
*AM 0146-521	01 46 06 −52 18.0	2.301 0.299	283.67 −62.90	246.1 −20.4	CLOSE GROUP of 5 Se 17/6 = 197-IG-1 = DN 0146-523	18 D 1.3	F152/J1862 (121,324)
AM 0146-271	01 46 42 −27 19.5	2.763 0.298	216.05 −77.19	272.6 −18.9	GROUP of 5 incl. MCG-5-5-16 = 477-G-5	18 D 9.0	F414/J0847 (274,321)
AM 0146-620	01 46 44 −62 02.2	1.943 0.298	291.52 −54.05	235.8 −20.1	SPIRAL w. FAINT OUTER ARM	10 D 0.7	F114/J1813 (224,69)
AM 0146-321	01 46 48 −32 14.8	2.694 0.298	237.69 −76.57	267.5 −19.5	SMALL RING (HI.S.BR.)	6 D 0.3	F354/J0832 (326,323)
AM 0147-271	01 47 13 −27 19.7	2.762 0.298	216.10 −77.08	272.6 −19.0	SPIRAL w. I/A COMP'N. MCG-5-5-17 = 477-G-8	1 S 0.8	F477/J3785 (213,54)
AM 0147-275	01 47 13 −27 56.8	2.753 0.298	218.87 −77.10	272.0 −19.1	COMPACT w. JET A 0147-27 = 414-IG-4	7,13 D 0.2	F477/J3785 (213,21)
AM 0147-643	01 47 24 −64 39.2	1.799 0.298	293.00 −51.59	233.0 −19.9	CLOSE PAIR	23 S 0.4	F080/J1004 (75,214)
AM 0147-350	01 47 31 −35 05.1	2.649 0.298	248.26 −75.34	264.5 −20.0	DISTURBED SPIRAL + COMP'N.[A82] NGC 698 = 353-G-51 (CHAIN of 5)	6,8,17 S 2.0	F353/J3596 (53,173)
AM 0147-274	01 47 33 −27 42.9	2.756 0.298	217.84 −77.02	272.3 −19.1	GALAXY w. PECULIAR ABSORPTION NGC 689 = MCG-5-5-19 = 414-G-5	14 D 0.6	F477/J3785 (209,33)
AM 0147-500	01 47 39 −50 09.6	2.348 0.298	280.76 −64.63	248.4 −20.7	CLOSE PAIR 197-IG-5 = B 0147-501	23 S 0.6	F197/J0938 (282,166)
AM 0147-280	01 47 40 −28 3.4	2.751 0.298	219.36 −77.00	271.9 −19.2	GROUP of 5 incl. MCG-5-5-20 = 414-IG-6	18 S 3.6	F414/J0847 (262,282)
AM 0147-561	01 47 40 −56 18.0	2.166 0.298	287.06 −59.24	241.9 −20.6	RING/SPIRAL + DIAMETRIC COMP'NS. 152-G-26	6,8 S 13.4	F152/J1862 (115,109)
AM 0147-323	01 47 53 −32 39.5	2.685 0.298	239.05 −76.22	267.1 −19.8	I/A DBL. in GROUP of 4	2,18 D 2.9	F354/J0832 (313,301)
AM 0148-341 [0148-3417]	01 48 17 −34 18.3	2.658 0.297	245.20 −75.54	265.3 −20.1	CLOSE PAIR (#1: SPIRAL + EDGE-ON)	23 S 1.3	F353/J3596 (43,215)
AM 0148-245	01 48 19 −24 59.3	2.790 0.297	206.13 −76.49	275.1 −18.9	LINE of 3	24 D 7.5	F477/J3785 (200,180)
AM 0148-341 [0148-3418]	01 48 22 −34 17.0	2.658 0.297	245.09 −75.53	265.4 −20.1	CLOSE PAIR (#2: SPIRAL + EDGE-ON)	23 S 1.3	F353/J3596 (42,216)
AM 0148-574	01 48 27 −57 41.3	2.110 0.297	288.03 −57.94	240.4 −20.6	SMALL GALAXY w. JET? 114-IG-11	7 D 0.4	F114/J1813 (220,303)
*AM 0148-472	01 48 28 −47 24.9	2.410 0.297	276.80 −66.84	251.4 −20.8	GALAXY w. FAINT OUTER ARMS ONLY 245-G-6 [A82] (PAIR inside)	10,12 S 1.3	F245/J3493 (211,49)
AM 0148-624	01 48 31 −62 45.1	1.890 0.297	291.65 −53.31	235.0 −20.2	I/A DBL. 80-IG-7	2 S 0.5	F114/J1813 (212,31)
AM 0148-271	01 48 35 −27 17.1	2.759 0.297	215.99 −76.77	272.7 −19.3	GALAXY w. BROKEN RING + COMPACT	13,16 D 0.7	F414/J0847 (252,324)
AM 0148-344	01 48 51 −34 43.3	2.650 0.297	246.50 −75.26	264.9 −20.2	COMPACT at END of CHAIN of 4 PKS 0149-348?	13,17 D 0.4	F354/J0832 (300,191)
*AM 0149-444	01 49 00 −44 42.1	2.468 0.297	272.21 −68.94	254.3 −20.9	LARGE (RESOLVED) DWARF 245-G-7	20 D 4.5	F245/J3493 (208,195)
AM 0149-254	01 49 22 −25 46.2	2.777 0.297	209.60 −76.41	274.4 −19.2	VERY CLOSE PAIR	23 S 0.2	F477/J3785 (187,138)
AM 0149-485	01 49 23 −48 51.5	2.370 0.297	278.52 −65.57	249.8 −21.0	GALAXY w. JET	7 D 0.3	F197/J0938 (270,237)
*AM 0149-334 [0150-3347]	01 49 46 −33 48.1	2.661 0.296	242.87 −75.45	265.9 −20.3	I/A TRIPLE (#1)	3 D 1.6	F353/J3596 (26,241)
*AM 0149-334 [0149-3348]	01 49 53 −33 46.5	2.661 0.296	242.74 −75.44	265.9 −20.3	I/A TRIPLE (#2)	3 D 1.6	F353/J3596 (24,243)
AM 0150-334 [0149-3346]	01 50 00 −33 47.7	2.660 0.296	242.78 −75.41	265.9 −20.4	I/A TRIPLE (#3)	3 D 1.6	F353/J3596 (23,242)
AM 0150-362	01 50 27 −36 21.2	2.617 0.296	251.39 −74.20	263.2 −20.7	I/A TRIPLE 354-IG-8	3 D 1.1	F354/J0832 (280,104)
AM 0150-355	01 50 38 −35 51.8	2.625 0.296	249.74 −74.41	263.7 −20.7	COMPACT	13 D 0.2	F354/J0832 (278,130)
AM 0150-640	01 50 44 −64 06.6	1.794 0.296	292.13 −51.96	233.5 −20.4	SMALL I/A DBL. in GROUP of 3	2,24 S 0.1	F080/J1004 (53,242)
AM 0150-355	01 50 59 −35 52.5	2.623 0.296	249.66 −74.34	263.7 −20.7	I/A DBL. w. TAIL	2,15 S 0.2	F354/J0832 (275,129)

Identification	R.A. (1950.0) Declination	Precession R.A./Dec.	Gal. Long. Gal. Lat.	S.G.L. S.G.B.	Description Cross-Identification	Coding Size (Min.)	Field/Plate (x, y)
AM 0151-494	01 51 14 −49 48.4	2.335 0.295	279.18 −64.61	248.8 −21.3	E + PEC. ABS'N. (2 APP. COMP'NS.) 197-G-10	8,14 D 0.9	F197/J0938 (252,186)
AM 0151-624 [0151-6240]	01 51 27 −62 42.7	1.863 0.295	291.11 −53.21	235.0 −20.6	PAIR (#1: CONDENSATION in DISK) 81-G-1	8,23 D 0.9	F081/J1978 (304,294)
AM 0151-624 [0151-6243]	01 51 35 −62 40.2	1.863 0.295	291.06 −53.25	235.0 −20.6	PAIR (#2: GAL. w. I/A COMP'N.) 81-G-1 + CONDENSATION in DISK	1,23 S 3.1	F080/J1004 (42,318)
AM 0151-333	01 51 59 −33 35.5	2.656 0.295	241.44 −75.10	266.2 −20.7	COMPACT (LUMPY)	13,16 D 0.4	F354/J0832 (266,252)
*AM 0152-565	01 52 21 −56 55.8	2.107 0.295	286.49 −58.39	241.2 −21.2	PAIR of E's (2 I/A + APP.COMP'NS) NGC 745 = Se 19/4 = 152-IG-32[DN + A82]	1,8,23 S 4.4	F153/J1882 (323,70)
AM 0152-355	01 52 28 −35 54.9	2.617 0.295	249.27 −74.06	263.7 −21.0	I/A DBL. in CHAIN of 4 PKS 0152-358	2,17 D 13.4	F354/J0832 (259,128)
AM 0152-282	01 52 31 −28 23.5	2.733 0.295	220.80 −75.94	271.7 −20.3	DBL. I/A COMPACTS 414-IG-9	2,13 S 0.3	F414/J0847 (205,264)
AM 0152-494	01 52 34 −49 47.3	2.328 0.295	278.75 −64.51	248.8 −21.5	E + I/A DBL. COMP'N.	2,8 S 1.2	F197/J0938 (240,188)
AM 0152-353	01 52 40 −35 35.0	2.622 0.294	248.13 −74.17	264.1 −21.1	I/A DBL.	2 S 0.2	F354/J0832 (257,145)
AM 0153-803	01 53 17 −80 38.9	−0.776 0.294	299.80 −36.34	216.2 −17.9	I/A DBL. (HI.S.BR.)	2 S 0.2	F013/J2551 (128,142)
*AM 0153-783	01 53 22 −78 30.3	−0.046 0.294	298.96 −38.37	218.4 −18.3	SPIRAL + 3 APP. COMP'NS. 13-G-24	8 S 1.6	F013/J2551 (117,257)
AM 0153-281	01 53 36 −28 14.9	2.732 0.294	220.21 −75.70	271.9 −20.5	COMPACT 414-G-12	13 D 0.4	F414/J0847 (192,272)
AM 0153-540	01 53 49 −54 07.6	2.193 0.294	283.45 −60.74	244.1 −21.5	CLOSE PAIR w. 3 COMP'NS. 152-IG-35	8,23 S 0.4	F153/J1882 (322,221)
AM 0153-693	01 53 54 −69 36.5	1.359 0.294	294.83 −46.74	227.7 −20.0	COMPACT	13 D 0.2	F052/J0930 (130,197)
AM 0154-501	01 54 10 −50 13.9	2.306 0.293	278.84 −63.99	248.3 −21.7	I/A QUARTET 197-IG-13	4 D 0.4	F197/J0938 (226,164)
*AM 0154-372	01 54 27 −37 28.5	2.582 0.293	253.35 −72.94	262.1 −21.6	LINEAR (GALAXIES + STARS?)	17,22 D 1.1	F297/J3693 (68,314)
AM 0154-253	01 54 36 −25 30.0	2.767 0.293	209.40 −75.20	274.8 −20.4	LOOSE ARC of 5	18 D 9.2	F477/J3785 (124,152)
*AM 0154-441	01 54 41 −44 12.8	2.449 0.293	269.22 −68.63	254.8 −21.9	SPIRAL w. ONE LONG ARM + COMPACT 245-G-10 = MCG-7-5-2 = B0154-442 [Lu]	8,10,13 S 4.5	F245/J3493 (153,221)
AM 0154-342	01 54 50 −34 26.0	2.633 0.293	243.60 −74.25	265.3 −21.4	COMPACT (LUMPY)	13,16 D 0.4	F354/J0832 (234,207)
AM 0154-353	01 54 56 −35 33.5	2.614 0.293	247.31 −73.77	264.1 −21.5	GALAXY w. LOOP (RING ?)	6,22 D 0.4	F354/J0832 (232,147)
*AM 0155-370	01 55 01 −37 05.1	2.587 0.293	252.01 −73.04	262.5 −21.6	CLOSE TRIPLE incl. COMPACT 354-G-15	13,24 D 0.7	F354/J0832 (230,65)
AM 0155-815	01 55 02 −81 53.0	−1.434 0.293	300.19 −35.14	214.9 −17.7	GALAXY w. FINE HALF RING	6 D 0.7	F003/J2565 (234,343)
AM 0155-775	01 55 30 −77 53.8	0.064 0.292	298.58 −38.92	219.0 −18.6	STAR CLUSTER 13-SC-25	22 D 1.3	F013/J2551 (108,289)
AM 0155-691	01 55 37 −69 14.7	1.369 0.292	294.43 −47.02	228.0 −20.2	GALAXY w. I/A COMP'N. 52-G-10	1 D 0.8	F052/J0930 (121,216)
AM 0155-353	01 55 51 −35 37.0	2.610 0.292	247.21 −73.58	264.1 −21.7	GALAXY w. JETS 354-G-20	7 D 0.5	F354/J0832 (222,144)
AM 0155-675	01 55 51 −67 57.5	1.475 0.292	293.68 −48.20	229.4 −20.4	I/A (CONTACT) DBL. ?	2,22 D 0.5	F080/J1004 (45,33)
AM 0155-580	01 55 56 −58 01.8	2.036 0.292	286.66 −57.19	239.9 −21.6	3-ARMED SPIRAL + APP.COMP'N. NGC 782 = 114-G-15	8,11 S 2.7	F114/J1813 (166,285)
AM 0156-643	01 56 02 −64 38.0	1.706 0.292	291.62 −51.25	232.9 −20.9	E w. I/A COMP'N. DN 0156-646	2 S 0.4	F080/J1004 (25,211)
AM 0156-551	01 56 04 −55 17.5	2.137 0.292	284.08 −59.57	242.9 −21.8	WIDE CHAIN of 6 DN 0156-553	17 D 13.4	F153/J1882 (301,160)
AM 0156-475	01 56 11 −47 58.5	2.353 0.292	275.06 −65.61	250.8 −22.1	GALAXY w. TAILS + MATERIAL B 0156-479 = Fa 715	15 D 0.5	F197/J0938 (210,286)
AM 0156-522	01 56 18 −52 21.5	2.231 0.292	280.84 −62.04	246.0 −22.0	GALAXY w. LOOP in I/A DBL.	2,15 D 0.6	F152/J1862 (38,317)
AM 0156-672	01 56 21 −67 25.7	1.511 0.292	293.31 −48.68	229.9 −20.5	DISTURBED SPIRAL	16 D 0.8	F052/J0930 (112,314)
AM 0156-335	01 56 27 −33 59.2	2.635 0.292	241.63 −74.09	265.8 −21.7	COMPACT in GROUP of 4	13,18 D 0.3	F354/J0832 (216,232)
AM 0156-542	01 56 29 −54 27.6	2.163 0.292	283.12 −60.25	243.8 −21.9	S0 + PEC. COMP'N. + COMP'N. 153-G-3	8 S 1.8	F153/J1882 (300,205)
AM 0156-624 [0157-6239]	01 56 47 −62 40.1	1.812 0.292	290.15 −53.00	235.0 −21.2	GROUP of 3 (#1)	24 D 2.5	F080/J1004 (10,315)
*AM 0156-562	01 56 49 −56 29.6	2.088 0.292	285.07 −58.48	241.6 −21.8	E + SPIRAL (I/A + APP.) B 0156-564 Se19/2 = 153-IG-4 = DN 0157-565 [A82]	1,8,23 S 1.3	F153/J1882 (292,96)

Identification	R.A. (1950.0) Declination	Precession R.A./Dec.	Gal. Long. Gal. Lat.	S.G.L. S.G.B.	Description Cross-Identification	Coding Size (Min.)	Field/Plate (x, y)
AM 0156-623 [0156-6240]	01 56 51 −62 39.9	1.812 0.292	290.14 −53.00	235.0 −21.2	GROUP of 3 (#2)	24 D 2.5	F080/J1004 (9,316)
AM 0157-813	01 57 03 −81 39.6	−1.383 0.291	300.02 −35.33	215.1 −17.8	GALAXY w. 9 APP. COMP'NS.	8 D 4.5	F014/J2559 (300,73)
AM 0157-623 [0156-6239]	01 57 08 −62 39.6	1.809 0.291	290.08 −52.99	235.0 −21.2	GROUP of 3 (#3)	24 D 2.5	F080/J1004 (7,316)
AM 0157-430	01 57 11 −43 04.0	2.462 0.291	266.08 −69.11	256.1 −22.3	ONE-ARMED SPIRAL + COMPACT	10,13 D 0.4	F298/J0860 (302,13)
AM 0157-522	01 57 12 −52 24.7	2.224 0.291	280.66 −61.92	246.0 −22.1	CLOSE PAIR (ONE w. DBL. NUCLEUS) 197-IG-15	23 S 0.6	F197/J0938 (199,48)
AM 0157-564	01 57 20 −56 40.4	2.078 0.291	285.12 −58.28	241.4 −21.9	SP. w. APP. COMP'N. 153-IG-5 (in 153-IG-4 Group)	8 S 1.2	F153/J1882 (287,86)
AM 0157-311	01 57 24 −31 15.8	2.675 0.291	231.68 −74.61	268.8 −21.7	GALAXY w. BROKEN RING	6,16 S 0.4	F414/J0847 (149,110)
AM 0157-624	01 57 26 −62 49.1	1.798 0.291	290.15 −52.83	234.8 −21.2	PAIR of HI.S.BR. GALAXIES	13,23 S 1.4	F114/J1813 (157,28)
AM 0157-233 [0157-2337]	01 57 47 −23 38.0	2.786 0.291	203.23 −74.05	276.9 −20.8	PAIR (#1)	23 D 0.4	F477/J3785 (83,251)
AM 0157-233 [0157-2338]	01 57 53 −23 37.3	2.786 0.291	203.21 −74.02	276.9 −20.8	PAIR (#2)	23 D 0.4	F477/J3785 (82,252)
AM 0157-385	01 57 58 −38 51.2	2.543 0.291	255.95 −71.62	260.6 −22.3	CLOSE PAIR	23 S 0.5	F297/J3693 (33,238)
AM 0157-680 [0158-6801]	01 57 58 −68 06.5	1.436 0.291	293.49 −47.99	229.2 −20.6	CLOSE PAIR (#1: E) NGC 802 = 52-G-13 = SS-3a	23 D 0.4	F052/J0930 (106,277)
AM 0158-310	01 58 04 −31 04.9	2.676 0.291	230.91 −74.50	269.0 −21.8	I/A DBL. w. COMP'N.	2,8 S 0.3	F414/J0847 (141,120)
AM 0158-560	01 58 05 −56 03.8	2.095 0.291	284.36 −58.76	242.0 −22.0	GALAXY w. APP. COMP'NS. NGC 795 = 153-G-8 (Group)	8 S 5.4	F153/J1882 (283,119)
*AM 0158-343	01 58 06 −34 33.7	2.619 0.291	243.13 −73.57	265.2 −22.1	GALAXY w. TAIL in GROUP of 3 354-G-25	15,24 D 13.4	F354/J0832 (197,201)
AM 0158-680 [0157-6806]	01 58 17 −68 01.6	1.439 0.290	293.41 −48.05	229.3 −20.6	CLOSE PAIR (#2: SPIRAL) 52-G-14 = SS-3b	23 D 0.6	F052/J0930 (104,281)
AM 0158-562	01 58 30 −56 26.3	2.078 0.290	284.63 −58.41	241.6 −22.0	CHAIN of KNOTS Se 19/1 = 153-IG-9	22 D 0.9	F153/J1882 (279,99)
AM 0158-543	01 58 33 −54 34.0	2.144 0.290	282.72 −60.00	243.6 −22.2	FAINT CLUSTER of GALAXIES	19 D 0.4	F152/J1862 (27,198)
AM 0158-345	01 58 44 −34 58.4	2.610 0.290	244.31 −73.30	264.8 −22.2	M51-TYPE	9 S 0.4	F354/J0832 (191,179)
AM 0158-783	01 58 55 −78 38.4	−0.224 0.290	298.68 −38.16	218.2 −18.6	SPIRAL + 3 APP. COMP'NS.(2 I/A) 13-G-26 (HI. + LO.S.BR.ARMS!)	1,8,10,20 S 4.3	F014/J2559 (339,231)
*AM 0159-383 [0159-3830]	01 59 00 −38 32.7	2.545 0.290	254.78 −71.61	261.0 −22.5	PAIR (#1: PECULIAR SPIRAL)	12,23 S 3.8	F298/J0860 (290,256)
AM 0159-231	01 59 01 −23 13.9	2.788 0.290	202.17 −73.65	277.4 −21.0	SMALL I/A DBL. w. TAIL	2 S 0.2	F477/J3785 (68,273)
AM 0159-251	01 59 13 −25 10.1	2.761 0.290	208.95 −74.12	275.3 −21.3	SPIRAL + APP.COMP'N. in LINE of 4 MCG-4-5-27 = 477-G-22	8,17 S 0.5	F477/J3785 (67,169)
*AM 0159-383 [0159-3832]	01 59 16 −38 30.7	2.544 0.290	254.60 −71.59	261.0 −22.6	PAIR (#2: E/S0) 298-G-1	23 S 3.8	F298/J0860 (287,258)
AM 0159-311	01 59 55 −31 15.9	2.668 0.289	231.34 −74.08	268.8 −22.2	GALAXY w. CONDENSATION in RIM MCG-5-5-28 = 414-G-26	12 D 0.6	F414/J0847 (120,109)
AM 0159-215	01 59 56 −21 59.9	2.803 0.289	198.39 −73.03	278.7 −21.0	DISTURBED SPIRAL + APP. COMP'N. [A82]	8,16 S 3.0	F477/J3785 (55,339)
*AM 0200-220	02 00 00 −22 00.3	2.803 0.289	198.41 −73.03	278.7 −21.0	KNOTTY SP. + COMPACT/STAR on ARM MCG-4-6-1 = 544-G-7	13,16 D 0.7	F478/J2708 (329,339)
AM 0200-583	02 00 08 −58 37.1	1.977 0.289	286.29 −56.41	239.2 −22.1	[N.S.] LINE of E + 2 SPIRALS Se 19/5 = 114-G-19 in G-21 Group	24 D 17.0	F114/J1813 (137,253)
AM 0200-713	02 00 09 −71 31.5	1.071 0.289	295.16 −44.77	225.6 −20.1	I/A DBL. (SMALL)	2 S 0.1	F053/J0923 (329,84)
AM 0200-581	02 00 11 −58 11.2	1.995 0.289	285.90 −56.78	239.7 −22.1	PECULIAR RING Se 19/3 = 114-I-20	6 D 0.4	F114/J1813 (136,276)
AM 0200-450	02 00 14 −45 1.3	2.403 0.289	268.83 −67.36	253.9 −22.9	ENVELOPED PAIR of E's 245-IG-11 (in GROUP of 3)	2,24 S 0.4	F245/J3493 (101,177)
AM 0200-533	02 00 32 −53 30.5	2.166 0.289	281.06 −60.73	244.7 −22.6	EDGE-ON w. COMPACT 153-G-11	8,13 S 0.5	F153/J1882 (271,257)
AM 0200-430	02 00 35 −43 2.1	2.447 0.289	264.85 −68.68	256.1 −22.9	SP. w. CONDENS. at BASE of ARMS 245-G-12	10,12 D 1.3	F298/J0860 (269,15)
AM 0200-601	02 00 53 −60 14.8	1.896 0.289	287.54 −54.93	237.5 −22.0	LINE of 3	24 D 4.5	F114/J1813 (133,166)
AM 0201-694	02 01 08 −69 41.3	1.253 0.288	294.03 −46.43	227.5 −20.6	I/A DBL. (LARGE) 52-IG-17 = Fa 717/8 [A82]	2 S 0.3	F052/J0930 (96,191)
AM 0201-802	02 01 14 −80 29.0	−0.949 0.288	299.35 −36.39	216.3 −18.2	SMALL I/A DBL.	2 S 0.4	F014/J2559 (308,137)

Identification	R.A. (1950.0) Declination	Precession R.A./Dec.	Gal. Long. Gal. Lat.	S.G.L. S.G.B.	Description Cross-Identification	Coding Size (Min.)	Field/Plate (x, y)
AM 0201-412	02 01 31 −41 20.4	2.479 0.288	260.86 −69.62	257.9 −23.1	RICH CLUSTER of GALAXIES	19 D 9.9	F298/J0860 (261,106)
AM 0201-721	02 01 35 −72 18.8	0.953 0.288	295.44 −44.00	224.7 −20.1	COMPACT	13 D 0.1	F052/J0930 (105,50)
AM 0201-233	02 01 38 −23 33.0	2.778 0.288	203.87 −73.18	277.1 −21.7	DISTURBED GALAXY w. DIAMETRICS NGC 808 = 478-G-1 + I/A COMP'NS.	1,8 S 1.3	F478/J2708 (306,256)
AM 0201-612	02 01 53 −61 27.9	1.827 0.288	288.35 −53.80	236.2 −21.9	SPIRAL w. PECULIAR ARM 114-G-22	10 D 0.7	F114/J1813 (128,100)
AM 0202-491	02 02 18 −49 10.3	2.286 0.288	274.98 −64.05	249.4 −23.1	COMPACT 197-G-19	13 D 0.2	F197/J0938 (156,221)
AM 0202-482	02 02 26 −48 24.8	2.306 0.287	273.82 −64.61	250.2 −23.1	DWARF	20 D 0.7	F197/J0938 (154,262)
AM 0202-365	02 02 33 −36 59.7	2.560 0.287	249.38 −71.76	262.7 −23.1	CHAIN of 5 GALAXIES	17 D 193.8	F298/J0860 (253,340)
AM 0202-680	02 02 35 −68 09.9	1.373 0.287	292.94 −47.76	229.0 −21.0	I/A DBL.	2 S 0.2	F052/J0930 (83,272)
AM 0202-674	02 02 39 −67 49.1	1.402 0.287	292.72 −48.08	229.4 −21.0	COMPACT	13 D 22.0	F052/J0930 (81,291)
AM 0202-522	02 02 42 −52 24.2	2.188 0.287	279.20 −61.45	245.9 −23.0	E w. APP. COMP'NS. 153-G-13 (in G-17 Group)	8 S 2.7	F153/J1882 (256,317)
AM 0202-795 [0203-7954]	02 02 43 −79 56.9	−0.773 0.287	299.04 −36.87	216.8 −18.4	I/A SPIRALS (#1: SPIRAL) 13-G-27 = B 0202-799	2 S 2.5	F014/J2559 (312,165)
AM 0202-770	02 02 43 −77 04.7	0.101 0.287	297.74 −39.55	219.8 −19.1	I/A NEST of 4 in GROUP of 6 30-IG-7	4,18 D 0.4	F030/J1949 (207,67)
*AM 0202-323	02 02 48 −32 31.5	2.638 0.287	235.31 −73.23	267.5 −22.9	I/A TRIPLE(incl. SP. w. PEC. ARM) MCG-5-6-4? = 354-G-33?	3,10,12 D 0.9	F354/J0832 (145,310)
AM 0202-354	02 02 57 −35 41.9	2.582 0.287	245.47 −72.23	264.1 −23.2	COMPACT	13 D 0.3	F354/J0832 (145,140)
AM 0203-235	02 03 01 −23 56.3	2.770 0.287	205.46 −72.98	276.8 −22.0	SMALL SPIRAL w. I/A COMP'N.	1 S 0.3	F477/J3785 (19,234)
*AM 0203-841	02 03 08 −84 13.6	−3.689 0.287	300.82 −32.84	212.5 −17.3	GALAXY w. BRIGHT INNER ARM 3-G-7 + 2 DISTURBED COMP'NS.	8,10 S 10.5	F003/J2565 (207,220)
AM 0203-795 [0202-7956]	02 03 09 −79 54.2	−0.768 0.287	299.00 −36.90	216.9 −18.5	I/A SPIRALS (#2: EDGE-ON) 13-G-28 = B 0202-799	2 S 2.5	F014/J2559 (312,168)
AM 0203-233	02 03 10 −23 38.5	2.774 0.287	204.52 −72.87	277.1 −22.0	DWARF	20 D 0.2	F478/J2708 (287,252)
AM 0203-552	02 03 20 −55 20.9	2.082 0.287	282.42 −58.97	242.7 −22.8	GRP.of 5 (3 I/A, 5 COMP'NS) Fa720 153-G-17= PKS0203-552= B 0203-554	2,3,8,10,18 D 9.0	F153/J1882 (245,159)
AM 0203-583	02 03 22 −58 39.3	1.948 0.287	285.67 −56.17	239.2 −22.5	COMPACT	13 D 0.3	F114/J1813 (114,251)
AM 0203-425	02 03 24 −42 55.4	2.436 0.287	263.70 −68.36	256.2 −23.4	KNOTTY IRR.	16 D 0.6	F298/J0860 (241,22)
AM 0203-325	02 03 33 −32 55.0	2.629 0.287	236.49 −72.99	267.1 −23.1	RING w. APP. COMP'N. 354-G-34	6,8 S 2.9	F415/J4607 (342,20)
AM 0203-711	02 03 34 −71 11.8	1.058 0.287	294.63 −44.96	225.8 −20.5	SP. w. PECULIAR ARM + APP. COMP'N	8,10 D 0.6	F052/J0930 (92,109)
AM 0203-600 [0203-6001]	02 03 41 −60 00.8	1.883 0.287	286.81 −54.97	237.7 −22.3	LO. and HI.S.BR. SPIRALS (#1) 114-G-25	23 D 0.3	F114/J1813 (114,178)
AM 0204-373	02 04 07 −37 34.2	2.543 0.286	250.56 −71.22	262.1 −23.5	SPIRAL w. I/A COMP'N. 298-G-7	1 D 2.2	F298/J0860 (236,309)
AM 0204-510	02 04 07 −51 2.5	2.221 0.286	277.08 −62.41	247.4 −23.3	RICH CLUSTER of GALAXIES 197-G-23 DN 0204-554	19 D 6.7	F197/J0938 (142,121)
AM 0204-633	02 04 08 −63 37.8	1.684 0.286	289.64 −51.77	233.8 −21.9	CLOSE TRIPLE (SUPERPOSED STAR)	24 D 0.9	F081/J1978 (224,249)
AM 0204-600 [0203-6000]	02 04 10 −60 06.2	1.875 0.286	286.80 −54.86	237.6 −22.4	LO. and HI.S.BR. SPIRALS (#2)	23 D 0.3	F114/J1813 (111,173)
AM 0204-284	02 04 20 −28 49.2	2.693 0.286	222.35 −73.34	271.6 −22.9	LO.S.BR. I/A DBL. in LINE of 3 414-IG-30	2,20,24 S 0.7	F415/J4607 (340,240)
AM 0204-283	02 04 21 −28 38.2	2.696 0.286	221.71 −73.34	271.8 −22.9	TIGHT GROUP of GALS.(7 EARLY)	18 D 1.9	F414/J0847 (65,249)
AM 0204-605	02 04 36 −60 50.6	1.834 0.286	287.35 −54.19	236.8 −22.3	GALAXY w. JET 114-IG-26	7 D 0.7	F114/J1813 (110,133)
AM 0204-445	02 04 37 −44 56.7	2.383 0.286	267.31 −66.86	254.0 −23.6	HEAVY-ARMED SPIRAL + I/A COMP'N.	1,7 D 0.4	F246/J2724 (325,180)
AM 0204-302	02 04 44 −30 28.5	2.666 0.286	228.07 −73.16	269.8 −23.2	SMALL I/A DBL.	2 S 0.3	F415/J4607 (333,151)
AM 0204-621	02 04 54 −62 12.1	1.759 0.286	288.41 −52.99	235.3 −22.2	LOOSE CHAIN of 5 + COMP'NS. 114-G-27	8,17 D 13.4	F114/J1813 (110,60)
AM 0205-371	02 05 03 −37 18.3	2.545 0.286	249.56 −71.18	262.4 −23.6	GALAXY w. PECULIAR ABS'N. + TAILS 354-G-38	14,15 D 0.7	F354/J0832 (123,53)
AM 0205-712	02 05 04 −71 21.4	1.018 0.285	294.56 −44.77	225.7 −20.6	CLOSE PAIR (COMPACT + M51-TYPE) 52-G-20/21 = B 0204-713	9,13,23 S 1.6	F052/J0930 (86,100)

Identification	R.A. (1950.0) Declination	Precession R.A./Dec.	Gal. Long. Gal. Lat.	S.G.L. S.G.B.	Description Cross-Identification	Coding Size (Min.)	Field/Plate (x, y)
AM 0205-571 [0205-5710]	02 05 06 −57 13.9	1.995 0.285	283.96 −57.26	240.7 −22.9	PAIR of SP. w. APP. COMP'N(#1) 153-G-22	8,23 D 0.4	F153/J1882 (229,59)
AM 0205-571 [0205-5713]	02 05 07 −57 10.0	1.998 0.285	283.89 −57.31	240.7 −22.9	PAIR of SP. w. APP. COMP'N(#2) 153-G-22	8,23 D 0.8	F153/J1882 (229,62)
AM 0205-611	02 05 07 −61 12.4	1.811 0.285	287.56 −53.85	236.4 −22.4	DWARF 114-G-28	20 D 1.1	F114/J1813 (107,113)
AM 0205-232	02 05 09 −23 22.3	2.773 0.285	204.09 −72.36	277.4 −22.4	I/A DBL. w. TAILS 478-IG-3	2,15 S 0.2	F478/J2708 (263,266)
AM 0205-540	02 05 11 −54 00.5	2.117 0.285	280.49 −59.93	244.1 −23.2	CLOSE TRIPLE (incl. I/A DBL.) 153-G-25	2,24 S 4.0	F153/J1882 (233,232)
AM 0205-292	02 05 46 −29 28.9	2.679 0.285	224.61 −73.01	270.9 −23.3	GROUP of 4 PKS 0205-229?	18 D 2.0	F415/J4607 (322,205)
AM 0205-464	02 05 52 −46 41.2	2.333 0.285	270.03 −65.49	252.1 −23.8	DBL. (COMPACT w. RING AROUND)	13,23 S 0.4	F246/J2724 (310,87)
AM 0205-285	02 05 53 −28 52.5	2.688 0.285	222.53 −73.00	271.5 −23.3	SP. w. PEC. ARMS + APP. COMP'N. MCG-5-6-7 = 414-G-32	8,10 S 1.7	F415/J4607 (322,238)
AM 0205-225	02 05 58 −22 52.0	2.779 0.285	202.71 −72.02	278.0 −22.6	GROUP of 4	18 D 5.4	F478/J2708 (253,294)
AM 0206-343	02 06 00 −34 35.3	2.592 0.285	241.36 −72.03	265.3 −23.7	I/A DBL. in GROUP of 3	2,24 S 0.5	F354/J0832 (110,199)
AM 0206-402 [0206-4019]	02 06 39 −40 22.1	2.477 0.284	257.00 −69.39	259.0 −24.0	PAIR (#1: SPIRAL) 298-G-11	23 S 3.1	F298/J0860 (209,159)
AM 0206-401 [0206-4022]	02 06 49 −40 19.8	2.477 0.284	256.86 −69.38	259.1 −24.1	PAIR (#2: IRR.) 298-G-12	16,23 S 3.1	F298/J0860 (207,161)
AM 0206-571	02 06 49 −57 18.2	1.979 0.284	283.67 −57.07	240.6 −23.1	COMPACT	13 D 0.3	F153/J1882 (217,55)
AM 0207-233	02 07 00 −23 39.0	2.765 0.284	205.36 −72.03	277.2 −22.9	SP. w. DWARF,COMPACT and I/A DBL. MCG-4-6-9 = 478-G-6	2,8,13,20 S 5.6	F478/J2708 (240,252)
*AM 0207-493	02 07 00 −49 31.0	2.249 0.284	274.18 −63.29	249.0 −23.8	SPIRAL + JET 197-IG-25 = B 0206-495 [A82]	7 D 0.9	F197/J0938 (115,202)
AM 0207-265	02 07 05 −26 56.3	2.716 0.284	216.02 −72.64	273.7 −23.3	CLOSE PAIR	23 S 0.9	F478/J2708 (238,75)
AM 0207-494	02 07 14 −49 43.0	2.242 0.284	274.40 −63.11	248.8 −23.9	I/A DBL.	2 S 0.3	F197/J0938 (113,191)
*AM 0207-360	02 07 21 −36 9.0	2.558 0.284	245.71 −71.23	263.6 −24.1	I/A DBL.	2 S 0.9	F354/J0832 (97,115)
AM 0207-374	02 07 37 −37 47.2	2.526 0.284	250.19 −70.51	261.8 −24.2	V. SMALL RING	6 D 0.2	F298/J0860 (199,298)
AM 0207-224	02 07 58 −22 40.0	2.777 0.283	202.58 −71.52	278.3 −23.0	SPIRAL w. COMP'NS. NGC 837 = 478-G-10	8 S 1.3	F478/J2708 (229,305)
AM 0208-323	02 08 02 −32 34.0	2.621 0.283	234.65 −72.15	267.6 −24.0	I/A DBL. in GROUP of 4	2,18 S 0.8	F354/J0832 (85,307)
AM 0208-522	02 08 02 −52 29.6	2.150 0.283	277.97 −60.89	245.7 −23.8	CLOSE TRIPLE, DIAMETRIC to SPIRAL	13,24 D 2.2	F197/J0938 (111,42)
AM 0208-503	02 08 12 −50 31.9	2.212 0.283	275.31 −62.39	247.9 −24.0	COMPACT	13 D 0.1	F197/J0938 (106,148)
AM 0208-604	02 08 19 −60 44.9	1.806 0.283	286.60 −54.05	236.8 −22.8	E + SPIRAL	23 S 1.2	F114/J1813 (85,137)
AM 0208-421	02 08 22 −42 19.5	2.427 0.283	260.92 −68.01	256.9 −24.3	IRR. 298-IG-14	16 D 1.1	F298/J0860 (192,54)
*AM 0208-223	02 08 28 −22 36.7	2.777 0.283	202.53 −71.39	278.4 −23.1	M51-TYPE w. ELONGATED COMP'N.	9 S 0.5	F478/J2708 (222,308)
AM 0208-371	02 08 35 −37 18.8	2.531 0.283	248.66 −70.54	262.4 −24.4	CLUSTER of GALAXIES PKS 0208-379?	19 D 9.0	F298/J0860 (189,323)
AM 0208-315	02 08 44 −31 50.2	2.631 0.283	232.22 −72.14	268.4 −24.1	GALAXY w. DIFFUSE TAIL 415-IG-3	15 D 0.4	F414/J0847 (20,76)
AM 0208-225	02 08 49 −22 52.5	2.773 0.283	203.39 −71.40	278.1 −23.2	GALAXY w. V. CLOSE I/A COMP'N.	1 S 0.1	F478/J2708 (218,293)
AM 0209-303	02 09 10 −30 36.8	2.651 0.282	228.22 −72.20	269.7 −24.1	GALAXY w. I/A COMP'NS. 415-IG-4	1 S 0.3	F415/J4607 (281,145)
AM 0209-471	02 09 20 −47 13.9	2.300 0.282	269.95 −64.68	251.5 −24.4	I/A DBL. PKS 0209-472?	2 S 0.2	F246/J2724 (278,59)
AM 0209-403	02 09 22 −40 31.5	2.462 0.282	256.59 −68.87	258.8 −24.5	cD CLUSTER of FAINT GALAXIES 298-G-17 = PKS 0209-405	19 D 7.2	F298/J0860 (182,150)
AM 0209-360	02 09 25 −36 04.5	2.552 0.282	245.02 −70.87	263.7 −24.5	DISTURBED ARMS + COMPACT + 6 APP. NGC 854 = 354-G-47	8,10,13 S 9.9	F355/J2713 (337,118)
AM 0209-483	02 09 29 −48 32.6	2.263 0.282	272.03 −63.73	250.0 −24.3	SPIRAL w. I/A COMP'N. B 0209-485	1 S 0.5	F197/J0938 (92,254)
AM 0209-270	02 09 53 −27 01.0	2.707 0.282	216.49 −72.03	273.7 −23.9	VERY CLOSE PAIR	23 S 0.1	F478/J2708 (205,71)
AM 0209-602	02 09 58 −60 21.3	1.812 0.282	285.95 −54.28	237.2 −23.1	I/A DBL.	2 S 0.2	F115/J1810 (325,154)

Identification	R.A. (1950.0) Declination	Precession R.A./Dec.	Gal. Long. Gal. Lat.	S.G.L. S.G.B.	Description Cross-Identification	Coding Size (Min.)	Field/Plate (x, y)
*AM 0210-270	02 10 07 −27 07.7	2.705 0.282	216.87 −71.99	273.5 −24.0	GROUP of 3	24 D 1.7	F478/J2708 (202,65)
*AM 0210-224	02 10 13 −22 42.4	2.772 0.282	203.21 −71.04	278.3 −23.5	3-ARMED SPIRAL w. I/A COMP'N. NGC 858 = 478-G-13 (GROUP of 4)	1,11,13,18 S 2.7	F478/J2708 (201,302)
AM 0210-503	02 10 14 −50 31.1	2.200 0.282	274.77 −62.19	247.9 −24.3	DISTURBED GALAXY 197-G-29 = B 0210-505	16 D 1.1	F197/J0938 (89,148)
AM 0210-445	02 10 19 −44 50.7	2.357 0.281	265.44 −66.17	254.1 −24.6	DWARF	20 D 0.6	F246/J2724 (271,187)
AM 0210-215	02 10 20 −21 57.7	2.783 0.281	201.09 −70.77	279.1 −23.4	GALAXY w. JET	7 D 0.2	F478/J2708 (199,342)
AM 0210-353	02 10 26 −35 34.5	2.558 0.281	243.36 −70.86	264.3 −24.7	HI.S.BR. LINEAR SHRED 354-G-48	22 D 1.3	F354/J0832 (63,145)
*AM 0210-270	02 10 37 −27 01.8	2.705 0.281	216.58 −71.87	273.7 −24.1	SPIRAL w. I/A COMP'N.	1 S 0.4	F415/J4607 (268,338)
AM 0210-565 [0211-5655]	02 10 41 −56 55.0	1.966 0.281	282.47 −57.10	240.9 −23.7	LINE of THREE (#1)	24 D 0.4	F153/J1882 (189,76)
AM 0210-512	02 10 53 −51 24.8	2.168 0.281	275.84 −61.45	246.9 −24.3	CLOSE PAIR (E's ?)	23 S 0.1	F198/J2634 (339,98)
AM 0210-601	02 10 53 −60 19.9	1.805 0.281	285.77 −54.24	237.2 −23.2	SMALL I/A DBL. (FILAMENT?) 114-IG-30	2,15 S 0.3	F114/J1813 (67,159)
AM 0211-732	02 11 01 −73 23.8	0.649 0.281	295.17 −42.72	223.4 −20.5	3 LO.S.BR. CIRCULAR ARMS + 30-G-11 BRIGHT NUCLEUS	8,10 S 2.3	F030/J1949 (183,265)
AM 0211-421	02 11 02 −42 16.0	2.416 0.281	260.03 −67.64	256.9 −24.8	COMPACT E w. LO.S.BR.COMP'N. NGC 862 = 298-G-20	8,20 S 1.3	F246/J2724 (267,326)
AM 0211-472	02 11 03 −47 22.5	2.287 0.281	269.72 −64.37	251.3 −24.6	CONSPICUOUS CLUSTER DN 0211-475	19 D 4.5	F246/J2724 (262,52)
AM 0211-565 [0210-5655]	02 11 15 −56 55.4	1.961 0.281	282.36 −57.05	240.9 −23.7	LINE of THREE (#2: COMPACT)	13,244 D 0.1	F153/J1882 (185,76)
AM 0211-472	02 11 22 −47 26.5	2.284 0.281	269.75 −64.29	251.2 −24.7	NUCLEATED DWARF ?	20 D 0.3	F246/J2724 (259,48)
AM 0211-271	02 11 24 −27 18.2	2.699 0.281	217.51 −71.72	273.4 −24.3	CLOSE PAIR (I/A TRIPLE?)	3,23 D 1.0	F478/J2708 (187,56)
AM 0211-682	02 11 30 −68 25.4	1.239 0.281	292.01 −47.17	228.6 −21.7	GALAXY w. I/A COMP'N.	1 S 0.2	F052/J0930 (40,254)
AM 0211-615	02 11 40 −61 58.7	1.709 0.280	287.09 −52.79	235.4 −23.0	WIDE I/A GROUP of 5 SPIRALS 114-G-31 Group	5,24 D 6.3	F115/J1810 (307,68)
AM 0211-480	02 11 49 −48 01.9	2.265 0.280	270.59 −63.82	250.6 −24.7	CLOSE PAIR	23 S 0.6	F198/J2634 (343,280)
AM 0211-565 [0211-5655]	02 11 50 −56 55.7	1.957 0.280	282.25 −57.00	240.9 −23.8	LINE of THREE (#3: 3-ARMED SP.)	24 D 0.4	F153/J1882 (180,76)
AM 0212-392	02 12 24 −39 25.1	2.473 0.280	253.12 −68.93	260.1 −25.1	MAGELLANIC DWARF 298-G-23	20 D 0.9	F298/J0860 (150,209)
AM 0212-412	02 12 27 −41 28.8	2.428 0.280	257.92 −67.86	257.8 −25.1	COMPACT w. ERUPTIVE MATERIAL 298-IG-24 = B 0212-414	13,15 D 0.6	F298/J0860 (151,99)
*AM 0212-250	02 12 33 −25 05.0	2.731 0.280	210.73 −71.14	275.8 −24.3	RING SPIRAL + COMP'NS. MCG-4-6-17 = 478-G-15 [A82]	8 S 3.2	F478/J2708 (173,175)
AM 0212-545 [0212-5458]	02 12 34 −54 54.8	2.033 0.280	279.86 −58.57	243.0 −24.2	LINE of 3(#1) 153-G-29	24 D 0.6	F153/J1882 (175,184)
AM 0212-220	02 12 36 −22 02.5	2.777 0.280	201.86 −70.30	279.1 −24.0	PAIR of E's	23 S 0.4	F478/J2708 (171,338)
AM 0212-545 [0212-5454]	02 12 44 −54 56.5	2.031 0.280	279.85 −58.53	243.0 −24.2	LINE of 3(#2) 153-G-30	24 D 1.5	F153/J1882 (174,182)
AM 0212-545 [0212-5456]	02 12 49 −54 58.1	2.029 0.279	279.86 −58.50	243.0 −24.2	LINE of 3(#3)	24 D 1.0	F153/J1882 (173,181)
AM 0212-245	02 12 51 −24 50.0	2.734 0.279	210.02 −71.03	276.1 −24.4	CLOSE PAIR in GROUP of 5 (incl. LARGE COMPACT)PKS 0212-248	13,18,23 D 2.2	F478/J2708 (169,188)
AM 0212-364	02 12 54 −36 44.1	2.527 0.279	246.08 −69.98	263.0 −25.2	CLUSTER of FAINT GALAXIES	19 D 9.0	F355/J2713 (299,84)
AM 0212-431	02 12 59 −43 14.3	2.384 0.279	261.53 −66.78	255.8 −25.2	COMPACT PN L02-2 **	13 D 0.3	F246/J2724 (247,274)
AM 0213-372	02 13 10 −37 24.0	2.512 0.279	247.80 −69.67	262.3 −25.3	GALAXY w. COMP'NS.	8 D 0.8	F298/J0860 (140,318)
AM 0213-505	02 13 24 −50 51.4	2.171 0.279	274.45 −61.60	247.5 −24.8	COMPACT (+ COMPACT ?)	13 D 0.4	F197/J0938 (63,128)
AM 0213-230	02 13 31 −23 00.2	2.761 0.279	204.77 −70.41	278.1 −24.3	CLOSE I/A DBL. 478-G-17	2 S 0.4	F478/J2708 (160,286)
AM 0213-441	02 13 35 −44 11.8	2.358 0.279	263.29 −66.12	254.8 −25.2	PAIR of COMPACTS w. TAIL	2,13,15 S 0.2	F246/J2724 (240,223)
*AM 0213-283	02 13 36 −28 36.4	2.672 0.279	221.70 −71.31	272.0 −24.9	SMALL M51-TYPE + 8 COMP'NS. 415-IG-14 [A82]	8,9 S 0.4	F415/J4607 (231,254)
AM 0213-220	02 13 39 −22 08.4	2.774 0.279	202.38 −70.11	279.1 −24.2	CHAIN of 4	17 D 2.3	F478/J2708 (158,333)

Identification	R.A. (1950.0) Declination	Precession R.A./Dec.	Gal. Long. Gal. Lat.	S.G.L. S.G.B.	Description Cross-Identification	Coding Size (Min.)	Field/Plate (x, y)
AM 0213-274	02 13 55 −27 46.9	2.685 0.279	219.15 −71.20	272.9 −24.9	PAIR of E's (EXTENDED MATERIAL)	23 S 0.8	F415/J4607 (228,298)
AM 0214-261	02 14 03 −26 10.9	2.710 0.278	214.24 −71.00	274.7 −24.8	VERY DISTURBED GALAXY w. LOOPS 478-G-19 + FILAMENTS	15 D 0.6	F478/J2708 (155,116)
AM 0214-475	02 14 13 −47 56.6	2.254 0.278	269.82 −63.60	250.6 −25.1	I/A DBL.	2 S 0.4	F198/J2634 (321,286)
AM 0214-234	02 14 47 −23 48.4	2.746 0.278	207.32 −70.35	277.3 −24.7	GALAXY w. LOOP and TAILS 478-G-20	15 D 0.6	F478/J2708 (145,243)
AM 0214-502	02 14 53 −50 26.6	2.175 0.278	273.49 −61.75	247.9 −25.0	PECULIAR SINGLE (ABS'N?)	14 D 0.4	F198/J2634 (308,152)
*AM 0215-331	02 15 03 −33 14.3	2.586 0.278	235.77 −70.57	266.9 −25.5	SPIRAL w. COMP'N. (M51-TYPE?) 355-G-2	1,9 S 0.6	F355/J2713 (280,273)
*AM 0215-505	02 15 13 −50 53.5	2.159 0.278	274.06 −61.39	247.4 −25.0	VERY COMPACT (JETS?)	7,13 D 0.2	F198/J2634 (304,128)
AM 0215-485	02 15 25 −48 56.4	2.219 0.277	271.10 −62.76	249.5 −25.3	RING GALAXY	6 D 0.4	F197/J0938 (40,230)
AM 0215-320	02 15 31 −32 03.6	2.607 0.277	232.22 −70.68	268.2 −25.6	SMALL M51-TYPE	9 S 0.2	F415/J4607 (207,69)
AM 0215-373	02 15 34 −37 34.2	2.500 0.277	247.71 −69.17	262.1 −25.7	DWARF	20 D 0.4	F355/J2713 (269,40)
AM 0215-372	02 15 36 −37 22.1	2.504 0.277	247.18 −69.24	262.3 −25.7	I/A DBL. (SMALL) w. COMP'N. 355-IG-3	2,8 S 0.2	F298/J0860 (114,319)
AM 0215-280	02 15 46 −28 02.8	2.676 0.277	220.03 −70.81	272.7 −25.3	ROUGH CHAIN of 6 incl. MCG-5-6-13 = 415-G-17	17 D 6.3	F415/J4607 (206,284)
*AM 0215-350	02 15 47 −35 01.7	2.550 0.277	240.80 −69.98	264.9 −25.7	IRR. + PEC. ARM (RESOLUTION) 355-G-4	10,16 D 1.1	F355/J2713 (269,177)
AM 0215-671	02 15 54 −67 13.8	1.294 0.277	290.63 −48.02	229.8 −22.4	I/A TRIPLE (VERY SMALL) 81-IG-4	3 D 0.5	F081/J1978 (157,57)
AM 0215-600	02 15 56 −60 05.9	1.774 0.277	284.65 −54.09	237.4 −23.8	2 E's + APP. COMP'N. NGC 888 = 115-G-2	8,23 S 8.4	F115/J1810 (287,171)
AM 0216-523	02 16 06 −52 32.7	2.097 0.277	276.12 −60.07	245.6 −25.0	GROUP of 9 GALAXIES DN 0216-525	18 D 4.5	F197/J0938 (45,37)
AM 0216-255	02 16 20 −25 59.2	2.708 0.277	213.90 −70.46	274.9 −25.3	SPIRAL w. MANY COMP'NS.(≈ 9) MCG-4-6-23 = 478-G-23	8 S 4.3	F478/J2708 (127,126)
AM 0216-224	02 16 21 −22 43.4	2.759 0.277	204.57 −69.70	278.5 −24.9	I/A TRIPLE MCG-4-6-24 = 478-G-24	3 D 1.1	F478/J2708 (125,301)
AM 0216-502	02 16 37 −50 28.3	2.164 0.276	273.12 −61.54	247.8 −25.3	SPIRAL w. ABS'N? + 2 COMPACTS Fa 724	8,13,14 D 0.8	F198/J2634 (293,151)
AM 0216-552	02 16 38 −55 22.4	1.987 0.276	279.52 −57.85	242.4 −24.7	SMALL I/A DBL. 153-IG-31 = B 0216-553	2 S 0.5	F153/J1882 (144,159)
*AM 0216-451	02 16 42 −45 18.7	2.314 0.276	264.58 −65.00	253.5 −25.7	SINGLE GALAXY w. JET + PUFF 246-IG-7 = B 0216-453	7 D 0.9	F246/J2724 (210,163)
AM 0216-264	02 16 46 −26 47.4	2.694 0.276	216.32 −70.48	274.1 −25.4	SMALL I/A TRIPLE	3 D 0.6	F478/J2708 (122,83)
AM 0216-481	02 16 56 −48 13.8	2.231 0.276	269.59 −63.07	250.3 −25.6	I/A DBL.	2 S 0.3	F197/J0938 (25,268)
AM 0217-340	02 17 06 −34 06.2	2.563 0.276	237.99 −69.96	266.0 −26.0	I/A DBL.	2 S 0.5	F355/J2713 (256,227)
*AM 0217-432	02 17 46 −43 29.0	2.356 0.275	260.77 −65.93	255.5 −26.0	SPIRAL w. JET 246-G-8	7 D 1.8	F246/J2724 (200,261)
AM 0217-483	02 17 49 −48 30.5	2.218 0.275	269.82 −62.77	250.0 −25.7	CHAIN of 7	17 D 8.1	F197/J0938 (18,252)
AM 0217-761	02 17 53 −76 18.9	−0.033 0.275	296.30 −39.90	220.3 −20.2	G.D. SP. w. DISTURBED COMP'N. 30-G-14	8 S 5.2	F030/J1949 (160,109)
AM 0217-650 [0218-6459]	02 17 56 −65 02.0	1.448 0.275	288.67 −49.81	232.0 −23.1	PAIR of SPIRALS (#1)	23 D 0.6	F081/J1978 (143,174)
AM 0217-413	02 17 59 −41 38.1	2.400 0.275	256.84 −66.90	257.6 −26.2	BRIGHT-RIMMED SPIRAL NGC 893 = Ag-11 = 298-G-29	12 D 1.1	F298/J0860 (96,89)
AM 0218-645 [0217-6502]	02 18 04 −64 59.8	1.450 0.275	288.62 −49.84	232.1 −23.1	PAIR of SPIRALS (#2)	23 D 0.5	F081/J1978 (143,176)
*AM 0218-321	02 18 16 −32 12.5	2.595 0.275	232.41 −70.08	268.1 −26.2	CLOSE PAIR	23 S 0.2	F415/J4607 (176,61)
AM 0218-554	02 18 24 −55 46.2	1.958 0.275	279.62 −57.38	242.0 −24.9	LINE of 3	24 D 1.6	F153/J1882 (131,138)
AM 0218-443	02 18 36 −44 31.2	2.326 0.275	262.61 −65.21	254.4 −26.1	HI.S.BR. IRR.	16 D 0.5	F246/J2724 (192,205)
*AM 0218-321	02 18 52 −32 10.3	2.594 0.275	232.25 −69.96	268.1 −26.3	OPEN SP.w. COMP'N. in ARM + COMPACT MCG-5-6-14 = 415-G-19	1,8,9,13,17 S 1.5	F415/J4607 (169,63)
*AM 0218-335	02 18 58 −33 57.1	2.560 0.274	237.31 −69.62	266.2 −26.4	TIGHT SP. + DIAM. COMP'NS (DISK) NGC 897 = 355-G-7	8 S 7.6	F355/J2713 (235,235)
*AM 0219-343	02 19 11 −34 33.0	2.548 0.274	238.94 −69.44	265.5 −26.4	E w. TILTED ABSORPTION 355-G-8	14 D 0.8	F355/J2713 (232,203)

Identification	R.A. (1950.0) Declination	Precession R.A./Dec.	Gal. Long. Gal. Lat.	S.G.L. S.G.B.	Description Cross-Identification	Coding Size (Min.)	Field/Plate (x, y)
AM 0219-421	02 19 13 −42 13.7	2.381 0.274	257.81 −66.40	256.9 −26.4	SPIRAL w. LINEAR MARK in DISK 298-G-31	12 D 1.6	F246/J2724 (185,328)
AM 0219-581	02 19 17 −58 16.4	1.839 0.274	282.22 −55.33	239.2 −24.6	DISTURBED GAL. w. DIAM. COMP'NS.	8,16 S 1.1	F115/J1810 (269,270)
AM 0219-390	02 19 26 −39 01.8	2.454 0.274	250.51 −67.88	260.5 −26.5	DWARF 298-G-33	20 D 0.7	F298/J0860 (76,228)
*AM 0219-273	02 19 27 −27 30.9	2.675 0.274	218.63 −69.96	273.3 −26.1	SPIRAL + I/A COMP'N. MCG-5-6-15 = 415-G-22 [A82]	1 S 1.8	F415/J4607 (162,313)
AM 0219-374	02 19 52 −37 40.9	2.482 0.274	247.10 −68.34	262.0 −26.6	SP. w. PEC. ARMS(JET?)+ 6 COMP'NS 298-G-36	7,8,10 D 1.3	F355/J2713 (223,35)
AM 0220-220	02 20 01 −22 01.7	2.763 0.274	203.47 −68.68	279.4 −25.7	LINE of 3 DISTURBED GALAXIES	24 D 1.4	F478/J2708 (79,338)
AM 0220-390	02 20 02 −39 00.8	2.452 0.274	250.34 −67.78	260.5 −26.6	LUMPY COMPACT 298-IG-37	13 D 0.3	F298/J0860 (70,229)
AM 0220-712	02 20 13 −71 24.5	0.792 0.273	293.08 −44.18	225.3 −21.7	I/A DBL. 53-IG-5	2 S 0.3	F053/J0923 (244,99)
AM 0220-592	02 20 17 −59 22.8	1.776 0.273	283.18 −54.36	238.0 −24.5	SP. w. PEC. ARMS + CONDENSATION 115-G-6 OFF NUCLEUS	12 D 0.7	F115/J1810 (259,211)
AM 0220-674	02 20 52 −67 41.9	1.195 0.273	290.37 −47.38	229.1 −22.7	I/A SPIRALS 53-IG-6	2 S 0.5	F053/J0923 (253,298)
AM 0220-370	02 20 54 −37 06.2	2.490 0.273	245.44 −68.35	262.6 −26.8	I/A DBL. (ONE VERY COMPACT) 355-IG-13	2,13 S 0.4	F355/J2713 (212,66)
AM 0221-402	02 21 00 −40 24.4	2.417 0.273	253.39 −67.01	258.9 −26.8	DWARF 298-G-39	20 D 0.3	F299/J3588 (333,154)
*AM 0221-330	02 21 07 −33 04.4	2.570 0.273	234.60 −69.35	267.2 −26.8	FILAMENT w. 3 KNOTS (I/A 3?) 355-IG-12	3,15 D 1.3	F355/J2713 (212,283)
AM 0221-672	02 21 15 −67 29.2	1.210 0.273	290.16 −47.54	229.4 −22.8	IRR.	16 D 0.5	F053/J0923 (252,310)
AM 0221-230	02 21 19 −23 01.5	2.745 0.272	206.33 −68.70	278.3 −26.1	LUMPY COMPACT	13 D 0.2	F479/J4497 (336,283)
AM 0221-291	02 21 20 −29 10.4	2.642 0.272	223.48 −69.63	271.5 −26.6	I/A TRIPLE	3 D 0.5	F415/J4607 (140,224)
AM 0221-332	02 21 27 −33 27.0	2.562 0.272	235.61 −69.22	266.7 −26.9	SMALL, FAINT GALAXY CLUSTER	19 D 3.6	F355/J2713 (208,262)
AM 0221-510	02 21 53 −51 04.5	2.113 0.272	272.77 −60.52	247.0 −26.1	I/A DBL. 198-IG-9	2 S 0.2	F198/J2634 (248,121)
AM 0222-273	02 22 01 −27 30.5	2.669 0.272	218.75 −69.39	273.4 −26.7	CLOSE DBL. (ONE COMPACT)	13,23 S 0.2	F415/J4607 (132,313)
AM 0222-253	02 22 30 −25 39.9	2.699 0.271	213.64 −69.04	275.4 −26.6	COMPACT GROUP of 4 V. COMPACT GAL	13,18 D 0.6	F478/J2708 (52,142)
*AM 0222-250	02 22 48 −25 01.2	2.709 0.271	211.91 −68.86	276.2 −26.6	SINGLE, BRIGHT-RIMMED SP. + COMP'N. NGC 922 = 478-G-28	7,8,16 D 1.2	F478/J2708 (48,176)
AM 0222-583	02 22 52 −58 37.1	1.794 0.271	281.93 −54.77	238.8 −25.0	E + I/A SPIRAL + COMP'NS. 115-G-8/9 = B 0223-586	2,8 S 2.2	F115/J1810 (243,252)
AM 0222-614	02 22 53 −61 48.1	1.618 0.271	285.11 −52.21	235.3 −24.3	CLOSE TRIPLE (E + 2 PECULIARS)	24 D 0.9	F115/J1810 (237,82)
AM 0222-295	02 22 59 −29 52.2	2.625 0.271	225.46 −69.26	270.8 −27.0	I/A DBL. w. MATERIAL(CHAIN of 10)	2,17 S 0.4	F415/J4607 (121,186)
AM 0222-324	02 22 59 −32 47.9	2.570 0.271	233.66 −69.01	267.5 −27.2	PAIR of I/A COMPACTS	2,13 S 0.3	F355/J2713 (191,297)
*AM 0223-403	02 23 13 −40 39.3	2.402 0.271	253.46 −66.52	258.6 −27.2	I/A DBL. w. TAIL Ag-12 = 299-IG-1 = B 0223-406	2,15 S 0.3	F299/J3588 (309,142)
*AM 0223-255	02 23 14 −25 51.8	2.694 0.271	214.26 −68.92	275.2 −26.8	SPIRAL in GROUP AM 0222-253 MCG-4-6-38 = 479-G-1	8 S 13.4	F478/J2708 (44,131)
*AM 0223-245	02 23 25 −24 59.6	2.708 0.271	211.91 −68.71	276.2 −26.8	COMPACT	13 D 0.1	F479/J4497 (308,178)
AM 0223-520	02 23 39 −52 06.8	2.066 0.271	273.85 −59.60	245.9 −26.2	GALAXY w. I/A COMP'N. 198-IG-10	1 S 0.1	F198/J2634 (231,65)
*AM 0223-245	02 23 42 −24 56.2	2.709 0.270	211.80 −68.64	276.3 −26.8	SBc w. I/A COMP'N. MCG-4-6-39 = 479-G-2	1 S 0.3	F478/J2708 (37,181)
AM 0223-614	02 23 43 −61 46.6	1.612 0.270	284.96 −52.17	235.3 −24.4	I/A DBL. w. TAIL (GROUP of 5) 115-IG-10 = B 0223-617	2,15,18 S 0.4	F115/J1810 (232,83)
AM 0223-231	02 23 44 −23 18.4	2.735 0.270	207.49 −68.25	278.1 −26.7	DWARF SPIRAL	20 D 0.9	F479/J4497 (306,269)
AM 0223-234	02 23 57 −23 42.9	2.728 0.270	208.59 −68.31	277.7 −26.8	CLOSE GROUP of 3	24 D 0.9	F478/J2708 (32,246)
AM 0224-594	02 24 05 −59 41.7	1.729 0.270	282.84 −53.82	237.6 −24.9	GALAXY w. DIAMETRIC COMP'NS.	8 S 1.0	F115/J1810 (233,195)
AM 0224-243	02 24 06 −24 31.0	2.715 0.270	210.73 −68.46	276.8 −26.9	RESOLVED SPIRAL + COMP'NS. A 0224-24 = 479-G-4	8 S 2.3	F478/J2708 (31,203)
AM 0224-250	02 24 08 −25 02.6	2.706 0.270	212.14 −68.56	276.2 −26.9	DISTURBED SINGLE 479-G-5 [A82]	16 D 0.7	F479/J4497 (299,176)

Identification	R.A. (1950.0) Declination	Precession R.A./Dec.	Gal. Long. Gal. Lat.	S.G.L. S.G.B.	Description Cross-Identification	Coding Size (Min.)	Field/Plate (x, y)
AM 0224-283	02 24 13 −28 34.5	2.645 0.270	221.83 −68.98	272.2 −27.2	I/A DBL.	2 S 0.3	F415/J4607 (106,256)
AM 0224-711	02 24 15 −71 12.7	0.761 0.270	292.55 −44.20	225.4 −22.1	I/A QUADRUPLE (SMALL) 53-IG-7	4 D 0.4	F053/J0923 (227,111)
AM 0224-471	02 24 16 −47 14.4	2.222 0.270	266.18 −62.78	251.2 −26.9	VERY CLOSE PAIR	23 S 0.2	F198/J2634 (232,327)
AM 0224-225	02 24 17 −22 52.1	2.741 0.270	206.46 −68.00	278.6 −26.7	GROUP of 3 (E's)	24 D 3.6	F478/J2708 (27,291)
*AM 0224-233 [0224-2340]	02 24 20 −23 38.5	2.729 0.270	208.46 −68.20	277.8 −26.8	PAIR (#1: E) + CHAIN of COMP'NS.	23 S 3.0	F478/J2708 (27,250)
AM 0224-234 [0224-2338]	02 24 34 −23 40.1	2.728 0.270	208.57 −68.16	277.7 −26.9	PAIR (#2: SPIRAL) + CHAIN of COMP'NS.	23 S 3.0	F478/J2708 (24,248)
AM 0224-602	02 24 37 −60 22.3	1.687 0.270	283.44 −53.24	236.8 −24.8	GROUP of 4	18 D 6.7	F115/J1810 (228,159)
AM 0224-231	02 24 39 −23 10.5	2.735 0.270	207.31 −68.01	278.3 −26.9	ROUGH CHAIN of 6	17 D 4.5	F478/J2708 (23,275)
AM 0224-230	02 24 43 −23 08.1	2.736 0.270	207.22 −67.98	278.3 −26.9	SPIRAL w. 7 APP. COMP'NS.	8 S 1.3	F479/J4497 (294,278)
AM 0224-490	02 24 44 −49 01.1	2.165 0.270	269.01 −61.58	249.3 −26.8	I/A DBL.	2 S 0.8	F198/J2634 (226,232)
AM 0224-725	02 24 48 −72 59.1	0.494 0.270	293.71 −42.63	223.6 −21.6	GALAXY w. TAIL	15 D 0.4	F053/J0923 (221,16)
AM 0225-221	02 25 00 −22 18.7	2.749 0.269	205.18 −67.68	279.3 −26.8	DISTANT CLUSTER of GALAXIES	19 D 8.1	F479/J4497 (292,323)
AM 0225-671	02 25 09 −67 15.9	1.186 0.269	289.52 −47.53	229.5 −23.3	RICH FAINT CLUSTER	19 D 11.2	F081/J1978 (109,53)
AM 0225-271	02 25 12 −27 19.4	2.664 0.269	218.42 −68.67	273.7 −27.4	DISTURBED SINGLE w. JET	7,16 D 0.4	F478/J2708 (22,52)
AM 0225-280	02 25 12 −28 09.2	2.650 0.269	220.70 −68.74	272.7 −27.4	ONE-ARMED SPIRAL	10 D 0.2	F415/J4607 (94,278)
AM 0225-280	02 25 22 −28 02.1	2.651 0.269	220.38 −68.70	272.9 −27.5	ONE-ARMED SP. + CONNECTED COMP'N.	1,10 S 0.8	F415/J4607 (92,285)
AM 0225-624	02 25 31 −62 42.1	1.537 0.269	285.55 −51.30	234.3 −24.4	E + SPIRAL 81-IG-11	2 S 0.2	F115/J1810 (219,34)
AM 0225-800	02 25 35 −80 08.9	−1.492 0.269	298.02 −36.29	216.3 −19.3	BRIGHT-RIMMED SPIRAL 14-G-1	10 D 0.6	F013/J2551 (53,157)
AM 0225-424	02 25 38 −42 42.7	2.341 0.269	257.35 −65.14	256.3 −27.5	I/A DBL. or I/A COMP'N.	1,22 S 0.8	F299/J3588 (282,32)
*AM 0226-505	02 26 08 −50 56.4	2.093 0.268	271.64 −60.13	247.1 −26.7	SP. + COMP'NS (BACKGROUND CLUSTER?) 198-G-11	8,19 S 1.1	F198/J2634 (212,129)
*AM 0226-320	02 26 11 −32 06.2	2.574 0.268	231.52 −68.43	268.3 −27.8	'PROLATE' w. RING MCG-5-7-1 = 415-G-26	6 D 0.6	F415/J4607 (86,66)
AM 0226-483	02 26 23 −48 36.5	2.169 0.268	267.98 −61.64	249.7 −27.1	I/A DBL.	2 S 0.4	F198/J2634 (211,254)
AM 0226-611	02 26 23 −61 10.3	1.626 0.268	283.95 −52.47	235.9 −24.9	COMPACT	13 D 0.2	F115/J1810 (216,116)
AM 0226-223	02 26 40 −22 34.3	2.741 0.268	206.14 −67.39	279.0 −27.2	SPIRAL w. I/A COMP'N. NGC 951 = 479-G-8	1,10 S 1.6	F479/J4497 (271,309)
AM 0226-540	02 26 47 −54 09.9	1.967 0.268	275.96 −57.79	243.5 −26.3	SMALL DWARF (PARTIALLY RESOLVED)	20 D 0.5	F154/J1878 (322,219)
AM 0226-441	02 26 55 −44 10.2	2.297 0.268	259.97 −64.18	254.6 −27.6	GALAXY w. FILAMENT to COMP'N. B 0226-441	1 S 1.0	F246/J2724 (112,223)
AM 0227-334	02 27 11 −33 43.2	2.539 0.267	235.75 −68.00	266.5 −28.1	CLUSTER of BRIGHT GALAXIES	19 D 9.0	F355/J2713 (144,248)
*AM 0227-264	02 27 14 −26 45.4	2.669 0.267	217.03 −68.16	274.3 −27.8	ONE-ARMED SPIRAL MCG-5-7-4 = 479-G-9	10 D 0.5	F479/J4497 (260,84)
AM 0227-233	02 27 28 −23 30.2	2.724 0.267	208.60 −67.47	278.0 −27.5	DWARF	20 D 0.1	F479/J4497 (260,259)
*AM 0227-484	02 27 29 −48 42.8	2.160 0.267	267.91 −61.43	249.5 −27.2	DBL.-RING-SP. w. I/A COMP'NS. 198-G-13 = Fa 728	1,6,10 D 1.1	F198/J2634 (202,248)
AM 0227-443	02 27 30 −44 35.3	2.283 0.267	260.65 −63.87	254.2 −27.7	TRIANGULAR GALAXY + COMP'N. 246-IG-17	2,16 D 0.3	F246/J2724 (107,200)
AM 0227-295	02 27 35 −29 51.5	2.613 0.267	225.39 −68.27	270.8 −28.0	SMALL I/A PAIR w. MATERIAL 415-G-29	2 S 0.3	F415/J4607 (68,186)
AM 0227-430 [0228-4315]	02 27 35 −43 01.8	2.324 0.267	257.57 −64.66	255.9 −27.8	GROUP of 8 (#1: S0) IC 1812 = 246-G-19	18 D 4.5	F246/J2724 (104,284)
AM 0227-471	02 27 44 −47 13.5	2.205 0.267	265.37 −62.32	251.2 −27.5	I/A DBL. (CHAIN of 5)	2,17 S 0.8	F246/J2724 (110,58)
AM 0227-485	02 27 50 −48 53.2	2.153 0.267	268.11 −61.28	249.3 −27.3	I/A DBL. ([S.E.] of Sa) Fa 729? near	2,8 S 0.4	F198/J2634 (198,239)
AM 0227-545	02 27 59 −54 58.7	1.926 0.267	276.76 −57.09	242.6 −26.4	PEC. SP. + 8 COMP'NS (incl. I/A 3) 153-IG-36 = B 0227-549	3,8,16 S 4.9	F153/J1882 (56,177)

Identification	R.A. (1950.0) Declination	Precession R.A./Dec.	Gal. Long. Gal. Lat.	S.G.L. S.G.B.	Description Cross-Identification	Coding Size (Min.)	Field/Plate (x, y)
AM 0228-543	02 28 23 −54 39.9	1.936 0.266	276.29 −57.27	242.9 −26.5	CLOSE I/A DBL. (ONE COMPACT)	2,13 S 0.1	F153/J1882 (52,194)
AM 0228-431 [0227-4301]	02 28 31 −43 15.2	2.314 0.266	257.82 −64.41	255.6 −28.0	GROUP of 8 (#2: SB(r)b) 246-G-21	18 D 4.5	F246/J2724 (95,271)
*AM 0228-342 [0228-3426]	02 28 32 −34 29.4	2.519 0.266	237.61 −67.57	265.6 −28.4	I/A SPIRALS + COMP'NS. (#1) IC 1811 = 355-G-20	2,8 D 0.5	F355/J2713 (129,206)
*AM 0228-342 [0228-3429]	02 28 43 −34 26.7	2.519 0.266	237.47 −67.55	265.6 −28.4	I/A SPIRALS + COMP'NS. (#2) IC 1813 = 355-G-22	2,8 D 0.3	F355/J2713 (127,209)
AM 0228-465	02 28 44 −46 56.6	2.209 0.266	264.67 −62.35	251.5 −27.7	ONE-ARMED SP. w. COMPACT at END	10,13 D 0.4	F198/J2634 (191,343)
AM 0228-510	02 28 47 −51 07.6	2.071 0.266	271.35 −59.70	246.8 −27.1	I/A DBL.	2 S 0.8	F198/J2634 (189,119)
AM 0228-234	02 28 51 −23 49.0	2.716 0.266	209.60 −67.24	277.7 −27.9	I/A DBL.	2 S 0.4	F479/J4497 (242,243)
AM 0229-231	02 29 11 −23 13.6	2.726 0.266	208.18 −67.02	278.3 −27.9	CURVED CHAIN of 7 E's	17 D 9.0	F479/J4497 (239,274)
AM 0229-604	02 29 31 −60 44.9	1.625 0.265	283.03 −52.58	236.3 −25.3	CLOSE PAIR (COMPACT + SPIRAL)	13,23 S 0.4	F115/J1810 (195,139)
AM 0229-580	02 29 40 −58 08.0	1.767 0.265	280.21 −54.58	239.1 −26.0	I/A GROUP (E + I/A COMP'NS.) 115-G-15 = PKS 0229-581 = Fa 222/3	1,18 S 2.2	F115/J1810 (196,279)
AM 0229-261	02 29 45 −26 16.3	2.672 0.265	215.96 −67.53	274.9 −28.3	I/A DBL. 479-G-13	2 S 0.5	F479/J4497 (230,111)
*AM 0229-444	02 29 45 −44 44.9	2.268 0.265	260.46 −63.44	253.9 −28.1	RING w. ABS'N in WIDE TRIPLE NGC 979 = 246-G-23 = UKS 0229-447	6,8,14,24 S 9.9	F246/J2724 (86,191)
*AM 0229-365	02 29 47 −36 53.5	2.463 0.265	243.45 −66.74	262.8 −28.6	3 DISTURBED ARMS in DISK IC 1816 = 355-G-25	10,11 D 1.2	F355/J2713 (117,77)
AM 0229-421	02 29 50 −42 16.0	2.334 0.265	255.54 −64.67	256.7 −28.3	GROUP of 3	24 D 2.0	F299/J3588 (241,57)
AM 0230-581	02 30 08 −58 14.2	1.758 0.265	280.24 −54.46	239.0 −26.0	SMALL RING w. OFF-SET NUCLEUS	6 D 0.4	F115/J1810 (193,274)
*AM 0230-351	02 30 12 −35 15.0	2.498 0.265	239.34 −67.08	264.7 −28.7	SPIRAL w. PEC. COMP'NS. 355-G-26	8 S 11.2	F355/J2713 (111,165)
AM 0230-295	02 30 22 −29 55.0	2.604 0.265	225.54 −67.66	270.8 −28.6	COMPACT 416-IG-1	13 D 0.4	F416/J1044 (300,180)
AM 0230-393	02 30 41 −39 31.1	2.399 0.264	249.44 −65.69	259.8 −28.7	DWARF IRR. NGC 986a = 299-G-6	16,20 D 1.1	F299/J3588 (234,205)
AM 0230-524 [0230-5242]	02 30 42 −52 43.9	2.000 0.264	273.26 −58.39	245.0 −27.2	I/A DBL.(#1: SPIRAL + DEBRIS) 154-IG-2 = B 0230-527	2,152 S 1.4	F198/J2634 (173,33)
AM 0230-524 [0230-5243]	02 30 50 −52 42.8	2.000 0.264	273.21 −58.39	245.0 −27.2	I/A DBL.(#2: ONE-ARMED)	2,10 S 1.4	F198/J2634 (172,34)
AM 0230-533 [0231-5331]	02 30 57 −53 33.9	1.965 0.264	274.35 −57.79	244.1 −27.0	CLOSE PAIR (#1)	23 D 0.3	F153/J1882 (28,252)
AM 0231-764	02 31 09 −76 49.2	−0.423 0.264	295.72 −39.09	219.5 −20.7	2 E's in ENVELOPE (CHAIN of 4) 30-IG-15	2,17 S 0.3	F031/J1947 (320,64)
AM 0231-533 [0231-5333]	02 31 11 −53 31.3	1.966 0.264	274.24 −57.80	244.1 −27.1	CLOSE PAIR (#2)	23 D 0.3	F153/J1882 (26,254)
AM 0231-434	02 31 32 −43 44.3	2.288 0.264	258.15 −63.69	255.0 −28.5	ONE-ARMED SPIRAL 246-G-25	10 D 0.9	F246/J2724 (67,244)
*AM 0231-391	02 31 34 −39 15.3	2.402 0.264	248.69 −65.63	260.1 −28.8	SP. w. PEC. ARMS + DIAMETRICS NGC 986 = 299-G-7	8,10 S 15.7	F299/J3588 (225,219)
AM 0231-302	02 31 38 −30 20.4	2.593 0.264	226.63 −67.38	270.3 −28.9	CLOSE PAIR of E's	23 S 0.4	F415/J4607 (21,159)
AM 0231-470	02 31 39 −47 07.3	2.189 0.263	264.35 −61.84	251.2 −28.1	CLOSE PAIR	23 S 0.3	F198/J2634 (164,334)
AM 0231-623	02 31 40 −62 34.5	1.491 0.263	284.53 −50.98	234.2 −25.1	COMPACTS CONNECTED by BRIDGE 81-I-12	2,13,13 S 0.5	F115/J1810 (181,41)
*AM 0231-335	02 31 49 −33 53.0	2.521 0.263	235.74 −67.02	266.3 −29.0	E + I/A DBL. COMP'N + TAIL GROUP of 6	1,2,15,18 S 0.3	F355/J2713 (92,238)
AM 0232-591	02 32 10 −59 13.2	1.691 0.263	280.99 −53.55	237.8 −26.0	I/A DBL. in GROUP of 7 115-IG-19	2,18 S 0.4	F115/J1810 (178,221)
AM 0232-401	02 32 24 −40 17.2	2.374 0.263	250.83 −65.09	259.0 −28.9	SMALL I/A TRIPLE 299-G-10	3 D 0.4	F299/J3588 (216,164)
AM 0232-460	02 32 52 −46 03.6	2.215 0.262	262.23 −62.27	252.4 −28.5	GALAXY w. JET	7 D 0.9	F246/J2724 (60,119)
*AM 0233-453	02 33 38 −45 37.0	2.225 0.262	261.27 −62.40	252.9 −28.7	GALAXY w. JET (DOG-LEG)	7 D 0.4	F246/J2724 (52,142)
AM 0233-503 [0235-5039]	02 33 51 −50 32.3	2.063 0.262	269.44 −59.48	247.3 −28.0	CHAIN of 4 (#1: SPIRAL) 198-G-20	15,17 S 19.9	F198/J2634 (146,150)
AM 0234-652	02 34 20 −65 28.6	1.247 0.261	286.87 −48.49	231.1 −24.6	COMPACT	13 D 0.3	F082/J1871 (300,149)
AM 0234-594	02 34 22 −59 46.3	1.643 0.261	281.24 −52.94	237.2 −26.2	GALAXY w. 3 APP. COMP'NS.	8 S 3.4	F115/J1810 (163,192)

Identification	R.A. (1950.0) Declination	Precession R.A./Dec.	Gal. Long. Gal. Lat.	S.G.L. S.G.B.	Description Cross-Identification	Coding Size (Min.)	Field/Plate (x, y)
AM 0234-325	02 34 26 −32 58.8	2.532 0.261	233.27 −66.61	267.3 −29.6	COMPACT	13 D 0.1	F355/J2713 (62,286)
AM 0234-503 [0233-5032]	02 34 26 −50 32.8	2.060 0.261	269.34 −59.40	247.3 −28.1	CHAIN of 4 (#2: SPIRAL) 198-G-22 = Fa 297	15,17 S 19.9	F198/J2634 (141,150)
AM 0235-503 [0234-5032]	02 35 00 −50 32.9	2.057 0.260	269.23 −59.33	247.3 −28.2	CHAIN of 4 (#3: SPIRAL)	15,17 S 19.9	F198/J2634 (137,150)
AM 0235-550	02 35 06 −55 04.8	1.875 0.260	275.56 −56.30	242.3 −27.3	SPIRAL w. APP. COMP'N. NGC 1031 = 154-G-5	8 S 2.3	F154/J1878 (255,174)
AM 0235-565	02 35 38 −56 59.4	1.781 0.260	277.85 −54.89	240.1 −27.0	PECULIAR SPIRAL (M51-TYPE?)	9 S 0.5	F115/J1810 (153,341)
AM 0235-424	02 35 43 −42 41.0	2.299 0.260	255.25 −63.52	256.2 −29.4	I/A DBL. (E's w. FILAMENT) 247-G-2	2 S 0.3	F299/J3588 (182,35)
AM 0235-225	02 35 48 −22 52.7	2.719 0.260	208.39 −65.46	278.9 −29.4	SMALL I/A TRIPLE	3 D 0.4	F479/J4497 (157,294)
AM 0235-503 [0235-5032]	02 35 49 −50 39.3	2.048 0.260	269.23 −59.16	247.1 −28.3	CHAIN of 4 (#4: EDGE-ON SPIRAL) 198-IG-23 = PKS 0236-505?	15,17 S 19.9	F198/J2634 (130,144)
AM 0235-793	02 35 54 −79 38.5	−1.524 0.260	297.23 −36.52	216.6 −19.9	I/A DBL. w. TAIL B 0235-796	2,15 S 0.3	F014/J2559 (237,195)
AM 0235-335	02 35 58 −33 55.3	2.508 0.260	235.50 −66.16	266.2 −29.9	I/A DBL. in WIDE PAIR 355-IG-31	2,23 S 0.4	F355/J2713 (46,235)
AM 0236-230	02 36 02 −23 03.8	2.715 0.260	208.85 −65.46	278.7 −29.4	CLOSE PAIR	23 S 0.2	F479/J4497 (154,284)
AM 0236-473	02 36 06 −47 38.9	2.150 0.259	264.35 −60.93	250.5 −28.8	COMPACT	13 D 0.2	F198/J2634 (124,305)
AM 0236-613	02 36 31 −61 33.0	1.517 0.259	282.79 −51.42	235.2 −26.0	EDGE-ON LATE SPIRAL w. RESOLUTION 115-IG-21 = UKS 0236-615	16 D 7.2	F115/J1810 (150,96)
AM 0236-310	02 36 38 −31 01.5	2.566 0.259	228.30 −66.29	269.6 −30.0	DISTURBED RING + 2 COMP'NS. 416-G-5	2,6,8 D 0.9	F416/J1044 (227,122)
AM 0236-530	02 36 45 −53 00.0	1.953 0.259	272.47 −57.54	244.5 −28.0	I/A DBL. (M51-TYPE)	2 S 0.6	F154/J1878 (246,286)
*AM 0236-273	02 36 51 −27 39.7	2.630 0.259	219.97 −66.13	273.4 −30.0	"RINGING" SPIRAL w. APP. COMP'NS IC 1830 = IC 1826 = 416-G-6	6,8,12 D 1.1	F479/J4497 (145,37)
AM 0236-582 [0236-5827]	02 36 51 −58 22.2	1.702 0.259	279.27 −53.77	238.6 −26.8	CONTACT PAIR (#1: COMPACT)	13,23 S 0.2	F115/J1810 (145,267)
AM 0236-582 [0236-5822]	02 36 58 −58 27.0	1.696 0.259	279.35 −53.71	238.5 −26.8	CONTACT PAIR (#2: DWARF) 115-G-22	20 D 0.7	F115/J1810 (144,262)
*AM 0237-525	02 37 07 −52 59.6	1.951 0.259	272.39 −57.50	244.5 −28.1	I/A DBL. + I/A COMP'N.	1,2,9 S 0.6	F198/J2634 (121,18)
AM 0237-611	02 37 17 −61 11.6	1.534 0.258	282.31 −51.63	235.5 −26.1	2 E's + INTEGRAL SPIRAL	24 D 2.2	F115/J1810 (145,115)
AM 0238-384	02 38 14 −38 41.7	2.391 0.257	246.40 −64.60	260.7 −30.2	ONE-ARMED SPIRAL + APP. COMP'N.	8,10 S 2.0	F299/J3588 (156,249)
AM 0238-795	02 38 27 −79 59.0	−1.750 0.257	297.32 −36.17	216.3 −19.9	EDGE-ON SP. w. 2 APP. COMP'NS. 14-G-02	8 S 1.1	F014/J2559 (229,178)
AM 0238-615	02 38 48 −61 52.7	1.476 0.257	282.80 −51.00	234.7 −26.1	COMPACT	13 D 0.1	F115/J1810 (136,78)
AM 0238-792	02 38 49 −79 25.0	−1.495 0.257	296.94 −36.65	216.8 −20.1	LO.S.BR. SPIRAL	20 D 0.5	F013/J2551 (12,187)
AM 0239-334	02 39 00 −33 48.2	2.501 0.257	234.99 −65.56	266.3 −30.5	LUMPY COMPACT	13 D 0.4	F355/J2713 (12,241)
AM 0239-614	02 39 16 −61 43.7	1.482 0.257	282.58 −51.07	234.9 −26.2	COMPACT w. FUZZ	13 D 0.1	F115/J1810 (133,86)
*AM 0239-273	02 39 30 −27 31.1	2.627 0.256	219.77 −65.53	273.6 −30.5	GALAXY w. APP.COMP'N.(KNOTS from MCG-5-7-14 = 416-G-8 NUCLEUS)	7,8 D 1.1	F479/J4497 (113,44)
AM 0239-590	02 39 38 −59 06.0	1.641 0.256	279.67 −53.00	237.7 −27.0	SPIRAL w. THIN OPPOSED JETS(?)	7 D 0.7	F115/J1810 (127,227)
AM 0239-242	02 39 51 −24 20.6	2.685 0.256	212.32 −64.93	277.3 −30.4	HI.S.BR. GALAXY + COMPACT COMP'N MCG-4-7-24 = 479-G-25	8,13 D 1.1	F479/J4497 (108,215)
AM 0240-242	02 40 17 −24 26.6	2.682 0.256	212.60 −64.85	277.2 −30.5	RING	6,15 D 0.4	F479/J4497 (102,209)
AM 0240-635	02 40 24 −63 51.1	1.320 0.255	284.57 −49.37	232.6 −25.7	SMALL I/A DBL. 82-IG-1	2 S 0.3	F082/J1871 (271,238)
AM 0240-254	02 40 35 −25 47.4	2.657 0.255	215.75 −65.05	275.7 −30.7	CLOSE TRIPLE	24 D 1.3	F479/J4497 (99,137)
AM 0240-600	02 40 56 −60 06.5	1.572 0.255	280.60 −52.15	236.6 −26.9	I/A DBL. 115-IG-25	2 S 0.2	F115/J1810 (120,173)
AM 0241-614	02 41 00 −61 49.9	1.461 0.255	282.44 −50.86	234.7 −26.4	SMALL I/A DBL. 115-IG-26	2 S 0.2	F115/J1810 (122,80)
AM 0241-223	02 41 14 −22 38.8	2.712 0.255	208.69 −64.19	279.3 −30.6	GALAXY w. I/A COMP'NS + TAILS 479-IG-28	1,15 D 0.4	F479/J4497 (90,306)
AM 0241-424	02 41 23 −42 47.1	2.272 0.255	254.45 −62.54	255.9 −30.4	DBL. incl. RING	6,23 S 0.5	F299/J3588 (126,29)

Identification	R.A. (1950.0) Declination	Precession R.A./Dec.	Gal. Long. Gal. Lat.	S.G.L. S.G.B.	Description Cross-Identification	Coding Size (Min.)	Field/Plate (x, y)
AM 0241-623	02 41 30 −62 39.1	1.400 0.254	283.22 −50.20	233.8 −26.2	E w. 4 DIAMETRIC COMP'NS.	8 D 1.3	F082/J1871 (268,303)
AM 0242-410	02 42 31 −41 08.2	2.313 0.253	250.96 −63.00	257.8 −30.8	PAIR	23 S 0.4	F299/J3588 (113,118)
*AM 0242-600	02 42 33 −60 09.1	1.557 0.253	280.41 −51.98	236.5 −27.1	SPIRAL w. I/A DBL. COMP'NS. NGC 1096 = 115-G-28 [A82]	1 D 2.5	F115/J1810 (109,170)
AM 0242-244	02 42 34 −24 43.6	2.672 0.253	213.51 −64.41	276.9 −31.1	E w. EXTENDED MATERIAL MCG-4-7-32 = 479-G-31	15 D 0.5	F479/J4497 (75,194)
AM 0242-513 [0242-5138]	02 42 40 −51 39.2	1.972 0.253	269.46 −57.70	245.8 −29.2	CLOSE PAIR of SPIRALS (#1:SB)	23 S 1.4	F198/J2634 (74,88)
AM 0242-474	02 42 43 −47 45.2	2.114 0.253	263.28 −59.94	250.2 −29.9	SP. w. DIAMETRIC I/A COMP'NS. 198-IG-29	1,9 S 0.3	F198/J2634 (65,298)
AM 0242-513 [0242-5139]	02 42 46 −51 38.1	1.973 0.253	269.41 −57.70	245.8 −29.2	CLOSE PAIR of SPIRALS (#2)	23 S 1.4	F198/J2634 (73,89)
AM 0242-590	02 42 52 −59 01.8	1.621 0.253	279.09 −52.76	237.6 −27.4	LO.S.BR. SPIRAL + PEC. COMP'NS 115-G-29	8 S 2.2	F115/J1810 (104,230)
AM 0242-432	02 42 57 −43 24.9	2.248 0.253	255.42 −62.01	255.1 −30.6	COMPACT	13 D 0.2	F247/J1929 (228,264)
*AM 0243-242	02 43 03 −24 20.5	2.678 0.253	212.70 −64.22	277.4 −31.2	I/A TRIPLE (BRIDGE w. CONDENS.)	3 D 0.7	F479/J4497 (69,214)
AM 0243-463	02 43 24 −46 39.8	2.147 0.253	261.29 −60.41	251.4 −30.2	RICH CLUSTER of FAINT GALAXIES R 56 = DN 0243-467	19 D 6.7	F247/J1929 (222,90)
AM 0243-250 [0243-2505]	02 43 27 −25 01.5	2.665 0.253	214.28 −64.28	276.6 −31.3	CLOSE PAIR of SPIRALS (#1) MCG-4-7-34 = 479-G-33	23 S 3.8	F479/J4497 (64,178)
AM 0243-281	02 43 27 −28 10.3	2.605 0.253	221.51 −64.73	272.9 −31.4	GALAXY w. I/A COMP'N. MCG-5-7-18 = IC 1845? = 416-G-15	1 S 2.2	F416/J1044 (147,275)
AM 0243-250 [0243-2501]	02 43 29 −25 05.7	2.664 0.253	214.44 −64.28	276.5 −31.3	CLOSE PAIR of SPIRALS (#2)	23 S 3.8	F479/J4497 (64,174)
AM 0243-653	02 43 44 −65 37.0	1.142 0.252	285.85 −47.78	230.6 −25.5	LINE of 5 (SEVERAL I/A) incl. Fa 226	2,17 D 3.6	F082/J1871 (247,145)
AM 0244-263 [0244-2627]	02 44 00 −26 31.0	2.636 0.252	217.72 −64.41	274.9 −31.5	PAIR of SPIRALS (#1) 479-G-35	23 S 4.5	F479/J4497 (59,98)
AM 0244-702	02 44 10 −70 20.7	0.617 0.252	289.98 −44.02	225.7 −23.9	I/A DBL. (in CHAIN of 4) 53-IG-12	2,17 D 1.1	F053/J0923 (140,158)
*AM 0244-302	02 44 11 −30 29.5	2.556 0.252	226.96 −64.68	270.2 −31.6	SPIRAL w. I/A COMP'N. (ARP 77) NGC 1097 = 416-G-20 = PKS 0244-304	1 D 9.0	F416/J1044 (140,151)
AM 0244-530	02 44 13 −53 04.7	1.905 0.252	271.25 −56.63	244.1 −29.1	M51-TYPE	9 S 0.2	F198/J2634 (64,11)
AM 0244-262 [0244-2631]	02 44 18 −26 27.6	2.636 0.252	217.61 −64.34	274.9 −31.6	PAIR of SPIRALS (#2) 479-G-37	23 S 4.5	F479/J4497 (55,101)
AM 0244-250	02 44 21 −25 04.4	2.662 0.252	214.48 −64.08	276.6 −31.5	GALAXY w. EXTENDED MATERIAL 479-G-38	15 D 0.9	F479/J4497 (53,175)
AM 0244-612	02 44 22 −61 29.7	1.456 0.252	281.62 −50.84	234.9 −26.9	PECULIAR M51-TYPE 115-IG-31	9 D 0.4	F115/J1810 (100,97)
*AM 0244-271	02 44 24 −27 11.0	2.622 0.252	219.28 −64.42	274.1 −31.6	SPIRAL w. 9 COMP'NS. MCG-5-7-25 = 479-G-39	8,13 S 3.1	F479/J4497 (54,62)
AM 0244-253	02 44 28 −25 33.2	2.653 0.252	215.57 −64.15	276.0 −31.6	SPIRAL + APP. COMP'N. MCG-4-7-38 = 479-G-40	8 S 2.5	F479/J4497 (52,149)
AM 0244-475	02 44 33 −47 54.2	2.100 0.252	263.21 −59.60	250.0 −30.2	HI.S.BR. (TRIPLE) NUCLEUS + TAIL	15 D 0.5	F247/J1929 (210,23)
AM 0244-531	02 44 35 −53 16.6	1.895 0.252	271.46 −56.46	243.9 −29.1	I/A DBL.	2 S 0.1	F154/J1878 (183,272)
AM 0244-225	02 44 38 −22 51.2	2.702 0.251	209.63 −63.50	279.2 −31.4	DISTURBED SINGLE MCG-4-7-39 = 479-G-42	16 D 1.0	F479/J4497 (48,294)
AM 0244-290	02 44 43 −29 08.9	2.582 0.251	223.84 −64.52	271.8 −31.7	COMPACT	13 D 0.2	F416/J1044 (133,223)
AM 0244-222 [0245-2228]	02 44 56 −22 25.2	2.710 0.251	208.75 −63.31	279.7 −31.4	CLOSE PAIR of SPIRALS (#1) MCG-4-7-41 = 546-G-19	23 S 3.8	F479/J4497 (44,317)
AM 0244-360	02 44 58 −36 02.3	2.432 0.251	239.69 −63.98	263.7 −31.7	DISTURBED SINGLE GALAXY 356-IG-10	16 D 0.5	F356/J4602 (212,123)
AM 0245-222 [0244-2225]	02 45 08 −22 28.6	2.708 0.251	208.90 −63.28	279.6 −31.5	CLOSE PAIR of SPIRALS (#2) NGC 1102 = MCG-4-7-40 = 546-G-20	23 S 3.8	F479/J4497 (41,314)
AM 0245-233	02 45 25 −23 30.5	2.689 0.251	211.15 −63.49	278.4 −31.6	DWARF	20 D 0.4	F479/J4497 (39,259)
AM 0246-472	02 46 35 −47 21.1	2.109 0.250	261.93 −59.59	250.5 −30.6	SMALL I/A TRIPLE 247-IG-4	3 D 0.5	F198/J2634 (29,318)
*AM 0246-365	02 46 44 −36 55.2	2.405 0.249	241.47 −63.45	262.6 −32.0	SPIRAL w. STRONG ARM + COMP'NS. 356-G-13	8,10 S 1.0	F300/J3536 (343,342)
AM 0246-694	02 46 49 −69 46.4	0.661 0.249	289.25 −44.34	226.2 −24.3	SPIRAL w. APP. COMP'N.(I/A ?) 53-IG-15	1 S 0.8	F053/J0923 (126,188)
AM 0247-421	02 47 02 −42 15.5	2.265 0.249	252.53 −61.79	256.4 −31.5	SMALL I/A DBL.	2 S 0.3	F299/J3588 (70,56)

Identification	R.A. (1950.0) Declination	Precession R.A./Dec.	Gal. Long. Gal. Lat.	S.G.L. S.G.B.	Description Cross-Identification	Coding Size (Min.)	Field/Plate (x, y)
*AM 0247-312	02 47 27 −31 23.6	2.529 0.249	229.03 −63.97	269.1 −32.3	E in CLUSTER + COMP'NS. IC 1860 = 416-G-31	8,19 D 5.4	F416/J1044 (103,102)
AM 0247-304	02 47 28 −30 47.3	2.542 0.249	227.65 −63.97	269.9 −32.3	COMPACT (FRINGED) MCG-5-7-34 = 416-G-32	13 D 0.7	F416/J1044 (102,134)
AM 0247-252	02 47 31 −25 25.8	2.649 0.249	215.59 −63.45	276.2 −32.2	SMALL I/A DBL.	2 S 0.2	F480/J3649 (282,155)
AM 0247-690	02 47 31 −69 07.6	0.734 0.249	288.62 −44.81	226.8 −24.6	SPIRAL w. CONDENSATION in ARM 53-G-16	10 D 0.7	F053/J0923 (121,223)
AM 0247-385	02 47 34 −38 58.5	2.351 0.249	245.79 −62.78	260.2 −32.0	SPIRAL + COMPACT COMP'N. 299-G-20	8,13 S 2.0	F300/J3536 (330,232)
AM 0247-713	02 47 44 −71 39.3	0.380 0.248	290.73 −42.80	224.3 −23.7	I/A TRIPLE (in GROUP of 4) 53-IG-17	3,18 D 0.9	F053/J0923 (127,87)
AM 0247-240	02 47 45 −24 04.9	2.674 0.248	212.68 −63.12	277.8 −32.2	PAIR + 3 COMP'NS.	8,23 S 1.3	F480/J3649 (281,227)
AM 0248-523	02 48 12 −52 35.5	1.903 0.248	269.89 −56.45	244.5 −29.8	I/A DBL. (DUSTY SPIRAL) 154-IG-16 = B 0248-525	2,14 S 0.5	F198/J2634 (30,35)
AM 0248-713	02 48 16 −71 36.0	0.382 0.248	290.64 −42.82	224.3 −23.7	CLOSE PAIR 53-IG-18	23 S 0.2	F053/J0923 (125,90)
AM 0248-470	02 48 18 −47 00.3	2.113 0.248	261.05 −59.51	250.8 −31.0	I/A DBL. 247-IG-6	2 S 0.2	F198/J2634 (12,336)
*AM 0248-350	02 48 28 −35 04.1	2.444 0.248	237.26 −63.44	264.8 −32.4	CHAIN of 5 356-IG-14	17 D 7.2	F356/J4602 (174,176)
AM 0248-231	02 48 38 −23 10.8	2.689 0.248	210.87 −62.70	278.9 −32.3	CLOSE PAIR	23 S 0.3	F480/J3649 (271,276)
AM 0248-494	02 48 55 −49 47.6	2.011 0.247	265.53 −57.97	247.6 −30.5	E + I/A COMP'N.(TAIL)	1,15 S 1.1	F198/J2634 (15,186)
AM 0248-270 [0249-2710]	02 48 58 −27 09.3	2.612 0.247	219.51 −63.41	274.2 −32.6	WIDE I/A DBL. (#1) MCG-5-7-41 = 480-G-5	2 S 2.2	F480/J3649 (262,63)
AM 0249-271 [0248-2709]	02 49 8 −27 10.5	2.612 0.247	219.57 −63.37	274.1 −32.7	WIDE I/A DBL. (#2) MCG-5-7-43 = 480-G-6	2 S 2.2	F480/J3649 (260,61)
AM 0249-250	02 49 13 −25 09.0	2.651 0.247	215.14 −63.02	276.6 −32.6	ELONGATED CLUSTER	19 D 8.1	F480/J3649 (261,170)
AM 0249-422	02 49 18 −42 21.8	2.253 0.247	252.40 −61.37	256.2 −31.9	GROUP of 3	24 D 0.9	F247/J1929 (166,321)
AM 0249-551	02 49 23 −55 10.8	1.779 0.247	273.28 −54.71	241.6 −29.3	SPIRAL w. I/A COMP'N. Fa 732 NGC 1136 = 154-G-19 [A82]	1 S 2.6	F154/J1878 (145,170)
AM 0250-405	02 50 07 −40 55.9	2.290 0.246	249.47 −61.72	257.8 −32.2	GROUP of 3(SP.w.I/A COMP'N.+TAIL)	1,15,24 D 1.6	F300/J3536 (300,128)
AM 0250-305 [0251-3104]	02 50 22 −30 58.6	2.530 0.246	228.07 −63.35	269.6 −33.0	LINE of 3 (#1: SPIRAL) MCG-5-7-44 = 416-G-40	24 D 10.8	F417/J3818 (336,125)
*AM 0250-310 [0250-3058]	02 50 48 −31 03.9	2.527 0.245	228.27 −63.26	269.5 −33.1	LINE of 3 (#2: COMPACT) 356-G-16	13,24 D 10.8	F417/J3818 (331,120)
AM 0251-601	02 51 00 −60 11.6	1.490 0.245	279.25 −51.21	236.1 −28.0	SPIRAL w. 2 APP. COMP'NS. 115-G-32 = B 0251-601	8 S 2.0	F116/J1814 (306,164)
AM 0251-580	02 51 02 −58 08.9	1.613 0.245	276.82 −52.60	238.3 −28.7	3-ARMED SPIRAL + PECULIAR COMP'N. 116-G-1	8,11 S 1.8	F115/J1810 (45,274)
AM 0251-345	02 51 08 −34 56.1	2.439 0.245	236.79 −62.92	264.9 −33.0	PAIR of E's + COMPACT PART	13,23 S 0.9	F356/J4602 (145,183)
AM 0251-310 [0250-3103]	02 51 22 −31 04.0	2.526 0.245	228.27 −63.13	269.5 −33.2	LINE of 3 (#3: SPIRAL) MCG-5-7-45 = 416-G-41	24 D 10.8	F417/J3818 (325,120)
AM 0251-374	02 51 38 −37 42.8	2.370 0.245	242.70 −62.33	261.6 −32.9	SMALL I/A DBL.	2 S 0.5	F299/J3588 (13,299)
AM 0251-342	02 51 44 −34 23.8	2.450 0.245	235.59 −62.87	265.5 −33.1	GALAXY w. DISTURBED COMP'N. IC 1864 = 356-G-17	8 S 7.6	F356/J4602 (138,211)
AM 0252-511 [0251-5113]	02 52 05 −51 16.0	1.937 0.244	267.29 −56.74	245.8 −30.7	CHAIN of FIVE (#1)	17 D 5.8	F199/J1892 (245,110)
AM 0252-555	02 52 06 −55 59.3	1.722 0.244	273.93 −53.90	240.6 −29.4	GALAXY w. I/A COMP'N.	1 S 0.4	F154/J1878 (126,126)
AM 0252-511 [0251-5116]	02 52 28 −51 15.0	1.935 0.244	267.21 −56.70	245.8 −30.7	CHAIN of FIVE (#2)	17 D 5.8	F199/J1892 (242,111)
AM 0252-511 [0251-5115]	02 52 34 −51 14.5	1.935 0.244	267.17 −56.69	245.9 −30.8	CHAIN of FIVE (#3)	17 D 5.8	F199/J1892 (241,112)
AM 0252-303	02 52 35 −30 37.6	2.532 0.244	227.31 −62.87	270.0 −33.4	GALAXY w. APP. COMP'NS. MCG-5-8-1 = 417-G-1	8 S 2.6	F417/J3818 (311,144)
AM 0252-511 [0251-5114]	02 52 44 −51 14.2	1.934 0.244	267.14 −56.67	245.9 −30.8	CHAIN of FIVE (#4)	17 D 5.8	F199/J1892 (240,112)
AM 0252-511 [0251-5114]	02 52 52 −51 13.6	1.934 0.243	267.10 −56.66	245.9 −30.8	CHAIN of FIVE (#5)	17 D 5.8	F199/J1892 (239,113)
*AM 0253-273	02 53 31 −27 37.6	2.593 0.243	220.81 −62.46	273.6 −33.7	SPIRAL w. RE-ENTRANT ARM MCG-5-8-4 = 417-G-3	10 D 1.6	F480/J3649 (208,38)
AM 0253-475	02 53 55 −47 54.4	2.055 0.242	261.67 −58.24	249.6 −31.7	I/A DBL. in CHAIN of 4	2,17 S 0.2	F199/J1892 (234,291)

Identification	R.A. (1950.0) Declination	Precession R.A./Dec.	Gal. Long. Gal. Lat.	S.G.L. S.G.B.	Description Cross-Identification	Coding Size (Min.)	Field/Plate (x, y)
AM 0254-604	02 54 07 −60 44.5	1.430 0.242	279.46 −50.55	235.3 −28.2	SMALL S0 + KNOTTY CHAIN 116-IG-2	22 S 0.3	F115/J1810 (34,134)
AM 0254-425	02 54 10 −42 59.6	2.215 0.242	252.93 −60.30	255.3 −32.7	I/A DBL.	2 S 0.5	F300/J3536 (257,18)
*AM 0254-322	02 54 18 −32 23.2	2.489 0.242	231.13 −62.49	267.9 −33.8	GALAXY w. APP. COMP'NS. MCG-5-8-6 = 417-G-6	8 S 0.5	F416/J1044 (26,46)
AM 0254-292	02 54 19 −29 28.9	2.552 0.242	224.84 −62.45	271.4 −33.8	M51-TYPE	9,15 S 0.3	F416/J1044 (21,203)
AM 0254-404	02 54 51 −40 43.0	2.278 0.241	248.46 −60.94	257.9 −33.1	I/A DBL. w. TAILS 300-IG-1 = B 0254-407	2,15 S 0.4	F300/J3536 (252,140)
*AM 0255-354	02 55 24 −35 46.0	2.406 0.241	238.31 −61.94	263.8 −33.8	SPIRAL w. PEC. ARMS + APP.COMP'NS 356-G-20	8,10 S 0.4	F356/J4602 (98,137)
*AM 0255-365	02 55 50 −36 55.0	2.376 0.240	240.68 −61.67	262.4 −33.8	DISTURBED SP. + APP. COMP'NS. 356-G-22	8,12 S 3.8	F356/J4602 (95,76)
AM 0256-231	02 56 08 −23 15.0	2.674 0.240	211.96 −61.05	278.9 −34.1	CLOSE TRIPLE of PECULIARS	24 D 1.2	F480/J3649 (178,273)
AM 0256-630	02 56 19 −63 04.3	1.243 0.240	281.76 −48.74	232.7 −27.7	DISTURBED SPIRAL w. 2 COMP'NS. 82-IG-6	8,16 S 1.2	F116/J1814 (262,12)
*AM 0256-321	02 56 44 −32 18.0	2.484 0.240	230.91 −61.97	268.0 −34.3	SPIRAL + IRR. COMP'N. NGC 1165 = 417-G-8	8,16 S 4.3	F417/J3818 (262,55)
AM 0256-364	02 56 56 −36 48.6	2.376 0.239	240.38 −61.47	262.5 −34.0	SPIRAL w. JET 356-IG-24	7 D 0.8	F357/J2770 (341,79)
AM 0257-372	02 57 03 −37 26.6	2.359 0.239	241.67 −61.33	261.8 −34.0	I/A DBL. 356-IG-25	2 S 0.4	F300/J3536 (232,316)
AM 0257-340	02 57 28 −34 02.8	2.442 0.239	234.59 −61.72	265.9 −34.3	CLOSE PAIR	23 S 0.3	F357/J2770 (341,228)
*AM 0257-370 [0258-3708]	02 57 59 −37 08.4	2.364 0.238	240.98 −61.21	262.1 −34.2	PAIR (#1: GALAXY w. PLUMES)	15,23 S 5.0	F300/J3536 (222,333)
AM 0258-650	02 58 16 −65 05.5	1.054 0.238	283.66 −47.16	230.5 −27.1	CLOSE PAIR of E'S	23 S 0.4	F082/J1871 (166,174)
AM 0258-370 [0257-3707]	02 58 22 −37 07.1	2.363 0.238	240.91 −61.14	262.1 −34.3	PAIR (#2: EDGE-ON) 356-G-26	23 S 5.0	F300/J3536 (218,334)
*AM 0258-743	02 58 22 −74 39.4	−0.346 0.238	292.36 −39.95	220.9 −23.1	DISTURBED GALAXY w. 3 COMP'NS 31-G-5	8,16 S 4.0	F031/J1947 (247,192)
AM 0300-402	03 00 06 −40 27.4	2.267 0.236	247.38 −60.06	258.1 −34.2	KNOTTY IRR.	16 D 0.5	F300/J3536 (199,155)
AM 0300-401	03 00 09 −40 11.4	2.274 0.236	246.86 −60.12	258.4 −34.2	S0 + DIAMETRIC COMP'NS. 300-G-5	8 S 3.7	F300/J3536 (198,169)
AM 0300-243	03 00 18 −24 39.2	2.639 0.236	215.22 −60.46	277.3 −35.1	SMALL SPIRAL w. LONG JET 480-IG-21	7 D 1.1	F480/J3649 (127,198)
AM 0300-781	03 00 21 −78 14.7	−1.474 0.236	295.04 −37.04	217.4 −21.6	SPIRAL + E COMP'N. (I/A) B 0300-782	2 S 0.6	F014/J2559 (179,274)
*AM 0300-230	03 00 25 −23 03.8	2.670 0.236	212.12 −60.05	279.2 −35.0	SPIRAL + I/A COMP'N. NGC 1187 = 480-G-23	1 S 3.8	F480/J3649 (125,283)
*AM 0300-423	03 00 40 −42 31.7	2.204 0.235	251.24 −59.34	255.6 −33.9	3-ARMED SPIRAL 247-G-14	11 D 0.7	F248/J3677 (331,309)
*AM 0300-222	03 00 53 −22 24.5	2.682 0.235	210.92 −59.77	280.0 −35.1	I/A DBL. (E + PEC. DISRUPTED SP.) MCG-4-8-18/19 = 547-G-2/3	2,17,23 S 1.6	F480/J3649 (119,319)
AM 0301-482	03 01 10 −48 29.1	2.000 0.235	261.55 −56.91	248.6 −32.8	I/A DBL. w. TAIL 199-IG-10	2,15 S 0.3	F199/J1892 (169,261)
AM 0301-360	03 01 27 −36 09.0	2.378 0.235	238.74 −60.68	263.2 −35.0	CLOSE PAIR	23 S 0.3	F357/J2770 (293,116)
AM 0301-252	03 01 40 −25 28.1	2.620 0.234	216.96 −60.32	276.3 −35.4	DWARF w. SOME STRUCTURE MCG-44-8-21 = 480-G-25	20 D 1.1	F480/J3649 (110,154)
AM 0301-245	03 01 43 −24 52.6	2.632 0.234	215.79 −60.19	277.0 −35.4	SP. w. COMPACT GRP. of 4 COMP'NS near MCG-4-8-22	8 S 1.3	F480/J3649 (110,186)
AM 0302-393	03 02 02 −39 38.3	2.283 0.234	245.61 −59.91	259.0 −34.6	GROUP of 4: (E + COMPACT + 2 SP.) IC 1875 = 300-G-6	18 D 4.9	F300/J3536 (179,199)
*AM 0302-504	03 02 03 −50 44.6	1.907 0.234	264.99 −55.69	246.0 −32.3	SPINDLE w. LOOP + JET + 4 COMP'NS 199-IG-12 in IC 1877 Group	7,8,15 D 2.9	F199/J1892 (162,139)
AM 0302-370	03 02 06 −37 08.2	2.351 0.234	240.68 −60.40	262.0 −35.0	DISTANT (ELONGATED) CLUSTER	19 D 3.4	F300/J3536 (178,333)
AM 0302-611	03 02 20 −61 19.8	1.328 0.234	279.09 −49.42	234.3 −29.0	I/A QUARTET w. CONDENS. in TAIL 116-IG-6/7 = Fa 14/15/16 = B 0302-613	4,13,15 D 2.0	F116/J1814 (229,107)
AM 0302-521 [0302-5221]	03 02 23 −52 19.4	1.837 0.234	267.32 −54.83	244.2 −31.9	PAIR w. I/A COMP'N. (#1) Fa 734 IC 1878 = 199-IG-13 = B 0302-523	1,23 S 2.0	F199/J1892 (159,55)
AM 0302-273	03 02 24 −27 39.3	2.573 0.234	221.39 −60.51	273.6 −35.6	DISK w. OFF-SET NUCLEUS IC 1876 = 417-G-13	12 D 0.9	F480/J3649 (102,37)
AM 0302-594	03 02 33 −59 40.8	1.440 0.234	277.12 −50.48	236.0 −29.6	GAL. w. 5 KNOTS in NUCLEUS + TAIL	15 D 0.7	F116/J1814 (230,196)
AM 0302-622	03 02 34 −62 26.3	1.243 0.233	280.32 −48.66	233.1 −28.6	VERY FAINT, VERY COMPACT GROUP PKS 0302-623 ?? QSO ?	22 D 0.7	F082/J1871 (138,316)

Identification	R.A. (1950.0) Declination	Precession R.A./Dec.	Gal. Long. Gal. Lat.	S.G.L. S.G.B.	Description Cross-Identification	Coding Size (Min.)	Field/Plate (x, y)
AM 0302-360	03 02 35 −36 08.1	2.375 0.233	238.65 −60.45	263.2 −35.2	CLOSE PAIR	23 S 0.4	F357/J2770 (281,117)
AM 0302-522 [0302-5219]	03 02 36 −52 21.2	1.835 0.233	267.33 −54.78	244.1 −32.0	PAIR w. I/A COMP'N. (#2) IC 1879 = 199-G-14 = Fa 736	1,23 S 2.0	F199/J1892 (157,53)
AM 0302-471	03 02 39 −47 18.4	2.037 0.233	259.41 −57.21	249.9 −33.3	COMPACT 247-G?-16	13 D 0.3	F199/J1892 (156,324)
AM 0302-274	03 02 48 −27 42.0	2.571 0.233	221.50 −60.43	273.6 −35.7	I/A DBL. MCG-5-8-14/15 = 417-IG-14	2 S 0.3	F480/J3649 (97,34)
AM 0302-273	03 02 59 −27 31.8	2.574 0.233	221.17 −60.37	273.8 −35.7	LINE of 3 incl. GALAXY w. FAINT MCG-5-8-16 = 417-G-15 RINGS	6,24 D 15.2	F417/J3818 (191,312)
AM 0303-742	03 03 12 −74 26.7	−0.368 0.233	291.86 −39.91	221.0 −23.5	NUCLEATED DWARF?	20 D 0.4	F031/J1947 (231,204)
AM 0303-490	03 03 21 −49 09.7	1.964 0.233	262.34 −56.27	247.7 −32.9	ELONGATED COMPACT 199-IG-15	13 D 0.4	F199/J1892 (150,224)
AM 0303-232	03 03 31 −23 27.1	2.657 0.233	213.22 −59.46	278.8 −35.8	CLOSE DBL. incl. COMPACT 480-G-29	13,23 S 0.8	F480/J3649 (87,262)
*AM 0304-391	03 04 11 −39 13.8	2.288 0.232	244.64 −59.60	259.4 −35.1	PECULIAR SPIRAL + I/A COMP'N. NGC 1217 = 300-G-10	1,12 S 0.8	F300/J3536 (157,221)
AM 0304-425	03 04 29 −42 52.4	2.179 0.231	251.44 −58.57	255.0 −34.6	CLOSE GROUP of 4	18 D 0.6	F300/J3536 (155,25)
AM 0304-470	03 04 29 −47 02.2	2.039 0.232	258.72 −57.04	250.1 −33.7	GALAXY w. LOOP and JET	7,15 D 0.7	F247/J1929 (29,66)
AM 0304-300	03 04 32 −30 01.9	2.516 0.231	226.26 −60.27	270.7 −36.0	GALAXY w. EXTENDED MATERIAL 417-G-17 + 2 APP. COMP'NS.	8,15 S 1.8	F417/J3818 (173,178)
AM 0304-255	03 04 36 −25 54.4	2.605 0.231	218.07 −59.76	275.8 −36.1	GALAXY w. MATERIAL (RING-LIKE ?) NGC 1210 = 480-G-31 = MC 0304-259	6,15 D 1.1	F481/J2583 (341,128)
AM 0304-470	03 04 41 −47 05.0	2.036 0.231	258.77 −56.99	250.1 −33.7	SMALL SPIRAL w. JET to COMP'N 248-G-3	7 D 0.8	F199/J1892 (137,336)
*AM 0305-313	03 05 09 −31 35.5	2.479 0.231	229.41 −60.19	268.7 −36.1	SPIRAL + APP. COMP'NS. MCG-5-8-18 = 417-G-18	8,10 S 6.5	F417/J3818 (166,94)
AM 0305-622	03 05 10 −62 23.5	1.226 0.231	279.97 −48.47	233.0 −28.9	CLOSE PAIR + GAL. w. I/A COMP'N	1,23,24 D 4.5	F116/J1814 (209,51)
AM 0305-512	03 05 12 −51 29.8	1.859 0.231	265.70 −54.88	245.0 −32.6	GALAXY w. I/A COMP'N. Fa 298?	1 D 0.4	F199/J1892 (135,99)
AM 0305-661	03 05 15 −66 15.8	0.876 0.231	284.12 −45.82	229.0 −27.3	CLOSE PAIR of EDGE-ON GALAXIES	23 S 0.3	F082/J1871 (129,111)
AM 0305-224	03 05 23 −22 42.5	2.668 0.231	212.04 −58.86	279.7 −36.2	ELONGATED COMPACT PKS 0305-226	13 D 0.4	F481/J2583 (336,300)
*AM 0305-824	03 05 27 −82 47.3	−4.569 0.230	298.17 −33.27	213.1 −19.5	EDGE-ON RING w. I/A COMP'N.	1,6 D 0.6	F003/J2565 (110,292)
AM 0305-622	03 05 41 −62 26.5	1.218 0.230	279.96 −48.39	233.0 −28.9	CLOSE PAIR (SPIRAL w. LONG ARM)	10,23 S 1.9	F082/J1871 (119,316)
*AM 0305-824	03 05 45 −82 47.2	−4.576 0.230	298.16 −33.27	213.1 −19.5	MULTIPLE-NUCLEUS GALAXY w. DISTURBED ENVELOPE (RING?)	9,16 D 0.5	F014/J2559 (169,30)
*AM 0305-665	03 05 51 −66 58.1	0.795 0.230	284.78 −45.28	228.3 −27.1	GROUP of 4 (EDGE-ON + COMPACT) NGC 1244 = Se 28/1 = 82-G-8 [A82]	8,13,18 S 0.4	F082/J1871 (128,73)
AM 0306-343	03 06 00 −34 36.9	2.404 0.230	235.46 −59.92	265.0 −36.0	GALAXY w. TAIL	15 D 0.3	F357/J2770 (245,200)
*AM 0306-231	03 06 04 −23 10.4	2.658 0.230	212.99 −58.83	279.2 −36.3	I/A CHAIN of 7 NGC 1230 = MCG-4-8-27 = 480-6-34	7,17 D 9.9	F480/J3649 (55,277)
*AM 0306-702	03 06 05 −70 25.9	0.346 0.230	288.11 −42.78	224.8 −25.6	ONE-ARMED SPIRAL (GAL. w. TAIL?) 54-IG-1 = B 0306-704	10,12,15 D 1.1	F054/J1155 (275,150)
AM 0306-670	03 06 21 −67 07.9	0.772 0.230	284.90 −45.14	228.1 −27.0	E w. DIAMETRIC COMPANIONS 82-G-9 = Fa 229	8 S 5.4	F082/J1871 (125,64)
AM 0306-424	03 06 27 −42 48.0	2.174 0.229	251.09 −58.25	255.0 −34.9	COMPACT	13 D 0.2	F248/J3677 (273,297)
AM 0306-324	03 06 58 −32 41.7	2.448 0.229	231.62 −59.81	267.3 −36.4	I/A DBL. in GROUP of 4 357-IG-6	2,18 S 0.4	F417/J3818 (146,34)
AM 0307-810	03 07 25 −81 00.7	−3.090 0.228	296.81 −34.65	214.7 −20.5	I/A DBL.	2 S 0.3	F014/J2559 (164,125)
*AM 0307-252	03 07 26 −25 26.5	2.609 0.228	217.41 −59.04	276.4 −36.7	'RI"RINGING" SP. w. 2 CLOSE COMP'NS IC 1895 = 481-G-1 (1 COMPACT)	1,8,13,22 S 1.8	F480/J3649 (40,155)
AM 0307-411	03 07 47 −41 13.3	2.218 0.228	248.07 −58.45	256.9 −35.5	LARGE Sd/DWARF 300-G-14	20 D 3.6	F300/J3536 (121,113)
AM 0307-705	03 07 50 −70 50.1	0.263 0.228	288.34 −42.39	224.3 −25.5	I/A TRIPLE 54-IG-3	3 D 0.6	F054/J1155 (265,129)
AM 0307-344	03 07 59 −34 40.8	2.397 0.228	235.52 −59.51	264.9 −36.4	GALAXY w. I/A COMP'N.	1 S 0.4	F357/J2770 (223,196)
AM 0307-504 [0308-5047]	03 07 59 −50 47.3	1.876 0.228	264.26 −54.85	245.6 −33.2	I/A DBL. (#1) 199-IG-20 = B 0307-507	2 S 0.5	F199/J1892 (111,136)
AM 0308-504 [0307-5047]	03 08 02 −50 47.4	1.875 0.228	264.25 −54.84	245.6 −33.2	I/A DBL. (#2) 199-IG-20 = B 0307-507	2 S 0.5	F199/J1892 (111,136)

Identification	R.A. (1950.0) Declination	Precession R.A./Dec.	Gal. Long. Gal. Lat.	S.G.L. S.G.B.	Description Cross-Identification	Coding Size (Min.)	Field/Plate (x, y)
AM 0308-771	03 08 05 −77 16.3	−1.254 0.228	293.88 −37.55	218.1 −22.4	CHAIN of 4 [N.S. 31-G-9?	17 D 9.0	F031/J1947 (207,54)
AM 0308-401	03 08 18 −40 11.5	2.247 0.227	246.12 −58.61	258.1 −35.7	DWARF	20 D 0.5	F300/J3536 (115,168)
AM 0309-251	03 09 08 −25 19.2	2.609 0.227	217.33 −58.64	276.5 −37.1	SMALL SP. w. KNOT at END of JET MCG-4-8-38 = 481-G-5	7,8 D 0.9	F480/J3649 (20,161)
AM 0309-473	03 09 11 −47 36.4	1.998 0.227	259.06 −56.08	249.2 −34.3	GROUP of 6	18 D 2.5	F248/J3677 (242,39)
AM 0309-321 [0309-3219]	03 09 14 −32 17.6	2.452 0.226	230.82 −59.33	267.8 −36.9	SMALL but WIDE I/A DBL.(#1)	2 S 1.6	F417/J3818 (120,56)
AM 0309-321 [0309-3217]	03 09 16 −32 19.4	2.452 0.226	230.88 −59.33	267.8 −36.9	SMALL but WIDE I/A DBL.(#2)	2 S 1.6	F417/J3818 (120,54)
AM 0309-473 [0309-4739]	03 09 19 −47 39.9	1.995 0.226	259.14 −56.04	249.1 −34.3	GROUP of 4 (#1)	18 D 3.1	F199/J1892 (95,304)
AM 0309-473 [0309-4739]	03 09 27 −47 37.4	1.996 0.226	259.06 −56.03	249.2 −34.3	GROUP of 4 (#2)	18 D 3.1	F199/J1892 (94,306)
AM 0309-473 [0309-4737]	03 09 29 −47 38.8	1.995 0.226	259.09 −56.02	249.2 −34.3	GROUP of 4 (#3)	18 D 3.1	F199/J1892 (94,305)
AM 0309-473 [0309-4738]	03 09 32 −47 39.7	1.994 0.226	259.11 −56.00	249.1 −34.3	GROUP of 4 (#4)	18 D 3.1	F199/J1892 (93,304)
AM 0309-222	03 09 50 −22 29.7	2.665 0.226	212.19 −57.81	280.1 −37.2	CLOSE GROUP of 5	18 D 1.6	F481/J2583 (281,313)
*AM 0310-252	03 10 03 −25 23.4	2.605 0.226	217.54 −58.45	276.5 −37.3	CHAIN of 5 (incl. RING) incl. MCG-4-8-42 = 481-G-7	6,17 D 4.0	F481/J2583 (275,157)
AM 0310-244 [0310-2444]	03 10 07 −24 48.3	2.618 0.226	216.45 −58.31	277.2 −37.3	CLOSE PAIR (#1: E) MCG-4-8-44 = 481-G-9	23 S 4.0	F481/J2583 (275,189)
AM 0310-244 [0310-2448]	03 10 14 −24 44.0	2.619 0.225	216.33 −58.27	277.3 −37.3	CLOSE PAIR (#2: EDGE-ON) MCG-4-8-48 = 481-G-11	23 S 4.0	F481/J2583 (274,193)
AM 0310-271	03 10 18 −27 14.5	2.565 0.225	221.05 −58.72	274.1 −37.4	I/A DBL. in GROUP of 12 481-IG-12 = R 66? = DN 0310-273	2,18 S 0.3	F417/J3818 (104,327)
AM 0310-424	03 10 18 −42 44.8	2.161 0.225	250.60 −57.59	254.9 −35.6	GALAXY w. TAILS 248-G-5	15 D 0.6	F300/J3536 (98,31)
AM 0310-644	03 10 41 −64 47.5	0.975 0.225	282.02 −46.42	230.2 −28.4	SMALL I/A DBL.	2 S 0.2	F082/J1871 (95,188)
*AM 0310-314 [0311-3150]	03 10 46 −31 40.3	2.463 0.225	229.61 −59.00	268.5 −37.3	PAIR (#1: SPIRAL) MCG-5-8-22 = 417-G-20	23 S 22.4	F417/J3818 (102,89)
AM 0310-473	03 10 54 −47 37.7	1.989 0.225	258.89 −55.81	249.1 −34.6	GROUP of 4	18 D 0.9	F248/J3677 (226,38)
*AM 0311-513	03 11 00 −51 31.9	1.828 0.225	264.99 −54.08	244.6 −33.4	I/A DBL. w. TAIL + MATERIAL 199-IG-23	2,15 S 0.2	F200/J2675 (338,92)
AM 0311-492	03 11 07 −49 22.3	1.920 0.224	261.66 −55.05	247.1 −34.1	DWARF	20 D 0.3	F200/J2675 (344,208)
AM 0311-473 [0311-4740]	03 11 11 −47 39.5	1.987 0.224	258.91 −55.75	249.0 −34.6	TIGHT GROUP of 4 (#1)	18 D 1.1	F199/J1892 (78,303)
AM 0311-315 [0310-3140]	03 11 13 −31 50.4	2.458 0.224	229.94 −58.91	268.3 −37.4	PAIR (#2: E) MCG-5-8-23 = 417-G-21	23 S 22.4	F417/J3818 (97,80)
AM 0311-473 [0311-4739]	03 11 14 −47 39.9	1.986 0.224	258.91 −55.74	249.0 −34.6	TIGHT GROUP of 4 (#2)	18 D 1.1	F199/J1892 (78,303)
AM 0311-474 [0311-4739]	03 11 14 −47 40.7	1.986 0.224	258.94 −55.73	249.0 −34.6	TIGHT GROUP of 4 (#3)	18 D 1.1	F199/J1892 (78,302)
AM 0311-474 [0311-4740]	03 11 17 −47 40.2	1.986 0.224	258.91 −55.73	249.0 −34.6	TIGHT GROUP of 4 (#4)	18 D 1.1	F199/J1892 (78,303)
*AM 0311-255	03 11 21 −25 54.7	2.592 0.224	218.62 −58.26	275.8 −37.6	BRIGHT-RIMMED SPIRAL + DIAMETRICS NGC 1255 = 481-G-13	8,12 S 9.9	F481/J2583 (259,130)
*AM 0311-252	03 11 30 −25 22.3	2.603 0.224	217.63 −58.13	276.5 −37.6	GALAXY w. WARPED PLANE + COMP'N. MCG-4-8-51 = 481-G-14 = UKS0311-253	8,12 D 3.6	F481/J2583 (258,159)
AM 0311-582	03 11 38 −58 20.3	1.465 0.224	274.33 −50.37	237.0 −31.2	SPIRAL w. 2 APP. COMP'NS.	8 S 3.8	F116/J1814 (168,269)
*AM 0311-573	03 11 48 −57 31.7	1.514 0.224	273.27 −50.82	237.9 −31.5	LARGE DISTURBED SPIRAL 116-G-12	16 D 3.6	F116/J1814 (167,312)
AM 0311-382	03 11 49 −38 29.3	2.284 0.224	242.67 −58.28	260.0 −36.7	2 E's in ENVELOPE + COMP'NS. 300-IG-19 = B 0311-384	2,8 S 0.4	F300/J3536 (76,259)
AM 0312-381	03 12 12 −38 10.8	2.292 0.223	242.07 −58.26	260.4 −36.8	DWARF 300-G-20	20 D 0.6	F300/J3536 (72,275)
AM 0312-305	03 12 58 −30 53.6	2.477 0.222	228.15 −58.50	269.5 −37.8	SPIRAL + COMPACT APP. COMP'N. IC 1904 = 417-G-22	8,13 S 0.9	F417/J3818 (76,130)
AM 0313-293	03 13 06 −29 31.7	2.508 0.222	225.55 −58.38	271.2 −37.9	GROUP of 7 E's	18 D 4.5	F417/J3818 (73,203)
AM 0313-321	03 13 10 −32 16.5	2.443 0.222	230.79 −58.50	267.7 −37.8	I/A DBL.	2 S 0.5	F357/J2770 (166,326)
AM 0313-581	03 13 18 −58 16.5	1.459 0.222	274.05 −50.23	237.0 −31.4	I/A E's in COMMON ENVELOPE 116-IG-13	2 S 0.4	F116/J1814 (156,272)

Identification	R.A. (1950.0) Declination	Precession R.A./Dec.	Gal. Long. Gal. Lat.	S.G.L. S.G.B.	Description Cross-Identification	Coding Size (Min.)	Field/Plate (x, y)
AM 0313-225	03 13 19 −22 59.8	2.649 0.222	213.49 −57.17	279.5 −38.0	I/A QUARTET 481-G-15	4 D 1.1	F481/J2583 (237,287)
AM 0313-765	03 13 30 −76 54.5	−1.221 0.222	293.29 −37.63	218.3 −22.9	GALAXY w. JETS or STRAIGHT ARMS	7 D 0.7	F014/J2559 (139,344)
*AM 0313-545	03 13 41 −54 59.8	1.646 0.222	269.64 −51.99	240.5 −32.7	I/A DBL. (SPIRAL + COMPACT) IC 1908 = 155-IG-13 = B 0313-550	2,13 D 1.3	F155/J1127 (210,177)
AM 0313-371	03 13 42 −37 18.9	2.311 0.222	240.36 −58.09	261.4 −37.3	SPIRAL + HI.S.BR. LINEAR COMP'N.	13,23 S 0.5	F301/J2765 (331,321)
AM 0314-242	03 14 00 −24 23.0	2.619 0.221	216.05 −57.36	277.7 −38.2	DWARF 481-G-16	20 D 0.9	F481/J2583 (228,212)
*AM 0314-631	03 14 03 −63 10.5	1.094 0.221	279.87 −47.19	231.7 −29.5	COMPACT [A82] ONE HI. + TWO LO.S.BR. COMP'NS	8,13 S 4.3	F082/J1871 (70,273)
AM 0314-762	03 14 48 −76 20.6	−1.058 0.220	292.75 −38.00	218.8 −23.2	I/A DBL.	2 S 0.5	F031/J1947 (187,104)
AM 0314-230	03 14 52 −23 03.0	2.646 0.220	213.76 −56.84	279.4 −38.4	GALAXY w. FAINT ARMS + 3 COMP'NS MCG-4-8-56 = 481-G-17	8,10 D 1.3	F481/J2583 (218,284)
AM 0314-493	03 14 54 −49 35.0	1.894 0.220	261.54 −54.40	246.6 −34.6	DWARF	20 D 0.4	F200/J2675 (311,198)
AM 0314-511 [0315-5106]	03 14 56 −51 11.0	1.825 0.220	263.98 −53.70	244.8 −34.1	CHAIN of 5 in CLUSTER (#1)	17,19 D 0.4	F199/J1892 (53,113)
AM 0314-510	03 14 59 −51 05.9	1.828 0.220	263.85 −53.73	244.8 −34.2	RICH CLUSTER	19 D 13.4	F200/J2675 (306,117)
AM 0315-510 [0314-5111]	03 15 08 −51 09.3	1.825 0.220	263.92 −53.68	244.8 −34.2	CHAIN of 5 in CLUSTER (#2)	17,19 D 0.4	F199/J1892 (52,114)
AM 0315-324	03 15 12 −32 45.5	2.426 0.220	231.72 −58.08	267.1 −38.1	SPIRAL w. PECULIAR ARM NGC 1288 = 357-G-13	10 D 1.8	F417/J3818 (53,29)
AM 0315-415	03 15 12 −41 53.2	2.171 0.220	248.62 −56.94	255.7 −36.7	DWARF	20 D 0.7	F300/J3536 (48,75)
AM 0315-511 [0315-5109]	03 15 16 −51 10.0	1.824 0.220	263.92 −53.66	244.8 −34.2	CHAIN of 5 in CLUSTER (#3)	17,19 D 0.4	F199/J1892 (51,114)
AM 0315-510 [0315-5110]	03 15 16 −51 09.3	1.824 0.220	263.90 −53.66	244.8 −34.2	CHAIN of 5 in CLUSTER (#4)	17,19 D 3.1	F199/J1892 (50,114)
AM 0315-712	03 15 18 −71 28.2	0.073 0.220	288.35 −41.51	223.4 −25.7	GALAXY w. MULTIPLE NUCLEUS	12 D 0.4	F054/J1155 (231,97)
AM 0315-322	03 15 21 −32 27.5	2.433 0.220	231.15 −58.04	267.5 −38.2	I/A DBL.	2 S 0.2	F357/J2770 (141,316)
AM 0315-510 [0315-5109]	03 15 21 −51 06.9	1.826 0.220	263.83 −53.67	244.8 −34.2	CHAIN of 5 in CLUSTER (#5)	17,19 D 3.1	F199/J1892 (50,116)
AM 0315-411	03 15 29 −41 17.6	2.189 0.220	247.55 −57.03	256.4 −36.9	Sa w. COMP'NS. (ONE DWARF) NGC 1291 = 301-G-2	8,20 S 4.0	F300/J3536 (43,107)
AM 0316-484	03 16 01 −48 41.1	1.926 0.219	259.99 −54.60	247.6 −35.1	DWARF (w. LINEAR FEATURE ?)	7,20 D 0.3	F200/J2675 (303,247)
AM 0316-350	03 16 02 −35 06.9	2.364 0.219	236.15 −57.83	264.1 −38.0	VERY CLOSE PAIR	23 S 0.1	F357/J2770 (135,173)
AM 0316-301	03 16 06 −30 16.8	2.484 0.219	227.06 −57.79	270.2 −38.5	GROUP of 3	24 D 4.0	F417/J3818 (39,162)
AM 0316-573	03 16 12 −57 38.3	1.481 0.219	272.89 −50.27	237.5 −32.0	DISRUPTED SINGLE B 0316-576	16 D 1.3	F116/J1814 (135,306)
*AM 0316-543	03 16 19 −54 33.7	1.654 0.219	268.72 −51.89	240.9 −33.2	CHAIN of 4 (I/A DBL. at CENTRE) IC 1908 = 155-IG-13	2,17 D 4.0	F155/J1127 (190,200)
AM 0316-260	03 16 23 −26 01.1	2.580 0.219	219.21 −57.17	275.7 −38.7	SPIRAL w. APP. COMP'N. MCG-4-8-57 = 481-G-18	8 S 4.0	F481/J2583 (198,125)
AM 0316-455	03 16 23 −45 52.8	2.032 0.219	255.40 −55.58	250.9 −35.9	CHAIN of 4 E's in CLUSTER	2,17,19 D 8.1	F248/J3677 (176,132)
AM 0316-371	03 16 30 −37 10.9	2.307 0.219	239.98 −57.55	261.5 −37.8	COMPACT	13 D 0.3	F301/J2765 (302,329)
AM 0316-235	03 16 33 −23 57.9	2.624 0.219	215.55 −56.70	278.3 −38.8	DWARF (SOME RESOLUTION) 481-G-19 = UKS 0316-239	20 D 1.8	F481/J2583 (197,235)
AM 0316-573	03 16 33 −57 37.8	1.479 0.219	272.84 −50.24	237.5 −32.0	CLOSE I/A DBL. 116-IG-15	2 S 0.3	F155/J1127 (187,36)
AM 0316-271	03 16 44 −27 18.3	2.551 0.218	221.58 −57.32	274.0 −38.8	COMPACT + SPIRAL	8,13 S 0.4	F418/J2654 (302,323)
AM 0317-325	03 17 19 −32 56.6	2.416 0.218	232.06 −57.63	266.8 −38.6	CLOSE TRIPLE	24 D 2.0	F357/J2770 (119,290)
*AM 0317-324	03 17 22 −32 49.7	2.419 0.218	231.86 −57.61	266.9 −38.6	IRR. (RESOLVED into KNOTS)	16 D 0.5	F418/J2654 (287,26)
AM 0317-621	03 17 26 −62 15.4	1.146 0.218	278.46 −47.45	232.5 −30.2	DWARF	20 D 0.4	F116/J1814 (133,58)
AM 0317-323	03 17 33 −32 38.7	2.423 0.217	231.51 −57.58	267.2 −38.6	EDGE-ON + APP. COMP'N. IC 1913 = 357-G-16	8,16 S 3.4	F417/J3818 (26,34)
AM 0317-664	03 17 39 −66 40.9	0.719 0.217	283.37 −44.63	228.0 −28.2	Sdm + HI.S.BR. COMP'N. + MATR'L. NGC 1313 = 82-G-11 = PKS 0317-666	8,13,15,16 D 13.4	F083/J1936 (298,84)

Identification	R.A. (1950.0) Declination	Precession R.A./Dec.	Gal. Long. Gal. Lat.	S.G.L. S.G.B.	Description Cross-Identification	Coding Size (Min.)	Field/Plate (x, y)
AM 0317-595	03 17 48 −59 55.7	1.320 0.217	275.61 −48.81	234.9 −31.3	CONDENS. at END of ONE-ARMED SP.	10,23 D 0.7	F116/J1814 (127,183)
AM 0318-582	03 18 02 −58 28.5	1.417 0.217	273.76 −49.61	236.5 −31.9	I/A DBL.	2 S 0.4	F116/J1814 (123,260)
AM 0318-255	03 18 10 −25 52.7	2.580 0.217	219.10 −56.75	275.8 −39.1	DWARF	20 D 0.4	F481/J2583 (177,132)
AM 0318-511	03 18 11 −51 13.8	1.807 0.217	263.67 −53.22	244.5 −34.6	SPIRAL w. ALIGNED I/A COMP'NS. 200-IG-4 [A82]	1 S 1.2	F200/J2675 (278,111)
AM 0318-292	03 18 21 −29 24.8	2.500 0.217	225.53 −57.23	271.3 −39.1	I/A DBL. E's	2 S 0.3	F418/J2654 (280,210)
*AM 0318-230	03 18 28 −23 06.6	2.639 0.216	214.26 −56.06	279.4 −39.2	GALAXY w. LOOPS MCG-4-9-1 = 481-G-22	15 D 0.7	F481/J2583 (174,281)
AM 0318-610	03 18 36 −61 07.4	1.227 0.216	276.99 −48.02	233.6 −30.8	V.SMALL RING w. OFF-SET NUCLEUS	6 D 0.2	F116/J1814 (124,118)
AM 0318-384	03 18 39 −38 42.9	2.257 0.216	242.69 −56.93	259.5 −38.0	CONTACT E's 301-IG-6	2 S 0.3	F301/J2765 (277,247)
AM 0318-481 [0318-4819]	03 18 41 −48 18.9	1.929 0.216	259.11 −54.34	247.8 −35.6	CLOSE TRIPLE (#1) 200-IG-6	24 D 0.7	F199/J1892 (13,265)
AM 0318-481 [0318-4818]	03 18 43 −48 18.7	1.929 0.216	259.10 −54.34	247.8 −35.6	CLOSE TRIPLE (#2) 200-IG-6	24 D 0.7	F199/J1892 (12,265)
AM 0318-481 [0318-4818]	03 18 44 −48 19.0	1.929 0.216	259.11 −54.33	247.8 −35.6	CLOSE TRIPLE (#3)	24 D 0.7	F199/J1892 (12,265)
AM 0318-491 [0318-4914]	03 18 51 −49 17.7	1.888 0.216	260.64 −53.93	246.7 −35.3	TRIPLE (#1: DWARF) IC 1916 = 200-G-8	20,24 S 1.3	F199/J1892 (14,212)
AM 0318-491 [0318-4917]	03 18 52 −49 13.6	1.891 0.216	260.53 −53.96	246.8 −35.4	TRIPLE (#2: COMPACT)	13,24 S 1.3	F199/J1892 (14,216)
AM 0318-491 [0318-4913]	03 18 53 −49 14.9	1.890 0.216	260.56 −53.95	246.7 −35.3	TRIPLE (#3: SPIRAL)	8,24 S 1.3	F199/J1892 (14,215)
*AM 0318-254	03 18 54 −25 41.4	2.583 0.216	218.83 −56.55	276.1 −39.3	LUMPY COMPACT / IRR. COMP'N. to NGC 1306?	13 D 0.5	F481/J2583 (168,142)
AM 0319-671	03 19 03 −67 10.9	0.649 0.216	283.77 −44.20	227.4 −28.1	COMPACT w. PLUME	13,15 D 0.4	F082/J1871 (59,57)
AM 0319-452	03 19 16 −45 26.0	2.037 0.216	254.36 −55.25	251.2 −36.5	GALAXY w. PECULIAR DUST 248-G-10	8,14 S 0.7	F248/J3677 (149,156)
AM 0319-272	03 19 17 −27 23.7	2.544 0.215	221.91 −56.77	273.9 −39.4	SMALL TIGHT GROUP of 4	18 D 0.7	F418/J2654 (272,318)
*AM 0319-542	03 19 17 −54 28.4	1.643 0.215	268.25 −51.56	240.8 −33.6	GROUP of 5 incl. RING 155-G-20	6,18 D 5.8	F155/J1127 (167,205)
AM 0319-421	03 19 34 −42 10.4	2.148 0.215	248.77 −56.09	255.2 −37.4	DWARF 301-G-7	20 D 0.5	F248/J3677 (144,331)
AM 0319-434	03 19 44 −43 45.8	2.094 0.215	251.50 −55.65	253.2 −37.1	EDGE-ON w. PLUMES + I/A COMP'N. 248-G-12 + 248-G-11	1,15 S 1.1	F248/J3677 (143,246)
AM 0319-454	03 19 45 −45 41.4	2.026 0.215	254.74 −55.08	250.9 −36.5	CLOSE PAIR of E's (GROUP of 4)	18,23 S 0.4	F248/J3677 (144,142)
AM 0319-662	03 19 59 −66 29.5	0.720 0.215	282.95 −44.58	228.0 −28.5	DWARF	20 D 0.5	F082/J1871 (51,93)
AM 0320-274	03 20 05 −27 46.2	2.534 0.215	222.63 −56.65	273.4 −39.5	GALAXY w. I/A COMP'N + TAIL (GROUP of 5)	1,15,18 S 0.3	F418/J2654 (262,298)
AM 0320-600	03 20 22 −60 01.7	1.296 0.214	275.46 −48.48	234.7 −31.5	I/A DBL.	2 S 0.1	F116/J1814 (110,177)
*AM 0320-495	03 20 24 −49 58.8	1.852 0.214	261.53 −53.43	245.8 −35.3	GALAXY w. CONDENSATION? in JET	7 D 0.5	F200/J2675 (262,179)
*AM 0320-372	03 20 46 −37 23.6	2.288 0.214	240.19 −56.69	261.0 −38.6	FORNAX A "RINGING" [PKS],[MC] NGC 1316 = ARP 154 = 357-G-22	1,15,22	F357/J2770 (85,50)
AM 0320-491	03 20 46 −49 15.4	1.882 0.214	260.37 −53.66	246.6 −35.6	COMPACT	13 D 0.3	F200/J2675 (260,218)
AM 0321-464	03 21 01 −46 44.7	1.981 0.214	256.35 −54.53	249.5 −36.5	I/A DBL. 248-IG-13	2 D 0.4	F248/J3677 (134,85)
AM 0321-425 [0321-4246]	03 21 09 −42 51.2	2.120 0.213	249.83 −55.64	254.2 −37.5	CLOSE PAIR (#1: E) MC 0321-429 = MC 0321-426	23 S 3.6	F248/J3677 (129,294)
AM 0321-291	03 21 12 −29 13.8	2.498 0.213	225.32 −56.59	271.5 −39.7	DWARF	20 D 0.4	F418/J2654 (247,220)
AM 0321-424 [0321-4251]	03 21 15 −42 46.8	2.122 0.213	249.69 −55.64	254.3 −37.6	CLOSE PAIR (#2: SPIRAL w. TAILS) 248-G-14	15,23 S 3.6	F248/J3677 (127,298)
AM 0321-294	03 21 17 −29 45.3	2.485 0.213	226.27 −56.63	270.8 −39.7	DWARF	20 D 0.4	F418/J2654 (246,192)
AM 0321-810 [0322-8106]	03 21 37 −81 05.5	−3.495 0.213	296.35 −34.23	214.3 −20.9	CLOSE DBL. w. COMP'NS. (#1:SP)	8,23 S 2.1	F014/J2559 (134,119)
AM 0321-370	03 21 42 −37 04.7	2.295 0.213	239.59 −56.53	261.4 −38.9	CLOSE TRIPLE (HI.S.BR. GALAXIES)	24 D 1.3	F301/J2765 (246,336)
AM 0321-871	03 21 48 −87 17.1	−0.311 0.213	301.07 −29.51	209.0 −17.3	IRR. in WIDE PAIR	16,23 D 1.6	F001/J3648 (292,87)

Identification	R.A. (1950.0) Declination	Precession R.A./Dec.	Gal. Long. Gal. Lat.	S.G.L. S.G.B.	Description Cross-Identification	Coding Size (Min.)	Field/Plate (x, y)
AM 0322-374	03 22 01 −37 41.4	2.276 0.212	240.68 −56.41	260.6 −38.8	RING in NUCLEUS w. BR. CONDENS. 301-IG-11	6,16 D 0.5	F358/J2698 (328,32)
AM 0322-573	03 22 07 −57 37.8	1.446 0.212	272.21 −49.61	237.1 −32.7	SPIRAL w. APP. COMP'N.	8 S 0.7	F116/J1814 (92,305)
AM 0322-575	03 22 20 −57 50.6	1.431 0.212	272.46 −49.48	236.9 −32.7	SPIRAL w. APP. COMP'N.	8 S 0.7	F116/J1814 (91,293)
AM 0322-670	03 22 20 −67 04.0	0.634 0.212	283.36 −44.03	227.3 −28.4	M51-TYPE w. ELONGATED COMP'N. 83-IG-2	9 S 0.3	F082/J1871 (42,61)
AM 0322-810 [0321-8105]	03 22 34 −81 06.3	−3.527 0.212	296.33 −34.20	214.3 −20.9	CLOSE DBL. w. COMP'NS.(#2: N)	8,23 S 2.1	F014/J2559 (132,118)
AM 0322-414	03 22 39 −41 48.7	2.149 0.212	247.92 −55.61	255.4 −38.1	COMPACT	13 D 0.2	F301/J2765 (233,82)
AM 0322-580	03 22 48 −58 04.9	1.413 0.212	272.72 −49.30	236.6 −32.6	GROUP of 6 (HI.S.BR. IRR. + E) 116-IG-17 = DN 0322-588?	8,16,18 D 4.3	F116/J1814 (89,280)
AM 0322-673	03 22 57 −67 38.7	0.558 0.211	283.92 −43.61	226.7 −28.2	COMPACT GALAXY	13 D 0.2	F054/J1155 (202,303)
AM 0323-420	03 23 13 −42 04.1	2.138 0.211	248.32 −55.45	255.1 −38.1	M51-TYPE in I/A SMALL CHAIN of 5 301-G-13	5,9,17 D 2.7	F248/J3677 (107,336)
*AM 0323-363	03 23 20 −36 32.5	2.305 0.211	238.57 −56.26	262.0 −39.3	I/A DBL.(KNOTS) NGC 1326a/b = 357-IG-28/29	2 S 2.2	F357/J2770 (57,95)
AM 0324-330	03 24 01 −33 04.2	2.397 0.210	232.32 −56.23	266.4 −39.9	E w. 2 FAINT JETS IC 1919 = 358-G-1	7 D 1.1	F357/J2770 (44,281)
AM 0324-334	03 24 17 −33 43.5	2.379 0.210	233.50 −56.17	265.6 −39.9	SMALL GALAXY w. I/A COMP'N.	1 S 0.2	F358/J2698 (310,246)
*AM 0324-524	03 24 18 −52 43.7	1.707 0.210	265.20 −51.72	242.4 −34.9	SPIRAL + M51-TYPE 155-G-27 = Fa 385	3,9,23 S 2.2	F200/J2675 (225,32)
AM 0324-362	03 24 34 −36 20.9	2.307 0.210	238.19 −56.02	262.2 −39.6	SMALL M51-TYPE	9 S 0.3	F358/J2698 (303,105)
AM 0324-342	03 24 47 −34 25.4	2.360 0.209	234.75 −56.07	264.6 −39.9	GALAXY w. JET (GROUP of 4) 358-IG-3	7,18 D 1.1	F357/J2770 (38,208)
AM 0325-364	03 25 06 −36 46.3	2.294 0.209	238.93 −55.88	261.6 −39.6	PAIR	23 S 0.5	F357/J2770 (38,82)
AM 0325-355	03 25 10 −35 51.4	2.320 0.209	237.30 −55.93	262.8 −39.8	DWARF/DEFECT ?	20 D 0.7	F357/J2770 (36,131)
AM 0325-640	03 25 35 −64 08.2	0.919 0.208	279.83 −45.59	230.1 −30.2	CLOSE PAIR	23 S 1.3	F083/J1936 (264,223)
AM 0326-485	03 26 14 −48 52.7	1.874 0.208	259.22 −52.97	246.7 −36.6	SPIRAL w. PEC. ARM (ONE-ARMED?)	10 D 0.9	F200/J2675 (213,239)
AM 0326-260	03 26 16 −26 03.2	2.561 0.208	220.04 −55.00	275.6 −41.0	COMPACT	13 D 0.1	F482/J4722 (344,120)
AM 0326-311	03 26 17 −31 14.5	2.438 0.208	229.09 −55.68	268.7 −40.6	S0 w. PLATEAUS (ARC of 4 [N.W.]) NGC 1340/4 = 418-G-5 = MC 0326-312	16,22 D 5.4	F418/J2654 (187,113)
AM 0326-335	03 26 30 −33 54.6	2.369 0.207	233.82 −55.72	265.2 −40.3	DWARF (NO RESOLUTION)	20 D 0.6	F358/J2698 (285,237)
AM 0326-662	03 26 41 −66 27.7	0.667 0.207	282.33 −44.08	227.7 −29.1	CLOSE PAIR in WIDE GROUP of 5 83-G-4 = B 0326-664	18,23 S 2.4	F083/J1936 (250,99)
AM 0326-594	03 26 45 −59 43.7	1.277 0.207	274.42 −47.98	234.6 −32.4	E w. FAINT JET	7 D 0.3	F116/J1814 (66,191)
AM 0326-460 [0327-4606]	03 26 53 −46 06.7	1.983 0.207	254.78 −53.77	249.9 −37.6	CLOSE PAIR (#1) PKS 0326-461	23 D 0.2	F249/J2672 (339,117)
AM 0327-460 [0326-4606]	03 27 03 −46 06.3	1.982 0.207	254.75 −53.75	249.9 −37.6	CLOSE PAIR (#2) PKS 0326-461	23 D 0.2	F249/J2672 (337,117)
AM 0327-732	03 27 10 −73 29.8	−0.470 0.207	289.45 −39.46	220.9 −25.4	GALAXY w. CONNECTED COMP'N.	1 S 1.0	F031/J1947 (143,256)
AM 0327-555	03 27 18 −55 53.4	1.522 0.206	269.33 −49.88	238.6 −34.1	RICH CLUSTER (cD at CENTRE)	19 D 9.0	F155/J1127 (107,128)
AM 0327-231	03 27 19 −23 12.3	2.623 0.206	215.35 −54.12	279.3 −41.2	DWARF	20 D 0.4	F482/J4722 (336,273)
AM 0327-594	03 27 22 −59 40.7	1.277 0.206	274.29 −47.95	234.6 −32.4	SMALL I/A DBL.	2 S 0.1	F116/J1814 (62,193)
*AM 0327-222	03 27 30 −22 27.0	2.639 0.206	214.13 −53.87	280.3 −41.2	GALAXY w. VERY CLOSE I/A COMP'N. NGC 1347 = ARP 39 = 548-G-27	1 S 0.7	F481/J2583 (62,316)
AM 0327-373	03 27 38 −37 38.8	2.262 0.206	240.40 −55.31	260.4 −39.9	COMPACT	13 D 0.2	F301/J2765 (183,306)
AM 0327-291	03 27 43 −29 12.3	2.485 0.206	225.56 −55.18	271.4 −41.1	I/A DBL. E's	2 S 0.2	F418/J2654 (171,222)
AM 0327-432	03 27 48 −43 29.4	2.075 0.206	250.40 −54.31	253.0 −38.5	LINE of 3 incl. V. COMPACT GALAXY	8,13,24 D 2.5	F249/J2672 (336,258)
*AM 0327-285	03 27 51 −28 56.8	2.491 0.206	225.12 −55.12	271.7 −41.2	I/A E + SPIRAL in GROUP of 3 MCG-5-9-9/8 = 418-G-7	2,24 S 0.4	F418/J2654 (169,236)
AM 0328-480	03 28 28 −48 02.0	1.901 0.205	257.69 −52.91	247.5 −37.2	LARGE COMPACT 200-G-29	13 D 0.5	F248/J3677 (68,14)

Identification	R.A. (1950.0) Declination	Precession R.A./Dec.	Gal. Long. Gal. Lat.	S.G.L. S.G.B.	Description Cross-Identification	Coding Size (Min.)	Field/Plate (x, y)
AM 0328-524	03 28 28 −52 41.1	1.690 0.205	264.71 −51.17	242.1 −35.5	VERY RICH CLUSTER of 8 GALAXIES	19 D 13.4	F200/J2675 (191,35)
AM 0328-320	03 28 32 −32 07.2	2.411 0.205	230.68 −55.25	267.5 −41.0	DWARF (NUCLEATED)	20 D 0.6	F418/J2654 (162,66)
*AM 0329-334	03 29 10 −33 47.4	2.366 0.204	233.62 −55.16	265.3 −40.9	LARGE SB w. APP. COMP'N. NGC 1350 = 358-G-13	8 S 4.3	F358/J2698 (255,244)
*AM 0329-502	03 29 13 −50 28.6	1.792 0.204	261.37 −51.94	244.6 −36.5	DBL. w. I/A COMP'NS.(DN 0329-505) NGC 1356 = Se 31/1 = 200-IG-31 [B]	1,2 S 0.9	F200/J2675 (186,154)
AM 0329-480	03 29 16 −48 08.8	1.893 0.204	257.80 −52.75	247.3 −37.3	CLOSE PAIR Fa 748/9 IC 1948/49 = 200-G-32/33	23 S 1.1	F200/J2675 (186,279)
AM 0329-321	03 29 17 −32 13.0	2.407 0.204	230.87 −55.09	267.3 −41.1	DWARF	20 D 0.7	F418/J2654 (153,60)
*AM 0329-362	03 29 18 −36 24.5	2.293 0.204	238.19 −55.07	261.9 −40.5	DWARF (BR. NUCLEUS) JET to KNOT	7,20 D 0.9	F358/J2698 (252,103)
AM 0329-431	03 29 23 −43 19.7	2.075 0.204	250.02 −54.07	253.1 −38.9	COMPACT PKS 0329-431?	13 D 0.1	F248/J3677 (49,266)
AM 0329-524	03 29 36 −52 42.0	1.684 0.204	264.61 −51.00	242.0 −35.7	RICH CLUSTER Se 31/5 = DN 0329-526	19 D 15.7	F155/J1127 (82,298)
AM 0329-324	03 29 42 −32 44.2	2.392 0.204	231.78 −55.03	266.6 −41.2	COMPACT	13 D 0.2	F358/J2698 (250,300)
AM 0329-313	03 29 53 −31 30.3	2.424 0.203	229.64 −54.93	268.3 −41.4	ISOLATED IRR. MCG-5-9-12 = 418-G-9	16 D 1.1	F418/J2654 (146,99)
*AM 0329-715	03 29 54 −71 55.4	−0.174 0.203	287.76 −40.36	222.3 −26.5	RING	6 D 0.5	F031/J1947 (128,340)
AM 0329-605	03 29 57 −60 52.8	1.170 0.203	275.54 −47.03	233.1 −32.2	E w. DIAMETRIC COMP'NS.	8,13 S 1.3	F117/J2635 (297,128)
*AM 0330-324	03 30 47 −32 49.6	2.387 0.202	231.95 −54.80	266.5 −41.4	GALAXY w. THIN CURVED TAIL(RING?)	6,15 D 0.5	F418/J2654 (136,27)
AM 0330-391	03 30 54 −39 19.3	2.202 0.202	243.17 −54.49	258.1 −40.2	COMPACT	13 D 0.3	F301/J2765 (149,216)
AM 0331-551	03 31 01 −55 19.4	1.536 0.202	268.17 −49.68	239.0 −34.8	CLOSE TRIPLE + 2	24 D 6.0	F156/J2661 (332,156)
AM 0331-260	03 31 07 −26 02.4	2.553 0.202	220.38 −53.93	275.5 −42.1	PLANETARY NEBULA (PKS 0331-260) NGC 1360 = PK 222-53-1 = 482-PN-7	25 D 6.7	F482/J4722 (286,122)
AM 0331-653	03 31 35 −65 31.1	0.733 0.201	280.86 −44.25	228.3 −30.0	CLOSE PAIR PKS 0331-654? = Fa 750	23 S 1.1	F083/J1936 (226,151)
AM 0331-323	03 31 40 −32 35.0	2.392 0.201	231.54 −54.61	266.8 −41.6	DWARF	20 D 0.4	F418/J2654 (126,40)
AM 0331-360	03 31 46 −36 01.8	2.298 0.201	237.49 −54.59	262.2 −41.0	PAIR of DWARFS in GROUP of 3	20,23,24 D 2.2	F358/J2698 (225,124)
AM 0331-425	03 31 50 −42 50.3	2.084 0.201	249.04 −53.74	253.6 −39.4	DISTURBED HI.S.BR. GAL. + COMP'N.	8,16 S 1.3	F249/J2672 (298,294)
AM 0331-432	03 31 54 −43 24.4	2.064 0.201	249.98 −53.61	252.9 −39.3	SPIRAL w. I/A COMP'N. 249-G-2 = B 0331-434	1 S 0.8	F249/J2672 (296,264)
*AM 0332-251	03 32 00 −25 18.9	2.568 0.201	219.25 −53.58	276.5 −42.3	I/A TRIPLE	3 D 0.4	F482/J4722 (276,161)
AM 0332-490	03 32 02 −49 07.8	1.840 0.201	259.07 −51.99	246.0 −37.4	I/A DBL. IC 1961 = 200-G-41	2 S 1.3	F200/J2675 (161,226)
AM 0332-642	03 32 07 −64 24.7	0.844 0.201	279.55 −44.85	229.4 −30.6	RICH CLUSTER PKS 0332-644 = DN 0322-644	19 D 11.2	F083/J1936 (225,210)
AM 0332-391	03 32 12 −39 11.0	2.203 0.201	242.88 −54.25	258.1 −40.5	GROUP of 5 + APP. COMP'NS. PKS 0332-39	8,18 D 4.5	F301/J2765 (135,223)
*AM 0332-305	03 32 15 −30 53.9	2.434 0.201	228.66 −54.38	269.0 −41.9	RING w. OFF-SET NUCLEUS (TRANS.) MCG-5-9-14 = 418-G-11	6 D 2.2	F418/J2654 (119,131)
AM 0332-421	03 32 15 −42 19.7	2.100 0.201	248.17 −53.76	254.2 −39.6	CLOSE PAIR of SPIRALS	23 S 2.1	F249/J2672 (294,322)
AM 0332-720	03 32 26 −72 00.1	−0.216 0.200	287.66 −40.16	222.1 −26.6	CONSPICUOUS CLUSTER DN 0333-720	19 D 13.4	F031/J1947 (118,336)
AM 0332-634	03 32 27 −63 40.4	0.913 0.200	278.66 −45.24	230.1 −31.0	GALAXY w. 4 DIAMETRIC COMP'NS.	8 D 5.4	F083/J1936 (224,250)
AM 0332-443	03 32 34 −44 39.3	2.017 0.200	251.97 −53.20	251.3 −39.0	CHAIN of 6 (2 I/A) incl. 249-IG-4 = B 0332-446	2,17 D 9.9	F249/J2672 (288,197)
AM 0332-360	03 32 47 −36 01.7	2.295 0.200	237.48 −54.39	262.2 −41.2	DWARF	20 D 4.9	F358/J2698 (214,124)
AM 0332-250	03 32 52 −25 6.1	2.572 0.200	218.98 −53.35	276.8 −42.5	SPIRAL w. APP. COMP'NS. NGC 1371 = NGC 1367 = 482-G-10	8 D 4.5	F482/J4722 (266,173)
AM 0332-324 [0333-3237]	03 32 58 −32 48.0	2.383 0.200	231.93 −54.34	266.4 −41.8	TRIPLE (#1: IRR.) MCG-5-9-15 = 358-G-20	16,24 S 5.4	F358/J2698 (213,297)
AM 0333-611	03 33 17 −61 16.0	1.119 0.200	275.70 −46.48	232.5 −32.3	DWARF (POSSIBLE RESOLUTION)	20 D 0.7	F117/J2635 (274,108)
AM 0333-324 [0332-3248]	03 33 21 −32 46.0	2.383 0.199	231.88 −54.26	266.5 −41.9	TRIPLE (#2: S0) MCG-5-9-16 = 358-G-22	24 S 5.4	F358/J2698 (209,299)

Identification	R.A. (1950.0) Declination	Precession R.A./Dec.	Gal. Long. Gal. Lat.	S.G.L. S.G.B.	Description Cross-Identification	Coding Size (Min.)	Field/Plate (x, y)
AM 0333-352	03 33 21 −35 23.5	2.312 0.199	236.38 −54.29	263.0 −41.5	CLOSE TRIPLE NGC 1374/75 = 358-G-23/24	24 D 6.7	F358/J2698 (208,158)
AM 0333-513	03 33 32 −51 36.9	1.719 0.199	262.65 −50.88	243.0 −36.7	DWARF (RESOLUTION into KNOTS) 200-G-45	20 D 0.9	F200/J2675 (150,93)
AM 0333-323 [0333-3246]	03 33 34 −32 37.9	2.386 0.199	231.65 −54.21	266.6 −42.0	TRIPLE (#3: E) MCG-5-9-17 = 358-G-25	24 S 5.4	F358/J2698 (207,306)
AM 0333-360	03 33 53 −36 09.3	2.289 0.199	237.68 −54.16	262.0 −41.4	DWARF	20 D 4.9	F358/J2698 (202,117)
AM 0334-483	03 34 05 −48 36.2	1.854 0.199	258.08 −51.85	246.4 −37.9	I/A DBL.	2 S 0.9	F200/J2675 (143,254)
AM 0334-254	03 34 17 −25 49.6	2.553 0.198	220.27 −53.19	275.8 −42.8	I/A DBL. 482-G-12	2 S 0.6	F482/J4722 (248,134)
AM 0334-452	03 34 17 −45 21.3	1.985 0.198	252.97 −52.74	250.3 −39.1	DISTURBED E at CENTRE of 5 IC 1969 = 249-G-5	18 D 8.3	F249/J2672 (271,160)
AM 0334-475	03 34 24 −47 52.0	1.884 0.198	256.92 −52.02	247.3 −38.2	I/A DBL. 200-IG-46	2 S 0.7	F249/J2672 (267,25)
AM 0334-350	03 34 33 −35 08.5	2.316 0.198	235.94 −54.05	263.3 −41.8	S0 w. I/A DBL. COMP'N. (VV 150) NGC 1380 = 358-G-28	1,2 S 4.4	F358/J2698 (195,172)
AM 0334-281	03 34 48 −28 15.1	2.494 0.198	224.31 −53.51	272.5 −42.7	RICH CLUSTER of FAINT GALAXIES	19 D 7.2	F418/J2654 (87,273)
AM 0334-395 [0334-3958]	03 34 48 −39 57.5	2.171 0.198	244.09 −53.66	257.0 −40.7	PAIR of E's + COMP'NS. (#1) PKS 0335-399	8,23 S 1.3	F301/J2765 (109,181)
AM 0334-440	03 34 49 −44 07.2	2.029 0.198	250.95 −52.94	251.8 −39.6	EDGE-ON w. I/A COMP'N. IC 1970 = 249-G-7	1 S 2.0	F249/J2672 (267,226)
AM 0334-395 [0334-3957]	03 34 56 −39 58.0	2.170 0.198	244.10 −53.64	257.0 −40.8	PAIR of E's + COMP'NS. (#2) PKS 0335-399	8,23 S 1.3	F301/J2765 (108,180)
AM 0334-460	03 34 58 −46 07.5	1.952 0.198	254.15 −52.42	249.3 −38.9	RING/SPIRAL	6 D 0.4	F249/J2672 (264,119)
AM 0335-252	03 35 1 −25 26.3	2.561 0.197	219.70 −52.95	276.3 −43.0	I/A TRIPLE (M51-TYPE ?) 482-G-15	3,9 D 2.8	F482/J4722 (239,155)
*AM 0335-244	03 35 20 −24 40.0	2.578 0.197	218.48 −52.70	277.4 −43.0	V. DISTURBED SPIRAL + I/A TRIPLE NGC 1385 = 482-G-16 = PKS 0335-246	3,12 S 3.1	F482/J4722 (236,197)
AM 0335-395	03 35 21 −39 58.9	2.168 0.197	244.11 −53.56	256.9 −40.8	CLOSE PAIR of E's	23 S 0.6	F301/J2765 (104,179)
AM 0335-434	03 35 36 −43 44.8	2.040 0.197	250.30 −52.88	252.2 −39.8	IRR . + JETS + LO.S.BR. COMP'N. 249-IG-8	7,8,16,20 S 1.2	F249/J2672 (260,246)
AM 0336-512 [0336-5124]	03 36 05 −51 24.6	1.718 0.196	262.12 −50.59	243.0 −37.1	CLOSE PAIR (#1) 200-G-53	23 D 0.8	F200/J2675 (128,103)
AM 0336-323	03 36 06 −32 35.1	2.382 0.196	231.62 −53.67	266.6 −42.5	PECULIAR SPIRAL + E + EDGE-ON	12,24 D 0.7	F419/J2576 (334,35)
AM 0336-512 [0336-5120]	03 36 07 −51 20.9	1.721 0.196	262.03 −50.61	243.1 −37.2	CLOSE PAIR (#2) 200-G-54 = Fa 753	23 D 0.6	F200/J2675 (128,107)
AM 0336-693	03 36 10 −69 34.2	0.170 0.196	284.91 −41.46	224.1 −28.2	GALAXY w. I/A COMP'N + MATERIAL	1,9 S 0.3	F054/J1155 (138,200)
AM 0336-232	03 36 12 −23 29.2	2.603 0.196	216.68 −52.22	279.0 −43.3	DWARF	20 D 0.1	F482/J4722 (226,260)
AM 0336-231	03 36 18 −23 11.4	2.610 0.196	216.23 −52.12	279.4 −43.3	DISTURBED E + 5 DWARFS + JET NGC 1395 = 482-G-19 = MC 0336-231	7,8,15,17,20 D 3.6	F482/J4722 (225,276)
AM 0336-353	03 36 35 −35 36.7	2.297 0.196	236.73 −53.63	262.5 −42.1	FORNAX CLUSTER CENTRE NGC 1399 = 358-G-45 = PKS 0336-35	19 D 4.5	F358/J2698 (173,147)
AM 0336-320	03 36 38 −32 03.8	2.395 0.196	230.75 −53.53	267.3 −42.7	CLOSE TRIPLE	24 D 1.3	F358/J2698 (172,337)
AM 0336-263	03 36 45 −26 30.0	2.533 0.195	221.55 −52.78	274.8 −43.3	SPIRAL + COMP'NS w. DIAMETRIC NGC 1398 = 482-G-22 GROUPS	8 D 5.4	F482/J4722 (218,98)
AM 0336-232	03 36 51 −23 24.6	2.604 0.195	216.63 −52.06	279.1 −43.4	DWARF	20 D 0.2	F482/J4722 (218,264)
AM 0336-304	03 36 54 −30 49.8	2.426 0.195	228.70 −53.38	268.9 −42.9	CLOSE PAIR (E + SPIRAL)	23 S 1.0	F419/J2576 (328,130)
AM 0337-334	03 37 14 −33 41.6	2.350 0.195	233.50 −53.48	265.0 −42.6	IRR. w. TAIL 358-IG-47	15,16 D 1.3	F358/J2698 (165,249)
AM 0337-353	03 37 19 −35 31.6	2.298 0.195	236.58 −53.48	262.6 −42.2	DWARF ELLIPTICAL	20 D 0.7	F358/J2698 (165,151)
*AM 0337-312	03 37 23 −31 28.8	2.408 0.195	229.80 −53.33	268.0 −42.9	ISOLATED WARPED PLANE (ABS'N.) NGC 1406 = 418-G-15	14,16 D 4.5	F419/J2576 (322,95)
AM 0337-355	03 37 26 −35 53.4	2.287 0.195	237.19 −53.45	262.1 −42.2	DWARF (V.SMALL NUCLEUS)	20 D 0.7	F358/J2698 (164,132)
AM 0337-390	03 37 30 −39 08.6	2.189 0.195	242.63 −53.24	257.9 −41.5	PECULIAR SPIRAL in GROUP of 4 301-G-21 = B 0337-391	12,18 D 2.2	F301/J2765 (80,224)
AM 0337-240	03 37 31 −24 00.3	2.589 0.195	217.63 −52.06	278.3 −43.6	DWARF 482-G-27	20 D 0.4	F482/J4722 (210,232)
AM 0337-315	03 37 33 −31 57.3	2.395 0.194	230.59 −53.33	267.4 −42.9	DWARF (RESOLUTION?)	20 D 0.7	F358/J2698 (161,343)

Identification	R.A. (1950.0) Declination	Precession R.A./Dec.	Gal. Long. Gal. Lat.	S.G.L. S.G.B.	Description Cross-Identification	Coding Size (Min.)	Field/Plate (x, y)
AM 0337-711	03 37 53 −71 13.4	−0.125 0.194	286.52 −40.33	222.5 −27.4	I/A DBL. (CONTACT) 54-IG-15	2 S 0.2	F054/J1155 (134,111)
AM 0337-303	03 37 55 −30 39.6	2.428 0.194	228.45 −53.14	269.1 −43.2	DWARF	20 D 0.2	F418/J2654 (53,143)
AM 0338-551	03 38 05 −55 14.7	1.505 0.194	267.39 −48.81	238.5 −35.7	I/A SPIRAL in CLUSTER	2,19 S 0.8	F156/J2661 (278,164)
AM 0338-375	03 38 14 −37 56.8	2.224 0.194	240.62 −53.20	259.4 −41.9	I/A SPIRALS + PECULIAR COMP'NS. 301-IG-22 = B 0338-379	2,8,16 S 0.7	F301/J2765 (71,288)
AM 0338-354	03 38 15 −35 47.5	2.288 0.194	237.02 −53.28	262.2 −42.4	DISTURBED IRR. NGC 1427a = 358-G-49	16 D 2.7	F358/J2698 (155,137)
AM 0338-270	03 38 23 −27 01.2	2.518 0.193	222.51 −52.52	274.1 −43.6	GALAXY w. I/A DBL. COMP'N. IC 1981 = 482-G-29	2,8 S 4.9	F418/J2654 (43,338)
AM 0338-285	03 38 24 −28 50.0	2.473 0.193	225.46 −52.81	271.6 −43.5	DBL. cD at CENTRE of CLUSTER PKS 0338-288	2,19 D 9.0	F418/J2654 (45,241)
AM 0338-315	03 38 26 −31 50.7	2.396 0.193	230.43 −53.14	267.5 −43.1	DWARF 419-G-1	20 D 0.6	F418/J2654 (49,79)
*AM 0338-320	03 38 29 −32 01.5	2.392 0.193	230.73 −53.14	267.2 −43.1	I/A DBL.	2 S 0.3	F419/J2576 (308,66)
AM 0338-265	03 38 35 −26 56.8	2.519 0.193	222.41 −52.46	274.2 −43.7	GALAXY w. I/A COMP'N. 482-G-32 (in PAIR w. NGC 1398)	1,8,23 S 1.6	F482/J4722 (195,75)
AM 0338-224	03 38 46 −22 43.5	2.616 0.193	215.75 −51.45	280.0 −43.9	DISTURBED w. DIAMETRIC COMP'NS. NGC 1415 = IC 1983 = 482-G-33	8 D 1.8	F482/J4722 (195,301)
*AM 0338-374	03 38 49 −37 40.5	2.231 0.193	240.16 −53.10	259.7 −42.1	LARGE COMPACT NGC 1419 = 301-G-23	13 D 0.5	F302/J2668 (340,301)
AM 0338-453	03 38 55 −45 31.1	1.962 0.193	252.91 −51.91	249.8 −39.8	IRR. DWARF (BR. KNOTS at CENTRE) IC 1986 = 249-G-13	20 D 1.6	F249/J2672 (227,152)
AM 0339-263	03 39 00 −26 39.7	2.525 0.193	221.97 −52.32	274.6 −43.8	DWARF (COMP'N to NGC 1398)	20 D 0.6	F482/J4722 (190,90)
AM 0339-652	03 39 00 −65 28.2	0.682 0.193	280.19 −43.65	227.9 −30.7	I/A DBL. (LONG-ARMED SPIRAL) 83-IG-8	2 S 0.2	F083/J1936 (185,154)
*AM 0339-484	03 39 04 −48 46.8	1.827 0.193	257.94 −51.01	245.8 −38.6	COMPACT	13 D 0.2	F200/J2675 (99,244)
AM 0339-235	03 39 05 −23 59.8	2.587 0.193	217.76 −51.71	278.3 −43.9	GALAXY w. APP. COMP'NS (DWARF) MCG-4-9-50 = 482-G-35	8,20 S 3.3	F482/J4722 (190,233)
*AM 0339-335	03 39 07 −33 56.4	2.338 0.193	233.93 −53.10	264.6 −42.9	GALAXY w. 2 COMP'NS. off MI.AX. 358-G-50	8 S 9.0	F358/J2698 (144,236)
AM 0339-571	03 39 08 −57 11.0	1.380 0.193	269.92 −47.83	236.4 −35.0	I/A DBL. w. TAIL (GROUP of 8) 156-IG-6	2,15,18 S 0.5	F156/J2661 (265,60)
AM 0339-270	03 39 12 −27 02.2	2.516 0.193	222.59 −52.34	274.0 −43.8	TIGHT GROUP of 4	18 D 0.9	F418/J2654 (33,337)
AM 0339-290	03 39 35 −29 01.2	2.467 0.192	225.83 −52.58	271.3 −43.7	SMALL I/A TRIPLE PKS 0339-291?	3 D 0.4	F419/J2576 (300,228)
AM 0339-300	03 39 35 −30 00.8	2.442 0.192	227.45 −52.71	269.9 −43.6	M51-TYPE	9 D 0.4	F419/J2576 (299,174)
AM 0339-483	03 39 35 −48 30.6	1.837 0.192	257.49 −51.01	246.1 −38.8	CLOSE TRIPLE (ALIGNED)	24 D 0.4	F200/J2675 (94,258)
AM 0339-381	03 39 47 −38 10.2	2.213 0.192	240.95 −52.87	259.0 −42.1	I/A DBL.	2,18 D 1.3	F301/J2765 (54,275)
AM 0339-625	03 39 52 −62 56.5	0.931 0.192	277.15 −44.94	230.3 −32.1	DBL. RING w. COMP'NS. 83-IG-9	1,6 S 0.3	F117/J2635 (230,20)
AM 0339-541	03 39 54 −54 10.2	1.558 0.192	265.74 −49.01	239.6 −36.5	GALAXY w. JETS 156-IG-7	7 D 0.4	F156/J2661 (267,222)
*AM 0339-335	03 39 57 −33 58.2	2.336 0.192	233.99 −52.93	264.5 −43.1	LONG CHAIN of 7 GALAXIES	17 D 28.7	F358/J2698 (135,234)
*AM 0340-280	03 40 03 −28 01.6	2.490 0.192	224.24 −52.33	272.7 −43.9	DISTURBED "BRIGHT-RIMMED" SPIRAL MCG-5-9-22 = 419-G-3	12 D 1.6	F482/J4722 (178,17)
AM 0340-244	03 40 05 −24 49.0	2.567 0.191	219.13 −51.69	277.1 −44.1	SMALL ONE-ARMED SPIRAL	10 D 0.5	F482/J4722 (178,189)
AM 0340-225	03 40 08 −22 55.0	2.610 0.191	216.18 −51.20	279.8 −44.2	LO.S.BR. SPIRAL + APP. COMP'N. MCG-4-9-53 = 482-G-36 = UKS0340-229	8,20 S 0.7	F482/J4722 (178,291)
AM 0340-381	03 40 08 −38 11.4	2.212 0.191	240.98 −52.80	258.9 −42.2	I/A DBL. + I/A TRIPLE 301-IG-25	2,3,18 D 4.9	F302/J2668 (325,274)
AM 0340-420	03 40 21 −42 04.4	2.084 0.191	247.32 −52.33	253.9 −41.2	SP. w. CONDENS.? + COMPACT COMP'N	8,12,13 S 2.2	F249/J2672 (214,337)
*AM 0340-472	03 40 30 −47 23.2	1.882 0.191	255.71 −51.18	247.4 −39.4	FACE-ON SB + VERY COMPACT COMP'N. NGC 1433 = 249-G-14 [A82]	8,13 S 6.7	F249/J2672 (212,51)
AM 0340-221	03 40 38 −22 16.0	2.624 0.191	215.24 −50.91	280.7 −44.3	GALAXY w. PECULIAR ISOPHOTES NGC 1426 = 549-G-1 (in CHAIN)	15,17 D 1.8	F482/J4722 (172,326)
AM 0341-432	03 41 01 −43 27.4	2.033 0.190	249.51 −51.97	252.1 −40.8	DISTURBED KNOTTY SINGLE GALAXY 249-G-15	16 D 0.7	F249/J2672 (207,262)
AM 0341-534	03 41 38 −53 47.7	1.571 0.190	265.06 −48.92	239.9 −36.9	2 cD in ENVELOPE (RICH CLUSTER) Se 31/7 = 156-G-8	2,19 D 17.9	F156/J2661 (254,243)

Identification	R.A. (1950.0) Declination	Precession R.A./Dec.	Gal. Long. Gal. Lat.	S.G.L. S.G.B.	Description Cross-Identification	Coding Size (Min.)	Field/Plate (x, y)
AM 0341-360	03 41 43 −36 00.7	2.273 0.190	237.37 −52.58	261.7 −43.0	RING-SPIRAL NGC 1437 = 358-G-58 = Lu0342-36	6 D 3.6	F358/J2698 (117,125)
AM 0341-250	03 41 50 −25 04.8	2.558 0.189	219.69 −51.37	276.7 −44.5	DWARF 482-G-15	20 D 0.4	F482/J4722 (157,175)
AM 0341-283	03 41 59 −28 33.2	2.474 0.189	225.20 −52.00	271.9 −44.3	I/A DBL.	2 S 0.4	F419/J2576 (272,253)
AM 0341-473	03 41 59 −47 39.0	1.865 0.189	256.01 −50.86	246.9 −39.5	SMALL GALAXY w. PLUMES and JETS	7,15 D 0.3	F249/J2672 (199,37)
AM 0342-444	03 42 52 −44 48.1	1.977 0.188	251.54 −51.39	250.3 −40.7	EDGE-ON w. 5 APP. COMP'NS. NGC1448 = NGC1457 = 249-G-16	8 D 9.0	F249/J2672 (190,190)
AM 0343-541	03 43 00 −54 13.7	1.540 0.188	265.55 −48.57	239.3 −36.8	I/A DBL. 156-IG-9	2 S 0.4	F156/J2661 (242,220)
AM 0343-362	03 43 07 −36 28.6	2.256 0.188	238.12 −52.29	261.0 −43.2	NUCLEATED DWARF	20 D 0.9	F358/J2698 (103,99)
AM 0343-435	03 43 11 −43 51.9	2.011 0.188	250.04 −51.52	251.4 −41.1	SMALL I/A QUARTET	4 D 1.8	F249/J2672 (186,240)
AM 0343-383	03 43 13 −38 32.3	2.193 0.188	241.48 −52.18	258.3 −42.7	SP. w. PEC. ARM (BROKEN RING?) 302-G-4	6,10 D 0.6	F302/J2668 (292,257)
AM 0343-471	03 43 43 −47 10.2	1.879 0.187	255.15 −50.70	247.4 −39.9	SMALL I/A DBL.	2 S 0.3	F249/J2672 (183,63)
AM 0343-495	03 43 47 −49 53.9	1.759 0.187	259.25 −49.94	244.1 −38.9	COMPACT	13 D 0.3	F200/J2675 (60,182)
AM 0344-363	03 44 02 −36 31.2	2.253 0.187	238.19 −52.10	260.9 −43.4	GALAXY w. I/A COMP'N. 358-G-61	1 S 2.0	F358/J2698 (93,97)
AM 0344-335	03 44 22 −33 54.9	2.327 0.186	233.95 −52.01	264.4 −44.0	EDGE-ON DWARF	20 D 1.0	F358/J2698 (86,237)
*AM 0344-365	03 44 22 −36 51.1	2.242 0.186	238.72 −52.03	260.4 −43.3	SB(r) + I/A DBL. COMP'N. NGC 1460 = 358-G-62	2,8 S 1.3	F358/J2698 (89,79)
AM 0344-571	03 44 38 −57 11.4	1.352 0.186	269.45 −47.16	235.9 −35.6	RICH CLUSTER 12 GALAXIES	19 D 9.0	F117/J2635 (203,329)
AM 0344-343	03 44 40 −34 32.4	2.309 0.186	234.97 −51.97	263.5 −43.9	GROUP of 7 PKS 0344-34	18 D 4.5	F358/J2698 (83,203)
AM 0344-370	03 44 43 −37 04.7	2.234 0.186	239.09 −51.95	260.1 −43.4	SUB-GROUP of 4 in GROUP of 5 358-G-64	18 D 4.5	F358/J2698 (86,67)
AM 0344-300	03 44 47 −30 05.7	2.430 0.186	227.80 −51.61	269.6 −44.7	M51-TYPE (2 ARMS to COMP'N) MCG-5-10-1 = 419-G-6	9,10,22 D 1.1	F419/J2576 (238,171)
AM 0345-854	03 45 03 −85 46.2	−0.199 0.185	299.59 −30.39	210.0 −18.5	GROUP of 4 (incl. COMPACT DBL.) PKS 0344-85?	13,18,23 D 1.8	F003/J2565 (101,124)
*AM 0345-335	03 45 08 −33 52.2	2.327 0.185	233.89 −51.85	264.4 −44.2	TRANSITION RING/Sc IC 1993 = 358-G-65	6 D 2.2	F358/J2698 (77,239)
AM 0345-363	03 45 08 −36 35.7	2.248 0.185	238.30 −51.88	260.7 −43.6	DWARF	20 D 0.3	F358/J2698 (81,93)
AM 0345-325	03 45 13 −32 55.9	2.353 0.185	232.37 −51.78	265.7 −44.3	DWARF (DEFECT?)	20 D 0.5	F419/J2576 (231,19)
*AM 0345-362	03 45 14 −36 28.8	2.251 0.185	238.12 −51.86	260.9 −43.6	DWARF PKS 0346-364	20 D 0.8	F358/J2698 (80,99)
AM 0345-441	03 45 14 −44 12.1	1.992 0.185	250.47 −51.09	250.9 −41.3	CLOSE PAIR	23 S 1.1	F249/J2672 (167,222)
AM 0345-480	03 45 25 −48 09.4	1.831 0.185	256.53 −50.17	246.0 −39.8	COMPACT	13 D 0.2	F201/J2725 (310,275)
AM 0345-450	03 45 41 −45 03.5	1.958 0.185	251.78 −50.85	249.8 −41.1	I/A DBL. in CHAIN of 4	2,17 D 2.2	F249/J2672 (163,176)
AM 0346-704	03 46 03 −70 45.1	−0.118 0.184	285.50 −40.08	222.5 −28.2	I/A DBL. 54-IG-18	2 S 0.3	F054/J1155 (97,134)
AM 0346-441	03 46 05 −44 12.6	1.989 0.184	250.44 −50.94	250.8 −41.4	CLOSE TRIPLE	24 D 2.0	F249/J2672 (159,221)
AM 0346-365	03 46 17 −36 53.5	2.236 0.184	238.78 −51.65	260.2 −43.7	DISTURBED SPIRAL 358-G-67	12 D 0.7	F359/J2709 (327,76)
AM 0346-634	03 46 18 −63 48.6	0.806 0.184	277.67 −43.88	229.0 −32.3	GALAXY w. I/A COMP'N.	1 S 0.5	F083/J1936 (142,243)
AM 0346-420	03 46 21 −42 00.8	2.068 0.184	246.97 −51.24	253.5 −42.2	CLOSE PAIR	23 S 0.3	F249/J2672 (154,339)
AM 0346-801	03 46 32 −80 15.6	−3.428 0.184	294.78 −34.11	214.3 −22.2	DISTURBED GALAXY w. APP. COMP'N. 15-G-01 (DWARF ?)	8,16,20 S 2.9	F014/J2559 (74,155)
AM 0346-570	03 46 36 −57 06.5	1.347 0.184	269.18 −46.95	235.8 −35.9	SMALL DBL. RING w. I/A COMP'N. IC 1999 = 156-G-14	1,6 D 0.5	F156/J2661 (211,66)
AM 0346-393	03 46 42 −39 34.7	2.150 0.183	243.08 −51.43	256.6 −43.0	DWARF 302-G-10	20 D 1.3	F302/J2668 (255,202)
AM 0346-454	03 46 45 −45 42.0	1.929 0.183	252.71 −50.53	248.9 −41.0	MEDIUM BRIGHT CLUSTER (15 GAL.)	19 D 7.2	F249/J2672 (154,141)
AM 0346-222	03 46 50 −22 23.5	2.613 0.183	216.06 −49.57	280.5 −45.7	DISTURBED SINGLE (POSSIBLE JET?) MCG-4-10-5 = 549-G-24	7,16 D 1.0	F482/J4722 (95,319)

Identification	R.A. (1950.0) Declination	Precession R.A./Dec.	Gal. Long. Gal. Lat.	S.G.L. S.G.B.	Description Cross-Identification	Coding Size (Min.)	Field/Plate (x, y)
AM 0347-483	03 47 12 −48 34.1	1.806 0.183	257.03 −49.78	245.4 −39.9	DWARF (RESOLUTION:) 201-G-2	20 D 0.9	F201/J2725 (293,254)
AM 0347-682	03 47 18 −68 22.5	0.252 0.183	282.84 −41.35	224.6 −29.7	PECULIAR GALAXY w. TAIL NGC 1473 = 54-G-19	15 D 1.3	F054/J1155 (81,261)
*AM 0347-370	03 47 21 −37 09.8	2.225 0.183	239.20 −51.42	259.8 −43.8	DWARF or FAINT STAR CLUSTER	20,22 D 0.4	F359/J2709 (315,61)
AM 0347-472	03 47 33 −47 22.0	1.857 0.182	255.20 −50.03	246.8 −40.5	SMALL I/A DBL. 249-IG-20	2 S 0.1	F201/J2725 (292,319)
AM 0347-442	03 47 39 −44 22.5	1.978 0.182	250.61 −50.63	250.4 −41.6	DWARF 249-G-21	20 D 0.9	F249/J2672 (144,212)
AM 0347-553	03 47 40 −55 35.0	1.439 0.182	267.03 −47.43	237.4 −36.8	SPIRAL + APP. COMP'N. 156-G-16	8 S 0.4	F156/J2661 (204,148)
AM 0347-754	03 47 57 −75 42.2	−1.323 0.182	290.43 −36.98	218.1 −25.2	GALAXY w. LOOP (ONE-ARMED SPIRAL)	15 D 0.4	F031/J1947 (78,132)
AM 0347-533	03 47 59 −53 33.4	1.556 0.182	264.20 −48.13	239.6 −37.8	I/A E's	2 S 0.3	F156/J2661 (204,257)
AM 0348-680	03 48 11 −68 06.1	0.284 0.182	282.48 −41.44	224.8 −29.9	CLOSE PAIR	23 S 0.7	F055/J1942 (336,268)
*AM 0348-502	03 48 56 −50 27.0	1.713 0.181	259.69 −48.99	243.0 −39.4	SP. w. I/A COMP'N. (GROUP of 3) 201-IG-4 = B 0348-504	1,24 S 0.4	F201/J2725 (274,154)
AM 0349-383	03 49 51 −38 36.1	2.174 0.180	241.46 −50.88	257.7 −43.9	DWARF (RESOLUTION) Ka-21 = 302-G-14 = UKS 0349-386	20 D 1.8	F302/J2668 (222,254)
AM 0349-714	03 49 58 −71 46.9	−0.350 0.179	286.35 −39.23	221.4 −27.8	DWARF SPIRAL 54-G-21 = UKS 0350-717	20 D 5.4	F054/J1155 (84,77)
*AM 0350-374	03 50 00 −37 45.3	2.200 0.179	240.12 −50.88	258.8 −44.2	GALAXY w. 4 JETS	7 D 0.9	F302/J2668 (221,300)
AM 0350-455	03 50 16 −45 53.7	1.910 0.179	252.81 −49.89	248.3 −41.5	CLOSE PAIR	23 S 0.7	F249/J2672 (121,130)
AM 0350-425	03 50 21 −42 53.4	2.025 0.179	248.19 −50.39	252.1 −42.6	CHAIN of 4	17 D 6.5	F249/J2672 (116,291)
AM 0350-321	03 50 25 −32 16.3	2.361 0.179	231.46 −50.64	266.3 −45.6	COMPACT	13 D 0.3	F359/J2709 (287,325)
AM 0350-444	03 50 31 −44 41.0	1.957 0.179	250.95 −50.08	249.8 −42.0	HI.S.BR. ELONGATED GALAXY NGC 1476 = 249-G-24	16 D 1.1	F249/J2672 (117,195)
AM 0351-600	03 51 03 −60 04.6	1.110 0.178	272.69 −45.16	232.4 −34.8	SPIRAL w. PECULIAR ARM IC 2010 = 117-G-11	10 D 0.9	F117/J2635 (159,174)
AM 0351-452	03 51 14 −45 21.0	1.929 0.178	251.93 −49.83	248.9 −41.9	CLOSE TRIPLE	24 D 2.0	F249/J2672 (111,159)
AM 0351-422	03 51 16 −42 25.3	2.040 0.178	247.42 −50.28	252.6 −42.9	CLOSE PAIR	23 S 1.3	F249/J2672 (106,316)
*AM 0351-290	03 51 41 −29 06.4	2.443 0.177	226.59 −49.98	270.7 −46.3	INTEGRAL SPIRAL + CONTACT COMP'N? 419-IG-10	1,12 D 0.8	F419/J2576 (158,225)
AM 0352-370	03 52 27 −37 07.1	2.214 0.176	239.10 −50.41	259.5 −44.8	SINGLE IRR. (GALAXY w. COMP'N.?) NGC 1484 = 359-G-6	8,16 D 2.2	F302/J2668 (195,334)
*AM 0352-360	03 52 37 −36 06.8	2.245 0.176	237.53 −50.38	260.8 −45.1	E (FAINT RING) + IRR. (MATERIAL) IC 2006 = 359-G-7	6,8,16 D 1.8	F359/J2709 (259,119)
AM 0352-445	03 52 52 −44 54.1	1.941 0.176	251.17 −49.63	249.3 −42.3	IRR. + CONDENS. + FNT.OBJECTS at END 249-G-27	16 D 1.1	F249/J2672 (95,183)
AM 0353-661	03 53 08 −66 10.0	0.500 0.176	279.94 −42.06	226.3 −31.5	WIDE GROUP of 3 (E + 2 SPIRALS) NGC 1490 = 83-G-11/12/13 = NGC 1503	24 D 17.5	F083/J1936 (108,115)
AM 0353-431	03 53 09 −43 11.2	2.006 0.176	248.53 −49.84	251.4 −43.0	SMALL E w. FAINT, WIDE JET	15 D 0.5	F249/J2672 (89,275)
AM 0353-535	03 53 15 −53 57.2	1.511 0.175	264.36 −47.26	238.7 −38.3	SMALL I/A QUARTET	4 D 1.3	F156/J2661 (162,236)
AM 0353-452	03 53 23 −45 27.9	1.917 0.175	252.00 −49.44	248.6 −42.2	I/A DBL. 249-IG-29 = B 0353-454	2 S 0.3	F249/J2672 (91,152)
AM 0353-704	03 53 23 −70 41.5	−0.173 0.175	284.98 −39.62	222.1 −28.7	I/A COMP'N. CONN. by FNT. FILAM:	1 S 1.1	F055/J1942 (295,133)
AM 0353-494	03 53 36 −49 45.6	1.728 0.175	258.37 −48.46	243.4 −40.3	DISTANT GLOBULAR 201-SC?-10	22 D 1.8	F201/J2725 (235,192)
AM 0353-355	03 53 57 −35 57.1	2.247 0.175	237.28 −50.10	260.9 −45.4	DWARF (near NEBULOSITY)	20 D 0.4	F359/J2709 (245,128)
*AM 0354-423	03 54 05 −42 30.6	2.029 0.174	247.46 −49.75	252.2 −43.4	IRR. w. LOOPS + I/A DBL. COMP'N. NGC 1487 = VV78 = 249-IG-31 = Ag-15 [B]	1,2,15,16 D 7.6	F302/J2668 (179,45)
AM 0354-462	03 54 19 −46 20.8	1.878 0.174	253.28 −49.12	247.4 −41.9	SMALL GALAXY w. JET	7 D 0.3	F250/J2681 (340,101)
AM 0354-721	03 54 24 −72 16.6	−0.495 0.174	286.61 −38.66	220.7 −27.8	SPIRAL w. I/A COMP'N. + TAIL 55-IG-2	1,15 S 0.2	F032/J1937 (300,318)
AM 0354-621	03 54 32 −62 19.7	0.898 0.174	275.25 −43.78	229.8 −33.9	COMPACT	13 D 0.3	F083/J1936 (89,320)
AM 0354-442	03 54 37 −44 27.7	1.953 0.174	250.42 −49.39	249.7 −42.8	GROUP of 7	18 D 2.7	F249/J2672 (77,206)

Identification	R.A. (1950.0) Declination	Precession R.A./Dec.	Gal. Long. Gal. Lat.	S.G.L. S.G.B.	Description Cross-Identification	Coding Size (Min.)	Field/Plate (x, y)
AM 0355-290	03 55 08 −29 01.2	2.439 0.173	226.64 −49.23	270.7 −47.1	EDGE-ON + APP. COMP'NS. MCG-5-10-6 = 419-G-12	8 S 5.4	F419/J2576 (117,229)
AM 0355-442	03 55 28 −44 24.0	1.953 0.173	250.29 −49.25	249.7 −42.9	CLOSE PAIR	23 S 1.0	F249/J2672 (69,209)
AM 0355-270	03 55 42 −27 09.9	2.486 0.172	223.90 −48.77	273.4 −47.5	LINE of 3	24 D 0.9	F483/J3814 (254,63)
AM 0355-593	03 55 44 −59 32.2	1.127 0.172	271.64 −44.85	232.5 −35.6	GAL. w. DUST LANE + OUTER MATR'L. IC 2017 = 117-G-15	14,16 D 0.9	F117/J2635 (127,202)
AM 0355-265	03 55 46 −26 59.3	2.490 0.172	223.64 −48.72	273.6 −47.5	LINE of 3	24 D 1.3	F419/J2576 (109,338)
AM 0355-463	03 55 49 −46 30.8	1.866 0.172	253.46 −48.83	247.0 −42.1	DWARF (LINEAR w. CONDENSATION) 249-G-32 = PKS 355-465 = [UKS]	20 D 1.1	F250/J2681 (326,93)
*AM 0355-603	03 55 55 −60 34.0	1.044 0.172	272.94 −44.40	231.5 −35.0	SPIRAL w. LONG ARM + COMP'NS. 117-G-16 = B 0355-605	8,10 D 1.8	F117/J2635 (127,147)
*AM 0355-462	03 55 56 −46 21.3	1.873 0.172	253.21 −48.84	247.2 −42.2	FACE-ON SP. + DIAMETRIC COMP'NS. NGC 1493 = 249-G-33	8 D 4.5	F249/J2672 (69,103)
AM 0356-522	03 56 00 −52 24.0	1.586 0.172	262.00 −47.36	240.1 −39.4	GALAXY w. LONG TAIL to COMP'N.	1,15 S 0.4	F156/J2661 (139,319)
AM 0356-353	03 56 20 −35 35.4	2.253 0.172	236.74 −49.61	261.3 −46.0	LARGE COMPACT ("AMERICA")	13 D 0.7	F359/J2709 (219,148)
AM 0356-425	03 56 31 −42 50.1	2.010 0.171	247.88 −49.27	251.6 −43.7	SMALL I/A DBL.	2 S 0.1	F249/J2672 (55,292)
AM 0356-402	03 56 42 −40 22.2	2.098 0.171	244.10 −49.46	254.8 −44.6	I/A DBL. + FUZZ 302-IG-23	2 S 0.2	F302/J2668 (152,159)
AM 0356-443	03 56 43 −44 36.5	1.941 0.171	250.56 −49.00	249.3 −43.0	HI.S.BR. ELONG. GAL. + 5 COMP'NS. NGC 1495 = 249-G-34	1,8,16 D 3.6	F249/J2672 (57,197)
AM 0356-300	03 56 47 −30 02.2	2.410 0.171	228.26 −49.03	269.1 −47.3	FAINT RICH CLUSTER	19 D 15.7	F419/J2576 (99,174)
AM 0356-463 [0357-4630]	03 56 51 −46 30.1	1.863 0.171	253.39 −48.66	246.9 −42.3	CLOSE PAIR in GROUP of 3 (#1)	24 S 0.7	F250/J2681 (317,94)
AM 0356-522	03 56 55 −52 28.4	1.578 0.171	262.04 −47.21	239.9 −39.5	CLUSTER of MEDIUM BR. GALAXIES NGC 1500 = 201-G-13	19 D 20.2	F156/J2661 (131,315)
AM 0357-463 [0356-4630]	03 57 01 −46 31.5	1.862 0.171	253.41 −48.62	246.9 −42.3	CLOSE PAIR in GROUP of 3 (#2)	23,24 D 2.2	F250/J2681 (315,92)
AM 0357-273	03 57 02 −27 31.0	2.475 0.171	224.50 −48.55	272.8 −47.7	CLOSE GROUP of 3	24 D 2.5	F419/J2576 (94,310)
AM 0357-463 [0357-4631]	03 57 03 −46 30.8	1.862 0.171	253.40 −48.62	246.9 −42.3	CLOSE PAIR in GROUP of 3 (#3)	24 D 2.2	F250/J2681 (315,93)
AM 0357-260	03 57 08 −26 02.1	2.512 0.171	222.33 −48.22	275.0 −47.9	GALAXY w. PUFF of MATERIAL 483-IG-5	15 D 0.4	F483/J3814 (238,124)
AM 0357-343	03 57 11 −34 32.3	2.282 0.171	235.13 −49.39	262.7 −46.4	DISTURBED + DIAMETRIC COMP'NS	8,10 D 10.8	F359/J2709 (210,204)
AM 0357-442 [0357-4419]	03 57 15 −44 20.5	1.950 0.170	250.13 −48.95	249.6 −43.2	CLOSE PAIR (#1)	23 S 1.0	F250/J2681 (318,210)
AM 0357-441 [0357-4420]	03 57 18 −44 19.8	1.950 0.170	250.12 −48.94	249.6 −43.2	CLOSE PAIR (#2)	23 S 1.0	F250/J2681 (317,210)
AM 0357-633	03 57 24 −63 37.0	0.756 0.170	276.62 −42.90	228.3 −33.4	I/A QUARTET (SMALL)	4 D 0.4	F084/J0959 (335,244)
AM 0357-251	03 57 30 −25 19.3	2.529 0.170	221.32 −47.97	276.1 −48.0	EDGE-ON SPIRAL w. APP. COMP'NS	8 S 4.3	F483/J3814 (233,162)
AM 0357-432	03 57 41 −43 28.8	1.982 0.170	248.82 −48.98	250.6 −43.6	SMALL I/A DBL.	2 S 0.3	F249/J2672 (45,257)
*AM 0357-460	03 57 41 −46 00.8	1.881 0.170	252.62 −48.60	247.5 −42.6	DWARF (SOME RESOLUTION) + COMP'NS 249-G-36 = UKS 357-460 = Fa392!?	8,20 D 2.0	F249/J2672 (52,121)
AM 0357-594	03 57 41 −59 43.3	1.103 0.170	271.74 −44.55	232.2 −35.7	GROUP of 6 (4+2) DN 0357-597 = Fa 18/19	18 D 4.0	F117/J2635 (114,192)
AM 0357-401	03 57 42 −40 16.8	2.098 0.170	243.95 −49.27	254.8 −44.8	I/A DBL. COMPACTS	2,13 S 0.2	F302/J2668 (141,164)
AM 0357-614	03 57 46 −61 41.3	0.937 0.170	274.22 −43.72	230.2 −34.6	CLOSE PAIR	23 S 0.8	F117/J2635 (117,86)
AM 0357-365	03 57 51 −36 50.9	2.211 0.170	238.68 −49.33	259.4 −45.9	GALAXY w. TAILS 359-G-13	15 D 0.9	F359/J2709 (203,80)
AM 0357-442 [0357-4426]	03 57 56 −44 25.5	1.945 0.170	250.23 −48.81	249.4 −43.3	CLOSE PAIR in GROUP of 6 (#1)	18,23 D 2.4	F250/J2681 (311,206)
AM 0357-442 [0357-4425]	03 57 57 −44 26.5	1.944 0.170	250.26 −48.81	249.4 −43.3	CLOSE PAIR in GROUP of 6 (#2)	18,23 D 2.4	F250/J2681 (311,205)
AM 0358-715	03 58 01 −71 53.9	−0.449 0.169	286.01 −38.64	220.8 −28.2	SMALL I/A DBL.	2 S 0.3	F055/J1942 (269,71)
AM 0358-835	03 58 09 −83 58.5	−7.850 0.169	297.82 −31.37	211.1 −20.0	S(r) + COMP'NS. (PEC. NUCLEUS) IC 2051 = 4-G-7	8,12 S 17.6	F004/J3821 (249,227)
AM 0358-465 [0359-4659]	03 58 25 −46 55.9	1.840 0.169	253.95 −48.31	246.3 −42.3	WIDE PAIR (#1)	23 S 10.3	F250/J2681 (301,71)

Identification	R.A. (1950.0) Declination	Precession R.A./Dec.	Gal. Long. Gal. Lat.	S.G.L. S.G.B.	Description Cross-Identification	Coding Size (Min.)	Field/Plate (x, y)
AM 0358-735	03 58 25 −73 50.3	−0.903 0.169	288.04 −37.54	219.1 −26.9	SP. w. 2 I/A COMP'NS (CHAIN of 7) B 0358-732?	1,17 D 13.0	F032/J1937 (275,236)
AM 0358-312	03 58 37 −31 20.7	2.371 0.169	230.31 −48.81	267.1 −47.4	DISTANT CLUSTER of GALAXIES	19 D 1.0	F420/J2703 (344,104)
AM 0358-445 [0358-4456]	03 58 39 −44 58.9	1.921 0.169	251.04 −48.61	248.6 −43.2	CLOSE PAIR (#1: SPIRAL)	23 D 0.4	F249/J2672 (40,176)
AM 0358-320	03 58 41 −32 09.3	2.348 0.169	231.54 −48.89	266.0 −47.3	I/A DBL. (ONE COMPACT) 419-IG-14	2,13 S 0.4	F419/J2576 (79,60)
AM 0358-275	03 58 43 −27 56.2	2.461 0.169	225.23 −48.26	272.1 −48.0	CHAIN of 4 (GROUP of 5) incl. 419-G-15	17,18 D 6.7	F419/J2576 (74,287)
AM 0358-623	03 58 47 −62 39.0	0.843 0.169	275.34 −43.19	229.2 −34.1	DWARF/IRR. (STRUCTURE)	16,208 D 0.7	F084/J0959 (331,296)
AM 0358-445 [0358-4458]	03 58 53 −44 56.9	1.921 0.168	250.98 −48.57	248.7 −43.2	CLOSE PAIR (#2: LUMPY NUCLEUS)	16,23 D 0.4	F249/J2672 (38,177)
AM 0358-520	03 58 56 −52 05.8	1.591 0.168	261.38 −47.02	240.2 −40.0	GROUP of 5 (4 + 1)	18 D 2.2	F156/J2661 (114,335)
AM 0359-465 [0358-4655]	03 59 12 −46 59.6	1.835 0.168	254.00 −48.17	246.1 −42.4	WIDE PAIR (#2) 250-G-1	23 S 10.3	F250/J2681 (294,68)
AM 0359-594	03 59 13 −59 46.5	1.091 0.168	271.70 −44.36	232.0 −35.8	CLOSE GROUP (4 + 2) Fa 20	18 D 0.9	F117/J2635 (104,189)
AM 0359-783	03 59 13 −78 33.0	−2.628 0.168	292.73 −34.74	215.3 −23.8	PAIR (Sa + COMPACT w. JET 15-G-04 + CONDENSATION)	7,13,23 S 4.9	F014/J2559 (23,238)
AM 0359-313 [0359-3135]	03 59 19 −31 36.7	2.362 0.168	230.74 −48.69	266.7 −47.5	E + SP. in GROUP of 5 (#1: E)	18,23 S 0.4	F420/J2703 (335,90)
*AM 0359-674	03 59 19 −67 47.1	0.245 0.168	281.39 −40.72	224.3 −31.0	M82-TYPE + FILAMENTS (2 COMP'NS.) NGC 1511 = 55-G-4 (GROUP of 3)[A82]	8,15,16,24 D 4.0	F055/J1942 (282,291)
AM 0359-313 [0359-3136]	03 59 21 −31 35.5	2.363 0.168	230.71 −48.69	266.7 −47.5	E + SP. in GROUP of 5 (#2: SP.)	18,23 S 0.4	F420/J2703 (335,91)
AM 0359-623	03 59 24 −62 38.7	0.840 0.168	275.29 −43.13	229.1 −34.1	I/A DBL (ONE VERY COMPACT)	2 S 0.3	F084/J0959 (328,297)
AM 0359-314	03 59 25 −31 49.6	2.356 0.168	231.07 −48.70	266.4 −47.5	COMPACT w. SMALL TAIL + ABS'N.	13,14,15 D 0.2	F420/J2703 (334,78)
AM 0359-661	03 59 30 −66 18.5	0.439 0.168	279.68 −41.43	225.7 −31.9	CLOSE PAIR	23 S 1.9	F083/J1936 (74,105)
AM 0359-525	03 59 38 −52 58.1	1.541 0.167	262.56 −46.67	239.1 −39.6	GROUP of 4(SP.w. OFF-SET NUCLEUS) Fa 762?	12,18 D 6.9	F156/J2661 (110,288)
AM 0359-291	03 59 53 −29 14.0	2.426 0.167	227.22 −48.24	270.2 −48.1	CLOSE GROUP of 4	18 D 2.2	F419/J2576 (62,217)
*AM 0400-513	04 00 04 −51 33.7	1.615 0.167	260.55 −47.00	240.7 −40.4	HI.S.BR. + I/A DBL. + PEC. COMP'N. 201-IG-15	2,8,12,13 S 3.1	F201/J2725 (179,96)
AM 0400-420	04 00 10 −42 04.7	2.028 0.167	246.64 −48.68	252.2 −44.6	ISOLATED IRR. w. STELLAR NUCLEUS 302-G-27 = B 0400-420	16 D 1.1	F250/J2681 (294,332)
*AM 0400-393	04 00 15 −39 36.0	2.115 0.167	242.88 −48.82	255.5 −45.5	I/A DBL. (TAILS) 302-IG-28 = B 0400-395	2 S 0.4	F302/J2668 (115,200)
AM 0400-423	04 00 52 −42 38.9	2.005 0.166	247.47 −48.50	251.4 −44.5	GALAXY w. JET	7 D 0.5	F302/J2668 (112,36)
AM 0400-441 [0401-4409]	04 00 52 −44 12.9	1.945 0.166	249.81 −48.32	249.4 −43.9	CLOSE PAIR (#1)	23 S 0.6	F250/J2681 (283,218)
AM 0401-632	04 01 03 −63 25.4	0.754 0.166	276.13 −42.62	228.2 −33.8	COMPACT	13 D 0.2	F084/J0959 (314,256)
AM 0401-301	04 01 07 −30 14.4	2.397 0.166	228.77 −48.13	268.6 −48.2	ARC of 5 GALAXIES	17 D 4.3	F419/J2576 (49,163)
AM 0401-641	04 01 09 −64 14.1	0.669 0.166	277.11 −42.25	227.5 −33.3	DISTORTED RING/SPIRAL in CLUSTER 83-G-14 = B 0401-642 = PKS 0401-64	2,6,8,16 D 3.1	F083/J1936 (56,215)
AM 0401-361	04 01 13 −36 16.9	2.221 0.165	237.84 −48.65	259.9 −46.8	GALAXY w. I/A COMP'N. 359-IG-14	1 S 0.4	F359/J2709 (166,110)
AM 0401-440 [0401-4412]	04 01 15 −44 09.5	1.946 0.165	249.72 −48.26	249.4 −44.0	CLOSE PAIR (#2)	23 S 0.6	F250/J2681 (280,221)
AM 0401-634	04 01 31 −63 48.3	0.712 0.165	276.57 −42.40	227.8 −33.6	I/A DBL. PECULIAR SPIRALS B 0401-637	2 S 0.6	F083/J1936 (52,238)
*AM 0401-425	04 01 32 −42 50.2	1.996 0.165	247.73 −48.36	251.1 −44.5	LUMPY COMPACT (I/A QUARTET?) Fa 395	4,13 D 0.3	F250/J2681 (279,292)
AM 0401-532	04 01 51 −53 27.1	1.505 0.165	263.09 −46.21	238.4 −39.6	CHAIN of 5	4,17 D 9.9	F156/J2661 (93,261)
AM 0401-414	04 01 58 −41 41.8	2.037 0.165	246.01 −48.38	252.5 −45.1	GALAXY w. 2 APP. COMP'NS. 302-G-29	8 S 0.5	F302/J2668 (100,87)
AM 0402-811	04 02 08 −81 12.1	−4.442 0.164	295.17 −33.02	213.1 −22.0	SB(r) + 4 APP. COMP'NS. 15-G-05	8,24 S 4.5	F014/J2559 (53,98)
*AM 0402-433	04 02 17 −43 30.0	1.969 0.164	248.70 −48.15	250.1 −44.4	SB w. FAINT CIRC. THIN ARMS + I/A NGC 1512 = 250-G-04 [A82]	1,10 S 5.6	F250/J2681 (271,257)
AM 0402-354	04 02 30 −35 47.0	2.234 0.164	237.10 −48.37	260.5 −47.1	CHAIN of 4 E's (2 COMPACTS)	13,13,17 D 6.3	F359/J2709 (152,137)

Identification	R.A. (1950.0) Declination	Precession R.A./Dec.	Gal. Long. Gal. Lat.	S.G.L. S.G.B.	Description Cross-Identification	Coding Size (Min.)	Field/Plate (x, y)
AM 0402-635	04 02 31 −63 55.5	0.694 0.164	276.64 −42.25	227.6 −33.6	I/A DBL. 84-IG-1	2 S 0.2	F084/J0959 (303,230)
AM 0402-440	04 02 32 −44 03.6	1.946 0.164	249.53 −48.04	249.4 −44.2	CLOSE DBL. Fa 396	23 S 0.3	F250/J2681 (268,227)
AM 0402-631	04 02 38 −63 14.8	0.763 0.164	275.81 −42.54	228.3 −34.1	GALAXY w. TAIL	15 D 0.4	F084/J0959 (305,266)
*AM 0402-541	04 02 49 −54 14.4	1.455 0.163	264.11 −45.84	237.4 −39.3	SPIRAL w. 7 COMP'NS. [A82] NGC 1515 = 156-G-36 = Fa 397	8 S 9.9	F156/J2661 (87,219)
AM 0402-631	04 02 51 −63 19.8	0.753 0.163	275.90 −42.48	228.2 −34.0	COMPACT	13 D 0.5	F084/J0959 (303,262)
AM 0402-480 [0403-4806]	04 02 53 −48 05.4	1.775 0.163	255.43 −47.35	244.4 −42.5	PAIR (#1: EARLY-TYPE, RINGED SP.)	23 S 2.2	F201/J2725 (153,282)
AM 0403-413	04 03 04 −41 37.0	2.037 0.163	245.87 −48.18	252.5 −45.3	FAINT CLUSTER	19 D 8.1	F302/J2668 (89,91)
*AM 0403-480 [0402-4805]	04 03 05 −48 06.7	1.773 0.163	255.45 −47.32	244.4 −42.5	PAIR (#2: LATE-TYPE SPIRAL) 201-G-18	23 S 2.2	F201/J2725 (151,281)
AM 0403-435 [0403-4400]	04 03 07 −43 59.2	1.948 0.163	249.40 −47.95	249.4 −44.3	I/A QUARTET in GROUP of 12 (#1)	4,18 D 1.8	F250/J2681 (262,231)
AM 0403-435 [0403-4359]	04 03 10 −43 59.4	1.947 0.163	249.40 −47.94	249.4 −44.3	I/A QUARTET in GROUP of 12 (#2)	4,18 D 1.8	F250/J2681 (262,231)
AM 0403-435 [0403-4359]	04 03 10 −43 59.1	1.947 0.163	249.40 −47.94	249.4 −44.3	I/A QUARTET in GROUP of 12 (#3) Fa 398?	4,18 D 1.8	F250/J2681 (262,231)
*AM 0403-555	04 03 10 −55 54.6	1.350 0.163	266.35 −45.27	235.6 −38.5	RING + I/A DBL. 156-G-37 = B 0403-559	2,6 D 1.3	F156/J2661 (87,129)
AM 0403-702	04 03 10 −70 23.5	−0.200 0.163	284.09 −39.09	221.8 −29.5	SMALL RING GALAXY Lo-2 = 55-G-9	6 D 0.5	F055/J1942 (253,153)
AM 0403-440 [0403-4359]	04 03 13 −44 00.1	1.947 0.163	249.42 −47.93	249.4 −44.3	I/A QUARTET in GROUP of 12 (#4)	4,18 D 1.8	F250/J2681 (261,230)
AM 0403-554	04 03 17 −55 41.5	1.364 0.163	266.05 −45.33	235.8 −38.6	CONTACT DBL.	2 S 0.1	F156/J2661 (86,141)
AM 0403-431 [0403-4326]	04 03 26 −43 17.6	1.974 0.163	248.36 −47.97	250.3 −44.7	LINE of 3 E's (#1)	24 D 3.1	F250/J2681 (260,268)
AM 0403-465	04 03 29 −46 59.6	1.822 0.163	253.81 −47.45	245.7 −43.1	GALAXY w. JET + TAIL	7,15 D 0.8	F201/J2725 (147,341)
AM 0403-431 [0403-4317]	04 03 30 −43 18.5	1.973 0.163	248.38 −47.95	250.3 −44.7	LINE of 3 E's (#2)	24 D 3.1	F250/J2681 (259,267)
*AM 0403-604	04 03 35 −60 49.3	0.982 0.162	272.74 −43.44	230.5 −35.6	M51-TYPE 117-IG-20 = B 0403-608	9 S 0.4	F118/J3762 (326,129)
AM 0403-432 [0403-4318]	04 03 36 −43 26.4	1.968 0.162	248.57 −47.92	250.1 −44.6	LINE of 3 E's (#3)	24 D 3.1	F250/J2681 (258,260)
AM 0404-720	04 04 17 −72 04.1	−0.540 0.162	285.86 −38.14	220.3 −28.5	SMALL I/A DBL.	2 S 0.2	F055/J1942 (242,64)
AM 0404-271	04 04 25 −27 17.4	2.470 0.161	224.66 −46.89	272.8 −49.4	V. SMALL RING / ONE-ARMED SPIRAL	6,10 D 0.3	F483/J3814 (150,57)
AM 0404-670	04 04 25 −67 01.6	0.314 0.161	280.20 −40.66	224.6 −31.8	DISTURBED SINGLE GAL. w. COMP'N. 84-IG-2	8,16 S 2.2	F055/J1942 (259,334)
AM 0404-624	04 04 42 −62 49.5	0.793 0.161	275.16 −42.51	228.5 −34.5	DWARF	20 D 0.4	F084/J0959 (294,290)
AM 0404-275	04 04 48 −27 57.6	2.452 0.161	225.65 −46.95	271.8 −49.3	I/A DBL. MCG-5-10-10/11 = 420-IG-1/2	2 S 0.6	F483/J3814 (145,21)
AM 0404-684 [0405-6840]	04 04 49 −68 40.0	0.075 0.161	282.06 −39.84	223.2 −30.8	CLOSE PAIR (#1)	23 S 2.4	F055/J1942 (251,246)
AM 0404-385	04 04 50 −38 55.1	2.128 0.161	241.81 −47.95	255.9 −46.6	COMPACT	13 D 0.6	F302/J2668 (66,235)
AM 0404-524	04 04 51 −52 48.3	1.530 0.161	262.01 −45.96	238.8 −40.4	HI.S.BR. NUCLEUS (OFF-SET ENVEL.) NGC1522 = 156-G-38 = MC0404-528 = Fa301	13,14 D 1.1	F201/J2725 (140,28)
AM 0405-684 [0404-6840]	04 05 08 −68 40.1	0.073 0.160	282.05 −39.81	223.1 −30.8	CLOSE PAIR (#2: SPIRAL)	23 S 2.4	F055/J1942 (250,246)
*AM 0405-371	04 05 23 −37 10.9	2.184 0.160	239.23 −47.84	258.3 −47.3	LINE of 3 (CLOSE)	22,24 D 0.5	F303/J4742 (336,327)
AM 0405-401	04 05 27 −40 18.5	2.078 0.160	243.88 −47.80	254.0 −46.2	GALAXY w. I/A COMP'N. w. TAIL Ag-16 = 302-IG-30	1,15 S 0.4	F302/J2668 (62,161)
AM 0405-654	04 05 40 −65 43.9	0.471 0.160	278.61 −41.14	225.7 −32.8	CLUSTER of GALAXIES	19 D 3.6	F084/J0959 (277,135)
AM 0405-433 [0406-4334]	04 05 55 −43 33.0	1.957 0.159	248.67 −47.49	249.7 −45.0	CLOSE GROUP of 5 (SOME I/A) (#1)	4,18 D 3.0	F250/J2681 (235,255)
AM 0405-552	04 05 56 −55 27.4	1.368 0.159	265.57 −45.04	235.8 −39.0	IRR. DWARF w. RESOLUTION IC 2032 = 156-G-42	16,20 D 1.3	F156/J2661 (65,152)
AM 0406-433 [0405-4333]	04 06 02 −43 33.8	1.956 0.159	248.69 −47.47	249.7 −45.0	CLOSE GROUP of 5 (SOME I/A) (#2) Fa 399?	4,18 D 3.0	F250/J2681 (234,254)
AM 0406-433 [0406-4333]	04 06 05 −43 34.9	1.955 0.159	248.72 −47.46	249.6 −45.0	CLOSE GROUP of 5 (SOME I/A) (#3)	4,18 D 3.0	F250/J2681 (234,253)

Identification	R.A. (1950.0) Declination	Precession R.A./Dec.	Gal. Long. Gal. Lat.	S.G.L. S.G.B.	Description Cross-Identification	Coding Size (Min.)	Field/Plate (x, y)
AM 0406-433 [0406-4334]	04 06 05 −43 34.4	1.956 0.159	248.70 −47.46	249.6 −45.0	CLOSE GROUP of 5 (SOME I/A) (#4)	4,18 D 3.0	F250/J2681 (234,253)
AM 0406-433 [0406-4334]	04 06 09 −43 34.4	1.956 0.159	248.70 −47.45	249.6 −45.0	CLOSE GROUP of 5 (SOME I/A) (#5) Fa 400?	4,18 D 3.0	F250/J2681 (233,253)
AM 0406-750	04 06 20 −75 07.0	−1.348 0.159	289.00 −36.38	217.7 −26.5	3 E's in ENVELOPE + COMP'NS. 32-IG-9	3,8 D 0.4	F032/J1937 (240,171)
AM 0406-310	04 06 30 −31 07.9	2.363 0.159	230.33 −47.11	266.9 −49.1	PAIR of E's (PKS 0406-311) near AM 0358-312	19,23 S 1.0	F420/J2703 (254,118)
AM 0406-305	04 06 35 −30 56.8	2.368 0.159	230.06 −47.07	267.2 −49.2	PAIR of E's	23 S 0.6	F420/J2703 (253,128)
AM 0406-431	04 06 40 −43 14.3	1.967 0.159	248.20 −47.38	250.0 −45.2	CLOSE PAIR of SPIRALS	23 S 1.3	F250/J2681 (228,272)
AM 0406-624	04 06 40 −62 45.1	0.790 0.159	274.94 −42.33	228.4 −34.7	GROUP of 3	24 D 4.9	F083/J1936 (16,292)
AM 0407-463 [0407-4637]	04 07 20 −46 36.2	1.827 0.158	253.09 −46.86	245.7 −43.8	CLOSE PAIR (#1)	23 S 1.3	F250/J2681 (220,91)
AM 0407-463 [0407-4636]	04 07 20 −46 37.4	1.827 0.158	253.11 −46.86	245.7 −43.8	CLOSE PAIR (#2)	23 S 1.3	F250/J2681 (220,90)
AM 0407-215	04 07 22 −21 54.1	2.599 0.158	217.41 −44.88	281.1 −50.5	SPIRAL w. APP. COMP'NS MCG-4-10-17 = 550-G-14	8 S 3.4	F483/J3814 (113,346)
*AM 0407-453	04 07 27 −45 39.6	1.867 0.158	251.71 −46.98	246.9 −44.3	LARGE COMPACT IC 2035 = 250-G-07	13 D 1.1	F250/J2681 (219,142)
*AM 0407-485	04 07 34 −48 51.1	1.724 0.157	256.30 −46.45	243.0 −42.8	EDGE-ON w. DIAMETRIC COMP'NS. 201-G-22	8 S 4.9	F201/J2725 (112,240)
*AM 0407-612	04 07 34 −61 24.2	0.911 0.157	273.21 −42.77	229.6 −35.7	V.DISTURBED SPIRAL + APP.COMP'N. 118-IG-4 = B 0407-613	8,12 S 0.8	F118/J3762 (297,100)
AM 0407-605	04 07 43 −60 57.5	0.950 0.157	272.64 −42.93	230.0 −36.0	E w. I/A COMP'NS. 118-IG-6	1,8 S 1.0	F118/J3762 (298,124)
AM 0407-553	04 07 47 −55 35.6	1.352 0.157	265.64 −44.75	235.4 −39.2	COMPACT	13 D 0.2	F156/J2661 (52,144)
AM 0407-560	04 07 52 −56 07.8	1.316 0.157	266.35 −44.58	234.9 −38.9	CLOSE PAIR IC 2038/39 = 157-G-1/2 = B 0407-561	23 S 1.3	F157/J1893 (300,115)
AM 0408-672	04 08 01 −67 28.2	0.229 0.157	280.51 −40.14	224.0 −31.8	VERY SMALL TIGHT GROUP 84-IG-5	18 D 0.4	F084/J0959 (259,42)
AM 0408-312	04 08 05 −31 23.4	2.353 0.157	230.77 −46.81	266.4 −49.4	DWARF 420-G-6 = UKS 0408-313	20 D 1.3	F420/J2703 (235,104)
AM 0408-625	04 08 07 −62 55.9	0.764 0.157	275.07 −42.11	228.1 −34.8	GALAXY w. STRONG ABS'N. in GROUP NGC 1534 = 84-G-6	14,18 D 1.0	F118/J3762 (288,18)
AM 0408-842	04 08 14 −84 29.9	−9.182 0.157	298.11 −30.85	210.6 −19.8	GALAXY w. LO.S.BR. COMP'N. 4-G-11	8 S 4.3	F003/J2565 (50,176)
AM 0408-373	04 08 16 −37 37.7	2.163 0.156	239.90 −47.27	257.4 −47.7	SMALL GALAXY w. TAILS	15 D 0.4	F303/J4742 (304,304)
AM 0408-814	04 08 41 −81 49.0	−5.143 0.156	295.58 −32.46	212.5 −21.7	SMALL I/A DBL.	2 S 0.2	F014/J2559 (50,63)
*AM 0408-561	04 08 48 −56 14.9	1.304 0.156	266.45 −44.42	234.6 −38.9	SB(r) w. I/A DBL. COMP'N. NGC 1533 = 157-G-3	1,2,12 D 2.2	F156/J2661 (46,109)
AM 0408-302	04 08 53 −30 27.9	2.378 0.156	229.48 −46.51	267.8 −49.8	SMALL I/A TRIPLET	3 D 1.0	F420/J2703 (227,154)
AM 0408-740	04 08 57 −74 02.2	−1.060 0.156	287.74 −36.82	218.4 −27.3	SPIRAL w. PECULIAR ARMS	10 D 0.6	F032/J1937 (235,229)
AM 0408-430 [0409-4303]	04 08 59 −43 05.7	1.967 0.156	247.93 −46.98	249.9 −45.7	CLOSE TRIPLE (#1)	24 S 4.0	F250/J2681 (206,279)
AM 0409-355	04 09 05 −35 57.4	2.215 0.155	237.46 −47.04	259.7 −48.4	IRR. w. JET 359-G-24	7,16 D 0.9	F360/J2771 (341,124)
AM 0409-430 [0408-4305]	04 09 07 −43 03.9	1.968 0.155	247.89 −46.95	250.0 −45.7	CLOSE TRIPLE (#2)	24 S 4.0	F250/J2681 (204,281)
AM 0409-474 [0409-4741]	04 09 14 −47 40.7	1.774 0.155	254.56 −46.38	244.2 −43.6	PECULIAR SPIRAL in ROW of 3 (#1)	12,24 D 4.5	F250/J2681 (202,33)
AM 0409-594	04 09 17 −59 46.8	1.042 0.155	271.04 −43.18	231.0 −36.8	I/A DBL. in WIDE GROUP	2,18 S 0.9	F117/J2635 (36,184)
AM 0409-430 [0409-4305]	04 09 18 −43 05.1	1.966 0.155	247.91 −46.92	249.9 −45.7	CLOSE TRIPLE (#3) 250-G-10	24 S 4.0	F250/J2681 (203,280)
AM 0409-415	04 09 23 −41 57.6	2.009 0.155	246.27 −46.98	251.4 −46.2	I/A DBL.	2 S 1.2	F250/J2681 (202,340)
AM 0409-474 [0409-4740]	04 09 26 −47 41.5	1.773 0.155	254.57 −46.34	244.2 −43.6	PECULIAR SPIRAL in ROW of 3 (#2)	12,24 D 4.5	F250/J2681 (200,33)
AM 0409-630	04 09 33 −63 08.9	0.735 0.155	275.25 −41.87	227.8 −34.7	I/A TRIPLET B 0409-631	3 D 1.1	F084/J0959 (264,274)
AM 0409-474 [0409-4741]	04 09 36 −47 41.8	1.772 0.155	254.57 −46.31	244.1 −43.6	PECULIAR SPIRAL in ROW of 3 (#3)	12,24 D 4.5	F250/J2681 (198,32)
AM 0409-424	04 09 37 −42 47.3	1.977 0.155	247.47 −46.88	250.3 −45.9	COMPACT w. SLIGHT HALO	13 D 0.1	F303/J4742 (281,28)

Identification	R.A. (1950.0) Declination	Precession R.A./Dec.	Gal. Long. Gal. Lat.	S.G.L. S.G.B.	Description Cross-Identification	Coding Size (Min.)	Field/Plate (x, y)
AM 0409-862	04 09 39 −86 28.4	−0.269 0.155	299.89 −29.61	209.1 −18.3	CONTACT I/A DBL.	2 S 0.2	F005/J2958 (315,43)
AM 0409-584	04 09 44 −58 47.4	1.118 0.155	269.74 −43.47	232.0 −37.5	I/A DBL. 118-G-7	2 S 0.3	F118/J3762 (292,241)
AM 0409-751	04 09 49 −75 16.8	−1.436 0.154	289.02 −36.11	217.4 −26.5	LINE of 3 (PKS 0409-75) 32-IG-10 = DN 0409-751	24 D 1.7	F032/J1937 (228,163)
*AM 0409-460	04 09 51 −46 08.1	1.840 0.154	252.32 −46.50	246.0 −44.4	HI.S.BR. GALAXY w. DIAM. COMP'NS. [A82]	8 D 0.6	F250/J2681 (197,116)
AM 0409-475	04 09 59 −47 50.2	1.764 0.154	254.75 −46.23	243.9 −43.6	SMALL I/A DBL.	2 S 0.3	F250/J2681 (195,25)
AM 0409-563	04 09 59 −56 36.8	1.275 0.154	266.87 −44.15	234.1 −38.8	IRR. or DISTURBED SPIRAL w. DUST NGC 1536 = 157-IG-5 = B0409-566	14,16 D 1.2	F157/J1893 (282,90)
*AM 0410-325	04 10 08 −32 59.7	2.303 0.154	233.17 −46.58	263.9 −49.4	DISTURBED SP. w. COMP'N. on BENT NGC 1531/32=39-G-26/27 [A82] ARM	1,8,10 S 1.3	F420/J2703 (211,18)
AM 0410-330	04 10 10 −33 00.4	2.303 0.154	233.19 −46.58	263.8 −49.4	SP. w. LINEAR ARM + COMP'NS.(I/A) NGC 1532 = 359-G-27	1,8,10 S 2.0	F359/J2709 (64,284)
AM 0410-560	04 10 13 −56 09.8	1.304 0.154	266.26 −44.26	234.6 −39.1	DWARF	20 D 0.2	F157/J1893 (282,114)
AM 0410-291 [0410-2911]	04 10 27 −29 11.1	2.410 0.154	227.73 −45.95	269.6 −50.4	WIDE I/A DBL. (#1: M51-TYPE) 420-IG-10 = MC 0410-291	2,9 S 2.2	F420/J2703 (209,223)
AM 0410-291 [0410-2911]	04 10 39 −29 11.7	2.410 0.153	227.76 −45.91	269.6 −50.4	WIDE I/A DBL. (#2: MATERIAL)	2 S 2.2	F420/J2703 (207,222)
AM 0410-422 [0410-4221]	04 10 39 −42 22.4	1.990 0.153	246.85 −46.72	250.7 −46.3	I/A DBL. (#1)	2 S 1.7	F250/J2681 (189,318)
AM 0410-511	04 10 40 −51 10.9	1.598 0.153	259.44 −45.48	239.9 −42.0	COMPACT w. INTEGRAL COMP'N.	8,13 S 0.7	F201/J2725 (89,114)
AM 0410-481	04 10 43 −48 17.7	1.741 0.153	255.38 −46.03	243.3 −43.5	I/A DBL. 20-IG-24	2 S 0.4	F201/J2725 (83,269)
AM 0410-421 [0410-4217]	04 10 45 −42 17.8	1.993 0.153	246.74 −46.71	250.8 −46.3	I/A DBL. (#1)	2 S 0.8	F250/J2681 (189,322)
AM 0410-422 [0410-4222]	04 10 45 −42 21.3	1.991 0.153	246.82 −46.70	250.7 −46.3	I/A DBL. (#2)	2 S 1.7	F250/J2681 (188,319)
AM 0410-421 [0410-4217]	04 10 49 −42 17.4	1.993 0.153	246.73 −46.70	250.8 −46.3	I/A DBL. (#2)	2 S 0.8	F250/J2681 (188,323)
AM 0410-502	04 10 49 −50 23.0	1.639 0.153	258.32 −45.63	240.8 −42.4	SPIRAL w. I/A COMP'N. 201-IG-25	1 S 0.6	F201/J2725 (87,157)
AM 0410-560	04 10 59 −56 00.1	1.312 0.153	266.00 −44.20	234.7 −39.3	DWARF	20 D 0.3	F157/J1893 (277,123)
AM 0411-591	04 11 02 −59 17.4	1.073 0.153	270.30 −43.15	231.3 −37.3	PAIR (ONE w. JET)	7,23 S 2.0	F118/J3762 (282,214)
AM 0411-324	04 11 04 −32 40.6	2.311 0.153	232.74 −46.35	264.3 −49.7	HI.S.BR. IRR. + PECULIAR ABS'N. IC 2040 = 359-G-30	13,14,16 D 1.3	F360/J2771 (326,301)
AM 0411-584 [0411-5752]	04 11 08 −58 41.4	1.120 0.153	269.52 −43.34	231.9 −37.7	IRR. COMP'N. to SB0 IC 2049 = 118-IG-9	16,20 D 0.9	F117/J2635 (18,241)
*AM 0411-680	04 11 14 −68 03.1	0.124 0.153	281.00 −39.61	223.2 −31.6	ONE-ARMED SPIRAL + 2 APP. COMP'NS	8,10 S 2.2	F055/J1942 (221,280)
AM 0411-343	04 11 16 −34 33.3	2.255 0.153	235.46 −46.50	261.5 −49.2	PEC. GAL. (SHARP EDGE + CONDENS.) 359-G-31 = PKS 0411-34	8,16 D 1.1	F360/J2771 (320,200)
AM 0411-541	04 11 23 −54 19.1	1.418 0.152	263.71 −44.62	236.4 −40.3	GROUP of 4 Fa 764	18 D 5.2	F156/J2661 (20,211)
AM 0411-422	04 11 30 −42 27.0	1.985 0.152	246.95 −46.56	250.5 −46.4	SPIRAL w. CLOSE I/A COMP'N. 303-G-2	1 S 0.4	F250/J2681 (181,314)
AM 0411-240	04 11 36 −24 09.3	2.540 0.152	220.84 −44.58	277.4 −51.3	I/A TRIPLE	3 D 1.3	F483/J3814 (62,224)
AM 0411-314	04 11 45 −31 46.2	2.336 0.152	231.47 −46.09	265.6 −50.1	LARGE E/S0 w. LINEAR STREAMERS NGC 1537 = MCG-5-11-5 = 420-G-12	15 D 1.3	F420/J2703 (193,84)
*AM 0411-575 [0411-5841]	04 11 45 −57 52.0	1.179 0.152	268.41 −43.53	232.7 −38.3	SB0 w. FAINT OUTER RING NGC 1543 = 118-G-10	6,8,12 D 2.5	F117/J2635 (9,285)
AM 0411-372	04 11 47 −37 24.9	2.163 0.152	239.62 −46.57	257.3 −48.4	CONDENS. + IRR. MATR'L. ONE SIDE	16 D 0.4	F360/J2771 (309,47)
*AM 0411-254	04 11 53 −25 43.7	2.500 0.152	223.01 −44.92	274.9 −51.2	ONE-ARMED SPIRAL + 2 COMP'NS. 483-G-14	8,10 D 1.3	F484/J5463 (326,138)
AM 0411-320	04 11 53 −32 07.4	2.326 0.152	231.98 −46.11	265.0 −50.0	LARGE COMPACT MCG-5-11-6 = 420-G-13	13 D 0.6	F420/J2703 (191,65)
AM 0411-822	04 11 57 −82 25.0	−5.868 0.152	296.08 −32.01	212.0 −21.4	RING around a STAR 15-G-8 = Fa 236 = STAR	12,21 D 0.9	F004/J3821 (243,313)
*AM 0412-381	04 12 12 −38 13.3	2.135 0.151	240.80 −46.51	256.1 −48.2	"N" GAL. (STRONG NUCLEUS + HALO) 303-G-5 + 5 CLOSE COMP'NS.	8,20,22 S 0.9	F303/J4742 (262,274)
AM 0412-540	04 12 16 −54 02.6	1.431 0.151	263.29 −44.57	236.6 −40.6	DWARF	20 D 0.4	F157/J1893 (272,229)
AM 0412-612 [0412-6122]	04 12 26 −61 22.1	0.890 0.151	272.86 −42.25	229.2 −36.1	I/A DBL. (#1) 118-IG-11	2,8 S 0.5	F118/J3762 (266,103)

Identification	R.A. (1950.0) Declination	Precession R.A./Dec.	Gal. Long. Gal. Lat.	S.G.L. S.G.B.	Description Cross-Identification	Coding Size (Min.)	Field/Plate (x, y)
AM 0412-390	04 12 27 −39 06.9	2.104 0.151	242.09 −46.47	254.9 −47.9	GROUP of 3	24 D 2.5	F303/J4742 (258,226)
AM 0412-422	04 12 29 −42 20.7	1.987 0.151	246.78 −46.39	250.5 −46.6	CLOSE PAIR	23 S 0.4	F250/J2681 (171,320)
AM 0412-474	04 12 29 −47 45.7	1.761 0.151	254.55 −45.83	243.7 −44.0	SMALL M51-TYPE	9 D 0.4	F250/J2681 (172,29)
AM 0412-612 [0412-6121]	04 12 29 −61 21.6	0.891 0.151	272.85 −42.24	229.2 −36.2	I/A DBL. (HI.S.BR.) + COMP'N.(#2) 118-IG-11	2,8,13 S 0.5	F118/J3762 (266,104)
AM 0412-660	04 12 37 −66 08.1	0.379 0.151	278.68 −40.33	224.8 −33.0	COMPACT ("SQUARE")	13 D 0.4	F084/J0959 (238,115)
*AM 0412-525	04 12 42 −52 56.3	1.494 0.151	261.76 −44.78	237.8 −41.3	CHAIN of 9	17 D 15.7	F202/J0941 (317,14)
AM 0412-613	04 12 58 −61 39.0	0.862 0.150	273.19 −42.08	228.9 −36.0	GROUP of 5 118-IG-13	18 D 2.2	F118/J3762 (262,89)
AM 0413-521	04 13 04 −52 15.8	1.531 0.150	260.82 −44.88	238.5 −41.7	I/A DBL.	2 S 0.2	F202/J0941 (316,51)
AM 0413-554	04 13 04 −55 49.8	1.315 0.150	265.65 −43.97	234.6 −39.6	NUCLEATED DWARF	20 D 0.3	F157/J1893 (261,133)
AM 0413-283	04 13 09 −28 36.6	2.422 0.150	227.08 −45.27	270.3 −51.1	I/A DBL. in LINE of 3 (HI.S.BR.) NGC 1540 = 420-IG-14	2,24 S 0.4	F420/J2703 (178,254)
AM 0413-244	04 13 25 −24 49.3	2.521 0.150	221.90 −44.36	276.3 −51.7	GROUP of 8	18 D 3.6	F483/J3814 (40,188)
AM 0413-561	04 13 32 −56 12.2	1.289 0.150	266.12 −43.80	234.2 −39.5	GAL. w. PEC. DUST (V.LARGE BULGE) NGC 1546 = 157-G-12	14 D 1.8	F157/J1893 (257,113)
AM 0413-590	04 13 33 −59 05.4	1.078 0.150	269.90 −42.91	231.3 −37.7	3 in a LINE incl. PECULIAR RING 118-IG-14	6,24 D 2.8	F118/J3762 (265,226)
AM 0413-380	04 13 38 −38 08.6	2.135 0.149	240.69 −46.22	256.1 −48.5	RING SPIRAL + E + COMP'NS.	6,8,23 S 1.1	F303/J4742 (247,278)
AM 0413-352	04 13 40 −35 28.0	2.222 0.149	236.83 −46.08	259.9 −49.4	PEC. SP. w. CONDENSATION in ARM 360-G-2	10,12 D 1.6	F360/J2771 (292,152)
AM 0413-424 [0413-4249]	04 13 42 −42 49.0	1.966 0.149	247.44 −46.13	249.8 −46.6	CLOSE GROUP of 3 (#1)	24 D 1.5	F250/J2681 (159,294)
*AM 0413-454	04 13 44 −45 43.6	1.847 0.149	251.62 −45.88	246.1 −45.2	GALAXY w. TAIL 250-IG-15	15 D 0.7	F250/J2681 (160,138)
AM 0413-424 [0413-4249]	04 13 47 −42 49.9	1.965 0.149	247.46 −46.12	249.7 −46.6	CLOSE GROUP of 3 (#2)	24 D 1.5	F250/J2681 (159,294)
AM 0413-424 [0413-4249]	04 13 47 −42 49.4	1.965 0.149	247.45 −46.12	249.7 −46.6	CLOSE GROUP of 3 (#3)	24 D 1.5	F250/J2681 (159,294)
AM 0413-560	04 13 48 −56 07.0	1.293 0.149	265.99 −43.79	234.2 −39.5	DWARF?	20 D 0.3	F157/J1893 (255,118)
AM 0413-404	04 13 53 −40 47.1	2.042 0.149	244.51 −46.18	252.4 −47.5	3 in a LINE incl. 303-G-6	24 D 2.2	F303/J4742 (241,137)
AM 0414-510	04 14 00 −51 04.2	1.593 0.149	259.13 −45.00	239.7 −42.5	GALAXY w. JET (LOOSE CLUSTER) 201-IG-26	7 D 0.6	F201/J2725 (61,119)
AM 0414-743 [0416-7435]	04 14 09 −74 39.2	−1.285 0.149	288.17 −36.20	217.7 −27.1	GROUP of 4 (#1: IRR.)	18 S 8.3	F032/J1937 (214,197)
*AM 0414-524	04 14 13 −52 47.7	1.497 0.149	261.49 −44.59	237.7 −41.5	CHAIN of 8 157-IG-14	17 D 8.5	F157/J1893 (259,296)
AM 0414-331	04 14 15 −33 11.1	2.291 0.149	233.59 −45.75	263.2 −50.2	DWARF w. RESOLUTION 360-G-3	20 D 0.9	F360/J2771 (289,275)
*AM 0414-524	04 14 27 −52 43.6	1.500 0.148	261.39 −44.57	237.8 −41.6	CONTACT E's in CLUSTER 157-IG-14	2,17,23 S 0.3	F201/J2725 (62,30)
AM 0414-744 [0414-7439]	04 14 35 −74 40.2	−1.294 0.148	288.17 −36.16	217.6 −27.2	GROUP of 4 (#2: S0)	18 S 8.3	F032/J1937 (213,196)
AM 0414-554 [0415-5553]	04 14 39 −55 42.8	1.317 0.148	265.41 −43.79	234.6 −39.9	CLOSE PAIR (#1) SS-4a NGC 1549 = 157-G-16 = MC 0414-557	23 D 1.8	F157/J1893 (250,140)
AM 0414-491	04 14 44 −49 18.7	1.680 0.148	256.65 −45.21	241.6 −43.5	CHAIN of 6 (3 RING GALAXIES) 201-G-28/29	6,17 D 30.5	F202/J0941 (311,210)
*AM 0414-492	04 14 46 −49 20.3	1.679 0.148	256.69 −45.20	241.6 −43.5	CLOSE PAIR (2 COMP'NS.) 201-IG-29	8,23 S 2.3	F201/J2725 (50,212)
AM 0414-574	04 14 57 −57 46.3	1.172 0.148	268.11 −43.15	232.4 −38.7	SB w. PECULIAR ARMS + COMP'NS. 118-G-15	8,10 D 1.1	F118/J3762 (258,297)
*AM 0415-475	04 15 03 −47 58.5	1.743 0.148	254.76 −45.37	243.2 −44.3	BROKEN RING w. OFF-SET NUCLEUS 202-G-1 (+DWARF)	6,8,15,20 D 1.6	F201/J2725 (44,285)
AM 0415-555 [0414-5542]	04 15 04 −55 53.9	1.303 0.148	265.63 −43.68	234.3 −39.8	CLOSE PAIR (#2) SS-4b NGC 1553 = 157-G-17 = MC 0415-559	23 D 2.2	F157/J1893 (246,130)
*AM 0415-390	04 15 05 −39 04.2	2.100 0.148	242.04 −45.96	254.6 −48.4	GALAXY w. TAILS, DEBRIS + COMP'N. 303-IG-7 = B 0415-390	8,15 D 0.5	F303/J4742 (231,229)
AM 0415-681	04 15 22 −68 15.4	0.066 0.147	281.01 −39.17	222.8 −31.8	GALAXY w. APP. COMP'N. 55-G-17	8 S 1.3	F055/J1942 (200,270)
AM 0415-743 [0414-7440]	04 15 30 −74 38.9	−1.296 0.147	288.11 −36.12	217.6 −27.2	GROUP of 4 (#3: ONE-ARMED SPIRAL)	10,18 S 8.3	F032/J1937 (209,198)

Identification	R.A. (1950.0) Declination	Precession R.A./Dec.	Gal. Long. Gal. Lat.	S.G.L. S.G.B.	Description Cross-Identification	Coding Size (Min.)	Field/Plate (x, y)
AM 0415-260	04 15 32 −26 04.0	2.486 0.147	223.75 −44.20	274.2 −52.0	GALAXY w. TAILS 484-G-3	15 D 0.9	F484/J5463 (281,121)
AM 0415-301	04 15 41 −30 11.0	2.375 0.147	229.42 −45.01	267.7 −51.3	I/A DBL. in LINE of 3	2,24 D 1.3	F420/J2703 (148,170)
AM 0415-502	04 15 58 −50 27.4	1.619 0.146	258.20 −44.81	240.1 −43.1	ALL LOOPS and TAILS	15 D 0.9	F201/J2725 (43,151)
AM 0416-420	04 16 10 −42 04.7	1.988 0.146	246.36 −45.70	250.4 −47.3	CLOSE GROUP of 3	24 D 2.2	F250/J2681 (134,334)
AM 0416-743 [0415-7438]	04 16 19 −74 35.3	−1.286 0.146	288.01 −36.11	217.6 −27.3	GROUP of 4 (#4: SPIRAL) 32-G-11	18 S 8.3	F032/J1937 (207,201)
AM 0416-270 [0416-2707]	04 16 22 −27 04.6	2.459 0.146	225.18 −44.25	272.6 −52.0	PAIR (#1)	23 S 2.7	F420/J2703 (139,336)
AM 0416-690 [0416-6902]	04 16 24 −69 03.4	−0.067 0.146	281.88 −38.73	222.0 −31.3	I/A SPIRALS (#1) B 0416-690	2 S 1.1	F055/J1942 (195,227)
AM 0416-424	04 16 25 −42 49.9	1.959 0.146	247.43 −45.64	249.4 −47.0	DISTURBED SINGLE	16 D 0.3	F303/J4742 (214,27)
*AM 0416-501	04 16 25 −50 16.9	1.626 0.146	257.94 −44.77	240.3 −43.2	DISTURBED GAL. w. CONDENS. at END NGC 1556 = 202-G-4	16 D 1.3	F201/J2725 (38,160)
AM 0416-270 [0416-2704]	04 16 28 −27 07.5	2.457 0.146	225.25 −44.24	272.5 −52.0	PAIR (#2)	23 S 2.7	F420/J2703 (138,334)
AM 0416-690 [0416-6903]	04 16 30 −69 02.6	−0.066 0.146	281.86 −38.73	222.0 −31.3	I/A SPIRALS (#2)	2 S 1.1	F055/J1942 (194,228)
*AM 0416-502	04 16 50 −50 29.0	1.615 0.145	258.20 −44.67	240.0 −43.2	I/A PAIR of COMPACTS 202-IG-6	2,13 S 0.3	F201/J2725 (36,149)
AM 0416-560	04 16 51 −56 02.9	1.286 0.145	265.74 −43.40	234.0 −39.9	EDGE-ON w. APP. COMP'NS. IC 2058 = 157-G-18	8 S 1.4	F157/J1893 (233,122)
AM 0416-555	04 16 59 −55 53.0	1.297 0.145	265.51 −43.43	234.1 −40.0	FAINT DWARF	20 D 0.4	F157/J1893 (232,131)
AM 0417-625	04 17 02 −62 54.5	0.720 0.145	274.52 −41.18	227.3 −35.5	LARGE IRR. NGC 1559 = 84-G-10	16 D 4.0	F084/J0959 (219,289)
AM 0417-435 [0417-4400]	04 17 07 −43 59.6	1.911 0.145	249.07 −45.44	247.8 −46.6	GROUP of 5 (#1)	18 D 2.5	F250/J2681 (127,231)
AM 0417-421	04 17 12 −42 19.7	1.976 0.145	246.70 −45.52	250.0 −47.4	CONTACT DBL. in CLUSTER	2,19 S 0.6	F250/J2681 (125,320)
AM 0417-440 [0417-4359]	04 17 12 −44 00.5	1.910 0.145	249.09 −45.42	247.8 −46.6	GROUP of 5 (#2)	18 D 2.5	F250/J2681 (126,230)
AM 0417-463	04 17 12 −46 37.3	1.799 0.145	252.79 −45.18	244.5 −45.3	CLOSE PAIR	23 S 1.0	F250/J2681 (129,90)
AM 0417-442 [0417-4421]	04 17 13 −44 20.1	1.897 0.145	249.55 −45.40	247.4 −46.4	CLOSE PAIR (#1)	23 S 3.4	F250/J2681 (126,212)
AM 0417-435 [0417-4359]	04 17 18 −43 58.8	1.911 0.145	249.05 −45.41	247.8 −46.6	GROUP of 5 (#3)	18 D 2.5	F250/J2681 (125,231)
AM 0417-435 [0417-4400]	04 17 18 −43 59.5	1.911 0.145	249.06 −45.41	247.8 −46.6	GROUP of 5 (#4)	18 D 2.5	F250/J2681 (125,231)
AM 0417-265 [0417-2651]	04 17 19 −26 54.9	2.462 0.145	225.03 −44.01	272.8 −52.3	CLOSE PAIR of SPIRALS (#1) MCG-4-11-7 = 484-G-5	23 S 3.6	F484/J5463 (259,76)
AM 0417-440 [0417-4358]	04 17 23 −44 00.8	1.909 0.145	249.09 −45.39	247.8 −46.6	GROUP of 5 (#5)	18 D 2.5	F250/J2681 (124,230)
AM 0417-265 [0417-2654]	04 17 24 −26 51.1	2.463 0.145	224.95 −43.98	272.9 −52.3	CLOSE PAIR of SPIRALS (#2) MCG-4-11-8 = 484-G-6	23 S 3.6	F484/J5463 (258,79)
AM 0417-442 [0417-4420]	04 17 28 −44 21.4	1.895 0.144	249.58 −45.35	247.3 −46.5	CLOSE PAIR (#2)	23 S 3.4	F250/J2681 (124,211)
AM 0417-662	04 17 34 −66 25.8	0.312 0.144	278.75 −39.76	224.1 −33.2	I/A DBL. (SPIRAL + GALAXY)	2 S 1.5	F084/J0959 (211,100)
*AM 0417-754	04 17 37 −75 47.9	−1.688 0.144	289.25 −35.42	216.6 −26.5	M51-TYPE	9 S 0.4	F032/J1937 (200,136)
AM 0417-571	04 17 48 −57 14.2	1.200 0.144	267.25 −42.95	232.7 −39.3	SMALL M51-TYPE	9 D 0.4	F118/J3762 (239,327)
*AM 0417-391	04 17 55 −39 17.5	2.086 0.144	242.37 −45.41	254.0 −48.8	HI.S.BR. RING (NO NUCLEUS) 303-IG-11 = B 0417-392 (5 COMP'NS	6,8 D 0.5	F303/J4742 (201,217)
AM 0417-532	04 17 56 −53 28.6	1.445 0.144	262.24 −43.89	236.6 −41.6	I/A DBL.	2 S 0.5	F157/J1893 (228,260)
*AM 0418-582	04 18 00 −58 22.5	1.115 0.144	268.73 −42.59	231.5 −38.6	E w. PEC. ABSORPTION (JET?) 118-G-19	7,14 D 0.3	F118/J3762 (236,266)
AM 0418-623 [0418-6231]	04 18 01 −62 31.1	0.755 0.144	273.98 −41.22	227.6 −35.9	3 I/A DBLS. in SMALL GROUP (#1)	2,18 S 1.2	F084/J0959 (213,310)
AM 0418-665	04 18 02 −66 52.5	0.249 0.144	279.25 −39.53	223.7 −32.9	GALAXY w. I/A + APP. COMP'N.	2,8 S 0.6	F055/J1942 (188,345)
AM 0418-623 [0418-6231]	04 18 12 −62 32.6	0.751 0.144	274.00 −41.19	227.5 −35.9	3 I/A DBLS. in SMALL GROUP (#2)	2,18 S 0.4	F084/J0959 (212,309)
*AM 0418-364	04 18 20 −36 48.4	2.170 0.143	238.84 −45.24	257.5 −49.8	DWARF w. 2 BRIGHT KNOTS 360-G-7	20 D 1.1	F360/J2771 (240,82)

Identification	R.A. (1950.0) Declination	Precession R.A./Dec.	Gal. Long. Gal. Lat.	S.G.L. S.G.B.	Description Cross-Identification	Coding Size (Min.)	Field/Plate (x, y)
AM 0418-625	04 18 20 −62 58.2	0.708 0.143	274.52 −41.02	227.1 −35.6	DWARF	20 D 0.3	F084/J0959 (211,286)
AM 0418-623 [0418-6232]	04 18 25 −62 31.5	0.752 0.143	273.97 −41.17	227.5 −35.9	3 I/A DBLS. in SMALL GROUP (#3)	2,18 S 0.2	F084/J0959 (211,310)
AM 0418-483	04 18 27 −48 30.0	1.709 0.143	255.38 −44.73	242.1 −44.5	COMPACT w. SLIGHT FUZZ	13 S 0.2	F201/J2725 (15,255)
AM 0418-450	04 18 44 −45 09.3	1.859 0.143	250.68 −45.06	246.1 −46.3	SPIRAL + APP. COMP'N. NGC 1558 = 250-G-17	8 S 7.6	F250/J2681 (113,168)
AM 0418-470	04 18 49 −47 03.5	1.775 0.143	253.36 −44.86	243.8 −45.3	I/A TRIPLE + 4 COMP'NS.	3,8 D 2.5	F250/J2681 (114,66)
AM 0418-465	04 18 52 −46 52.8	1.783 0.143	253.11 −44.87	244.0 −45.4	CLOSE PAIR	23 S 0.3	F250/J2681 (114,76)
AM 0418-513	04 18 55 −51 33.5	1.551 0.143	259.59 −44.15	238.5 −42.8	I/A 3 (CLUSTER of COMPACTS)	3,13,18 D 1.3	F202/J0941 (269,91)
AM 0419-573	04 19 00 −57 30.9	1.175 0.142	267.56 −42.71	232.3 −39.2	DWARF	20 D 0.2	F157/J1893 (215,44)
AM 0419-480	04 19 22 −48 02.3	1.728 0.142	254.71 −44.64	242.5 −44.9	I/A DBL. (SMALL)	2 S 0.2	F202/J0941 (272,280)
*AM 0419-482	04 19 44 −48 22.4	1.711 0.141	255.17 −44.54	242.1 −44.7	E + PECULIAR SP. (PKS 0419-484?) NGC 1567 = 202-IG-10 + G-9 = SS-5	12,23 S 3.6	F202/J0941 (269,262)
AM 0419-680 [0419-6809]	04 19 49 −68 08.7	0.055 0.141	280.65 −38.85	222.5 −32.1	CLOSE PAIR (#1)	23 S 1.2	F055/J1942 (178,276)
AM 0419-625	04 19 53 −62 59.5	0.698 0.141	274.47 −40.85	227.0 −35.7	DWARF 84-G-13	20 D 0.9	F084/J0959 (201,285)
AM 0419-680 [0419-6808]	04 19 54 −68 09.7	0.052 0.141	280.67 −38.83	222.5 −32.1	CLOSE PAIR (#2)	23 S 1.2	F055/J1942 (178,275)
AM 0420-420	04 20 28 −42 01.6	1.981 0.141	246.24 −44.92	250.0 −48.0	CLOSE PAIR	23 S 1.9	F250/J2681 (92,336)
AM 0420-434	04 20 31 −43 44.8	1.913 0.140	248.67 −44.84	247.7 −47.2	E w. PLUMES + APP. COMP'NS. NGC 1751 = 250-G-19 = MC 0420-437	8,15 S 2.7	F250/J2681 (94,243)
AM 0420-440	04 20 35 −44 05.4	1.899 0.140	249.15 −44.81	247.3 −47.1	ISOLATED RING	6 D 0.8	F250/J2681 (94,225)
AM 0420-452 [0421-4523]	04 20 48 −45 23.5	1.844 0.140	250.97 −44.68	245.6 −46.5	CLOSE TRIPLE (#1)	24 D 3.1	F250/J2681 (94,155)
AM 0420-452 [0420-4523]	04 20 55 −45 25.2	1.842 0.140	251.01 −44.66	245.5 −46.5	CLOSE TRIPLE (#2)	24 D 3.1	F250/J2681 (93,153)
*AM 0420-570 [0421-5705]	04 20 58 −57 05.6	1.198 0.140	266.90 −42.57	232.5 −39.7	DWARF near NGC 1574 = 157-G-22	20 D 0.2	F157/J1893 (201,67)
*AM 0421-404	04 21 00 −40 43.1	2.029 0.140	244.39 −44.84	251.7 −48.7	SB w. BENT BAR + COMPACT COMP'N. NGC 1572 = 303-G-14	8,12,13 D 1.1	F303/J4742 (169,141)
AM 0421-570 [0420-5705]	04 21 00 −57 05.5	1.198 0.140	266.90 −42.57	232.5 −39.7	S0 w. SHARP EDGE + 2 DWARFS NGC 1574 = 157-G-22	8,12,20 D 2.7	F118/J3762 (216,335)
AM 0421-452 [0420-4525]	04 21 02 −45 23.0	1.843 0.140	250.96 −44.64	245.6 −46.5	CLOSE TRIPLE (#3)	24 D 3.1	F250/J2681 (91,155)
AM 0421 272	04 21 22 −27 20.0	2.445 0.139	225.87 −43.23	271.8 −53.1	SPIRAL w. CLOSE COMP'N. 484-G-13	8 S 0.6	F420/J2703 (80,322)
AM 0421-562	04 21 27 −56 27.0	1.242 0.139	266.04 −42.68	233.0 −40.2	I/A DBL. Se 37/2 = 157-IG-24	2 S 0.3	F157/J1893 (198,101)
AM 0421-443	04 21 37 −44 30.8	1.879 0.139	249.73 −44.60	246.6 −47.0	GALAXY w. APP. COMP'N. PKS 0421-444?	8 S 0.9	F250/J2681 (84,202)
*AM 0421-492	04 21 50 −49 23.3	1.656 0.139	256.50 −44.05	240.6 −44.5	CLOSE GROUP of 4 S0's 202-G-13	18 D 1.9	F202/J0941 (248,208)
AM 0421-283	04 21 57 −28 38.9	2.409 0.139	227.67 −43.39	269.6 −52.9	CLOSE PAIR of E's	23 S 0.3	F420/J2703 (74,251)
AM 0421-681	04 21 58 −68 15.5	0.025 0.139	280.68 −38.62	222.2 −32.2	CLOSE I/A DBL.	2 S 0.2	F055/J1942 (168,270)
AM 0422-471 [0422-4712]	04 22 16 −47 12.5	1.759 0.138	253.48 −44.26	243.2 −45.7	CONTACT DBL. in GROUP of 3 (#1)	23,24 D 0.2	F250/J2681 (83,57)
AM 0422-280	04 22 17 −28 03.3	2.425 0.138	226.90 −43.19	270.6 −53.1	EDGE-ON + I/A COMP'N. 420-G-18	1 S 1.2	F420/J2703 (69,283)
AM 0422-471 [0422-4710]	04 22 18 −47 10.8	1.760 0.138	253.44 −44.26	243.2 −45.7	GROUP of 3 (#2)	24 D 3.1	F250/J2681 (83,58)
AM 0422-670	04 22 18 −67 03.0	0.200 0.138	279.24 −39.08	223.2 −33.1	SMALL I/A TRIPLE 84-IG-16 = B 0422-670	3 D 0.4	F055/J1942 (165,335)
*AM 0422-275	04 22 21 −27 52.9	2.429 0.138	226.67 −43.14	270.9 −53.2	M51-TYPE in GROUP of 3 420-G-19	9,24 D 3.6	F420/J2703 (69,292)
AM 0422-375	04 22 26 −37 56.3	2.124 0.138	240.50 −44.48	255.4 −50.1	GROUP of 5	18 D 5.4	F360/J2771 (195,22)
AM 0422-471 [0422-4712]	04 22 28 −47 12.5	1.759 0.138	253.47 −44.22	243.1 −45.8	GROUP of 3 (#3)	24 D 3.1	F250/J2681 (81,57)
AM 0422-654	04 22 28 −65 45.4	0.371 0.138	277.69 −39.57	224.3 −34.0	I/A DBL. (SMALL)	2 S 0.3	F084/J0959 (185,137)

Identification	R.A. (1950.0) Declination	Precession R.A./Dec.	Gal. Long. Gal. Lat.	S.G.L. S.G.B.	Description Cross-Identification	Coding Size (Min.)	Field/Plate (x, y)
*AM 0422-514	04 22 29 −51 42.6	1.532 0.138	259.65 −43.58	237.9 −43.2	FUZZY SPIRAL w. HI.S.BR. JET NGC 1578 = 202-G-14 = Fa 771	7 D 1.1	F202/J0941 (239,84)
*AM 0422-514	04 22 29 −51 42.2	1.532 0.138	259.65 −43.57	237.9 −43.2	COMPACT	13 D 0.2	F202/J0941 (239,84)
AM 0422-542	04 22 36 −54 22.5	1.374 0.138	263.23 −43.02	235.1 −41.6	DWARF	20 D 0.3	F157/J1893 (190,213)
AM 0422-343	04 22 42 −34 38.1	2.232 0.138	235.90 −44.17	260.2 −51.4	PAIR (ONE w. APP. COMP'N.) MC 0422-345?	8,23 S 2.2	F360/J2771 (194,199)
AM 0422-473	04 22 53 −47 38.4	1.737 0.137	254.06 −44.11	242.6 −45.6	GALAXY w. FAINT RING 202-G-15 = MC 0422-476	6 D 2.7	F202/J0941 (241,302)
AM 0422-485	04 22 55 −48 52.6	1.678 0.137	255.76 −43.95	241.1 −44.9	E w. DIAMETRIC JETS + PEC. COMP'N	7,8 D 0.2	F202/J0941 (239,236)
AM 0423-292	04 23 01 −29 20.7	2.388 0.137	228.68 −43.29	268.4 −53.0	ONE-ARMED SPIRAL 420-G-20	10 D 0.8	F420/J2703 (62,214)
AM 0423-264 [0423-2642]	04 23 04 −26 48.7	2.457 0.137	225.29 −42.74	272.6 −53.5	PAIR of IRR. (#1) MCG-4-11-13 = 484-G-15	2,23 S 8.5	F484/J5463 (190,82)
AM 0423-441	04 23 06 −44 11.4	1.889 0.137	249.25 −44.36	246.8 −47.4	CLOSE PAIR	23 S 1.5	F250/J2681 (70,219)
AM 0423-671	04 23 11 −67 14.8	0.168 0.137	279.43 −38.92	223.0 −33.0	COMPACT w. MATERIAL	13 D 0.5	F084/J0959 (181,57)
AM 0423-264 [0423-2648]	04 23 36 −26 42.0	2.460 0.136	225.19 −42.60	272.7 −53.7	PAIR of IRR. (#2) 484-G-17	2,23 S 8.5	F484/J5463 (184,88)
AM 0423-580	04 23 40 −58 05.7	1.114 0.136	268.08 −41.95	231.2 −39.3	SPIRAL w. APP. COMP'NS. IC 2070 = 118-G-23	8 S 2.2	F118/J3762 (196,281)
AM 0423-733	04 23 42 −73 39.3	−1.086 0.136	286.70 −36.12	217.9 −28.3	CHAIN of 5	17 D 12.5	F032/J1937 (181,252)
AM 0423-465 [0423-4656]	04 23 50 −46 55.0	1.768 0.136	253.03 −44.02	243.3 −46.1	GROUP of 3 (#1)	24 D 1.9	F250/J2681 (68,72)
AM 0423-465 [0423-4655]	04 23 55 −46 55.1	1.768 0.136	253.03 −44.01	243.3 −46.1	GROUP of 3 (#2)	24 D 1.9	F250/J2681 (67,72)
AM 0423-445 [0424-4455]	04 23 57 −44 53.7	1.857 0.136	250.22 −44.16	245.8 −47.2	GROUP of 4 (#1)	18,24 S 4.5	F250/J2681 (63,181)
AM 0423-284	04 23 58 −28 43.8	2.404 0.136	227.91 −42.97	269.3 −53.3	CLUSTER of GALAXIES	19 D 5.4	F420/J2703 (50,246)
AM 0423-465 [0423-4655]	04 23 58 −46 56.1	1.767 0.136	253.06 −44.00	243.3 −46.1	GROUP of 3 (#3)	24 D 1.9	F250/J2681 (67,71)
AM 0424-445 [0423-4453]	04 24 05 −44 54.6	1.856 0.136	250.24 −44.14	245.8 −47.2	GROUP of 4 (#2: in CLOSE PAIR)	13,18,24 S 1.1	F250/J2681 (62,180)
AM 0424-445 [0424-4454]	04 24 06 −44 53.8	1.857 0.136	250.23 −44.14	245.8 −47.2	GROUP of 4 (#3: IN CLOSE PAIR)	13,18,24 S 1.1	F250/J2681 (61,180)
AM 0424-605	04 24 08 −60 54.7	0.879 0.136	271.65 −41.07	228.4 −37.5	CLOSE PAIR of E's in a COMMON ENVELOPE	2 S 0.1	F118/J3762 (191,130)
AM 0424-425	04 24 12 −42 58.3	1.936 0.136	247.54 −44.21	248.2 −48.2	CHAIN of 4	17 D 4.5	F251/J1045 (332,282)
AM 0424-445 [0424-4453]	04 24 12 −44 55.4	1.855 0.136	250.26 −44.12	245.7 −47.2	GROUP of 4 (#4)	24 S 4.5	F250/J2681 (61,179)
AM 0424-372 [0424-3720]	04 24 17 −37 22.4	2.140 0.135	239.74 −44.09	255.9 −50.7	CLOSE PAIR (JET?) (#1)	7,23 D 0.3	F360/J2771 (176,52)
AM 0424-433 [0424-4328]	04 24 18 −43 30.4	1.914 0.135	248.29 −44.17	247.5 −48.0	WIDE PAIR w. DIFFUSE ENVEL. (#1)	2 S 3.0	F250/J2681 (56,255)
AM 0424-433	04 24 20 −43 30.4	1.914 0.135	248.29 −44.17	247.5 −48.0	CHAIN of 5	17 D 4.9	F251/J1045 (329,254)
AM 0424-493	04 24 21 −49 34.1	1.640 0.135	256.67 −43.63	240.1 −44.7	DISTURBED GALAXY (SMALL)	16 D 0.5	F202/J0941 (226,199)
AM 0424-685 [0424-6852]	04 24 24 −68 51.8	−0.086 0.135	281.26 −38.16	221.6 −31.9	CLOSE PAIR in GROUP of 4 (#1)	18,23 S 0.4	F055/J1942 (156,238)
AM 0424-474	04 24 26 −47 40.2	1.732 0.135	254.06 −43.84	242.3 −45.8	I/A DBL. (E + SPIRAL)	2 S 0.3	F202/J0941 (227,301)
AM 0424-372 [0424-3722]	04 24 29 −37 20.7	2.141 0.135	239.70 −44.05	256.0 −50.8	CLOSE PAIR (JET?) (#2)	23 D 0.3	F360/J2771 (174,54)
AM 0424-432 [0424-4330]	04 24 29 −43 28.5	1.915 0.135	248.24 −44.14	247.5 −48.0	WIDE PAIR w. DIFFUSE ENVEL. (#2)	2 S 3.0	F250/J2681 (55,257)
AM 0424-685 [0424-6851]	04 24 29 −68 52.1	−0.088 0.135	281.26 −38.16	221.5 −31.9	CLOSE PAIR in GROUP of 4 (#2)	18,23 S 0.4	F055/J1942 (156,237)
AM 0424-744	04 24 31 −74 44.4	−1.405 0.135	287.85 −35.56	217.0 −27.5	DWARF (NUCLEATED)	20 D 0.4	F032/J1937 (177,193)
AM 0424-565	04 24 33 −56 58.4	1.194 0.135	266.57 −42.14	232.2 −40.2	DWARF Se 37/3 = 157-G-27	20 D 0.3	F157/J1893 (175,73)
AM 0424-673 [0425-6735]	04 24 39 −67 34.3	0.112 0.135	279.74 −38.66	222.6 −32.9	CLOSE PAIR (#1)	23 S 2.8	F055/J1942 (154,307)
AM 0424-565	04 24 59 −56 50.4	1.201 0.135	266.38 −42.11	232.2 −40.3	DWARF	20 D 0.3	F157/J1893 (172,81)

Identification	R.A. (1950.0) Declination	Precession R.A./Dec.	Gal. Long. Gal. Lat.	S.G.L. S.G.B.	Description Cross-Identification	Coding Size (Min.)	Field/Plate (x, y)
*AM 0425-421	04 25 00 −42 12.9	1.964 0.135	246.48 −44.08	249.1 −48.7	2 SPIRALS MCG-7-10-4 = B 0424-422 IC 2068 = 303-G-17 = Fa 408	8,23 S 8.1	F303/J4742 (129,60)
AM 0425-673 [0424-6734]	04 25 02 −67 35.4	0.107 0.134	279.75 −38.62	222.5 −32.9	CLOSE PAIR (#2)	23 S 2.8	F055/J1942 (152,306)
*AM 0425-421	04 25 03 −42 13.4	1.963 0.134	246.50 −44.07	249.1 −48.7	RING + I/A COMP'N. in LOOSE GROUP IC 2068 = 303-G-17	1,6,18 S 1.7	F250/J2681 (46,324)
AM 0425-412	04 25 21 −41 20.7	1.996 0.134	245.27 −44.02	250.2 −49.2	COMPACT	13 D 0.2	F303/J4742 (125,107)
AM 0425-670 [0425-6709]	04 25 23 −67 09.8	0.167 0.134	279.22 −38.76	222.9 −33.2	GROUP of 3 (#1)	24 D 1.3	F055/J1942 (149,329)
AM 0425-670 [0425-6709]	04 25 25 −67 09.1	0.168 0.134	279.21 −38.76	222.9 −33.2	GROUP of 3 (#2)	24 D 1.3	F055/J1942 (149,329)
AM 0425-333	04 25 31 −33 38.2	2.259 0.134	234.62 −43.47	261.4 −52.3	DWARF w. RESOLUTION 360-G-10	20 D 1.3	F360/J2771 (163,253)
AM 0425-670 [0425-6709]	04 25 36 −67 09.7	0.166 0.134	279.21 −38.74	222.8 −33.2	GROUP of 3 (#3)	24 D 1.3	F055/J1942 (148,329)
*AM 0425-491	04 25 42 −49 13.9	1.653 0.134	256.17 −43.45	240.3 −45.1	CHAIN of 10 (incl. I/A DBL.) 202-IG-21 = B0425-492 = PKS0425-491	2,17 D 9.0	F202/J0941 (215,217)
AM 0425-473	04 25 44 −47 39.7	1.729 0.134	254.02 −43.63	242.2 −46.0	I/A DBL. 202-G-22	2 S 0.7	F251/J1045 (306,31)
*AM 0425-612	04 25 44 −61 27.8	0.821 0.134	272.26 −40.72	227.8 −37.3	SPIRAL w. DOUBLE ARMS 118-G-24	10 D 0.7	F118/J3762 (181,101)
AM 0425-463	04 25 46 −46 31.8	1.781 0.133	252.46 −43.73	243.5 −46.6	RICH DISTANT CLUSTER	19 D 9.0	F250/J2681 (50,92)
AM 0425-334	04 25 51 −33 43.9	2.255 0.133	234.76 −43.41	261.2 −52.3	ONE-ARMED SPIRAL (PAIR w. DWARF) 360-IG-11	10,20,23 D 1.1	F360/J2771 (159,248)
AM 0426-424	04 26 00 −42 45.4	1.940 0.133	247.23 −43.88	248.3 −48.6	DISTURBED (BRIGHT-RIMMED) SPIRAL B 0426-248	10,12 D 0.7	F303/J4742 (120,31)
*AM 0426-425	04 26 07 −42 53.7	1.934 0.133	247.42 −43.86	248.1 −48.5	E + I/A COMP'N. + JETS 251-G-1 (CHAIN of 5)	1,7,17 S 0.4	F251/J1045 (313,287)
AM 0426-551	04 26 8 −55 17.6	1.304 0.133	264.30 −42.32	233.7 −41.4	DWARF	20 D 0.2	F157/J1893 (163,164)
AM 0426-584 [0426-5846]	04 26 11 −58 47.2	1.050 0.133	268.84 −41.45	230.2 −39.1	PAIR (#1)	23 S 1.7	F118/J3762 (178,244)
AM 0426-481	04 26 22 −48 17.5	1.697 0.133	254.86 −43.46	241.3 −45.7	I/A DBL.	2 S 0.2	F202/J0941 (209,268)
AM 0426-584 [0426-5847]	04 26 24 −58 46.1	1.051 0.133	268.81 −41.43	230.2 −39.2	PAIR (#2) 118-G-25	23 S 1.7	F118/J3762 (176,245)
AM 0426-323	04 26 25 −32 33.0	2.290 0.133	233.18 −43.13	263.0 −52.8	CLOSE GROUP of 3 MCG-5-11-8 = 320-G-12	24 D 0.7	F420/J2703 (27,41)
AM 0426-372	04 26 26 −37 22.8	2.136 0.133	239.80 −43.67	255.6 −51.1	CLOSE PAIR of SPIRALS	23 S 1.1	F303/J4742 (111,319)
AM 0426-550 [0426-5509]	04 26 29 −55 08.2	1.313 0.133	264.08 −42.31	233.8 −41.6	CLOSE PAIR (#1) NGC 1596 = 157-G-31 = SS-6a	16,23 D 2.0	F157/J1893 (160,172)
AM 0426-480 [0427-4754]	04 26 31 −48 02.2	1.709 0.132	254.51 −43.46	241.6 −45.9	E w. DIAMETRIC PEC. SPIRALS (#1) 202-G-23 = 'CARAFFE'	12,24 S 10.8	F250/J2681 (46,11)
AM 0426-733	04 26 34 −73 32.1	−1.077 0.132	286.45 −35.99	217.8 −28.5	CLOSE GROUP of 3	24 D 2.0	F032/J1937 (170,258)
AM 0426 625	04 26 38 −62 53.4	0.677 0.132	273.99 −40.16	226.4 −36.3	2 E's in COMMON ENVELOPE	23 S 1.0	F084/J0959 (160,290)
AM 0426-484	04 26 39 −48 49.7	1.670 0.132	255.59 −43.35	240.7 −45.4	CLOSE PAIR (incl. RING)	6,23 S 0.5	F202/J0941 (207,239)
AM 0426-621	04 26 40 −62 16.9	0.738 0.132	273.24 −40.35	226.9 −36.8	COMPACT	13 D 0.2	F084/J0959 (159,323)
AM 0426-550 [0426-5508]	04 26 49 −55 09.9	1.310 0.132	264.10 −42.25	233.7 −41.6	CLOSE PAIR (#2: IRR.) NGC 1602 = 157-G-32 = SS-6b	16,23 D 1.1	F157/J1893 (158,170)
AM 0426-430	04 26 50 −43 09.2	1.923 0.132	247.78 −43.72	247.6 −48.5	I/A DBL.	2 S 0.3	F251/J1045 (305,274)
AM 0426-475 [0426-4802]	04 26 52 −47 56.5	1.713 0.132	254.37 −43.41	241.7 −46.0	E w. DIAMETRIC PEC. SPIRALS (#2) NGC 1595 = 202-G-25 [A82]	12,24 S 10.8	F250/J2681 (43,16)
*AM 0427-475 [0426-4756]	04 27 03 −47 54.3	1.714 0.132	254.32 −43.38	241.7 −46.0	E w. DIAMETRIC PEC. SPIRALS (#3) NGC 1598 = 202-G-26 [A82]	12,24 S 10.8	F250/J2681 (41,18)
AM 0427-474	04 27 08 −47 43.9	1.722 0.132	254.08 −43.38	241.9 −46.1	COMPACT Fa 409	13 D 0.2	F202/J0941 (203,298)
AM 0427-344	04 27 14 −34 40.9	2.223 0.132	236.11 −43.25	259.6 −52.3	LINE of 3	24 D 5.2	F360/J2771 (144,197)
*AM 0427-373	04 27 18 −37 35.2	2.128 0.131	240.09 −43.50	255.2 −51.2	SPIRAL w. 2 I/A COMP'NS. MCG-6-10-10 = 303-G-21 = B 0427-375	1 S 1.5	F360/J2771 (143,41)
AM 0427-821	04 27 34 −82 11.2	−5.882 0.131	295.51 −31.71	211.7 −21.9	DWARF 15-G-10	20 D 0.5	F014/J2559 (23,32)
*AM 0427-273	04 27 35 −27 31.9	2.432 0.131	226.56 −41.93	271.0 −54.4	I/A DBL. w. CONNECTING FILAMENT MCG-5-11-11 = 421-IG-2	2,22,25 D 1.6	F484/J5463 (136,43)

Identification	R.A. (1950.0) Declination	Precession R.A./Dec.	Gal. Long. Gal. Lat.	S.G.L. S.G.B.	Description Cross-Identification	Coding Size (Min.)	Field/Plate (x, y)
AM 0427-265	04 27 39 −26 53.0	2.450 0.131	225.72 −41.76	272.1 −54.5	3 in a LINE MCG-4-11-16 = 484-G-26	24 D 8.1	F484/J5463 (135,78)
AM 0427-480	04 27 42 −48 08.2	1.701 0.131	254.62 −43.25	241.3 −46.0	PECULIAR SPIRAL 202-G-27	12 D 1.1	F202/J0941 (198,276)
AM 0427-454	04 27 47 −45 42.0	1.813 0.131	251.28 −43.44	244.3 −47.3	I/A TRIPLE	3 D 0.5	F251/J1045 (291,137)
AM 0427-615	04 27 53 −61 53.3	0.772 0.131	272.69 −40.34	227.2 −37.2	COMPACT	13 D 0.1	F084/J0959 (151,344)
AM 0427-535	04 27 57 −53 56.0	1.385 0.131	262.43 −42.35	234.9 −42.5	RICH CLUSTER (PKS 0427-53) [B] IC 2082 = Se 37/1 = 157-IG-35	19 D 9.0	F157/J1893 (148,236)
AM 0428-390	04 28 16 −39 03.2	2.075 0.130	242.12 −43.40	253.0 −50.7	SMALL, ONE-ARMED SPIRAL	10 D 0.3	F303/J4742 (93,229)
AM 0428-605	04 28 32 −60 56.5	0.858 0.130	271.47 −40.56	227.9 −37.9	VERY CLOSE PAIR of SPIRALS PKS 0428-611?	23 S 0.4	F118/J3762 (163,129)
AM 0428-484	04 28 37 −48 46.8	1.668 0.130	255.48 −43.03	240.4 −45.7	DWARF 202-G-28	20 D 0.8	F202/J0941 (189,242)
AM 0428-450	04 28 48 −45 08.2	1.835 0.129	250.49 −43.29	244.8 −47.8	RING + I/A DBL. COMP'N.	2,6 S 0.3	F251/J1045 (282,168)
AM 0429-444	04 29 12 −44 43.1	1.853 0.129	249.91 −43.24	245.3 −48.1	CLOSE PAIR	23 S 0.4	F251/J1045 (279,191)
AM 0429-545	04 29 19 −54 52.6	1.320 0.129	263.62 −41.96	233.7 −42.0	DWARF	20 D 0.4	F157/J1893 (138,186)
AM 0429-505	04 29 20 −50 51.2	1.559 0.129	258.26 −42.65	238.0 −44.6	CLOSE CONTACT TRIPLE (SMALL)	3 D 0.5	F202/J0941 (183,131)
AM 0429-442	04 29 27 −44 20.6	1.868 0.129	249.40 −43.21	245.7 −48.3	DWARF 251-G-8 = Ka-23	20 D 0.4	F251/J1045 (278,211)
AM 0429-485	04 29 30 −48 53.9	1.659 0.128	255.61 −42.87	240.2 −45.8	DWARF (RESOLUTION) 202-G-29	20 D 0.7	F202/J0941 (181,236)
AM 0429-295	04 29 32 −29 51.3	2.365 0.128	229.73 −42.01	267.0 −54.3	CHAIN of 4 in WIDE GROUP of 7	17,18 D 3.8	F421/J4737 (255,186)
AM 0429-583	04 29 45 −58 34.0	1.054 0.128	268.40 −41.06	230.0 −39.6	SPIRAL w. CONDENSATION in ARM 118-G-28 = B 0429-585 + COMP'N.	8,10 D 0.7	F118/J3762 (153,256)
*AM 0430-285	04 30 13 −28 57.8	2.390 0.128	228.60 −41.68	268.4 −54.6	M51-TYPE (GRAND DESIGN) 421-G-8	9 S 0.8	F421/J4737 (247,234)
AM 0430-520	04 30 15 −52 08.1	1.485 0.127	259.95 −42.32	236.5 −43.9	SPIRAL + I/A COMP'N. 202-G-31	1 S 2.2	F202/J0941 (175,62)
*AM 0430-303	04 30 18 −30 30.1	2.345 0.127	230.63 −41.97	265.8 −54.2	M51-TYPE 421-G-9	9 S 0.4	F421/J4737 (245,152)
AM 0430-605	04 30 19 −60 53.1	0.855 0.127	271.32 −40.36	227.8 −38.1	DISTURBED IRR. + JET and CONDENSATION	7,16 D 1.0	F118/J3762 (151,132)
AM 0430-423	04 30 23 −42 34.1	1.938 0.127	246.96 −43.08	247.9 −49.4	I/A DBL.	2 S 0.1	F251/J1045 (271,306)
AM 0430-454	04 30 27 −45 47.4	1.803 0.127	251.37 −42.97	243.8 −47.7	E (CENTRE of CHAIN of 7) Fa 415	17 D 11.2	F251/J1045 (266,133)
AM 0430-544	04 30 31 −54 42.2	1.328 0.127	263.35 −41.83	233.7 −42.3	BRIGHT GALAXY (PEC. DUST in ARMS) NGC 1617 = 157-G-41 (3 COMP'NS)	8,14 D 3.1	F157/J1893 (129,195)
AM 0430-613	04 30 34 −61 35.1	0.789 0.127	272.18 −40.13	227.2 −37.6	RICH ELONGATED CLUSTER (DBL. cD) Se 40/6 = 118-IG-30 = B 0430-615	2,19 S 0.2	F118/J3762 (150,94)
AM 0430-621	04 30 37 −62 18.6	0.718 0.127	273.09 −39.91	226.5 −37.1	GALAXY w. OFF-SET NUCLEUS	22 D 0.5	F084/J0959 (135,321)
AM 0430-632	04 30 38 −63 25.5	0.603 0.127	274.46 −39.56	225.5 −36.3	DISTURBED GALAXY w. LOOPS 84-G-21 = Fa 775	15,16 D 1.0	F084/J0959 (136,261)
AM 0430-392	04 30 41 −39 23.9	2.058 0.127	242.63 −42.95	252.2 −51.0	VERY FAINT, POOR STAR CLUSTER	22 D 1.8	F304/J2655 (338,208)
AM 0430-322	04 30 48 −32 20.5	2.290 0.127	233.09 −42.19	262.8 −53.8	SMALL SPIRAL w. TAIL	15 D 0.6	F360/J2771 (103,322)
*AM 0431-434	04 31 06 −43 49.8	1.886 0.126	248.68 −42.93	246.1 −48.8	BRIGHT, 3-ARMED SP. (GROUP of 5) NGC 1616 = Ag-20 = 251-G-10	11,12,18 D 1.3	F251/J1045 (263,239)
AM 0431-624	04 31 09 −62 45.4	0.671 0.126	273.62 −39.72	226.1 −36.8	I/A DBL.	2 S 0.2	F084/J0959 (132,297)
AM 0431-490	04 31 30 −49 01.7	1.648 0.126	255.74 −42.53	239.8 −46.0	I/A DBL. (in CHAIN of 4)	2,17 S 0.4	F202/J0941 (164,229)
AM 0432-625	04 32 01 −62 58.1	0.645 0.125	273.84 −39.56	225.8 −36.7	GALAXY w. BROKEN RING	6 D 1.3	F084/J0959 (127,285)
AM 0432-710 [0434-7104]	04 32 54 −71 00.5	−0.532 0.124	283.36 −36.63	219.3 −30.8	CHAIN of 5 in WING of LMC (#1)	17 D 7.6	F055/J1942 (121,121)
AM 0433-710 [0432-7100]	04 33 02 −71 01.5	−0.536 0.124	283.37 −36.61	219.2 −30.8	CHAIN of 5 in WING of LMC (#2)	17 D 7.6	F055/J1942 (120,120)
AM 0433-260	04 33 13 −26 05.8	2.465 0.123	225.11 −40.36	273.1 −55.9	CHAIN of 4 [N.S.] MCG-4-11-21 = 484-G-34	17 D 10.8	F484/J5463 (68,120)
AM 0433-400	04 33 15 −40 08.2	2.026 0.123	243.66 −42.49	250.8 −51.1	PAIR of I/A DBL. 304-IG-1	2,2,18 S 1.6	F304/J2655 (310,169)

Identification	R.A. (1950.0) Declination	Precession R.A./Dec.	Gal. Long. Gal. Lat.	S.G.L. S.G.B.	Description Cross-Identification	Coding Size (Min.)	Field/Plate (x, y)
AM 0433-710 [0433-7101]	04 33 27 −71 02.2	−0.541 0.123	283.37 −36.58	219.2 −30.8	CHAIN of 5 in WING of LMC (#3)	17 D 7.6	F055/J1942 (119,120)
AM 0433-654	04 33 40 −65 48.6	0.307 0.123	277.23 −38.48	223.2 −34.8	DWARF	20 D 0.7	F084/J0959 (123,133)
AM 0433-710 [0433-7102]	04 33 43 −71 03.0	−0.546 0.123	283.38 −36.55	219.2 −30.8	CHAIN of 5 in WING of LMC (#4)	17 D 7.6	F055/J1942 (117,119)
AM 0433-393	04 33 48 −39 36.9	2.044 0.123	242.96 −42.36	251.4 −51.4	GAL. w. FAINT OUTER RING or LOOP	6 D 1.3	F304/J2655 (306,197)
AM 0434-710 [0433-7103]	04 34 02 −71 04.9	−0.554 0.122	283.40 −36.52	219.1 −30.8	CHAIN of 5 in WING of LMC (#5)	17 D 7.6	F055/J1942 (116,117)
AM 0434-350	04 34 06 −35 04.1	2.200 0.122	236.86 −41.90	258.1 −53.4	LINE of 3 (COMPACT w. TAIL)	13,15,24 D 4.5	F360/J2771 (68,175)
AM 0434-360	04 34 14 −36 05.9	2.166 0.122	238.24 −41.99	256.5 −53.0	SPIRAL + M51-TYPE 360-G-17/18	9,23 S 1.4	F360/J2771 (68,120)
AM 0434-223	04 34 28 −22 32.7	2.557 0.122	220.78 −39.09	279.3 −56.7	E w. JET in GROUP of 4 484-G-37 = PKS 0434-225	7,18 D 0.5	F485/J5480 (322,310)
AM 0435-401	04 35 01 −40 18.1	2.017 0.121	243.90 −42.16	250.3 −51.3	CLOSE I/A DBL. 304-IG-5	2 S 0.5	F303/J4742 (25,160)
*AM 0435-394	04 35 07 −39 44.8	2.037 0.121	243.16 −42.12	251.1 −51.6	SPIRAL w. OFF-SET ARM 304-G-6	10 D 0.8	F304/J2655 (292,191)
AM 0435-465	04 35 07 −46 59.7	1.738 0.121	252.93 −42.10	241.6 −47.7	GALAXY w. I/A COMP'N. 251-G-13	1 S 0.7	F202/J0941 (130,338)
AM 0435-585	04 35 18 −58 57.7	1.003 0.121	268.67 −40.26	229.0 −39.9	RESOLVED DWARF Reticulum = Se 40/3 = 118-G-31	20 D 1.4	F118/J3762 (115,234)
AM 0435-383	04 35 27 −38 31.3	2.081 0.120	241.52 −41.97	252.8 −52.2	I/A DBL.	2 S 0.2	F304/J2655 (290,257)
AM 0435-443	04 35 27 −44 33.2	1.846 0.120	249.64 −42.14	244.6 −49.1	IRR.	16 D 0.6	F251/J1045 (220,201)
AM 0435-383	04 35 32 −38 37.7	2.077 0.120	241.66 −41.96	252.6 −52.2	ONE-ARMED SPIRAL	10 D 0.7	F304/J2655 (289,251)
AM 0435-524	04 35 41 −52 40.3	1.438 0.120	260.49 −41.42	235.2 −44.2	E + SPIRAL (I/A)	2 S 1.8	F202/J0941 (131,33)
AM 0435-443	04 35 43 −44 37.4	1.843 0.120	249.73 −42.09	244.4 −49.1	DISTURBED COMPACT w. FAINT PLUME Fa 419	13,16 D 0.3	F251/J1045 (217,197)
AM 0435-521	04 35 44 −52 16.8	1.461 0.120	259.97 −41.47	235.6 −44.5	DWARF 202-G-41	20 D 1.3	F202/J0941 (130,54)
AM 0435-322	04 35 51 −32 28.5	2.279 0.120	233.51 −41.16	262.0 −54.7	IRR. w. PLUME off END 421-G-12	7,15,16 D 1.0	F360/J2771 (46,314)
AM 0436-593	04 36 05 −59 37.6	0.945 0.120	269.48 −40.01	228.3 −39.5	CONTACT E's (PKS 0436-596) 118-IG-32 = DN 0436-596	2 S 0.1	F118/J3762 (111,198)
AM 0436-560	04 36 22 −56 01.1	1.221 0.119	264.85 −40.76	231.7 −42.0	IRR. 157-G-44	16 D 0.8	F157/J1893 (86,123)
*AM 0436-472	04 36 22 −47 21.7	1.717 0.119	253.41 −41.86	241.0 −47.6	I/A DBL. w. DISTORTED RING 251-G-14	2 S 0.7	F251/J1045 (209,50)
AM 0436-465	04 36 27 −46 51.6	1.741 0.119	252.73 −41.88	241.6 −47.9	COMPACT	13 D 0.1	F202/J0941 (118,345)
AM 0436-513	04 36 29 −51 31.3	1.502 0.119	258.95 −41.46	236.3 −45.0	E + 6 COMP'NS. 202-G-43	8 D 1.1	F202/J0941 (123,94)
AM 0436-644	04 36 32 −64 40.5	0.435 0.119	275.73 −38.56	223.9 −35.8	I/A DBL. 84-IG-27 = Fa 779	2 S 0.3	F084/J0959 (104,193)
AM 0436-461	04 36 34 −46 18.9	1.766 0.119	252.00 −41.88	242.2 −48.3	DWARF	20 D 0.4	F251/J1045 (208,106)
AM 0436-342	04 36 54 −34 21.1	2.219 0.118	236.01 −41.23	258.8 −54.2	V.SMALL RING w. ABS'N. (DEFECT?)	6,14 D 0.5	F360/J2771 (37,213)
AM 0437-241	04 37 03 −24 16.9	2.509 0.118	223.15 −39.04	276.0 −57.1	CLOSE PAIR of SPIRALS MCG-4-12-3/4 = 485-G-3/4	23 S 0.9	F484/J5463 (20,216)
AM 0437-273	04 37 16 −27 38.0	2.419 0.118	227.35 −39.87	270.0 −56.5	CONTACT PAIR of COMPACTS	2,13 S 0.1	F485/J5480 (283,37)
AM 0437-621	04 37 17 −62 19.4	0.690 0.118	272.81 −39.17	225.8 −37.6	DISTURBED GALAXY w. MATERIAL	16 D 0.5	F084/J0959 (93,319)
AM 0437-344	04 37 25 −34 43.8	2.206 0.118	236.53 −41.18	258.2 −54.2	PECULIAR SPIRAL w. APP. COMP'N.	8,12 S 1.8	F360/J2771 (31,192)
AM 0437-691 [0437-6914]	04 37 27 −69 18.1	−0.236 0.118	281.21 −36.92	220.2 −32.3	PAIR of SP. in WING of LMC (#1) 55-G-29	23 S 4.3	F055/J1942 (95,211)
AM 0437-691 [0437-6918]	04 37 28 −69 14.4	−0.225 0.118	281.13 −36.94	220.2 −32.4	PAIR of SP. in WING of LMC (#2) 55-G-28	23 S 4.3	F055/J1942 (94,215)
AM 0437-644	04 37 30 −64 45.1	0.422 0.118	275.78 −38.44	223.7 −35.8	RING GALAXY (DEFORMED)	6 D 0.5	F084/J0959 (99,188)
AM 0437-780	04 37 37 −78 08.0	−2.879 0.117	291.07 −33.31	214.0 −25.4	GALAXY w. TAIL 15-IG-11 = B 0437-781	15 D 1.1	F015/J3673 (158,279)
AM 0437-314	04 37 45 −31 48.3	2.297 0.117	232.73 −40.64	262.8 −55.3	RING w. OFF-SET NUCLEUS	6 D 0.4	F421/J4737 (160,82)

Identification	R.A. (1950.0) Declination	Precession R.A./Dec.	Gal. Long. Gal. Lat.	S.G.L. S.G.B.	Description Cross-Identification	Coding Size (Min.)	Field/Plate (x, y)
AM 0437-365	04 37 49 −36 56.8	2.132 0.117	239.47 −41.37	254.7 −53.3	WIDE PAIR of I/A SPIRALS	2 S 9.0	F360/J2771 (31,73)
AM 0437-370	04 37 53 −37 08.7	2.125 0.117	239.74 −41.37	254.4 −53.3	PAIR + COMP'N. 361-G-2	8,23 S 2.2	F361/J3738 (287,63)
*AM 0438-503	04 38 02 −50 37.6	1.548 0.117	257.73 −41.32	237.0 −45.8	I/A DBL. w. TAILS B 0438-506	2,6,15 S 0.4	F202/J0941 (109,142)
AM 0438-641	04 38 03 −64 12.7	0.483 0.117	275.10 −38.54	224.1 −36.3	I/A SPIRALS 84-IG-31	2 S 0.4	F084/J0959 (94,217)
AM 0438-354	04 38 09 −35 43.7	2.173 0.117	237.87 −41.16	256.5 −53.9	I/A DBL. (E + SPIRAL) GROUP of 5 361-G-3 = PKS 0438-357	2,8,24 D 0.9	F361/J3738 (287,139)
AM 0438-541	04 38 14 −54 18.5	1.330 0.117	262.56 −40.80	233.1 −43.4	I/A TRIPLE 157-G-47	3 D 2.5	F158/J2819 (326,211)
AM 0438-444	04 38 15 −44 44.1	1.833 0.117	249.87 −41.63	243.9 −49.4	E + COMP'N. CONNECTED by FILAMENT 251-G-21	1 S 1.3	F251/J1045 (193,191)
AM 0438-472	04 38 21 −47 24.6	1.711 0.116	253.45 −41.52	240.6 −47.8	SPIRAL w. PECULIAR ARMS	10 D 0.4	F251/J1045 (192,48)
AM 0438-473	04 38 24 −47 37.6	1.700 0.116	253.74 −41.50	240.4 −47.7	DISTURBED (RESOLVED DWARF?) 202-IG-48	16,20 D 0.6	F251/J1045 (191,36)
AM 0438-661	04 38 24 −66 19.6	0.217 0.116	277.65 −37.85	222.4 −34.7	RING GALAXY (DEFORMED) Fa 304	6 D 0.6	F085/J1113 (336,97)
*AM 0438-530	04 38 38 −53 05.9	1.405 0.116	260.97 −40.92	234.3 −44.2	SPIRAL + I/A DBL.(CHAIN of 4) 157-IG-48/49	1,2,17 D 26.9	F202/J0941 (107,9)
AM 0438-445	04 38 48 −44 59.4	1.820 0.116	250.21 −41.53	243.5 −49.3	M51-TYPE (CHAIN of 4) 251-G-20	9,17 S 0.3	F251/J1045 (188,178)
AM 0438-481	04 38 54 −48 17.0	1.667 0.116	254.60 −41.38	239.5 −47.4	DWARF 202-G-49	20 D 0.7	F202/J0941 (98,268)
AM 0438-511	04 38 56 −51 11.3	1.515 0.116	258.45 −41.11	236.3 −45.5	LINEAR DWARF (?) 202-G-50	20 D 1.0	F202/J0941 (102,112)
AM 0439-372	04 39 03 −37 26.9	2.113 0.116	240.17 −41.17	253.8 −53.3	SPIRAL w. DISTURBED COMP'N. 361-G-6	8 S 1.3	F360/J2771 (18,46)
AM 0439-373	04 39 14 −37 39.9	2.105 0.115	240.46 −41.16	253.4 −53.3	SPIRAL w. I/A COMP'N. (RING?) 304-G-11	1,6 S 0.6	F304/J2655 (252,304)
AM 0439-393	04 39 15 −39 35.5	2.036 0.115	243.02 −41.31	250.6 −52.3	DISTURBED SINGLE 304-IG-12 = B 0439-395	16 D 0.8	F304/J2655 (249,200)
AM 0439-451	04 39 28 −45 16.8	1.806 0.115	250.60 −41.41	243.0 −49.3	LARGE CLOUD of FAINT GALAXIES	19 D 8.1	F251/J1045 (182,162)
AM 0439-585	04 39 29 −58 50.5	0.998 0.115	268.35 −39.77	228.6 −40.3	IRR. w. HI.S.BR. CENTRE Se 40/4 = 118-G-34	16 D 0.6	F118/J3762 (86,239)
AM 0439-621	04 39 31 −62 15.1	0.688 0.115	272.63 −38.93	225.6 −37.8	DISTURBED GALAXY	16 D 0.4	F084/J0959 (79,322)
AM 0439-525	04 39 33 −52 50.8	1.417 0.115	260.61 −40.82	234.4 −44.5	PECULIAR SPIRAL + I/A COMP'N. 157-IG-50 = B0439-528(CENTRE of 6)	1,18 S 3.8	F203/J2836 (343,19)
AM 0439-631	04 39 57 −63 18.2	0.577 0.114	273.91 −38.60	224.7 −37.1	I/A DBL. (SMALL)	2 S 0.3	F085/J1113 (345,259)
AM 0440-215	04 40 17 −21 53.0	2.569 0.114	220.52 −37.60	280.3 −58.1	I/A DBL.	2 S 0.2	F485/J5480 (250,346)
AM 0440-415	04 40 17 −41 59.8	1.942 0.114	246.23 −41.24	247.2 −51.2	SPIRAL w. I/A COMP'N.	1 S 0.7	F251/J1045 (174,338)
AM 0440-453	04 40 40 −45 31.0	1.793 0.113	250.91 −41.20	242.6 −49.3	I/A DBL. (IRR. + COMPACT)	2,13,16 S 0.4	F251/J1045 (170,149)
AM 0440-645	04 40 52 −64 55.3	0.387 0.113	275.85 −38.04	223.3 −35.9	GALAXY w. TAIL	15 D 0.3	F085/J1113 (330,173)
AM 0440-554	04 40 54 −55 43.6	1.228 0.113	264.32 −40.19	231.3 −42.7	GROUP of 6	18 D 4.3	F158/J2819 (301,136)
AM 0440-442	04 40 55 −44 21.1	1.844 0.113	249.36 −41.16	244.0 −50.0	DISTURBED E in CHAIN of 7 PKS 0440-443	16,17 D 14.3	F251/J1045 (168,212)
AM 0440-494	04 40 55 −49 42.3	1.590 0.113	256.45 −40.94	237.6 −46.7	PECULIAR SPIRAL + 2 APP. COMP'NS. B 0440-497	8,12 D 0.5	F202/J0941 (82,191)
AM 0440-433	04 40 56 −43 36.9	1.875 0.113	248.38 −41.16	244.9 −50.4	I/A DBL. 251-IG-25	2 S 0.6	F251/J1045 (167,251)
AM 0440-621	04 40 58 −62 13.5	0.686 0.113	272.54 −38.78	225.5 −38.0	DISTURBED GALAXY w. MATERIAL	15,16 D 0.4	F084/J0959 (70,323)
AM 0441-403	04 41 02 −40 36.7	1.995 0.113	244.40 −41.03	248.9 −52.1	CLOSE PAIR	23 S 1.6	F304/J2655 (230,146)
AM 0441-473	04 41 15 −47 30.1	1.700 0.113	253.53 −41.03	240.1 −48.2	SPIRAL + I/A COMPACT	1,13 S 0.2	F251/J1045 (165,43)
AM 0441-633	04 41 21 −63 38.2	0.535 0.112	274.27 −38.36	224.3 −36.9	DWARF	20 D 0.5	F085/J1113 (335,242)
AM 0441-614	04 41 23 −61 49.5	0.724 0.112	272.03 −38.83	225.8 −38.3	SMALL I/A DBL. + COMP'NS. 118-IG-36?	2,8 S 0.1	F084/J0959 (66,344)
AM 0441-334	04 41 31 −33 43.8	2.233 0.112	235.40 −40.19	259.2 −55.4	I/A TRIPLE 361-G-7	3 D 1.8	F361/J3738 (252,247)

Identification	R.A. (1950.0) Declination	Precession R.A./Dec.	Gal. Long. Gal. Lat.	S.G.L. S.G.B.	Description Cross-Identification	Coding Size (Min.)	Field/Plate (x, y)
AM 0441-473	04 41 44 −47 31.9	1.698 0.112	253.57 −40.95	240.0 −48.2	E + I/A COMP'N. (CENTRE of GROUP)	1,18 S 0.2	F251/J1045 (161,41)
AM 0441-624	04 41 48 −62 46.9	0.625 0.112	273.20 −38.54	224.9 −37.6	COMPACT	13 D 0.2	F085/J1113 (337,288)
AM 0442-393	04 42 11 −39 31.3	2.034 0.111	242.98 −40.74	250.3 −52.9	CLOSE GROUP of 4 304-IG-14	18 D 1.0	F304/J2655 (219,205)
AM 0442-413 [0443-4140]	04 42 24 −41 33.2	1.956 0.111	245.66 −40.83	247.4 −51.8	TRIPLE (#1: SPIRAL in PAIR) NGC 1658 = 304-G-16	8,23,24 S 2.7	F304/J2655 (215,96)
AM 0442-622	04 42 33 −62 25.3	0.660 0.111	272.73 −38.55	225.1 −37.9	E + JET (LO.S.BR.)	7 D 0.3	F084/J0959 (61,311)
AM 0442-622	04 42 33 −62 25.2	0.660 0.111	272.72 −38.55	225.1 −37.9	GALAXY w. MATERIAL 84-G-39 DBL. NUCLEUS	7,22 D 0.9	F085/J1113 (334,308)
AM 0442-413 [0442-4133]	04 42 34 −41 35.3	1.954 0.111	245.71 −40.80	247.3 −51.8	TRIPLE (#2: SPIRAL in PAIR) NGC 1660 = 304-G-18	8,23,24 S 2.7	F304/J2655 (214,94)
AM 0442-580	04 42 35 −58 01.4	1.054 0.111	267.21 −39.54	229.0 −41.2	SMALL M51-TYPE	9 S 0.1	F118/J3762 (61,282)
AM 0443-625	04 43 13 −62 59.1	0.598 0.110	273.39 −38.33	224.6 −37.6	I/A TRIPLET (SMALL)	3 D 0.6	F085/J1113 (327,278)
AM 0443-414 [0442-4135]	04 43 31 −41 40.0	1.950 0.109	245.82 −40.62	247.1 −51.9	TRIPLE (#3: SB0) 304-G-19	24 S 16.1	F304/J2655 (204,90)
*AM 0443-360	04 43 40 −36 00.7	2.155 0.109	238.43 −40.09	255.2 −54.8	9 GALAXIES in GROUP (3 I/A DBLS.)	2,18 D 6.7	F361/J3738 (226,125)
*AM 0443-624	04 43 40 −62 42.3	0.626 0.109	273.03 −38.35	224.8 −37.8	ONE-ARMED SPIRAL w. JETS 84-G-40 (3 'STARS'[S.])	7,10 D 0.6	F085/J1113 (326,293)
AM 0443-301	04 43 43 −30 13.0	2.338 0.109	231.03 −39.08	264.8 −57.1	COMPACT	13 D 0.2	F421/J4737 (90,167)
AM 0443-542	04 43 56 −54 23.9	1.309 0.109	262.52 −39.96	232.2 −43.9	CONTACT PAIR	2 S 0.1	F158/J2819 (281,209)
AM 0444-581	04 44 06 −58 10.1	1.038 0.109	267.34 −39.31	228.6 −41.2	SMALL I/A TRIPLE 118-IG-41	3 D 0.4	F119/J3787 (319,274)
AM 0444-621	04 44 30 −62 18.8	0.664 0.108	272.52 −38.36	225.0 −38.2	I/A TRIPLET 119-IG-1	3 D 1.1	F085/J1113 (323,315)
AM 0444-444	04 44 36 −44 49.8	1.816 0.108	249.99 −40.51	242.8 −50.3	E (CENTRE of ARC of 13 GALAXIES) NGC 1668 = 251-G-30	17 D 6.3	F251/J1045 (133,186)
AM 0444-592	04 44 56 −59 20.2	0.939 0.107	268.79 −38.98	227.5 −40.4	B.M. SPIRAL + COMP'NS. NGC 1672 = 118-G-43 = PKS 0444-593	8 D 6.3	F118/J3762 (50,210)
AM 0445-490	04 45 08 −49 07.9	1.611 0.107	255.64 −40.30	237.6 −47.6	I/A DBL.	2 S 0.4	F203/J2836 (307,221)
AM 0445-274	04 45 10 −27 46.0	2.407 0.107	228.07 −38.20	269.0 −58.1	CLOSE PAIR of E's (COMPACTS)	23 S 0.5	F421/J4737 (70,298)
AM 0445-360	04 45 10 −36 00.1	2.154 0.107	238.47 −39.78	255.0 −55.1	ELONG. DWARF (STRUCTURE, RESOL.?) 361-G-9	20 D 1.8	F361/J3738 (210,126)
AM 0445-445	04 45 10 −44 52.3	1.813 0.107	250.05 −40.41	242.6 −50.3	I/A DBL (E + SPIRAL)	2 S 0.4	F251/J1045 (127,184)
AM 0445-525	04 45 11 −52 59.3	1.394 0.107	260.66 −39.96	233.5 −45.0	RING + STELLAR NUCLEUS (DEFECT?)	6 D 0.4	F202/J0941 (54,13)
AM 0445-622	04 45 12 −62 22.6	0.655 0.107	272.57 −38.26	224.9 −38.2	DWARF	20 D 0.7	F085/J1113 (318,312)
AM 0445-474	04 45 15 −47 43.0	1.681 0.107	253.78 −40.35	239.2 −48.6	I/A DBL.	2 S 0.2	F203/J2836 (310,297)
AM 0445-635	04 45 15 −63 53.3	0.491 0.107	274.43 −37.88	223.6 −37.0	RING GALAXY (near BRIGHT STAR)	6 D 0.4	F085/J1113 (310,231)
*AM 0445-572	04 45 23 −57 25.9	1.091 0.107	266.36 −39.28	229.1 −41.9	ONE-ARMED SP. w. COMPACT COMP'N. Se 40/1 = 158-G-3 = UKS 0445-574	10,13,16 S 5.6	F119/J3787 (313,314)
AM 0445-483	04 45 27 −48 38.6	1.635 0.107	254.99 −40.27	238.1 −48.0	I/A DBL. (SMALL) 203-IG-1	2 S 0.1	F202/J0941 (40,246)
AM 0445-622	04 45 44 −62 29.2	0.641 0.106	272.69 −38.18	224.7 −38.1	I/A DBL. (in GROUP of 3) 119-IG-1	2,244 S 0.6	F084/J0959 (41,306)
AM 0445-444	04 45 50 −44 46.4	1.816 0.106	249.92 −40.29	242.6 −50.5	3-ARMED SPIRAL at END of 13 GALS. 251-G-32	11,17 D 0.8	F251/J1045 (121,189)
AM 0446-623	04 46 02 −62 33.0	0.634 0.106	272.76 −38.13	224.6 −38.1	LOOSE GROUP of DISTURBED I/A GAL 85-G-1	16,18 D 13.4	F118/J3762 (55,38)
AM 0446-622	04 46 58 −62 27.5	0.640 0.105	272.61 −38.05	224.6 −38.2	GALAXY w. SHARP ABS'N. + DIAMETRIC 119-G-2 COMP'NS.	8,13,14,17 S 1.3	F118/J3762 (49,42)
AM 0446-634	04 46 59 −63 49.1	0.492 0.105	274.28 −37.71	223.5 −37.2	I/A DBL. in GROUP of 8	2,18 D 3.0	F085/J1113 (300,235)
AM 0447-550 [0447-5458]	04 47 16 −55 00.5	1.260 0.104	263.22 −39.40	231.2 −43.8	PAIR of E's (#1)	23 S 2.2	F158/J2819 (254,177)
AM 0447-593 [0447-5930]	04 47 19 −59 30.4	0.918 0.104	268.92 −38.65	227.0 −40.5	I/A DBL. (#1: LO.S.BR.) 119-G-3	2 S 1.3	F118/J3762 (34,200)
AM 0447-545 [0447-5500]	04 47 23 −54 58.1	1.262 0.104	263.16 −39.39	231.2 −43.9	PAIR of E's (#2)	23 S 2.2	F158/J2819 (253,180)

Identification	R.A. (1950.0) Declination	Precession R.A./Dec.	Gal. Long. Gal. Lat.	S.G.L. S.G.B.	Description Cross-Identification	Coding Size (Min.)	Field/Plate (x, y)
AM 0447-593 [0447-5930]	04 47 32 −59 30.1	0.917 0.104	268.91 −38.62	227.0 −40.5	I/A DBL. (#2: HI.S.BR.) Se 40/8 = 119-G-4	2 S 1.3	F118/J3762 (33,200)
AM 0447-595	04 47 37 −59 52.9	0.884 0.104	269.39 −38.53	226.7 −40.3	DISTURBED SPIRAL + COMPACT [N.E.] NGC 1688 = 119-G-6	13,16 D 2.0	F118/J3762 (34,180)
AM 0447-633	04 47 49 −63 33.1	0.519 0.103	273.92 −37.69	223.6 −37.5	GALAXY w. PECULIAR ARM + COMP'N. 85-IG-5 = B 0447-635	8,10,12 D 0.9	F085/J1113 (297,250)
AM 0448-320	04 48 00 −32 02.8	2.278 0.103	233.57 −38.55	261.0 −57.3	LARGE Sd/DWARF NGC 1679 = 422-G-1	8,10,20 D 2.7	F421/J4737 (43,67)
AM 0448-574	04 48 02 −57 44.6	1.060 0.103	226.68 −38.88	228.5 −41.9	CIRCULAR-ARMED SPIRAL + 119-G-8 = Fa 239 CONDENS. in ARM	10 D 0.7	F158/J2819 (243,31)
AM 0448-235	04 48 03 −23 58.6	2.508 0.103	223.70 −36.54	275.8 −59.6	VERY CLOSE I/A DBL. MCG-4-12-17 = 485-G-19	2 S 0.1	F485/J5480 (154,234)
AM 0448-420	04 48 06 −42 09.4	1.922 0.103	246.52 −39.80	245.6 −52.4	SPIRAL w. 2 APP. COMP'NS. 304-G-21	8 S 2.9	F251/J1045 (96,329)
AM 0448-492	04 48 12 −49 22.6	1.591 0.103	255.92 −39.79	236.8 −47.8	I/A DBL. + MATERIAL 203-IG-5 (in GROUP of 7)	2,18 S 0.2	F203/J2836 (280,209)
*AM 0448-622	04 48 18 −62 28.9	0.633 0.103	272.59 −37.89	224.4 −38.3	M51-TYPE (E + SPIRAL)	2,91 S 0.4	F085/J1113 (298,308)
AM 0448-594	04 48 24 −59 42.7	0.896 0.103	269.15 −38.47	226.7 −40.5	COMPACT	13 D 0.2	F119/J3787 (285,193)
AM 0448-543	04 48 31 −54 37.6	1.282 0.102	262.70 −39.27	231.3 −44.2	PAIR	23 S 0.4	F158/J2819 (245,198)
AM 0448-604	04 48 33 −60 46.2	0.801 0.102	270.46 −38.24	225.8 −39.7	SMALL I/A TRIPLE w. LINEAR 119-IG-10 = B 0448-607 ELEMENT	3,7 D 0.6	F119/J3787 (281,136)
*AM 0448-622	04 48 39 −62 26.6	0.636 0.102	272.53 −37.86	224.4 −38.4	LARGE, LUMPY COMPACT in GROUP 119-IG-11 = PKS 448-62?	13,18 D 0.4	F084/J0959 (23,307)
AM 0448-452	04 48 53 −45 27.6	1.780 0.102	250.82 −39.76	241.3 −50.5	I/A TRIPLE	3 D 0.8	F251/J1045 (93,151)
AM 0449-551	04 49 10 −55 12.8	1.241 0.102	263.43 −39.11	230.7 −43.9	GROUP of 3 incl. Fa 423?	24 D 6.6	F158/J2819 (239,167)
AM 0449-424	04 49 19 −42 42.8	1.898 0.101	247.26 −39.60	244.7 −52.2	DWARF	20 D 0.4	F304/J2655 (146,34)
AM 0449-472	04 49 26 −47 27.7	1.685 0.101	253.42 −39.65	238.8 −49.3	I/A DBL. (SMALL) in GROUP of 5	2,18 S 0.3	F251/J1045 (91,44)
AM 0449-482	04 49 30 −48 24.3	1.639 0.101	254.65 −39.61	237.7 −48.7	SMALL I/A DBL. 203-IG-8	2 S 0.4	F203/J2836 (270,261)
AM 0449-625	04 49 32 −62 54.8	0.583 0.101	273.08 −37.65	223.9 −38.1	I/A DBL. (SMALL)	2 S 0.4	F084/J0959 (20,281)
AM 0449-612 [0449-6125]	04 49 49 −61 26.7	0.732 0.101	271.26 −37.95	225.1 −39.2	SMALL PAIR of I/A SPIRALS (#1) 119-G-14?	2 S 1.4	F118/J3762 (26,95)
AM 0450-612 [0449-6126]	04 50 01 −61 25.9	0.733 0.100	271.24 −37.93	225.1 −39.3	SMALL PAIR of I/A SPIRALS (#2) 119-G-13?	2 S 1.4	F118/J3762 (25,95)
AM 0450-462	04 50 21 −46 29.7	1.730 0.100	252.17 −39.50	239.8 −50.0	GALAXY w. I/A COMP'N.	1 S 0.2	F251/J1045 (81,95)
AM 0450-423	04 50 30 −42 30.4	1.904 0.100	247.00 −39.37	244.7 −52.5	SPIRAL w. ASYMM. ARM + APP.COMP'N 304-G-27 = B 0450-424	8,10 S 0.8	F251/J1045 (73,309)
AM 0450-635	04 50 38 −63 52.5	0.473 0.100	274.22 −37.31	223.1 −37.4	I/A DBL. w. LOOP	2 S 0.1	F085/J1113 (279,234)
*AM 0450-614	04 50 51 −61 43.6	0.701 0.099	271.58 −37.77	224.7 −39.1	DISTURBED DWARF + I/A COMPACT 119-G-16	1,13,20 S 1.3	F118/J3762 (21,79)
AM 0450-634	04 50 56 −63 46.9	0.482 0.099	274.10 −37.30	223.1 −37.5	I/A TRIPLE	3 D 1.2	F085/J1113 (277,239)
AM 0450-620	04 50 57 −62 05.6	0.664 0.099	272.03 −37.68	224.4 −38.8	SMALL I/A DBL. w. PLUME	2 S 0.1	F084/J0959 (7,324)
AM 0451-453	04 51 36 −45 34.9	1.770 0.098	250.99 −39.28	240.6 −50.8	SPIRAL + I/A COMP'N. (CHAIN of 9) 251-G-39 + 251-IG-40 = B 0451-455	1,17 S 0.6	F251/J1045 (68,144)
*AM 0452-630	04 52 04 −63 03.9	0.558 0.098	273.19 −37.34	223.5 −38.1	SPIRAL + I/A COMP'N. Fa 240 NGC 1706 = 85-G-7 = B 0452-630	1,9 S 1.0	F084/J0959 (6,271)
*AM 0452-594	04 52 09 −59 49.4	0.875 0.097	269.17 −37.99	226.2 −40.7	LARGE SPIRAL w. APP. COMP'NS. NGC 1703 = 119-G-199 (incl. RING)	6,8 S 8.1	F119/J3787 (259,188)
AM 0452-360	04 52 11 −36 00.3	2.144 0.097	238.74 −38.38	253.8 −56.3	SPIRAL w. I/A COMP'N. 361-G-17	1 S 0.8	F361/J3738 (134,125)
AM 0452-500	04 52 22 −50 08.6	1.542 0.097	256.87 −39.08	235.3 −47.8	CLOSE PAIR in LINE of 3	24 D 4.5	F203/J2836 (243,169)
AM 0452-435	04 52 32 −43 57.6	1.840 0.097	248.90 −39.07	242.5 −51.9	CLOSE PAIR 251-G-41	23 S 3.4	F252/J1933 (329,231)
AM 0452-640	04 52 39 −64 04.6	0.442 0.097	274.40 −37.04	222.7 −37.4	I/A DBL. (SMALL)	2 S 0.2	F084/J0959 (8,217)
AM 0452-260	04 52 58 −26 06.8	2.446 0.096	226.64 −36.09	271.2 −60.2	SP. (LARGE NUCLEUS) + I/A COMP'N. MCG-4-12-21 = 485-G-23	1,12 S 0.2	F485/J5480 (95,119)
AM 0453-632	04 53 05 −63 28.9	0.509 0.096	273.66 −37.13	223.1 −37.9	COMPACT	13 D 0.2	F085/J1113 (265,256)

Identification	R.A. (1950.0) Declination	Precession R.A./Dec.	Gal. Long. Gal. Lat.	S.G.L. S.G.B.	Description Cross-Identification	Coding Size (Min.)	Field/Plate (x, y)
*AM 0453-372	04 53 07 −37 24.1	2.095 0.096	240.53 −38.39	251.5 −55.8	SPIRAL w. DIAMETRIC + I/A COMP'N. 361-G-19 (FILAMENT)	1,8,13 S 6.9	F361/J3738 (125,50)
AM 0453-385	04 53 07 −38 50.9	2.042 0.096	242.37 −38.57	249.3 −55.0	I/A DBL. 304-IG-29 = B 0453-388	2 S 0.1	F304/J2655 (105,241)
AM 0453-532	04 53 08 −53 26.4	1.348 0.096	261.08 −38.73	231.8 −45.5	DISTURBED GAL. w. ABS'N. and JETS NGC 1705 = 158-G-13	7,14,16 D 2.2	F158/J2819 (210,263)
AM 0453-472	04 53 45 −47 23.9	1.680 0.095	253.33 −38.92	238.1 −49.9	CONTACT DBL. + GALAXY w. TAIL (in GROUP of 6)	15,18,23 S 0.2	F203/J2836 (234,316)
AM 0453-640	04 53 53 −64 02.7	0.442 0.095	274.33 −36.92	222.6 −37.5	SPIRAL + I/A COMP'N.	2 S 0.4	F085/J1113 (259,226)
AM 0454-424	04 54 04 −42 46.8	1.888 0.095	247.41 −38.73	243.7 −52.9	SPIRAL + APP. COMP'N.	8 S 0.9	F252/J1933 (317,295)
*AM 0454-561	04 54 45 −56 19.1	1.149 0.094	264.71 −38.19	228.9 −43.6	IRR. w. STAR-LIKE CONDENSATIONS 158-G-15	16 D 1.0	F158/J2819 (196,108)
AM 0454-460	04 54 57 −46 07.0	1.739 0.094	251.69 −38.71	239.4 −50.9	SP.w. FAINT CONN. to DBL. COMP'N.	1 S 0.9	F251/J1045 (37,114)
AM 0454-471	04 54 59 −47 16.0	1.685 0.093	253.16 −38.71	238.0 −50.1	E w. JET + CONDENSATION (SMALL)	7 D 0.3	F251/J1045 (40,52)
AM 0455-283	04 55 21 −28 34.0	2.374 0.093	229.74 −36.23	266.2 −60.0	CLOSE TRIPLE (near LARGE SPIRAL) 422-G-13	24 D 2.0	F422/J3765 (222,256)
AM 0455-425	04 55 27 −42 53.2	1.881 0.093	247.56 −38.49	243.3 −53.0	I/A DBL.(w. RESOLVED DWARF) [PKS] Ag-22 = MCG-7-11-2 = 252-IG-1 [B]	2,20 S 0.5	F251/J1045 (24,287)
AM 0455-424	04 55 30 −42 40.7	1.890 0.093	247.30 −38.46	243.6 −53.1	LINE of 3	24 D 4.4	F252/J1933 (303,301)
AM 0455-465	04 55 36 −46 50.2	1.704 0.093	252.61 −38.61	238.4 −50.5	RING w. I/A COMP'N. (in GROUP)	1,6,18 S 0.4	F251/J1045 (33,75)
AM 0455-510	04 55 37 −51 02.8	1.485 0.093	257.99 −38.52	233.8 −47.5	SMALL I/A DBL. (JET ?)	2,7 S 0.4	F203/J2836 (214,121)
AM 0455-471	04 55 38 −47 14.0	1.685 0.093	253.12 −38.60	238.0 −50.2	SPIRAL w. I/A COMP'N. 252-G-2	1 S 0.4	F203/J2836 (217,326)
AM 0456-312	04 56 27 −31 29.9	2.286 0.091	233.35 −36.68	260.6 −59.2	SMALL TRIPLE	24 D 0.9	F422/J3765 (208,99)
AM 0456-250	04 56 32 −25 04.5	2.472 0.091	225.69 −35.03	272.9 −61.3	CLOSE I/A DBL.	2 S 0.2	F486/J4784 (319,174)
AM 0456-275	04 56 37 −27 54.9	2.392 0.091	229.05 −35.79	267.2 −60.5	DISTURBED (GALAXY + APP. COMP'N.) MCG-5-12-12 = 422-G-15	8,16 S 3.4	F485/J5480 (53,21)
AM 0456-505	04 56 38 −50 57.7	1.488 0.091	257.88 −38.36	233.7 −47.7	GALAXY w. I/A COMP'N. 203-IG-14 = B 0456-509	1 S 0.5	F203/J2836 (205,125)
AM 0456-224	04 56 46 −22 44.9	2.534 0.091	223.02 −34.27	277.6 −61.8	SPIRAL w. VERY FAINT ARMS 485-G-29 + EXTENSIONS	10 D 1.1	F486/J4784 (319,299)
*AM 0456-264	04 56 46 −26 46.6	2.424 0.091	227.71 −35.46	269.4 −60.9	GAL. w. LONG THIN ARM (M51-TYPE) 486-G-1	9,10 S 0.5	F485/J5480 (50,83)
AM 0456-464	04 56 49 −46 43.9	1.707 0.091	252.48 −38.40	238.3 −50.7	GROUP of 19 GALAXIES	18 D 22.4	F251/J1045 (22,80)
AM 0456-340	04 56 53 −34 03.1	2.204 0.091	236.51 −37.10	256.1 −58.1	GROUP of 3	24 D 0.9	F361/J3738 (80,229)
AM 0457-541 [0457-5414]	04 57 02 −54 13.1	1.289 0.091	262.01 −38.09	230.5 −45.4	PAIR of I/A DBLS. (#1)	2 S 1.3	F158/J2819 (179,221)
AM 0457-541 [0457-5413]	04 57 07 −54 14.2	1.287 0.090	262.03 −38.07	230.4 −45.4	PAIR of I/A DBLS. (#2)	2 S 1.3	F158/J2819 (178,220)
AM 0457-382	04 57 16 −38 28.9	2.051 0.090	242.03 −37.72	249.1 −55.9	I/A DBL. (SMALL) 304-IG-30 = B 0457-384	2 S 0.5	F305/J2776 (334,257)
AM 0457-224	04 57 28 −22 46.3	2.533 0.090	223.11 −34.12	277.5 −61.9	DWARF MCG-4-12-27	20 D 1.3	F486/J4784 (310,298)
AM 0457-635	04 57 40 −63 59.9	0.434 0.090	274.16 −36.52	222.2 −37.8	IRR. or DISTURBED GALAXY 85-G-24	16 D 1.3	F085/J1113 (237,230)
AM 0457-440	04 57 46 −44 02.0	1.829 0.090	249.05 −38.14	241.4 −52.6	CLOSE PAIR	23 S 0.3	F251/J1045 (5,224)
*AM 0457-620	04 57 50 −62 06.3	0.642 0.089	271.84 −36.89	223.6 −39.3	E w. JETS NGC 1765 = 119-G-24	7 D 0.8	F085/J1113 (240,331)
AM 0458-325	04 58 06 −32 58.2	2.238 0.089	235.24 −36.65	257.7 −58.8	DWARF/PLANETARY NEBULA ? (PECULIAR CIRCULAR RIM)	20,25	F422/J3765 (188,20)
AM 0458-455	04 58 13 −45 53.0	1.745 0.089	251.41 −38.13	239.1 −51.4	CONTACT I/A DBL. w. TAIL	2,15 S 0.1	F252/J1933 (270,130)
AM 0458-584	04 58 20 −58 43.9	0.952 0.089	267.65 −37.39	226.3 −42.0	SPIRAL + COMPACT COMP'N. 119-G-25	8,13 S 2.0	F119/J3787 (218,247)
AM 0458-461	04 58 35 −46 14.7	1.727 0.088	251.87 −38.08	238.6 −51.2	LOOSE CHAIN of 5	17 D 11.2	F252/J1933 (266,111)
*AM 0458-250	04 58 38 −25 08.8	2.468 0.088	225.95 −34.59	272.5 −61.7	M51-TYPE + PECULIAR COMP'N. MCG-4-12-31 = 486-G-7	8,9 S 1.4	F485/J5480 (26,170)
AM 0459-223	04 59 00 −22 39.4	2.535 0.088	223.12 −33.75	277.6 −62.3	CONTACT I/A DBL. in GROUP	2,18 S 0.4	F485/J5480 (18,303)

Identification	R.A. (1950.0) Declination	Precession R.A./Dec.	Gal. Long. Gal. Lat.	S.G.L. S.G.B.	Description Cross-Identification	Coding Size (Min.)	Field/Plate (x, y)
AM 0459-753	04 59 04 −75 30.7	−1.916 0.088	287.61 −33.22	214.5 −28.2	I/A TRIPLE in GROUP of 4 33-G-4/5 = PKS 0458-75	3,18 D 2.9	F032/J1937 (62,143)
AM 0459-384	04 59 08 −38 44.9	2.038 0.088	242.42 −37.39	248.3 −56.0	CLUSTER of 10 GALAXIES NGC 1759 = 305-G-1 = B 0459-387?	19 D 6.7	F304/J2655 (42,244)
AM 0459-231	04 59 25 −23 11.2	2.521 0.087	223.76 −33.83	276.5 −62.3	GROUP of 5 incl. I/A DBL.	2,18 D 9.9	F485/J5480 (14,275)
AM 0459-265	04 59 26 −26 54.1	2.419 0.087	228.06 −34.92	268.8 −61.4	PECULIAR SPIRAL w. CONDENSATION 486-G-9 on END of ARM	9,12 D 0.6	F485/J5480 (18,75)
AM 0459-394	04 59 42 −39 40.7	2.003 0.087	243.60 −37.40	246.9 −55.6	DWARF	20 D 0.6	F304/J2655 (38,194)
*AM 0459-340	04 59 51 −34 06.2	2.199 0.087	236.71 −36.51	255.5 −58.6	GAL. w. I/A COMP'N. in GROUP of 3 361-IG-25	1,24 S 1.1	F362/J5542 (314,225)
*AM 0500-620	05 00 00 −62 09.0	0.631 0.086	271.83 −36.63	223.3 −39.4	M51-TYPE 119-IG-27 = Fa 241/2	9 S 0.4	F085/J1113 (226,329)
*AM 0500-590	05 00 11 −59 06.6	0.915 0.086	268.08 −37.10	225.7 −41.9	ONE-ARMED SPIRAL w. CONNECTED 119-IG-28 = B 0500-591 COMP'N.	9,10 S 0.7	F119/J3787 (205,227)
AM 0500-444	05 00 53 −44 48.3	1.790 0.085	250.06 −37.62	239.9 −52.5	I/A DBL. (END of CHAIN of 8) 252-G-5/6	2,17 S 0.9	F252/J1933 (247,189)
AM 0501-225	05 01 04 −22 53.7	2.527 0.085	223.57 −33.37	276.9 −62.7	CLOSE PAIR MCG-4-12-38/39 = 486-G-17/19	23 S 1.1	F486/J4784 (265,292)
*AM 0501-632	05 01 04 −63 21.9	0.497 0.085	273.29 −36.28	222.3 −38.5	GALAXY w. JET 85-G-30	7 D 1.3	F085/J1113 (218,264)
*AM 0501-761	05 01 10 −76 14.1	−2.202 0.085	288.37 −32.86	213.9 −27.7	E w. JET + CONDENSATION 33-IG-8	7 D 1.0	F032/J1937 (61,104)
AM 0501-252	05 01 17 −25 29.5	2.457 0.085	226.56 −34.12	271.4 −62.2	DISTURBED (RESOLVED) w. JETS? MCG-4-12-41 = 486-G-21 (GRP of 4)	7,16,18 D 1.0	F486/J4784 (260,152)
AM 0501-395	05 01 23 −39 50.2	1.995 0.084	243.85 −37.10	246.3 −55.8	PLANETARY NEBULA or COMPACT	13,25 D 0.3	F304/J2655 (21,185)
AM 0501-342	05 01 36 −34 29.5	2.185 0.084	237.27 −36.23	254.5 −58.7	I/A DBL. (ONE COMPACT)	2,13 S 0.4	F361/J3738 (29,204)
AM 0501-232	05 01 40 −23 29.5	2.511 0.084	224.30 −33.43	275.6 −62.7	CHAIN of 10	17 D 14.8	F486/J4784 (257,260)
AM 0502-601	05 02 38 −60 14.6	0.809 0.083	269.43 −36.63	224.5 −41.1	GROUP of 4	24 D 2.7	F119/J3787 (188,166)
AM 0503-775	05 03 01 −77 50.0	−2.933 0.082	290.12 −32.22	212.9 −26.3	COMPACT	13 D 0.1	F033/J1958 (253,21)
AM 0503-251	05 03 06 −25 19.7	2.460 0.082	226.52 −33.68	271.5 −62.6	I/A DBL. 486-IG-26	2 S 0.3	F486/J4784 (238,162)
AM 0503-473	05 03 32 −47 38.5	1.652 0.081	253.66 −37.27	236.0 −50.9	CLOSE PAIR in GROUP of 3 incl. Fa 424?	8,23,24 S 0.5	F252/J1933 (219,37)
AM 0503-380	05 03 33 −38 02.3	2.059 0.081	241.69 −36.44	248.5 −57.2	DISTURBED SPIRAL + 5 COMP'NS. NGC 1792 = 305-G-6 = Lu 0503-38	8,10 D 5.8	F305/J2776 (269,283)
AM 0503-553 [0505-5542]	05 03 36 −55 35.8	1.181 0.081	263.64 −37.03	228.2 −44.9	WIDE PAIR of SPIRALS (#1) 158-G-17	23 S 11.2	F158/J2819 (129,147)
AM 0503-471	05 03 56 −47 18.5	1.668 0.081	253.24 −37.19	236.3 −51.2	I/A DBL.	2 S 1.1	F203/J2836 (141,321)
AM 0504-474	05 04 05 −47 41.6	1.649 0.081	253.73 −37.18	235.9 −50.9	CLOSE PAIR	23 S 0.6	F203/J2836 (140,301)
AM 0504-511	05 04 07 −51 10.3	1.462 0.081	258.09 −37.18	232.2 −48.4	GALAXY w. LONG DIAMETRIC TAILS	15 D 1.1	F203/J2836 (142,114)
AM 0504-493 [0504-4941]	05 04 08 −49 38.0	1.548 0.081	256.16 −37.19	233.8 −49.5	CLOSE PAIR of SPIRALS (#1) IC 1803 = 203-G-19 = Fa 305	23 D 0.7	F203/J2836 (141,197)
AM 0504-493 [0504-4938]	05 04 17 −49 39.8	1.546 0.080	256.20 −37.17	233.7 −49.5	CLOSE PAIR of SPIRALS (#2) 203-G-19	23 D 0.5	F203/J2836 (140,195)
AM 0504-320	05 04 34 −32 01.0	2.262 0.080	234.45 −35.11	258.2 −60.5	IRR.(DUST) w. HALO + COMP'NS. NGC 1800 = 422-G-30	8,14,16 D 0.7	F362/J5542 (264,339)
AM 0504-281	05 04 40 −28 17.5	2.375 0.080	230.06 −34.18	265.2 −62.0	CLOSE PAIR in GROUP of 5 422-IG-31	18,23 S 0.3	F422/J3765 (113,271)
AM 0504-274	05 04 50 −27 43.2	2.391 0.080	229.41 −33.99	266.4 −62.3	WIDE PAIR of PECULIAR GALAXIES MCG-5-13-6 = 422-G-33 (RESOLUTION)	23 D 24.6	F486/J4784 (216,33)
AM 0505-554 [0503-5535]	05 05 11 −55 42.0	1.170 0.079	263.75 −36.80	227.8 −45.0	WIDE PAIR of SPIRALS (#2) 158-G-18	23 S 11.2	F158/J2819 (117,141)
AM 0505-230	05 05 29 −23 08.7	2.518 0.079	224.24 −32.49	276.0 −63.7	I/A DBL. (E + SPIRAL) 486-G-29/30	2 S 1.3	F486/J4784 (210,279)
AM 0505-374	05 05 39 −37 46.4	2.067 0.078	241.45 −35.99	248.4 −57.6	I/A PAIR of SPIRALS	2 S 0.3	F362/J5542 (246,30)
*AM 0505-373	05 05 58 −37 34.7	2.073 0.078	241.22 −35.90	248.7 −57.8	SB w. PECULIAR ABS'N + 11 COMP'NS NGC 1808 = 305-G-8	8,14 D 4.0	F362/J5542 (243,40)
AM 0506-221	05 06 02 −22 11.4	2.543 0.078	223.22 −32.06	278.0 −64.0	E + DIAMETRIC COMP'NS. + OTHERS 553-G-9 (in GROUP of 5)	8,18 S 3.6	F486/J4784 (204,330)
AM 0506-410	05 06 09 −41 03.5	1.941 0.078	245.49 −36.34	243.6 −55.7	CLOSE PAIR	23 S 0.4	F305/J2776 (238,122)

Identification	R.A. (1950.0) Declination	Precession R.A./Dec.	Gal. Long. Gal. Lat.	S.G.L. S.G.B.	Description Cross-Identification	Coding Size (Min.)	Field/Plate (x, y)
AM 0506-594	05 06 15 −59 47.4	0.841 0.078	268.79 −36.24	224.3 −41.8	ELONGATED IRR. NGC 1824 = 119-G-36	16 D 2.2	F119/J3787 (164,191)
AM 0506-382	05 06 25 −38 22.4	2.044 0.077	242.21 −35.93	247.4 −57.4	LARGE Sd/DWARF w. RESOLUTION 305-G-9 = UKS 0506-383	20 D 4.0	F305/J2776 (238,266)
AM 0506-750	05 06 27 −75 01.4	−1.785 0.077	286.88 −32.93	214.3 −28.9	I/A DBL. 33-IG-12	2 S 0.4	F033/J1958 (259,172)
AM 0506-374	05 06 42 −37 43.4	2.067 0.077	241.43 −35.77	248.3 −57.8	I/A DBL. 305-IG-10 = B 0506-377	2 S 0.3	F362/J5542 (235,33)
AM 0506-292	05 06 45 −29 20.5	2.342 0.077	231.43 −34.01	262.8 −62.1	I/A DBL. (#1: EDGE-ON SPIRAL) NGC 1811 = 422-G-37	2 D 1.6	F422/J3765 (89,215)
AM 0506-291	05 06 55 −29 19.4	2.343 0.077	231.42 −33.97	262.8 −62.1	I/A DBL. (#2: Sa) NGC 1812 = 422-G-39	2 D 1.0	F422/J3765 (87,216)
AM 0507-512	05 07 15 −51 25.7	1.442 0.076	258.40 −36.69	231.4 −48.5	M51-TYPE + RING w. OFF-SET NUCLEUS 203-G-21	6,8,9 S 3.1	F203/J2836 (116,100)
AM 0507-375	05 07 18 −37 53.4	2.061 0.076	241.65 −35.69	247.9 −57.8	7 in a CHAIN	17 D 8.5	F362/J5542 (229,24)
AM 0507-630	05 07 18 −63 03.2	0.514 0.076	272.76 −35.65	221.8 −39.2	DWARF 85-G-47	20 D 0.7	F119/J3787 (160,15)
AM 0507-611	05 07 21 −61 15.3	0.701 0.076	270.56 −35.91	223.1 −40.6	SPIRAL + INTEGRAL-SPIRAL (I/A) NGC 1796a? = 119-IG-37	2 S 0.2	F119/J3787 (158,112)
AM 0507-375	05 07 35 −37 55.2	2.059 0.076	241.70 −35.64	247.8 −57.9	CHAIN of 8 305-G-12	17 D 11.2	F305/J2776 (226,291)
AM 0507-220	05 07 39 −22 05.9	2.545 0.076	223.26 −31.68	278.1 −64.4	GROUP of 3 SPIRALS	24 D 4.5	F486/J4784 (184,335)
AM 0507-254	05 07 46 −25 40.0	2.448 0.075	227.28 −32.78	270.2 −63.5	GROUP of 3 SPIRALS 486-G-32	18,24 D 13.0	F486/J4784 (182,144)
AM 0507-222	05 07 49 −22 21.5	2.538 0.075	223.57 −31.72	277.5 −64.3	OFF-CENTER NUCLEUS and BROKEN MCG-4-13-2 = 553-G-15 ARMS	10,12 D 0.9	F486/J4784 (182,321)
AM 0508-370	05 08 20 −37 01.5	2.091 0.075	240.65 −35.34	249.0 −58.5	DIFFRACTION SPIKE on NUCLEUS NGC 1827 = 362-G-6	16 D 3.1	F305/J2776 (219,339)
AM 0508-373	05 08 21 −37 38.9	2.068 0.075	241.40 −35.44	248.0 −58.2	I/A DBL. E + PECULIAR SPIRAL	2,12 S 0.7	F305/J2776 (218,305)
AM 0508-330	05 08 36 −33 04.6	2.225 0.074	235.94 −34.52	255.5 −60.7	SMALL DWARF (STRUCTURE) 362-G-7	20 D 0.5	F362/J5542 (218,282)
AM 0509-290	05 09 25 −29 03.1	2.349 0.073	231.28 −33.38	262.9 −62.7	E + RING + 5 COMP'NS. 422-G-43	6,8,23 D 2.7	F423/J2710 (327,225)
AM 0509-572	05 09 48 −57 24.8	1.033 0.073	265.81 −36.03	225.7 −44.0	I/A DBL. 158-IG-20 near PKS 0509-573 = QSO	2 S 0.5	F158/J2819 (86,48)
*AM 0510-330	05 10 07 −33 02.2	2.225 0.072	235.97 −34.20	255.2 −61.0	LARGE LO.S.BR. Sd (RESOLUTION) Ka-26 = 362-G-9 = UKS 0510-330	20 D 4.0	F362/J5542 (201,285)
AM 0510-423	05 10 07 −42 30.8	1.877 0.072	247.39 −35.77	240.8 −55.3	CLOSE PAIR of SPIRALS	23 S 0.7	F305/J2776 (198,44)
AM 0510-762	05 10 07 −76 24.6	−2.323 0.072	288.37 −32.30	213.3 −27.8	GALAXY w. TAIL 33-G-16	15 D 1.1	F033/J1958 (240,99)
AM 0510-260	05 10 19 −26 06.7	2.433 0.072	227.98 −32.36	268.8 −64.0	SPIRAL w. 8 COMP'NS 486-G-34	8 S 4.0	F486/J4784 (151,120)
AM 0510-233	05 10 48 −23 30.2	2.505 0.071	225.10 −31.45	274.5 −64.8	GROUP of 3	24 D 3.1	F486/J4784 (145,260)
*AM 0511-572	05 11 25 −57 27.5	1.026 0.070	265.84 −35.81	225.4 −44.1	SPIRAL w. WIDE DIAM. COMP'NS. NGC 1853 = 158-G-22 (I/A DBL.)	2,8 S 10.8	F119/J3787 (125,315)
*AM 0511-303	05 11 37 −30 32.3	2.302 0.070	233.14 −33.30	259.5 −62.5	E w. JET to COMP'N. w. TAIL PKS 0511-30	7,8,15 S 0.5	F423/J2710 (299,146)
AM 0511-775	05 11 39 −77 51.3	−2.997 0.070	289.97 −31.78	212.5 −26.5	GALAXY w. PECULIAR COMP'N. 15-G-19	8 S 2.4	F033/J1958 (229,22)
AM 0512-485	05 12 11 −48 59.1	1.570 0.069	255.38 −35.87	232.9 −50.9	CLOSE PAIR	23 S 0.5	F203/J2836 (70,229)
AM 0512-611	05 12 13 −61 13.3	0.693 0.069	270.43 −35.34	222.4 −41.0	QUARTET	18 D 1.6	F119/J3787 (127,113)
AM 0512-465	05 12 42 −46 54.8	1.675 0.068	252.83 −35.69	235.0 −52.5	SMALL I/A DBL.	2 S 0.2	F252/J1933 (136,76)
AM 0512-451	05 12 46 −45 18.6	1.751 0.068	250.87 −35.56	236.8 −53.7	CONTACT TRIPLE 252-IG-13	3 D 0.3	F252/J1933 (134,162)
AM 0513-511	05 13 01 −51 12.9	1.445 0.068	258.13 −35.79	230.6 −49.2	I/A E's in COMMON ENVELOPE	2 S 0.3	F204/J2662 (323,112)
AM 0513-271	05 13 10 −27 11.3	2.401 0.068	229.42 −32.06	266.0 −64.2	VERY DISTANT CLUSTER of GALAXIES	19 D 1.1	F486/J4784 (117,62)
AM 0513-412	05 13 25 −41 26.8	1.918 0.067	246.18 −35.03	241.5 −56.5	IRR. SINGLE GALAXY 305-G-17	16 D 2.2	F305/J2776 (165,102)
AM 0513-553	05 13 27 −55 34.4	1.164 0.067	263.50 −35.64	226.6 −45.8	COMPACT	13 D 0.1	F159/J2675 (305,145)
AM 0513-284	05 13 40 −28 45.7	2.355 0.067	231.24 −32.40	262.6 −63.7	S(r) w. FAINT OUTER ARMS + COMP'N	8,10 S 1.1	F423/J2710 (277,241)

Identification	R.A. (1950.0) Declination	Precession R.A./Dec.	Gal. Long. Gal. Lat.	S.G.L. S.G.B.	Description Cross-Identification	Coding Size (Min.)	Field/Plate (x, y)
AM 0513-224	05 13 48 −22 47.9	2.523 0.067	224.59 −30.57	275.8 −65.6	E w. PECULIAR ABS'N. + SPIRAL	14,23 S 3.2	F486/J4784 (108,298)
AM 0513-620	05 13 56 −62 03.5	0.605 0.067	271.42 −35.03	221.6 −40.4	DISTURBED GALAXY	16 D 0.4	F085/J1113 (139,335)
AM 0514-523	05 14 00 −52 37.5	1.359 0.067	259.87 −35.64	229.1 −48.2	SMALL I/A DBL.	2 S 0.7	F203/J2836 (63,33)
AM 0514-621 [0514-6213]	05 14 00 −62 13.4	0.587 0.067	271.62 −35.00	221.5 −40.3	CLOSE TRIPLE (#1)	24 D 1.0	F085/J1113 (138,326)
AM 0514-621 [0514-6213]	05 14 00 −62 17.2	0.581 0.067	271.69 −35.00	221.5 −40.2	CLOSE TRIPLE (#2) 119-G-47 = Fa 243	24 D 1.1	F085/J1113 (139,322)
AM 0514-621 [0514-6217]	05 14 27 −62 13.9	0.586 0.066	271.62 −34.95	221.4 −40.3	CLOSE TRIPLE (#3)	24 D 0.5	F085/J1113 (136,325)
AM 0514-273 [0514-2728]	05 14 46 −27 31.3	2.390 0.065	229.92 −31.82	265.0 −64.4	CLOSE PAIR (#1: E)	23 D 0.1	F486/J4784 (98,44)
*AM 0514-370	05 14 53 −37 09.1	2.080 0.065	241.08 −34.08	247.3 −59.5	GALAXY w. I/A COMP'N. 362-G-11	1 S 4.5	F362/J5542 (148,64)
AM 0514-360	05 14 54 −36 07.5	2.117 0.065	239.86 −33.88	248.9 −60.1	NGC 2685 TYPE (JET ?) (SMALL)	7,16 D 0.5	F362/J5542 (148,119)
AM 0514-272 [0514-2731]	05 14 55 −27 28.5	2.392 0.065	229.88 −31.78	265.1 −64.4	CLOSE PAIR (#2: SPIRAL)	23 D 0.4	F486/J4784 (96,46)
*AM 0514-370	05 14 55 −37 09.5	2.080 0.065	241.09 −34.07	247.2 −59.5	EDGE-ON next to CHAIN of 4 362-G-11	8,17 D 6.7	F305/J2776 (149,332)
AM 0515-540	05 15 03 −54 09.7	1.259 0.065	261.76 −35.46	227.5 −47.1	DISTURBED GAL. w. LOOPS and TAILS 159-G-3 = MC 0515-541 (in PAIR)	1,2,15,23 D 1.8	F158/J2819 (38,220)
AM 0515-234	05 15 22 −23 47.9	2.495 0.065	225.82 −30.56	273.2 −65.7	M51-TYPE MCG-4-13-7 = 486-G-49	9 S 0.3	F486/J4784 (89,244)
AM 0515-292	05 15 43 −29 26.8	2.333 0.064	232.16 −32.16	260.8 −63.8	GALAXY w. TAILS 423-IG-3	15 D 0.8	F423/J2710 (253,205)
AM 0516-433	05 16 22 −43 37.1	1.823 0.063	248.89 −34.76	238.0 −55.3	COMPACT	13 D 0.1	F252/J1933 (98,253)
AM 0516-621	05 16 32 −62 12.5	0.584 0.063	271.56 −34.71	221.2 −40.4	COMPACT PKS 0516-621?	13 D 0.3	F085/J1113 (123,326)
AM 0516-364	05 16 49 −36 41.5	2.095 0.063	240.62 −33.61	247.5 −60.1	CLOSE PAIR (E + SPIRAL)	23 S 0.4	F362/J5542 (127,88)
AM 0516-512	05 16 54 −51 28.9	1.424 0.062	258.47 −35.18	229.6 −49.4	I/A DBL. in CLOSE TRIPLE	2,24 S 1.6	F203/J2836 (36,93)
*AM 0517-370	05 17 17 −37 08.4	2.079 0.062	241.17 −33.61	246.7 −59.9	E + 4 COMP'NS. (OFF-SET NUCLEUS) IC 2122 = 362-G-14	8,20,22 S 5.4	F305/J2776 (123,333)
AM 0517-324	05 17 42 −32 42.4	2.230 0.061	236.03 −32.57	254.0 −62.5	PEC. ABS'N. + TAIL in WIDE PAIR MCG-5-13-17 = 362-G-18	14,15,23 D 0.6	F362/J5542 (115,302)
AM 0517-250	05 17 46 −25 08.8	2.456 0.061	227.51 −30.46	269.7 −65.9	CLOSE PAIR near HALO of cD cD = IC 2121 = 486-G-53 [MC]	23 S 0.3	F486/J4784 (60,171)
AM 0517-621	05 17 47 −62 15.1	0.576 0.061	271.59 −34.56	221.0 −40.5	RING GALAXY	6 D 0.6	F085/J1113 (115,323)
AM 0517-260	05 17 54 −26 03.3	2.431 0.061	228.52 −30.72	267.6 −65.6	SMALL I/A DBL.	2 S 0.2	F487/J3827 (325,120)
*AM 0517-321	05 17 56 −32 11.6	2.246 0.061	235.45 −32.40	254.9 −62.9	SP. w. IRR. COMP'N. or CONDENS. NGC 1879 = 423-G-6	1,8,16 S 5.8	F423/J2710 (226,58)
AM 0517-632	05 17 58 −63 28.8	0.441 0.061	273.06 −34.40	220.1 −39.4	I/A DBL. in CHAIN of 4	2,17 D 1.8	F085/J1113 (116,257)
AM 0518-242	05 18 09 −24 27.3	2.476 0.061	226.78 −30.16	271.3 −66.1	GALAXY w. MATERIAL	15 D 0.7	F487/J3827 (324,206)
AM 0518-622	05 18 10 −62 22.3	0.563 0.061	271.73 −34.51	220.9 −40.4	STAR w. APPARENT NEBULA	21 D 0.2	F085/J1113 (113,317)
AM 0518-615	05 18 27 −61 53.6	0.613 0.060	271.14 −34.52	221.1 −40.8	CLOSE PAIR (SMALL)	23 S 0.2	F085/J1113 (110,342)
AM 0518-572	05 18 28 −57 27.8	1.013 0.060	265.77 −34.86	224.3 −44.6	DBL. E in CHAIN of 4	17,23 S 0.5	F119/J3787 (74,312)
AM 0518-265	05 18 34 −26 50.0	2.408 0.060	229.44 −30.81	265.7 −65.4	GALAXY + I/A COMP'NS. 486-G-57	1 S 1.2	F487/J3827 (315,78)
AM 0518-261	05 18 40 −26 12.8	2.426 0.060	228.76 −30.60	267.1 −65.7	DWARF	20 D 0.2	F487/J3827 (315,112)
AM 0519-414	05 19 03 −41 49.3	1.897 0.059	246.80 −34.04	239.6 −57.0	SPIRAL + APP. COMP'N. 305-G-20 = B 0519-418	8 S 0.9	F305/J2776 (109,81)
AM 0519-370	05 19 20 −37 00.5	2.082 0.059	241.11 −33.18	246.4 −60.3	IRR. w. BRIGHT KNOTS Lu 0519-37 = 362-G-19	16 D 2.7	F305/J2776 (101,339)
AM 0519-261	05 19 29 −26 16.5	2.424 0.059	228.89 −30.45	266.8 −65.8	I/A TRIPLE 487-G-1	3 D 1.3	F487/J3827 (305,109)
AM 0519-620	05 19 30 −62 00.4	0.599 0.059	271.27 −34.39	220.9 −40.8	I/A DBL.(SMALL)	2 S 0.1	F085/J1113 (103,336)
AM 0519-235	05 19 42 −23 51.2	2.491 0.058	226.25 −29.63	272.4 −66.7	EDGE-ON + 2 APP. + PEC. COMP'N. NGC 1886 = 487-G-2	8 S 7.2	F486/J4784 (35,240)

Identification	R.A. (1950.0) Declination	Precession R.A./Dec.	Gal. Long. Gal. Lat.	S.G.L. S.G.B.	Description Cross-Identification	Coding Size (Min.)	Field/Plate (x, y)
AM 0519-574	05 19 42 −57 48.2	0.984 0.058	266.17 −34.68	223.8 −44.4	NEST of 6 119-G-56	18 D 1.6	F119/J3787 (66,293)
AM 0519-611	05 19 46 −61 18.4	0.670 0.058	270.42 −34.42	221.3 −41.4	PAIR of I/A SPIRALS (#1) 119-IG-54 + COMP'NS.	2,8 S 1.9	F119/J3787 (78,106)
AM 0519-271	05 19 48 −27 12.8	2.397 0.058	229.95 −30.66	264.6 −65.5	GALAXY w. JET (in PAIR) 487-IG-31	7,23 S 0.8	F487/J3827 (300,58)
AM 0519-612	05 19 48 −61 20.6	0.666 0.058	270.46 −34.42	221.3 −41.4	PAIR of I/A SPIRALS (#2) 119-IG-55 + COMP'NS.	2,8 S 1.9	F119/J3787 (78,104)
AM 0520-222	05 20 14 −22 23.7	2.531 0.058	224.72 −29.03	276.0 −67.1	SPIRAL w. I/A COMP'N. (M51-TYPE?) MCG-4-13-14 = 553-G-37	1,9 S 0.3	F487/J3827 (302,317)
AM 0520-612	05 20 22 −61 26.9	0.654 0.057	270.58 −34.34	221.2 −41.3	I/A DBL. 119-IG-57	2 S 0.2	F119/J3787 (75,98)
AM 0520-363	05 20 41 −36 32.6	2.097 0.057	240.64 −32.82	246.7 −60.8	E + SPIRAL w. LONG ARM	10,23 S 1.8	F363/J2794 (345,93)
AM 0520-870	05 20 42 −87 02.6	−0.373 0.057	299.88 −28.46	207.9 −18.4	DWARF/DEFECT	20 D 1.1	F005/J2958 (253,51)
AM 0520-272	05 20 44 −27 27.8	2.389 0.057	230.30 −30.54	263.9 −65.6	GALAXY + APP. COMP'N. 487-G-6	8 S 1.0	F486/J4784 (27,46)
*AM 0520-390	05 20 59 −39 06.9	2.003 0.057	243.64 −33.25	242.8 −59.1	RING w. I/A COMP'NS MCG-7-12-1 = 305-G-21 = B0521-391	1,6,23 D 0.9	F305/J2776 (86,227)
AM 0520-265 [0521-2652]	05 20 58 −26 56.1	2.404 0.057	229.74 −30.33	265.0 −65.9	CLOSE PAIR (#1)	23 D 0.4	F487/J3827 (286,74)
AM 0521-434	05 21 02 −43 48.1	1.810 0.057	249.22 −33.94	236.7 −55.8	RING/SPIRAL w. I/A COMP'N.	1,6 S 0.4	F253/J2715 (324,240)
*AM 0521-505	05 21 04 −50 53.2	1.454 0.056	257.77 −34.51	229.3 −50.2	M51-TYPE 204-G-3	9 S 0.3	F204/J2662 (255,132)
AM 0521-265 [0520-2656]	05 21 05 −26 52.4	2.406 0.056	229.68 −30.29	265.1 −65.9	CLOSE PAIR (#2)	23 D 0.4	F487/J3827 (285,77)
AM 0521-385	05 21 21 −38 55.1	2.010 0.056	243.44 −33.15	243.0 −59.3	RING w. PUFF (in PAIR) 305-G-22	1,6,23 D 0.8	F305/J2776 (82,236)
AM 0521-325	05 21 34 −32 53.3	2.221 0.056	236.47 −31.83	252.7 −63.1	SMALL SPIRAL + MANY CLOSE COMP'NS 362-G-22 = PKS 0521-325	8 S 0.8	F362/J5542 (71,292)
AM 0521-343	05 21 35 −34 37.3	2.164 0.056	238.45 −32.23	249.7 −62.1	DWARF (RESOLUTION)	20 D 0.5	F363/J2794 (338,197)
AM 0521-623	05 21 40 −62 39.4	0.525 0.056	272.02 −34.07	220.2 −40.3	CLOSE PAIR	23 S 0.4	F085/J1113 (92,300)
AM 0521-641	05 21 40 −64 14.9	0.342 0.056	273.93 −33.90	219.2 −39.0	PLANETARY NEBULA(?)	25 D 0.4	F085/J1113 (97,215)
AM 0521-424	05 21 49 −42 45.1	1.855 0.055	248.00 −33.67	237.8 −56.6	PEC. SP. w. FILAM. to I/A COMP'N. MCG-7-12-2 = 252-G-22 = B0521-427	1,15 S 0.7	F305/J2776 (82,30)
*AM 0522-611	05 22 03 −61 19.9	0.663 0.055	270.42 −34.15	221.0 −41.5	I/A DBL. (in GROUP of 3) 119-G-60?	2,24 S 2.5	F120/J1894 (325,98)
*AM 0522-774	05 22 05 −77 46.6	−3.010 0.055	289.70 −31.28	212.0 −26.9	GALAXY w. ABS'N. + APP. COMP'NS. NGC 1956 = 16-G-2	1,14 D 2.2	F033/J1958 (200,28)
AM 0522-614	05 22 06 −61 43.8	0.623 0.055	270.90 −34.11	220.7 −41.2	CLOSE PAIR	23 S 5.8	F120/J1894 (322,77)
AM 0522-620	05 22 30 −62 00.6	0.593 0.054	271.23 −34.04	220.5 −41.0	COMPACT	13 D 0.1	F085/J1113 (85,335)
*AM 0522-574	05 22 34 −57 48.9	0.978 0.054	266.17 −34.30	223.4 −44.6	GROUP of 6	18 D 1.8	F120/J1894 (338,287)
*AM 0522-423	05 22 35 −42 38.4	1.859 0.054	247.89 −33.51	237.7 −56.8	SMALL M51-TYPE B 0522-426	9 S 0.7	F253/J2715 (312,303)
AM 0522-495	05 22 40 −49 57.0	1.505 0.054	256.65 −34.21	229.8 −51.2	SPIRAL w. ONE STRONG ARM + APP. 204-G-4 = Fa 306 COMP'NS.	8,10 S 1.8	F204/J2662 (242,182)
AM 0522-431	05 22 43 −43 10.1	1.836 0.054	248.52 −33.56	237.1 −56.4	DISTURBED GALAXY w. I/A COMP'N.	1 S 1.1	F252/J1933 (35,275)
AM 0522-870	05 22 50 −87 05.3	−0.381 0.054	299.92 −28.42	207.9 −18.3	DWARF w. RESOLUTION 4-G-17	20 D 1.3	F003/J2565 (69,22)
AM 0522-612	05 22 53 −61 27.0	0.650 0.054	270.55 −34.04	220.8 −41.5	I/A DBL.	2 S 0.3	F120/J1894 (319,92)
AM 0523-233	05 23 00 −23 31.3	2.499 0.054	226.18 −28.81	272.7 −67.5	VERY CLOSE I/A DBL.	2 S 0.2	F487/J3827 (266,258)
*AM 0523-400	05 23 11 −40 04.1	1.964 0.053	244.87 −33.00	240.9 −58.8	M51-TYPE 305-IG-23	9 S 0.5	F306/J2669 (329,170)
AM 0523-332	05 23 19 −33 28.9	2.201 0.053	237.25 −31.62	251.2 −63.1	I/A DBL. or IRR. 362-G-24	2,16 S 0.1	F363/J2794 (321,259)
AM 0523-244	05 23 33 −24 43.2	2.466 0.053	227.52 −29.09	269.7 −67.2	GALAXY w. JET + I/A COMP'N. (M51-TYPE ?)	1,7,9 S 0.6	F487/J3827 (258,193)
AM 0523-313	05 23 42 −31 35.4	2.262 0.053	235.13 −31.06	254.6 −64.2	16 MEDIUM-BRIGHT GALAXIES MCG-5-13-18 = 423-G-13 = PKS 0523-315	19 D 15.7	F423/J2710 (160,90)
AM 0523-483	05 23 46 −48 38.2	1.575 0.053	255.08 −33.95	230.9 −52.3	EDGE-ON w. 2 COMP'NS in WIDE PAIR Fa 795?	8,23 S 0.6	F204/J2662 (233,253)

Identification	R.A. (1950.0) Declination	Precession R.A./Dec.	Gal. Long. Gal. Lat.	S.G.L. S.G.B.	Description Cross-Identification	Coding Size (Min.)	Field/Plate (x, y)
AM 0523-460	05 23 53 −46 03.1	1.704 0.052	251.99 −33.70	233.5 −54.4	SPIRAL w. PEC. ARMS in GROUP of 3 253-G-3 = B 0523-460 + DWARF	10,20,24 D 6.3	F253/J2715 (292,121)
AM 0523-251	05 23 57 −25 18.2	2.449 0.052	228.19 −29.19	268.2 −67.1	SPIRAL w. TAIL	10,15 D 0.5	F487/J3827 (253,162)
AM 0525-780 [0526-7804]	05 25 04 −78 07.2	−3.204 0.051	290.04 −31.04	211.6 −26.6	CLOSE GROUP of 4 (#1: E:)	18 D 6.3	F033/J1958 (191,9)
AM 0525-511	05 25 40 −51 17.1	1.425 0.050	258.29 −33.81	228.0 −50.3	E + 3 APP. COMP'NS.	8 S 2.2	F204/J2662 (216,111)
AM 0525-454	05 25 44 −45 49.4	1.713 0.050	251.75 −33.35	233.3 −54.8	I/A TRIPLE	3 D 2.2	F253/J2715 (275,133)
AM 0525-371	05 25 45 −37 15.6	2.068 0.050	241.71 −31.98	244.3 −61.1	IRR. at CENTRE of ARC of 8 incl. 363-G-2	16,17 D 5.8	F363/J2794 (289,57)
AM 0525-780 [0525-7807]	05 25 47 −78 03.8	−3.176 0.050	289.96 −31.02	211.6 −26.7	CLOSE GROUP of 4 (#2: SPIRAL)	18 D 6.3	F033/J1958 (189,12)
AM 0525-621	05 25 53 −62 11.0	0.569 0.050	271.40 −33.63	219.9 −41.0	I/A DBL.	2 S 0.3	F085/J1113 (64,324)
AM 0525-371	05 25 56 −37 17.6	2.067 0.049	241.76 −31.95	244.2 −61.1	ARC of 7 (I/A DBL. at CENTRE) incl. 363-G-2	2,17 D 7.8	F306/J2669 (306,320)
AM 0526-780 [0525-7803]	05 26 12 −78 06.1	−3.199 0.049	290.00 −30.99	211.6 −26.6	CLOSE GROUP of 4 (#3: E)	18 D 6.3	F033/J1958 (188,10)
AM 0526-393	05 26 23 −39 37.0	1.979 0.049	244.47 −32.31	240.7 −59.5	I/A DBL. w. LOOPS + TAILS 306-IG-1	2,15 S 0.4	F306/J2669 (297,195)
*AM 0526-634	05 26 26 −63 48.1	0.387 0.049	273.33 −33.43	218.9 −39.6	E w. ABSORPTION LANE NGC 1947 = 85-G-87	14 D 1.8	F086/J1020 (328,234)
AM 0526-244	05 26 31 −24 41.3	2.465 0.049	227.74 −28.44	269.1 −67.9	LINEAR DWARF 487-G-14	20 D 0.4	F487/J3827 (222,196)
AM 0526-780 [0526-7806]	05 26 43 −78 04.2	−3.183 0.048	289.95 −30.97	211.6 −26.7	CLOSE GROUP of 4 (#4: IRR.)	18 D 6.3	F033/J1958 (186,12)
AM 0527-284	05 27 00 −28 42.7	2.349 0.048	232.15 −29.58	259.6 −66.3	DWARF:	20 D 0.4	F423/J2710 (120,244)
AM 0527-622	05 27 20 −62 27.6	0.537 0.047	271.71 −33.44	219.6 −40.8	I/A DBL.	2 S 0.2	F085/J1113 (56,309)
*AM 0527-421	05 27 41 −42 16.6	1.870 0.047	247.62 −32.52	236.9 −57.7	GALAXY w. 9 APP. COMP'NS. 306-G-4	8 S 1.8	F305/J2776 (23,54)
AM 0527-621	05 27 42 −62 11.2	0.566 0.047	271.38 −33.42	219.7 −41.1	I/A TRIPLE	3 D 0.3	F085/J1113 (53,323)
AM 0528-442	05 28 04 −44 22.5	1.778 0.046	250.10 −32.76	234.3 −56.1	GALAXY w. LOOP	15 D 0.3	F253/J2715 (255,212)
AM 0528-873	05 28 19 −87 37.3	−0.480 0.046	300.47 −28.18	207.6 −17.9	DWARF 1-G-2	20 D 0.9	F001/J3648 (53,197)
AM 0528-245	05 28 26 −24 54.7	2.458 0.046	228.14 −28.10	268.2 −68.2	DWARF 487-G-17	20 D 0.5	F487/J3827 (198,184)
AM 0528-332	05 28 49 −33 25.4	2.199 0.045	237.51 −30.49	249.7 −64.0	I/A SPIRALS + COMP'N. 363-G-3	2 S 2.0	F363/J2794 (260,263)
*AM 0529-565	05 29 15 −56 53.8	1.042 0.045	265.04 −33.41	222.9 −45.8	I/A TRIPLE w. TAIL 159-G-16	3,15 D 1.1	F159/J2675 (185,77)
*AM 0529-442	05 29 38 −44 27.7	1.773 0.044	250.25 −32.50	233.8 −56.2	I/A DBL.	1,2 S 0.6	F253/J2715 (240,207)
AM 0530-622	05 30 00 −62 25.2	0.537 0.044	271.64 −33.14	219.2 −41.0	I/A TRIPLE (CHAIN) 120-G-3 = B 0530-624	3 D 2.0	F086/J1020 (313,310)
*AM 0530-421	05 30 06 −42 12.3	1.872 0.043	247.63 −32.07	236.3 −58.1	RING w. ONE VERY LONG THIN ARM 306-G-9	6,8 D 0.8	F253/J2715 (238,329)
AM 0530-620	05 30 16 −62 07.2	0.569 0.043	271.28 −33.12	219.4 −41.3	CLOSE TRIPLE	24 D 0.2	F086/J1020 (313,326)
AM 0530-465	05 30 19 −46 58.9	1.653 0.043	253.23 −32.70	231.0 −54.3	E + DISTORTED SPIRAL	23 S 6.3	F253/J2715 (231,72)
AM 0530-245	05 30 43 −24 57.6	2.456 0.043	228.38 −27.63	267.5 −68.7	DWARF w. CONDENSATIONS	20 D 0.2	F487/J3827 (171,182)
AM 0530-293	05 30 43 −29 39.7	2.318 0.043	233.46 −29.08	256.5 −66.5	ASYMM. SP. w. DISTURBED COMP'N. MCG-5-14-2 = 423-G-19	8,16 S 6.7	F424/J2726 (346,195)
*AM 0531-270	05 31 27 −27 01.2	2.397 0.041	230.65 −28.14	262.3 −68.0	SP. w. DIFFRACTION SPIKE NUCLEUS 487-G-22	12,22 D 0.9	F424/J2726 (342,337)
*AM 0531-362	05 31 29 −36 26.1	2.094 0.041	241.06 −30.68	243.9 −62.5	EDGE-ON + PEC. DUST + 6 COMP'NS. IC 2135 = IC 2136 = 363-G-7	8,13,14 D 2.9	F363/J2794 (228,102)
AM 0531-232	05 31 54 −23 20.4	2.501 0.041	226.77 −26.83	271.5 −69.5	E + 5 COMP'NS. (GROUP of 5) NGC 1979 = 487-G-24	8,18 S 5.4	F487/J3827 (157,269)
*AM 0532-524	05 32 05 −52 40.5	1.334 0.041	260.01 −32.89	225.6 −49.7	GALAXY w. VERY PECULIAR ABS'N. 159-G-19 + COMP'NS.	8,14 D 0.8	F204/J2662 (164,36)
AM 0532-640	05 32 05 −64 08.8	0.336 0.041	273.68 −32.78	218.0 −39.6	GROUP of 8 DN 0532-642	18 D 4.0	F085/J1113 (35,216)
*AM 0532-282	05 32 09 −28 29.8	2.353 0.040	232.30 −28.44	258.7 −67.4	SP. w. I/A COMP'N. in GROUP of 6 MCG-5-14-3 = 423-G-20	1,18 S 1.4	F424/J2726 (331,258)

Identification	R.A. (1950.0) Declination	Precession R.A./Dec.	Gal. Long. Gal. Lat.	S.G.L. S.G.B.	Description Cross-Identification	Coding Size (Min.)	Field/Plate (x, y)
AM 0532-304	05 32 29	2.281	234.87	253.6	TRIPLE (#1: E:)	24	F423/J2710
[0532-3055]	−30 49.9	0.040	−29.05	−66.2	NGC 1989 = 423-G-21	S 12.1	(59,129)
AM 0532-275	05 32 32	2.372	231.65	260.0	CLOSE PAIR of SPIRALS (#1)	23	F487/J3827
[0532-2750]	−27 52.2	0.040	−28.17	−67.8		D 0.5	(148,26)
AM 0532-304	05 32 34	2.285	234.75	253.8	TRIPLE (#2: EDGE-ON S0)	24	F423/J2710
[0532-3049]	−30 43.1	0.040	−29.00	−66.3	MCG-5-14-5 = 423-G-22	S 12.1	(57,135)
AM 0532-305	05 32 37	2.278	234.99	253.3	TRIPLE (#3: BROAD-ARMED SPIRAL)	24	F423/J2710
[0532-3043]	−30 55.8	0.040	−29.05	−66.2	NGC 1992 = 423-G-23	S 12.1	(57,124)
AM 0532-275	05 32 42	2.372	231.63	260.0	CLOSE PAIR of SPIRALS (#2)	23	F487/J3827
[0532-2752]	−27 50.5	0.040	−28.12	−67.8		D 0.5	(146,27)
AM 0532-291	05 32 46	2.330	233.18	256.8	E w. HALO + COMPACT COMP'N.	8,13	F423/J2710
	−29 15.8	0.040	−28.54	−67.1	MCG-5-14-6 = 423-G-24	S 1.6	(53,213)
AM 0532-341	05 32 52	2.171	238.63	247.1	DWARF	20	F363/J2794
	−34 12.6	0.039	−29.87	−64.2	363-G-8	D 0.9	(214,222)
AM 0532-734	05 32 59	−1.489	284.96	213.1	SHRED or EDGE-ON GALAXY	22	F033/J1958
	−73 47.1	0.039	−31.50	−30.8	33-G-22	D 2.0	(166,242)
AM 0533-324	05 33 09	2.218	237.07	249.6	TRIPLE	24	F424/J2726
	−32 47.1	0.039	−29.45	−65.1		D 1.6	(313,28)
AM 0533-242	05 33 19	2.471	227.99	268.4	DWARF	20	F487/J3827
	−24 23.1	0.039	−26.88	−69.4	487-G-28	D 0.7	(139,213)
AM 0533-381	05 33 21	2.030	243.12	240.7	CHAIN of 4	17	F306/J2669
	−38 10.2	0.039	−30.70	−61.5		D 8.5	(226,274)
AM 0533-235	05 33 23	2.486	227.46	269.8	I/A E's + COMP'NS.	2,8	F487/J3827
	−23 52.3	0.039	−26.69	−69.6		S 0.8	(138,240)
AM 0533-474	05 33 27	1.613	254.18	229.5	I/A DBL. w. TAIL	2,15	F204/J2662
	−47 43.5	0.039	−32.26	−54.0	204-IG-18	S 0.4	(146,301)
AM 0533-505	05 33 51	1.435	258.02	226.6	PAIR of SPIRALS + 2 APP. COMP'NS	8,23	F204/J2662
	−50 58.5	0.038	−32.51	−51.3	NGC 2007/8 = 204-G-19/20	S 2.6	(147,127)
AM 0534-373	05 34 23	2.054	242.43	241.4	CLOSE DBL.	23	F306/J2669
	−37 30.7	0.037	−30.36	−62.1		S 0.2	(216,310)
AM 0534-320	05 34 25	2.243	236.30	250.6	CLOSE GROUP of 6 (MOSTLY I/A)	18	F424/J2726
	−32 01.0	0.037	−28.98	−65.8	424-G-2	D 4.5	(300,70)
AM 0534-521	05 34 36	1.363	259.45	225.4	TRIPLE	24	F159/J2675
	−52 10.4	0.037	−32.48	−50.3	204-G-21 = Fa 796	D 9.9	(141,331)
AM 0535-382	05 35 06	2.020	243.48	239.8	CLOSE PAIR (#1: E)	23	F306/J2669
[0535-3823]	−38 24.6	0.036	−30.41	−61.5		S 2.0	(208,262)
AM 0535-382	05 35 09	2.021	243.45	239.9	CLOSE PAIR (#2: SPIRAL)	23	F306/J2669
[0535-3824]	−38 23.0	0.036	−30.40	−61.6		S 2.0	(207,263)
AM 0535-804	05 35 10	−5.081	292.86	210.1	I/A DBL.	2	F016/J3763
	−80 45.0	0.036	−29.95	−24.3		S 0.6	(233,136)
AM 0535-262	05 35 18	2.412	230.35	262.6	EDGE-ON + 2 APP. COMP'NS.	8	F487/J3827
	−26 27.4	0.036	−27.14	−69.0	487-G-30	S 4.9	(115,102)
AM 0535-521	05 35 18	1.361	259.48	225.3	DWARF	20	F204/J2662
	−52 11.9	0.036	−32.37	−50.3	204-G-22	D 0.6	(137,60)
AM 0535-385	05 35 24	1.999	244.12	239.0	CLOSE I/A DBL.	2	F306/J2669
	−38 57.9	0.036	−30.48	−61.2		S 0.2	(204,232)
*AM 0535-422	05 35 46	1.858	248.10	234.5	DISTURBED SPIRAL w. I/A COMP'N.	1	F306/J2669
	−42 26.3	0.035	−31.08	−58.5	306-G-12 = B 0535-424	S 0.6	(199,46)
*AM 0536-502	05 36 13	1.465	257.42	226.5	SMALL M51-TYPE (DBL. ?)	9	F204/J2662
	−50 26.0	0.035	−32.09	−51.9	204-IG-24	S 0.3	(126,155)
AM 0536-641	05 36 56	0.319	273.74	217.3	I/A DBL.	2	F086/J1020
	−64 14.2	0.034	−32.25	−39.7		D 0.7	(264,216)
AM 0536-433	05 36 58	1.811	249.38	232.9	I/A DBL. in GROUP of 3	2,10,24	F253/J2715
	−43 30.4	0.033	−31.05	−57.8	B 0536-435	D 1.3	(170,259)
*AM 0537-292	05 37 25	2.324	233.67	255.0	SPIRAL w. I/A COMP'N.	1	F424/J2726
	−29 24.0	0.033	−27.61	−67.9	MCG-5-14-9 = 424-G-5	S 0.7	(268,211)
AM 0537-262	05 37 58	2.410	230.60	261.7	COMPACT	13	F487/J3827
	−26 29.3	0.032	−26.59	−69.5		D 0.3	(83,100)
AM 0538-220	05 38 05	2.535	225.95	274.0	GALAXY w. I/A COMP'N.	1	F487/J3827
	−22 01.5	0.032	−25.03	−71.2	MCG-4-14-8 = 554-G-23 [A82]	S 0.4	(80,340)
AM 0538-295	05 38 19	2.308	234.27	253.7	GROUP of 8 (I/A DBL.)	2,18	F424/J2726
	−29 53.9	0.032	−27.58	−67.8		D 4.0	(257,184)
AM 0538-425	05 38 24	1.835	248.79	233.2	I/A DBL.	2	F306/J2669
	−42 56.6	0.031	−30.69	−58.4		S 0.4	(173,19)
AM 0538-405	05 38 31	1.922	246.42	235.6	S0 w. MANY COMP'NS.	8	F306/J2669
	−40 51.7	0.031	−30.28	−60.1	306-G-17	S 8.7	(172,130)
AM 0538-272	05 38 41	2.381	231.71	259.1	SMALL I/A DBL. w. ABS'N.	2,14	F424/J2726
	−27 28.7	0.031	−26.75	−69.2	487-IG-31	S 0.5	(255,314)
AM 0539-553	05 39 05	1.132	263.50	222.1	SPIRAL w. 3 DIAMETRIC COMP'NS.	8	F159/J2675
	−55 34.0	0.030	−32.02	−47.6	159-G-23	S 11.2	(110,148)
AM 0539-271	05 39 27	2.388	231.51	259.4	PAIR (#1: S0)	2	F424/J2726
[0539-2715]	−27 14.0	0.030	−26.51	−69.4		S 1.4	(246,327)

Identification	R.A. (1950.0) Declination	Precession R.A./Dec.	Gal. Long. Gal. Lat.	S.G.L. S.G.B.	Description Cross-Identification	Coding Size (Min.)	Field/Plate (x, y)
AM 0539-271 [0539-2714]	05 39 29 −27 15.1	2.388 0.030	231.53 −26.51	259.3 −69.4	PAIR (#2: LONG-ARMED SPIRAL) 487-G-33	2 S 1.4	F424/J2726 (245,326)
AM 0539-235	05 39 34 −23 58.9	2.481 0.030	228.11 −25.40	268.0 −70.9	DWARF (DBL. ?)	20 D 0.6	F487/J3827 (63,234)
AM 0539-270	05 39 34 −27 09.1	2.391 0.030	231.43 −26.46	259.6 −69.5	COMPACT	13 D 0.2	F487/J3827 (64,64)
AM 0539-483	05 39 41 −48 38.0	1.562 0.030	255.39 −31.33	227.2 −53.8	LARGE COMPACT	13 D 0.2	F204/J2662 (92,250)
*AM 0539-225	05 39 57 −22 58.3	2.509 0.029	227.10 −24.97	270.8 −71.4	EDGE-ON w. 5 APP. COMP'NS. MCG-4-14-11 = 487-G-35	8 S 2.2	F488/J3843 (328,286)
AM 0540-260	05 40 04 −26 08.4	2.420 0.029	230.41 −26.03	262.0 −70.1	GROUP of 4	18 D 1.3	F487/J3827 (57,118)
AM 0540-253	05 40 12 −25 33.7	2.436 0.029	229.81 −25.81	263.4 −70.4	CLOSE PAIR (S0/E + S(r))	23 S 1.3	F488/J3843 (322,147)
AM 0540-382	05 40 16 −38 28.9	2.015 0.029	243.82 −29.44	238.1 −62.2	PECULIAR SPIRAL 306-G-22	12 D 1.1	F306/J2669 (153,258)
AM 0540-264	05 40 32 −26 48.6	2.400 0.028	231.15 −26.15	260.1 −69.8	CLOSE PAIR of SPIRALS 487-G-37	23 S 0.8	F487/J3827 (52,82)
AM 0540-485	05 40 38 −48 57.9	1.543 0.028	255.80 −31.22	226.7 −53.6	DISTURBED GALAXY w. PLUME	15,16 D 0.5	F204/J2662 (85,232)
AM 0540-514	05 40 38 −51 45.9	1.383 0.028	259.06 −31.52	224.5 −51.1	CONTACT DBL. w. MATERIAL 204-IG-28	2,15 S 0.2	F204/J2662 (92,82)
AM 0540-240	05 40 56 −24 05.1	2.478 0.028	228.34 −25.15	267.3 −71.2	SMALL I/A DBL.	2 S 0.4	F487/J3827 (46,229)
*AM 0541-294 [0541-3006]	05 41 00 −29 45.2	2.312 0.028	234.31 −26.98	253.1 −68.3	GROUP (#1: MERGER + TAIL) MCG-5-14-10 = 424-G-10	15,16,24 D 0.8	F424/J2726 (226,192)
AM 0541-255	05 41 07 −25 57.6	2.425 0.027	230.31 −25.74	262.1 −70.4	GALAXY (w. JETS ?) in CHAIN of 4	7,17 D 3.6	F487/J3827 (45,128)
AM 0541-861 [0544-8621]	05 41 15 −86 12.9	−0.284 0.027	298.88 −28.43	207.9 −19.2	PAIR (#1: PECULIAR SPIRAL)	12,23 D 0.5	F005/J2958 (258,98)
AM 0541-300 [0541-2945]	05 41 20 −30 06.1	2.301 0.027	234.71 −27.01	252.2 −68.2	GROUP (#2: SB + DIAMETRICS) NGC 2049 = 424-G-11	8,24 S 3.8	F424/J2726 (222,174)
AM 0541-274	05 41 34 −27 40.3	2.375 0.027	232.14 −26.21	257.6 −69.6	E w. I/A COMP'N. + PEC. KNOTTY MCG-5-14-12 = 424-G-12	1,23 S 5.4	F487/J3827 (40,36)
AM 0541-524	05 41 59 −52 43.8	1.323 0.026	260.20 −31.40	223.5 −50.4	DWARF (RESOLUTION) 159-G-25 = UKS 0542-547	20 D 0.9	F204/J2662 (83,30)
AM 0542-331	05 42 05 −33 17.9	2.198 0.026	238.21 −27.78	245.7 −66.2	IRR. w. KNOTS 363-G-18	16 D 0.6	F363/J2794 (111,270)
AM 0542-481	05 42 05 −48 17.7	1.578 0.026	255.06 −30.89	226.9 −54.3	I/A DBL. in GROUP	2,18 S 0.1	F204/J2662 (70,267)
AM 0542-532	05 42 20 −53 20.7	1.283 0.026	260.93 −31.40	223.0 −49.8	GALAXY w. LOOPS	15 D 0.9	F159/J2675 (80,266)
AM 0542-453	05 42 34 −45 31.5	1.716 0.025	251.89 −30.40	229.3 −56.7	DWARF 253-G-19	20 D 0.6	F253/J2715 (117,151)
AM 0542-292	05 42 36 −29 21.0	2.324 0.025	234.00 −26.52	253.4 −68.8	DWARF 424-G-16	20 D 0.9	F424/J2726 (207,214)
AM 0542-875	05 42 41 −87 50.9	−0.540 0.025	300.66 −27.98	207.4 −17.7	GALAXY w. JET to STAR	7 D 0.8	F005/J2958 (223,18)
AM 0542-403	05 42 46 −40 37.5	1.930 0.025	246.34 −29.44	234.6 −60.8	CLOSE PAIR	23 S 0.5	F306/J2669 (128,143)
AM 0542-483	05 42 50 −48 39.9	1.558 0.025	255.50 −30.82	226.4 −54.0	CHAIN of 6	17 D 7.2	F205/J3868 (332,246)
AM 0543-324 [0543-3242]	05 43 06 −32 41.0	2.218 0.025	237.61 −27.41	246.4 −66.8	PAIR (#1: E)	23 S 1.7	F424/J2726 (201,35)
AM 0543-324 [0534-3241]	05 43 08 −32 42.0	2.218 0.025	237.63 −27.41	246.4 −66.8	PAIR (#2: SPIRAL)	23 S 1.7	F424/J2726 (200,34)
AM 0543-234	05 43 25 −23 42.6	2.488 0.024	228.17 −24.48	267.7 −71.8	DWARF	20 D 0.4	F487/J3827 (15,248)
AM 0543-515	05 43 30 −51 59.5	1.368 0.024	259.37 −31.10	223.7 −51.1	DWARF/IRR. SINGLE 204-G-34	16,20 D 0.7	F159/J2675 (67,338)
AM 0543-281	05 43 31 −28 16.6	2.356 0.024	232.93 −25.99	255.5 −69.6	SMALL SPIRAL w. REVERSING ARM	10 D 0.8	F424/J2726 (197,272)
AM 0543-274	05 43 36 −27 48.0	2.370 0.024	232.44 −25.82	256.6 −69.9	CHAIN of 5	17 D 5.5	F424/J2726 (196,297)
AM 0543-480	05 43 42 −48 06.2	1.588 0.024	254.88 −30.60	226.7 −54.6	CLOSE PAIR in DISTANT GROUP 204-G-35	18,23 S 0.4	F204/J2662 (55,277)
AM 0543-475	05 43 58 −47 58.1	1.594 0.023	254.73 −30.54	226.7 −54.7	RICH DISTANT CLUSTER PKS 0543-479	19 D 9.0	F205/J3868 (323,284)
*AM 0544-393	05 44 12 −39 31.0	1.974 0.023	245.17 −28.93	235.5 −61.8	BULL'S-EYE + ELLIPTICAL OUTER RING 306-G-25	6 D 1.2	F306/J2669 (113,202)
AM 0544-862 [0541-8612]	05 44 35 −86 21.2	−0.297 0.022	299.02 −28.34	207.8 −19.1	PAIR (#2: SPIRAL)	23 D 1.1	F005/J2958 (252,93)

Identification	R.A. (1950.0) Declination	Precession R.A./Dec.	Gal. Long. Gal. Lat.	S.G.L. S.G.B.	Description Cross-Identification	Coding Size (Min.)	Field/Plate (x, y)
AM 0544-484	05 44 54 −48 43.4	1.554 0.022	255.62 −30.49	225.9 −54.1	SPIRAL w. I/A COMP'N. 204-G-37 = B 0544-487	1 S 0.7	F205/J3868 (313,244)
*AM 0545-341	05 45 13 −34 16.0	2.165 0.022	239.46 −27.42	242.8 −66.0	LATE-TYPE SP. w. FAINT OUTER ARMS NGC 2090 = 363-G-23	10 D 5.4	F363/J2794 (77,217)
*AM 0545-520	05 45 16 −52 06.4	1.360 0.021	259.54 −30.85	223.2 −51.2	DWARF w. STRONG KNOTS (MATERIAL) NGC 2101 = 205-G-1 = UKS 0545-521 [B]	2,15,20 D 1.3	F204/J2662 (55,61)
AM 0545-820	05 45 26 −82 00.4	−6.422 0.021	294.18 −29.30	209.2 −23.3	I/A DBL.(#1: ONE-ARMED SPIRAL)	2 S 8.5	F004/J3821 (73,327)
*AM 0545-434	05 45 29 −43 46.2	1.796 0.021	250.00 −29.58	230.2 −58.5	BROKEN RING w. 2 I/A COMP'NS. 253-G-25 = B 0545-437	1,6 S 0.4	F253/J2715 (87,244)
AM 0545-643	05 45 34 −64 39.0	0.259 0.021	274.18 −31.30	216.0 −39.7	CLOSE PAIR (in CHAIN of FIVE)	23,17 D 3.4	F086/J1020 (213,196)
AM 0545-820 [0546-8208]	05 45 35 −82 00.8	−6.431 0.021	294.18 −29.29	209.2 −23.3	I/A DBL. (#2: IRR.) 16-G-9	2 S 9.0	F015/J3673 (41,48)
*AM 0545-355	05 45 37 −35 50.3	2.110 0.021	241.19 −27.77	240.2 −64.9	RING GALAXY w. COMP'N. 363-G-24	6,8,9 S 0.6	F363/J2794 (75,133)
*AM 0545-453	05 45 46 −45 30.3	1.716 0.021	251.97 −29.84	228.5 −57.0	I/A DBL.("SACRED MUSHROOM"-TYPE) 253-IG-26 = B 0545-455	2 S 0.9	F253/J2715 (87,151)
*AM 0545-513	05 45 56 −51 34.2	1.392 0.020	258.93 −30.68	223.5 −51.7	DISTURBED GAL. w. 2 APP. COMP'NS. NGC 2104 = 205-G-2	8 S 4.0	F204/J2662 (47,90)
*AM 0546-820 [0545-8200]	05 46 26 −82 08.9	−6.590 0.020	294.32 −29.23	209.1 −23.2	I/A DBL. (#2: BR.-RIMMED SPIRAL) NGC 2144 = 16-G-10	2,10 S 8.5	F015/J3673 (41,41)
AM 0546-325	05 46 38 −32 59.4	2.207 0.019	238.18 −26.79	244.5 −67.1	E + 3 APP. COMP'NS. (I/A DBL.) MCG-5-14-17 = 363-G-27	2,8 S 1.8	F364/J4743 (330,285)
*AM 0546-253	05 46 39 −25 31.7	2.436 0.019	230.32 −24.43	261.4 −71.7	CLUSTER w. 8 PAIRS (MUCH I/A) 488-G-28/29 (4 E + S(r)'s)	19,23	F488/J3843 (244,150)
*AM 0546-324	05 46 43 −32 46.1	2.214 0.019	237.94 −26.70	244.9 −67.3	3-ARMED SPIRAL + COMP'N. MCG-5-14-18 = 363-G-28	8,11 S 0.6	F364/J4743 (329,297)
*AM 0546-324	05 46 44 −32 40.7	2.217 0.019	237.85 −26.67	245.1 −67.3	I/A COMPACT E's (MATERIAL) PKS 0546-329?	2,13 S 0.6	F363/J2794 (58,302)
*AM 0546-242	05 46 45 −24 23.7	2.468 0.019	229.17 −24.01	264.6 −72.2	LINE of 3 (incl. SP. w. PEC. ARMS) MCG-4-14-34 = 488-G-30	8,10,24 D 2.0	F488/J3843 (243,211)
*AM 0546-472	05 46 47 −47 25.3	1.621 0.019	254.19 −29.99	226.5 −55.4	GAL. w. FAINT ABS'N + 15 COMP'NS. 253-G-27	8,14 S 2.7	F204/J2662 (25,312)
*AM 0547-474	05 47 04 −47 45.1	1.604 0.019	254.57 −29.99	226.1 −55.2	GAL. w. LOOPS + TAILS + 3 COMP'NS 205-IG-3 = B 0547-477	2,8,15,18 D 1.1	F204/J2662 (24,294)
*AM 0547-244	05 47 24 −24 47.1	2.457 0.018	229.62 −24.01	263.3 −72.2	SMALL GALAXY w. I/A COMP'N. 488-G-34	1 S 0.8	F488/J3843 (235,190)
AM 0547-330 [0548-3302]	05 47 40 −33 04.6	2.204 0.018	238.34 −26.60	244.0 −67.2	PAIR (#1: ONE-ARMED SPIRAL) 363-G-31	10,23 S 5.4	F363/J2794 (48,280)
*AM 0547-471	05 47 42 −47 10.9	1.633 0.018	253.94 −29.79	226.5 −55.7	SPIRAL w. PEC. ARMS (to RING?) 253-G-29	6,8,10 D 0.9	F254/J4748 (328,59)
AM 0547-324	05 47 47 −32 43.7	2.216 0.018	237.98 −26.48	244.6 −67.5	I/A QUARTET	4 D 1.3	F364/J4743 (317,299)
*AM 0548-330 [0547-3304]	05 48 03 −33 02.6	2.205 0.017	238.33 −26.52	243.9 −67.3	PAIR (#2: S(r)) 363-G-2	6,23 S 5.4	F363/J2794 (43,282)
AM 0548-393	05 48 37 −39 34.1	1.970 0.017	245.45 −28.12	234.0 −62.3	SMALL SPIRAL w. 3 I/A COMP'NS.	8 S 3.1	F307/J2767 (339,196)
AM 0548-463	05 48 45 −46 37.8	1.660 0.016	253.35 −29.53	226.6 −56.3	GALAXY w. DIAMETRIC COMP'NS	8 S 2.5	F253/J2715 (62,90)
AM 0549-451	05 49 02 −45 15.4	1.727 0.016	251.81 −29.23	227.8 −57.5	DISTURBED SINGLE	16 D 0.5	F253/J2715 (56,163)
*AM 0549-344	05 49 8 −34 47.8	2.146 0.016	240.28 −26.79	240.5 −66.1	CHAIN of 6 incl. 364-G-4	17 D 8.5	F364/J4743 (299,188)
AM 0549-385	05 49 17 −38 55.8	1.995 0.016	244.78 −27.84	234.5 −62.9	GALAXY w. I/A COMP'N. 306-G-30	1 S 0.6	F307/J2767 (334,231)
AM 0549-404	05 49 27 −40 40.1	1.926 0.015	246.71 −28.21	232.4 −61.5	CLOSE PAIR	23 S 0.5	F307/J2767 (326,138)
*AM 0549-345	05 49 28 −34 53.4	2.142 0.015	240.40 −26.75	240.2 −66.1	PAIR + 2 COMP'NS. (I/A?)	2,8 S 0.7	F364/J4743 (295,184)
AM 0549-463	05 49 54 −46 35.7	1.662 0.015	253.35 −29.33	226.4 −56.4	GALAXY w. JET 253-G-31	7 D 0.5	F253/J2715 (51,91)
AM 0549-510	05 49 58 −51 03.5	1.421 0.015	258.43 −30.00	222.9 −52.4	GROUP of 3 w. APP. COMP'NS. 205-G-5	8,24 D 2.9	F205/J3868 (264,121)
AM 0550-820	05 50 06 −82 03.5	−6.495 0.014	294.20 −29.13	209.0 −23.3	WIDE PAIR I/A SPIRALS	2 S 7.6	F005/J2958 (328,311)
AM 0550-503	05 50 09 −50 36.0	1.448 0.014	257.91 −29.91	223.2 −52.8	CLOSE I/A DBL.(ABS'N. over ONE ?) NGC 2115 = 205-G-6 = B 0550-505	2,14 S 0.5	F205/J3868 (263,146)
AM 0550-302	05 50 14 −30 27.8	2.288 0.014	235.75 −25.30	248.0 −69.4	DISTURBED SP. FNT. TAIL to COMP'N 424-G-34	1 S 1.3	F424/J2726 (119,154)
*AM 0550-342	05 50 31 −34 21.4	2.160 0.014	239.90 −26.40	240.7 −66.6	M51-TYPE 364-G-7	9 S 0.7	F363/J2794 (18,211)

Identification	R.A. (1950.0) Declination	Precession R.A./Dec.	Gal. Long. Gal. Lat.	S.G.L. S.G.B.	Description Cross-Identification	Coding Size (Min.)	Field/Plate (x, y)
AM 0551-801	05 51 04 −80 19.3	−4.755 0.013	292.23 −29.39	209.5 −25.0	DWARF w. OFF-SET NUCLEUS 16-IG-12	16,20 D 0.5	F016/J3763 (199,162)
AM 0551-395	05 51 06 −39 55.9	1.956 0.013	245.98 −27.74	232.7 −62.2	2 E's in COMMON ENVELOPE	2 S 0.4	F307/J2767 (312,178)
AM 0551-392 [0551-3931]	05 51 11 −39 29.8	1.973 0.013	245.51 −27.62	233.2 −62.6	PAIR (#1: Sc)	23 S 5.6	F306/J2669 (41,201)
AM 0551-342	05 51 18 −34 24.2	2.159 0.013	240.00 −26.26	240.3 −66.7	GROUP of 5	18 D 4.5	F364/J4743 (276,210)
AM 0551-393 [0551-3929]	05 51 34 −39 31.8	1.971 0.012	245.56 −27.55	233.0 −62.6	PAIR (#2: Sc)	23 S 5.6	F306/J2669 (37,199)
*AM 0552-324	05 52 00 −32 45.0	2.214 0.012	238.29 −25.64	242.8 −68.0	ONE-ARMED SPIRAL in PAIR 364-G-11/12	10,23 S 1.6	F364/J4743 (270,299)
*AM 0552-332	05 52 02 −33 21.2	2.194 0.012	238.93 −25.81	241.7 −67.6	CHAIN of 6	17 D 9.0	F364/J4743 (269,267)
AM 0552-463 [0552-4629]	05 52 32 −46 30.5	1.666 0.011	253.34 −28.87	225.7 −56.7	PAIR (#1) 254-G-2	23 S 1.2	F254/J4748 (285,97)
AM 0552-462 [0552-4630]	05 52 33 −46 29.0	1.667 0.011	253.32 −28.86	225.7 −56.7	PAIR (#2: RING) 254-G-3	6,23 S 1.2	F254/J4748 (285,98)
AM 0552-410 [0552-4105]	05 52 42 −41 03.9	1.910 0.011	247.30 −27.71	230.8 −61.5	I/A DBL. (#1: SPIRAL)	2 S 1.8	F307/J2767 (292,118)
AM 0552-263	05 52 48 −26 34.1	2.405 0.010	231.91 −23.48	256.2 −72.3	I/A DBL. 488-G-42	2 S 0.8	F488/J3843 (169,95)
AM 0552-515 [0553-5159]	05 52 49 −51 59.6	1.364 0.010	259.56 −29.68	221.6 −51.7	PAIR (#1)	23 D 0.7	F205/J3868 (238,72)
AM 0552-410 [0552-4103]	05 52 51 −41 05.2	1.909 0.010	247.34 −27.68	230.8 −61.4	I/A DBL. (#2: E)	2 S 1.8	F307/J2767 (291,117)
AM 0552-350	05 52 57 −35 08.9	2.133 0.010	240.90 −26.14	238.4 −66.3	GALAXY w. FAINT OUTER STRUCTURE 2 APP. COMP'NS.	8,10 S 2.4	F364/J4743 (257,171)
AM 0553-225	05 53 01 −22 54.3	2.509 0.010	228.22 −22.14	267.1 −74.2	PLANETARY NEBULA	25 D 1.8	F488/J3843 (167,292)
AM 0553-515 [0552-5159]	05 53 06 −51 59.0	1.365 0.010	259.56 −29.64	221.6 −51.8	PAIR (#2) 205-G-7 = Fa 799	23 D 0.6	F205/J3868 (236,72)
*AM 0553-394	05 53 11 −39 46.5	1.961 0.010	245.91 −27.31	232.1 −62.6	I/A DBL. in GROUP of 3 307-IG-2	2,24 D 2.7	F306/J2669 (20,186)
*AM 0553-295	05 53 15 −29 56.6	2.304 0.010	235.43 −24.51	247.8 −70.2	RESOLVED LO.S.BR. Sd 424-G-39	20 D 5.4	F424/J2726 (83,181)
AM 0553-421	05 53 16 −42 13.4	1.861 0.010	248.61 −27.87	229.4 −60.5	COMPACT	13 D 0.1	F254/J4748 (286,327)
AM 0553-351	05 53 58 −35 14.8	2.129 0.009	241.07 −25.97	237.9 −66.4	SPIRAL + 6 APP. COMP'NS. 364-G-15	8 S 1.0	F364/J4743 (245,165)
AM 0554-514	05 54 02 −51 41.6	1.382 0.009	259.25 −29.45	221.5 −52.1	GROUP of 4 incl. SPIRAL (SOME I/A)	18 D 3.1	F205/J3868 (229,88)
AM 0554-234 [0554-2351]	05 54 24 −23 49.7	2.483 0.008	229.27 −22.18	263.6 −74.0	PAIR of PECULIAR GALAXIES (#1)	16,23 S 4.3	F488/J3843 (150,242)
AM 0554-380	05 54 25 −38 09.2	2.024 0.008	244.22 −26.67	233.7 −64.1	SPIRAL w. APP. COMP'N. 307-G-5	8 S 1.1	F307/J2767 (282,275)
AM 0554-235 [0554-2349]	05 54 28 −23 51.0	2.483 0.008	229.30 −22.17	263.5 −74.0	PAIR of PECULIAR GALAXIES (#2)	16,23 S 4.3	F488/J3843 (149,241)
*AM 0554-340	05 54 38 −34 06.4	2.168 0.008	239.90 −25.51	239.4 −67.4	3-ARMED SPIRAL + APP. COMP'N. 364-G-17	8,11 D 1.6	F364/J4743 (239,227)
AM 0554-332	05 54 39 −33 26.3	2.191 0.008	239.20 −25.31	240.5 −67.9	IRR./DWARF 364-G?-16	16,20 D 0.7	F364/J4743 (239,263)
AM 0555-612 [0555-6124]	05 55 13 −61 27.7	0.617 0.007	270.48 −30.18	216.1 −43.0	PAIR of E's (#1)	23 S 2.6	F121/J3822 (337,92)
AM 0555-590	05 55 19 −59 06.8	0.840 0.007	267.78 −30.04	217.1 −45.3	DWARF	20 D 0.5	F120/J1894 (106,223)
AM 0555-612 [0555-6127]	05 55 23 −61 24.9	0.622 0.007	270.43 −30.16	216.1 −43.1	PAIR of E's (#2) 120-G-23	23 S 2.6	F121/J3822 (336,94)
AM 0555-643	05 55 28 −64 38.4	0.255 0.007	274.16 −30.24	214.7 −40.0	I/A DBL. (SMALL)	2 S 0.1	F086/J1020 (156,197)
AM 0555-372	05 55 29 −37 29.0	2.049 0.007	243.56 −26.29	234.1 −64.8	E w. 6 APP.COMP'NS. 364-G-18 = PKS 0555-374	8 S 2.9	F364/J4743 (227,46)
AM 0555-590	05 55 36 −59 04.5	0.843 0.006	267.73 −30.00	217.1 −45.3	DWARF	20 D 0.6	F120/J1894 (104,225)
*AM 0555-231	05 55 48 −23 11.1	2.501 0.006	228.75 −21.65	265.2 −74.6	SP. in PAIR + PECULIAR COMP'NS. IC 2152 = 488-G-47	8,23 S 2.9	F488/J3843 (132,277)
AM 0556-455	05 56 26 −45 51.6	1.696 0.005	252.77 −28.08	225.1 −57.6	SB w. IRR. OUTER RING (COMP'N ?) 254-G-6	8,10 D 0.7	F254/J4748 (250,132)
AM 0556-345	05 56 31 −34 53.6	2.141 0.005	240.86 −25.37	237.4 −67.0	GALAXY w. I/A COMP'NS.(DUST) 364-G-19 WIDE GROUP of 4	1,14,18 S 0.5	F364/J4743 (218,185)
AM 0556-330	05 56 32 −33 07.9	2.201 0.005	239.01 −24.84	240.2 −68.4	COMPACT	13 D 0.6	F364/J4743 (218,279)

Identification	R.A. (1950.0) Declination	Precession R.A./Dec.	Gal. Long. Gal. Lat.	S.G.L. S.G.B.	Description Cross-Identification	Coding Size (Min.)	Field/Plate (x, y)
AM 0556-425	05 56 34 −42 51.4	1.834 0.005	249.46 −27.42	227.7 −60.3	DOUBLE I/A PAIR in GROUP of 8 254-IG-7	2,2,18 D 3.4	F254/J4748 (252,294)
AM 0556-361	05 56 42 −36 18.9	2.091 0.005	242.39 −25.74	235.2 −65.9	GALAXY w. JET	7 D 0.4	F364/J4743 (215,108)
AM 0556-232	05 56 48 −23 20.5	2.497 0.005	229.00 −21.49	264.2 −74.7	SP. w. APP.COMP'NS.(in WIDE PAIR) MCG-4-15-3 = 488-G-51	8,23 S 2.2	F488/J3843 (120,268)
AM 0556-471	05 56 53 −47 18.3	1.625 0.005	254.39 −28.28	223.9 −56.3	PAIR of GALAXIES (ONE COMPACT)	13,23 S 0.4	F254/J4748 (244,55)
AM 0557-423	05 57 24 −42 39.5	1.842 0.004	249.28 −27.23	227.6 −60.5	COMPACT	13 D 0.2	F307/J2767 (242,35)
AM 0557-415	05 57 27 −41 57.6	1.872 0.004	248.52 −27.05	228.3 −61.1	ONE-ARMED SPIRAL + 2 COMP'NS.	8,10 S 0.4	F307/J2767 (243,72)
AM 0557-481	05 57 39 −48 19.7	1.572 0.003	255.56 −28.34	222.9 −55.4	GALAXY w. CLOSE I/A COMP'N. 205-G-12 + MATERIAL	1 S 0.3	F205/J3868 (200,269)
AM 0557-765	05 57 45 −76 55.6	−2.680 0.003	288.32 −29.51	210.2 −28.3	SPIRAL w. PECULIAR ARM 33-G-32	10 D 2.0	F033/J1958 (93,69)
AM 0557-522	05 57 54 −52 28.8	1.333 0.003	260.24 −28.97	220.2 −51.6	2 SPIRALS + GALAXY w. Fa 801 PECULIAR DUST	14,24 D 2.6	F205/J3868 (196,46)
*AM 0558-335	05 58 15 −33 55.3	2.175 0.003	239.96 −24.74	238.1 −68.0	I/A DBL. + TAIL IC 2153 = 364-IG-22	2,15 D 1.6	F364/J4743 (199,237)
AM 0558-330	05 58 17 −33 01.0	2.205 0.002	239.01 −24.46	239.6 −68.7	GALAXY w. 3 APP. COMP'NS 364-G-23 (+COMPACTS?)	8,13 S 7.2	F364/J4743 (199,286)
AM 0558-473	05 58 32 −47 31.2	1.614 0.002	254.69 −28.05	223.2 −56.2	SPIRAL w. JET	7 D 0.6	F254/J4748 (229,44)
AM 0558-244	05 58 33 −24 42.0	2.459 0.002	230.51 −21.61	259.1 −74.4	CLOSE PAIR	23 S 1.5	F488/J3843 (99,195)
AM 0558-285	05 58 38 −28 59.6	2.333 0.002	234.87 −23.09	247.5 −71.7	DWARF SPIRAL w. ELONGATED NUCLEUS DDO 233 = MCG-5-15-2 = 425-G-2 [UKS]	12,20 D 1.8	F425/J5596 (289,231)
*AM 0558-522	05 58 44 −52 24.3	1.338 0.002	260.17 −28.84	220.0 −51.7	SPIRAL + 5 PECULIAR COMP'NS. 205-G-14	8 S 3.1	F205/J3868 (189,50)
*AM 0558-754	05 58 48 −75 40.8	−2.161 0.002	286.88 −29.57	210.5 −29.5	PLANETARY/DISTANT SPIRAL 33-PN-33 = K1-27 = Fa 246	25 D 0.6	F033/J1958 (82,135)
AM 0558-600	05 58 59 −60 04.7	0.752 0.001	268.92 −29.64	216.1 −44.5	THIN-ARMED SPIRAL w. DIAMETRICS 120-G-26 (COMP'N. on END of ARM)	8,9,10 S 0.8	F121/J3822 (320,168)
AM 0559-281 [0559-2815]	05 59 01 −28 17.1	2.354 0.001	234.18 −22.78	248.9 −72.3	PAIR (#1: E)	23 S 1.8	F425/J5596 (285,269)
AM 0559-234	05 59 03 −23 40.5	2.487 0.001	229.53 −21.13	262.2 −75.0	DISTURBED SINGLE w. LONG TAIL NGC 2139 = IC 2154 = 488-G-54	15,16 D 1.8	F488/J3843 (93,250)
AM 0559-281 [0559-2817]	05 59 08 −28 15.8	2.355 0.001	234.17 −22.75	248.9 −72.3	PAIR (#2: SPIRAL)	23 S 1.8	F425/J5596 (284,270)
AM 0559-323	05 59 34 −32 37.2	2.218 0.001	238.69 −24.08	239.7 −69.2	I/A DBL. PKS 0559-326	2 S 0.7	F364/J4743 (184,307)
AM 0559-284	05 59 36 −28 41.8	2.342 0.001	234.65 −22.79	247.7 −72.1	I/A DBL.	2 S 0.1	F425/J5596 (278,247)
AM 0559-385	05 59 45 −38 59.9	1.991 0.000	245.44 −25.89	230.6 −63.9	GROUP of 9 (TRIPLE, E + RING)	1,18,23,24 D 4.5	F307/J2767 (224,232)
AM 0600-583	06 00 27 −58 35.6	0.885 −0.001	267.24 −29.33	216.5 −46.0	CHAIN of 5 DN 0600-586	17 D 4.5	F121/J3822 (317,248)
AM 0600-395	06 00 32 −39 56.7	1.954 −0.001	246.50 −25.99	229.3 −63.2	NEST of 6 (CENTRE WIDE CLUSTER) PKS 0600-399	18,19 D 2.7	F307/J2767 (215,181)
AM 0601-544	06 01 14 −54 48.4	1.179 −0.002	262.95 −28.80	218.1 −49.6	CLOSE TRIPLE	24 D 1.6	F160/J3854 (193,190)
AM 0601-645	06 01 14 −64 56.7	0.215 −0.002	274.53 −29.64	213.8 −39.9	CHAIN of 6 GALAXIES 86-G-52 Group	17 D 16.8	F086/J1020 (124,179)
AM 0601-583	06 01 22 −58 36.5	0.884 −0.002	267.27 −29.21	216.3 −46.0	SB w. APP. COMP'N. 121-G-2	8 S 1.3	F121/J3822 (310,248)
AM 0601-603	06 01 25 −60 31.6	0.709 −0.002	269.46 −29.37	215.5 −44.2	CLUSTER of STARS (LMC?) 121-SC-3	22 D 1.6	F121/J3822 (302,146)
AM 0601-263	06 01 38 −26 39.7	2.402 −0.002	232.75 −21.67	251.8 −73.8	E + 5 APP. COMP'NS. 488-G-59	8 S 2.0	F488/J3843 (63,89)
*AM 0601-320	06 01 48 −32 08.6	2.234 −0.003	238.35 −23.48	239.5 −69.8	GALAXY w. I/A COMP'N. (8 APP.) MCG-5-15-3 = 425-G-4	1,8,9 S 0.9	F364/J4743 (159,333)
*AM 0602-600	06 02 10 −60 07.0	0.748 −0.003	269.00 −29.25	215.5 −44.6	M51-TYPE	9 S 1.1	F120/J1894 (62,167)
AM 0602-303	06 02 13 −30 35.7	2.283 −0.003	236.79 −22.89	242.2 −71.1	DISTURBED w. KNOTS and TAIL 425-G-5	15,16 D 0.7	F425/J5596 (246,146)
AM 0602-575	06 02 32 −57 58.9	0.937 −0.004	266.57 −29.00	216.4 −46.6	FAINT RING GALAXY	6 D 0.8	F120/J1894 (53,281)
AM 0602-353	06 02 36 −35 34.9	2.117 −0.004	241.99 −24.38	233.7 −67.1	CLOSE PAIR of SPIRALS (3-ARMED)	11,23 S 0.8	F364/J4743 (151,148)
AM 0603-332	06 03 10 −33 24.2	2.192 −0.005	239.76 −23.61	236.7 −69.0	TRIPLE (ABSORPTION)	14,24 D 3.4	F364/J4743 (144,265)

Identification	R.A. (1950.0) Declination	Precession R.A./Dec.	Gal. Long. Gal. Lat.	S.G.L. S.G.B.	Description Cross-Identification	Coding Size (Min.)	Field/Plate (x, y)
AM 0603-634	06 03 38 −63 41.9	0.370 −0.005	273.11 −29.32	213.9 −41.2	SPIRAL w. COMP'N. in DISK 86-G-56	8 D 1.3	F086/J1020 (107,246)
*AM 0603-330	06 03 54 −33 04.6	2.202 −0.006	239.49 −23.36	236.9 −69.4	LARGE DWARF (RESOLUTION) 364-G?-29 = UKS 0603-330	20 D 2.0	F364/J4743 (135,282)
*AM 0604-451	06 04 01 −45 10.6	1.729 −0.006	252.33 −26.64	223.4 −58.7	SPIRAL w. SPLIT ARM + COMP'N. 254-G-13	8,10 D 0.7	F254/J4748 (179,170)
AM 0604-693	06 04 08 −69 34.7	−0.515 −0.006	279.86 −29.43	211.9 −35.5	CLOSE PAIR NGC 2187 = 57-G-68 = DN 0604-695	23 S 1.1	F057/J1023 (176,199)
AM 0604-275	06 04 38 −27 52.3	2.367 −0.007	234.22 −21.47	247.0 −73.5	LINEAR DWARF 425-G-8	20 D 0.9	F425/J5596 (219,292)
AM 0604-751	06 04 52 −75 18.3	−2.020 −0.007	286.43 −29.21	210.2 −30.0	SHRED or EDGE-ON GALAXY 34-G-2	22 D 1.6	F033/J1958 (59,152)
AM 0604-421	06 04 54 −42 19.0	1.857 −0.007	249.28 −25.80	225.4 −61.4	CONTACT TRIPLE (WIDE GROUP of 6)	3,18 D 0.2	F307/J2767 (168,55)
AM 0605-450	06 05 22 −45 02.3	1.736 −0.008	252.24 −26.37	223.1 −59.0	CONTACT PAIR (SPIKE) 254-IG-18	2 S 0.1	F254/J4748 (166,177)
AM 0605-385	06 05 31 −38 57.1	1.993 −0.008	245.72 −24.80	228.4 −64.5	DISTURBED GALAXY w. PLUMES 307-G-19	15 D 0.8	F307/J2767 (164,236)
AM 0605-341	06 05 32 −34 11.6	2.166 −0.008	240.75 −23.39	234.4 −68.6	IRR. + COMPACT	13,16 S 0.8	F364/J4743 (118,222)
AM 0605-492	06 05 42 −49 29.2	1.510 −0.008	257.13 −27.26	220.0 −54.8	CHAIN of 4 in GRP. of 8 (PEC.SP.) PKS 0605-494	12,17,18 D 4.3	F205/J3868 (129,207)
AM 0605-332	06 05 58 −33 28.8	2.190 −0.009	240.04 −23.08	235.3 −69.3	I/A DBL. 364-G-31	2 S 0.9	F364/J4743 (113,260)
*AM 0606-770	06 06 01 −77 08.9	−2.781 −0.009	288.53 −29.03	209.6 −28.2	CLOSE PAIR (INCIPIENT M51-TYPE)	9,23 S 0.8	F033/J1958 (70,53)
AM 0606-472	06 06 25 −47 24.4	1.620 −0.009	254.87 −26.72	221.1 −56.8	SPIRAL + APP. GROUP of 5 COMP'NS 254-G-22 = Fa 805	8 S 4.3	F205/J3868 (121,318)
*AM 0607-444	06 07 50 −44 42.7	1.751 −0.011	252.00 −25.87	222.5 −59.4	SPIRAL w. I/A COMP'N. 254-G-28	1 S 0.3	F254/J4748 (143,195)
*AM 0608-364	06 08 06 −36 40.0	2.079 −0.012	243.49 −23.65	229.9 −66.8	SMALL I/A DBL.	2 S 0.2	F364/J4743 (92,89)
AM 0608-333	06 08 08 −33 38.0	2.185 −0.012	240.35 −22.70	233.9 −69.4	I/A DBL. off of AM 0608-332 364-G-35/36	2 S 1.1	F364/J4743 (88,252)
*AM 0608-473	06 08 17 −47 36.7	1.610 −0.012	255.17 −26.46	220.4 −56.7	E + I/A COMP'N. + MATERIAL + COMPACT 205-G-27	1,2,13,17 S 0.4	F205/J3868 (104,307)
*AM 0608-340	06 08 21 −34 05.4	2.169 −0.012	240.84 −22.80	233.2 −69.0	HI.S.BR. IRR. NGC 2188 = 364-G-37 = Lu 0608-34	8,16 D 3.4	F364/J4743 (86,227)
*AM 0608-473	06 08 23 −47 32.3	1.614 −0.012	255.09 −26.43	220.4 −56.8	COMPACT	13 D 0.1	F254/J4748 (140,43)
*AM 0608-332	06 08 33 −33 25.0	2.192 −0.012	240.16 −22.55	234.1 −69.6	CHAIN of 10	17 D 15.7	F364/J4743 (84,263)
AM 0608-752	06 08 34 −75 21.3	−2.036 −0.012	286.47 −28.98	209.9 −30.0	OFF-SET NUCLEUS (RINGS + COMP'NS) IC 2164 = 34-G-5 = Fa 248	8,12 D 1.1	F033/J1958 (47,147)
AM 0608-513	06 08 35 −51 34.9	1.390 −0.013	259.54 −27.21	218.1 −53.0	GROUP of 3 SPIRALS	24 D 4.9	F205/J3868 (107,94)
AM 0608-654	06 08 41 −65 43.5	0.113 −0.013	275.46 −28.89	212.5 −39.4	CLOSE TRIPLE 86-G-62 = PKS 0608-65?	24 D 2.0	F087/J1171 (327,130)
AM 0608-631	06 08 52 −63 12.3	0.430 −0.013	272.59 −28.71	213.3 −41.8	CLOSE PAIR	23 S 2.1	F087/J1171 (341,265)
AM 0609-323	06 09 09 −32 33.5	2.221 −0.013	239.33 −22.15	235.1 −70.4	GAL.w. BROKEN RING + COMP'N. (HALO) 364-G-40	6,8 S 0.4	F364/J4743 (76,309)
AM 0609-361	06 09 09 −36 12.7	2.096 −0.013	243.08 −23.31	230.0 −67.3	GALAXY w. 3 APP. COMP'NS. 364-G-41	8 S 1.1	F364/J4743 (80,113)
AM 0609-473	06 09 11 −47 39.7	1.608 −0.013	255.26 −26.32	220.1 −56.7	OUTLYING SPIRAL + I/A DBL.COMP'N 205-G-29	1 S 0.9	F205/J3868 (96,304)
AM 0609-331	06 09 16 −33 17.5	2.196 −0.014	240.09 −22.37	233.9 −69.8	I/A DBL. 364-IG-42	2 S 0.4	F365/J2882 (344,268)
AM 0609-314 [0609-3145]	06 09 37 −31 44.2	2.247 −0.014	238.53 −21.78	236.2 −71.2	PAIR (#1: E)	23 S 0.7	F425/J5596 (161,85)
AM 0609-314 [0609-3144]	06 09 38 −31 45.0	2.247 −0.014	238.54 −21.78	236.2 −71.1	PAIR (#2: SPIRAL)	23 S 0.7	F425/J5596 (160,84)
AM 0609-330	06 09 49 −33 06.7	2.203 −0.014	239.94 −22.20	233.9 −70.0	GALAXY w. PECULIAR DUST LANE? (PROLATE?)	14 D 0.4	F364/J4743 (69,279)
AM 0610-623	06 10 06 −62 31.5	0.506 −0.015	271.84 −28.51	213.4 −42.5	E w. JET NGC 2205 = 86-G-63 = MC 0610-625	7 D 0.7	F086/J1020 (64,306)
*AM 0610-443	06 10 12 −44 38.2	1.755 −0.015	252.04 −25.44	221.8 −59.6	I/A QUINTET w. INTEGRAL-SIGN 254-G-32/34 SPIRAL	5,12 D 3.6	F254/J4748 (120,198)
AM 0610-430	06 10 57 −43 01.2	1.828 −0.016	250.35 −24.90	222.7 −61.2	DISTURBED SPIRAL + APP. COMP'N. 254-G-35	8,12 S 2.0	F307/J2767 (108,18)
AM 0611-450	06 11 05 −45 03.6	1.736 −0.016	252.53 −25.40	221.2 −59.3	2 E's in COMMON ENVELOPE 254-IG-13	2 S 0.5	F254/J4748 (112,175)

Identification	R.A. (1950.0) Declination	Precession R.A./Dec.	Gal. Long. Gal. Lat.	S.G.L. S.G.B.	Description Cross-Identification	Coding Size (Min.)	Field/Plate (x, y)
AM 0611-345	06 11 07 −34 58.9	2.139 −0.016	241.95 −22.54	230.6 −68.5	I/A DBL. + SHREDS	2 S 0.9	F365/J2882 (321,178)
AM 0611-614 [0611-6143]	06 11 29 −61 43.6	0.592 −0.017	270.95 −28.27	213.4 −43.3	PAIR (#1)	23 S 0.8	F121/J3822 (232,85)
AM 0611-430	06 11 32 −43 09.8	1.822 −0.017	250.53 −24.84	222.4 −61.1	RING w. OFF-SET NUCLEUS in PAIR 254-G-38	6,23 D 0.6	F254/J4748 (105,277)
AM 0611-614 [0611-6143]	06 11 35 −61 43.1	0.593 −0.017	270.94 −28.26	213.4 −43.3	PAIR (#2)	23 S 0.8	F121/J3822 (232,86)
AM 0611-433	06 11 47 −43 38.8	1.800 −0.017	251.06 −24.92	221.9 −60.7	TRIPLE (SP.w. PEC.ARMS + COMPACT) NGC 2200/01 = 254-G-39/40	10,13,24 D 6.5	F254/J4748 (104,251)
AM 0611-492	06 11 52 −49 25.3	1.515 −0.017	257.28 −26.27	218.4 −55.2	SMALL I/A DBL.	2 S 0.5	F205/J3868 (76,209)
AM 0611-700	06 11 54 −70 02.1	−0.600 −0.017	280.39 −28.76	211.0 −35.2	CLOSE PAIR 57-IG-72	23 S 0.8	F057/J1023 (141,174)
AM 0612-362	06 12 05 −36 23.6	2.090 −0.018	243.47 −22.80	228.4 −67.4	COMPACT	13 D 0.3	F364/J4743 (48,103)
AM 0612-373	06 12 05 −37 39.6	2.043 −0.018	244.78 −23.19	227.0 −66.2	GALAXY w. TAIL and DUST 307-IG-25	2,14,15 D 1.3	F364/J4743 (51,34)
AM 0612-332	06 12 18 −33 29.0	2.191 −0.018	240.50 −21.83	232.0 −70.0	EDGE-ON + 2 APP. COMP'NS 365-G-1	8 S 0.6	F365/J2882 (310,259)
*AM 0612-411	06 12 25 −41 10.1	1.906 −0.018	248.47 −24.14	223.6 −63.0	COMPACT w. DISK ("N" GALAXY ?) DIFFRACTION SPIKE !?	13 D 0.7	F307/J2767 (93,117)
AM 0612-425	06 12 32 −42 50.4	1.836 −0.018	250.24 −24.58	222.3 −61.5	TRIPLE (COMPACT w. HI.S.BR. RING) incl. 254-G-43? + E + SPIRAL)	13,24 D 4.5	F307/J2767 (92,27)
AM 0612-254	06 12 34 −25 46.6	2.429 −0.018	232.83 −19.06	247.9 −76.2	I/A DBL. in GROUP of 3 MCG-4-15-15 = 489-G-20	2,24 S 1.0	F489/J3797 (196,138)
AM 0612-513	06 12 53 −51 31.8	1.394 −0.019	259.62 −26.54	217.1 −53.2	SMALL DWARF (RESOLUTION) 205-G?-35	20 D 0.9	F206/J2820 (322,93)
AM 0612-593 [0613-5938]	06 12 54 −59 37.6	0.797 −0.019	268.61 −27.86	213.9 −45.4	PAIR (#1)	23 S 0.9	F121/J3822 (228,198)
AM 0613-593 [0612-5937]	06 13 02 −59 38.0	0.796 −0.019	268.62 −27.84	213.8 −45.4	PAIR (#2)	23 S 0.9	F121/J3822 (227,198)
AM 0613-675	06 13 16 −67 53.0	−0.209 −0.019	277.95 −28.56	211.4 −37.4	CHAIN of 4	17 D 11.6	F057/J1023 (130,289)
AM 0613-621	06 13 17 −62 16.9	0.535 −0.019	271.60 −28.12	213.0 −42.8	I/A DBL. (DWARF)	2,20 S 0.4	F087/J1171 (319,317)
AM 0613-263	06 13 19 −26 33.7	2.406 −0.019	233.66 −19.20	245.1 −75.7	DWARF (SOME RESOLUTION) 489-G-22	20 D 1.4	F489/J3797 (187,96)
AM 0613-641	06 13 20 −64 18.2	0.301 −0.019	273.89 −28.31	212.3 −40.9	COMPACT	13 D 0.2	F086/J1020 (52,209)
AM 0613-440	06 13 39 −44 04.5	1.782 −0.020	251.61 −24.71	221.0 −60.4	GROUP of 3	24 D 1.3	F254/J4748 (86,228)
*AM 0614-264	06 14 00 −26 44.9	2.401 −0.020	233.91 −19.12	244.2 −75.7	SPIRAL w. I/A COMP'N. NGC 2206 = 489-G-26	1 S 2.3	F489/J3797 (179,86)
AM 0614-573	06 14 00 −57 35.2	0.973 −0.020	266.35 −27.43	214.4 −47.4	TRIPLE COMPACT	13,24 D 1.3	F161/J1122 (335,32)
AM 0614-370	06 14 23 −37 08.7	2.063 −0.021	244.40 −22.60	226.5 −66.9	EDGE-ON w. DUST + 5 APP. COMP'NS. CHAIN of 5	8,14 S 2.0	F364/J4743 (25,61)
*AM 0615-574	06 15 05 −57 42.7	0.963 −0.022	266.51 −27.41	214.1 −47.3	DWARF (SOME RESOLUTION) 121-G-20	20 D 0.7	F160/J3854 (92,32)
AM 0615-481	06 15 34 −48 18.0	1.577 −0.023	256.22 −25.42	217.9 −56.5	E + SPIRAL 206-G-4 = Fa 810	23 S 3.8	F206/J2820 (308,268)
AM 0616-324	06 16 11 −32 47.5	2.214 −0.024	240.10 −20.83	230.9 −71.0	GALAXY w. 2 LARGE COMP'NS. 365-G-3	8 S 3.3	F365/J2882 (267,297)
AM 0616-350	06 16 39 −35 08.0	2.135 −0.024	242.50 −21.52	227.6 −68.9	GROUP of 5	18 D 2.7	F365/J2882 (260,172)
AM 0616-484	06 16 55 −48 43.8	1.555 −0.025	256.74 −25.30	217.3 −56.1	CHAIN of 4 PKS 0616-48	17 D 3.1	F206/J2820 (295,245)
AM 0616-485	06 16 56 −48 55.1	1.545 −0.025	256.94 −25.35	217.2 −55.9	GROUP of 4	18 D 2.2	F205/J3868 (30,234)
AM 0616-351	06 16 59 −35 17.1	2.130 −0.025	242.68 −21.51	227.2 −68.8	M51-TYPE ? 365-G-5	9 S 0.5	F365/J2882 (256,164)
AM 0617-243	06 17 06 −24 33.4	2.464 −0.025	232.05 −17.65	248.5 −77.8	GROUP of 4 incl. 489-G-36	18 D 9.2	F489/J3797 (141,203)
AM 0617-274	06 17 11 −27 42.0	2.374 −0.025	235.12 −18.82	239.6 −75.4	COMPACT	13 D 0.1	F426/J3890 (344,300)
*AM 0617-302	06 17 13 −30 23.1	2.292 −0.025	237.77 −19.79	234.1 −73.2	I/A DBL. w. TAILS	2,15 S 0.4	F426/J3890 (339,156)
AM 0617-624	06 17 27 −62 44.7	0.487 −0.025	272.18 −27.69	212.2 −42.5	I/A DBL. (SMALL)	2 S 0.6	F087/J1171 (291,294)
AM 0618-371 [0618-3712]	06 18 15 −37 10.5	2.063 −0.027	244.69 −21.87	224.6 −67.2	PAIR (#1: CONTACT I/A DBL.) 365-IG-6	2,23 S 0.2	F307/J2767 (29,331)

Identification	R.A. (1950.0) Declination	Precession R.A./Dec.	Gal. Long. Gal. Lat.	S.G.L. S.G.B.	Description Cross-Identification	Coding Size (Min.)	Field/Plate (x, y)
AM 0618-431	06 18 19 −43 17.6	1.818 −0.027	251.03 −23.68	219.9 −61.4	COMPACT w. BROKEN RING 255-G-1	13 D 0.3	F254/J4748 (39,268)
*AM 0618-223	06 18 22 −22 33.1	2.520 −0.027	230.23 −16.60	255.0 −79.3	SPIRAL w. 4 COMP'NS. MCG-4-15-25 = 489-G-39	8 S 3.7	F489/J3797 (126,311)
AM 0618-363	06 18 23 −36 33.1	2.086 −0.027	244.06 −21.65	225.2 −67.7	I/A DBL.(RING + FILAM. to COMP'N)	2 S 0.4	F365/J2882 (240,96)
AM 0618-371 [0618-3710]	06 18 29 −37 12.2	2.062 −0.027	244.74 −21.84	224.5 −67.1	PAIR (#2: SPIRAL) PKS 0618-371	23 S 0.2	F307/J2767 (27,329)
AM 0618-275	06 18 47 −27 53.2	2.368 −0.027	235.44 −18.56	238.0 −75.5	V.LO.S.BR. 2-ARMED SPIRAL 425-G-18	10 D 1.1	F425/J5596 (51,290)
*AM 0619-573 [0619-5728]	06 19 27 −57 32.9	0.980 −0.028	266.44 −26.71	213.3 −47.6	I/A TRIPLE (#1: SPIRAL) NGC 2221 = 121-IG-24	3 D 5.4	F160/J3854 (60,39)
*AM 0619-573 [0619-5732]	06 19 30 −57 30.9	0.982 −0.028	266.40 −26.69	213.3 −47.7	I/A TRIPLE (#2: SPIRAL) NGC 2222 = 121-IG-25	3 D 5.4	F160/J3854 (60,41)
AM 0619-652	06 19 37 −65 25.6	0.162 −0.029	275.22 −27.75	211.2 −39.9	I/A DBL.	2 S 0.2	F087/J1171 (268,151)
*AM 0619-271	06 19 41 −27 12.7	2.389 −0.029	234.86 −18.12	238.9 −76.1	SB (DUST + KNOTS in OUTER CIRC. NGC 2217 = 489-G-42 ARMS)	8,10,13,14 D 3.1	F425/J5596 (40,326)
AM 0619-273	06 19 44 −27 32.2	2.379 −0.029	235.18 −18.24	238.1 −75.9	DWARF 426-G-1	20 D 1.6	F425/J5596 (40,309)
AM 0619-572 [0619-5730]	06 19 44 −57 28.3	0.986 −0.029	266.36 −26.66	213.3 −47.7	I/A TRIPLE (#3: SB) 161-G-1	3 D 5.4	F160/J3854 (58,43)
AM 0620-645	06 20 00 −64 59.9	0.219 −0.029	274.75 −27.66	211.2 −40.3	I/A TRIPLE (of GROUP to [S.E.]) near 87-G-4	3 D 0.7	F086/J1020 (18,168)
AM 0620-344	06 20 10 −34 40.0	2.153 −0.029	242.28 −20.69	226.3 −69.6	M51-TYPE ? + DIAMETRIC COMP'NS. 365-G-9	8,9 S 1.1	F365/J2882 (221,197)
AM 0620-363	06 20 13 −36 31.8	2.087 −0.029	244.17 −21.30	224.3 −67.9	GAL.w. LONG ARM + 4 APP. COMP'NS 365-G-10 (DIAMETRICS)	8,10 S 2.9	F365/J2882 (220,97)
AM 0620-531	06 20 18 −53 11.0	1.295 −0.030	261.69 −25.77	214.6 −51.9	I/A DBL.	2 S 0.4	F161/J1122 (303,271)
AM 0620-594	06 20 59 −59 42.8	0.795 −0.031	268.87 −26.86	212.4 −45.5	LARGE SB w. COMP'N. or KNOT 121-G-26 + PEC. FEATURE in RIM	8,10 S 1.3	F121/J3822 (173,195)
AM 0621-645	06 21 15 −64 58.1	0.224 −0.031	274.73 −27.53	211.1 −40.4	DBL. E (CENTRE of LARGE GROUP) NGC 2229/30/33/35 = 87-G-9+	18,23 D 26.9	F087/J1171 (260,176)
AM 0621-273	06 21 49 −27 35.5	2.378 −0.032	235.42 −17.83	236.4 −76.1	DWARF	20 D 0.2	F425/J5596 (15,305)
AM 0621-353	06 21 52 −35 36.7	2.120 −0.032	243.36 −20.68	224.4 −68.9	DISTURBED SPIRAL + 6 COMP'NS. 365-G-11	1,8 S 1.6	F365/J2882 (202,147)
AM 0621-700	06 21 52 −70 04.6	−0.596 −0.032	280.48 −27.92	209.9 −35.4	I/A TRIPLE 57-G-78	3 D 1.5	F057/J1023 (95,170)
AM 0622-350	06 22 07 −35 00.1	2.142 −0.032	242.76 −20.42	224.8 −69.5	CONTACT DBL. E + 2 APP. COMP'NS. 365-IG-13	2,8 S 1.8	F365/J2882 (200,179)
AM 0622-350	06 22 09 −35 05.0	2.139 −0.032	242.85 −20.44	224.7 −69.4	GALAXY w. DISTORTED ARMS (+ I/A COMP'N.)	1,10 S 1.8	F365/J2882 (199,175)
AM 0622-385	06 22 19 −38 53.2	2.001 −0.032	246.71 −21.66	221.3 −65.8	TRIPLE w. 3 APP. COMP'NS.	8,24 D 0.4	F308/J4792 (260,238)
*AM 0622-224	06 22 30 −22 48.6	2.514 −0.033	230.88 −15.83	250.4 −79.9	SPIRAL w. APP. COMP'NS NGC 2223 = 489-G-49	8 S 4.0	F489/J3797 (74,297)
AM 0622-643	06 22 33 −64 32.5	0.280 −0.033	274.27 −27.34	211.0 −40.8	CHAIN of 5 87-G-15	17 D 7.8	F087/J1171 (254,199)
AM 0622-625 [0622-6258]	06 22 34 −62 56.3	0.471 −0.033	272.48 −27.14	211.4 −42.4	PAIR (#1: HI.S.BR.) 87-IG-14	23 S 2.7	F121/J3822 (162,22)
AM 0622-625 [0622-6256]	06 22 52 −62 58.8	0.466 −0.033	272.53 −27.11	211.3 −42.4	PAIR (#2: LO.S.BR.)	23 S 2.7	F121/J3822 (160,19)
*AM 0623-605	06 23 00 −60 56.9	0.680 −0.033	270.28 −26.80	211.8 −44.4	SMALL M51-TYPE w. COMP'N w. TAIL 121-IG-28 [A82]	8,9 S 1.3	F121/J3822 (159,129)
AM 0623-371	06 23 07 −37 19.8	2.059 −0.034	245.18 −21.01	222.1 −67.3	LOOSE CHAIN of 13 (M82-TYPE)	13,17 D 20.2	F365/J2882 (189,54)
AM 0623-651	06 23 10 −65 19.3	0.180 −0.034	275.15 −27.37	210.7 −40.1	GROUP of 5 incl. 87-IG-16	18 D 6.7	F087/J1171 (249,158)
AM 0623-541	06 23 15 −54 16.7	1.225 −0.034	262.97 −25.57	213.5 −50.9	CLOSE PAIR	23 S 0.3	F161/J1122 (277,213)
AM 0623-584	06 23 29 −58 40.4	0.890 −0.034	267.78 −26.37	212.3 −46.6	SMALL 'S'-SHAPED GALAXY 121-IG-31	9,12 S 0.3	F121/J3822 (156,251)
AM 0623-321	06 23 39 −32 16.6	2.234 −0.034	240.15 −19.17	227.1 −72.1	DISTURBED GALAXY in CLUSTER 426-G-7 + MANY COMP'NS.	8,14 D 0.9	F426/J3890 (262,57)
AM 0624-620 [0624-6207]	06 24 04 −62 06.8	0.562 −0.035	271.59 −26.85	211.3 −43.3	PAIR (#1)	23 S 0.9	F121/J3822 (153,66)
AM 0624-620 [0624-6206]	06 24 08 −62 07.8	0.561 −0.035	271.61 −26.85	211.3 −43.2	PAIR (#2)	23 S 0.9	F121/J3822 (152,65)
AM 0624-534	06 24 15 −53 43.6	1.263 −0.035	262.41 −25.31	213.5 −51.5	CLOSE PAIR	23 S 0.3	F161/J1122 (270,243)

Identification	R.A. (1950.0) Declination	Precession R.A./Dec.	Gal. Long. Gal. Lat.	S.G.L. S.G.B.	Description Cross-Identification	Coding Size (Min.)	Field/Plate (x, y)
AM 0624-261	06 24 16 −26 14.3	2.418 −0.035	234.32 −16.80	237.6 −77.5	IRR. 489-G?-56	22 D 0.7	F490/J5469 (320,111)
AM 0624-631	06 24 40 −63 17.9	0.432 −0.036	272.92 −26.95	211.0 −42.1	GROUP of 4 (incl. COMPACT) 87-G-17	13,18 D 4.5	F087/J1171 (245,267)
AM 0624-344	06 24 41 −34 48.4	2.150 −0.036	242.76 −19.86	223.6 −69.8	SPIRAL w. 3 APP. + I/A COMP'N. 365-G-20	1,2,8 S 2.2	F365/J2882 (172,190)
AM 0624-283	06 24 42 −28 38.0	2.348 −0.036	236.68 −17.63	232.0 −75.5	GALAXY w. TAIL MCG-5-16-3 = 426-G-10	15 D 0.8	F426/J3890 (254,252)
AM 0625-252	06 25 00 −25 21.0	2.444 −0.036	233.54 −16.31	239.4 −78.3	RING/PLANETARY NEBULA PK 233-16-1 = 490-PN?-1	6 D 0.5	F489/J3797 (46,160)
AM 0625-621	06 25 00 −62 13.2	0.552 −0.036	271.73 −26.76	211.1 −43.2	I/A QUARTET (SMALL)	4 D 0.5	F087/J1171 (246,325)
AM 0625-540	06 25 02 −54 00.2	1.245 −0.036	262.74 −25.25	213.2 −51.3	COMPACT CLUSTER of GALAXIES 161-IG-6 = DN 0625-540	19 D 5.4	F161/J1122 (263,229)
AM 0625-533	06 25 14 −53 39.7	1.268 −0.037	262.38 −25.15	213.3 −51.6	I/A DBL. 161-IG-7 = PKS 0625-53	2 S 0.4	F161/J1122 (262,247)
AM 0625-490	06 25 16 −49 05.6	1.540 −0.037	257.49 −24.06	214.8 −56.1	I/A DBL. w. LONG TAIL	2,15 S 0.4	F206/J2820 (221,228)
AM 0625-615	06 25 18 −61 57.4	0.580 −0.037	271.44 −26.68	211.2 −43.4	DWARF (DEFECT ?)	20 D 0.6	F087/J1171 (244,339)
AM 0625-540	06 25 33 −54 08.1	1.236 −0.037	262.90 −25.21	213.1 −51.1	CLOSE PAIR	23 S 0.2	F161/J1122 (259,222)
AM 0625-355	06 25 40 −35 59.1	2.109 −0.037	244.01 −20.08	222.0 −68.8	HI.S.BR. IRR. + SP. w. PEC. ARMS	10,16,23 S 2.8	F365/J2882 (161,127)
AM 0625-652	06 25 47 −65 27.5	0.166 −0.038	275.34 −27.11	210.4 −40.0	I/A TRIPLE 87-G-19	3 D 0.9	F087/J1171 (234,151)
AM 0625-630	06 25 57 −63 06.4	0.456 −0.038	272.73 −26.78	210.8 −42.3	I/A TRIPLE	3 D 0.5	F087/J1171 (238,277)
*AM 0626-470	06 26 00 −47 08.8	1.642 −0.038	255.47 −23.44	215.3 −58.0	I/A TRIPLE 255-IG-7	3 D 0.7	F206/J2820 (216,333)
AM 0626-610	06 26 22 −61 09.8	0.663 −0.038	270.59 −26.43	211.2 −44.2	QUARTET 121-G-32	18 D 2.6	F121/J3822 (137,117)
AM 0626-355	06 26 37 −35 56.3	2.111 −0.039	244.03 −19.88	221.5 −68.9	I/A DBL.	2 S 0.2	F365/J2882 (151,129)
AM 0626-542	06 26 37 −54 25.1	1.218 −0.039	263.24 −25.12	212.7 −50.9	LARGE CLUSTER CENTERED on cD 161-G-8	19 D 5.4	F161/J1122 (250,207)
AM 0626-634	06 26 38 −63 48.3	0.376 −0.039	273.51 −26.80	210.6 −41.6	GROUP of FOUR (CLOSE) 87-G-21	18 D 1.3	F087/J1171 (232,240)
AM 0627-472 [0627-4719]	06 27 05 −47 20.1	1.633 −0.039	255.73 −23.31	214.9 −57.9	GROUP of 3 (#1)	24 D 0.2	F255/J2826 (224,54)
AM 0627-484	06 27 10 −48 44.0	1.561 −0.039	257.20 −23.67	214.4 −56.5	SPIRAL w. 5 COMP'NS. Fa 815	8 D 1.8	F206/J2820 (204,247)
AM 0627-262	06 27 13 −26 27.6	2.413 −0.040	234.81 −16.28	234.5 −77.7	RING MCG-4-16-5 = 490-G-6	6 D 0.7	F489/J3797 (20,100)
AM 0627-472 [0627-4720]	06 27 16 −47 22.1	1.632 −0.040	255.77 −23.29	214.8 −57.8	GROUP of 3 (#2)	24 D 0.3	F255/J2826 (222,52)
AM 0627-674	06 27 16 −67 49.9	−0.183 −0.040	278.01 −27.24	209.7 −37.6	GALAXY w. PECULIAR ARM	10 D 0.5	F087/J1171 (221,24)
AM 0627-535	06 27 20 −53 51.4	1.257 −0.040	262.66 −24.89	212.7 −51.5	COMPACT (DBL.) 161-IG-9	13,13,23 S 0.8	F161/J1122 (245,237)
AM 0627-263	06 27 22 −26 36.7	2.408 −0.040	234.97 −16.31	234.0 −77.5	E + SPIRAL	23 S 1.2	F489/J3797 (18,92)
AM 0627-471 [0627-4722]	06 27 27 −47 19.9	1.634 −0.040	255.74 −23.25	214.8 −57.9	GROUP of 3 (#3)	24 D 0.5	F255/J2826 (221,54)
AM 0628-264	06 28 04 −26 48.0	2.403 −0.041	235.21 −16.24	232.9 −77.5	I/A DBL. (E + SPIRAL)	2 S 0.6	F489/J3797 (10,81)
AM 0628-325	06 28 10 −32 52.1	2.216 −0.041	241.11 −18.50	223.5 −71.9	GALAXY w. PEC. DUST + I/A COMP'N.	1,14 S 0.9	F426/J3890 (210,25)
AM 0628-321	06 28 29 −32 14.6	2.237 −0.041	240.52 −18.21	224.0 −72.5	SPIRAL w. PECULIAR DOUBLE ARM 426-G-18	10 D 0.7	F426/J3890 (207,59)
*AM 0628-632	06 28 38 −63 27.8	0.419 −0.042	273.17 −26.53	210.3 −42.0	RING GALAXY 87-G-22 = Fa 252	6 D 0.6	F087/J1171 (221,259)
AM 0629-660	06 29 02 −66 09.4	0.074 −0.042	276.16 −26.87	209.8 −39.3	I/A DBL. 87-IG-23 = PKS 0629-661	2 S 0.2	F087/J1171 (215,114)
AM 0629-305	06 29 12 −30 59.5	2.277 −0.042	239.36 −17.61	225.0 −73.7	CLOSE PAIR of SPIRALS + COMPACT MCG-5-16-9/10 = 426-G-20	13,23 S 0.5	F426/J3890 (200,126)
AM 0629-693	06 29 19 −69 39.9	−0.503 −0.043	280.08 −27.24	209.2 −35.9	3 CLOSE SPIRALS (ALIGNED)	24 D 2.1	F057/J1023 (59,189)
AM 0630-522	06 30 01 −52 23.2	1.354 −0.044	261.20 −24.16	212.5 −53.0	DWARF 206-G-16 = UKS 0630-523	20 D 0.9	F161/J1122 (226,317)
AM 0630-672	06 30 10 −67 24.3	−0.109 −0.044	277.57 −26.92	209.5 −38.1	DWARF	20 D 0.5	F058/J1021 (310,309)

Identification	R.A. (1950.0) Declination	Precession R.A./Dec.	Gal. Long. Gal. Lat.	S.G.L. S.G.B.	Description Cross-Identification	Coding Size (Min.)	Field/Plate (x, y)
AM 0630-353	06 30 21 −35 35.2	2.125 −0.044	243.96 −19.05	219.8 −69.4	I/A DBL. 365-G-26	2 S 0.4	F365/J2882 (110,147)
AM 0630-332	06 30 31 −33 28.6	2.197 −0.044	241.90 −18.26	221.4 −71.5	E + 2 APP. COMP'NS.	8 S 1.8	F365/J2882 (106,261)
AM 0631-622	06 31 33 −62 21.8	0.546 −0.046	272.02 −26.03	210.1 −43.1	3-ARMED SP. + COMP'N. (SMALL)	8,11 D 0.3	F087/J1171 (205,318)
AM 0631-341	06 31 49 −34 13.8	2.173 −0.046	242.74 −18.29	220.0 −70.8	I/A PAIR in WIDE GROUP of 5 365-G-28	2,18 S 1.1	F365/J2882 (92,220)
AM 0632-254	06 32 06 −25 41.3	2.437 −0.047	234.52 −14.97	231.5 −78.9	PLANETARY NEBULA MCG-4-16-11 = 490-G?-15	25 D 0.5	F490/J5469 (226,142)
AM 0632-673	06 32 28 −67 36.6	−0.137 −0.047	277.83 −26.72	209.2 −37.9	BARRED SPIRAL w. PECULIAR ARM 58-G-4	10 D 0.7	F058/J1021 (297,299)
AM 0632-310	06 32 35 −31 03.8	2.276 −0.047	239.71 −16.96	222.5 −73.9	DWARF	20 D 0.7	F426/J3890 (161,122)
AM 0632-305	06 32 53 −30 56.2	2.281 −0.048	239.61 −16.86	222.4 −74.0	DWARF 426-G-25	20 D 0.6	F426/J3890 (157,129)
*AM 0632-625	06 32 54 −62 57.1	0.483 −0.048	272.69 −25.98	209.8 −42.6	CONTACT DBL. w. MATERIAL 87-G-28 = MC 0632-629	2,14 S 0.1	F121/J3822 (99,20)
AM 0633-352	06 33 27 −35 26.6	2.132 −0.049	244.06 −18.41	218.1 −69.7	SPIRAL w. DIAMETRIC JETS 365-G-33	7 D 0.6	F365/J2882 (76,154)
AM 0633-391	06 33 59 −39 12.9	1.995 −0.049	247.84 −19.62	215.7 −66.1	SPIRAL w. I/A COMP'N. + COMPACT	1,13 S 2.7	F308/J4792 (138,221)
AM 0634-292	06 34 18 −29 22.3	2.329 −0.050	238.23 −15.97	223.1 −75.6	CLOSE PAIR of E's	23 S 0.8	F426/J3890 (141,213)
AM 0634-292	06 34 22 −29 27.8	2.327 −0.050	238.33 −16.00	222.9 −75.5	RING (PAIR NEAR-BY)	6 D 0.5	F426/J3890 (140,208)
AM 0634-343	06 34 48 −34 34.3	2.163 −0.051	243.31 −17.83	217.9 −70.6	GALAXY w. I/A COMP'N.	1 S 0.9	F365/J2882 (60,201)
AM 0634-350	06 34 58 −35 01.7	2.147 −0.051	243.77 −17.97	217.5 −70.2	BROAD ONE-ARMED SP. + PEC. COMP'N. 365-G-35	8,10 S 4.0	F365/J2882 (59,176)
AM 0635-372	06 35 42 −37 22.6	2.065 −0.052	246.13 −18.67	215.8 −67.9	GROUP of 6 incl. 366-G-1	18 D 4.5	F366/J2716 (310,49)
AM 0635-240	06 35 55 −24 09.5	2.481 −0.052	233.44 −13.56	230.9 −80.6	SP. in WIDE PAIR + 2 APP. COMP'NS 490-G-18	8,23 S 5.4	F490/J5469 (180,224)
AM 0636-580	06 36 05 −58 03.0	0.957 −0.052	267.46 −24.63	210.0 −47.5	DWARF ?	20 D 0.4	F121/J3822 (66,282)
*AM 0636-662	06 36 38 −66 29.4	0.041 −0.053	276.65 −26.16	208.8 −39.1	SB w. PROTO-ARM 87-G-32	10 D 1.6	F087/J1171 (173,97)
AM 0636-593	06 36 49 −59 36.3	0.825 −0.053	269.16 −24.88	209.6 −45.9	LINEAR FEATURE or GALAXY w. JET 121-G-41 (in WIDE PAIR w. PEC.)	7,23 D 0.7	F121/J3822 (65,198)
AM 0637-365	06 37 20 −36 52.4	2.084 −0.054	245.76 −18.19	215.1 −68.5	CLOSE QUARTET	18 D 2.7	F366/J2716 (294,77)
AM 0637-271	06 37 45 −27 12.2	2.396 −0.055	236.48 −14.42	222.9 −77.9	REFLECTION NEBULA VV 134 = 490-G?-22	21 D 1.8	F426/J3890 (99,329)
AM 0637-404	06 37 56 −40 40.4	1.941 −0.055	249.56 −19.41	213.3 −64.8	DWARF	20 D 0.4	F308/J4792 (99,142)
AM 0637-373	06 37 55 −37 35.9	2.058 −0.055	246.52 −18.33	214.5 −67.8	CLOSE TRIPLE (MATERIAL) in WIDE GROUP of 10 GALAXIES	18,24 S 0.7	F308/J4792 (95,307)
AM 0638-255	06 38 09 −25 50.8	2.435 −0.055	235.24 −13.79	224.6 −79.2	SMALL GALAXY w. JETS MCG-4-16-15 = 490-IG-26	7 D 1.0	F490/J5469 (153,134)
*AM 0638-270	06 38 23 −27 02.5	2.401 −0.056	236.38 −14.23	222.4 −78.1	REFLECTION NEBULA + GROUP of 6 MCG-5-16-13 = 490-G-28	8,21	F490/J5469 (150,70)
AM 0638-675	06 38 52 −67 50.5	−0.160 −0.056	278.18 −26.16	208.4 −37.8	DWARF w. 3 COMP'NS.	8,20 D 0.6	F087/J1171 (162,24)
AM 0638-322	06 38 59 −32 26.0	2.236 −0.057	241.57 −16.23	216.6 −72.9	GALAXY w. APP. COMP'N. NGC 2267 = 426-G-29	8 S 0.9	F426/J3890 (88,48)
AM 0639-344	06 39 50 −34 41.9	2.162 −0.058	243.83 −16.92	214.8 −70.7	GROUP of 3 incl. 366-G-8	24 D 5.2	F366/J2716 (269,194)
AM 0639-630	06 39 51 −63 03.4	0.484 −0.058	272.97 −25.22	208.7 −42.5	I/A DBL.	2 S 0.4	F087/J1171 (153,281)
*AM 0639-582	06 39 56 −58 28.2	0.928 −0.058	268.04 −24.23	209.2 −47.1	OPEN SPIRAL w. APP. COMP'N. 121-IG-1/2	1,8,9 S 2.8	F121/J3822 (40,258)
AM 0640-630	06 40 17 −63 02.5	0.487 −0.058	272.96 −25.17	208.6 −42.6	DBL. E CONN. w. LINEAR FILAMENT	2,15 S 0.6	F087/J1171 (151,282)
AM 0640-505	06 40 37 −50 57.1	1.452 −0.059	260.16 −22.19	210.0 −54.6	DWARF (RESOLVED) 206-GA-20 = "CARINA DWARF"	20 D 30.	F206/J2820 (89,127)
AM 0640-272	06 40 42 −27 24.5	2.391 −0.059	236.94 −13.90	219.6 −77.9	E w. PECULIAR COMP'NS.	8 S 8.1	F427/J3900 (339,316)
AM 0640-750	06 40 56 −75 03.7	−1.855 −0.059	286.21 −26.90	207.6 −30.6	GAL. w. I/A COMP'N.(PEC. NUCLEUS) 034-IG-10	1 D 1.1	F034/J2020 (159,173)
AM 0640-601	06 40 57 −60 12.5	0.777 −0.059	269.93 −24.50	208.8 −45.4	GALAXY w. I/A COMP'N. 122-IG-3 (in WIDE PAIR)	1,23 S 0.3	F121/J3822 (40,164)

Identification	R.A. (1950.0) Declination	Precession R.A./Dec.	Gal. Long. Gal. Lat.	S.G.L. S.G.B.	Description Cross-Identification	Coding Size (Min.)	Field/Plate (x, y)
AM 0640-812	06 40 59 −81 27.3	−5.676 −0.059	293.37 −27.36	207.2 −24.2	SMALL NEBULA w. OFF-SET CONDENS.	16 D 0.4	F016/J3763 (98,94)
AM 0641-522	06 41 05 −52 29.7	1.360 −0.060	261.79 −22.56	209.7 −53.1	COMPACT	13 D 0.2	F161/J1122 (135,311)
AM 0641-325	06 41 07 −32 59.3	2.220 −0.060	242.29 −16.02	214.8 −72.5	LINE of 3 incl. 366-G-10	24 D 10.8	F366/J2716 (257,287)
AM 0641-411	06 41 52 −41 16.6	1.920 −0.061	250.43 −18.92	211.4 −64.3	GALAXY w. DUST LANE	14 D 0.3	F308/J4792 (60,109)
AM 0642-405 [0642-4059]	06 42 02 −40 56.7	1.933 −0.061	250.12 −18.77	211.4 −64.6	CLOSE PAIR (#1: M51-TYPE)	1,23 S 1.8	F308/J4792 (57,127)
AM 0642-405 [0642-4056]	06 42 14 −40 59.1	1.932 −0.061	250.17 −18.75	211.3 −64.6	CLOSE PAIR (#2: SB(r))	1,23 S 1.8	F308/J4792 (56,125)
AM 0642-275	06 42 17 −27 50.9	2.379 −0.061	237.50 −13.76	217.5 −77.6	GROUP of 5	18 D 3.6	F427/J3900 (319,293)
*AM 0642-645	06 42 20 −64 56.2	0.265 −0.061	275.07 −25.32	208.2 −40.7	TRIPLE NUCLEUS w. BROKEN RING	6,22 D 0.4	F087/J1171 (141,180)
*AM 0642-801	06 42 33 −80 11.9	−4.527 −0.062	291.97 −27.23	207.2 −25.4	BROKEN RING + TAIL + COMP'N. 16-G-16 (ONE-ARMED SPIRAL?)	6,8,15 D 1.1	F016/J3763 (83,160)
AM 0642-273	06 42 49 −27 35.2	2.387 −0.062	237.31 −13.55	217.2 −77.8	SPIRAL + 5 APP. COMP'NS. NGC 2280 = 427-G-2	8 D 5.4	F490/J5469 (98,40)
AM 0642-641 [0643-6411]	06 42 50 −64 12.6	0.356 −0.062	274.29 −25.12	208.2 −41.4	CLOSE PAIR (#1: EDGE-ON 87-G-36 + 2 COMP'NS.)	8,23 D 0.6	F087/J1171 (137,219)
AM 0643-272	06 43 07 −27 21.3	2.394 −0.062	237.12 −13.39	217.1 −78.1	DWARF	20 D 0.4	F490/J5469 (94,52)
*AM 0643-563	06 43 15 −56 38.1	1.080 −0.063	266.21 −23.35	208.7 −49.0	M51-TYPE (ASYMMETRIC ARMS) 161-IG-24	9 S 0.5	F161/J1122 (123,88)
AM 0643-641 [0642-6412]	06 43 35 −64 11.9	0.359 −0.063	274.30 −25.04	208.0 −41.4	CLOSE PAIR (#2: SPIRAL) 87-G-38	23 D 0.8	F087/J1171 (133,219)
AM 0643-462	06 43 38 −46 23.7	1.696 −0.063	255.67 −20.33	209.8 −59.2	RING GALAXY + COMP'N. 255-G-18 = Lu 0643-46	1,6 D 1.0	F255/J2826 (72,103)
AM 0643-365	06 43 45 −36 57.2	2.086 −0.063	246.33 −17.02	211.6 −68.6	E w. DIAMETRIC COMP'NS. 366-G-16	8 S 2.7	F366/J2716 (224,74)
AM 0643-311	06 43 52 −31 10.5	2.279 −0.064	240.79 −14.78	213.8 −74.3	GALAXY w. PLUMES MCG-5-16-21 = 427-G-6	15 D 0.7	F426/J3890 (31,115)
AM 0644-624	06 44 01 −62 46.3	0.524 −0.064	272.77 −24.70	208.1 −42.8	M51-TYPE (SMALL)	9 S 0.2	F087/J1171 (128,296)
AM 0644-280	06 44 14 −28 00.4	2.376 −0.064	237.83 −13.43	215.4 −77.5	DWARF	20 D 0.5	F426/J3890 (23,285)
AM 0644-665	06 44 16 −66 57.4	−0.009 −0.064	277.31 −25.50	207.8 −38.7	PAIR of I/A DBLS. (RICH FIELD)	2 S 2.7	F087/J1171 (133,71)
AM 0644-260	06 44 20 −26 03.3	2.432 −0.064	236.02 −12.61	217.0 −79.4	SPIRAL w. CONDENSATION in DISK MCG-4-16-19 = 490-G-14	12 S 0.4	F490/J5469 (78,122)
*AM 0644-741	06 44 25 −74 11.0	−1.555 −0.064	285.26 −26.58	207.4 −31.4	LARGE RING LINDSAY-SHAPLEY RING = 54-IG-11	6 D 2.0	F034/J2020 (146,220)
AM 0644-234	06 44 33 −23 41.5	2.498 −0.065	233.85 −11.57	219.8 −81.7	SMALL ONE-ARMED SPIRAL	10 D 0.5	F490/J5469 (74,249)
AM 0644-723	06 44 36 −72 32.2	−1.093 −0.065	283.44 −26.35	207.4 −33.1	SPIRAL w. PECULIAR ARM 34-G-12	10 D 2.0	F057/J1023 (14,27)
AM 0645-430	06 45 12 −43 04.1	1.848 −0.065	252.44 −18.96	209.7 −62.5	LINE of 3	24 D 2.2	F255/J2826 (50,280)
AM 0645-642	06 45 31 −64 27.6	0.332 −0.066	274.63 −24.89	207.8 −41.2	CLOSE TRIPLE	24 D 1.8	F087/J1171 (122,205)
*AM 0645-264	06 45 39 −26 41.2	2.415 −0.066	236.73 −12.60	214.9 −78.8	I/A DBL. SP. w. COMP'N. in DISK NGC 2292/3 = VV178 = 490-IG-48/9	1,2 S 0.8	F490/J5469 (63,88)
*AM 0646-645	06 46 34 −64 54.4	0.279 −0.067	275.13 −24.87	207.6 −40.7	I/A DBL. w. JET 87-IG-41	2,7 S 0.4	F087/J1171 (117,181)
*AM 0646-320	06 46 52 −32 02.2	2.254 −0.068	241.87 −14.54	211.1 −73.6	STAR w. ASYMM. JET (EDGE-ON?)	7 D 1.6	F366/J2716 (192,338)
AM 0646-235	06 46 59 −23 56.2	2.492 −0.068	234.31 −11.17	215.8 −81.6	SMALL LOOSE GROUP of STARS 490-SC-50	22 D 3.1	F491/J3951 (315,235)
AM 0647-313 [0647-3132]	06 47 04 −31 31.5	2.271 −0.068	241.40 −14.29	211.1 −74.1	PAIR + COMP'NS. (#1: SPIRAL)	8,23 S 1.3	F427/J3900 (258,97)
AM 0647-313 [0647-3131]	06 47 07 −31 32.8	2.270 −0.068	241.43 −14.29	211.1 −74.0	PAIR + COMP'NS. (#2: E)	8,23 S 1.3	F427/J3900 (258,95)
AM 0647-653	06 47 26 −65 38.7	0.185 −0.069	275.95 −24.93	207.4 −40.0	DIFFUSE GROUP of 6 87-G-42	18 D 9.4	F087/J1171 (114,141)
AM 0647-325	06 47 32 −32 58.0	2.225 −0.069	242.81 −14.78	210.4 −72.6	CLOSE PAIR PKS 0647-329	23 S 0.4	F366/J2716 (185,288)
AM 0647-691	06 47 48 −69 13.4	−0.372 −0.069	279.84 −25.58	207.3 −36.4	I/A DBL. (E + SPIRAL)	2 S 0.4	F058/J1021 (215,218)
AM 0648-641 [0648-6416]	06 48 20 −64 13.1	0.368 −0.070	274.44 −24.54	207.4 −41.4	CLOSE PAIR (#1: S0) NGC 2305 = 87-G-44 = SS-7a	23 D 1.1	F087/J1171 (105,217)

Identification	R.A. (1950.0) Declination	Precession R.A./Dec.	Gal. Long. Gal. Lat.	S.G.L. S.G.B.	Description Cross-Identification	Coding Size (Min.)	Field/Plate (x, y)
AM 0648-674	06 48 31 −67 45.5	−0.121 −0.070	278.26 −25.24	207.2 −37.9	I/A DBL. (COMPACT near-by)	2,13 S 0.1	F087/J1171 (113,27)
AM 0648-222	06 48 33 −22 22.4	2.535 −0.070	233.03 −10.18	214.9 −83.2	PLANETARY NEBULA ?	25 D 0.5	F491/J3951 (297,319)
AM 0648-641 [0648-6413]	06 48 33 −64 16.8	0.361 −0.070	274.51 −24.53	207.3 −41.4	CLOSE PAIR (#2: SB(r)) NGC 2307 = 87-G-45 = SS-7b	23 D 1.4	F087/J1171 (104,214)
AM 0648-674	06 48 50 −67 45.7	−0.121 −0.071	278.27 −25.22	207.2 −37.9	I/A DBL. (CLOSE,in a PAIR)	2 D 0.2	F058/J1021 (213,296)
AM 0649-351	06 49 06 −35 17.4	2.149 −0.071	245.16 −15.39	208.9 −70.3	I/A SPIRALS 366-IG-25	2 S 0.4	F366/J2716 (167,164)
AM 0649-674	06 49 07 −67 44.5	−0.117 −0.071	278.26 −25.19	207.2 −37.9	COMPACT (between 2 I/A DBL.)	13 D 0.1	F058/J1021 (211,297)
AM 0649-674	06 49 21 −67 44.9	−0.117 −0.071	278.27 −25.16	207.1 −37.9	COMPACT	13 D 0.1	F058/J1021 (210,297)
AM 0649-674	06 49 23 −67 41.9	−0.109 −0.071	278.21 −25.15	207.1 −37.9	CLOSE I/A DBL. (in a PAIR)	2 D 0.2	F058/J1021 (210,300)
AM 0649-554	06 49 24 −55 46.0	1.155 −0.071	265.55 −22.30	207.4 −49.9	E + COMP'N. in COMMON ENVELOPE 161-IG-27 [PKS] (in GROUP of 4)	1,18 S 1.1	F162/J3880 (327,132)
AM 0649-640	06 49 25 −64 04.0	0.389 −0.071	274.30 −24.39	207.2 −41.6	IRR. (NGC 520-TYPE) 87-IG-46	16 D 1.1	F087/J1171 (98,225)
AM 0649-520	06 49 31 −52 05.3	1.397 −0.072	261.77 −21.20	207.6 −53.6	DWARF 207-G-7 = UKS 0649-520	20 D 1.6	F161/J1122 (65,331)
AM 0649-624	06 49 33 −62 45.5	0.538 −0.072	272.91 −24.08	207.2 −42.9	IRR. (Sd or DWARF) 87-G-47	16 D 0.8	F087/J1171 (94,295)
AM 0649-570	06 49 35 −57 03.5	1.059 −0.072	266.90 −22.63	207.4 −48.6	GAL.w. I/A COMP'N. + APP.COMP'NS. 161-G-28	1,8 S 0.4	F122/J3850 (241,336)
AM 0651-674	06 51 08 −67 42.9	−0.106 −0.074	278.27 −24.99	206.9 −37.9	COMPACT + MATERIAL	13 D 0.2	F058/J1021 (201,299)
AM 0651-640	06 51 10 −64 04.4	0.393 −0.074	274.36 −24.21	206.9 −41.6	CHAIN of 4 87-G-48 included	17 D 8.5	F087/J1171 (88,224)
AM 0651-292	06 51 22 −29 29.5	2.336 −0.074	239.88 −12.62	208.0 −76.1	PAIR (S0 + SPIRAL) 427-G-16	23 S 2.0	F427/J3900 (210,206)
AM 0652-405	06 52 03 −40 50.2	1.948 −0.075	250.73 −16.97	207.0 −64.8	ISOLATED S0 NGC 2310 = 309-G-7	16 D 0.2	F309/J1253 (219,134)
*AM 0652-622	06 52 10 −62 26.6	0.579 −0.075	272.65 −23.71	206.8 −43.2	3-ARMED SPIRAL	11 D 0.6	F087/J1171 (76,311)
AM 0652-365	06 52 20 −36 54.5	2.096 −0.076	246.97 −15.42	206.9 −68.7	CHAIN of 4 incl. COMPACT	13,17 D 0.3	F309/J1253 (219,345)
*AM 0654-365	06 54 13 −36 52.3	2.099 −0.078	247.08 −15.06	205.9 −68.8	ONE-ARMED SPIRAL + I/A COMP'N. 366-G-28	1 S 2.7	F366/J2716 (112,78)
AM 0654-634	06 54 40 −63 41.1	0.448 −0.079	274.04 −23.74	206.4 −42.0	3-ARMED SPIRAL (3rd ARM FUZZY) 87-G-55	11 D 1.6	F087/J1171 (66,243)
AM 0654-281	06 54 48 −28 17.3	2.374 −0.079	239.09 −11.43	204.6 −77.3	I/A DBL. w. TAIL (M51-TYPE ?) [A82]	2,9,14,15 D 0.5	F427/J3900 (170,271)
*AM 0655-652	06 55 14 −65 25.7	0.236 −0.080	275.91 −24.10	206.4 −40.2	SPIRAL w. APP. COMP'NS. 87-G-54 (CHAIN of 7)	8,17 D 1.8	F088/J1115 (315,148)
*AM 0655-671	06 55 32 −67 16.6	−0.024 −0.080	277.89 −24.49	206.4 −38.4	I/A DBL. [A82]	2 S 0.5	F087/J1171 (76,51)
AM 0655-293 [0655-2932]	06 55 48 −29 34.0	2.337 −0.081	240.37 −11.77	204.0 −76.1	SMALL PAIR of SPIRALS (#1)	23 S 1.7	F427/J3900 (158,203)
AM 0655-293 [0655-2934]	06 55 50 −29 32.1	2.338 −0.081	240.34 −11.75	203.9 −76.1	SMALL PAIR of SPIRALS (#2)	23 S 1.7	F427/J3900 (158,204)
AM 0655-522	06 55 50 −52 27.0	1.386 −0.081	262.46 −20.40	205.9 −53.2	PAIR 207-G-10	23 S 4.3	F162/J3880 (288,313)
*AM 0655-652	06 55 55 −65 23.9	0.242 −0.081	275.90 −24.02	206.3 −40.2	I/A DBL.	2 S 0.4	F087/J1171 (66,151)
AM 0656-671	06 56 04 −67 14.9	−0.018 −0.081	277.88 −24.43	206.3 −38.4	GALAXY w. PECULIAR ABSORPTION 87-G-57 + 3 COMP'NS.	8,14 D 0.9	F088/J1115 (301,51)
AM 0656-454	06 56 07 −45 44.5	1.743 −0.081	255.80 −18.07	205.5 −59.9	GAL. w. APP.COMP'N. (PEC.NUCLEUS) 256-G-11	8 D 1.1	F256/J2727 (215,139)
AM 0656-704	06 56 22 −70 40.8	−0.623 −0.081	281.58 −25.13	206.4 −35.0	SP. w. FORKED ARM + I/A COMP'N. 58-G-21 = Fa 262	1,10 S 2.2	F058/J1021 (175,140)
AM 0656-623	06 56 25 −62 36.9	0.572 −0.081	272.97 −23.28	206.1 −43.0	CLOSE PAIR	23 S 0.5	F087/J1171 (51,300)
AM 0656-283	06 56 47 −28 30.4	2.370 −0.082	239.48 −11.13	202.7 −77.1	CLOSE PAIR	23 S 0.8	F427/J3900 (147,259)
AM 0656-370	06 56 54 −37 04.9	2.094 −0.082	247.50 −14.65	204.4 −68.5	CLOSE TRIPLE (COMMON ENVEL.)	3	F309/J1253 (170,336)
*AM 0656-745	06 56 56 −74 57.8	−1.748 −0.082	286.25 −25.87	206.4 −30.7	BARRED BULL'S-EYE (OUTER RING)	6 D 0.7	F034/J2020 (104,176)
AM 0657-393	06 57 16 −39 38.0	2.001 −0.083	249.96 −15.59	204.6 −66.0	CHAIN of 5 incl. 309-G-12	6,17 D 0.4	F309/J1253 (166,199)

Identification	R.A. (1950.0) Declination	Precession R.A./Dec.	Gal. Long. Gal. Lat.	S.G.L. S.G.B.	Description Cross-Identification	Coding Size (Min.)	Field/Plate (x, y)
AM 0657-703	06 57 29 −70 33.8	−0.595 −0.083	281.47 −25.01	206.2 −35.1	DWARF (VERY LO.S.BR.)	20 D 0.4	F058/J1021 (170,146)
AM 0658-590	06 58 19 −59 03.5	0.916 −0.084	269.33 −22.08	205.7 −46.6	I/A DBL. w. LONG TAIL 122-IG-11	2,15 D 3.1	F122/J3850 (177,230)
AM 0658-374	06 58 58 −37 41.7	2.075 −0.085	248.25 −14.52	203.4 −67.9	HI.S.BR. NUCLEUS + LO.S.BR. ARMS 309-G-15 (APP. COMP'NS.)	8,10,12 D 1.2	F309/J1253 (148,303)
AM 0659-755	06 59 11 −75 51.3	−2.053 −0.085	287.26 −25.88	206.3 −29.8	ONE-ARMED SB w. APP. COMP'N. 34-G-14	8,10 D 1.1	F035/J1951 (333,114)
AM 0659-680	06 59 45 −68 05.7	−0.137 −0.086	278.87 −24.28	205.9 −37.5	I/A DBL.	2 S 0.8	F058/J1021 (158,279)
AM 0700-673	07 00 02 −67 32.1	−0.048 −0.086	278.28 −24.13	205.8 −38.1	I/A DBL.	2 S 0.1	F088/J1115 (279,37)
AM 0700-283	07 00 41 −28 37.4	2.369 −0.087	239.96 −10.41	199.0 −76.9	E w. DIAMETRIC COMP'NS.	8 S 6.0	F427/J3900 (101,253)
AM 0701-674	07 01 08 −67 44.4	−0.076 −0.088	278.52 −24.07	205.7 −37.9	I/A DBL. (SMALL)	2 S 0.4	F058/J1021 (150,298)
AM 0701-695	07 01 55 −69 56.2	−0.452 −0.089	280.89 −24.51	205.7 −35.7	COMPACT	13 D 0.2	F058/J1021 (149,180)
AM 0702-423	07 02 57 −42 38.7	1.887 −0.090	253.26 −15.85	202.8 −62.9	GALAXY w. PECULIAR DUST 256-G-14	14 D 1.4	F309/J1253 (114,36)
AM 0702-772 [0703-7724]	07 02 41 −77 25.5	−2.693 −0.090	289.01 −25.94	206.1 −28.2	CLOSE PAIR (#1)	23 D 0.2	F017/J1954 (258,313)
AM 0702-601 [0702-6012]	07 02 43 −60 10.8	0.829 −0.090	270.66 −21.88	205.0 −45.4	PAIR (#1)	23 S 1.2	F122/J3850 (148,170)
AM 0702-665	07 02 43 −66 54.1	0.057 −0.090	277.67 −23.72	205.5 −38.7	GROUP of 5	18 D 2.2	F087/J1171 (36,67)
AM 0702-601 [0702-6010]	07 02 48 −60 12.1	0.827 −0.090	270.69 −21.88	204.9 −45.4	PAIR (#2)	23 S 1.2	F122/J3850 (147,168)
AM 0703-321	07 03 18 −32 19.3	2.260 −0.091	243.61 −11.49	199.0 −73.2	SMALL DBL.	2 S 0.3	F427/J3900 (73,54)
AM 0703-772 [0702-7725]	07 03 20 −77 24.5	−2.681 −0.091	289.00 −25.90	206.1 −28.2	CLOSE PAIR (#2) 34-G-15	23 D 0.7	F017/J1954 (257,314)
AM 0704-582	07 04 31 −58 26.8	0.983 −0.093	268.96 −21.13	204.5 −47.1	V.LO.S.BR. DWARF(SOME RESOLUTION)	20 D 1.6	F122/J3850 (134,262)
AM 0705-695	07 05 13 −69 59.6	−0.449 −0.094	281.02 −24.25	205.4 −35.6	I/A DBL. (MESSY)	2 S 0.6	F058/J1021 (134,176)
AM 0705-280	07 05 26 −28 08.7	2.387 −0.094	239.99 −9.26	194.0 −77.2	SPIRAL w. 5 APP. COMP'NS.	8 S 5.2	F427/J3900 (44,278)
AM 0705-293 [0706-2935]	07 05 59 −29 34.3	2.346 −0.095	241.34 −9.78	194.9 −75.7	CLOSE PAIR (E + SPIRAL) (#1)	23 D 0.3	F428/J2901 (308,200)
AM 0706-293 [0705-2934]	07 06 04 −29 35.6	2.346 −0.095	241.37 −9.77	194.9 −75.7	CLOSE PAIR (E + SPIRAL) (#2)	23 D 0.5	F428/J2901 (307,199)
AM 0707-663	07 07 05 −66 34.0	0.122 −0.096	277.44 −23.22	204.9 −39.0	I/A TRIPLE (in GROUP of 6) 88-IG-3	3,18 D 0.3	F087/J1171 (11,83)
AM 0707-273	07 07 45 −27 30.4	2.408 −0.097	239.64 −8.51	191.0 −77.6	I/A DBL. MCG-5-17-9/10 = 491-G-20/21	2 S 0.9	F492/J3819 (325,42)
AM 0708-381	07 08 33 −38 12.1	2.068 −0.098	249.53 −13.00	198.7 −67.2	CLOSE PAIR	23 S 0.1	F309/J1253 (47,274)
AM 0709-380	07 09 05 −38 08.3	2.071 −0.099	249.51 −12.88	198.4 −67.2	CLOSE PAIR of SPIRALS 310-G-1/2	23 S 2.2	F310/J2837 (316,276)
AM 0709-263	07 09 41 −26 37.5	2.434 −0.100	239.04 −7.73	187.7 −78.3	SPIRAL w. I/A COMP'N. 492-G-2	1 D 1.3	F491/J3951 (39,91)
AM 0711-602	07 11 08 −60 22.4	0.836 −0.102	271.21 −20.95	203.5 −45.1	GROUP of 5 incl. 122-G-15/16 = Fa 26	18 D 6.3	F122/J3850 (92,157)
AM 0712-513	07 12 06 −51 37.2	1.469 −0.103	262.56 −17.76	2 1.6 −53.8	I/A DBL. w. CONDENSATION in ARM 207-G-26/27	2,6,8 S 1.0	F208/J3844 (328,88)
AM 0712-643	07 12 47 −64 36.0	0.400 −0.104	275.58 −22.09	203.9 −40.9	I/A DBL. w. TAIL 88-IG-6	2 S 0.6	F088/J1115 (218,198)
AM 0712-694	07 12 54 −69 41.4	−0.356 −0.104	280.88 −23.54	204.6 −35.9	SMALL SPIRAL w. DBL. NUCLEUS	12 D 0.7	F058/J1021 (98,191)
AM 0713-275	07 13 09 −27 53.0	2.402 −0.105	240.52 −7.62	186.5 −76.9	I/A DBL. 428-G-8	2 S 0.4	F428/J2901 (225,293)
AM 0713-715	07 13 10 −71 51.4	−0.798 −0.105	283.17 −24.08	204.8 −33.7	I/A DBL.	2 S 0.6	F058/J1021 (105,74)
AM 0713-483	07 13 21 −48 39.1	1.632 −0.105	259.77 −16.45	200.5 −56.7	NEBULOSITY 207-N-28	21	F207/J2805 (61,252)
AM 0713-691	07 13 21 −69 10.2	−0.260 −0.105	280.35 −23.36	204.4 −36.4	DISTURBED GALAXY w. I/A COMP'N.	1,16 S 0.2	F058/J1021 (94,218)
AM 0713-465	07 13 22 −46 52.1	1.719 −0.105	258.07 −15.74	200.0 −58.5	PECULIAR PLANETARY NEBULA Lo-3 = 256-PN-19	25 D 1.1	F256/J2727 (56,77)
AM 0714-720	07 14 30 −72 00.9	−0.827 −0.107	283.37 −24.02	204.7 −33.5	CHAIN of 4	17 D 1.3	F058/J1021 (100,66)

Identification	R.A. (1950.0) Declination	Precession R.A./Dec.	Gal. Long. Gal. Lat.	S.G.L. S.G.B.	Description Cross-Identification	Coding Size (Min.)	Field/Plate (x, y)
AM 0714-571	07 14 58 −57 15.2	1.106 −0.107	268.26 −19.42	202.2 −48.2	DISTURBED SINGLE w. MATERIAL 162-G-17 = Lu 0714-57	16 D 1.3	F123/J3886 (334,319)
AM 0715-715	07 15 24 −71 51.7	−0.786 −0.108	283.23 −23.91	204.6 −33.7	COMPACT	13 D 0.1	F058/J1021 (95,73)
AM 0716-621	07 16 04 −62 14.7	0.673 −0.109	273.31 −21.00	203.0 −43.2	SPIRAL + APP. COMP'N. NGC 2369 = 122-G-18	8 S 4.9	F123/J3886 (302,53)
AM 0716-495	07 16 26 −49 59.8	1.569 −0.109	261.27 −16.51	200.0 −55.3	CLOSE PAIR 208-IG-2	23 S 1.3	F208/J3844 (296,177)
AM 0717-571	07 17 40 −57 19.1	1.109 −0.111	268.46 −19.11	2 1.7 −48.1	DWARF (RESOLVED?)	20 D 1.1	F122/J3850 (37,318)
AM 0717-441	07 17 41 −44 11.5	1.848 −0.111	255.86 −13.95	197.5 −61.0	I/A DBL. in GROUP of 3 257-G-6/7	2,24 D 1.2	F257/J1315 (279,221)
AM 0718-625 [0719-6258]	07 18 06 −62 51.0	0.618 −0.112	274.00 −20.98	202.8 −42.6	PAIR (#1: SPIRAL) NGC 2369a = 88-G-8	23 S 9.5	F122/J3850 (56,22)
AM 0718-711	07 18 09 −71 19.3	−0.651 −0.112	282.72 −23.56	204.3 −34.2	COMPACT w. TAIL + CONDENSATION	13,15 D 0.4	F058/J1021 (81,101)
AM 0718-585 [0718-5858]	07 18 15 −58 54.9	0.985 −0.112	270.07 −19.61	2 1.9 −46.5	PAIR (#1) 123-IG-2	23 D 0.5	F123/J3886 (303,232)
AM 0718-585 [0718-5854]	07 18 20 −58 58.8	0.980 −0.112	270.14 −19.62	2 1.9 −46.4	PAIR (#2)	23 D 0.3	F123/J3886 (302,229)
*AM 0718-274	07 18 21 −27 41.4	2.412 −0.112	240.88 −6.51	181.5 −76.6	GAL. w. ABS'N or JET (GALACTIC?) 428-*N?-21	7,14,22 D 1.1	F428/J2901 (163,303)
AM 0719-625 [0718-6251]	07 19 26 −62 58.1	0.610 −0.113	274.18 −20.87	202.6 −42.4	PAIR (#2: RING w. ELONG. NUCLEUS) NGC 2381 = 88-G-10 = Fa 266	6,23 S 9.5	F122/J3850 (48,15)
AM 0719-685	07 19 49 −68 56.3	−0.188 −0.114	280.29 −22.74	203.7 −36.5	IRR.	16 D 0.4	F058/J1021 (62,228)
AM 0719-615	07 19 54 −61 58.3	0.714 −0.114	273.19 −20.49	202.3 −43.4	GALAXY w. APP. COMP'N. 123-G-5	8 S 6.7	F088/J1115 (177,340)
AM 0719-685	07 19 55 −68 50.5	−0.171 −0.114	280.19 −22.70	203.7 −36.6	DWARF	20 D 0.5	F058/J1021 (61,233)
AM 0720-723	07 20 36 −72 37.2	−0.934 −0.115	284.14 −23.74	204.2 −32.9	DWARF (VERY LO.S.BR.) UKS 0720-726	20 D 0.7	F059/J1164 (286,30)
AM 0720-292	07 20 56 −29 24.3	2.367 −0.116	242.68 −6.80	182.7 −74.8	STAR CLUSTER 428-SC-24	22	F428/J2901 (133,211)
AM 0721-685	07 21 31 −68 54.6	−0.174 −0.116	280.31 −22.58	203.5 −36.5	BR.GAL. (SPIRALS off both MI.AX.) NGC 2397a = 58-G-30	8 D 1.8	F058/J1021 (53,229)
AM 0722-684	07 22 03 −68 44.3	−0.143 −0.117	280.15 −22.48	203.4 −36.7	PECULIAR GALAXY w. JET NGC 2397a = 58-G-31	7,16 D 0.6	F059/J1164 (302,238)
AM 0722-615	07 22 14 −61 55.7	0.728 −0.117	273.25 −20.21	2 1.9 −43.4	PAIR of SPIRALS 123-G-9	23 S 13.4	F123/J3886 (265,73)
AM 0722-591 [0722-5913]	07 22 16 −59 12.9	0.973 −0.117	270.56 −19.24	2 1.3 −46.1	PAIR (#1)	23 D 0.3	F123/J3886 (274,218)
AM 0722-364	07 22 26 −36 42.7	2.141 −0.118	249.38 −9.88	190.7 −67.9	CHAIN of 5 (E at CENTRE) incl. 367-G-24	17 D 17.9	F368/J2845 (325,84)
AM 0722-591 [0722-5912]	07 22 26 −59 13.7	0.973 −0.118	270.58 −19.22	2 1.3 −46.1	PAIR (#2)	23 D 0.3	F123/J3886 (273,217)
AM 0723-381	07 23 22 −38 18.9	2.086 −0.119	250.92 −10.43	191.6 −66.3	GALAXY w. I/A COMP'N. 310-G-14	1 S 0.9	F310/J2837 (165,270)
AM 0725-511	07 25 56 −51 16.0	1.523 −0.122	263.11 −15.64	197.9 −53.8	GALAXY + 2 APP. COMP'NS. 208-G-15	8 S 1.2	F208/J3844 (213,111)
AM 0726-453	07 26 05 −45 34.9	1.805 −0.123	257.80 −13.20	195.3 −59.3	POSSIBLE LARGE DWARF 257-G?-17	20 D 2.2	F257/J1315 (197,148)
AM 0726-374	07 26 38 −37 42.9	2.113 −0.123	250.68 −9.59	189.6 −66.7	PLANETARY NEBULA	25	F367/J2743 (23,31)
*AM 0727-621	07 27 34 −62 15.1	0.717 −0.125	273.81 −19.75	2 1.2 −43.0	SPIRAL + DBL. I/A COMP'N. SS-8 IC 2200/a = Se 55/1 = 123-G-11/12	2 S 1.0	F123/J3886 (230,57)
AM 0727-672	07 27 52 −67 28.5	0.085 −0.125	279.05 −21.56	202.5 −37.9	GALAXY w. APP. COMP'N. IC 2202 = 88-G-16	8 S 5.8	F088/J1115 (136,43)
AM 0728-585	07 28 35 −58 58.4	1.016 −0.126	270.65 −18.39	200.1 −46.2	CLOSE PAIR in GROUP of 4	23,24 S 0.7	F123/J3886 (231,233)
*AM 0728-664	07 28 50 −66 47.6	0.188 −0.126	278.39 −21.24	202.2 −38.5	SPIRAL w. I/A + APP. COMP'NS. 88-G-17 [A82]	1,8 S 1.3	F088/J1115 (130,80)
AM 0729-554	07 29 14 −55 49.5	1.253 −0.127	267.65 −17.07	198.9 −49.2	CHAIN of 4	17 D 6.0	F163/J2850 (275,133)
AM 0729-620	07 29 36 −62 09.1	0.735 −0.127	273.81 −19.49	200.8 −43.0	SPIRAL + 2 APP. COMP'NS. NGC 2417 = 123-G-15	1,2,8 S 5.8	F123/J3886 (218,63)
*AM 0729-660	07 29 46 −66 07.0	0.284 −0.128	277.75 −20.92	2 1.9 −39.2	M51-TYPE in GROUP of 4 88-G-20	9,18 S 1.4	F088/J1115 (124,116)
AM 0729-745	07 29 55 −74 56.9	−1.517 −0.128	286.78 −23.78	203.9 −30.5	DWARF + COMP'N. 035-G-09	8,20 D 1.3	F035/J1951 (237,177)
*AM 0731-680	07 31 19 −68 05.0	0.013 −0.130	279.78 −21.46	202.2 −37.2	DWARF (FILAMENTARY STRUCTURE) 59-G-1	20,22 D 1.1	F059/J1164 (260,277)

Identification	R.A. (1950.0) Declination	Precession R.A./Dec.	Gal. Long. Gal. Lat.	S.G.L. S.G.B.	Description Cross-Identification	Coding Size (Min.)	Field/Plate (x, y)
AM 0733-273	07 33 22	2.432	242.39	170.2	LOOSE CLUSTER of STARS	22	F493/J3891
	−27 35.8	−0.132	−3.54	−74.8	429-SC-3	D 5.4	(281,39)
AM 0733-235	07 33 28	2.531	239.14	159.0	PAIR of REFLECTION NEBULOSITIES	21	F492/J3819
	−23 51.7	−0.132	−1.69	−77.4		S 1.6	(14,239)
AM 0734-250	07 34 31	2.498	240.38	162.6	NEBULOSITY w. STAR	21	F493/J3891
	−25 09.3	−0.134	−2.12	−76.4		D 11.6	(270,170)
AM 0734-720	07 34 38	−0.718	283.97	202.9	ONE-ARMED SPIRAL	10	F058/J1021
	−72 06.6	−0.134	−22.57	−33.2	59-G-4	D 0.4	(18,51)
AM 0735-694	07 35 05	−0.227	281.51	202.2	DWARF (near NGC 2434)	20	F059/J1164
	−69 40.3	−0.135	−21.70	−35.6	59-G-6	D 0.9	(237,193)
AM 0736-521	07 36 22	1.502	264.67	195.7	CLOSE PAIR	23	F208/J3844
	−52 10.5	−0.136	−14.57	−52.5	208-G-30	S 1.3	(127,62)
AM 0736-521	07 36 22	1.502	264.68	195.7	EDGE-ON w. I/A COMP'N.	1	F163/J2850
	−52 11.2	−0.136	−14.58	−52.4	208-G-31	S 1.2	(225,330)
*AM 0736-692	07 36 31	−0.173	281.29	202.0	SPIRAL w. PEC. ARMS (7 COMP'NS.)	10,8	F059/J1164
	−69 24.7	−0.137	−21.49	−35.8	NGC 2442 = 2443?? = 59-G-8	D 4.5	(230,207)
AM 0736-703	07 36 34	−0.387	282.48	202.3	LO.S.BR. ONE-ARMED SPIRAL	22,10	F059/J1164
	−70 35.4	−0.137	−21.90	−34.6	59-G-9 (DBL. NUCLEUS)	D 1.2	(227,144)
AM 0736-550	07 36 48	1.328	267.40	197.0	CLOSE PAIR (#1)	23	F163/J2850
[0737-5504]	−55 04.2	−0.137	−15.78	−49.7	163-G-10	S 1.2	(219,175)
*AM 0736-334	07 36 54	2.260	248.12	179.9	LOW DENSITY GLOBULAR (DWARF?)	20,22	F368/J2845
	−33 43.4	−0.137	−5.86	−69.4	368-SC-7	D 4.5	(169,248)
AM 0737-550	07 37 00	1.328	267.42	196.9	CLOSE PAIR (#2)	23	F163/J2850
[0736-5504]	−55 04.7	−0.137	−15.75	−49.7	163-G-11	S 1.6	(218,175)
AM 0737-764	07 37 06	−2.122	288.86	203.9	SB w. I/A COMP'N.	1	F035/J1951
	−76 48.4	−0.137	−23.93	−28.6	035-G-11	S 0.7	(207,79)
AM 0737-691	07 37 18	−0.136	281.14	2 1.9	V.LO.S.BR. DWARF	20	F059/J1164
	−69 13.6	−0.138	−21.36	−36.0		D 0.7	(227,218)
*AM 0738-332	07 38 22	2.271	248.01	178.7	4 CONDENSATIONS in DIFFUSE NEBULA	18,22	F368/J2845
	−33 25.5	−0.139	−5.45	−69.4		D 0.5	(153,264)
AM 0738-692	07 38 22	−0.153	281.31	2 1.8	PECULIAR GALAXY w. I/A COMP'N	12,1	F059/J1164
	−69 21.9	−0.139	−21.32	−35.8	59-G-11	S 0.7	(222,210)
AM 0738-683	07 38 35	−0.033	280.60	2 1.6	DOUBLE SPIRAL w. WIDE ARMS	22	F059/J1164
	−68 38.9	−0.139	−21.04	−36.5	59-G-12	D 1.8	(222,249)
AM 0738-571	07 38 53	1.184	269.62	197.5	LINE of 3 COMPACTS	13,24	F163/J2850
	−57 17.4	−0.140	−16.46	−47.5	163-IG-12	D 1.0	(202,56)
AM 0739-675	07 39 27	0.083	279.92	2 1.2	I/A DBL.(SMALL)(SPIRAL + GALAXY)	2	F059/J1164
	−67 55.7	−0.140	−20.70	−37.2	59-IG-14	S 0.5	(219,288)
AM 0739-835	07 39 37	−8.378	296.45	205.4	CLOSE TRIPLE (incl. PEC. SPIRAL)	12,24	F005/J2958
	−83 57.6	−0.141	−25.91	−21.6	5-G-10	D 4.0	(140,233)
*AM 0741-761	07 41 28	−1.889	288.46	203.4	DWARF + COMP'N.	8,20	F035/J1951
	−76 19.5	−0.143	−23.54	−29.0	035-G-12	D 1.3	(195,105)
AM 0742-651	07 42 20	0.465	277.37	200.0	SPIRAL w. APP. COMP'NS.	8	F088/J1115
	−65 12.0	−0.144	−19.37	−39.7	89-G-1	S 4.5	(51,161)
AM 0742-252	07 42 59	2.502	241.53	158.4	NEBULOSITY w. ABS'N.	14,21	F493/J3891
	−25 22.7	−0.145	−0.57	−74.7		D 4.5	(167,159)
AM 0744-675	07 44 08	0.116	280.08	200.7	SPIRAL w. 3 APP. COMP'NS.	8	F088/J1115
	−67 54.3	−0.147	−20.28	−37.1		S 1.3	(55,15)
AM 0746-384	07 46 11	2.113	253.47	182.4	I/A DBL. in GROUP of 3	2,24	F311/J1291
	−38 46.2	−0.149	−6.73	−64.1	311-G-13	S 0.6	(199,245)
AM 0747-272	07 47 39	2.453	243.87	161.9	PLANETARY NEBULA	25	F429/J0400
	−27 29.0	−0.151	−0.74	−72.6	493-PN-19	D 0.7	(89,314)
AM 0748-541	07 48 07	1.415	267.47	194.2	DISTURBED GALAXY w. I/A COMP'N.	1	F163/J2850
	−54 19.6	−0.152	−13.98	−49.8	163-IG-19 = UKS 0748-543	S 1.7	(131,215)
*AM 0748-665	07 48 25	0.274	279.36	199.9	GALAXY w. IRREGULAR TAIL	15	F088/J1115
	−66 58.9	−0.152	−19.53	−37.9	89-IG-4	D 1.2	(28,62)
AM 0748-524	07 48 36	1.506	266.13	193.2	GROUP of 5	18	F163/J2850
	−52 48.5	−0.152	−13.20	−51.2		D 6.7	(126,296)
AM 0748-325	07 48 40	2.302	248.71	173.1	REFLECTION NEBULA	21	F430/J0490
	−32 58.6	−0.153	−3.36	−68.5		D 0.8	(338,15)
AM 0750-490	07 50 21	1.706	262.94	190.4	CLUSTER of 7 GALAXIES	19	F209/J1320
	−49 06.5	−0.155	−11.18	−54.5		D 4.5	(264,228)
AM 0752-282	07 52 41	2.434	245.30	162.1	CLOSE PAIR	23	F430/J0490
	−28 28.8	−0.158	−0.30	−71.1		S 0.2	(298,257)
AM 0753-233	07 53 02	2.561	241.09	148.1	CLOSE PAIR (ONE COMPACT)	13,23	F493/J3891
	−23 30.3	−0.158	2.36	−73.8	PK 241+2-1 = 494-PN-2	S 3.2	(44,259)
AM 0754-521	07 54 34	1.563	266.00	191.5	LARGE E w. MATERIAL	15	F163/J2850
	−52 10.2	−0.160	−12.10	−51.5	NGC 2502 = 209-G-8 = MC 0754-521	D 2.4	(75,329)
AM 0755-785	07 55 54	−2.925	291.46	203.4	I/A DBL. (RING ?) + I/A DBL.	2,6	F017/J1954
	−78 58.4	−0.162	−23.74	−26.2	17-IG-8	S 0.2	(113,231)
AM 0755-585	07 55 54	1.129	272.25	195.3	NEBULOSITY	21	F124/J2902
	−58 59.5	−0.162	−15.22	−45.1	IC 2220 = 124-RN?-3	D 4.0	(305,229)

Identification	R.A. (1950.0) Declination	Precession R.A./Dec.	Gal. Long. Gal. Lat.	S.G.L. S.G.B.	Description Cross-Identification	Coding Size (Min.)	Field/Plate (x, y)
AM 0756-525	07 56 19 −52 59.1	1.524 −0.162	266.86 −12.27	191.7 −50.6	PAIR of SPIRALS	23 S 5.0	F164/J2906 (331,282)
AM 0756-494	07 56 51 −49 43.2	1.697 −0.163	264.02 −10.57	189.2 −53.5	EDGE-ON w. 2 COMP'NS. 209-G-9	8 D 6.7	F209/J1320 (207,197)
AM 0759-451	07 59 11 −45 18.8	1.901 −0.166	260.39 −7.98	184.9 −57.2	REFLECTION NEBULA 258-N-3	21 D 1.7	F258/J2802 (149,162)
AM 0759-670	07 59 44 −67 01.6	0.342 −0.167	279.91 −18.56	198.6 −37.5	RING GALAXY (PLANETARY NEBULA?) 89-G-8	6 D 0.4	F059/J1164 (115,335)
AM 0802-721	08 02 09 −72 15.4	−0.523 −0.170	284.98 −20.68	200.5 −32.5	I/A DBL.	2 S 0.1	F060/J1172 (331,44)
AM 0802-721	08 02 48 −72 16.4	−0.521 −0.171	285.02 −20.65	200.5 −32.4	PECULIAR SPIRAL w. 5 COMP'NS. 59-G-24	12,8 D 2.0	F060/J1172 (328,44)
AM 0803-701	08 03 38 −70 13.4	−0.115 −0.172	283.09 −19.68	199.6 −34.3	SPIRAL + COMP'N. (near V.BR.STAR) 59-G-25 = PKS 0803-702?	8 S 2.9	F060/J1172 (341,154)
AM 0804-272	08 04 09 −27 22.7	2.480 −0.172	245.71 2.43	155.0 −69.7	OBSCURED SP. + COMP'N. w. PEC.DUST NGC 2566 = 495-G-3	8,14 S 4.5	F494/J3855 (178,52)
AM 0804-765	08 04 54 −76 55.9	−1.848 −0.173	289.62 −22.51	202.2 −28.0	DWARF w. CONDENSATION 35-G-20	20 D 0.7	F018/J2959 (345,325)
AM 0806-354	08 06 20 −35 49.2	2.252 −0.175	253.05 −1.77	171.0 −64.0	NEBULOSITY or Sb NUCLEUS	14,22 D 1.1	F369/J2855 (110,135)
AM 0807-613	08 07 01 −61 37.3	0.970 −0.176	275.32 −15.30	194.9 −42.2	GALAXY w. I/A COMP'N. 124-IG-12	1 S 0.3	F124/J2902 (224,92)
AM 0807-611	08 07 02 −61 14.9	1.003 −0.176	274.99 −15.11	194.7 −42.5	DWARF	20 D 0.4	F124/J2902 (224,112)
AM 0807-285	08 07 37 −28 58.9	2.445 −0.177	247.47 2.20	157.6 −68.2	STAR + NEBULOSITY 430-*N?-22	21 D 0.8	F430/J0490 (122,231)
AM 0808-490	08 8 47 −49 03.0	1.770 −0.178	264.46 −8.56	185.9 −53.2	NEBULOSITY (MANY STARS) NGC 2547 = 209-SC-18	21 D 11.2	F209/J1320 (102,232)
AM 0810-742	08 10 44 −74 22.0	−0.945 −0.180	287.30 −21.07	200.7 −30.3	DWARF 035-G-21	20 D 1.3	F035/J1951 (92,206)
AM 0811-342	08 11 09 −34 25.8	2.303 −0.181	252.43 −0.18	167.2 −64.3	NEBULOSITY 369-*N-8	21 D 2.2	F370/J2873 (322,207)
AM 0813-284	08 13 40 −28 42.2	2.462 −0.184	247.97 3.46	155.1 −67.2	DWARF	20 D 0.5	F431/J0486 (322,245)
*AM 0814-760	08 14 45 −76 08.3	−1.432 −0.185	289.12 −21.64	2 1.2 −28.5	SMALL RING w. COMP'N. off MI.AX. 36-IG-1	6,8 S 0.3	F036/J2048 (283,109)
AM 0814-674	08 14 50 −67 47.2	0.352 −0.185	281.33 −17.64	197.3 −36.2	DWARF	20 D 0.7	F060/J1172 (304,290)
AM 0814-540	08 14 51 −54 01.7	1.542 −0.185	269.23 −10.45	188.6 −48.4	SMALL, BRIGHT I/A DBL. 164-IG-6	2 S 0.3	F164/J2906 (181,231)
AM 0814-231	08 14 52 −23 16.3	2.595 −0.185	243.59 6.71	141.5 −69.2	SMALL I/A DBL. 494-G-39?	2 S 1.0	F495/J2986 (317,271)
AM 0815-271	08 15 01 −27 18.0	2.500 −0.186	246.97 4.49	151.6 −67.6	SPIRAL w. PEC. ARM + APP. COMP'N. NGC 2559 = 494-G-41	8,10 D 3.4	F430/J0490 (33,320)
AM 0815-675	08 15 08 −67 52.8	0.341 −0.186	281.43 −17.66	197.3 −36.1	DISTORTED GALAXY	16,22 D 0.6	F060/J1172 (302,285)
AM 0815-420	08 15 41 −42 04.9	2.072 −0.186	259.28 −3.72	176.9 −58.1	2 BRIGHT STARS w. NEBULOSITY	21 D 0.4	F312/J1255 (160,65)
*AM 0815-295	08 15 42 −29 58.5	2.434 −0.186	249.28 3.11	157.3 −66.2	LARGE SB w. APP. COMP'N. DB-431-4 = 431-G-2	8 S 2.3	F431/J0486 (297,177)
AM 0816-415	08 16 21 −41 57.3	2.078 −0.187	259.24 −3.55	176.5 −58.1	REFLECTION NEBULA	21 D 1.8	F312/J1255 (153,72)
AM 0816-252	08 16 37 −25 20.8	2.549 −0.188	245.55 5.88	146.4 −68.1	LARGE SB + 4 APP. COMP'NS NGC 2566 = 495-G-3	8 S 6.7	F495/J2986 (293,160)
AM 0816-480	08 16 41 −48 07.4	1.840 −0.188	264.40 −6.96	183.2 −53.2	LINEAR NEBULOSITY	21	F209/J1320 (30,279)
AM 0816-714	08 16 45 −71 41.8	−0.268 −0.188	285.01 −19.42	199.0 −32.6	ONE-ARMED SPIRAL w. LARGE NUCLEUS 60-IG-3	10 D 0.2	F059/J1164 (56,79)
AM 0817-322	08 17 30 −32 28.0	2.371 −0.189	251.55 2.02	161.7 −64.5	E? w. ASYMM. NEBULA (GALACTIC?)	22 D 0.5	F431/J0486 (273,44)
AM 0818-672	08 18 03 −67 22.5	0.432 −0.189	281.12 −17.17	196.7 −36.4	GROUP of 3 (2 E's + SPIRAL)	24 D 5.8	F090/J2170 (334,42)
AM 0818-501	08 18 10 −50 10.7	1.753 −0.189	266.26 −7.91	184.8 −51.4	NEBULOSITY	21 D 2.2	F209/J1320 (23,168)
AM 0819-500	08 19 33 −50 04.4	1.764 −0.191	266.30 −7.67	184.4 −51.4	NEBULOSITY	21	F210/J2806 (269,167)
AM 0819-650	08 19 49 −65 08.7	0.709 −0.191	279.23 −15.86	195.3 −38.4	CLOSE PAIR 89-G-18/19	23 S 1.9	F090/J2170 (338,162)
AM 0820-525	08 20 03 −52 51.7	1.628 −0.192	268.67 −9.17	186.7 −49.0	CLOSE PAIR (SMALL) 164-G-8	23 S 0.9	F209/J1320 (16,23)
AM 0820-401	08 20 31 −40 16.4	2.147 −0.192	258.30 −1.95	173.3 −58.8	REFLECTION NEBULA	21 D 1.1	F312/J1255 (110,162)

Identification	R.A. (1950.0) Declination	Precession R.A./Dec.	Gal. Long. Gal. Lat.	S.G.L. S.G.B.	Description Cross-Identification	Coding Size (Min.)	Field/Plate (x, y)
AM 0820-401	08 20 46 −40 14.8	2.149 −0.193	258.31 −1.89	173.2 −58.8	REFLECTION NEBULA	21 D 0.9	F312/J1255 (107,163)
AM 0820-751	08 20 48 −75 16.2	−1.080 −0.193	288.49 −20.91	200.5 −29.2	CONTACT I/A DBL. 36-IG-3	2 S 0.1	F036/J2048 (269,158)
*AM 0821-783	08 21 12 −78 32.1	−2.303 −0.193	291.59 −22.44	21.9 −26.2	SPIRAL w. 3 APP. COMP'NS. 18-G-2	8 D 1.6	F018/J2959 (282,250)
AM 0824-802	08 24 15 −80 24.6	−3.318 −0.197	293.45 −23.19	202.6 −24.4	GALAXY w. PECULIAR DUST 18-G-6	14 D 1.3	F018/J2959 (259,152)
AM 0824-691	08 24 46 −69 10.2	0.239 −0.197	283.07 −17.57	197.0 −34.6	VERY SMALL GROUP of FIVE	18 D 0.5	F060/J1172 (249,220)
*AM 0824-774 [0824-7739]	08 24 46 −77 41.3	−1.867 −0.197	290.88 −21.88	21.4 −26.9	CHAIN of 3 (#1: S(r)) 18-G-7	24 D 5.2	F018/J2959 (280,296)
AM 0824-773 [0825-7737]	08 24 53 −77 39.4	−1.852 −0.197	290.85 −21.85	21.3 −26.9	CHAIN of 3 (#2: DISTURBED SP.) 18-G-8	24 D 5.2	F018/J2959 (280,298)
AM 0825-773 [0824-7741]	08 25 42 −77 37.1	−1.824 −0.198	290.84 −21.80	21.3 −26.9	CHAIN of 3 (#3: SPIRAL) 18-G-9	24 D 5.2	F018/J2959 (278,300)
AM 0826-273	08 26 42 −27 33.2	2.514 −0.199	248.67 6.47	149.3 −65.2	GALAXY w. I/A COMP'N. DB431-9 = 431-G-11	1 S 0.3	F431/J0486 (168,308)
AM 0826-684	08 26 58 −68 40.4	0.331 −0.200	282.74 −17.13	196.5 −34.9	M51-TYPE (SMALL)	9 D 0.6	F060/J1172 (240,247)
AM 0827-845	08 27 17 −84 58.8	−9.104 −0.200	297.95 −25.21	204.6 −20.2	PAIR (#1: KNOTTY IRR.) 6-G-1	6,9,16,23 D 0.7	F006/J2945 (259,168)
AM 0828-844	08 28 08 −84 48.8	−8.675 −0.201	297.80 −25.12	204.5 −20.4	PAIR (#2: SPIRAL) 6-G-2	6,9,16,23 D 1.1	F006/J2945 (260,177)
AM 0830-235	08 30 30 −23 52.2	2.605 −0.204	246.15 9.31	140.3 −65.6	SPIRAL + CLOSE I/A COMP'N. MCG-4-21-1 = 495-G-16	1 S 0.4	F495/J2986 (124,240)
AM 0830-593	08 30 36 −59 36.9	1.269 −0.204	275.18 −11.78	190.2 −42.5	SPIRAL w. APP. COMP'N. 124-G-18 = Fa 275	8 S 1.9	F124/J2902 (67,197)
AM 0831-224	08 31 11 −22 48.1	2.629 −0.205	245.36 10.06	137.7 −65.7	EDGE-ON SB w. 5 APP. COMP'NS. NGC 2613 = 495-G-18	8 S 6.5	F495/J2986 (115,297)
AM 0832-693	08 32 55 −69 32.4	0.260 −0.207	283.80 −17.16	196.4 −33.9	I/A DBL. in SMALL GROUP of 4	2,18 S 0.5	F060/J1172 (210,202)
AM 0833-315	08 33 36 −31 58.4	2.420 −0.208	253.15 5.08	156.7 −61.9	LARGE IRR. DB 431-15 = 431-G-18	16 D 2.1	F370/J2873 (70,341)
AM 0834-261	08 34 06 −26 14.1	2.558 −0.208	248.57 8.58	145.1 −64.1	PECULIAR DBL. NUCLEUS (DUST?) PK 248+8-1 = 495-G-21	14,22 D 1.1	F495/J2986 (81,113)
AM 0834-690	08 34 39 −69 00.9	0.353 −0.209	283.43 −16.74	196.0 −34.3	COMPACT w. MATERIAL	13 D 0.2	F060/J1172 (202,230)
AM 0836-271	08 36 53 −27 13.8	2.540 −0.211	249.75 8.49	146.7 −63.2	I/A SPIRAL (M51-TYPE) 431-G-18	2,9 S 0.6	F431/J0486 (47,324)
AM 0836-680	08 36 56 −68 05.2	0.501 −0.211	282.75 −16.05	195.2 −35.0	I/A TRIPLET (SMALL)	3 D 0.5	F060/J1172 (192,280)
AM 0838-674	08 38 19 −67 45.1	0.556 −0.213	282.55 −15.75	194.8 −35.2	SPIRAL + APP. COMP'N. 60-G-10	8 S 1.6	F060/J1172 (185,298)
AM 0838-321	08 38 39 −32 11.9	2.425 −0.213	253.97 5.79	156.0 −60.8	PLANETARY NEBULA PK 254+5-1 = 432-PN-1	25 D 0.3	F370/J2873 (13,328)
AM 0839-745	08 39 20 −74 58.6	−0.750 −0.214	288.87 −19.74	199.1 −28.9	VERY OBSCURED SPIRAL + COMP'NS. 36-G-6 = UKS 0839-749	8 S 3.4	F036/J2048 (207,178)
AM 0839-252	08 39 49 −25 20.7	2.588 −0.215	248.63 10.15	142.3 −63.2	I/A DBL.	2 S 0.6	F496/J3861 (278,160)
AM 0839-694	08 39 54 −69 46.3	0.295 −0.215	284.36 −16.79	196.0 −33.4	DWARF	20 D 0.4	F060/J1172 (177,190)
AM 0840-693	08 40 16 −69 37.6	0.320 −0.215	284.25 −16.68	195.8 −33.5	DWARF	20 D 0.5	F060/J1172 (175,198)
AM 0842-675	08 42 37 −67 54.0	0.577 −0.218	282.91 −15.50	194.5 −34.8	SPIRAL w. I/A COMP'N. 60-G-12	1 S 0.5	F060/J1172 (163,290)
AM 0843-721	08 43 36 −72 16.4	−0.086 −0.219	286.69 −17.97	197.2 −31.1	LUMPY COMPACT in CHAIN of 7	13,17 S 15.7	F060/J1172 (162,55)
AM 0844-594	08 44 02 −59 46.1	1.343 −0.219	276.33 −10.51	188.5 −41.4	NEBULOSITY	21	F125/J4755 (233,191)
AM 0844-333	08 44 22 −33 30.1	2.407 −0.220	255.75 5.93	157.2 −59.1	WIDE PAIR (E + SPIRAL) 371-G-15/16	23 D 24.6	F371/J3997 (220,259)
AM 0845-275 [0845-2759]	08 45 15 −27 56.7	2.541 −0.221	251.47 9.52	146.8 −61.2	PAIR of SPIRALS (#1) 432-G-10	8,23 S 1.7	F496/J3861 (212,21)
AM 0845-252 [0846-2525]	08 45 23 −25 23.9	2.597 −0.221	249.45 11.11	141.7 −62.0	SPIRALS in LINE of 3 (#1) MCG-4-21-13 = 496-G-19	23,24 S 1.7	F496/J3861 (211,158)
AM 0845-275 [0845-2756]	08 45 24 −27 59.2	2.540 −0.221	251.52 9.53	146.9 −61.2	PAIR of SPIRALS (#2) 432-G-11	8,23 S 1.7	F496/J3861 (210,19)
AM 0846-252 [0846-2524]	08 46 00 −25 25.3	2.597 −0.221	249.55 11.21	141.6 −61.8	SPIRALS in LINE of 3 (#2) MCG-4-21-12 = 496-G-18 (in PAIR)	23,24 S 1.7	F496/J3861 (203,157)
AM 0846-252 [0845-2523]	08 46 08 −25 24.7	2.598 −0.221	249.56 11.24	141.6 −61.8	SPIRALS in LINE of 3 (#3) (in PAIR)	23,24 S 1.7	F496/J3861 (202,157)

Identification	R.A. (1950.0) Declination	Precession R.A./Dec.	Gal. Long. Gal. Lat.	S.G.L. S.G.B.	Description Cross-Identification	Coding Size (Min.)	Field/Plate (x, y)
AM 0847-685	08 47 02 −68 55.7	0.487 −0.222	284.03 −15.78	194.8 −33.8	I/A TRIPLE (VERY SMALL)	3 D 0.3	F060/J1172 (142,234)
AM 0847-700	08 47 17 −70 01.2	0.335 −0.223	284.95 −16.41	195.5 −32.8	SPIRAL w. 2 APP. + I/A COMP'N. 60-IG-14	1,8 S 0.5	F060/J1172 (143,176)
AM 0848-484	08 48 24 −48 44.5	1.943 −0.224	268.05 −3.12	177.4 −49.3	NEBULOSITY	21	F211/J2243 (282,238)
AM 0849-272	08 49 24 −27 21.4	2.562 −0.225	251.58 10.61	145.0 −60.5	COMP'N. on END of SPIRAL ARM	8,9 S 0.3	F496/J3861 (163,53)
AM 0849-841	08 49 55 −84 19.2	−6.832 −0.226	297.61 −24.42	203.8 −20.6	GROUP of 5 6-G-4	18 D 4.0	F005/J2958 (53,189)
AM 0850-701	08 50 05 −70 19.6	0.319 −0.226	285.36 −16.41	195.4 −32.4	I/A DBL. in GROUP of 14	2,18 S 0.1	F060/J1172 (131,159)
AM 0850-244	08 50 07 −24 40.6	2.621 −0.226	249.54 12.40	139.7 −61.1	PAIR	23 S 0.7	F496/J3861 (153,197)
AM 0850-482	08 50 16 −48 26.2	1.963 −0.226	268.02 −2.68	176.8 −49.3	NEBULOSITY	21	F211/J2243 (266,255)
AM 0851-351	08 51 04 −35 17.1	2.379 −0.227	258.01 5.87	158.9 −57.1	LOOSE STAR CLUSTER 371-SC-25	22 D 2.6	F371/J3997 (146,164)
AM 0851-313	08 51 10 −31 32.6	2.471 −0.227	255.12 8.27	152.5 −58.7	I/A DBL. PKS 0851-314?	2 S 0.4	F432/J0430 (154,97)
AM 0851-683	08 51 23 −68 34.2	0.578 −0.227	283.98 −15.25	194.1 −33.8	CLOSE TRIPLE	24 D 1.1	F060/J1172 (120,253)
AM 0852-685	08 52 07 −68 50.3	0.550 −0.228	284.24 −15.36	194.3 −33.6	SMALL IRR. ("V"-SHAPED) 60-IG-16	16 D 0.7	F060/J1172 (118,238)
AM 0852-244 [0853-2454]	08 52 58 −24 41.7	2.625 −0.229	249.97 12.90	139.4 −60.4	LINE of 3 (#1: S0)	24 D 11.0	F496/J3861 (118,195)
AM 0853-244 [0852-2441]	08 53 09 −24 49.3	2.623 −0.229	250.09 12.85	139.6 −60.4	LINE of 3 (#2: SPIRAL)	24 D 11.0	F496/J3861 (116,189)
AM 0853-245 [0853-2449]	08 53 09 −24 54.1	2.622 −0.229	250.16 12.80	139.8 −60.4	LINE of 3 (#3: LO.S.BR. SPIRAL)	24 D 11.0	F496/J3861 (116,184)
AM 0854-242 [0854-2433]	08 54 49 −24 28.9	2.633 −0.231	250.07 13.36	138.8 −60.1	E + SP. w. 5 COMP'NS (#1: E) NGC 2717 = 496-G-21 (I/A DBL.)	2,8,23 S 4.0	F496/J3861 (96,207)
AM 0854-243 [0854-2428]	08 54 55 −24 33.3	2.632 −0.231	250.14 13.33	139.0 −60.0	E + SP. w. 5 COMP'NS.(#2: SP.) MCG-4-21-16 = 496-G-22 (I/A DBL.)	2,8,23 S 4.0	F496/J3861 (95,203)
AM 0855-284	08 55 39 −28 46.0	2.545 −0.232	253.58 10.79	146.9 −58.8	PLANETARY NEBULA PK 253 + 10-1 = 432-PN-14	25 D 2.0	F432/J0430 (101,245)
AM 0855-674	08 55 47 −67 40.8	0.730 −0.232	283.51 −14.38	193.1 −34.3	LONG-ARMED SB + COMP'N. 60-G-18 = UKS 0856-688 = Fa 278	8,10 S 4.3	F090/J2170 (139,35)
*AM 0857-685	08 57 01 −68 52.0	0.598 −0.233	284.54 −15.03	193.9 −33.3	SB(ONE-ARMED) + 6 APP. COMP'NS. 60-G-19	10,8 D 3.6	F060/J1172 (94,235)
AM 0857-651	08 57 25 −65 11.1	1.007 −0.234	281.63 −12.68	191.1 −36.2	SPIRAL w. APP. COMP'N. 90-G-11	8 S 2.0	F090/J2170 (126,169)
AM 0857-663	08 57 30 −66 32.5	0.872 −0.234	282.71 −13.53	192.1 −35.1	EDGE-ON SPIRAL w. 2 APP. COMP'NS. 90-G-12	8 S 3.6	F090/J2170 (128,96)
AM 0858-335	08 58 17 −33 52.8	2.434 −0.234	257.90 7.92	155.4 −56.4	PEC. DWARF or PLANETARY NEBULA	20,25 D 2.5	F372/J2982 (332,237)
AM 0858-730	08 58 43 −73 07.4	−0.058 −0.235	288.11 −17.58	196.7 −29.8	RING w. ELONG. NUCLEUS + COMP'N. 36-G-15 = Fa 279	6,8 D 0.9	F036/J2048 (135,277)
AM 0858-274	08 58 51 −27 40.6	2.575 −0.235	253.18 12.03	144.5 −58.4	SPIRAL w. HEAVY ARM to KNOT 432-G-17	9 D 0.6	F497/J3845 (312,34)
AM 0859-471	08 59 18 −47 15.5	2.048 −0.235	268.11 −0.77	173.9 −49.0	NEBULOSITY	21	F211/J2243 (186,320)
AM 0859-215	08 59 22 −21 56.6	2.691 −0.236	248.70 15.77	133.5 −59.4	COMP'N. APPEARS STELLAR w. PLUME MCG-4-22-1? = 564-G-7 (M51-TYPE)	9,13 S 0.2	F496/J3861 (37,342)
AM 0859-240	08 59 37 −24 02.6	2.651 −0.236	250.43 14.48	137.6 −59.1	CONTACT E's	2 S 0.2	F497/J3845 (307,229)
AM 0900-640	09 00 43 −64 05.5	1.134 −0.237	281.01 −11.70	189.9 −36.8	I/A DBL. (M51-TYPE?) 90-IG-14	2,9 S 0.3	F090/J2170 (105,226)
*AM 0902-680	09 02 01 −68 01.8	0.752 −0.238	284.18 −14.15	192.8 −33.7	EDGE-ON SPIRAL w. SQUARE NUCLEUS 60-G-24 (+ I/A DBL.)	1,12 D 3.1	F061/J1173 (327,274)
AM 0902-442	09 02 25 −44 21.0	2.159 −0.239	266.31 1.58	169.8 −50.3	PLANETARY NEBULA PK 266 + 1-2 = 260-PN-16	25	F261/J1362 (348,209)
AM 0902-734	09 02 25 −73 47.1	−0.141 −0.239	288.82 −17.78	196.9 −29.1	SMALL LUMPY NUCLEUS w. JETS(?) 36-IG-16	7,16 D 0.7	F036/J2048 (122,241)
AM 0902-674	09 02 38 −67 46.6	0.788 −0.239	284.02 −13.94	192.6 −33.8	DISTURBED GALAXY 60-G-25	16 D 0.9	F090/J2170 (105,28)
AM 0902-253	09 02 59 −25 34.6	2.627 −0.239	252.15 14.09	140.2 −58.0	DWARF w. STRUCTURE 497-G-7	20 D 0.8	F497/J3845 (264,148)
AM 0903-473	09 03 29 −47 35.5	2.055 −0.240	268.83 −0.46	173.6 −48.2	NEBULOSITY (NOVA SHELL?)	21	F211/J2243 (148,302)
*AM 0903-715	09 03 41 −71 51.0	0.238 −0.240	287.31 −16.49	195.4 −30.6	I/A TRIPLE (in GROUP of 6) 60-IG-26/27 = "MADORE'S MONSTER"	2,3,18 D 0.9	F060/J1172 (78,73)

Identification	R.A. (1950.0) Declination	Precession R.A./Dec.	Gal. Long. Gal. Lat.	S.G.L. S.G.B.	Description Cross-Identification	Coding Size (Min.)	Field/Plate (x, y)
AM 0904-274	09 04 18 −27 49.4	2.584 −0.241	254.10 12.85	144.1 −57.2	LO.S.BR. SPIRAL + APP. COMP'N. 433-G-2	8 S 7.2	F433/J2965 (271,295)
AM 0905-274	09 05 20 −27 49.3	2.586 −0.242	254.25 13.03	144.0 −57.0	I/A DBL.	2 S 1.9	F433/J2965 (271,295)
*AM 0905-232	09 05 27 −23 25.1	2.674 −0.242	250.82 15.91	136.0 −57.9	SPIRAL w. I/A COMP'N. + COMPACT NGC 2772 = 497-G-14	1,13 S 0.7	F497/J3845 (235,264)
AM 0905-683	09 05 30 −68 32.7	0.728 −0.242	284.80 −14.24	192.9 −33.1	CHAIN of 4	17 D 7.2	F060/J1172 (51,249)
AM 0906-511	09 06 33 −51 19.8	1.927 −0.243	271.92 −2.63	177.3 −45.4	NEBULOSITY	21	F211/J2243 (124,101)
AM 0906-694	09 06 35 −69 44.3	0.588 −0.243	285.80 −14.94	193.7 −32.1	COMPACT w. DIFFRACTION SPIKE IC 2448 = PK 285-14-1 = 61-PN-1	13,21 D 0.5	F061/J1173 (295,185)
AM 0906-753	09 06 52 −75 36.5	−0.497 −0.243	290.51 −18.70	197.8 −27.4	SPIRAL + I/A COMP'N. + APP. COMP'N 36-G-19 [A82]	1,8 S 2.5	F036/J2048 (114,142)
AM 0906-231	09 06 57 −23 10.0	2.681 −0.243	250.85 16.34	135.4 −57.6	DWARF	20 D 0.2	F497/J3845 (217,277)
AM 0907-721	09 07 11 −72 14.5	0.217 −0.243	287.81 −16.53	195.5 −30.1	I/A COMPACT w. TAIL	13,15 D 1.6	F060/J1172 (66,51)
AM 0907-482	09 07 27 −48 23.0	2.045 −0.244	269.87 −0.51	173.9 −47.2	NEBULOSITY	21	F211/J2243 (113,259)
AM 0907-224	09 07 31 −22 48.2	2.689 −0.244	250.65 16.67	134.7 −57.5	CENTRALLY CONDENSED DWARF DDO 56 = MCG-4-22-3 = 497-G-17	20 D 0.9	F497/J3845 (210,297)
AM 0908-325	09 08 01 −32 56.7	2.483 −0.244	258.55 10.06	152.4 −54.9	DWARF SPIRAL (OBSCURED?) MCG-5-22-1 = 372-G-8	20 D 2.0	F372/J2982 (224,289)
AM 0908-674	09 08 21 −67 43.7	0.851 −0.245	284.34 −13.50	192.0 −33.5	GALAXY w. I/A COMP'N. NGC 2788 = 61-G-2	1 S 1.3	F061/J1173 (297,293)
AM 0909-303 [0910-3042]	09 09 13 −30 39.9	2.536 −0.245	257.01 11.77	148.5 −55.4	PAIR (#1: IRR.) MCG-5-22-2 = 433-G-7	8 S 0.5	F433/J2965 (211,144)
AM 0909-452	09 09 17 −45 23.8	2.155 −0.246	267.90 1.76	170.0 −48.7	3 REFLECTION NEBULAE	21 S 9.0	F261/J1362 (280,156)
AM 0910-304 [0909-3039]	09 10 04 −30 42.4	2.537 −0.246	257.16 11.88	148.5 −55.2	PAIR (#2: S0) MCG-5-22-3 = 433-G-8	8 S 0.5	F433/J2965 (201,141)
*AM 0910-235	09 10 05 −23 58.1	2.672 −0.246	251.97 16.36	136.7 −56.7	S0(LENS) + PEC. COMP'N. + 3 DWARFS NGC 2784 = 497-G-23 (+NUCLEATED)	1,8,20 S 3.4	F497/J3845 (178,235)
AM 0910-421	09 10 31 −42 13.5	2.256 −0.247	265.75 4.10	165.8 −50.3	PLANETARY NEBULA NGC 2792 = PK 265+4-1 = 314-PN-6	25	F261/J1362 (274,327)
AM 0911-790	09 11 35 −79 01.9	−1.548 −0.248	293.49 −20.68	199.9 −24.5	SMOOTH-ARMED, ISOLATED SPIRAL 18-G-15	10 D 1.0	F018/J2959 (150,231)
AM 0911-314	09 11 36 −31 41.2	2.520 −0.248	258.13 11.46	149.9 −54.6	I/A PAIR (ONE w. TAIL)	2,15 S 2.2	F433/J2965 (183,89)
AM 0912-242	09 12 12 −24 22.7	2.668 −0.248	252.62 16.45	137.3 −56.2	COMPACT	13 D 0.1	F497/J3845 (152,213)
AM 0912-254	09 12 14 −25 47.3	2.642 −0.248	253.72 15.52	139.8 −56.0	PAIR	23 S 0.5	F497/J3845 (152,137)
AM 0912-232	09 12 19 −23 20.8	2.688 −0.249	251.83 17.15	135.5 −56.3	DWARF 497-G-25	20 D 0.2	F497/J3845 (151,268)
AM 0912-241	09 12 42 −24 18.6	2.671 −0.249	252.65 16.58	137.2 −56.1	DWARF (ELONGATED)	20 D 0.7	F497/J3845 (146,216)
AM 0912-280	09 12 54 −28 03.3	2.599 −0.249	255.57 14.12	143.7 −55.3	SPIRAL w. 3 APP. COMP'NS. MCG-5-22-4 = 433-G-10	8 S 4.3	F497/J3845 (145,15)
AM 0913-690	09 13 07 −69 07.9	0.741 −0.249	285.72 −14.11	192.7 −32.2	SPIRAL (w. STRAIGHT JET ?) 61-G-3	7 D 3.1	F060/J1172 (18,213)
AM 0913-251	09 13 11 −25 10.8	2.655 −0.249	253.40 16.09	138.7 −55.9	E w. COMPACT COMP'N.	8,13 S 0.8	F497/J3845 (141,169)
AM 0913-252	09 13 16 −25 21.5	2.652 −0.249	253.55 15.99	139.0 −55.8	COMPACT w. APP. COMP'NS. 497-G-27	8,13 S 2.6	F497/J3845 (140,160)
AM 0913-242	09 13 54 −24 22.0	2.672 −0.250	252.88 16.75	137.2 −55.8	GROUP of 3 497-G-31	24 D 1.8	F497/J3845 (132,213)
*AM 0914-362	09 14 00 −36 25.2	2.420 −0.250	261.98 8.59	157.0 −52.4	HI.S.BR. GAL. or P.N. in CLUSTER NGC 2118=PK 268+8-1=372-PN-13	13,25 D 1.3	F372/J2982 (158,103)
*AM 0914-232	09 14 06 −23 25.7	2.690 −0.250	252.17 17.41	135.6 −55.9	Sb w. STRONG ARM + CHAIN of 4 NGC 2815 = 497-G-32 COMP'NS.	8,10,17 S 10.8	F497/J3845 (129,263)
AM 0914-240	09 14 32 −24 07.6	2.678 −0.251	252.79 17.02	136.8 −55.7	PAIR 497-G-33	23 S 0.8	F497/J3845 (124,226)
AM 0915-241	09 15 11 −24 11.2	2.678 −0.251	252.94 17.09	136.8 −55.6	PAIR	23 S 0.5	F497/J3845 (116,223)
AM 0915-311	09 15 24 −31 14.6	2.540 −0.251	258.36 12.36	148.8 −54.0	VERY CLOSE, SMALL I/A DBL.	2 S 0.1	F433/J2965 (140,113)
*AM 0915-220	09 15 33 −22 08.9	2.715 −0.252	251.40 18.50	133.2 −55.7	3-ARMED SPIRAL NGC 2835 = 564-G-35	11 D 5.4	F497/J3845 (111,332)
AM 0915-321 [0916-3214]	09 15 38 −32 11.3	2.520 −0.252	259.10 11.75	150.3 −53.6	3 in CHAIN of 9 (#1: E)	17 D 24.2	F433/J2965 (137,62)

Identification	R.A. (1950.0) Declination	Precession R.A./Dec.	Gal. Long. Gal. Lat.	S.G.L. S.G.B.	Description Cross-Identification	Coding Size (Min.)	Field/Plate (x, y)
AM 0915-242	09 15 48 −24 26.2	2.674 −0.252	253.24 17.03	137.2 −55.4	DISTURBED GALAXY w. I/A COMP'N. 497-IG-38	1 S 0.4	F497/J3845 (108,209)
AM 0916-321 [0915-3211]	09 16 08 −32 16.5	2.520 −0.252	259.24 11.76	150.4 −53.5	3 in CHAIN of 9 (#2: EDGE-ON) 433-G-12	17 D 24.2	F433/J2965 (132,57)
AM 0916-241	09 16 26 −24 12.7	2.680 −0.252	253.16 17.29	136.8 −55.3	PAIR	23 S 1.5	F497/J3845 (101,221)
AM 0916-321 [0916-3216]	09 16 44 −32 14.0	2.522 −0.253	259.30 11.89	150.2 −53.4	3 in CHAIN of 9 (#3: DWARF) 433-G-13	17 D 24.2	F433/J2965 (125,59)
AM 0917-291	09 17 53 −29 18.5	2.586 −0.254	257.29 14.08	145.4 −54.0	SMALL I/A DBL.	2 S 0.8	F433/J2965 (110,216)
AM 0918-645	09 18 33 −64 59.0	1.219 −0.255	282.99 −10.89	188.9 −34.9	CHAIN of 4 91-G-8 included	17 D 11.2	F091/J2147 (256,178)
AM 0919-475	09 19 57 −47 51.3	2.124 −0.256	270.94 1.35	171.5 −45.8	REFLECTION NEBULA	21	F261/J1362 (179,26)
AM 0921-264	09 21 08 −26 42.2	2.644 −0.257	255.83 16.38	140.8 −53.8	PAIR (incl. HI.S.BR. IRR.)	16,23 S 5.4	F497/J3845 (46,86)
*AM 0921-225	09 21 14 −22 56.8	2.712 −0.257	252.95 18.95	134.4 −54.3	"RINGING" w. INTERNAL FILAMENTS NGC 2865 = 498-G-1 = MCG 0921-229	8,16 D 2.0	F497/J3845 (41,288)
AM 0921-252	09 21 33 −25 25.2	2.668 −0.257	254.92 17.33	138.6 −53.9	cD + 4 COMP'NS. MGC-4-22-13 = 498-G-4	8 S 3.7	F498/J3856 (306,156)
AM 0922-252	09 22 06 −25 23.4	2.670 −0.258	254.98 17.44	138.5 −53.8	COMP'N. GAL. w. PECULIAR ABS'N.	14 D 0.4	F497/J3845 (33,157)
AM 0922-633	09 22 16 −63 35.9	1.364 −0.258	282.28 −9.63	187.3 −35.6	E w. 6 APP. COMP'NS. NGC 2887 = 91-G-9	8 D 2.8	F091/J2147 (238,253)
AM 0922-320	09 22 25 −32 04.2	2.542 −0.258	260.03 12.86	149.4 −52.3	DWARF ? (near STAR) 433-G-20	20 D 0.7	F372/J2982 (61,335)
AM 0922-245	09 22 26 −24 52.6	2.680 −0.258	254.64 17.84	137.6 −53.8	SPIRAL w. STRONG ARM + COMP'N. A 0922-24 = MCG-4-23-1 = 498-G-5	8,10 S 1.2	F497/J3845 (28,184)
*AM 0923-335	09 23 13 −33 53.2	2.506 −0.259	261.47 11.71	152.1 −51.6	DISTURBED IRR. w. PEC. ABS'N. NGC 2883 = 372-G-24	14,16 D 2.2	F372/J2982 (54,237)
AM 0923-275	09 23 33 −27 59.5	2.625 −0.259	257.18 15.88	142.7 −53.1	SMALL I/A DBL. 434-IG-1	2 S 0.4	F497/J3845 (19,17)
AM 0925-314	09 25 16 −31 47.8	2.555 −0.261	260.27 13.49	148.7 −51.8	MAGELLANIC DWARF (RESOLUTION:) 434-G-5	20 D 1.1	F433/J2965 (27,81)
*AM 0926-355	09 26 18 −35 56.9	2.471 −0.262	263.41 10.69	154.8 −50.3	GROUP of 4 incl. NUCLEAR JET 373-G-3	7,18 D 4.9	F373/J2916 (263,128)
AM 0926-243	09 26 37 −24 36.9	2.693 −0.262	255.13 18.71	137.0 −52.9	I/A DBL. w. TAILS	2,15 S 1.2	F498/J3856 (245,200)
*AM 0926-762	09 26 37 −76 24.1	−0.353 −0.262	291.97 −18.34	197.3 −26.1	IRR. + APP. COMP'N. (JETS?) NGC 2915 = 37-G-3	7,8,16 S 7.2	F036/J2048 (55,92)
AM 0927-603	09 27 07 −60 34.8	1.607 −0.262	280.56 −7.07	184.1 −37.2	CLOSE TRIPLE 126-G-14 included	24 D 10.1	F126/J2979 (198,148)
AM 0927-301	09 27 16 −30 11.7	2.592 −0.263	259.41 14.92	146.0 −51.8	DWARF	20 D 1.7	F434/J4001 (269,168)
AM 0928-302 [0929-3005]	09 28 06 −30 23.2	2.591 −0.263	259.68 14.91	146.2 −51.6	GROUP of 5 (#1: EDGE-ON S0) MGC-5-23-2 = 434-G-7	18 D 1.7	F434/J4001 (259,157)
AM 0928-300 [0928-3023]	09 28 08 −30 09.7	2.595 −0.263	259.52 15.08	145.8 −51.6	GROUP of 5 (#2: E/S0) NGC 2904 = 434-G-6	18 D 0.9	F434/J4001 (259,170)
AM 0928-331	09 28 22 −33 18.0	2.534 −0.264	261.83 12.88	150.7 −50.7	I/A QUINTET	5 D 2.9	F373/J2916 (242,270)
AM 0928-343	09 28 22 −34 35.9	2.507 −0.264	262.76 11.95	152.6 −50.3	DWARF/DEFECT/OBSCURED GALAXY ?	20 D 0.4	F373/J2916 (241,201)
AM 0928-560	09 28 49 −56 04.4	1.856 −0.264	277.61 −3.64	179.5 −39.9	PLANETARY NEBULA vdB-2 = 166-PN-21	25 D	F166/J2251 (121,121)
AM 0928-300 [0928-3009]	09 28 57 −30 08.1	2.598 −0.264	259.63 15.22	145.7 −51.5	GROUP of 5 (#3: E) MGC-5-23-4 = 434-G-9	18 D 0.8	F434/J4001 (249,171)
AM 0929-295 [0934-3009]	09 29 02 −29 59.5	2.601 −0.264	259.54 15.34	145.5 −51.5	GROUP of 5 (#4: EDGE-ON) 434-G-10	18 D 0.9	F434/J4001 (248,179)
AM 0929-300 [0929-2959]	09 29 27 −30 05.1	2.600 −0.264	259.67 15.33	145.6 −51.4	GROUP of 5 (#5: SPIRAL) MGC-5-23-5 = 434-G-11	18 D 0.9	F434/J4001 (243,174)
AM 0930-330	09 30 38 −33 01.6	2.546 −0.266	261.98 13.41	150.0 −50.4	NUCLEATED DWARF 373-G-7	20 D 1.1	F434/J4001 (227,16)
AM 0931-324	09 31 14 −32 48.8	2.552 −0.266	261.92 13.65	149.7 −50.3	EDGE-ON w. DISTORTED RING COMP'N MGC-5-23-6 = 373-G-8	6,8 S 2.2	F434/J4001 (220,28)
AM 0931-253	09 31 18 −25 34.0	2.687 −0.266	256.63 18.81	138.4 −51.7	CONTACT SPIRALS 498-G-13	2 S 0.5	F498/J3856 (188,149)
AM 0933-290	09 33 26 −29 00.1	2.631 −0.268	259.52 16.71	143.7 −50.7	SMALL TRIPLE	24 D 1.0	F434/J4001 (198,233)
AM 0933-244	09 33 42 −24 45.1	2.706 −0.268	256.42 19.77	137.0 −51.3	SPIRAL + I/A COMP'N. MGC-4-23-9 = 498-G-15	1 S 1.4	F498/J3856 (158,193)
AM 0935-221	09 35 30 −22 10.8	2.752 −0.270	254.77 21.87	132.8 −51.1	SPIRAL w. APP. COMP'N. MGC-4-23-11 = 565-G-29	8 S 1.4	F498/J3856 (136,331)

Identification	R.A. (1950.0) Declination	Precession R.A./Dec.	Gal. Long. Gal. Lat.	S.G.L. S.G.B.	Description Cross-Identification	Coding Size (Min.)	Field/Plate (x, y)
AM 0935-320	09 35 49 −32 03.5	2.581 −0.270	262.11 14.86	148.1 −49.6	DWARF 434-G-17	20 D 0.7	F434/J4001 (169,69)
AM 0936-271	09 36 00 −27 16.4	2.668 −0.270	258.69 18.35	140.8 −50.5	SMALL I/A DBL.	2 S 0.1	F498/J3856 (131,57)
AM 0936-305	09 36 47 −30 54.2	2.605 −0.271	261.44 15.84	146.3 −49.6	I/A QUARTET	4 D 2.7	F434/J4001 (158,131)
AM 0937-240	09 37 41 −24 01.6	2.726 −0.272	256.55 20.93	135.7 −50.5	CLOSE I/A DBL. in GROUP of 5	2,18 S 0.6	F498/J3856 (110,232)
AM 0937-245	09 37 51 −24 50.1	2.714 −0.272	257.19 20.38	136.9 −50.4	PECULIAR-ARMED SPIRAL + COMP'N. MGC-4-23-12 = 498-G-20	8,10 S 1.8	F498/J3856 (108,188)
AM 0937-304	09 37 51 −30 48.9	2.610 −0.272	261.55 16.07	146.1 −49.4	S0 + 12 APP. COMP'NS. 434-G-18 (incl. PEC. SPIRAL)	8,12 D 4.0	F434/J4001 (146,136)
AM 0938-320	09 38 35 −32 00.3	2.590 −0.272	262.51 15.30	147.8 −49.0	DWARF (SOME RESOLUTION) 434-G-19	20 D 0.9	F434/J4001 (137,72)
AM 0939-233 [0939-2338]	09 39 44 −23 37.0	2.737 −0.273	256.60 21.55	135.0 −50.0	PAIR of SPIRALS (#1)	23 D 0.4	F498/J3856 (84,253)
AM 0939-233 [0939-2337]	09 39 56 −23 38.4	2.737 −0.274	256.65 21.56	135.0 −50.0	PAIR of SPIRALS (#2)	23 D 0.4	F498/J3856 (82,252)
AM 0941-323	09 41 25 −32 30.8	2.589 −0.275	263.33 15.33	148.4 −48.3	DWARF w. STRUCTURE + APP. COMP'N 373-G-20	8,20 D 1.3	F373/J2916 (95,312)
AM 0941-315	09 41 55 −31 56.0	2.601 −0.275	263.00 15.83	147.5 −48.3	DISTURBED DWARF 434-G-27	20 D 0.9	F434/J4001 (99,76)
AM 0942-311	09 42 20 −31 19.6	2.613 −0.276	262.65 16.34	146.6 −48.4	PECULIAR DWARF/SPIRAL? 434-G-30	20 D 0.6	F434/J4001 (94,108)
AM 0942-313 [0942-3135]	09 42 23 −31 33.6	2.609 −0.276	262.82 16.17	146.9 −48.3	PAIR of IRR. (#1) IC 2507 = 434-G-31	16,23 D 1.3	F434/J4001 (94,96)
AM 0942-290	09 42 25 −29 05.4	2.653 −0.276	261.08 18.00	143.2 −48.8	BRIGHT-RIMMED SPIRAL	10,12 D 1.1	F434/J4001 (92,228)
AM 0942-313 [0942-3133]	09 42 35 −31 35.8	2.609 −0.276	262.88 16.18	146.9 −48.3	PAIR of IRR. (#2) MGC-5-23-10 = 434-G-33	16,23 D 2.0	F434/J4001 (92,94)
AM 0942-315	09 42 53 −31 56.7	2.604 −0.276	263.17 15.96	147.4 −48.1	I/A DBL.	2 S 0.4	F373/J2916 (78,342)
*AM 0943-333	09 43 08 −33 33.2	2.575 −0.276	264.33 14.80	149.8 −47.7	GALAXY w. JET and CONDENSATIONS 373-IG-22	7 D 0.9	F373/J2916 (77,256)
AM 0943-300 [0944-3012]	09 43 18 −30 06.8	2.638 −0.276	261.95 17.38	144.7 −48.4	PAIR (#1: DWARF/EDGE-ON) MGC-5-23-11 = 434-G-34	20,23 D 2.9	F434/J4001 (83,173)
AM 0943-305	09 43 27 −30 57.5	2.623 −0.276	262.58 16.77	146.0 −48.2	PARENT to MANY DWARFS NGC 2997 = 434-G-35	8 D 7.2	F434/J4001 (81,128)
AM 0944-301 [0943-3006]	09 44 07 −30 12.2	2.638 −0.277	262.16 17.43	144.8 −48.2	PAIR (#2: SPIRAL w. PEC. ARMS) NGC 3001 = 434-G-38	10,23 D 4.3	F434/J4001 (73,169)
*AM 0944-332	09 44 38 −33 22.2	2.583 −0.277	264.44 15.14	149.4 −47.5	GALAXY w. 7 COMP'NS. ALIGNED MCG-5-23-15 = 373-G-26 [N.S.]	8,17 S 22.4	F373/J2916 (60,265)
AM 0944-243	09 44 56 −24 36.5	2.733 −0.278	258.26 21.66	136.4 −48.8	DISTURBED SP. w. COMPACT COMP'N. MCG-4-23-17 = 499-G-5	8,13,16 S 1.5	F499/J5653 (290,199)
AM 0945-304	09 45 28 −30 43.2	2.633 −0.278	262.74 17.24	145.5 −47.9	GROUP of 3 MCG-5-23-16 = 436-G-40	24 D 4.9	F435/J1299 (322,137)
AM 0945-484	09 45 32 −48 44.3	2.230 −0.278	274.69 3.55	169.5 −41.7	PLANETARY NEBULA PK 274+3 = 213-PN-1	25 D 4.5	F213/J2269 (307,244)
AM 0945-311	09 45 33 −31 16.5	2.624 −0.278	263.15 16.84	146.3 −47.7	DISRUPTED DWARF 434-G-41	20 D 1.8	F434/J4001 (57,111)
AM 0945-323 [0949-3250]	09 45 34 −32 36.2	2.600 −0.278	264.07 15.85	148.2 −47.4	GROUP of 5 (#1: SPIRAL) IC 2510 = 374-G-29	17 D 0.7	F434/J4001 (58,30)
AM 0946-254 [0946-2548]	09 46 25 −25 48.8	2.717 −0.279	259.41 21.01	138.2 −48.3	GROUP of 3 w. COMP'NS. (#1)	8,24 D 0.3	F499/J5653 (271,135)
AM 0946-254 [0946-2548]	09 46 39 −25 44.4	2.719 −0.279	259.40 21.10	138.0 −48.3	GROUP of 3 w. COMP'NS. (#2) 499-G-7	8,24 D 0.6	F499/J5653 (268,139)
AM 0946-254 [0946-2544]	09 46 40 −25 48.5	2.718 −0.279	259.45 21.05	138.1 −48.3	GROUP of 3 w. COMP'NS. (#3)	8,24 D 0.5	F499/J5653 (268,135)
AM 0947-422	09 47 07 −42 23.1	2.405 −0.279	270.82 8.62	161.5 −44.2	PEC. SPIRAL (LARGEST of 5 GALS.) 316-G-01	12,18 D 0.7	F316/J1368 (285,47)
AM 0947-323 [0945-3236]	09 47 15 −32 36.4	2.605 −0.280	264.35 16.08	148.1 −47.1	GROUP of 5 (#2: EDGE-ON Sb) IC 2511/12 = 374-G-49 = PKS 0947-324?	17 D 2.7	F434/J4001 (39,39)
AM 0947-474	09 47 23 −47 41.0	2.271 −0.280	274.26 4.56	168.0 −41.9	JET = STAR! + COMP'N. 213-G-2	7,8 D 6.7	F213/J2269 (293,301)
AM 0947-323 [0947-3236]	09 47 51 −32 39.1	2.606 −0.280	264.48 16.13	148.1 −47.0	GROUP of 5 (#3: EDGE-ON Sb) IC 2513/14? = 374-G-50	17 D 3.0	F434/J4001 (32,36)
AM 0948-303	09 48 19 −30 38.4	2.643 −0.280	263.17 17.71	145.2 −47.3	DWARF	20 D 0.7	F435/J1299 (289,142)
AM 0948-283	09 48 36 −28 30.6	2.679 −0.281	261.72 19.34	142.1 −47.5	3 in a LINE	24 D 4.0	F435/J1299 (288,256)
AM 0949-323 [0947-3239]	09 49 04 −32 31.2	2.612 −0.281	264.59 16.40	147.9 −46.7	GROUP of 5 (#4: Sc) NGC 3038 = 374-G-2	17 D 2.2	F434/J4001 (18,43)

Identification	R.A. (1950.0) Declination	Precession R.A./Dec.	Gal. Long. Gal. Lat.	S.G.L. S.G.B.	Description Cross-Identification	Coding Size (Min.)	Field/Plate (x, y)
AM 0949-264	09 49 09 −26 46.9	2.709 −0.281	260.59 20.71	139.5 −47.6	IRR. w. KNOTS NGC 3037 = 499-G-10	16 D 0.8	F499/J5653 (237,83)
AM 0949-320	09 49 20 −32 00.2	2.622 −0.281	264.28 16.82	147.1 −46.8	SMALL I/A DBL.	2 S 0.2	F434/J4001 (15,71)
AM 0949-372	09 49 26 −37 22.7	2.522 −0.281	267.90 12.75	154.6 −45.4	SMALL SB(r) w. CONNECTED COMP'N.	1 S 0.6	F316/J1368 (268,317)
AM 0949-525	09 49 29 −52 56.4	2.119 −0.281	277.83 0.69	173.9 −39.1	CLUSTER ? 167-SC-8	22	F167/J3872 (216,290)
AM 0949-325 [0949-3231]	09 49 46 −32 50.4	2.609 −0.282	264.92 16.25	148.3 −46.5	GROUP of 5 (#5: EDGE-ON Sc) MGC-5-24-2 = 374-G-3	17 D 2.0	F434/J4001 (11,26)
AM 0950-264	09 50 06 −26 42.4	2.712 −0.282	260.71 20.91	139.4 −47.4	DWARF w. ELONGATED NUCLEUS 499-G-12	20 D 0.6	F499/J5653 (226,87)
AM 0950-291	09 50 06 −29 12.1	2.672 −0.282	262.47 19.04	143.0 −47.1	GALAXY w. I/A COMP'N. + HALO 435-G-3 = MC 0950-291	1 D 3.1	F435/J1299 (269,219)
AM 0952-325	09 52 12 −32 52.5	2.616 −0.283	265.35 16.55	148.2 −46.0	GALAXY w. I/A COMP'N. in TRIPLE 374-G-8	1,24 S 1.0	F374/J3862 (267,293)
AM 0952-280	09 52 18 −28 04.0	2.696 −0.283	262.06 20.21	141.3 −46.8	E w. I/A DBL. [N.] NGC 3056 = 435-G-7	2,8 S 12.1	F435/J1299 (244,281)
AM 0952-334 [0952-3400]	09 52 24 −33 47.3	2.601 −0.284	266.00 15.88	149.4 −45.8	PAIR (#1: E)	23 S 4.6	F374/J3862 (264,244)
AM 0952-273	09 52 42 −27 36.8	2.704 −0.284	261.81 20.61	140.6 −46.8	DISTURBED SINGLE w. TAILS	15 D 0.7	F435/J1299 (240,305)
AM 0952-340 [0952-3347]	09 52 47 −34 00.7	2.598 −0.284	266.21 15.76	149.7 −45.7	PAIR (#2: SPIRAL)	23 S 4.6	F374/J3862 (259,232)
AM 0952-325	09 52 58 −32 54.0	2.618 −0.284	265.50 16.64	148.1 −45.9	SPIRAL w. I/A COMP'N. IC 2522/23 = 374-G-10/11	1,2,8 S 3.7	F374/J3862 (258,291)
AM 0953-675	09 53 23 −67 52.6	1.348 −0.284	287.59 −10.67	188.7 −30.4	DWARF	20 D 0.6	F091/J2147 (71,21)
AM 0953-330	09 53 49 −33 03.3	2.618 −0.285	265.74 16.64	148.3 −45.6	DWARF	20 D 0.3	F374/J3862 (249,283)
AM 0953-373	09 53 59 −37 32.1	2.537 −0.285	268.72 13.20	154.5 −44.5	SPIRAL w. 3 APP. COMP'NS. 316-G-4	8 S 2.2	F374/J3862 (243,43)
AM 0954-280	09 54 00 −28 01.2	2.702 −0.285	262.33 20.50	141.1 −46.4	I/A DBL.	2 S 1.7	F435/J1299 (224,283)
AM 0954-365	09 54 34 −36 52.8	2.551 −0.285	268.39 13.78	153.5 −44.6	COMPACT	13 D 0.4	F374/J3862 (238,78)
AM 0954-793	09 54 54 −79 37.3	−0.714 −0.285	295.43 −19.66	198.7 −22.8	VERY SMALL I/A TRIPLE (JET?)	3,7 S 0.2	F019/J2028 (264,193)
AM 0955-412	09 55 18 −41 21.1	2.464 −0.286	271.36 10.38	159.4 −43.1	GAL. w. PEC. ABS.[S.W.] of BR.ST. 316-G-7	14 D 0.8	F316/J1368 (204,104)
AM 0955-323 [0955-3237]	09 55 38 −32 35.9	2.632 −0.286	265.74 17.23	147.6 −45.4	PAIR (#1: E) 374-G-13	23 S 3.2	F374/J3862 (228,308)
AM 0955-323 [0955-3235]	09 55 49 −32 37.7	2.632 −0.286	265.79 17.23	147.6 −45.3	PAIR (#2: SBd) 374-G-14	23 S 3.2	F374/J3862 (226,306)
AM 0956-375	09 56 04 −37 59.4	2.536 −0.286	269.34 13.10	154.9 −44.0	I/A DWARFS 316-IG-09	2,20,20 S 0.4	F316/J1368 (197,285)
AM 0956-263	09 56 05 −26 30.9	2.730 −0.286	261.65 21.94	138.9 −46.1	CONTACT E's	2 S 0.2	F499/J5653 (154,98)
AM 0956-264	09 56 08 −26 41.1	2.727 −0.286	261.78 21.81	139.1 −46.1	SB + SP. + E (STAR on 3-ARMED SP.) E = NGC 3078 = 449-G-27	11,24 D 13.4	F499/J5653 (154,89)
*AM 0956-282	09 56 29 −28 23.0	2.703 −0.287	263.03 20.58	141.6 −45.8	PECULIAR SPIRAL w. JET	7,12 D 1.8	F435/J1299 (195,264)
AM 0956-300 [0956-3000]	09 56 40 −30 07.2	2.676 −0.287	264.25 19.28	144.0 −45.6	CLOSE PAIR (#1) NGC 3082 = 435-G-18	23 D 1.0	F435/J1299 (192,171)
AM 0956-343	09 56 46 −34 33.7	2.602 −0.287	267.23 15.86	150.2 −44.7	CLOSE PAIR	23 S 0.6	F374/J3862 (215,203)
*AM 0956-265	09 56 50 −26 53.8	2.726 −0.287	262.05 21.76	139.4 −45.9	3-ARMED SPIRAL w. CONDENS. in ARM NCG 3084 = 499-G-29 [A82]	11 D 2.0	F435/J1299 (191,344)
AM 0956-300 [0956-3007]	09 56 50 −30 00.6	2.678 −0.287	264.21 19.38	143.9 −45.5	CLOSE PAIR (#2) MCG-5-24-12 = 435-G-19	23 D 2.3	F435/J1299 (190,177)
AM 0956-335 [0959-3352]	09 56 59 −33 59.1	2.612 −0.287	266.89 16.34	149.4 −44.8	CHAIN of 3 (#1: E) NGC 3087 = 374-G-15 = SS-9a	24 S 22.4	F374/J3862 (213,234)
AM 0957-275	09 57 06 −27 53.7	2.712 −0.287	262.80 21.04	140.8 −45.8	COMPACT w. DIFFUSE MATERIAL off 435-IG-20 = To 0957-278 ONE SIDE	13,15 D 0.2	F499/J5653 (142,24)
AM 0957-223	09 57 10 −22 35.2	2.789 −0.287	259.01 25.04	133.2 −46.1	ISOLATED SO(r) w. FNT OUTER ARMS NGC 3081 = IC 2529 = 499-G-31	12 D 2.2	F499/J5653 (140,309)
AM 0957-280	09 57 20 −28 05.8	2.709 −0.287	262.98 20.92	141.1 −45.7	3-ARMED SPIRAL w. PEC. COMPACT NGC 3089 = 435-G-24	8,10,11,13 D 2.2	F435/J1299 (185,279)
AM 0957-292	09 57 40 −29 22.7	2.690 −0.288	263.92 19.98	142.9 −45.5	EDGE-ON SP. w. I/A DBL. + COMPACT IC 2531 = 435-G-25	2,8,13 D 8.1	F435/J1299 (181,211)
AM 0957-311 [1000-3124]	09 57 50 −31 18.8	2.660 −0.288	265.26 18.52	145.6 −45.1	3 PECULIAR GALAXIES ALIGNED (#1) NGC 3095 = 435-G-26	24 D 2.2	F435/J1299 (179,107)

Identification	R.A. (1950.0) Declination	Precession R.A./Dec.	Gal. Long. Gal. Lat.	S.G.L. S.G.B.	Description Cross-Identification	Coding Size (Min.)	Field/Plate (x, y)
AM 0957-335 [0956-3359]	09 57 54 −33 59.2	2.616 −0.288	267.04 16.46	149.4 −44.6	CHAIN of 3 (#2: SB,BROKEN RING) IC 2532 = 374-G-16 = SS-9b	6,24 S 22.4	F374/J3862 (203,234)
AM 0958-250 [0958-2511]	09 58 08 −25 07.8	2.755 −0.288	261.04 23.28	136.9 −45.7	LINE of 3 (#1)	24 D 0.4	F499/J5653 (129,172)
AM 0958-250 [0958-2507]	09 58 16 −25 09.0	2.755 −0.288	261.08 23.29	136.9 −45.7	LINE of 3 (#2)	24 D 0.4	F499/J5653 (127,171)
AM 0958-310	09 58 18 −31 00.5	2.666 −0.288	265.14 18.82	145.2 −45.1	E w. ARC of APP. COMP'NS. I 2533 = 435-G-27	8,17 D 2.2	F435/J1299 (173,123)
*AM 0958-312 [0957-3118]	09 58 27 −31 25.2	2.660 −0.288	265.44 18.52	145.8 −45.0	3 PECULIAR GALAXIES ALIGNED (#2) NGC 3100 = NGC 3103 = 435-G-30	24 D 1.8	F435/J1299 (172,101)
AM 0958-355	09 58 32 −35 59.2	2.583 −0.288	268.45 14.98	152.0 −44.1	CLOSE PAIR (S0/S + SPIRAL)	23 S 0.8	F374/J3862 (195,126)
AM 0958-251 [0958-2509]	09 58 39 −25 11.6	2.756 −0.288	261.18 23.31	137.0 −45.6	LINE of 3 (#3)	24 D 0.5	F499/J5653 (123,169)
AM 0958-362	09 58 42 −36 22.7	2.576 −0.288	268.73 14.69	152.6 −43.9	OPEN SPIRAL + I/A COMP'N	1 S 1.5	F374/J3862 (193,105)
AM 0958-672	09 58 44 −67 24.2	1.453 −0.288	287.70 −9.99	187.9 −30.2	DWARF	20 S 1.2	F061/J1173 (39,309)
AM 0958-543	09 58 55 −54 32.9	2.127 −0.288	279.92 0.28	174.8 −37.1	CLUSTER of STARS NGC 3105 = 167-SC-14	22 D 1.3	F167/J3872 (141,203)
AM 0959-380	09 59 04 −38 03.1	2.546 −0.289	269.86 13.42	154.8 −43.4	SMALL I/A DBL.	2 S 1.0	F316/J1368 (165,282)
AM 0959-335 [0957-3359]	09 59 18 −33 52.2	2.622 −0.289	267.20 16.73	149.1 −44.4	CHAIN of 3 (#3: Sa) IC 2534 = 374-G-19	24 S 22.4	F374/J3862 (187,240)
AM 0959-545	09 59 43 −54 52.3	2.122 −0.289	280.21 0.09	175.1 −36.9	CLUSTER? CONDENSATION ? 167-SC-15	22 D 2.7	F167/J3872 (135,186)
*AM 1000-255	10 00 01 −25 59.5	2.747 −0.289	262.00 22.91	138.1 −45.3	SMALL HI.S.BR. GALAXY w. JET	7 D 0.4	F499/J5653 (107,126)
AM 1000-572	10 00 11 −57 29.7	2.026 −0.289	281.83 −1.98	177.9 −35.5	REFLECTION NEBULAE? FUZZY	21 S 1.8	F167/J3872 (134,45)
AM 1000-312 [0958-3125]	10 00 16 −31 26.1	2.666 −0.289	265.77 18.75	145.7 −44.6	3 PECULIAR GALAXIES ALIGNED (#3) NGC 3108 = 435-G-32	24 D 0.5	F435/J1299 (151,100)
AM 1000-254	10 00 18 −25 49.4	2.751 −0.289	261.94 23.08	137.8 −45.2	STELLAR NUCLEUS w. THIN ARM	22 D 0.4	F499/J5653 (103,135)
AM 1000-404	10 00 19 −40 44.9	2.499 −0.289	271.74 11.44	158.2 −42.4	DWARF (OFF-SET NUCLEUS)	20 D 0.5	F316/J1368 (153,137)
*AM 1000-255	10 00 50 −25 55.0	2.750 −0.290	262.10 23.08	137.9 −45.1	RESOLVED DWARF NGC 3109 = DDO 236 = 499-G-36	20 D 17.9	F499/J5653 (97,130)
AM 1001-363 [1001-3629]	10 01 15 −36 32.7	2.583 −0.290	269.25 14.88	152.6 −43.4	PAIR of SPIRALS (#1: SPIRAL)	23 S 3.2	F374/J3862 (166,96)
AM 1001-370	10 01 16 −37 08.9	2.572 −0.290	269.64 14.40	153.4 −43.2	SB w. CONDENS.(STAR?) in GRP of 5 374-G-25 [A82]	12,18 D 0.8	F316/J1368 (142,330)
AM 1001-362 [1001-3632]	10 01 18 −36 29.0	2.584 −0.290	269.22 14.93	152.5 −43.4	PAIR of SPIRALS (#2: SPIRAL)	23 S 3.2	F374/J3862 (165,100)
AM 1001-264	10 01 25 −26 47.0	2.739 −0.290	262.82 22.51	139.1 −44.9	LO.S.BR. SPIRAL w. KNOTS + COMP'N MCG-4-24-14 = 499-G-37	8,12 D 2.7	F499/J5653 (90,83)
AM 1001-270	10 01 29 −27 05.8	2.735 −0.290	263.05 22.27	139.6 −44.9	DWARF?	20 D 1.1	F499/J5653 (90,66)
AM 1001-590	10 01 31 −59 02.2	1.973 −0.290	282.89 −3.11	179.4 −34.6	NEBULOSITY	21	F127/J2252 (224,230)
AM 1001-272	10 01 35 −27 20.4	2.732 −0.290	263.23 22.10	139.9 −44.8	SPIRAL w. 4 DIAMETRIC COMP'NS. IC 2537 = 499-G-39	1,2,8 S 3.6	F499/J5653 (89,53)
AM 1001-270	10 01 48 −27 05.9	2.736 −0.291	263.11 22.32	139.6 −44.8	DWARF (SOME RESOLUTION)	20 D 1.8	F435/J1299 (131,333)
AM 1001-250 [1002-2508]	10 01 51 −25 07.3	2.764 −0.291	261.73 23.84	136.8 −44.9	PAIR (incl. S(r)) (#1)	23 D 0.3	F499/J5653 (84,172)
*AM 1002-310	10 02 00 −31 07.4	2.676 −0.291	265.87 19.23	145.2 −44.3	GALAXY w. DIAMETRIC COMP'NS. + IC 2539 = 435-G-34	8 S 4.9	F435/J1299 (131,117)
AM 1002-250 [1001-2507]	10 02 05 −25 08.3	2.765 −0.291	261.79 23.86	136.8 −44.9	PAIR (incl. S(r)) (#2)	23 D 0.4	F499/J5653 (81,171)
AM 1002-281	10 02 09 −28 12.1	2.721 −0.291	263.93 21.52	141.1 −44.6	GALAXY w. APP. COMP'N. NGC 3113 = 435-G-35	8 S 17.0	F435/J1299 (128,273)
AM 1002-411	10 02 19 −41 10.3	2.499 −0.291	272.31 11.33	158.6 −41.9	EDGE-ON w. 6 COMP'NS. [S.E] 316-G-21	8 D 2.6	F316/J1368 (133,114)
AM 1002-671	10 02 22 −67 12.3	1.512 −0.291	287.86 −9.62	187.5 −30.1	DWARF NGC 3136a = 92-G-7	20 S 1.4	F061/J1173 (19,317)
AM 1002-370	10 02 25 −37 05.3	2.577 −0.291	269.79 14.59	153.3 −43.0	DISTURBED SINGLE GALAXY PKS 1002-370 = 374-G-28	16 D 1.0	F316/J1368 (129,333)
AM 1002-292	10 02 42 −29 23.1	2.705 −0.291	264.83 20.68	142.8 −44.4	COMPACT	13 D 0.1	F435/J1299 (122,210)
AM 1002-582	10 02 48 −58 25.3	2.009 −0.291	282.66 −2.52	178.7 −34.8	V.BR. NEBULOSITY 127-EN-4	21	F127/J2252 (216,264)

Identification	R.A. (1950.0) Declination	Precession R.A./Dec.	Gal. Long. Gal. Lat.	S.G.L. S.G.B.	Description Cross-Identification	Coding Size (Min.)	Field/Plate (x, y)
AM 1002-482	10 02 50 −48 29.9	2.334 −0.291	276.80 5.50	167.6 −39.3	DWARF	20 D 5.4	F213/J2269 (154,260)
AM 1003-215	10 03 21 −21 51.3	2.812 −0.292	259.67 26.54	132.2 −44.6	RING w. 2 CONDENSATIONS on RIM (NUCLEUS?)	6 D 0.4	F500/J2983 (338,346)
AM 1003-412	10 03 43 −41 28.3	2.499 −0.292	272.71 11.25	158.9 −41.6	LO.S.BR. DWARF (STRUCTURE) 316-G-22	20 D 1.3	F316/J1368 (119,98)
AM 1003-440	10 03 43 −44 07.0	2.444 −0.292	274.31 9.12	162.2 −40.7	DWARF?/PLANETARY NEBULA ? Lo-4 = 263-PN-2	20,25 D 0.7	F263/J1447 (297,225)
AM 1003-410	10 03 48 −41 04.8	2.508 −0.292	272.48 11.57	158.4 −41.7	N GALAXY or PLANETARY NEBULA	22,25 D 0.1	F316/J1368 (118,119)
AM 1003-435	10 03 50 −43 58.8	2.447 −0.292	274.24 9.24	162.0 −40.8	I/A DBL. (MERGER) 263-IG-3 + 263-IG-1	2 S 0.7	F263/J1447 (297,232)
AM 1003-333	10 03 53 −33 38.5	2.642 −0.292	267.84 17.50	148.5 −43.5	NUCLEUS w. BROAD PECULIAR ARM + IC 2545? = 374-IG-32 COUNTER ARM	10,12 D 0.7	F374/J3862 (136,252)
AM 1004-393	10 04 02 −39 30.1	2.539 −0.292	271.55 12.86	156.3 −42.1	CLOSE PAIR (E + SPIRAL)	23 S 1.1	F316/J1368 (114,203)
AM 1004-673	10 04 02 −67 30.1	1.510 −0.292	288.17 −9.76	187.6 −29.8	SPIRAL w. I/A COMP'N. 62-G-2	1 S 0.3	F061/J1173 (12,300)
AM 1004-722	10 04 08 −72 26.4	1.028 −0.292	291.19 −13.71	192.2 −26.9	CLOSE PAIR (E + SPIRAL)	23 S 0.4	F061/J1173 (47,37)
AM 1004-294	10 04 17 −29 41.4	2.705 −0.292	265.32 20.66	143.1 −44.0	BULGING COMPACT NGC 3125 = 435-G-41 = To 1004-296	13 D 0.9	F435/J1299 (104,193)
AM 1004-424	10 04 54 −42 49.4	2.477 −0.293	273.71 10.29	160.5 −41.0	GALAXY + APP. COMP'N. 263-G-6/7 = Fa 426	8 S 1.1	F316/J1368 (109,25)
AM 1004-401	10 04 57 −40 11.4	2.530 −0.293	272.12 12.41	157.1 −41.7	PLANETARY w. DIFFRACTION SPIKES NGC 3132 = PK272+12.1 = 316-PN-27	25 D 1.4	F316/J1368 (106,166)
AM 1005-260	10 05 02 −26 09.4	2.758 −0.293	263.06 23.50	138.2 −44.1	I/A QUARTET	4 D 4.7	F499/J5653 (46,116)
AM 1005-394	10 05 11 −39 42.1	2.540 −0.293	271.86 12.83	156.5 −41.8	EDGE-ON or JETS 316-G-28	7 S 0.3	F316/J1368 (103,192)
AM 1005-280	10 05 27 −28 00.0	2.733 −0.293	264.40 22.13	140.7 −43.9	COMPACT	13 D 0.4	F435/J1299 (88,284)
*AM 1005-345	10 05 43 −34 59.0	2.626 −0.293	269.01 16.67	150.2 −42.8	SB + I/A COMP'N. IC 2548 = 374-G-37	1,11 S 3.2	F374/J3862 (116,180)
AM 1005-571	10 05 57 −57 18.3	2.080 −0.293	282.34 −1.36	177.2 −35.0	REFLECTION NEBULA?	21 D 1.0	F167/J3872 (92,54)
AM 1006-404	10 06 11 −40 44.4	2.525 −0.294	272.65 12.11	157.7 −41.4	DWARF or OBSCURED SPIRAL	20 D 0.4	F316/J1368 (94,136)
AM 1006-263	10 06 50 −26 30.1	2.757 −0.294	263.64 23.49	138.6 −43.7	SMALL SPIRAL + 3 CLOSE COMP'NS.	8 S 0.9	F500/J2983 (289,98)
*AM 1006-284	10 06 51 −28 49.3	2.725 −0.294	265.21 21.68	141.8 −43.5	Sc w. I/A COMP'NS. off MI.AX. NGC 3137 = 435-G-47 = PKS 1006-286?	1 S 5.4	F435/J1299 (73,239)
AM 1006-404	10 06 52 −40 42.8	2.528 −0.294	272.74 12.21	157.7 −41.2	SMALL HI.S.BR. DWARF	20 D 0.3	F316/J1368 (87,138)
*AM 1006-380	10 06 55 −38 09.7	2.576 −0.294	271.19 14.27	154.4 −41.9	DISTURBED RING + E COMP'N. 316-IG-32	6,23 S 0.9	F316/J1368 (83,274)
AM 1007-664	10 07 30 −66 46.9	1.605 −0.295	288.02 −8.98	186.7 −29.9	VERY DISTORTED GALAXY in TRIPLE NGC 3136b = 92-G-13 = Se 72/1	16,24 D 14.1	F062/J2049 (268,349)
AM 1007-394	10 07 33 −39 41.4	2.551 −0.295	272.22 13.11	156.3 −41.4	GALAXY w. I/A COMP'N. 316-G-34	1 S 0.4	F316/J1368 (78,192)
AM 1007-375 [1008-3753]	10 07 59 −37 53.2	2.585 −0.295	271.20 14.62	153.9 −41.8	PAIR of DISTURBED SP. (#1: SP?) 316-G-38	16,23 S 5.6	F374/J3862 (95,23)
AM 1008-374 [1007-3753]	10 08 12 −37 47.2	2.588 −0.295	271.17 14.72	153.8 −41.7	PAIR of DISTURBED SP. (#2: SP.) 316-G-40	16,23 S 5.6	F374/J3862 (92,29)
AM 1008-253	10 08 16 −25 34.4	2.774 −0.295	263.27 24.41	137.3 −43.4	LO.S.BR. SPIRAL + HI.S.BR. COMP'N MCG-4-24-19 = 500-G-6	8 S 15.2	F500/J2983 (273,148)
AM 1008-394	10 08 17 −39 45.9	2.552 −0.295	272.39 13.14	156.3 −41.2	[N.S.]SHRED (EDGE-ON PEC.SP?) 316-G-41	22 D 0.8	F316/J1368 (71,188)
AM 1008-385	10 08 26 −38 53.7	2.569 −0.295	271.89 13.85	155.2 −41.4	SB(r) + FAINT RING + MATERIAL 316-G-42	12 D 0.5	F316/J1368 (68,235)
*AM 1008-392	10 08 56 −39 26.8	2.561 −0.296	272.30 13.47	155.9 −41.2	SP. + RING + PEC. ABS'N. + COMP'N. 316-G-44	6,8,14 D 1.0	F316/J1368 (64,205)
AM 1009-371	10 09 23 −37 11.2	2.603 −0.296	271.00 15.34	152.9 −41.7	DWARF	20 D 0.4	F316/J1368 (55,326)
AM 1009-393	10 09 24 −39 37.8	2.560 −0.296	272.48 13.37	156.1 −41.1	CLOSE PAIR	23 S 0.9	F316/J1368 (59,195)
AM 1009-375	10 09 35 −37 51.0	2.592 −0.296	271.44 14.83	153.8 −41.5	DWARF	20 D 0.4	F316/J1368 (54,290)
AM 1009-803	10 09 54 −80 36.5	−0.660 −0.296	296.62 −20.03	199.1 −21.6	PLANETARY NEBULA NGC 3195 = PK 296-20-1 = 19-PN-2	25 D 1.0	F019/J2028 (223,145)
*AM 1010-342	10 10 24 −34 28.9	2.651 −0.297	269.50 17.65	149.3 −42.0	SPIRAL w. 3 APP. COMP'NS. IC 2556 = 374-G-42	8 S 2.5	F374/J3862 (64,205)

Identification	R.A. (1950.0) Declination	Precession R.A./Dec.	Gal. Long. Gal. Lat.	S.G.L. S.G.B.	Description Cross-Identification	Coding Size (Min.)	Field/Plate (x, y)
AM 1010-262 [1010-2634]	10 10 28 −26 29.7	2.767 −0.297	264.33 24.00	138.6 −42.9	I/A DBL. w. TAILS (#1)	2,15 D 0.5	F500/J2983 (245,99)
AM 1010-445	10 10 29 −44 59.3	2.459 −0.297	275.82 9.12	162.7 −39.3	ASYMMETRIC SPIRAL in WIDE PAIR 263-IG-16	23 S 17.9	F263/J1447 (232,180)
AM 1010-263 [1010-2629]	10 10 38 −26 34.7	2.766 −0.297	264.41 23.96	138.7 −42.9	I/A DBL. w. TAILS (#2) 500-IG-11	2,15 D 0.6	F500/J2983 (243,94)
AM 1011-354	10 11 07 −35 44.1	2.633 −0.297	270.40 16.72	150.9 −41.6	PAIR 374-G-44	23 S 3.7	F374/J3862 (58,138)
AM 1011-432 [1011-4334]	10 11 25 −43 28.1	2.495 −0.297	275.07 10.46	160.8 −39.6	CLOSE PAIR (#1: EDGE-ON SPIRAL) 263-G-18	23 S 5.4	F263/J1447 (224,261)
AM 1011-433 [1011-4328]	10 11 25 −43 34.2	2.493 −0.297	275.13 10.38	160.9 −39.6	CLOSE PAIR (#2: Sc) 263-G-19	23 S 5.4	F263/J1447 (224,256)
AM 1011-343	10 11 27 −34 36.5	2.653 −0.297	269.76 17.67	149.4 −41.8	I/A DBL. (E + COMPACT) 374-IG-45	2 S 0.5	F375/J1179 (315,196)
AM 1011-345	10 11 48 −34 53.3	2.649 −0.297	269.99 17.49	149.8 −41.6	PAIR of CONTACT E's in GROUP 374-G-46	2,18 S 0.3	F374/J3862 (49,183)
AM 1011-284	10 11 52 −28 41.6	2.741 −0.298	266.06 22.46	141.5 −42.4	RING GALAXY	6 D 0.5	F435/J1299 (13,245)
AM 1012-272	10 12 17 −27 26.3	2.759 −0.298	265.31 23.51	139.8 −42.4	SPIRAL + APP. COMP'N. NGC 3173 = 500-G-16	8 S 9.9	F500/J2983 (223,48)
AM 1012-283	10 12 23 −28 37.5	2.743 −0.298	266.11 22.58	141.4 −42.3	SPIRAL w. APP. COMP'N. NGC 3175 = 436-G-3	8 S 5.4	F435/J1299 (7,248)
AM 1012-443	10 12 35 −44 35.9	2.478 −0.298	275.90 9.65	162.1 −39.1	DWARF IRR. 263-G-21	20 D 1.1	F263/J1447 (212,201)
AM 1012-375	10 12 36 −37 56.3	2.603 −0.298	271.99 15.11	153.7 −40.9	DISTURBED (ISOLATED) GALAXY	16 D 1.3	F316/J1368 (22,285)
AM 1013-271	10 13 05 −27 10.3	2.765 −0.298	265.28 23.83	139.4 −42.3	CLOSE PAIR (incl. BULLS-EYE) 500-G-19	6,23 S 11.2	F436/J0530 (271,329)
AM 1013-335	10 13 40 −33 51.9	2.672 −0.299	269.69 18.54	148.4 −41.4	DWARF 375-G-3	20 D 0.8	F375/J1179 (292,236)
AM 1013-411	10 13 53 −41 18.3	2.549 −0.299	274.19 12.49	157.9 −39.8	GALAXY w. I/A COMP'N. 317-IG-7	1 S 0.3	F317/J1363 (278,105)
AM 1013-394	10 13 57 −39 44.3	2.577 −0.299	273.28 13.79	156.0 −40.2	DWARF	20 D 0.5	F317/J1363 (280,190)
AM 1014-312	10 14 53 −31 24.3	2.712 −0.299	268.37 20.68	145.1 −41.5	COMPACT 436-G-8	13 D 0.3	F436/J0530 (247,102)
AM 1014-331	10 14 54 −33 10.7	2.686 −0.299	269.48 19.25	147.4 −41.3	DWARF (+ STAR?)	20 D 0.4	F375/J1179 (279,274)
AM 1014-483	10 14 55 −48 38.0	2.401 −0.300	278.53 6.54	166.8 −37.4	ISOLATED G.D. SPIRAL 213-G-11 = UKS 1014-486?	22 D 4.5	F213/J2269 (47,250)
AM 1015-334	10 15 36 −33 40.4	2.681 −0.300	269.92 18.93	148.0 −41.1	I/A DBL. 375-IG-6	2 S 0.6	F375/J1179 (270,247)
AM 1016-332	10 16 12 −33 28.4	2.686 −0.300	269.90 19.17	147.7 −41.0	DWARF	20 D 0.5	F375/J1179 (264,258)
AM 1016-622	10 16 14 −62 25.4	1.954 −0.300	286.31 −4.86	181.8 −31.4	P.N. w. FAINT OUTER RING NGC 3211 = PK 286-4-1 = 127-PN-15	25	F127/J2252 (128,49)
AM 1016-370	10 16 45 −37 08.8	2.632 −0.301	272.22 16.23	152.5 −40.2	GALAXY w. I/A COMP'N.	1 S 0.6	F317/J1363 (254,329)
AM 1016-264	10 16 47 −26 44.1	2.780 −0.301	265.71 24.68	138.8 −41.5	SPIRAL + COMPACT COMP'N. MCG-4-25-1 = 500-G-22	8,13 S 0.4	F500/J2983 (169,87)
AM 1016-405	10 16 47 −40 57.4	2.568 −0.301	274.45 13.09	157.3 −39.4	COMPACT GALAXY	13 D 0.4	F317/J1363 (249,125)
AM 1017-733	10 17 08 −73 34.8	1.106 −0.301	292.67 −14.08	192.6 −25.4	GROUP of 4	18 D 4.9	F038/J2058 (341,240)
AM 1017-262	10 17 14 −26 26.8	2.785 −0.301	265.61 24.97	138.4 −41.4	EDGE-ON + 3 COMP'NS (DBL.I/A) NGC 3202 = 500-G-24	2,8 S 2.9	F500/J2983 (164,102)
AM 1017-253	10 17 22 −25 33.9	2.796 −0.301	265.05 25.69	137.2 −41.4	MULTI-ARMED SPIRAL + APP. COMP'N NGC 3208 = 500-G-25	8,10 S 5.4	F500/J2983 (162,149)
AM 1017-271	10 17 26 −27 12.5	2.776 −0.301	266.15 24.39	139.4 −41.3	SMALL M51-TYPE (I/A SPIRALS ?) 500-IG-26	2,9 S 0.4	F500/J2983 (162,61)
AM 1017-411	10 17 45 −41 19.9	2.566 −0.301	274.82 12.88	157.7 −39.1	GALAXY w. I/A COMP'N. (LO.S.BR.)	1 S 1.2	F317/J1363 (239,105)
*AM 1018-283	10 18 03 −28 37.3	2.759 −0.301	267.19 23.33	141.3 −41.1	I/A DBL. SPIRALS 436-IG-11	2 S 0.9	F436/J0530 (212,252)
AM 1019-220	10 19 35 −22 01.2	2.844 −0.302	263.07 28.81	132.5 −40.9	SB w. 2 DIAMETRIC COMP'NS. NGC 3233 = 568-G-1	8 S 2.0	F500/J2983 (135,340)
AM 1019-705	10 19 53 −70 50.2	1.447 −0.303	291.29 −11.69	189.9 −26.8	CLOSE I/A DOUBLE	2 S 0.1	F062/J2049 (198,134)
AM 1020-332	10 20 31 −33 28.3	2.701 −0.303	270.68 19.69	147.6 −40.1	CLOSE PAIR	23 S 0.7	F375/J1179 (216,259)
AM 1021-222	10 21 17 −22 25.8	2.843 −0.303	263.71 28.72	133.1 −40.5	LUMPY COMPACT	13 D 0.2	F500/J2983 (114,318)

Identification	R.A. (1950.0) Declination	Precession R.A./Dec.	Gal. Long. Gal. Lat.	S.G.L. S.G.B.	Description Cross-Identification	Coding Size (Min.)	Field/Plate (x, y)
AM 1021-291	10 21 19 −29 13.3	2.761 −0.304	268.20 23.26	142.0 −40.3	GALAXY w. JETS (in LINE of 3)	7,24 D 0.8	F436/J0530 (173,219)
AM 1021-285	10 21 21 −28 58.2	2.764 −0.304	268.05 23.47	141.7 −40.4	I/A DBL. in GROUP of 3 To 1021-289	2,24 S 0.4	F436/J0530 (173,233)
AM 1021-282	10 21 57 −28 28.6	2.772 −0.304	267.86 23.95	141.0 −40.3	I/A DBL. or DISTURBED SINGLE? 436-IG-15	2 D 0.9	F436/J0530 (166,259)
AM 1022-330	10 22 07 −33 05.3	2.712 −0.304	270.74 20.19	147.0 −39.8	DWARF 375-G-19	20 D 0.7	F375/J1179 (198,279)
AM 1022-261	10 22 29 −26 19.7	2.800 −0.304	266.58 25.76	138.2 −40.2	CLUSTER of 7 GALAXIES + FAINTER MEMBERS	19 D 4.0	F500/J2983 (101,108)
AM 1022-354	10 22 51 −35 43.8	2.678 −0.304	272.44 18.09	150.4 −39.3	DWARF	20 D 0.5	F375/J1179 (189,138)
*AM 1023-291 [1023-2920]	10 23 23 −29 16.7	2.767 −0.305	268.64 23.48	142.0 −39.9	GALAXY w. I/A COMP'NS. (in PAIR) 436-G-18	1,2,23 S 0.6	F436/J0530 (149,216)
*AM 1023-292 [1023-2916]	10 23 29 −29 20.8	2.766 −0.305	268.70 23.44	142.1 −39.9	SPIRAL w. I/A COMP'NS. (in PAIR) 436-G-19	1,2,23 S 1.6	F436/J0530 (148,213)
AM 1023-323	10 23 42 −32 39.2	2.724 −0.305	270.77 20.74	146.4 −39.5	SPIRAL w. COMPLEX NUCLEUS IC 2576 = 375-G-23	12,22 D 0.9	F375/J1179 (180,303)
AM 1023-432	10 23 42 −43 29.8	2.556 −0.305	276.98 11.67	160.0 −37.5	LATE-TYPE SP. w. OFF-SET NUCLEUS NGC3256a = Se 77/4 = 263-G-34	16 D 1.3	F263/J1447 (104,259)
AM 1024-363	10 24 29 −36 36.4	2.671 −0.305	273.24 17.54	151.4 −38.8	DWARF 375-G-25	20 D 0.7	F375/J1179 (171,91)
AM 1024-234	10 24 33 −23 49.9	2.834 −0.305	265.35 28.05	134.9 −39.8	SPIRAL w. FAINT OUTER RING MCG-4-25-8 = 500-G-41	6 D 1.3	F501/J2987 (344,239)
*AM 1024-342	10 24 35 −34 24.9	2.703 −0.305	271.98 19.38	148.6 −39.1	E w. JETS + APP. COMP'NS.	7,8 D 0.7	F375/J1179 (170,208)
AM 1024-283	10 24 46 −28 35.6	2.779 −0.306	268.48 24.21	141.1 −39.6	GALAXY w. LOOP	15,16 D 0.4	F436/J0530 (133,253)
AM 1024-330	10 24 51 −33 02.9	2.722 −0.306	271.22 20.55	146.9 −39.2	GAL. w. CONDENS. at END of PLUME 375-G-27	9,15 D 0.6	F375/J1179 (167,282)
AM 1024-485	10 24 51 −48 53.3	2.456 −0.306	280.05 7.21	166.4 −35.7	CHAIN of 4 214-IG-5	17 D 5.8	F214/J2278 (225,238)
AM 1024-323	10 24 54 −32 38.4	2.728 −0.306	270.99 20.89	146.4 −39.3	DWARF (PECULIAR STRUCTURE)	20 D 0.6	F375/J1179 (166,304)
AM 1025-370	10 25 00 −37 09.6	2.665 −0.306	273.65 17.14	152.1 −38.6	KNOTTY RING (NO NUCLEUS ?)	6 D 0.4	F375/J1179 (166,61)
AM 1025-333	10 25 14 −33 35.1	2.717 −0.306	271.61 20.15	147.6 −39.1	DWARF	20 D 0.5	F375/J1179 (163,253)
AM 1025-293	10 25 20 −29 32.5	2.769 −0.306	269.18 23.50	142.4 −39.5	LO.S.BR. GAL. w. HI.S.BR. COMP'N. 436-G-23	8 S 0.9	F436/J0530 (126,202)
*AM 1025-401	10 25 33 −40 10.7	2.621 −0.306	275.44 14.66	155.9 −37.9	"ASTERISK" ("JETS" in DISK) NGC 3250b = 317-G-29	7 D 2.0	F317/J1363 (160,167)
*AM 1025-433	10 25 44 −43 39.0	2.563 −0.306	277.38 11.74	160.1 −37.1	I/A DBL. w. TAIL (MERGER) [PKS] NGC3256 = VV 65 = Se77/5 = 263-IG-38	2,15 D 3.6	F263/J1447 (85,250)
AM 1025-335	10 25 49 −33 52.0	2.715 −0.306	271.89 19.98	147.9 −38.9	I/A DBL. + MATERIAL	2 D 0.7	F375/J1179 (156,238)
AM 1026-304	10 26 25 −30 46.9	2.757 −0.307	270.16 22.61	143.9 −39.1	GALAXY w. HI.S.BR. BAR 436-G-26	16 D 0.5	F436/J0530 (115,135)
AM 1026-312	10 26 36 −31 21.3	2.750 −0.307	270.54 22.16	144.7 −39.0	DISTURBED SPIRAL in PAIR + COMP'N IC 2580? = 436-G-27	1,8,23 D 3.1	F436/J0530 (113,104)
*AM 1026-442	10 26 53 −44 24.1	2.556 −0.307	277.96 11.21	160.9 −36.7	SB(rs)c w. BR. ARM + 4 I/A COMP'NS NGC 3261 = Se 77/7 = 263-G-40	1,10 S 2.9	F263/J1447 (75,209)
AM 1026-433	10 26 57 −43 35.5	2.571 −0.307	277.54 11.91	160.0 −36.9	DISRUPTED SPIRAL NGC 3256c = 263-G-41 = Fa 431	2 D 1.3	F263/J1447 (73,253)
*AM 1027-435	10 27 05 −43 52.3	2.566 −0.307	277.71 11.68	160.3 −36.8	SPIRAL w. STRAIGHT ARM + COMP'N. NGC 3263 in Se 77/6 = 263-IG-43	8,15 S 3.1	F263/J1447 (72,238)
AM 1027-393	10 27 21 −39 34.8	2.639 −0.307	275.41 15.35	155.0 −37.7	HI.S.BR. GALAXY w. LO.S.BR. ARMS 317-G-36	12 D 1.3	F317/J1363 (142,199)
AM 1027-380	10 27 29 −38 05.4	2.662 −0.307	274.61 16.62	153.2 −38.0	RING GALAXY ? w. STRONG NUCLEUS 317-G-38	6 D 1.2	F375/J1179 (140,11)
AM 1027-350	10 27 44 −35 04.2	2.706 −0.307	272.94 19.19	149.4 −38.4	CHAIN of 17 (DN 1027-353) incl. NGC 3267/68/69/71/73	17 D 4.5	F375/J1179 (136,173)
AM 1028-344	10 28 03 −34 40.9	2.712 −0.307	272.77 19.55	148.9 −38.4	V.LO.S.BR. DWARF	20 D 0.4	F375/J1179 (132,194)
AM 1028-460	10 28 24 −46 00.6	2.534 −0.308	279.05 9.98	162.8 −36.0	NUCLEATED DWARF 263-G-47	20 D 1.1	F263/J1447 (64,123)
*AM 1028-362	10 28 37 −36 28.7	2.690 −0.308	273.90 18.10	151.1 −38.0	SB + THIN ARM NGC 3275 = 375-G-50	10,12 D 1.6	F375/J1179 (127,97)
AM 1028-364	10 28 47 −36 43.0	2.687 −0.308	274.06 17.92	151.4 −37.9	GALAXY w. EXTENDED MATERIAL 375-G-52	15 D 0.8	F375/J1179 (125,84)
AM 1029-455	10 29 04 −45 59.8	2.538 −0.308	279.14 10.05	162.7 −35.9	S0(TILTED ABS'N) + LINEAR COMPACT Se 77/8 = 263-G-48	8,13,14 S 2.3	F263/J1447 (57,123)

Identification	R.A. (1950.0) Declination	Precession R.A./Dec.	Gal. Long. Gal. Lat.	S.G.L. S.G.B.	Description Cross-Identification	Coding Size (Min.)	Field/Plate (x, y)
AM 1029-294	10 29 07 −29 41.7	2.779 −0.308	270.02 23.84	142.5 −38.6	SPIRAL w. DISTURBED COMP'N. 436-G-31	8 S 3.6	F436/J0530 (82,193)
AM 1029-375	10 29 22 −37 51.0	2.673 −0.308	274.80 17.02	152.8 −37.6	I/A DBL.	2 S 0.2	F317/J1363 (119,292)
AM 1029-300	10 29 31 −30 07.6	2.775 −0.308	270.36 23.53	143.0 −38.5	CHAIN of 5 IC 2588 = 436-G-33	17 D 6.7	F436/J0530 (78,170)
*AM 1029-343	10 29 36 −34 35.8	2.719 −0.308	273.01 19.79	148.7 −38.1	TIGHTLY WOUND SPIRAL w. PEC. ARM NGC 3281 = 375-G-55	10 D 3.1	F375/J1179 (115,198)
*AM 1030-271 [1030-2715]	10 30 07 −27 16.4	2.810 −0.309	268.73 25.97	139.4 −38.5	CLOSE TRIPLE incl. COMPACT (#1) MCG-4-25-14 = 501-G-5	13,24 D 0.3	F436/J0530 (68,323)
*AM 1030-271 [1030-2716]	10 30 21 −27 16.0	2.811 −0.309	268.77 26.01	139.4 −38.4	CLOSE TRIPLE incl. COMPACT (#2) MCG-4-25-14 = 501-G-7	13,24 D 0.8	F436/J0530 (65,323)
*AM 1030-271 [1030-2716]	10 30 28 −27 15.7	2.811 −0.309	268.79 26.02	139.4 −38.4	CLOSE TRIPLE incl. COMPACT (#3) NGC 3285a = MCG-4-25-15 = 501-G-8	13,24 D 0.8	F436/J0530 (63,323)
AM 1031-265	10 31 02 −26 56.4	2.816 −0.309	268.71 26.36	139.0 −38.3	PECULIAR SPIRAL w. I/A COMP'N. 501-G-12	1 S 1.2	F436/J0530 (56,341)
AM 1031-602	10 31 10 −60 29.2	2.181 −0.309	286.80 −2.27	178.9 −30.6	NEBULOSITY	21	F127/J2252 (26,147)
*AM 1031-271	10 31 14 −27 11.8	2.814 −0.309	268.91 26.17	139.3 −38.2	DISTURBED SPIRAL w. DIAMETRICS NGC 3285 = 501-G-15 (COMPACTS)	8,12,13,14 D 13.4	F501/J2987 (259,61)
AM 1032-315	10 32 02 −31 55.7	2.761 −0.310	271.93 22.31	145.3 −37.8	DISTURBED SPIRAL MCG-5-25-15 = 436-G-40	12,16 D 1.2	F375/J1179 (85,341)
AM 1032-285	10 32 11 −28 55.5	2.797 −0.310	270.17 24.85	141.5 −38.0	PLANETARY NEBULA Ko-28 = 436-PN-41	25 D 1.0	F436/J0530 (46,234)
AM 1032-272	10 32 16 −27 23.6	2.814 −0.310	269.25 26.14	139.5 −38.0	SP.w.THIN ARMS + I/A COMP'NS. NGC 3285b = MCG-4-25-22= 501-G-18	1,10 S 30.2	F437/J0515 (317,315)
*AM 1032-281	10 32 29 −28 19.5	2.805 −0.310	269.86 25.39	140.7 −37.9	THIN-ARMED SPIRAL + 6 COMP'NS. MCG-5-25-17=436-G-46= To 1032-283	2,8,18 S 145.6	F437/J0515 (313,265)
AM 1032-425	10 32 29 −42 58.3	2.609 −0.310	278.09 12.96	158.9 −36.1	DWARF Se 77/2 = 264-G-6	20 D 0.5	F317/J1363 (93,16)
AM 1032-274	10 32 38 −27 49.1	2.811 −0.310	269.58 25.83	140.1 −37.9	SP.w. PEC.BR. ARM + APP. COMP'NS. MCG-5-25-18 = 437-G-2	8,10 D 24.6	F437/J0515 (312,292)
AM 1032-291	10 32 42 −29 14.9	2.795 −0.310	270.46 24.64	141.9 −37.9	IRR.	16 D 0.4	F436/J0530 (40,216)
AM 1033-242	10 33 00 −24 29.5	2.847 −0.310	267.57 28.63	135.8 −37.9	LO.S.BR.(SP.+COMP'N.)+ DIAMETRICS	8,20 S 9.0	F501/J2987 (240,206)
AM 1033-363	10 33 01 −36 37.1	2.705 −0.310	274.77 18.45	151.1 −37.1	LARGE COMPACT 375-G-69	13 D 0.6	F375/J1179 (79,89)
*AM 1033-280	10 33 02 −28 03.4	2.809 −0.310	269.81 25.68	140.4 −37.8	SP. w. THIN ARMS + 5 I/A COMP'NS. MCG-5-25-19 = 437-G-4	8,10 S 190.4	F437/J0515 (307,280)
AM 1033-320	10 33 28 −32 06.0	2.764 −0.310	272.30 22.33	145.5 −37.5	EDGE-ON SP. w. THIN ABS'N (PAIR) NGC 3302 = 437-G-7	8,14,23 D 33.6	F437/J0515 (297,62)
*AM 1033-280	10 33 31 −28 01.8	2.811 −0.311	269.89 25.76	140.3 −37.7	CHAIN of 6 (SPINDLE) KLEM 13 = DN 1033-282	17 D 22.4	F501/J2987 (232,16)
*AM 1033-365	10 33 55 −36 58.6	2.704 −0.311	275.12 18.24	151.5 −36.9	LARGE DWARF (RESOLVED) Ka-30 = 375-G-71 = UKS 1033-369	20 D 2.2	F376/J5018 (330,70)
AM 1034-401	10 34 05 −40 18.9	2.658 −0.311	276.95 15.39	155.6 −36.3	COMPACT GALAXY (LUMPY)	13,16 D 0.2	F317/J1363 (73,158)
AM 1034-271	10 34 14 −27 15.4	2.821 −0.311	269.57 26.49	139.3 −37.6	CLUSTER (CORE=DBL. E) DN 1034-272 NGC 3309 = (ABELL1060) = 501-G-36	19 S 2.0	F436/J0530 (19,323)
AM 1034-271	10 34 21 −27 16.2	2.822 −0.311	269.60 26.50	139.4 −37.6	CLUSTER (DISTURBED SP.,M82-TYPE) NGC 3311 = 501-G-38 (PEC. DUST)	14,16,19 D 17.9	F501/J2987 (222,57)
AM 1034-324	10 34 27 −32 47.4	2.759 −0.311	272.88 21.86	146.3 −37.3	STAR + NEBULOSITY	21 D 0.6	F375/J1179 (59,294)
AM 1034-273	10 34 30 −27 39.7	2.818 −0.311	269.87 26.19	139.8 −37.5	CLOSE PAIR MCG-5-25-22/24 = 437-G-11/13	23 S 0.7	F501/J2987 (220,36)
AM 1034-271	10 34 41 −27 18.5	2.822 −0.311	269.69 26.51	139.4 −37.5	LARGE DISTURBED SP. w. PEC. DUST NGC 3312 = IC 629 = 501-G-43	12,14 D 89.6	F437/J0515 (288,320)
AM 1034-272	10 34 52 −27 25.2	2.821 −0.311	269.80 26.43	139.5 −37.4	SPINDLE GALAXY = PROJECTION? NGC 3314 = 501-IG-46	22 D 1.8	F436/J0530 (11,314)
AM 1034-371	10 34 52 −37 10.2	2.705 −0.311	275.40 18.17	151.7 −36.7	DISTURBED GALAXY	16 D 0.4	F375/J1179 (60,59)
*AM 1034-423	10 34 58 −42 35.4	2.628 −0.311	278.29 13.52	158.3 −35.7	CHAIN of 4 (I/A DBL. CENTRAL) 264-IG-13	2,17 D 9.6	F318/J1380 (326,35)
AM 1035-271	10 35 00 −27 17.6	2.823 −0.311	269.75 26.56	139.4 −37.4	PAIR (KNOTTY IRR. + 'V'-SHAPED) 501-G-49	16,23 S 28.0	F437/J0515 (285,321)
AM 1035-412 [1035-4112]	10 35 03 −41 21.9	2.647 −0.311	277.67 14.58	156.9 −35.9	PAIR of SPIRALS + COMP'N. (#1) NGC 3318 = 317-G-52	8,23 D 1.8	F317/J1363 (65,101)
*AM 1035-250	10 35 06 −25 04.0	2.847 −0.311	268.38 28.42	136.6 −37.4	SB + 5 COMP'NS (E+ SP+ PECULIAR) NGC 3313 = MCG-4-25-44 = 501-G-5	8,23 S 13.4	F501/J2987 (214,176)
*AM 1035-260	10 35 07 −26 03.4	2.837 −0.311	269.01 27.60	137.8 −37.4	SP. w. DISTURBED ARMS + 6 COMP'NS MCG-4-25-45 = 501-G-51 (+COMPACT)	8,10,13 S 5.8	F501/J2987 (214,122)

Identification	R.A. (1950.0) Declination	Precession R.A./Dec.	Gal. Long. Gal. Lat.	S.G.L. S.G.B.	Description Cross-Identification	Coding Size (Min.)	Field/Plate (x, y)
AM 1035-271	10 35 16 −27 19.9	2.824 −0.311	269.83 26.56	139.4 −37.4	GALAXY w. I/A COMP'N. NGC 3316 = 501-G-54	1 S 10.1	F437/J0515 (281,319)
AM 1035-270	10 35 17 −27 07.8	2.826 −0.311	269.71 26.73	139.2 −37.4	SMALL SP. w. PEC. ABS'N. in DISK MCG-4-25-47 = 501-G-52	14 D 22.4	F437/J0515 (281,330)
*AM 1035-411 [1035-4121]	10 35 17 −41 12.1	2.650 −0.311	277.62 14.75	156.6 −35.9	PAIR of SPIRALS + COMP'N.(#2) NGC 3318b = 317-G-53	8,23 D 1.0	F317/J1363 (62,110)
AM 1035-264	10 35 25 −26 49.3	2.829 −0.312	269.55 27.00	138.8 −37.3	E w. 3 APP. COMP'NS. IC 2597 = 501-G-58 = PKS 1035-268	8 S 2.2	F501/J2987 (210,81)
AM 1035-243	10 35 52 −24 39.7	2.853 −0.312	268.29 28.86	136.1 −37.2	HI.S.BR. IRR. near CLUSTER of E's 501-IG-64	16,19 D 0.4	F501/J2987 (205,197)
AM 1035-375	10 35 58 −37 50.0	2.701 −0.312	275.96 17.72	152.5 −36.4	DWARF SPIRAL w. I/A COMP'N. 317-G-54 = UKS 1035-378	1,20 S 1.9	F376/J5018 (306,25)
AM 1036-272	10 36 12 −27 28.6	2.825 −0.312	270.11 26.55	139.6 −37.1	DISTURBED SPIRAL MCG-4-25-52 = 501-G-65	12 D 33.6	F437/J0515 (270,312)
AM 1036-300	10 36 54 −30 02.1	2.799 −0.312	271.78 24.47	142.8 −36.9	EDGE-ON + DISTURBED SP. + RING-SP. MCG-5-25-31 = 437-G-30	6,12,24 D 302.4	F437/J0515 (259,174)
AM 1037-233 [1037-2318]	10 37 09 −23 39.7	2.866 −0.312	267.93 29.85	134.8 −36.9	WIDE GROUP of 3 (#1: S0) NGC 3335 = MCG-4-25-55 = 501-G-7	24 D 8.5	F501/J2987 (190,251)
AM 1037-233 [1037-2339]	10 37 44 −23 33.6	2.868 −0.313	268.00 30.01	134.7 −36.7	WIDE GROUP of 3 (#2: SPIRAL) NGC 3331 = 501-G-72	24 D 8.5	F501/J2987 (183,257)
*AM 1037-404	10 37 47 −40 43.7	2.669 −0.313	277.80 15.39	156.0 −35.5	HI.S.BR. SP. + NEAR-BY "STAR"	12 D 0.2	F318/J1380 (301,136)
AM 1037-231 [1037-2333]	10 37 54 −23 18.6	2.871 −0.313	267.87 30.24	134.4 −36.7	WIDE GROUP of 3 (#3: EDGE-ON)	24 D 8.5	F501/J2987 (181,270)
AM 1037-273	10 37 57 −27 30.8	2.829 −0.313	270.50 26.73	139.6 −36.8	SPIRAL + 15 COMP'NS. (CLOSE PAIR) NGC 3336 = MCG-5-25-36 = 437-G-36	8,23 D 56.0	F437/J0515 (249,310)
AM 1038-360	10 38 03 −36 08.9	2.731 −0.313	275.43 19.38	150.4 −36.2	IRR./Sd w. 3 APP. COMP'NS. NGC 3347a = 376-G-4 (2 DWARFS)	8,16,20 S 4.3	F376/J5018 (287,116)
*AM 1038-465	10 38 03 −46 50.5	2.574 −0.313	280.93 10.10	163.2 −34.2	NO NUCLEUS INTEGRAL + DWARF 264-G-29	12,20 D 1.3	F264/J2980 (234,80)
AM 1038-460	10 38 04 −46 04.3	2.588 −0.313	280.55 10.77	162.3 −34.4	TRIPLE E in CLUSTER of 24 264-IG-30 = KLEM 15 = DN 1038-459	3,19 D 0.8	F264/J2980 (234,121)
AM 1038-273	10 38 10 −27 36.6	2.829 −0.313	270.60 26.67	139.8 −36.7	CLOSE PAIR (END of CHAIN) incl. To 1039-275?	17,23 S 53.8	F437/J0515 (247,305)
AM 1038-481	10 38 11 −48 18.6	2.549 −0.313	281.68 8.83	164.9 −33.8	OBSCURED SB w. APP. COMP'N. 214-G-17 = UKS 1038-483	8 D 2.5	F214/J2278 (106,269)
AM 1039-230	10 39 02 −23 07.2	2.876 −0.313	268.00 30.54	134.2 −36.4	LO.S.BR. SPIRAL + APP. COMP'NS. MCG-4-25-58 = NGC 3355? = 501-G-79	8,13,20 S 3.1	F501/J2987 (167,280)
AM 1039-313	10 39 05 −31 32.5	2.790 −0.313	273.08 23.44	144.7 −36.4	CLOSE PAIR of PECULIAR SPIRALS 376-G-6? = KLEM 16 = DN 1039-361	2,12,23 S 75.0	F437/J0515 (233,94)
AM 1039-365	10 39 27 −36 52.9	2.728 −0.314	276.08 18.89	151.2 −35.8	SUB-GROUP of 3 in GROUP of 4	18,24 D 10.3	F376/J5018 (271,77)
*AM 1039-283	10 39 33 −28 31.0	2.823 −0.314	271.43 26.07	140.9 −36.4	I/A TRIPLE + PAIR (PEC. SP.) 437-G-44 ?	3,10,23 D 201.6	F437/J0515 (230,256)
AM 1039-325	10 39 44 −32 59.0	2.776 −0.314	274.02 22.27	146.4 −36.1	EDGE-ON S0 + APP. COMP'N. MCG-5-25-39 = 376-G-9	8 S 34.7	F437/J0515 (225,16)
AM 1039-365	10 39 53 −36 53.5	2.730 −0.314	276.16 18.92	151.2 −35.7	DWARF (RESOLVED) 376-G-11	20 D 0.9	F375/J1179 (6,72)
AM 1039-625	10 39 57 −62 54.9	2.179 −0.314	288.90 −3.88	181.0 −28.7	NEBULOSITY	22	F093/J2065 (302,286)
AM 1040-281	10 40 22 −28 12.9	2.829 −0.314	271.42 26.42	140.5 −36.2	CLOSE I/A DBL.	2 S 4.5	F437/J0515 (220,272)
AM 1040-250 [1041-2505]	10 40 37 −25 06.8	2.860 −0.314	269.60 29.07	136.7 −36.1	PAIR (#1: E)	23 D 12.3	F501/J2987 (147,173)
AM 1040-472	10 40 42 −47 21.2	2.581 −0.314	281.58 9.87	163.7 −33.6	GALAXY w. APP. COMP'NS. 264-G-35	8 D 1.2	F214/J2278 (82,320)
AM 1040-303	10 40 43 −30 31.2	2.806 −0.314	272.83 24.49	143.4 −36.1	SP. + DIAMETRIC COMP'NS(ARC of 11)	8,18 S 28.0	F437/J0515 (215,149)
AM 1040-360	10 40 46 −36 06.0	2.743 −0.314	275.90 19.69	150.2 −35.6	COMPACT I/A w. SP.(in LINE of 3) NGC 3354 = 376-G-14 = To1040-360?	1,13,24 D 13.4	F376/J5018 (257,120)
AM 1041-250 [1040-2506]	10 41 03 −25 02.9	2.862 −0.314	269.66 29.18	136.6 −36.0	PAIR (#2: SPIRAL) 501-G-83?	23 S 12.3	F501/J2987 (142,177)
AM 1041-250	10 41 20 −25 04.2	2.863 −0.315	269.73 29.19	136.6 −36.0	DWARF (?)	20 D 0.7	F501/J2987 (139,176)
AM 1041-240	10 41 23 −24 06.2	2.872 −0.315	269.15 30.01	135.4 −35.9	SPIRAL w. PECULIAR ARMS MCG-4-26-4 = 501-G-86	10 D 1.6	F501/J2987 (138,227)
AM 1041-222 [1041-2233]	10 41 24 −22 27.8	2.887 −0.315	268.11 31.39	133.4 −35.8	GAL. + I/A COMP'N. in GRP.of 3(#1)	1,24 S 0.4	F501/J2987 (137,316)
AM 1041-223 [1041-2233]	10 41 24 −22 33.3	2.886 −0.315	268.16 31.31	133.5 −35.8	GAL. + I/A COMP'N. in GRP.of 3(#2) MCG-4-26-5 = 501-G-85	1,24 S 0.4	F501/J2987 (137,311)
*AM 1041-395	10 41 46 −39 51.5	2.700 −0.315	278.04 16.52	154.8 −34.9	HI.S.BR. SP. w. 6 APP.COMP'NS. 318-G-6	8 D 0.6	F318/J1380 (262,184)

Identification	R.A. (1950.0) Declination	Precession R.A./Dec.	Gal. Long. Gal. Lat.	S.G.L. S.G.B.	Description Cross-Identification	Coding Size (Min.)	Field/Plate (x, y)
AM 1041-223	10 41 52	2.887	268.28	133.5	GAL. + I/A COMP'N. in GRP. of 3(#3)	1,24	F501/J2987
[1041-2227]	−22 33.7	−0.315	31.37	−35.7	MCG-4-26-6 = 501-G-88	S 0.4	(132,310)
AM 1041-342	10 41 56	2.767	275.24	148.2	I/A TRIPLE	3	F376/J5018
	−34 27.3	−0.315	21.24	−35.6		D 1.8	(246,208)
AM 1042-361	10 42 06	2.747	276.20	150.3	DWARF	20	F376/J5018
	−36 10.9	−0.315	19.76	−35.4	376-G-18	D 0.5	(243,116)
AM 1042-402	10 42 18	2.696	278.38	155.3	GROUP of 4	18	F318/J1380
	−40 20.7	−0.315	16.15	−34.8	incl. 318-G-7/9	D 5.8	(255,158)
AM 1042-361	10 42 53	2.751	276.33	150.3	DWARF	20	F376/J5018
	−36 10.0	−0.315	19.85	−35.2		D 0.4	(234,117)
AM 1043-421	10 43 11	2.674	279.48	157.6	SMALL I/A DBL. in LINE of 3	2,24	F318/J1380
	−42 15.4	−0.315	14.55	−34.3		D 1.3	(245,56)
AM 1043-224	10 43 12	2.889	268.70	133.8	SMALL HI.S.BR. IRR. + JET	7,16	F501/J2987
	−22 44.8	−0.315	31.38	−35.4	501-IG-92	D 0.4	(115,300)
AM 1043-233	10 43 25	2.882	269.23	134.7	SMALL I/A DBL.	2	F501/J2987
	−23 30.9	−0.316	30.76	−35.4		S 0.7	(113,259)
AM 1043-285	10 43 42	2.831	272.57	141.5	COMPACT	13	F437/J0515
	−28 59.9	−0.316	26.13	−35.5		D 7.8	(181,231)
*AM 1044-451	10 44 21	2.636	281.15	161.1	IRR. w. CONDENSATIONS	16	F264/J2980
	−45 16.6	−0.316	11.99	−33.5		D 0.4	(176,165)
AM 1044-241	10 44 55	2.880	269.97	135.6	SPIRAL w. APP. COMP'NS.	8	F501/J2987
	−24 10.3	−0.316	30.39	−35.1	NGC 3383 = 501-G-97	S 2.5	(95,224)
AM 1045-365	10 45 06	2.750	277.16	151.2	BIFURCATED COMPACT	13,16	F318/J1380
	−36 57.8	−0.316	19.37	−34.7		D 0.4	(229,340)
*AM 1045-245	10 45 59	2.876	270.66	136.5	PECULIAR "TRUNCATED" SP. + COMP'N	6,8,10,12	F501/J2987
	−24 53.9	−0.317	29.90	−34.9	NGC 3393 = 501-G-100	S 3.8	(82,184)
AM 1046-720	10 46 11	1.759	293.81	190.1	GALAXY w. JET?/BAR in DISK	7	F062/J2049
	−72 09.9	−0.317	−11.77	−24.3	Fa 285	D 0.8	(88,59)
*AM 1046-310	10 46 52	2.821	274.37	143.9	SB.w.THIN SPLIT ARMS + I/A COMP'N.	1,10	F437/J0515
	−31 02.8	−0.317	24.71	−34.7	MCG-5-26-8 = 437-G-65	S 33.6	(144,121)
AM 1048-624	10 48 32	2.278	289.69	180.4	NEBULOSITY	22	F093/J2065
	−62 44.2	−0.318	−3.27	−27.9			(250,299)
AM 1048-232	10 48 33	2.896	270.32	134.7	PECULIAR SPIRAL w. APP. COMP'N.	8,12	F502/J2984
	−23 23.8	−0.318	31.48	−34.3	MCG-4-26-12 = 501-G-102	S 6.0	(321,264)
AM 1049-362	10 49 13	2.774	277.64	150.4	DWARF	20	F376/J5018
	−36 23.6	−0.318	20.27	−33.9		D 0.5	(166,105)
AM 1049-322	10 49 55	2.818	275.72	145.6	DOUBLE-ARMED SB in LINE of 3	10,24	F437/J0515
	−32 24.3	−0.319	23.84	−34.0	MCG-5-26-9 = 437-G-67	D 89.6	(110,47)
AM 1051-290	10 51 10	2.853	274.22	141.6	DISTURBED S0 + 2 COMP'NS (GROUP)	8,18	F437/J0515
	−29 07.9	−0.319	26.83	−33.8	MCG-5-26-11 = 437-G-69	S 50.4	(93,223)
AM 1051-325	10 51 46	2.820	276.33	146.1	3rd in CHAIN w. AM1049-322	24	F376/J5018
	−32 51.1	−0.319	23.64	−33.6	MCG-5-26-12 = 376-G-26	D 26.9	(137,295)
AM 1052-301	10 52 58	2.849	275.18	142.9	SPIRAL w. 3 ALIGNED STARS(?)	8	F438/J0536
	−30 12.0	−0.320	26.09	−33.4	MCG-5-26-13 = 437-G-71	S 1.3	(339,164)
AM 1053-280	10 53 38	2.869	274.18	140.4	I/A QUARTET	4	F437/J0515
	−28 05.3	−0.320	28.01	−33.3	PKS 1053-282	D 58.2	(63,278)
AM 1054-273	10 54 13	2.875	274.06	139.8	GALAXY w. SPUR to [N.E.]	7	F438/J0536
	−27 37.9	−0.320	28.47	−33.1		D 0.6	(328,302)
AM 1054-624	10 54 14	2.339	290.28	180.1	NEBULOSITY	22	F093/J2065
	−62 44.5	−0.320	−2.99	−27.2			(215,300)
*AM 1054-325	10 54 43	2.830	276.95	146.1	V.KNOTTY SP. I/A w. DIFFUSE SP.	2,12,15	F437/J0515
	−32 53.4	−0.321	23.89	−33.0	376-G-28	S 38.1	(56,20)
AM 1055-250	10 55 12	2.898	272.89	136.9	E + SPIRAL + APP. COMP'N. (#1)	8,23	F502/J2984
[1055-2514]	−25 09.6	−0.321	30.74	−32.8	MCG-4-26-15 = 502-G-5	D 0.6	(237,170)
AM 1055-251	10 55 15	2.897	272.95	137.0	E + SPIRAL + APP. COMP'N. (#2)	8,23	F502/J2984
[1055-2514]	−25 14.2	−0.321	30.67	−32.8	MCG-4-26-16 = 502-G-5	D 0.6	(237,166)
AM 1055-350	10 55 15	2.811	278.19	148.8	SMALL I/A DBL.	2	F376/J5018
	−35 09.9	−0.321	21.92	−32.8		S 0.5	(100,171)
AM 1055-391	10 55 33	2.771	280.17	153.5	"X"-SHAPED DWARF	16,20	F318/J1380
	−39 10.2	−0.321	18.37	−32.4	318-G-24	D 3.4	(119,221)
AM 1055-260	10 55 48	2.893	273.53	137.9	I/A SPIRALS (#1)	2	F502/J2984
[1055-2601]	−26 02.5	−0.321	30.03	−32.7	502-G-7	D 0.3	(229,123)
AM 1055-260	10 55 52	2.893	273.53	137.9	I/A SPIRALS (#2)	2	F502/J2984
[1055-2602]	−26 01.3	−0.321	30.06	−32.7		D 0.4	(229,124)
AM 1055-362	10 55 57	2.801	278.93	150.2	SMALL M51-TYPE	9	F376/J5018
	−36 24.0	−0.321	20.89	−32.6		S 0.3	(93,104)
AM 1056-273	10 56 03	2.880	274.47	139.8	TIGHT GROUP of GALAXIES	18	F438/J0536
	−27 38.5	−0.321	28.66	−32.7		D 0.4	(306,302)
AM 1056-463	10 56 15	2.686	283.63	162.1	I/A DBL. (CLOSE) CHAIN of 9	2,17	F265/J1369
	−46 33.8	−0.321	11.77	−31.2		S 0.1	(322,89)
AM 1056-692	10 56 29	2.101	293.33	186.9	PAIR (#1)	23	F062/J2049
[1057-6926]	−69 23.1	−0.321	−8.91	−24.7	63-G-1	S 7.2	(27,203)

Identification	R.A. (1950.0) Declination	Precession R.A./Dec.	Gal. Long. Gal. Lat.	S.G.L. S.G.B.	Description Cross-Identification	Coding Size (Min.)	Field/Plate (x, y)
AM 1056-281	10 56 36 −28 12.5	2.877 −0.321	274.90 28.22	140.5 −32.6	SP. w. FAINT CIRCULAR OUTER RING NGC 3483 = 438-G-1	10 D 56.0	F437/J0515 (28,271)
AM 1056-431	10 56 47 −43 10.1	2.732 −0.321	282.21 14.88	158.2 −31.7	SP. in GROUP of 4 (COMPACT NEAR)	12,13,18 D 0.8	F265/J1369 (326,272)
*AM 1056-430	10 56 49 −43 06.1	2.733 −0.321	282.19 14.94	158.1 −31.7	LINE of 3 SPIRALS 264-G-58	24 D 12.1	F318/J1380 (110,10)
AM 1057-692 [1056-6923]	10 57 32 −69 26.6	2.114 −0.322	293.44 −8.92	186.9 −24.5	PAIR (#2) 63-G-2	23 S 7.2	F062/J2049 (22,199)
AM 1057-282	10 57 58 −28 29.6	2.879 −0.322	275.35 28.11	140.9 −32.3	FAINT RING w. STELLAR NUCLEUS	6 D 0.4	F438/J0536 (283,257)
AM 1058-243	10 58 23 −24 38.9	2.910 −0.322	273.34 31.53	136.3 −32.1	SMALL SPIRAL w. I/A COMP'N.	1 S 0.3	F502/J2984 (199,198)
AM 1059-593	10 59 13 −59 34.5	2.477 −0.322	289.50 0.14	176.5 −27.7	COMETARY NEBULA NGC 3503 = 128-EN-28	22	F129/J2244 (348,194)
AM 1100-253	11 00 01 −25 35.9	2.907 −0.323	274.25 30.88	137.5 −31.8	SPIRAL + 3 APP. COMP'NS. 502-G-10	8 S 1.0	F502/J2984 (179,147)
AM 1100-255	11 00 05 −25 53.9	2.905 −0.323	274.43 30.62	137.8 −31.8	DISTURBED SPIRAL MCG-4-26-18 = 502-IG-11	16 D 0.7	F502/J2984 (178,131)
AM 1100-392	11 00 18 −39 20.8	2.791 −0.323	281.12 18.62	153.6 −31.5	SMALL DWARF (DEFECT?)	20 D 0.7	F318/J1380 (70,211)
AM 1100-244	11 00 27 −24 43.9	2.915 −0.323	273.87 31.68	136.5 −31.6	GROUP of 7 (TRIPLE and PEC. SP.)	12,18,24 D 8.1	F502/J2984 (174,194)
AM 1100-254	11 00 36 −25 45.0	2.908 −0.323	274.47 30.80	137.7 −31.7	SMALL I/A DBL.	2 S 0.1	F502/J2984 (172,139)
AM 1100-224 [1101-2258]	11 00 55 −22 49.2	2.930 −0.323	272.89 33.42	134.2 −31.4	PAIR of SPIRALS + COMP'NS. (#1)	8,23 D 3.4	F502/J2984 (168,297)
AM 1101-354	11 01 07 −35 48.3	2.828 −0.323	279.65 21.88	149.5 −31.6	GALAXY w. RING + FAINT TAIL 377-G-2	6,15 S 0.9	F377/J3993 (298,134)
AM 1101-271	11 01 12 −27 17.3	2.898 −0.323	275.43 29.51	139.5 −31.6	GALAXY w. 2 OBJECTS to [E](SMALL)	8 D 0.6	F438/J0536 (245,322)
AM 1101-225 [1100-2249]	11 01 20 −22 58.5	2.930 −0.323	273.08 33.32	134.4 −31.3	PAIR of SPIRALS + COMP'NS. (#2)	8,23 D 2.0	F502/J2984 (163,288)
AM 1101-322	11 01 36 −32 20.6	2.860 −0.323	278.10 25.04	145.4 −31.6	COMPACT I/A DBL.	2 S 0.2	F438/J0536 (238,51)
AM 1102-313	11 02 04 −31 31.0	2.868 −0.323	277.78 25.82	144.4 −31.5	I/A DBL.	2 S 0.6	F438/J0536 (233,95)
AM 1102-225	11 02 26 −22 51.9	2.933 −0.324	273.28 33.54	134.3 −31.0	ONE-ARMED SPIRAL	10 D 0.5	F502/J2984 (150,294)
AM 1102-233	11 02 48 −23 34.4	2.929 −0.324	273.77 32.96	135.2 −31.0	DISTURBED SPIRAL w. 4 COMP'NS. 502-IG-15	8,16 S 1.5	F502/J2984 (145,256)
AM 1104-362 [1104-3624]	11 04 04 −36 22.2	2.835 −0.324	280.49 21.63	150.1 −30.9	PAIR (#1) 377-G-6	23 D 0.9	F377/J3993 (265,104)
AM 1104-362 [1104-3622]	11 04 05 −36 24.4	2.835 −0.324	280.51 21.59	150.1 −30.9	PAIR (#2: LINEAR) 377-G-7	16,23 D 1.1	F377/J3993 (264,102)
AM 1104-290	11 04 06 −29 02.8	2.894 −0.324	277.00 28.23	141.6 −31.0	FAINT, LINKED GALAXIES	17 D 0.9	F438/J0536 (210,228)
*AM 1104-365	11 04 47 −36 54.0	2.834 −0.324	280.87 21.20	150.7 −30.8	SB w. 2 DWARF COMP'NS NGC 3533 = NGC 3557a = 377-G-11	8 S 5.4	F377/J3993 (256,76)
AM 1104-224	11 04 53 −22 44.4	2.940 −0.324	273.81 33.92	134.2 −30.5	E + DIAMETRIC COMP'NS.	8 S 2.6	F502/J2984 (119,301)
AM 1104-252	11 04 55 −25 22.7	2.922 −0.324	275.27 31.59	137.3 −30.7	DWARF 502-G-18?	20 D 0.7	F502/J2984 (119,159)
AM 1105-365	11 05 16 −36 51.6	2.836 −0.325	280.95 21.28	150.6 −30.7	PEC. DWARF w. JET from NUCLEUS	7,20 D 0.6	F377/J3993 (251,78)
AM 1105-700	11 05 25 −70 09.0	2.200 −0.325	294.36 −9.29	187.4 −23.7	PAIR of SPIRALS w. PECULIAR ARMS	10,23 S 2.9	F063/J2071 (228,170)
AM 1105-461	11 05 33 −46 15.1	2.745 −0.325	284.99 12.72	161.5 −29.7	IRR. 265-G-07	16 D 4.0	F265/J1369 (236,109)
AM 1105-215	11 05 46 −21 51.9	2.947 −0.325	273.53 34.79	133.3 −30.2	SMALL I/A DBL.	2 S 0.3	F502/J2984 (108,348)
AM 1105-412	11 05 49 −41 27.5	2.797 −0.325	283.05 17.13	156.0 −30.2	DISTURBED GALAXY in GROUP of 3	16,24 D 2.2	F318/J1380 (18,95)
AM 1106-280	11 06 33 −28 06.0	2.908 −0.325	277.07 29.32	140.5 −30.4	DWARF + HI.S.BR. PEC. SPIRAL [E] 438-G-5 = UKS 1106-280	8,20 S 2.9	F438/J0536 (182,279)
AM 1106-223	11 06 53 −22 36.1	2.945 −0.325	274.23 34.26	134.1 −30.0	SPIRAL + APP. COMP'N. MCG-4-27-1 = 502-G-20	8 S 1.9	F502/J2984 (94,308)
AM 1107-232	11 07 27 −23 27.3	2.941 −0.325	274.84 33.56	135.1 −29.9	STRONG-ARMED SPIRAL w. COMP'NS. IC 2627 = 502-G-21	8,10 S 10.8	F502/J2984 (88,262)
AM 1107-280	11 07 34 −28 06.5	2.911 −0.325	277.30 29.41	140.5 −30.2	GALAXY w. BRIGHT RIM	16 D 0.3	F438/J0536 (170,278)
*AM 1107-371	11 07 35 −37 16.2	2.843 −0.325	281.58 21.10	151.1 −30.2	LARGE E w. 8 DISTURBED COMP'NS NGC 3557 = 377-G-16= PKS 1107-372	8 S 9.0	F377/J3993 (226,57)

Identification	R.A. (1950.0) Declination	Precession R.A./Dec.	Gal. Long. Gal. Lat.	S.G.L. S.G.B.	Description Cross-Identification	Coding Size (Min.)	Field/Plate (x, y)
AM 1107-363	11 07 56 −36 36.4	2.849 −0.325	281.35 21.73	150.3 −30.1	LINEAR DWARF (SOME RESOLUTION) 377-G-17	20 D 0.9	F377/J3993 (223,93)
AM 1107-464	11 07 58 −46 47.1	2.753 −0.325	285.59 12.39	162.0 −29.2	I/A DBL. (CLOSE) CHAIN of 6 265-IG-13	2,17 S 0.2	F265/J1369 (214,81)
AM 1108-371	11 08 14 −37 16.3	2.845 −0.326	281.71 21.15	151.1 −30.0	GALAXY w. JETS in GROUP of 3 NGC 3564 = 377-G-18 = KLEM 18 [DN]	7,24 D 1.6	F319/J1312 (263,325)
*AM 1108-300	11 08 15 −30 04.5	2.900 −0.326	278.42 27.70	142.8 −30.1	M51-TYPE 438-G-8 [A82]	9 S 1.0	F438/J0536 (162,173)
AM 1108-350	11 08 21 −35 04.8	2.863 −0.326	280.76 23.16	148.6 −30.1	SPIRAL w. APP. COMP'N. 377-G-19	8 S 1.1	F377/J3993 (219,175)
AM 1108-362 [1108-3620]	11 08 21 −36 23.5	2.853 −0.326	281.34 21.96	150.1 −30.1	PAIR (#1)	23 D 0.4	F377/J3993 (218,104)
AM 1108-472	11 08 27 −47 28.0	2.748 −0.326	285.94 11.80	162.7 −29.0	DWARF	20 D 0.4	F265/J1369 (209,44)
AM 1108-362 [1108-3623]	11 08 28 −36 20.9	2.854 −0.326	281.35 22.01	150.0 −30.0	PAIR (#2: IRR.)	16,23 D 0.2	F377/J3993 (217,107)
AM 1108-463 [1108-4638]	11 08 28 −46 38.6	2.758 −0.326	285.62 12.56	161.8 −29.1	I/A DBL.(GROUP of 6) (#1: E)	2,18 S 0.4	F265/J1369 (209,88)
AM 1108-463 [1108-4638]	11 08 29 −46 38.2	2.758 −0.326	285.62 12.56	161.8 −29.1	I/A DBL.(GROUP of 6) (#2: SP)	2,18 S 0.4	F265/J1369 (209,89)
*AM 1108-270	11 08 45 −27 06.8	2.921 −0.326	277.08 30.42	139.4 −29.9	NGC 7603-TYPE	9 D 0.4	F438/J0536 (155,332)
AM 1108-465	11 08 54 −46 50.0	2.758 −0.326	285.76 12.41	162.0 −29.0	DISTURBED GALAXY 265-G-15	16 D 0.6	F265/J1369 (205,78)
*AM 1108-363	11 08 56 −36 36.2	2.854 −0.326	281.55 21.82	150.3 −29.9	EDGE-ON SPIRAL w. STRONG ABS'N. NGC 3573 = 377-G-22	8,14 D 1.8	F377/J3993 (212,93)
AM 1108-271	11 08 57 −27 10.0	2.922 −0.326	277.15 30.39	139.4 −29.9	CLOSE I/A DBL. (M51-TYPE)	2,9 S 0.4	F503/J2330 (333,59)
AM 1108-471	11 08 58 −47 16.3	2.754 −0.326	285.95 12.01	162.5 −28.9	DWARF	20 D 0.4	F265/J1369 (204,55)
AM 1109-232	11 09 13 −23 26.6	2.946 −0.326	275.27 33.75	135.2 −29.5	IRR. w. ADJACENT KNOTS (ARC?)	15,16 D 0.4	F503/J2330 (334,259)
AM 1109-473 [1109-4738]	11 09 34 −47 38.9	2.753 −0.326	286.19 11.70	162.9 −28.8	I/A PAIR of DWARFS (#1) 215-IG-33	2,20 S 1.6	F265/J1369 (199,35)
AM 1109-220 [1109-2203]	11 09 35 −22 02.1	2.955 −0.326	274.58 35.05	133.6 −29.3	PAIR of E's (#1) PKS 1109-220	23 D 0.3	F502/J2984 (60,338)
AM 1109-220 [1109-2202]	11 09 38 −22 03.2	2.955 −0.326	274.60 35.04	133.6 −29.3	PAIR of E's (#2) PKS 1109-220	23 D 0.3	F502/J2984 (60,337)
AM 1109-473 [1109-4738]	11 09 43 −47 38.4	2.754 −0.326	286.21 11.72	162.9 −28.8	I/A PAIR of DWARFS (#2) 215-G-34	2,20 S 1.6	F265/J1369 (197,35)
AM 1109-375	11 09 53 −37 51.3	2.848 −0.326	282.28 20.75	151.7 −29.7	DWARF (LINEAR) 319-G-7	20 D 0.7	F377/J3993 (201,26)
AM 1109-610	11 09 54 −61 01.9	2.550 −0.326	291.29 −0.69	177.7 −26.0	'PRETTY PICTURE' NGC 3581 = 129-EN-9	22	F129/J2244 (271,121)
AM 1110-251	11 10 31 −25 18.9	2.938 −0.326	276.57 32.21	137.3 −29.4	CLOSE GROUP of 3 (incl. RING)	6,24 D 7.2	F503/J2330 (316,159)
AM 1110-262	11 10 50 −26 29.1	2.931 −0.326	277.24 31.19	138.7 −29.4	E + MANY LO.S.BR. COMP'NS. NGC 3585 = 502-G-25	8 D 4.5	F503/J2330 (311,96)
AM 1111-272	11 11 01 −27 22.9	2.926 −0.326	277.74 30.39	139.7 −29.4	SMALL I/A DBL.	2 S 0.2	F502/J2984 (47,51)
AM 1111-474	11 11 10 −47 46.3	2.762 −0.326	286.49 11.69	163.0 −28.5	DWARF w. STRUCTURE (8 COMP'NS)	8,20 D 0.3	F265/J1369 (184,28)
AM 1111-320	11 11 13 −32 03.8	2.896 −0.326	280.00 26.16	145.1 −29.5	I/A DBL.	2 S 0.4	F438/J0536 (128,66)
AM 1111-685	11 11 13 −68 59.5	2.338 −0.326	294.38 −8.03	186.0 −23.6	GALAXY w. APP. COMP'N. 63-G-11	8 S 3.4	F063/J2071 (203,233)
AM 1111-294	11 11 34 −29 41.0	2.913 −0.327	278.97 28.36	142.3 −29.4	GROUP of 4 DISTURBED GALAXIES incl. 438-G-14	17 S 7.8	F438/J0536 (123,193)
AM 1111-474	11 11 35 −47 40.6	2.765 −0.327	286.52 11.80	162.9 −28.4	PLANETARY NEBULA Lo-5 = 215-PN-35	25 D 2.4	F216/J2283 (346,299)
AM 1111-364	11 11 44 −36 42.1	2.865 −0.327	282.15 21.95	150.4 −29.4	DWARF (COMP'N. to AM 1108-363)	20 D 0.4	F377/J3993 (182,88)
AM 1112-250	11 12 08 −25 03.6	2.943 −0.327	276.83 32.60	137.1 −29.0	DWARF	20 D 0.3	F503/J2330 (296,173)
AM 1112-232	11 12 14 −23 27.2	2.953 −0.327	276.02 34.05	135.2 −28.8	DISTURBED GALAXY w. PLUMES NGC 3597 = 503-G-3	15 D 1.3	F502/J2984 (29,262)
AM 1112-450	11 12 35 −45 07.9	2.797 −0.327	285.71 14.23	160.0 −28.6	I/A DBL.(SMALL)	2 S 0.3	F265/J1369 (171,170)
AM 1112-471	11 12 40 −47 12.1	2.777 −0.327	286.51 12.31	162.3 −28.3	DWARF w. STRUCTURE (COMPLEX) 265-G-19	20 D 1.1	F265/J1369 (171,59)
AM 1112-272	11 12 48 −27 23.1	2.932 −0.327	278.15 30.56	139.7 −29.0	E + FAINT IRREGULAR RING 503-G-5	15 D 2.2	F438/J0536 (107,317)

Identification	R.A. (1950.0) Declination	Precession R.A./Dec.	Gal. Long. Gal. Lat.	S.G.L. S.G.B.	Description Cross-Identification	Coding Size (Min.)	Field/Plate (x, y)
AM 1113-270	11 13 04 −27 00.0	2.935 −0.327	278.03 30.93	139.3 −28.9	GALAXY w. FAN to [NW] (V.SMALL)	15 D 0.3	F438/J0536 (104,337)
AM 1113-334 [1113-3333]	11 13 38 −33 41.9	2.894 −0.327	281.25 24.87	147.0 −29.0	PAIR (#1: SPIRAL) A 1113-33 = 377-G-31 = SS-10b	23 S 7.2	F377/J3993 (161,249)
AM 1113-333 [1113-3341]	11 13 51 −33 33.0	2.896 −0.327	281.23 25.02	146.8 −29.0	PAIR (#2: E) SS-10a NGC 3606 = MCG-5-27-4 = 377-G-32	23 S 7.2	F377/J3993 (159,257)
AM 1113-444	11 13 55 −44 42.0	2.809 −0.327	285.77 14.72	159.5 −28.4	SPIRAL w. PECULIAR ARM	10 D 0.7	F265/J1369 (158,193)
AM 1114-283	11 14 00 −28 35.2	2.928 −0.327	279.01 29.57	141.1 −28.8	HI.S.BR. GALAXY w. I/A COMP'N. To 1114-285	1 S 0.2	F438/J0536 (94,252)
AM 1114-232	11 14 06 −23 23.5	2.958 −0.327	276.45 34.30	135.2 −28.4	CONTACT I/A DBL.	2 S 0.1	F503/J2330 (274,263)
AM 1114-234 [1114-2343]	11 14 06 −23 42.4	2.956 −0.327	276.62 34.01	135.6 −28.4	CLOSE PAIR (#1)	23 D 0.6	F503/J2330 (273,246)
AM 1114-322	11 14 20 −32 22.5	2.905 −0.327	280.81 26.14	145.4 −28.9	DWARF	20 D 0.4	F377/J3993 (154,320)
AM 1114-234 [1114-2342]	11 14 22 −23 43.1	2.957 −0.327	276.69 34.03	135.6 −28.4	CLOSE PAIR (#2)	23 D 0.3	F503/J2330 (270,246)
AM 1114-344	11 14 39 −34 40.9	2.891 −0.327	281.88 24.04	148.1 −28.8	SP. w. 2 APP. COMP'NS (SPIRALS) 377-G-34 in a LINE	8 S 7.4	F377/J3993 (150,196)
AM 1114-222	11 14 55 −22 26.8	2.965 −0.328	276.16 35.23	134.2 −28.1	I/A DBL. in GROUP of 3 570-G-15	2,24 D 1.8	F503/J2330 (264,314)
AM 1115-254 [1115-2551]	11 15 17 −25 46.2	2.948 −0.328	277.95 32.26	137.9 −28.4	CLOSE PAIR (#1: E) 503-G-11	23 S 6.3	F503/J2330 (258,136)
AM 1115-255 [1115-2546]	11 15 23 −25 51.7	2.948 −0.328	278.02 32.18	138.0 −28.3	CLOSE PAIR (#2: LO.S.BR. SPIRAL) NGC 3617 = 503-G-12	23 D 0.3	F503/J2330 (256,131)
AM 1115-282	11 15 35 −28 26.8	2.934 −0.328	279.31 29.84	141.0 −28.5	SB(r) w. BRANCHED ARMS	10 D 0.7	F438/J0536 (75,259)
AM 1115-323	11 15 51 −32 32.1	2.910 −0.328	281.21 26.12	145.6 −28.6	MULTI-ARMED SPIRAL (LACEY ABS'N.) NGC 3621 = 377-G-37= PKS 1115-325	14 D 9.9	F377/J3993 (136,312)
AM 1115-461	11 15 53 −46 18.5	2.805 −0.328	286.70 13.35	161.3 −27.9	I/A DBL. (in GROUP of 4) PKS 1116-46	2,18 S 0.5	F265/J1369 (141,106)
AM 1116-274	11 16 08 −27 43.3	2.939 −0.328	279.10 30.56	140.2 −28.3	RING GALAXY	6 D 0.8	F438/J0536 (68,298)
*AM 1116-290	11 16 27 −29 08.8	2.932 −0.328	279.83 29.28	141.8 −28.3	GALAXY w. I/A + APP.COMP'NS. MCG-5-27-9 = 438-G-20/21	1,8 S 0.8	F439/J1377 (336,221)
AM 1116-324	11 16 50 −32 48.3	2.912 −0.328	281.54 25.95	145.9 −28.4	PECULIAR GALAXY + I/A COMP'N	1 S 0.6	F438/J0536 (66,25)
AM 1117-230	11 17 17 −23 05.1	2.967 −0.328	277.09 34.88	135.0 −27.7	4-ARMED SPIRAL	10 D 0.5	F503/J2330 (235,280)
AM 1118-272	11 18 06 −27 20.9	2.947 −0.328	279.39 31.07	139.8 −27.8	RINGED SPIRAL w. BAR (SMALL) 503-G-15	12 D 1.0	F438/J0536 (44,318)
AM 1118-281	11 18 12 −28 12.0	2.943 −0.328	279.80 30.30	140.7 −27.9	BRIGHT STAR w. NEBULOSITY [W] 438-G-22?	21 D 0.7	F438/J0536 (44,272)
AM 1118-425 [1118-4258]	11 18 24 −42 59.8	2.848 −0.329	285.91 16.60	157.5 −27.8	I/A DBL.(RING + SPIRAL) (#1)	2,6 S 1.3	F265/J1369 (114,284)
AM 1118-290	11 18 26 −29 07.8	2.939 −0.329	280.28 29.47	141.8 −27.9	EDGE-ON Sa w. ASYMM. OUTER ENV. MCG-5-27-10 = 438-G-23	12 D 0.8	F438/J0536 (42,222)
AM 1118-425 [1118-4259]	11 18 27 −42 58.6	2.849 −0.329	285.91 16.62	157.5 −27.7	I/A DBL.(RING + SPIRAL) (#2) 265-G-23	2,6 S 1.3	F265/J1369 (113,285)
AM 1118-461	11 18 41 −46 13.0	2.823 −0.329	287.14 13.60	161.1 −27.4	CLOSE PAIR of SPIRALS 265-G-26	23 S 0.9	F265/J1369 (114,111)
*AM 1118-350	11 18 56 −35 07.1	2.906 −0.329	282.95 23.97	148.6 −27.9	DISTURBED GAL. w. I/A DBL.COMP'N.	1 S 0.5	F377/J3993 (103,173)
AM 1118-322	11 18 59 −32 29.3	2.922 −0.329	281.87 26.41	145.6 −27.9	SPIRAL + HI.S.BR. COMP'N. 439-G-1	8 S 0.8	F377/J3993 (101,314)
AM 1119-335	11 19 30 −33 51.6	2.915 −0.329	282.55 25.18	147.1 −27.8	SMALL GALAXY w. FAINT JET	7 D 0.2	F377/J3993 (96,240)
AM 1119-373 [1119-3747]	11 19 30 −37 38.1	2.892 −0.329	284.07 21.67	151.4 −27.8	CLOSE PAIR (#1) 319-G-15	23 D 0.5	F319/J1312 (142,306)
AM 1119-374 [1119-3738]	11 19 42 −37 47.3	2.892 −0.329	284.17 21.55	151.6 −27.8	CLOSE PAIR (#2) 319-G-16	23 D 1.1	F319/J1312 (140,298)
AM 1119-233	11 19 53 −23 33.6	2.972 −0.329	278.00 34.70	135.6 −27.1	COMPACT	13 D 0.4	F503/J2330 (202,254)
AM 1121-450	11 21 17 −45 05.6	2.848 −0.329	287.18 14.81	159.8 −27.1	COMPACT (between 2 BR.STARS)	13 D 0.1	F265/J1369 (89,171)
AM 1121-342	11 21 18 −34 22.2	2.919 −0.329	283.15 24.85	147.7 −27.5	PAIR	23 S 1.1	F377/J3993 (77,212)
AM 1121-344	11 21 54 −34 44.4	2.920 −0.329	283.42 24.55	148.1 −27.3	GALAXY w. OUTER MATERIAL 377-G-41 + 2 COMP'NS.(I/A)	1,8,15 S 2.4	F377/J3993 (70,192)
AM 1122-374	11 22 25 −37 40.8	2.905 −0.330	284.67 21.84	151.5 −27.2	3 in a LINE	24 D 1.0	F319/J1312 (111,304)

Identification	R.A. (1950.0) Declination	Precession R.A./Dec.	Gal. Long. Gal. Lat.	S.G.L. S.G.B.	Description Cross-Identification	Coding Size (Min.)	Field/Plate (x, y)
AM 1122-262	11 22 45 −26 27.9	2.965 −0.330	280.10 32.29	138.9 −26.7	SPIRAL w. APP. COMP'N. NGC 3673 = 503-G-16	8 S 4.0	F503/J2330 (168,99)
AM 1123-283	11 23 21 −28 37.1	2.957 −0.330	281.20 30.35	141.3 −26.8	I/A TRIPLE To 1123-286 = 4389-IG-6	3 D 0.3	F439/J1377 (255,251)
AM 1123-363	11 23 21 −36 32.2	2.915 −0.330	284.42 22.97	150.2 −27.0	NUCLEATED DWARF 377-G-45	20 D 0.5	F377/J3993 (56,96)
AM 1123-374 [1123-3738]	11 23 31 −37 46.9	2.909 −0.330	284.93 21.82	151.6 −27.0	CLOSE PAIR of SPIRALS (#1) 319-G-19	23 D 0.5	F319/J1312 (100,298)
AM 1123-241	11 23 32 −24 14.9	2.978 −0.330	279.26 34.40	136.4 −26.4	GALAXY w. 4 APP. COMP'NS.	8 S 0.7	F503/J2330 (158,218)
AM 1123-535	11 23 47 −53 57.4	2.784 −0.330	290.57 6.59	169.6 −25.6	GALAXY w. APP. COMP'N. 170-G-3	8 S 0.1	F170/J2852 (253,233)
AM 1123-331	11 23 49 −33 15.0	2.935 −0.330	283.24 26.08	146.5 −26.9	PAIR of E's	23 S 1.5	F377/J3993 (47,272)
AM 1123-373 [1123-3746]	11 23 50 −37 38.0	2.911 −0.330	284.94 21.98	151.4 −26.9	CLOSE PAIR of SPIRALS (#2) 319-G-20	23 D 0.5	F319/J1312 (96,306)
AM 1124-340	11 24 19 −34 05.4	2.933 −0.330	283.68 25.33	147.4 −26.8	PLANETARY NEBULA To 1124-340 Ko 283+25-1 = 378-PN-1	25 D 2.6	F377/J3993 (43,227)
*AM 1124-272	11 24 24 −27 25.8	2.966 −0.330	280.93 31.54	140.0 −26.5	S0 w. SPIKE along MI.AX. 503-G-17	7 D 0.8	F439/J1377 (244,315)
AM 1124-273	11 24 49 −27 32.9	2.966 −0.330	281.08 31.47	140.1 −26.4	COMPACT (LUMPY)	13 D 0.2	F439/J1377 (239,308)
*AM 1125-285	11 25 03 −28 54.1	2.961 −0.330	281.72 30.23	141.6 −26.4	PEC. RING I/A w. PEC. SPIRAL 439-G-10 = To 1124-289?	2,6,12 S 5.2	F439/J1377 (235,236)
AM 1125-374	11 25 10 −37 40.9	2.917 −0.330	285.22 22.03	151.4 −26.7	SMALL GALAXY w. I/A COMP'N.	1 S 0.2	F377/J3993 (39,34)
AM 1125-412	11 25 30 −41 20.1	2.897 −0.330	286.60 18.61	155.5 −26.5	GALAXY w. I/A COMP'N. 319-G-22	1 S 0.5	F319/J1312 (83,107)
AM 1125-594	11 25 39 −59 40.9	2.732 −0.330	292.65 1.25	175.7 −24.4	NEBULOSITY NGC 3699 = PK 292+11 = 129-PN-21	25 D	F129/J2244 (168,196)
AM 1126-290	11 26 18 −29 01.1	2.964 −0.330	282.07 30.22	141.8 −26.2	COMPACT GALAXY	13 D 0.3	F439/J1377 (220,230)
*AM 1126-291	11 26 19 −29 16.4	2.963 −0.330	282.18 29.98	142.1 −26.2	COMPACT (DOUBLE)	13,13,23 S 0.5	F439/J1377 (220,216)
AM 1126-341	11 26 50 −34 10.0	2.942 −0.331	284.25 25.44	147.5 −26.3	I/A DBL. (ONE w. TAIL)	2,15 S 0.4	F377/J3993 (15,222)
AM 1127-365	11 27 21 −36 56.3	2.930 −0.331	285.39 22.87	150.6 −26.2	DISTURBED SINGLE w. PEC. ABS'N. 378-G-7	14,16 D 1.3	F319/J1312 (58,342)
AM 1127-633	11 27 27 −63 38.5	2.691 −0.331	294.09 −2.44	179.9 −23.4	NEBULOSITY	22 D	F093/J2065 (16,244)
*AM 1127-351	11 27 36 −35 15.7	2.940 −0.331	284.82 24.47	148.7 −26.2	HI.S.BR. SPIRAL + APP. COMP'NS To 1127-352 = 378-G-8	8,12 S 1.0	F378/J1317 (271,164)
AM 1128-340	11 28 06 −34 08.4	2.947 −0.331	284.51 25.56	147.5 −26.0	COMPACT	13 D 0.2	F378/J1317 (267,224)
AM 1128-421	11 28 09 −42 17.6	2.905 −0.331	287.42 17.87	156.6 −26.0	I/A DBL.	2 S 0.4	F265/J1369 (16,318)
AM 1128-393	11 28 21 −39 34.2	2.921 −0.331	286.53 20.46	153.5 −26.0	COMPACT	13 D 0.1	F320/J1448 (320,200)
AM 1128-420	11 28 34 −42 04.7	2.908 −0.331	287.43 18.10	156.3 −25.9	GALAXY w. TAIL	15 D 0.5	F265/J1369 (11,330)
AM 1129-441	11 29 00 −44 17.2	2.897 −0.331	288.24 16.03	158.8 −25.8	SPIRAL w. I/A COMP'N.	1 S 0.2	F265/J1369 (13,211)
*AM 1129-300	11 29 03 −30 02.0	2.969 −0.331	283.13 29.48	143.0 −25.6	GAL. w. LARGE COMPACT (= IC 2913) NGC 3717 = 439-G-15 off MI.AX.	8,13 S 7.4	F439/J1377 (188,175)
AM 1129-350	11 29 11 −35 06.2	2.947 −0.331	285.10 24.73	148.6 −25.8	I/A DBL.	2 S 0.3	F378/J1317 (254,173)
AM 1130-332	11 30 19 −33 22.3	2.959 −0.331	284.70 26.44	146.7 −25.6	GROUP of 5 incl. I/A DBL.	2,18 D 3.6	F378/J1317 (243,266)
AM 1130-272	11 30 25 −27 21.8	2.984 −0.331	282.37 32.09	140.0 −25.1	DISTURBED SINGLE GALAXY	16 D 0.5	F439/J1377 (172,319)
AM 1131-521	11 31 14 −52 15.6	2.857 −0.331	291.11 8.55	167.5 −24.7	GALAXY + DIAMETRIC COMP'NS.	8	F170/J2852 (197,326)
AM 1131-263	11 31 59 −26 35.9	2.991 −0.332	282.43 32.93	139.2 −24.7	LUMPY COMPACT (JET ?) MCG-4-27-13 = 503-G-25	7,13 D 0.6	F503/J2330 (57,90)
AM 1132-495	11 32 06 −49 55.9	2.880 −0.332	290.54 10.82	165.0 −24.8	CLOSE PAIR incl. 216-G-16	23 D 2.0	F216/J2283 (161,183)
AM 1132-270	11 32 23 −27 07.7	2.991 −0.332	282.75 32.46	139.8 −24.7	HI.S.BR. SB(r) + OUTER MATERIAL	15,16 D 0.6	F503/J2330 (53,61)
AM 1132-231	11 32 31 −23 13.8	3.004 −0.332	281.13 36.12	135.6 −24.2	SMALL SP. w. 2 APP. COMP'NS. 503-G-26	8,12 D 0.5	F504/J2927 (321,272)
AM 1132-450	11 32 42 −45 06.2	2.914 −0.332	289.15 15.45	159.7 −25.1	SP. w. I/A COMP'N. (FILAMENT) 266-IG-5	1 S 1.0	F266/J2870 (248,173)

Identification	R.A. (1950.0) Declination	Precession R.A./Dec.	Gal. Long. Gal. Lat.	S.G.L. S.G.B.	Description Cross-Identification	Coding Size (Min.)	Field/Plate (x, y)
*AM 1132-324	11 32 51 −32 45.4	2.971 −0.332	285.04 27.19	146.0 −25.0	ISOLATED RING GALAXY	6 D 0.7	F439/J1377 (145,29)
AM 1133-245	11 33 01 −24 52.3	3.000 −0.332	281.97 34.63	137.4 −24.3	LO.S.BR. RING (I/A COMP'NS.) 504-G-1	1,6 D 1.3	F503/J2330 (42,183)
AM 1133-374	11 33 04 −37 40.8	2.952 −0.332	286.82 22.54	151.5 −25.1	I/A SP. + COMP'NS off ENDS (#1) NGC 3742 = 320-G-6	2,8 D 1.6	F320/J1448 (274,303)
AM 1133-282	11 33 17 −28 21.6	2.989 −0.332	283.47 31.37	141.2 −24.6	CLOSE TRIPLE	24 D 0.5	F439/J1377 (138,265)
AM 1133-281	11 33 23 −28 13.1	2.990 −0.332	283.44 31.51	141.0 −24.5	I/A TRIPLE	3 D 0.6	F439/J1377 (137,273)
AM 1133-374	11 33 25 −37 43.4	2.953 −0.332	286.90 22.52	151.5 −25.0	I/A SP. + COMP'NS off ENDS (#2) NGC 3749 = 320-G-8	2,8 D 1.6	F320/J1448 (271,300)
AM 1133-274	11 33 38 −27 41.8	2.992 −0.332	283.29 32.02	140.5 −24.4	ONE-ARMED SPIRAL	10 D 0.7	F503/J2330 (38,31)
*AM 1134-322	11 34 31 −32 26.4	2.979 −0.332	285.30 27.61	145.7 −24.6	COMPACT (PAIR)	13,23 D 0.2	F378/J1317 (197,317)
*AM 1134-323	11 34 39 −32 33.1	2.979 −0.332	285.37 27.51	145.8 −24.6	I/A TRIPLE 378-G-13 = PKS 1134-325	3 D 3.1	F378/J1317 (195,311)
*AM 1135-284	11 35 12 −28 48.0	2.994 −0.332	284.10 31.10	141.7 −24.2	RING GALAXY w. OFF-SET NUCLEUS	6 D 0.4	F439/J1377 (116,241)
AM 1135-502	11 35 58 −50 26.4	2.904 −0.332	291.29 10.51	165.5 −24.2	CHAIN of 4 NGC 3778 = 216-G-26	17 D 17.0	F216/J2283 (128,155)
AM 1136-522	11 36 29 −52 28.2	2.895 −0.332	291.95 8.58	167.7 −23.9	GALAXY + APP. COMP'N.	8	F170/J2852 (154,316)
AM 1136-392	11 36 38 −39 20.0	2.962 −0.332	288.08 21.18	153.3 −24.4	COMPACT w. FUZZ	13 D 0.2	F320/J1448 (235,215)
AM 1137-415	11 37 54 −41 56.4	2.958 −0.332	289.13 18.75	156.1 −24.2	CLOSE PAIR E's (I/A?)	23 S 1.2	F266/J2870 (200,344)
AM 1138-641	11 38 18 −64 13.4	2.812 −0.333	295.41 −2.66	180.3 −22.1	NEBULOUS STAR	21	F094/J3898 (213,221)
AM 1138-441	11 38 29 −44 12.5	2.951 −0.333	289.91 16.61	158.6 −24.1	CHAIN of 5 PEC. SP. in GROUP 7 266-G-15	12,17,18 D 16.6	F266/J2870 (194,222)
AM 1139-621	11 39 15 −62 12.4	2.844 −0.333	294.97 −0.68	178.1 −22.3	PLANETARY NEBULA PK 294-01 = 129-PN-29	25	F130/J2280 (321,54)
AM 1140-310	11 40 20 −31 09.6	3.004 −0.333	286.18 29.21	144.4 −23.3	I/A DBL.	2 S 0.9	F439/J1377 (58,113)
AM 1141-325	11 41 25 −32 58.5	3.003 −0.333	287.04 27.54	146.4 −23.2	CLOSE PAIR (E + SPIRAL)	23 S 1.9	F378/J1317 (119,288)
AM 1142-294	11 42 04 −29 48.1	3.013 −0.333	286.13 30.61	143.0 −22.8	CONTACT I/A DBL. 440-IG-2	2 S 0.3	F440/J3271 (306,188)
*AM 1142-272	11 42 28 −27 24.8	3.020 −0.333	285.38 32.92	140.4 −22.5	RING in CHAIN of 4	6,17 D 0.4	F439/J1377 (28,314)
AM 1142-501 [1143-5018]	11 42 28 −50 15.4	2.950 −0.333	292.26 10.97	165.2 −23.2	PAIR of SPIRALS (#1) 216-G-37	23 S 4.9	F217/J2194 (330,161)
AM 1142-432 [1143-4318]	11 42 48 −43 22.7	2.978 −0.333	290.46 17.62	157.7 −23.3	CLOSE PAIR (#1: S(r)c) 266-G-21	23 S 5.8	F266/J2870 (152,266)
AM 1143-431 [1143-4322]	11 43 01 −43 18.0	2.980 −0.333	290.47 17.71	157.6 −23.3	CLOSE PAIR (#2: SB(r)c) 266-G-23	23 S 5.8	F266/J2870 (150,270)
AM 1143-501 [1142-5015]	11 43 01 −50 18.9	2.954 −0.333	292.36 10.93	165.3 −23.1	PAIR of SPIRALS (#2) 216-G-39	23 S 4.9	F217/J2194 (325,158)
AM 1143-280	11 43 11 −28 05.3	3.021 −0.333	285.80 32.32	141.1 −22.4	DISTURBED DWARF w. COMPACT 440-G-4	13,20 D 2.2	F440/J3271 (296,281)
*AM 1144-273	11 44 16 −27 38.7	3.025 −0.333	285.92 32.82	140.7 −22.1	SB w. WARPED PLANE (ISOLATED) NGC 3885 = 440-G-7	12,14 D 2.2	F440/J3271 (283,305)
AM 1144-323	11 44 30 −32 30.9	3.015 −0.333	287.60 28.17	145.9 −22.5	CHAIN of 4	17 D 0.9	F439/J1377 (13,39)
AM 1144-424	11 44 54 −42 42.6	2.992 −0.333	290.66 18.37	157.0 −22.9	GALAXY w. I/A COMP'N. 266-G-25	1 S 1.2	F266/J2870 (131,302)
AM 1145-340	11 45 53 −34 05.2	3.017 −0.333	288.39 26.73	147.7 −22.4	I/A DBL.	2 S 0.1	F378/J1317 (70,228)
AM 1146-270	11 46 14 −27 06.0	3.032 −0.333	286.23 33.47	140.2 −21.6	COMPACT w. TAIL or JET MCG-4-28-3 = 504-G-17	7,13,15 D 1.0	F504/J2927 (152,67)
AM 1146-280	11 46 14 −28 00.8	3.030 −0.333	286.54 32.59	141.1 −21.7	LO.S.BR. GALAXY w. APP. COMP'NS. MCG-5-28-7 = 440-G-11	8,20 S 6.3	F440/J3271 (260,285)
AM 1146-371 [1146-3714]	11 46 20 −37 13.9	3.013 −0.333	289.42 23.73	151.1 −22.5	PAIR (#1: LO.S.BR. SPIRAL) 378-G-23 = Lu 1146-37	20,23 S 1.9	F379/J1532 (326,57)
AM 1146-371 [1146-3713]	11 46 32 −37 14.3	3.013 −0.333	289.46 23.73	151.1 −22.4	PAIR (#2: HI.S.BR. SPIRAL) NGC 3903 = 378-G-24 = Lu 1146-37	12,233 S 1.9	F379/J1532 (323,57)
AM 1146-500	11 46 52 −50 08.6	2.981 −0.333	292.92 11.25	165.0 −22.5	CLOSE PAIR 217-G-4/5	23 D 2.0	F216/J2283 (34,168)
AM 1146-332	11 46 58 −33 21.8	3.023 −0.333	288.42 27.49	146.9 −22.1	SP. w. 2 COMP'NS + DISTURBED GAL. To 1146-333 = 378-G-25	8,16,23 D 4.0	F379/J1532 (325,265)

Identification	R.A. (1950.0) Declination	Precession R.A./Dec.	Gal. Long. Gal. Lat.	S.G.L. S.G.B.	Description Cross-Identification	Coding Size (Min.)	Field/Plate (x, y)
*AM 1147-321	11 47 04 −32 18.9	3.025 −0.333	288.13 28.51	145.8 −22.0	GROUP of 6	18 D 2.2	F379/J1532 (326,321)
AM 1147-371	11 47 09 −37 15.9	3.016 −0.333	289.60 23.74	151.1 −22.3	DWARF (STRUCTURE)	20 D 0.5	F379/J1532 (317,56)
*AM 1147-383	11 47 19 −38 30.5	3.014 −0.333	289.98 22.54	152.4 −22.3	SP. w. DIAMETRIC COMP'NS. + MORE 320-G-26	8,14 S 10.3	F320/J1448 (124,259)
AM 1147-565	11 47 48 −56 54.3	2.964 −0.334	294.69 4.72	172.3 −21.9	PLANETARY NEBULA NGC 3918 = PK 294+4-1 = 170-PN-13	25	F170/J2852 (69,77)
AM 1148-750	11 48 03 −75 05.7	2.812 −0.334	299.02 −12.95	191.6 −19.2	LO.S.BR. GAL. w. 4 APP. COMP'NS. 39-G-2 = UKS 1148-750	8 S 3.4	F039/J2124 (241,172)
AM 1148-283	11 48 30 −28 31.7	3.037 −0.334	287.28 32.24	141.8 −21.3	"RINGING" S0 w. APP. COMP'NS. NGC 3923 = 440-G-17 = MC 1148-285	8,22 S 6.7	F440/J3271 (232,258)
AM 1149-364	11 49 09 −36 40.6	3.026 −0.334	289.86 24.41	150.5 −21.9	DWARF	20 D 0.3	F379/J1532 (296,88)
AM 1149-420	11 49 57 −42 00.8	3.020 −0.334	291.43 19.28	156.2 −22.0	SMALL RING w. BRIGHT EDGE (P.N.?) 320-PN?-28 = PK 291+19-1	6,25 D 0.4	F320/J1448 (100,71)
*AM 1150-312	11 50 20 −31 22.0	3.039 −0.334	288.61 29.60	144.8 −21.2	CHAIN of 5	17 D 1.8	F440/J3271 (210,106)
AM 1150-321 [1147-3218]	11 50 36 −32 17.4	3.038 −0.334	288.94 28.73	145.8 −21.2	E + SPIRAL (2 APP. COMP'NS.) (#1) MCG-5-28-14 = 440-G-25 = PKS 1151-324	8,23,24 S 4.5	F379/J1532 (286,324)
AM 1150-503	11 50 36 −50 34.4	3.006 −0.334	293.62 10.98	165.5 −21.8	PLANETARY NEBULA BDS = 217-PN-11	25 D 1.2	F217/J2194 (259,147)
AM 1150-322 [1150-3217]	11 50 47 −32 22.5	3.039 −0.334	289.00 28.65	145.9 −21.2	E + SPIRAL (2 APP. COMP'NS.) (#2) 440-G-26 = PKS 1151-324	8,23,24 S 4.5	F379/J1532 (283,319)
AM 1151-535 [1153-5252]	11 51 12 −53 50.1	3.003 −0.334	294.45 7.82	169.0 −21.6	PAIR of SPIRALS (#1) 171-G-1	23 S 3.7	F170/J2852 (36,240)
AM 1151-225	11 51 26 −22 53.4	3.052 −0.334	286.15 37.84	135.9 −19.8	GALAXY w. PECULIAR ABSORPTION NGC 3955 = 504-G-26	14 D 2.2	F504/J2927 (88,292)
AM 1151-535 [1153-5250]	11 51 46 −53 53.0	3.007 −0.334	294.55 7.79	169.0 −21.5	PAIR of SPIRALS (#2) 171-G-2	23 S 3.7	F170/J2852 (32,238)
AM 1152-453	11 52 05 −45 35.4	3.026 −0.334	292.70 15.88	160.1 −21.6	CLOSE PAIR of SPIRALS 266-G-31	23 S 1.3	F266/J2870 (66,146)
AM 1152-331	11 52 09 −33 16.9	3.043 −0.334	289.58 27.85	146.9 −21.0	DWARF 379-G-7	20 D 0.4	F379/J1532 (267,271)
AM 1152-311	11 52 20 −31 17.4	3.046 −0.334	289.06 29.79	144.8 −20.8	GALAXY w. COMP'N. in ENVELOPE	1 S 0.2	F440/J3271 (187,110)
AM 1152-372	11 52 42 −37 25.2	3.040 −0.334	290.81 23.86	151.3 −21.2	RINGING GALAXY w. PEC. NUCLEUS IC 2977 = 379-G-9 = MC 1152-374	16,22 D 1.8	F379/J1532 (257,49)
*AM 1152-421	11 52 50 −42 12.6	3.035 −0.334	292.03 19.21	156.5 −21.4	BROKEN RING 320-G-34	6 D 0.7	F267/J1560 (331,325)
AM 1154-320 [1154-3210]	11 54 06 −32 06.6	3.051 −0.334	289.71 29.08	145.7 −20.5	PAIR of I/A DBLS. (#1) 440-G-32	2,23 S 3.1	F379/J1532 (246,334)
AM 1154-340	11 54 10 −34 08.8	3.050 −0.334	290.27 27.11	147.9 −20.7	DWARF("S"-SHAPED) + COMPACT + SP. 379-G?-13	13,20,23 D 1.1	F379/J1532 (244,225)
AM 1154-321 [1154-3206]	11 54 13 −32 10.3	3.052 −0.334	289.75 29.03	145.8 −20.5	PAIR of I/A DBLS. (#2) 440-G-33	2,23 S 3.1	F379/J1532 (245,331)
*AM 1154-375	11 54 13 −37 55.1	3.047 −0.334	291.25 23.44	151.9 −21.0	3-ARMED SPIRAL (ONE ENDS on 320-G-35 CONDENSATION)	10,11 D 2.7	F321/J1310 (322,287)
AM 1154-330	11 54 57 −33 00.5	3.054 −0.334	290.15 28.25	146.7 −20.4	SPIRAL + I/A COMP'N. 379-G-15	1 S 0.7	F379/J1532 (236,286)
*AM 1155-443	11 55 17 −44 37.3	3.046 −0.334	293.05 16.95	159.1 −21.1	I/A QUARTET (DISRUPTED GALAXY?) 267-IG-1	4,16 D 0.8	F267/J1560 (301,197)
AM 1155-364	11 55 36 −36 44.6	3.054 −0.334	291.25 24.65	150.7 −20.6	DWARF	20 D 0.2	F379/J1532 (227,85)
AM 1156-474	11 56 39 −47 47.9	3.052 −0.334	293.99 13.90	162.5 −20.9	GALAXY w. PECULIAR ABSORPTION 217-G-18	14 D 0.5	F267/J1560 (281,27)
AM 1157-445	11 57 23 −44 50.9	3.058 −0.334	293.48 16.81	159.3 −20.7	DISTURBED SINGLE GALAXY	16 D 0.4	F267/J1560 (280,186)
AM 1157-242 [1158-2417]	11 57 58 −24 26.6	3.068 −0.334	288.47 36.73	137.8 −18.6	PAIR (#1: LO.S.BR. SPIRAL) MCG-4-28-9 = 505-G-2	20,23 S 10.5	F505/J2913 (277,208)
AM 1158-471	11 58 07 −47 16.4	3.061 −0.334	294.13 14.46	161.9 −20.6	PLANETARY NEBULA Lo-6 = 267-PN-7 = PK 294+14-1	25 D 1.0	F267/J1560 (269,56)
*AM 1158-345	11 58 26 −34 55.0	3.067 −0.334	291.43 26.56	148.8 −19.9	I/A SPIRAL + LINEAR GALAXY 379-G-20 + IG-21 = Fa 307 + 451	2,23 S 1.1	F379/J1532 (196,184)
AM 1158-241 [1157-2426]	11 58 35 −24 17.6	3.069 −0.334	288.60 36.91	137.6 −18.4	PAIR (#2: LINEAR GALAXY) MCG-4-29-1 = 505-G-3	23 S 10.5	F505/J2913 (270,216)
AM 1158-422 [1158-4223]	11 58 35 −42 28.2	3.066 −0.334	293.19 19.18	156.8 −20.4	CLOSE PAIR (#1) 321-G-2	23 D 0.5	F321/J1310 (270,44)
*AM 1158-333	11 58 45 −33 36.2	3.068 −0.334	291.18 27.86	147.4 −19.7	GAL. w. COMP'N. + JET in ENVELOPE 379-G-22 = To 1158-335	7,8 D 0.7	F379/J1532 (193,254)
AM 1158-422 [1158-4228]	11 58 47 −42 23.4	3.067 −0.334	293.21 19.27	156.7 −20.3	CLOSE PAIR (#2) 321-G-3	23 D 0.7	F321/J1310 (268,48)

Identification	R.A. (1950.0) Declination	Precession R.A./Dec.	Gal. Long. Gal. Lat.	S.G.L. S.G.B.	Description Cross-Identification	Coding Size (Min.)	Field/Plate (x, y)
AM 1158-451	11 58 47 −45 17.0	3.066 −0.334	293.83 16.43	159.8 −20.5	THREE-ARMED SPIRAL	11 D 0.6	F267/J1560 (266,163)
*AM 1159-530	11 59 01 −53 04.9	3.065 −0.334	295.45 8.80	168.1 −20.5	RING w. MATERIAL in WIDE PAIR 171-IG-5 = Fa 452	6,7,15,23 D 1.2	F171/J2253 (236,281)
AM 1159-235	11 59 15 −23 50.6	3.071 −0.334	288.65 37.38	137.2 −18.2	SMALL CONTACT DBL. (MATERIAL) 505-IG-5	2 D 0.5	F505/J2913 (262,240)
AM 1159-295	11 59 24 −29 57.5	3.071 −0.334	290.41 31.44	143.6 −19.1	DISTURBED SINGLE DWARF 440-G-39	20 D 1.1	F440/J3271 (105,181)
AM 1159-291	11 59 39 −29 12.4	3.072 −0.334	290.27 32.19	142.8 −18.9	ONE-ARMED SPIRAL MCG-5-29-2? = 440-G-42	10,23 S 1.8	F440/J3271 (101,221)
AM 1200-221	12 00 14 −22 16.3	3.074 −0.334	288.44 38.96	135.6 −17.7	GROUP of 3	24 D 2.2	F505/J2913 (252,325)
AM 1200-295	12 00 22 −29 51.0	3.074 −0.334	290.62 31.59	143.5 −18.9	CLOSE TRIPLE	24 D 0.8	F440/J3271 (94,187)
AM 1200-251	12 00 55 −25 11.4	3.076 −0.334	289.50 36.16	138.7 −18.1	PECULIAR SHREDS (DWARF?) 505-G-7	20,22 D 1.1	F505/J2913 (241,168)
AM 1201-224	12 01 31 −22 40.7	3.077 −0.334	288.94 38.64	136.1 −17.5	SMALL DWARF	20 D 0.3	F505/J2913 (236,303)
AM 1201-291	12 01 38 −29 17.3	3.078 −0.334	290.79 32.20	142.9 −18.5	LO.S.BR.GAL. + COMPACT (GRP. of 3) 440-G-45	13,20,24 D 2.2	F440/J3271 (78,217)
AM 1201-670	12 01 39 −67 01.9	3.096 −0.334	298.39 −4.84	182.9 −19.4	PLANETARY NEBULA NGC 4071 = PK 298-4-1 = 94-PN-12	25	F064/J2055 (213,338)
*AM 1202-275	12 02 07 −27 50.3	3.080 −0.334	290.54 33.64	141.5 −18.2	DISTURBED DWARF in WIDE PAIR MCG-5-29-5 = 440-G-46	20,23 D 2.0	F440/J3271 (71,294)
AM 1202-495	12 02 14 −49 58.4	3.089 −0.334	295.35 11.94	164.8 −20.0	SPIRAL + APP. COMP'N. 217-G-22	8 S 0.4	F217/J2194 (160,181)
AM 1202-432	12 02 17 −43 27.1	3.086 −0.334	294.10 18.35	157.9 −19.8	E + LO.S.BR. SP. + DWARF (ALIGNED) 267-G-13 = Fa 155	20,24 S 22.4	F267/J1560 (235,262)
AM 1202-352	12 02 36 −35 29.5	3.084 −0.334	292.49 26.17	149.5 −19.1	3 DWARFS PKS 1202-355?	20,20,20,24 S 5.8	F379/J1532 (151,153)
AM 1202-432	12 02 48 −43 29.6	3.089 −0.334	294.20 18.33	157.9 −19.7	DWARF 267-G-14	20 D 0.9	F267/J1560 (230,260)
AM 1203-220	12 03 14 −22 02.9	3.081 −0.334	289.25 39.34	135.5 −17.0	DWARF	20 D 0.4	F505/J2913 (215,338)
AM 1203-223	12 03 33 −22 34.3	3.082 −0.334	289.49 38.85	136.1 −17.0	LO.S.BR. SPIRAL + 2 COMP'NS. MCG-4-29-6 = VV 49 = 505-G-13	8 S 4.5	F505/J2913 (210,310)
AM 1203-354	12 03 41 −35 42.0	3.088 −0.334	292.78 26.01	149.7 −18.9	CLOSE PAIR (E + SPIRAL) 379-G-26	23 S 1.2	F379/J1532 (139,141)
AM 1203-441	12 03 47 −44 10.1	3.094 −0.334	294.51 17.70	158.7 −19.5	SPIRAL w. PECULIAR ARM 267-G-16	10,12 D 0.8	F267/J1560 (220,224)
AM 1203-241	12 03 48 −24 10.7	3.083 −0.334	290.02 37.29	137.7 −17.3	GROUP of 3	24 D 1.3	F505/J2913 (206,223)
AM 1203-275	12 03 51 −27 53.7	3.085 −0.334	291.01 33.67	141.6 −17.8	LINE of 3 incl. RING	6,24 D 4.0	F505/J2913 (203,24)
*AM 1204-280	12 04 08 −28 03.2	3.086 −0.334	291.12 33.52	141.7 −17.8	INTEGRAL-SIGN SP. in GROUP of 4 MCG-5-29-15 = 440-G-55	12,18 D 4.0	F440/J3271 (47,282)
*AM 1204-292	12 04 08 −29 29.2	3.087 −0.334	291.47 32.12	143.2 −18.0	E + DIFFUSE SPIRAL + 6 COMP'NS. NGC 4105/6 = 440-G-54/6	2,8 S 1.2	F441/J0562 (315,203)
*AM 1204-314	12 04 16 −31 40.2	3.088 −0.334	292.01 29.99	145.5 −18.3	I/A DBL. MCG-5-29-16/17 = 440-IG-58	2 S 0.2	F440/J3271 (50,88)
AM 1204-272	12 04 30 −27 25.1	3.087 −0.334	291.06 34.16	141.1 −17.6	E w. 9 COMP'NS. MCG-4-29-7 = 505-G-14	8 S 3.1	F505/J2913 (196,49)
*AM 1204-395	12 04 34 −39 55.8	3.095 −0.334	293.84 21.89	154.2 −19.1	PECULIAR SPIRAL + APP. COMP'N. NGC 4112 = 321-G-6 = To 1204-399	8,12 S 2.5	F321/J1310 (212,181)
AM 1204-294	12 04 38 −29 44.8	3.088 −0.334	291.66 31.89	143.5 −17.9	EDGE-ON SPIRAL + I/A COMP'N. IC 3005 = 441-G-2	1 S 2.1	F441/J0562 (309,189)
AM 1205-314 [1205-3143]	12 05 08 −31 44.6	3.092 −0.334	292.24 29.95	145.6 −18.1	PAIR of SPIRALS (#1) MCG-5-29-19 = 441-G-5	23 D 0.6	F441/J0562 (301,82)
AM 1205-341	12 05 08 −34 12.4	3.093 −0.334	292.78 27.54	148.2 −18.4	I/A TRIPLE in GROUP of 5	3,18,23 D 1.6	F379/J1532 (122,221)
*AM 1205-513	12 05 08 −51 34.4	3.111 −0.334	296.10 10.45	166.5 −19.6	SUPERNOVA SHELL ? 217-SNR-25	22	F217/J2194 (136,95)
*AM 1205-300	12 05 23 −30 03.6	3.091 −0.334	291.92 31.61	143.9 −17.8	RING/SB(r) TRANSITION IC 3010 = 441-G-6	6 D 1.6	F441/J0562 (300,173)
AM 1205-460	12 05 29 −46 00.9	3.106 −0.334	295.17 15.94	160.6 −19.3	CLOSE PAIR (BR. and FAINT RINGS)	6,23 S 3.1	F267/J1560 (203,125)
AM 1205-314 [1205-3144]	12 05 39 −31 43.9	3.093 −0.334	292.36 29.99	145.6 −18.0	PAIR of SPIRALS (#2) MCG-5-29-21 = 441-G-7	23 D 1.9	F441/J0562 (295,83)
AM 1206-311	12 06 26 −31 14.5	3.096 −0.334	292.44 30.50	145.1 −17.8	LINE of THREE IC 3015 = 441-G-9	24 D 13.0	F441/J0562 (286,109)
AM 1206-470 [1207-4704]	12 06 48 −47 02.3	3.116 −0.334	295.58 14.97	161.7 −19.1	CLOSE PAIR (#1: SPIRAL)	23 D 0.5	F267/J1560 (190,70)

Identification	R.A. (1950.0) Declination	Precession R.A./Dec.	Gal. Long. Gal. Lat.	S.G.L. S.G.B.	Description Cross-Identification	Coding Size (Min.)	Field/Plate (x, y)
AM 1206-240	12 06 52 −24 00.1	3.091 −0.334	290.83 37.62	137.7 −16.5	SPIRAL w. APP. COMP'N. MCG-4-29-9 = 505-G-22	8 S 1.4	F505/J2913 (169,233)
AM 1207-272	12 07 03 −27 24.8	3.094 −0.334	291.72 34.28	141.2 −17.1	CLOSE DBL.	23 S 0.6	F441/J0562 (283,315)
AM 1207-470 [1207-4702]	12 07 12 −47 04.2	3.118 −0.334	295.66 14.95	161.8 −19.1	CLOSE PAIR (#2: SPIRAL)	23 D 0.3	F267/J1560 (186,68)
AM 1207-294 [1207-2949]	12 07 38 −29 48.2	3.098 −0.334	292.42 31.96	143.7 −17.3	I/A DBL. (#1) MCG-5-29-26 = 441-G-14	2 D 1.0	F441/J0562 (274,187)
*AM 1207-292	12 07 39 −29 27.7	3.098 −0.334	292.35 32.30	143.3 −17.2	SPIRAL w. JET-LIKE ARM IC 764 = 441-G-13	10 D 5.6	F441/J0562 (274,205)
AM 1207-321 [1207-3214]	12 07 42 −32 16.9	3.101 −0.334	292.97 29.53	146.3 −17.6	I/A DBL. (#1)	2 D 0.5	F441/J0562 (271,54)
AM 1207-321 [1207-3216]	12 07 46 −32 14.1	3.102 −0.334	292.98 29.58	146.2 −17.6	I/A DBL. (#2)	2 D 0.3	F441/J0562 (270,56)
AM 1207-294 [1207-2948]	12 07 51 −29 48.8	3.099 −0.334	292.48 31.96	143.7 −17.3	I/A DBL. (#2)	2 D 0.3	F441/J0562 (271,186)
*AM 1208-334	12 08 03 −33 46.8	3.104 −0.334	293.36 28.07	147.8 −17.8	PECULIAR SP. w. 7 APP. COMP'NS. 379-G-31 = MC 1208-337	8 S 3.8	F379/J1532 (89,244)
*AM 1208-273	12 08 08 −27 32.2	3.098 −0.334	292.04 34.21	141.4 −16.8	SPIRAL w. I/A COMP'N. MCG-5-29-28 = 441-G-16	1 S 1.1	F441/J0562 (270,309)
AM 1208-361	12 08 16 −36 18.4	3.108 −0.334	293.91 25.59	150.5 −18.0	DWARF	20 D 0.4	F379/J1532 (90,108)
*AM 1208-305	12 08 33 −30 50.7	3.103 −0.334	292.87 30.97	144.8 −17.3	DISTURBED GALAXY + ELONG. COMP'N. MCG-5-29-29 = 441-G-17	8 S 1.7	F441/J0562 (262,131)
AM 1209-521	12 09 45 −52 11.0	3.146 −0.334	296.91 9.97	167.2 −18.9	SPIRAL w. PECULIAR ARMS 217-G-30 = Fa 308	10 D 1.3	F218/J4216 (345,56)
*AM 1209-414	12 09 52 −41 45.7	3.124 −0.334	295.24 20.26	156.2 −18.3	PAIR of E's w. NARROW BRIDGE	2 S 0.4	F321/J1310 (158,83)
AM 1210-321	12 10 02 −32 18.3	3.110 −0.334	293.54 29.59	146.3 −17.2	SPIRAL + APP. COMP'N. MCG-5-29-30 = 441-G-19	8 S 0.7	F380/J1540 (338,321)
AM 1210-283	12 10 06 −28 33.3	3.105 −0.334	292.78 33.29	142.5 −16.6	SMALL SPIRAL w. JET to COMP'N.	7,8 D 0.9	F441/J0562 (246,254)
AM 1210-251	12 10 17 −25 14.3	3.101 −0.334	292.09 36.55	139.1 −16.0	CONTACT DBL. in GROUP of 3	2,24 S 0.2	F505/J2913 (127,167)
*AM 1210-341	12 10 30 −34 12.0	3.115 −0.334	294.01 27.74	148.3 −17.3	I/A DBL. SPIRALS 379-G-32/33	2 S 1.3	F379/J1532 (62,220)
*AM 1210-341	12 10 32 −34 12.8	3.115 −0.334	294.02 27.73	148.3 −17.3	GROUP of 5 (PAIR of SP. + COMPACT) incl. AM 1210-3412.0	13,18 D 10.8	F380/J1540 (329,219)
AM 1211-341	12 11 02 −34 12.8	3.117 −0.334	294.14 27.75	148.4 −17.2	LARGE COMPACT 379-IG-35 = To 1211-342	13 D 0.4	F379/J1532 (57,219)
*AM 1211-375	12 11 11 −37 57.2	3.124 −0.334	294.85 24.06	152.3 −17.6	DWARF (SOME RESOLUTION) 321-G-14	20 D 1.3	F321/J1310 (143,287)
AM 1211-465 [1211-4657]	12 11 13 −46 59.4	3.143 −0.334	296.34 15.13	161.7 −18.4	I/A SPIRALS (#1) 267-G-29	2 D 0.6	F267/J1560 (149,73)
AM 1211-253	12 11 21 −25 32.2	3.105 −0.334	292.45 36.31	139.4 −15.8	CONTACT DBL. in GROUP of 3	2,24 S 0.1	F505/J2913 (114,151)
AM 1211-465 [1211-4659]	12 11 33 −46 57.2	3.145 −0.334	296.39 15.18	161.7 −18.3	I/A SPIRALS (#2) 267-G-30	2 D 0.7	F267/J1560 (146,75)
AM 1212-351	12 12 08 −35 13.6	3.123 −0.334	294.57 26.78	149.4 −17.1	CLOSE PAIR + COMP'N. 380-G-1 = Fa 309	8,23 S 10.3	F379/J1532 (46,165)
AM 1212-363	12 12 37 −36 32.8	3.128 −0.333	294.91 25.50	150.8 −17.2	DWARF	20 D 0.3	F379/J1532 (43,94)
AM 1212-405	12 12 40 −40 50.4	3.137 −0.333	295.64 21.25	155.3 −17.6	DWARF 321-G-15	20 D 0.8	F321/J1310 (129,132)
AM 1212-424	12 12 40 −42 44.7	3.141 −0.333	295.94 19.37	157.3 −17.8	CHAIN of 4 (CENTRAL E) 267-G-34	17 D 13.9	F267/J1560 (134,300)
AM 1212-344	12 12 47 −34 46.6	3.125 −0.333	294.64 27.25	149.0 −16.9	IRR. 380-G-5	16 D 0.7	F379/J1532 (38,189)
*AM 1212-375 [1213-3749]	12 12 48 −37 51.8	3.131 −0.333	295.18 24.20	152.2 −17.3	PAIR of SP. (#1: DWARF BETWEEN) 321-G-16	20,23 S 4.9	F379/J1532 (44,23)
AM 1213-311	12 13 00 −31 17.1	3.119 −0.333	294.06 30.70	145.4 −16.4	DWARF 441-G-23	20 D 0.9	F441/J0562 (211,108)
*AM 1213-343	12 13 07 −34 37.2	3.126 −0.333	294.69 27.41	148.8 −16.8	3-ARMED SPIRAL A 1213-34 = 380-G-7	11 D 0.9	F379/J1532 (34,197)
AM 1213-323	12 13 09 −32 38.8	3.122 −0.333	294.34 29.36	146.8 −16.5	CLOSE PAIR (SPIRALS)	23 S 0.8	F380/J1540 (302,304)
AM 1213-220	12 13 22 −22 08.8	3.105 −0.333	292.26 39.72	136.1 −14.7	SHRED (PECULIAR FILAMENTS)	20,22 D 0.5	F505/J2913 (89,333)
AM 1213-374 [1212-3751]	12 13 36 −37 48.9	3.135 −0.333	295.34 24.27	152.2 −17.1	PAIR of SP. (#2: DWARF BETWEEN) 321-G-18 = To 1213-378	23 S 4.9	F379/J1532 (35,25)
AM 1213-430	12 13 48 −43 02.1	3.148 −0.333	296.20 19.11	157.6 −17.6	LARGE SPIRAL w. EXTENDED WINGS NGC 4219 = Ag-34 = 267-G-37	12 D 4.5	F267/J1560 (123,285)

Identification	R.A. (1950.0) Declination	Precession R.A./Dec.	Gal. Long. Gal. Lat.	S.G.L. S.G.B.	Description Cross-Identification	Coding Size (Min.)	Field/Plate (x, y)
AM 1214-371	12 14 20 −37 11.2	3.136 −0.333	295.39 24.91	151.5 −16.9	Sd/DWARF w. PECULIAR ARM+ COMP'NS 380-G-8	8,10,20 D 3.1	F321/J1310 (109,328)
AM 1214-255	12 14 22 −25 56.0	3.114 −0.333	293.36 36.03	140.0 −15.2	I/A DBL. w. TAILS MCG-4-29-16 = VV 333b = 505-IG-31	2,15 S 0.5	F505/J2913 (78,129)
AM 1214-505	12 14 25 −50 54.6	3.176 −0.333	297.45 11.33	165.9 −18.1	GROUP of 5 217-G-32	18 D 7.2	F218/J4216 (311,127)
AM 1214-324	12 14 36 −32 43.8	3.128 −0.333	294.70 29.33	146.9 −16.3	DWARF	20 D 0.4	F379/J1532 (14,298)
AM 1214-504	12 14 52 −50 46.9	3.179 −0.333	297.51 11.47	165.7 −18.0	I/A DBL.	2 S 0.9	F218/J4216 (307,134)
AM 1215-793	12 15 16 −79 31.2	3.554 −0.333	301.40 −17.01	195.9 −17.1	DWARF IC 3104 = 20-G-4 = UKS 1215-794	20 D 0.1	F020/J2059 (142,204)
AM 1215-421	12 15 42 −42 14.7	3.156 −0.333	296.45 19.94	156.8 −17.2	CLOSE PAIR (E + IRR./DEFECT?)	23 S 3.0	F267/J1560 (103,327)
AM 1215-243	12 15 57 −24 38.8	3.116 −0.333	293.54 37.36	138.7 −14.6	PAIR	23 S 0.6	F505/J2913 (58,198)
AM 1216-431	12 16 10 −43 13.3	3.162 −0.333	296.69 18.99	157.8 −17.2	DISTURBED GALAXY + APP. COMP'N. 267-IG-39	8,16 D 0.7	F267/J1560 (100,274)
AM 1216-350	12 16 49 −35 02.2	3.142 −0.333	295.60 27.11	149.4 −16.1	DWARF 380-G-13	20 D 0.7	F380/J1540 (258,176)
AM 1217-633	12 17 07 −63 31.1	3.273 −0.333	299.45 −1.13	179.1 −17.9	NEBULOSITY	21 D	F095/J2356 (251,257)
*AM 1217-354	12 17 20 −35 40.9	3.146 −0.333	295.82 26.49	150.1 −16.1	I/A DBL. (GALAXY + "S"-SHAPE) 380-IG-15 = PKS 1217-357?	2 S 0.5	F380/J1540 (252,142)
AM 1217-254	12 17 33 −25 47.3	3.122 −0.333	294.21 36.29	139.9 −14.5	SPIRAL w. COMP'NS. in PAIR w. E MCG-4-29-19 = 506-G-2	8,23 S 7.2	F505/J2913 (39,137)
AM 1217-624	12 17 48 −62 42.1	3.274 −0.333	299.43 −0.31	178.3 −17.9	5 NEBULOSITIES PK 299-0-2	21	F095/J2356 (249,301)
AM 1218-343	12 18 14 −34 30.5	3.146 −0.333	295.85 27.68	148.9 −15.8	GALAXY w. LO.S.BR. KNOTS 380-G-17	16 D 0.8	F380/J1540 (243,205)
AM 1219-383	12 19 12 −38 30.6	3.162 −0.333	296.64 23.74	153.0 −16.1	BRIGHT STAR on NEBULOSITY 321-G?-26	21 D 0.7	F322/J1370 (332,254)
*AM 1219-430	12 19 18 −43 03.4	3.178 −0.333	297.26 19.23	157.7 −16.6	GAL.w.I/A COMP'N(LONG ARM)[Ag-36] Se 90/3 = 267-IG-41 = Fa 156/7 [Lu]	1 S 1.6	F267/J1560 (69,282)
AM 1219-455	12 19 23 −45 55.8	3.190 −0.333	297.64 16.37	160.7 −16.9	GALAXY w. APP. COMP'NS. 267-G-43	8 S 1.3	F267/J1560 (73,128)
AM 1220-630	12 20 08 −63 00.7	3.303 −0.333	299.73 −0.58	178.6 −17.6	2 NEBULOSITIES PK 299-1-1	21 D	F095/J2356 (234,285)
AM 1220-335	12 20 16 −33 54.7	3.152 −0.333	296.23 28.32	148.3 −15.3	IRR. + APP. COMP'N. To 1220-339 = 380-G-21	8,16 S 4.0	F380/J1540 (221,237)
AM 1220-445	12 20 47 −44 55.1	3.194 −0.333	297.77 17.41	159.7 −16.6	FAINT RING	6 D 0.5	F267/J1560 (58,182)
AM 1221-595	12 21 09 −59 56.6	3.286 −0.333	299.52 2.48	175.4 −17.4	PLANETARY NEBULA PK 299+2+1 = 131-PN-1	25	F130/J2280 (51,178)
AM 1221-415	12 21 10 −41 54.9	3.184 −0.333	297.48 20.40	156.6 −16.2	SPIRAL + APP. COMP'N. 322-G-1	8 S 1.0	F321/J1310 (45,72)
AM 1221-351	12 21 33 −35 16.1	3.162 −0.333	296.72 27.01	149.8 −15.2	COMPACT	13 D 0.1	F380/J1540 (206,165)
AM 1221-403	12 21 33 −40 39.6	3.181 −0.333	297.40 21.65	155.3 −16.0	M51-TYPE (very small)	9 S 0.3	F321/J1310 (39,139)
AM 1221-445	12 21 38 −44 54.7	3.199 −0.333	297.93 17.43	159.7 −16.4	CLOSE PAIR (HI.S.BR. GAL. + SP.) 267-G-45	23 S 0.8	F268/J2229 (316,180)
AM 1221-420	12 21 48 −42 00.7	3.187 −0.332	297.62 20.32	156.7 −16.1	STAR w. NEBULOSITY	21 D 0.7	F321/J1310 (39,66)
AM 1221-350	12 21 58 −35 07.7	3.163 −0.332	296.79 27.16	149.6 −15.1	ISOLATED IRR. 380-G-25	16 D 1.6	F380/J1540 (202,172)
AM 1222-254	12 22 20 −25 45.3	3.136 −0.332	295.53 36.47	140.1 −13.4	MDM.S.BR. DISK + HI.S.BR. NUCLEUS IC 3289 = 506-G-7	16,22 D 0.9	F506/J2908 (248,138)
AM 1222-500	12 22 24 −50 01.4	3.229 −0.332	298.63 12.36	165.0 −16.8	3 GALAXIES in COMMON ENVELOPE 218-IG-4	3 D 0.5	F218/J4216 (245,177)
AM 1222-392	12 22 34 −39 29.3	3.181 −0.332	297.47 22.84	154.1 −15.6	E + SB (CLOSE) SS-11 IC 3290/NGC 4373 = 322-G-4/6	23 S 2.0	F321/J1310 (26,201)
AM 1222-471	12 22 38 −47 11.8	3.215 −0.332	298.37 15.18	162.1 −16.5	cD + PECULIAR COMP'N. 268-G-4	8 S 5.8	F268/J2229 (301,58)
AM 1223-385	12 23 30 −38 50.9	3.183 −0.332	297.59 23.50	153.5 −15.3	CHAIN of 4 GALAXIES 322-G-9 = To 1223-388	17 D 15.7	F321/J1310 (14,235)
AM 1224-360	12 24 14 −36 08.8	3.176 −0.332	297.43 26.20	150.7 −14.8	IRR./DWARF 380-G-29	16,20 D 1.8	F380/J1540 (176,118)
AM 1224-245	12 24 41 −24 55.1	3.140 −0.332	296.06 37.37	139.4 −12.7	CLOSE PAIR	23 S 1.0	F506/J2908 (220,184)
*AM 1224-331	12 24 45 −33 14.8	3.167 −0.332	297.20 29.09	147.8 −14.2	SPIRAL + I/A COMP'N. MCG-5-30-1? = 380-G-30 = To1224-332	1 S 0.7	F380/J1540 (171,273)

Identification	R.A. (1950.0) Declination	Precession R.A./Dec.	Gal. Long. Gal. Lat.	S.G.L. S.G.B.	Description Cross-Identification	Coding Size (Min.)	Field/Plate (x, y)
AM 1224-340 [1225-3408]	12 24 45 −34 06.8	3.171 −0.332	297.30 28.23	148.7 −14.4	PAIR of HI.S.BR. GALAXIES (#1) To 1224-340 = 380-IG-31	23 D 0.3	F380/J1540 (171,227)
*AM 1224-390	12 24 55 −39 03.5	3.191 −0.332	297.91 23.32	153.7 −15.1	E w. JETS in OUTER REGIONS IC 3370 = 322-G-14	7 D 1.8	F322/J1370 (271,226)
AM 1224-405	12 24 56 −40 52.6	3.199 −0.332	298.11 21.51	155.6 −15.3	DWARF 322-G-13	20 D 0.6	F322/J1370 (269,129)
AM 1224-374	12 24 58 −37 49.7	3.186 −0.332	297.79 24.54	152.5 −14.9	DWARF 322-G-15	20 D 0.7	F322/J1370 (272,292)
AM 1225-340 [1224-3406]	12 25 04 −34 08.4	3.172 −0.332	297.38 28.21	148.7 −14.3	PAIR of HI.S.BR. GALAXIES (#2) To 1225-341 = 380-IG-33	23 D 0.5	F380/J1540 (168,225)
AM 1225-370	12 25 19 −37 03.2	3.184 −0.332	297.78 25.32	151.7 −14.7	DWARF-SHRED (RESOLUTION?) 380-G-34	20 D 2.0	F380/J1540 (165,69)
AM 1226-291	12 26 28 −29 14.6	3.159 −0.332	297.13 33.11	143.8 −13.2	CLOSE PAIR	23 S 1.7	F441/J0562 (54,216)
AM 1226-402	12 26 28 −40 24.4	3.204 −0.332	298.38 22.00	155.2 −15.0	CLOSE PAIR of DWARFS 322-G-19/20	20,23 S 1.8	F322/J1370 (254,154)
*AM 1226-225	12 26 44 −22 53.6	3.139 −0.332	296.36 39.43	137.5 −11.9	DISTURBED SPIRAL w. 5 off MI.AX. NGC 4462 = 506-G-13	2,8,16 S 14.6	F506/J2908 (195,292)
AM 1228-435	12 28 10 −43 57.6	3.231 −0.331	299.05 18.49	158.9 −15.1	PLANETARY NEBULA PK 299+18-1 Ko-23 = 268-PN-12 = Lu 1228-43	25 D 0.9	F268/J2229 (255,234)
AM 1228-330	12 28 25 −33 07.1	3.181 −0.331	298.06 29.30	147.8 −13.5	CLOSE PAIR of E's	23 S 1.3	F380/J1540 (130,280)
AM 1228-344	12 28 30 −34 41.6	3.188 −0.331	298.24 27.73	149.4 −13.7	GALAXY w. HI.S.BR. KNOTS	16 D 0.7	F380/J1540 (130,196)
AM 1228-353 [1228-3539]	12 28 30 −35 35.9	3.192 −0.331	298.33 26.83	150.3 −13.9	PAIR of SPIRALS (#1)	23 D 0.3	F380/J1540 (130,147)
AM 1228-353 [1228-3535]	12 28 31 −35 39.1	3.192 −0.331	298.34 26.78	150.4 −13.9	PAIR of SPIRALS (#2)	23 D 0.5	F380/J1540 (130,144)
AM 1228-260	12 28 39 −26 00.6	3.154 −0.331	297.32 36.38	140.7 −12.1	SPIRAL w. I/A COMP'N. MCG-4-30-3 = 506-G-15	1 S 0.4	F506/J2908 (171,125)
*AM 1229-512	12 29 09 −51 28.3	3.286 −0.331	299.85 11.02	166.6 −15.8	EDGE-ON or NUCLEUS w. JETS 218-L-8	7 D 2.5	F218/J4216 (186,101)
AM 1229-332	12 29 34 −33 26.7	3.187 −0.331	298.37 28.99	148.2 −13.3	CLOSE PAIR (ONE w. BR. KNOTS)	23 S 0.8	F380/J1540 (117,262)
AM 1230-323 [1230-3235]	12 30 27 −32 36.3	3.186 −0.331	298.49 29.84	147.4 −13.0	I/A DBL. (#1)	2 D 0.4	F380/J1540 (107,307)
AM 1230-323 [1230-3236]	12 30 32 −32 35.0	3.186 −0.331	298.51 29.86	147.4 −12.9	I/A DBL. (#2)	2 D 0.5	F380/J1540 (106,308)
AM 1230-272	12 30 46 −27 22.7	3.166 −0.331	298.05 35.06	142.1 −11.9	SMALL MAGELLANIC TYPE MCG-4-30-4 = 506-G-17	16 D 1.0	F506/J2908 (146,52)
AM 1231-345 [1231-3451]	12 31 07 −34 50.0	3.199 −0.331	298.86 27.64	149.6 −13.2	PAIR of SPIRALS (#1) 380-G-42	6,23 D 0.8	F380/J1540 (101,188)
AM 1231-345 [1231-3450]	12 31 26 −34 51.5	3.200 −0.331	298.93 27.62	149.7 −13.1	PAIR of SPIRALS (#2) 380-G-44	6,23 D 0.7	F380/J1540 (97,186)
AM 1232-363	12 32 17 −36 35.3	3.212 −0.331	299.27 25.90	151.5 −13.3	SPIRAL + 4 APP. COMP'NS. 380-G-46 = To 1231-365?	8 S 2.2	F380/J1540 (90,93)
AM 1232-365	12 32 22 −36 59.4	3.215 −0.331	299.32 25.50	151.9 −13.3	CLOSE PAIR	23 S 1.3	F380/J1540 (89,72)
AM 1232-321	12 32 46 −32 15.0	3.193 −0.331	299.03 30.24	147.1 −12.4	SMALL I/A DBL. 442-IG-8	2 S 0.1	F380/J1540 (80,326)
AM 1233-424 [1233-4246]	12 33 08 −42 46.8	3.251 −0.331	299.91 19.74	157.8 −14.1	PAIR (#1: DISTURBED E) 268-G-23	23 D 0.4	F268/J2229 (208,298)
AM 1233-424 [1233-4246]	12 33 16 −42 44.8	3.252 −0.330	299.94 19.74	157.8 −14.1	PAIR (#2: DBL. RING + COMP'N.) 268-G-24	6,8,23 D 1.0	F268/J2229 (207,298)
*AM 1234-415	12 34 04 −41 51.9	3.250 −0.330	300.03 20.66	156.9 −13.8	SB(r) (CONDENSATION in BAR) 322-G-31	12 D 1.3	F322/J1370 (176,77)
AM 1235-432	12 35 00 −43 20.8	3.265 −0.330	300.31 19.19	158.4 −13.8	SB(r) w. FAINT OUTER RING NGC 4573 = 268-G-26 = Fa 455	6,12 D 2.0	F268/J2229 (189,268)
AM 1235-325 [1236-3253]	12 35 49 −32 50.6	3.207 −0.330	299.81 29.69	147.8 −11.9	CLOSE PAIR (both RESOLVED) (#1) 380-G-52	23 D 0.4	F380/J1540 (46,294)
AM 1235-284	12 35 52 −28 40.2	3.187 −0.330	299.53 33.85	143.7 −11.0	CLOSE PAIR	23 S 1.9	F442/J1333 (213,248)
AM 1236-325 [1235-3250]	12 36 03 −32 53.3	3.208 −0.330	299.87 29.65	147.9 −11.8	CLOSE PAIR (both RESOLVED) (#2) 381-G-2	23 D 0.5	F380/J1540 (44,291)
AM 1236-270	12 36 14 −27 02.1	3.180 −0.330	299.51 35.49	142.0 −10.6	SPIRAL w. CHAIN of 4 COMP'NS. MCG-4-30-7 = 506-G-27	8,17 D 7.6	F506/J2908 (81,69)
AM 1236-343	12 36 29 −34 30.3	3.219 −0.330	300.08 28.04	149.5 −12.1	S0 w. V.LO.S.BR. ARMS + JET 381-G-4 = Fa 832	7,15 D 0.5	F381/J2339 (307,204)
*AM 1237-364	12 37 50 −36 41.6	3.237 −0.329	300.52 25.87	151.8 −12.2	M51-TYPE or LONG ARM + COMP'N 381-G-5	8,9,10 S 1.1	F381/J2339 (288,87)
AM 1237-231	12 37 51 −23 17.7	3.168 −0.329	299.69 39.25	138.4 −9.4	EDGE-ON or GAL. w. JETS (in PAIR)	7,23 D 0.5	F507/J2340 (329,269)

Identification	R.A. (1950.0) Declination	Precession R.A./Dec.	Gal. Long. Gal. Lat.	S.G.L. S.G.B.	Description Cross-Identification	Coding Size (Min.)	Field/Plate (x, y)
AM 1237-225	12 37 56 −22 59.5	3.166 −0.329	299.70 39.55	138.1 −9.4	I/A DBL. 506-IG-34	2 S 0.3	F507/J2340 (328,286)
AM 1238-340	12 38 00 −34 05.1	3.222 −0.329	300.41 28.47	149.2 −11.7	IRR. w. I/A COMP'N.	1,16 S 0.2	F380/J1540 (24,226)
*AM 1238-362	12 38 10 −36 28.8	3.237 −0.329	300.58 26.08	151.6 −12.1	HI.S.BR. SP. + BARRED RING + GAL. IC 3639 = 381-G-8 = Fa 312/13 [To]	6,12,24 S 1.6	F381/J2339 (285,98)
AM 1238-404	12 38 10 −40 42.1	3.264 −0.329	300.79 21.87	155.8 −12.8	GALAXY + 2 COMP'NS. NGC 4603 = 322-G-52	8 D 2.9	F322/J1370 (134,139)
AM 1238-254	12 38 24 −25 41.1	3.180 −0.329	300.02 36.87	140.8 −9.9	PECULIAR SPIRAL w. 3 COMP'NS. MCG-4-30-11 = 506-IG-35	8,12 S 3.4	F507/J2340 (320,141)
AM 1238-405	12 38 29 −40 53.2	3.266 −0.329	300.86 21.68	156.0 −12.8	V.LO.S.BR. DWARF	20 D 0.8	F322/J1370 (131,129)
AM 1239-393	12 39 14 −39 32.7	3.261 −0.329	300.96 23.03	154.7 −12.4	I/A DBL. w. MATERIAL 322-IG-54	2 S 0.3	F322/J1370 (123,201)
AM 1239-421	12 39 22 −42 11.4	3.280 −0.329	301.10 20.39	157.4 −12.9	N GAL. or PLANETARY NEBULA	22,25 D 0.4	F322/J1370 (123,59)
AM 1239-250	12 39 30 −25 09.4	3.181 −0.329	300.30 37.41	140.3 −9.5	GALAXY w. JET + CONDENSATION 507-G-2	7 S 0.8	F506/J2908 (39,170)
*AM 1239-402	12 39 51 −40 28.1	3.270 −0.329	301.12 22.11	155.6 −12.5	SPIRAL w. THIN, OFF-CENTRE ARMS NGC 4622 = Fa-33 = 322-G-57 [To?]	10 D 1.3	F322/J1370 (117,151)
AM 1240-233	12 40 01 −23 30.1	3.174 −0.329	300.35 39.06	138.7 −9.0	DISTURBED SINGLE w. FILAMENT MCG-4-30-12 = 507-G-4	15,16 D 0.4	F507/J2340 (302,259)
AM 1240-423	12 40 17 −42 39.1	3.288 −0.329	301.30 19.93	157.9 −12.8	DISTURBED EDGE-ON + 5 COMP'NS 268-G-34 = Fa 456 (+ Fa 457?)	8 S 6.0	F268/J2229 (138,305)
AM 1240-404	12 40 57 −40 41.7	3.277 −0.329	301.36 21.89	155.9 −12.3	DWARF 322-G-63	20 D 0.5	F322/J1370 (106,139)
*AM 1241-402	12 41 03 −40 26.6	3.276 −0.329	301.37 22.14	155.7 −12.3	I/A DBL. + MATERIAL Ag-38? NGC 4622a/b = 322-IG-64 = Fa 33?	2 S 0.4	F322/J1370 (105,152)
AM 1241-290	12 41 04 −29 06.4	3.206 −0.329	300.93 33.47	144.3 −10.0	COMPACT w. JET	7,13 D 0.1	F442/J1333 (152,225)
AM 1241-335	12 41 22 −33 55.8	3.234 −0.329	301.20 28.66	149.1 −11.0	GAL. w. I/A COMP'NS. (LINE of 3) 381-G-12 = MC 1241-339 (+ Fa 458)	1,24 S 1.6	F381/J2339 (253,236)
*AM 1241-402	12 41 33 −40 27.4	3.278 −0.329	301.47 22.13	155.7 −12.2	SB (CENTRE of CHAIN of 5) NGC 4650 = 322-G-67	17 S 2.0	F322/J1370 (100,151)
AM 1241-541	12 41 58 −54 14.9	3.411 −0.328	301.96 8.35	169.7 −14.2	PLANETARY NEBULA 172-?-7	25	F172/J2384 (156,222)
AM 1242-402	12 42 00 −40 26.4	3.281 −0.328	301.56 22.15	155.7 −12.1	SPINDLE w. ERUPTION NGC 4650a = Ag-38 = 320-IG-69	16 D 1.8	F322/J1370 (95,152)
AM 1242-301	12 42 48 −30 13.7	3.218 −0.328	301.42 32.36	145.5 −9.9	I/A DBL.	2 S 0.4	F442/J1333 (132,165)
AM 1243-262	12 43 01 −26 21.7	3.197 −0.328	301.34 36.23	141.7 −9.0	PECULIAR SPIRAL w. CLOSE COMP'N. MCG-4-30-15 = 507-G-8	8,12 S 1.6	F507/J2340 (263,106)
AM 1243-333	12 43 18 −33 33.9	3.239 −0.328	301.64 29.03	148.9 −10.5	LARGE DWARF (SOME RESOLUTION) 381-G-20 = UKS 1243-336	20 D 1.8	F381/J2339 (232,256)
AM 1243-422	12 43 35 −42 25.0	3.304 −0.328	301.94 20.18	157.7 −12.1	LUMPY COMPACT	13,16 D 0.3	F268/J2229 (105,318)
AM 1243-412	12 43 38 −41 29.3	3.297 −0.328	301.92 21.11	156.8 −12.0	IRR. w. ABSORPTION NGC 4672	16,14 D 1.1	F323/J 642 (341,94)
AM 1243-490	12 43 57 −49 00.5	3.366 −0.328	302.15 13.60	164.4 −13.2	GROUP of 3 218-G-19/16/17	24 D 8.5	F218/J4216 (57,230)
AM 1244-405	12 44 12 −40 53.9	3.295 −0.328	302.03 21.70	156.2 −11.7	DWARF 322-G-79	20 D 1.0	F322/J1370 (73,127)
AM 1245-255	12 45 00 −25 55.6	3.200 −0.328	301.88 36.68	141.4 −8.5	3 SPIRALS (STRONG ARM) + COMPACT MCG-4-30-16 = VV290a = 507-G-11	10,13,18 S 2.7	F507/J2340 (240,129)
AM 1245-354	12 45 04 −35 48.0	3.261 −0.328	302.11 26.81	151.2 −10.6	CLOSE PAIR 381-G-24	23 S 0.6	F381/J2339 (211,136)
AM 1245-215	12 45 19 −21 59.7	3.179 −0.327	301.88 40.61	137.5 −7.5	SPIRAL w. 3 COMP'NS.(ONE COMPACT) MCG-4-30-19 = 574-G-33	8,13 S 3.6	F507/J2340 (238,340)
AM 1246-410	12 46 01 −41 02.2	3.305 −0.327	302.40 21.57	156.4 −11.4	CENTAURUS CLUSTER = KLEM 19/20 NGC 4696 = 322-G-91 = DN 1246-410	18 D 1.8	F322/J1370 (55,119)
AM 1246-251	12 46 48 −25 19.1	3.201 −0.327	302.38 37.29	140.9 −7.9	LINEAR DWARF	20 D 0.4	F507/J2340 (218,162)
AM 1247-245	12 47 01 −24 53.2	3.199 −0.327	302.43 37.72	140.5 −7.8	SB + DWARF COMP'N. 507-G-16	8,20 S 2.7	F507/J2340 (216,185)
AM 1247-233	12 47 05 −23 35.5	3.192 −0.327	302.44 39.02	139.2 −7.5	SPIRAL w. PECULIAR ARMS 507-G-17	10 D 0.5	F507/J2340 (215,255)
*AM 1247-253	12 47 21 −25 38.8	3.205 −0.327	302.54 36.96	141.2 −7.9	RING SP.(LARGE HALO) + 2 COMPACTS IC 3813 = 507-G-19 + DWARF	6,8,13,20 S 2.0	F507/J2340 (211,144)
AM 1247-410	12 47 23 −41 09.3	3.313 −0.327	302.68 21.46	156.6 −11.2	E w. BRIDGE to COMP'N. NGC 4709	1 S 1.2	F323/J 642 (303,114)
AM 1247-440	12 47 32 −44 08.8	3.340 −0.327	302.72 18.47	159.6 −11.7	DISTURBED DWARF 268-G-46 = PKS 1247-442?	20 D 2.0	F269/J1553 (338,222)

Identification	R.A. (1950.0) Declination	Precession R.A./Dec.	Gal. Long. Gal. Lat.	S.G.L. S.G.B.	Description Cross-Identification	Coding Size (Min.)	Field/Plate (x, y)
AM 1247-403	12 47 36 −40 34.4	3.309 −0.327	302.71 22.04	156.0 −11.1	DISTURBED GALAXY (near BR.STAR) 323-G-7	16 D 0.9	F322/J1370 (38,143)
AM 1247-421	12 47 55 −42 16.6	3.325 −0.327	·302.79 20.34	157.7 −11.3	3-ARM SP. w. FILAMENT to CONDENS. 323-G-10	10,11 D 1.6	F268/J2229 (61,324)
AM 1248-223	12 48 02 −22 37.1	3.189 −0.327	302.71 39.99	138.3 −7.0	SPIRAL w. I/A + APP. COMP'N. 507-G-22	1,8 S 0.8	F507/J2340 (204,307)
AM 1248-221	12 48 30 −22 15.9	3.188 −0.327	302.85 40.35	138.0 −6.8	GROUP/CLUSTER	17,18 D 1.6	F507/J2340 (198,326)
AM 1248-451	12 48 43 −45 16.0	3.358 −0.326	302.95 17.35	160.8 −11.7	COMPACT	13 D 0.3	F269/J1553 (324,162)
AM 1248-261	12 48 51 −26 10.9	3.212 −0.326	302.96 36.43	141.8 −7.7	CENTRE of EXTENDED CLUSTER MCG-4-30-29 = 507-G-25	8,19 D 1.8	F507/J2340 (193,116)
AM 1248-273	12 48 51 −27 30.9	3.220 −0.326	302.96 35.10	143.2 −8.0	SPIRAL + 5 COMP'NS (I/A DBL.SP.) IC 3829 = 442-G-24	2,8 S 2.0	F507/J2340 (193,44)
AM 1248-275	12 48 53 −27 58.4	3.223 −0.326	302.97 34.64	143.6 −8.1	SPIRAL + 3 APP. COMP'NS 442-G-25	8 S 3.1	F507/J2340 (193,20)
AM 1248-254	12 48 54 −25 49.4	3.210 −0.326	302.97 36.79	141.5 −7.6	DISRUPTED DWARF ? 507-G-26	20 S 2.2	F507/J2340 (193,135)
AM 1248-255 [1249-2549]	12 48 56 −25 50.9	3.210 −0.326	302.98 36.76	141.5 −7.6	PAIR + LO.S.BR. KNOTS(DWARF?)(#1) A 1248-25 = MCG-4-30-30 = 507-G-27	20,23 S 2.2	F507/J2340 (192,134)
AM 1249-254 [1248-2550]	12 49 04 −25 49.2	3.210 −0.326	303.02 36.79	141.5 −7.5	PAIR (#2) MCG-4-30-31 = 507-G-28	23 S 2.2	F507/J2340 (191,135)
AM 1249-462	12 49 48 −46 23.0	3.375 −0.326	303.15 16.23	161.9 −11.8	NUCLEUS in SIDE of RING + COMP'N.	6,8 S 1.3	F268/J2229 (51,103)
*AM 1249-313	12 49 58 −31 37.0	3.251 −0.326	303.24 30.99	147.2 −8.7	DWARF 442-G-28	20 D 1.3	F442/J1333 (51,89)
AM 1250-422	12 50 01 −42 23.2	3.337 −0.326	303.20 20.22	157.9 −11.0	GALAXY w. PECULIAR ABSORPTION 323-G-29	14 D 1.0	F322/J1370 (18,45)
AM 1250-401	12 50 06 −40 13.2	3.318 −0.326	303.23 22.39	155.8 −10.5	SPIRAL w. BRANCHED ARM	10 D 1.2	F323/J 642 (277,165)
AM 1250-410	12 50 09 −41 06.2	3.326 −0.326	303.23 21.51	156.7 −10.7	SPIRAL w. I/A COMP'N.	1 S 0.7	F323/J 642 (275,117)
AM 1250-234	12 50 18 −23 46.8	3.201 −0.326	303.38 38.83	139.6 −6.8	CLOSE PAIR + 5 APP. COMP'NS. 507-G-36 (E + EDGE-ON SPIRAL)	8,23 S 3.1	F507/J2340 (176,245)
*AM 1250-271	12 50 28 −27 11.5	3.223 −0.326	303.40 35.42	142.9 −7.6	DISTURBED SINGLE GALAXY MCG-4-31-1 = 507-G-37	16 D 1.2	F443/J1469 (315,326)
AM 1250-220	12 50 49 −22 04.6	3.192 −0.326	303.56 40.53	137.9 −6.2	Sd/DWARF 575-G-19	20 D 1.0	F507/J2340 (169,336)
AM 1250-250	12 50 53 −25 06.7	3.211 −0.326	303.54 37.50	140.9 −7.0	DISTURBED SINGLE GALAXY MCG-4-31-4 = 507-G-41	16 D 0.5	F507/J2340 (168,173)
AM 1251-425	12 51 23 −42 52.1	3.349 −0.326	303.47 19.74	158.5 −10.8	DWARF (RESOLUTION:) 269-G-3	20 D 1.1	F268/J2229 (28,291)
AM 1251-240 [1251-2409]	12 51 26 −24 09.4	3.206 −0.326	303.71 38.45	140.0 −6.6	PAIR of IRR. (#1)	16,23 S 1.6	F507/J2340 (162,224)
AM 1251-240 [1251-2409]	12 51 33 −24 09.9	3.207 −0.326	303.75 38.44	140.0 −6.6	PAIR of IRR. (#2)	16,23 S 1.6	F507/J2340 (160,224)
*AM 1251-313	12 51 53 −31 36.6	3.258 −0.325	303.72 31.00	147.3 −8.3	DWARF (V.LO.S.BR.) 443-G-6	20 D 1.8	F443/J1469 (293,89)
AM 1251-283	12 51 58 −28 36.5	3.237 −0.325	303.79 34.00	144.4 −7.6	RING + COMPACT in CHAIN of 4	6,13,17 D 0.5	F442/J1333 (23,250)
AM 1252-285	12 52 07 −28 53.8	3.239 −0.325	303.82 33.71	144.7 −7.6	CHAIN of 4 KLEM 21 = DN 1252-287	17 D 13.4	F443/J1469 (293,235)
AM 1252-230	12 52 10 −23 06.8	3.202 −0.325	303.94 39.49	139.0 −6.2	CHAIN of 6	17 D 6.7	F507/J2340 (153,280)
AM 1252-280	12 52 11 −28 04.2	3.234 −0.325	303.85 34.54	143.9 −7.4	DWARF 443-G-9	20 D 0.4	F442/J1333 (20,278)
AM 1252-453	12 52 11 −45 37.8	3.381 −0.325	303.58 16.98	161.2 −11.2	GROUP of 4 (incl. 3 PEC. SPIRALS) incl. 269-G-6 = Fa 459	18 D 9.0	F269/J1553 (290,144)
*AM 1252-264	12 52 15 −26 40.9	3.225 −0.325	303.90 35.93	142.5 −7.1	SMALL I/A TRIPLE in LINE of 3	3,24 D 1.8	F507/J2340 (152,89)
*AM 1252-443	12 52 32 −44 32.7	3.372 −0.325	303.66 18.06	160.2 −10.9	HI.S.BR.SP.(I/A COMP'N. + COMPACT) Se 94/1 = 269-G-8 = Fa 460 = SS-13	1,13 S 0.9	F269/J1553 (289,202)
AM 1252-484	12 52 35 −48 48.3	3.420 −0.325	303.61 13.80	164.4 −11.7	GALAXY w. JETS on BOTH SIDES 219-G-7	7 D 0.9	F219/J2247 (245,217)
AM 1252-441	12 52 39 −44 14.8	3.369 −0.325	303.69 18.36	159.9 −10.9	WIDE PAIR (BOTH w. FILAMENTS)	23 S 8.5	F268/J2229 (19,217)
AM 1252-445	12 52 39 −44 57.5	3.377 −0.325	303.68 17.65	160.6 −11.0	IRR. DWARF w. STRUCTURE Se 94/2 = 269-G-9 = UKS 1252-449	20 D 1.3	F269/J1553 (287,180)
AM 1252-434	12 52 45 −43 44.6	3.365 −0.325	303.72 18.86	159.4 −10.7	PECULIAR S(r) w. I/A COMP'N. 269-G-10 = Fa 461	1,12 S 1.3	F268/J2229 (17,244)
AM 1252-251	12 52 48 −25 16.0	3.217 −0.325	304.08 37.34	141.2 −6.6	GALAXY w. HI. + LO.S.BR. ARMS 507-G-44	10 D 0.5	F507/J2340 (145,165)

Identification	R.A. (1950.0) Declination	Precession R.A./Dec.	Gal. Long. Gal. Lat.	S.G.L. S.G.B.	Description Cross-Identification	Coding Size (Min.)	Field/Plate (x, y)
*AM 1252-300	12 52 55 −30 04.7	3.250 −0.325	304.01 32.53	145.9 −7.7	I/A E's in CHAIN of 7 443-G-11	2,3,7,17 S 1.0	F443/J1469 (282,172)
AM 1253-262	12 53 01 −26 28.1	3.226) −0.325	304.11 36.14	142.3 −6.8	"RINGING" GALAXY + 4 COMP'NS. MC 1252-266	8,16,22 S 3.6	F507/J2340 (143,100)
AM 1253-240	12 53 17 −24 01.1	3.210 −0.325	304.25 38.58	140.0 −6.2	COMPACT	13 D 0.2	F507/J2340 (139,232)
AM 1253-291	12 53 29 −29 14.0	3.246 −0.325	304.17 33.37	145.1 −7.4	SP. w. PECULIAR APP. COMP'N. NGC 4806 = 443-G-12	8 S 9.9	F443/J1469 (277,217)
AM 1253-252	12 53 37 −25 29.4	3.221 −0.325	304.31 37.11	141.4 −6.5	GALAXY w. 3 DIAMETRIC COMP'NS. MCG-4-31-8 = 507-G-48	8 S 3.6	F507/J2340 (135,153)
AM 1254-463	12 54 06 −46 37.6	3.404 −0.325	303.91 15.98	162.3 −11.1	CLOSE PAIR 269-G-12/14 = Fa 316/18	23 S 4.5	F269/J1553 (271,91)
AM 1254-413	12 54 05 −41 31.9	3.351 −0.325	304.04 21.04	157.3 −10.0	I/A DBL. NGC 4811/12 = 323-G-47/8	2 S 1.0	F323/J 642 (234,93)
*AM 1254-321	12 54 26 −32 14.7	3.271 −0.325	304.33 30.35	148.1 −7.9	DUSTY COMPACT + COMP'NS. + PLUME MCG-5-31-4/5 = 443-G-15	8,10,13,14,15 D 2.0	F381/J2339 (107,327)
AM 1254-245	12 54 44 −24 59.2	3.220 −0.325	304.64 37.61	141.0 −6.1	CLOSE PAIR LO.S.BR. GALAXIES 507-G-50/52	20,23 S 0.9	F507/J2340 (122,180)
AM 1254-393	12 54 59 −39 31.3	3.335 −0.324	304.26 23.08	155.3 −9.5	S0 + PEC. COMP'NS to [W.]	8 S 8.4	F323/J 642 (227,203)
AM 1255-241	12 55 00 −24 12.6	3.216 −0.324	304.75 38.38	140.3 −5.8	LINE of 3 (incl. M51-TYPE) 507-G-57	9,24 D 3.1	F507/J2340 (118,221)
AM 1255-235	12 55 01 −23 55.0	3.214 −0.324	304.76 38.67	140.0 −5.8	CLOSE PAIR (SP. w. HI.S.BR. ARM) MCG-4-31-11 = 507-G-56	10,23 S 2.2	F507/J2340 (118,237)
*AM 1255-455	12 55 17 −45 59.9	3.404 −0.324	304.14 16.60	161.7 −10.8	SP. w. I/A COMP'N. + APP. COMP'NS NGC 4835 in Se 94/5 =269-G-19[DN]	1,8 S 1.4	F269/J1553 (261,125)
AM 1255-430	12 55 19 −43 03.3	3.371 −0.324	304.23 19.54	158.8 −10.1	DISTURBED GALAXY w. PLUMES 269-IG-20	15 D 1.3	F269/J1553 (264,283)
AM 1255-503	12 55 21 −50 37.1	3.462 −0.324	304.03 11.98	166.3 −11.6	CLOSE PAIR (incl. M51-TYPE) 219-G-15	9,23 S 2.2	F219/J2247 (219,121)
AM 1255-223	12 55 30 −22 36.4	3.206 −0.324	304.96 39.98	138.7 −5.3	E + 2 APP. COMP'NS IC 3927 = 507-G-58	8 S 1.3	F507/J2340 (111,307)
AM 1255-325	12 55 33 −32 57.4	3.281 −0.324	304.58 29.64	148.8 −7.9	DWARF (SOME RESOLUTION) 381-G-31	20 D 1.1	F443/J1469 (250,18)
AM 1255-430	12 55 33 −43 05.0	3.373 −0.324	304.27 19.51	158.8 −10.1	GALAXY w. STREAMERS	15 D 1.6	F323/J0642 (220,12)
AM 1255-315 [1256-3150]	12 55 59 −31 51.2	3.274 −0.324	304.73 30.74	147.8 −7.5	CLOSE PAIR (#1)	23 D 0.4	F443/J1469 (246,77)
*AM 1256-433 [1256-4334]	12 56 01 −43 36.3	3.381 −0.324	304.34 18.99	159.4 −10.1	PAIR (#1: E) 269-IG-22	23 S 2.2	F269/J1553 (256,254)
*AM 1256-433 [1256-4336]	12 56 09 −43 34.1	3.381 −0.324	304.37 19.03	159.3 −10.1	PAIR (#2: PEC.SP. w. I/A COMP'NS) 269-IG-23	1,23 S 2.2	F269/J1553 (255,256)
AM 1256-315 [1255-3151]	12 56 22 −31 50.9	3.275 −0.324	304.82 30.74	147.8 −7.4	CLOSE PAIR (#2)	23 D 0.4	F443/J1469 (241,78)
AM 1256-485	12 56 33 −48 52.9	3.447 −0.324	304.28 13.71	164.6 −11.1	DWARF 219-G-16	20 D 1.8	F219/J2247 (210,214)
AM 1256-285 [1256-2859]	12 56 36 −28 54.8	3.253 −0.324	305.00 33.67	144.9 −6.7	CLOSE PAIR (#1: E) 443-G-19	23 D 0.3	F443/J1469 (240,235)
AM 1256-285 [1256-2854]	12 56 53 −28 59.5	3.255 −0.324	305.07 33.59	145.0 −6.6	CLOSE PAIR (#2: PECULIAR)	23 D 0.3	F443/J1469 (237,231)
AM 1257-475	12 57 01 −47 59.1	3.438 −0.324	304.39 14.61	163.8 −10.9	DWARF 219-G-17	20 D 0.5	F269/J1553 (243,19)
*AM 1257-291	12 57 02 −29 19.7	3.258 −0.324	305.10 33.25	145.4 −6.7	SP. w. HI.S.BR. FILAMENT in DISK MCG-5-31-9 = 443-G-21 = [To?]	8,12,16 D 1.9	F443/J1469 (235,213)
AM 1257-251	12 57 05 −25 19.7	3.229 −0.324	305.30 37.25	141.5 −5.7	MULTIPLE NUCLEUS + MATERIAL PKS 1257-253	12 D 0.7	F507/J2340 (93,161)
AM 1257-501	12 57 14 −50 11.1	3.469 −0.324	304.35 12.41	166.0 −11.2	CLOSE PAIR 219-G-18/19	23 S 1.5	F219/J2247 (203,144)
AM 1257-485	12 57 19 −48 58.0	3.453 −0.324	304.41 13.62	164.8 −11.0	GALAXY w. JET	7 D 0.7	F219/J2247 (203,210)
AM 1257-340	12 57 20 −34 09.7	3.297 −0.324	304.96 28.42	150.1 −7.8	SPIRAL w. I/A COMP'N. in ARM	1 S 0.4	F382/J1530 (342,221)
AM 1257-332	12 57 24 −33 27.8	3.292 −0.324	305.01 29.12	149.4 −7.6	SPIRAL w. I/A COMP'N.	1 S 0.2	F382/J1530 (343,259)
*AM 1257-222	12 57 38 −22 25.7	3.210 −0.323	305.61 40.14	138.7 −4.8	GALAXY w. 2 I/A COMP'NS. MCG-4-31-17 = 575-G-44	1 S 0.9	F507/J2340 (84,317)
AM 1257-435	12 57 38 −43 56.9	3.393 −0.323	304.64 18.64	159.8 −9.9	CLOSE PAIR MC 1257-439	23 S 1.2	F269/J1553 (240,235)
AM 1257-450	12 57 39 −45 07.0	3.407 −0.323	304.60 17.47	160.9 −10.2	E + DISTURBED DWARF 269-G-30/31	20,23 S 2.7	F269/J1553 (239,173)
AM 1257-421	12 57 40 −42 13.1	3.375 −0.323	304.71 20.37	158.1 −9.5	DWARF	20 D 0.4	F269/J1553 (242,328)

Identification	R.A. (1950.0) Declination	Precession R.A./Dec.	Gal. Long. Gal. Lat.	S.G.L. S.G.B.	Description Cross-Identification	Coding Size (Min.)	Field/Plate (x, y)
AM 1258-321	12 58 16 −32 10.4	3.284 −0.323	305.28 30.40	148.2 −7.1	ARC of 4 (incl. 2 E's + SPIRAL) MCG-5-31-12 = 443-G-24 [PKS]	17 D 9.0	F382/J1530 (335,329)
AM 1258-352	12 58 20 −35 21.0	3.312 −0.323	305.14 27.22	151.3 −7.9	E w. VERY FAINT OUTER RING 381-G-47	6 D 1.1	F382/J1530 (329,158)
AM 1258-254	12 58 22 −25 49.5	3.236 −0.323	305.63 36.74	142.0 −5.5	DWARF + IRR. (HII RING) 507-G-65	16,20 S 7.2	F507/J2340 (78,134)
AM 1258-453	12 58 32 −45 32.2	3.417 −0.323	304.75 17.04	161.4 −10.1	SMALL I/A DWARF w. STRUCTURE 269-G-33	20 D 0.7	F269/J1553 (231,150)
*AM 1258-303	12 58 45 −30 36.0	3.273 −0.323	305.47 31.97	146.7 −6.6	GROUP of 4 incl. 2 V.LO.S.BR. SP. NGC 4903/05 = 443-G-30/34	10,18 D 10.3	F443/J1469 (215,145)
AM 1258-320	12 58 48 −32 01.2	3.285 −0.323	305.41 30.55	148.1 −7.0	GROUP of 5 (E + SPIRAL) IC 3986? = 443-G-32	18 D 7.6	F443/J1469 (213,69)
AM 1258-234	12 58 58 −23 43.2	3.222 −0.323	305.93 38.84	140.0 −4.8	E + 10 COMP'NS. MCG-4-31-20 = 507-G-66	8 S 2.0	F507/J2340 (69,247)
AM 1259-452	12 59 10 −45 28.8	3.420 −0.323	304.87 17.10	161.3 −10.0	SMALL I/A DBL 269-IG-34	2 S 0.4	F269/J1553 (225,153)
*AM 1259-423	12 59 12 −42 30.1	3.386 −0.323	305.00 20.07	158.4 −9.3	SPIRAL w. FAINT OUTER ARMS NGC 4909 = 269-G-35 = Fa 462	10 D 1.3	F269/J1553 (226,313)
AM 1259-360	12 59 19 −36 09.9	3.323 −0.323	305.33 26.40	152.2 −7.9	COMPACT 381-G-49	13 D 0.1	F382/J1530 (317,114)
*AM 1259-272	12 59 21 −27 25.2	3.251 −0.323	305.81 35.14	143.6 −5.7	SPIRAL w. DIAMETRIC COMP'NS.	8 S 0.4	F443/J1469 (209,316)
AM 1259-323	12 59 21 −32 31.0	3.291 −0.323	305.52 30.05	148.6 −7.0	SPIRAL w. PEC. ARM in GROUP of 3 MCG-5-31-199 = 381-G-50	3,10 D 2.7	F381/J2339 (52,312)
AM 1259-493	12 59 26 −49 35.6	3.476 −0.323	304.74 12.98	165.5 −10.8	GALAXY w. APP. COMP'N. 219-G-21	8 S 4.3	F219/J2247 (184,176)
*AM 1259-322	12 59 29 −32 29.6	3.291 −0.323	305.55 30.07	148.6 −6.9	I/A TRIPLE incl. SP. w. PEC. ARMS MCG-5-31-21 = 443-G-37	1,3,10 D 3.6	F443/J1469 (206,43)
AM 1259-250 [1259-2504]	12 59 30 −25 06.6	3.234 −0.323	306.00 37.44	141.4 −5.1	PAIR (#1) 507-G-68	23 S 1.6	F507/J2340 (64,172)
AM 1259-484	12 59 30 −48 43.7	3.464 −0.323	304.79 13.85	164.6 −10.6	SPIRAL w. 2 COMP'NS along MI.AX.	8 S 1.5	F219/J2247 (184,223)
AM 1259-250 [1259-2506]	12 59 32 −25 04.9	3.234 −0.323	306.01 37.47	141.4 −5.1	PAIR (#2)	23 S 1.6	F507/J2340 (63,174)
AM 1259-361	12 59 43 −36 11.0	3.325 −0.323	305.41 26.38	152.2 −7.8	SHRED or EDGE-ON	22 D 0.7	F382/J1530 (312,114)
AM 1259-381	12 59 43 −38 19.9	3.345 −0.323	305.31 24.23	154.3 −8.3	PECULIAR SPIRAL w. I/A COMP'N.	1 S 0.6	F323/J 642 (178,267)
AM 1259-340	12 59 52 −34 03.0	3.306 −0.323	305.56 28.51	150.1 −7.2	SPIRAL w. I/A COMP'N. 381-G-52	9 S 1.0	F381/J2339 (48,229)
*AM 1300-233	13 00 10 −23 39.3	3.225 −0.323	306.29 38.89	140.0 −4.6	IRR. MCG-4-31-23 = 507-G-70	16 D 1.3	F507/J2340 (54,250)
AM 1300-412	13 00 17 −41 26.9	3.380 −0.323	305.27 21.11	157.4 −8.9	HI.S.BR. RING GALAXY (SMALL)	6 D 0.3	F323/J 642 (173,100)
AM 1300-243	13 00 20 −24 32.8	3.232 −0.322	306.27 37.99	140.9 −4.8	LO.S.BR. ONE-ARMED SPIRAL	10,20 D 0.5	F507/J2340 (53,202)
AM 1300-461	13 00 42 −46 17.8	3.439 −0.322	305.11 16.27	162.2 −9.9	FAINT ELONGATED DWARF 269-G-37 (NO STRUCTURE)	20 D 0.9	F269/J1553 (210,110)
AM 1300-424	13 00 58 −42 41.6	3.397 −0.322	305.34 19.86	158.7 −9.1	TRIPLE in COMMON ENVELOPE 269-G-38	3 D 0.5	F269/J1553 (209,303)
AM 1301-301	13 01 31 −30 15.3	3.280 −0.322	306.20 32.28	146.5 −6.0	E + ABS'N. in GROUP of 4 E's + SP NGC 4936 = 443-G-47 = MC 1301-302	14,18 D 1.3	F443/J1469 (183,163)
AM 1301-282	13 01 39 −28 26.7	3.265 −0.322	306.36 34.08	144.8 −5.5	E + SPIRAL	23 S 2.7	F443/J1469 (181,261)
AM 1301-341	13 01 49 −34 13.0	3.315 −0.322	306.01 28.32	150.4 −6.9	SPIRAL w. I/A COMP'N.	1 S 0.4	F382/J1530 (292,220)
AM 1302-350	13 02 34 −35 04.1	3.326 −0.322	306.13 27.46	151.3 −7.0	SPIRAL w. PEC. ARMS + COMP'N. NGC 4947 = IC 3974 = 382-G-5	8,10 S 9.9	F382/J1530 (283,174)
AM 1302-484	13 02 35 −48 43.3	3.484 −0.322	305.31 13.83	164.7 −10.1	LARGE EDGE-ON SPIRAL + COMP'NS. NGC 4945 = 219-G-24	8 D 22.4	F219/J2247 (157,223)
AM 1302-232 [1302-2324]	13 02 43 −23 24.2	3.229 −0.322	307.06 39.10	139.9 −3.9	PAIR (#1: E)	23 S 1.0	F507/J2340 (22,263)
AM 1302-232 [1302-2324]	13 02 48 −23 24.6	3.230 −0.322	307.08 39.09	139.9 −3.9	PAIR (#2: SPIRAL)	23 S 1.0	F507/J2340 (21,263)
AM 1302-281 [1303-2809]	13 02 52 −28 11.5	3.267 −0.322	306.70 34.32	144.6 −5.2	CLOSE PAIR of SPIRALS (#1) MCG-5-31-32 = 443-G-59 = To1303-281	23 D 1.4	F443/J1469 (167,274)
AM 1303-280 [1302-2811]	13 03 03 −28 09.3	3.267 −0.321	306.75 34.36	144.6 −5.1	CLOSE PAIR of SPIRALS (#2) MCG-5-31-33 = 443-G-61 = To1303-281	23 D 0.6	F443/J1469 (164,276)
AM 1303-372	13 03 12 −37 20.2	3.351 −0.321	306.12 25.19	153.5 −7.4	FIVE I/A GALAXIES (COMMON ENV.)	5 D 1.7	F323/J 642 (141,320)
AM 1303-292	13 03 20 −29 29.2	3.279 −0.321	306.72 33.02	145.9 −5.4	E + APP. COMP'NS. NGC 4955 = 443-G-62	8 S 2.2	F443/J1469 (161,205)

Identification	R.A. (1950.0) Declination	Precession R.A./Dec.	Gal. Long. Gal. Lat.	S.G.L. S.G.B.	Description Cross-Identification	Coding Size (Min.)	Field/Plate (x, y)
*AM 1303-371	13 03 20 −37 19.0	3.351 −0.321	306.15 25.21	153.5 −7.4	DISRUPTED GALAXY w. 6 COMP'NS. NGC 4953 = 382-G-8	8,16,17 D 2.9	F381/J2339 (15,53)
*AM 1303-340	13 03 22 −34 02.7	3.319 −0.321	306.39 28.47	150.3 −6.5	E w. DIAMETRIC COMP'N.	8 S 0.7	F382/J1530 (275,229)
AM 1303-490	13 03 32 −49 05.5	3.495 −0.321	305.45 13.45	165.1 −10.0	SHRED 219-G-27	22 D 1.1	F219/J2247 (148,203)
AM 1303-235	13 03 41 −23 53.8	3.235 −0.321	307.30 38.59	140.5 −3.9	HI.S.BR. RING w. HALO + COMP'N. MCG-4-31-27 = 508-G-3	6,8 D 0.7	F507/J2340 (11,236)
AM 1303-380	13 03 50 −38 00.5	3.360 −0.321	306.21 24.51	154.2 −7.4	SPIRAL w. MULTIPLE NUCLEUS 323-IG-79	12 D 0.3	F381/J2339 (11,15)
*AM 1303-371	13 03 53 −37 18.4	3.353 −0.321	306.27 25.21	153.5 −7.3	SPIRAL w. PECULIAR ARM	10 D 1.0	F323/J 642 (133,321)
*AM 1304-333	13 04 21 −33 35.7	3.319 −0.321	306.65 28.91	149.9 −6.2	M51-TYPE 382-G-10	9 S 0.3	F382/J1530 (264,254)
AM 1304-275	13 04 25 −27 57.8	3.270 −0.321	307.13 34.53	144.4 −4.8	SPIRAL w. APP. COMP'N. NGC 4965 = 443-G-70	8 S 5.4	F443/J1469 (148,286)
AM 1304-241	13 04 28 −24 17.2	3.240 −0.321	307.49 38.19	140.9 −3.8	DWARF	20 D 0.3	F508/J2347 (271,217)
AM 1305-223	13 05 03 −22 35.4	3.229 −0.321	307.83 39.87	139.3 −3.2	EDGE-ON w. 5 COMP'NS incl. DWARF MCG-4-31-35 = 508-G-11	8,20 D 2.2	F508/J2347 (266,308)
AM 1305-620	13 05 06 −62 01.1	3.778 −0.321	304.89 0.54	178.1 −12.3	NEBULOSITY	21	F132/J2097 (260,69)
AM 1305-280 [1305-2801]	13 05 13 −28 05.1	3.273 −0.321	307.34 34.39	144.6 −4.6	DWARF (#1)	20,23 S 3.6	F508/J2347 (259,13)
AM 1305-365	13 05 27 −36 52.8	3.355 −0.320	306.65 25.62	153.2 −6.8	DWARF 382-G-12	20 D 0.4	F382/J1530 (250,78)
AM 1305-280 [1305-2805]	13 05 29 −28 01.8	3.273 −0.320	307.41 34.44	144.6 −4.6	DWARF (#2)	20,23 S 3.6	F508/J2347 (256,16)
AM 1305-315 [1306-3151]	13 05 50 −31 53.0	3.308 −0.320	307.15 30.59	148.3 −5.5	CLOSE PAIR (#1) 443-G-72	23 D 0.5	F443/J1469 (133,76)
AM 1305-501	13 05 53 −50 18.5	3.529 −0.320	305.76 12.21	166.4 −9.9	PAIR of LO.S.BR. DISTURBED GALS. 219-G-30	23 S 5.6	F219/J2247 (129,138)
AM 1306-315 [1305-3153]	13 06 15 −31 51.4	3.310 −0.320	307.26 30.61	148.3 −5.4	CLOSE PAIR (#2) 443-G-73	23 D 0.3	F443/J1469 (129,77)
*AM 1306-282	13 06 25 −28 22.7	3.279 −0.320	307.63 34.08	145.0 −4.5	DISTURBED SINGLE GALAXY NGC 4980 = 443-G-75	16 D 1.6	F443/J1469 (125,264)
AM 1306-374 [1306-3748]	13 06 31 −37 47.2	3.369 −0.320	306.81 24.70	154.1 −6.9	CLOSE PAIR (#1)	23 D 0.3	F382/J1530 (238,29)
AM 1306-240	13 06 36 −24 07.0	3.244 −0.320	308.12 38.32	140.9 −3.3	DWARF IRR. w. KNOTS MCG-4-31-37 = 508-G-15	20 D 0.9	F508/J2347 (246,226)
AM 1306-474	13 06 45 −47 41.4	3.494 −0.320	306.09 14.82	163.8 −9.2	PAIR 219-IG?-32/35	23 S 4.0	F269/J1553 (155,35)
AM 1306-374 [1306-3747]	13 06 46 −37 48.1	3.371 −0.320	306.86 24.68	154.1 −6.8	CLOSE PAIR (#2) 323-G-87	23 D 0.5	F382/J1530 (235,28)
AM 1306-265	13 06 53 −26 52.4	3.268 −0.320	307.90 35.57	143.5 −4.0	TYPE II IRR. (RESOLUTION?)	16,209 D 0.5	F508/J2347 (240,78)
AM 1307-231	13 07 04 −23 16.7	3.239 −0.320	308.35 39.15	140.1 −2.9	DWARF	20 D 0.4	F508/J2347 (240,271)
AM 1307-230	13 07 06 −23 07.2	3.238 −0.320	308.38 39.30	139.9 −2.9	"RINGING" GALAXY MC 1307-231 NGC 4993 = NGC 4994 = 508-G-18	16,22 D 1.3	F508/J2347 (240,280)
*AM 1307-425	13 07 08 −42 56.7	3.432 −0.320	306.52 19.54	159.2 −8.0	LINEAR NUCLEUS w. KNOTTY CLUMPS Se 102/2 = 269-IG-56 (FILAMENT)	2,8,16 D 2.7	F269/J1553 (148,289)
AM 1307-235	13 07 10 −23 58.5	3.245 −0.320	308.30 38.45	140.8 −3.1	DWARF IRR. w. KNOTS MCG-4-31-40 = 508-G-19	20 D 1.6	F508/J2347 (239,234)
*AM 1307-461	13 07 10 −46 10.2	3.475 −0.320	306.28 16.33	162.3 −8.8	INNER RING w. OUTER SPIRALS Se 94/4 = 269-G-57	6,10 D 2.7	F269/J1553 (150,116)
AM 1307-263	13 07 23 −26 30.8	3.266 −0.320	308.08 35.92	143.2 −3.7	DWARF (RESOLUTION)	20 D 0.5	F508/J2347 (234,98)
*AM 1307-464	13 07 39 −46 43.6	3.486 −0.320	306.32 15.77	162.9 −8.8	E w. SHARP ABSORPTION PATCH Se 94/6 = 269-G-58	14 D 2.2	F269/J1553 (146,86)
AM 1307-735	13 07 54 −73 52.6	4.423 −0.319	304.34 −11.31	190.1 −14.0	NEBULA w. DIFFRACTION SPIKES	21 D 1.0	F040/J2072 (195,239)
AM 1308-233	13 08 03 −23 36.0	3.244 −0.319	308.60 38.81	140.5 −2.8	SPIRAL + APP. COMP'N. MCG-4-31-41 = 508-G-24	8 S 4.0	F508/J2347 (228,254)
AM 1308-253	13 08 25 −25 38.1	3.262 −0.319	308.47 36.77	142.4 −3.3	ONE-ARMED SP w. HI. + LO.S.BR. ARMS 508-G-25 (BROKEN RING ?)	6,10 D 0.5	F508/J2347 (222,145)
AM 1308-274	13 08 26 −27 44.8	3.280 −0.319	308.23 34.67	144.5 −3.9	PAIR of IRR. MCG-5-31-38 = 443-G-80	16,23 S 7.6	F508/J2347 (221,32)
AM 1309-425 [1311-4241]	13 09 59 −42 50.2	3.445 −0.319	307.08 19.61	159.2 −7.5	CLOSE PAIR (#1: E)	23 D 0.9	F324/J1381 (335,21)
AM 1310-322	13 10 07 −32 25.3	3.329 −0.318	308.15 29.98	149.1 −4.8	IRR. + COMP'N. MCG-5-31-39 = 443-G-83	8,16 S 2.7	F382/J1530 (200,317)

Identification	R.A. (1950.0) Declination	Precession R.A./Dec.	Gal. Long. Gal. Lat.	S.G.L. S.G.B.	Description Cross-Identification	Coding Size (Min.)	Field/Plate (x, y)
AM 1310-443	13 10 14 −44 37.4	3.471 −0.318	306.97 17.83	161.0 −7.9	LARGE V.LO.S.BR. DWARF 269-G?-66	20 D 1.6	F269/J1553 (120,199)
AM 1310-425	13 10 17 −42 59.5	3.449 −0.318	307.13 19.45	159.4 −7.5	CLOSE PAIR in LARGE GROUP 269-G-67/68 = Fa 464	18,23 S 0.9	F269/J1553 (117,286)
AM 1310-362	13 10 25 −36 27.2	3.371 −0.318	307.79 25.96	153.0 −5.8	E w. PEC. LINEAR COMP'N. + DUST 382-G-16	8,14 S 2.5	F382/J1530 (197,101)
AM 1310-393	13 10 33 −39 33.2	3.407 −0.318	307.51 22.87	156.0 −6.6	CLOSE I/A DBL. w. MATERIAL 323-IG-94	2 S 0.1	F324/J1381 (337,198)
AM 1310-430	13 10 33 −43 07.0	3.452 −0.318	307.16 19.32	159.5 −7.5	GAL.w. I/A COMP'N in LARGER GROUP Se 102/3 = 269-G-70 = Fa 465	1,18 S 0.4	F269/J1553 (115,279)
AM 1310-484	13 10 53 −48 45.0	3.537 −0.318	306.71 13.70	165.0 −8.8	SPIRAL w. PECULIAR COMP'N. 219-G-39/41	8 S 2.5	F219/J2247 (83,220)
AM 1310-373 [1311-3731]	13 10 57 −37 33.4	3.386 −0.318	307.79 24.85	154.1 −6.0	CLOSE PAIR (#1)	23 D 0.3	F324/J1381 (337,305)
AM 1311-315	13 11 11 −31 50.0	3.327 −0.318	308.47 30.54	148.6 −4.4	CLOSE PAIR	23 S 1.0	F443/J1469 (72,78)
AM 1311-373 [1310-3733]	13 11 14 −37 31.9	3.387 −0.318	307.86 24.87	154.1 −5.9	CLOSE PAIR (#2: BULLS-EYE)	6,23 D 0.3	F324/J1381 (334,307)
AM 1311-280	13 11 18 −28 01.9	3.291 −0.318	308.96 34.32	144.9 −3.3	DWARF?	20 D 0.4	F508/J2347 (187,16)
AM 1311-424 [1311-4250]	13 11 19 −42 41.7	3.450 −0.318	307.35 19.73	159.1 −7.2	CLOSE PAIR (#2: SPIRAL)	23 D 1.0	F324/J1381 (322,30)
*AM 1311-455	13 11 26 −45 50.8	3.495 −0.318	307.08 16.59	162.2 −8.0	CONTACT SPIRALS 269-IG-74	2 S 0.8	F269/J1553 (110,133)
AM 1311-450	13 11 40 −45 03.0	3.485 −0.318	307.19 17.38	161.4 −7.7	SMALL DWARF	20 D 0.7	F269/J1553 (107,175)
AM 1311-730	13 11 43 −73 08.1	4.430 −0.318	304.67 −10.60	189.4 −13.6	CLOSE DOUBLE	23 S 0.5	F040/J2072 (181,279)
AM 1311-265	13 11 48 −26 57.7	3.282 −0.318	309.23 35.37	143.9 −2.9	CLOSE I/A DBL. (LO.S.BR.)	2 S 0.1	F443/J1469 (60,339)
AM 1311-345	13 11 55 −34 55.0	3.361 −0.318	308.30 27.46	151.6 −5.1	SMALL SPIRAL w. APP. COMP'N.	8 S 0.5	F382/J1530 (180,183)
AM 1311-365	13 11 55 −36 54.7	3.383 −0.318	308.07 25.47	153.5 −5.6	CHAIN of 4 w. M51-TYPE at END	9,17 D 13.4	F324/J1381 (328,340)
AM 1311-653	13 11 55 −65 39.5	3.985 −0.318	305.36 −3.15	181.9 −12.3	FAINT GLOBULAR CLUSTER AL-1 = 96-SC-4	22	F096/J2079 (180,144)
AM 1312-225	13 12 14 −22 53.0	3.248 −0.318	309.93 39.42	140.0 −1.7	DWARF + APP. COMP'N. 508-G-30 = UKS 1312-228	8,20 D 1.6	F508/J2347 (177,293)
*AM 1312-365	13 12 16 −36 52.9	3.384 −0.318	308.15 25.50	153.5 −5.5	THIN-ARMED M51-TYPE in CHAIN of 5 382-G-23 (in AM 1311-3654)	9,17 S 0.4	F382/J1530 (177,78)
AM 1312-254	13 12 20 −25 41.0	3.272 −0.318	309.55 36.63	142.7 −2.5	LINE of 3 (I/A ?)	3 D 1.3	F508/J2347 (175,143)
AM 1312-392	13 12 27 −39 27.6	3.415 −0.317	307.91 22.93	156.0 −6.2	COMPACT 323-G-100	13 D 0.1	F324/J1381 (317,204)
AM 1312-315	13 12 28 −31 59.1	3.332 −0.317	308.77 30.36	148.8 −4.2	PECULIAR SPIRAL w. APP. COMP'N. 443-G-85 (in GROUP ?)	8,12 S 2.0	F443/J1469 (58,69)
*AM 1312-344	13 12 34 −34 40.2	3.361 −0.317	308.47 27.69	151.4 −4.9	E w. LONG TAIL + COMP'N. 382-G-24	8,15 S 1.9	F382/J1530 (173,197)
AM 1312-322	13 12 41 −32 26.9	3.338 −0.317	308.76 29.90	149.3 −4.2	CONTACT E's in WIDE GROUP	2,18 S 0.1	F382/J1530 (171,316)
AM 1312-280	13 12 43 −28 01.3	3.295 −0.317	309.34 34.30	145.0 −3.0	E + SPIRAL + SPIRAL NGC 5048/51 = MCG-5-31-40	24 D 13.4	F443/J1469 (50,282)
AM 1312-222	13 12 59 −22 28.3	3.246 −0.317	310.22 39.80	139.7 −1.4	GAL. w. OUTER RING (3 COMP'NS.) 576-G-18	6,8 S 2.7	F508/J2347 (168,315)
AM 1313-221	13 13 03 −22 13.9	3.244 −0.317	310.28 40.04	139.5 −1.3	DWARF	20 D 0.3	F508/J2347 (167,328)
*AM 1313-370	13 13 15 −37 04.1	3.390 −0.317	308.35 25.29	153.8 −5.4	I/A TRIPLE (SPIRAL w. PEC. ARMS) 382-G-27	3,10 D 1.8	F382/J1530 (166,68)
*AM 1313-444	13 13 44 −44 41.5	3.491 −0.317	307.61 17.70	161.2 −7.3	COMPACT	13 D 0.3	F269/J1553 (86,194)
AM 1314-273	13 14 00 −27 37.4	3.295 −0.317	309.74 34.66	144.7 −2.6	Sd/DWARF + I/A COMP'N (SHRED) 444-G-2	20 D 1.6	F508/J2347 (155,38)
AM 1314-350	13 14 08 −35 08.5	3.372 −0.317	308.78 27.19	151.9 −4.7	DWARF w. STRUCTURE 382-G?-29	20 D 0.7	F382/J1530 (156,171)
AM 1314-370	13 14 30 −37 04.0	3.395 −0.317	308.62 25.27	153.8 −5.1	CHAIN of 6 (near EDGE-ON) incl. 382-G-31	17 D 8.5	F382/J1530 (153,68)
AM 1314-370	13 14 36 −37 00.2	3.395 −0.316	308.65 25.33	153.8 −5.1	GALAXY + COMP'NS	8 D 2.6	F323/J 642 (18,334)
AM 1314-454	13 14 46 −45 45.6	3.513 −0.316	307.69 16.62	162.3 −7.4	DWARF/DEFECT?	20 D 0.4	F270/J1313 (341,134)
AM 1314-390 [1315-3905]	13 14 50 −39 04.6	3.421 −0.316	308.46 23.26	155.8 −5.6	CLOSE TRIPLE. (#1)	24 D 0.2	F324/J1381 (293,225)

Identification	R.A. (1950.0) Declination	Precession R.A./Dec.	Gal. Long. Gal. Lat.	S.G.L. S.G.B.	Description Cross-Identification	Coding Size (Min.)	Field/Plate (x, y)
AM 1314-622	13 14 53 −62 25.8	3.894 −0.316	305.99 0.03	178.7 −11.3	NEBULOSITY 132-EN-3	21	F132/J2097 (198,49)
AM 1314-463	13 14 55 −46 34.4	3.526 −0.316	307.63 15.81	163.1 −7.6	DWARF (SEMI-CIRCULAR) 269-G-79	20 D 0.6	F270/J1313 (337,91)
AM 1315-390 [1314-3904]	13 15 07 −39 01.7	3.422 −0.316	308.52 23.31	155.8 −5.6	CLOSE TRIPLE. (#2: PEC. SPIRAL) 324-G-2	12,24 D 0.4	F324/J1381 (290,228)
AM 1315-390 [1315-3901]	13 15 12 −39 05.0	3.423 −0.316	308.53 23.25	155.8 −5.6	CLOSE TRIPLE. (#3) 324-G-3	24 D 0.3	F324/J1381 (289,225)
*AM 1315-263	13 15 20 −26 34.3	3.289 −0.316	310.25 35.67	143.8 −2.1	"SQUARE" E GAL. w. 5 APP. COMP'NS NGC 5061 = 508-G-38	8,22 D 1.8	F508/J2347 (139,95)
*AM 1315-350	13 15 37 −35 05.4	3.377 −0.316	309.12 27.21	152.0 −4.4	SP.w.THIN, LO.S.BR.DISTURBED ARMS NGC 5063 = 382-G-36 (GROUP of 3)	1,10,13,23,24 S 5.4	F382/J1530 (140,174)
AM 1315-621	13 15 52 −62 18.9	3.901 −0.316	306.12 0.13	178.6 −11.2	NEBULOSITY 132-EN-4	21	F132/J2097 (192,55)
AM 1315-365	13 15 53 −36 57.7	3.400 −0.316	308.94 25.34	153.8 −4.9	CHAIN of 5 GALAXIES	17 D 4.8	F323/J 642 (4,336)
AM 1316-473	13 16 02 −47 38.7	3.550 −0.316	307.70 14.72	164.2 −7.7	SPIRAL w. APP. COMP'N. NGC 5064 = 220-G-2	8 S 1.3	F269/J1553 (71,35)
AM 1316-234	13 16 05 −23 40.7	3.264 −0.316	310.93 38.52	141.0 −1.1	DWARF	20 D 0.9	F508/J2347 (129,250)
AM 1316-342	13 16 05 −34 20.7	3.371 −0.316	309.33 27.94	151.3 −4.1	I/A DBL. 382-G-38	2 S 0.7	F382/J1530 (134,214)
AM 1316-284	13 16 13 −28 41.4	3.312 −0.316	310.16 33.54	145.9 −2.5	I/A DBL. 444-IG-8	2 S 0.4	F444/J1484 (281,246)
AM 1316-270	13 16 14 −27 04.3	3.296 −0.316	310.42 35.15	144.3 −2.0	NUCLEATED DWARF	20 D 0.2	F508/J2347 (128,68)
AM 1316-425	13 16 16 −42 59.0	3.480 −0.316	308.28 19.35	159.6 −6.4	SPIRAL w. I/A COMP'N. 269-G-83	1 S 0.1	F270/J1313 (334,284)
AM 1316-365	13 16 18 −36 58.3	3.402 −0.316	309.03 25.32	153.8 −4.8	GROUP of 10 (incl. I/A DBL.)	2,18 D 6.3	F382/J1530 (133,73)
AM 1316-240	13 16 19 −24 09.6	3.269 −0.316	310.92 38.03	141.5 −1.2	E + EDGE-ON 508-G-43	1 S 1.1	F508/J2347 (126,224)
AM 1316-771	13 16 20 −77 18.0	5.011 −0.316	304.55 −14.77	193.7 −14.0	EDGE-ON SPIRAL w. 3 COMP'NS. 40-G-7	8 D 0.1	F021/J2083 (220,323)
AM 1316-432	13 16 29 −43 23.1	3.487 −0.316	308.28 18.95	160.0 −6.5	GALAXY w. APP. COMP'NS. NGC 5090a = 269-G-84	8 S 1.8	F270/J1313 (330,263)
AM 1316-341	13 16 32 −34 16.5	3.371 −0.316	309.44 27.99	151.2 −4.0	PECULIAR SB in WIDE GROUP of 3	10,24 S 5.8	F382/J1530 (129,217)
AM 1316-330	13 16 34 −33 00.9	3.358 −0.316	309.62 29.24	150.0 −3.6	DWARF (SOME RESOLUTION) 382-G-40 near To 1316-331	20 D 0.6	F444/J1484 (273,14)
*AM 1316-241	13 16 48 −24 13.4	3.271 −0.315	311.05 37.96	141.6 −1.1	E in CONTACT w. EDGE-ON w. DUST 508-IG-45	2,14 S 0.4	F508/J2347 (121,221)
AM 1316-263	13 16 50 −26 32.3	3.293 −0.315	310.67 35.66	143.8 −1.7	DWARF	20 D 0.4	F508/J2347 (121,97)
AM 1316-364	13 16 52 −36 47.6	3.402 −0.315	309.18 25.49	153.7 −4.6	CHAIN of 5	17 D 10.8	F382/J1530 (127,82)
AM 1316-463	13 16 53 −46 34.0	3.538 −0.315	307.97 15.78	163.1 −7.3	SPIRAL w. 3 DIAMETRIC COMP'NS. 269-G-85	8 S 3.6	F219/J2247 (24,336)
AM 1316-330	13 16 57 −33 06.2	3.360 −0.315	309.70 29.15	150.1 −3.6	GROUP of 16 (I/A) incl. 382-G-43	18 D 13.4	F382/J1530 (123,280)
AM 1316-444	13 16 59 −44 48.9	3.511 −0.315	308.20 17.52	161.5 −6.8	CLOSE PAIR	23 S 4.7	F270/J1313 (322,186)
AM 1317-470	13 17 01 −47 01.1	3.546 −0.315	307.95 15.33	163.6 −7.3	SPIRAL in GROUP of 4 269-G-85 (SP. w. I/A COMP'N.)	1,18 D 2.2	F269/J1553 (61,68)
AM 1317-233	13 17 04 −23 32.5	3.265 −0.315	311.24 38.62	141.0 −0.8	GROUP of 5 508-G-40	18 D 2.5	F508/J2347 (117,258)
*AM 1317-270	13 17 05 −27 08.7	3.299 −0.315	310.64 35.05	144.4 −1.8	EDGE-ON Sa w. I/A COMP'N. + DUST NGC 5078 = 508-G-48	1,14,23 S 2.7	F444/J1484 (273,329)
AM 1317-220	13 17 09 −22 00.8	3.251 −0.315	311.54 40.13	139.5 −0.4	DISTURBED SPIRAL MCG-4-32-3 = 576-G-32	12 D 1.1	F508/J2347 (116,340)
AM 1317-472	13 17 18 −47 25.1	3.554 −0.315	307.95 14.92	164.0 −7.4	IRR./DWARF 269-G-87	20 D 0.6	F270/J1313 (313,47)
AM 1317-354	13 17 25 −35 46.7	3.392 −0.315	309.44 26.48	152.7 −4.2	LARGE DWARF w. RESOLUTION 382-G-45 = UKS 1317-357	20 D 3.1	F382/J1530 (120,137)
AM 1317-433	13 17 25 −43 36.2	3.495 −0.315	308.43 18.71	160.3 −6.4	WIDE GROUP of 5 (+ 5 DWARFS) NGC 5090b = 269-G-88	8,17,20 S 6.3	F270/J1313 (321,251)
AM 1317-241	13 17 34 −24 10.6	3.272 −0.315	311.27 37.98	141.6 −0.9	SPIRAL w. COMPACT in DISK NGC 5085 = 508-G-50	8,13 S 1.4	F508/J2347 (111,223)
AM 1317-303	13 17 40 −30 38.4	3.336 −0.315	310.24 31.57	147.8 −2.7	I/A DBL. (INTEGRAL + GALAXY) 440-G-10	2,12 S 0.9	F444/J1484 (262,142)
AM 1317-404	13 17 41 −40 46.6	3.456 −0.315	308.83 21.51	157.6 −5.6	CLOSE GROUP of 4 in ENVELOPE PKS 1317-407	18 D 0.8	F324/J1381 (261,135)

Identification	R.A. (1950.0) Declination	Precession R.A./Dec.	Gal. Long. Gal. Lat.	S.G.L. S.G.B.	Description Cross-Identification	Coding Size (Min.)	Field/Plate (x, y)
AM 1317-425	13 17 52 −42 51.1	3.486 −0.315	308.61 19.45	159.6 −6.1	IRR. w. KNOTS	16 D 0.5	F270/J1313 (318,292)
AM 1317-560	13 17 52 −56 02.5	3.734 −0.315	307.05 6.34	172.5 −9.5	NEBULOSITY 173-EN-10	21	F173/J (135,126)
AM 1317-252	13 17 58 −25 22.0	3.284 −0.315	311.18 36.79	142.8 −1.1	SMALL M51-TYPE	9 D 0.4	F508/J2347 (107,159)
AM 1318-772	13 18 09 −77 29.1	5.086 −0.315	304.63 −14.96	193.9 −13.9	GALAXY w. APP. COMP'NS. 40-IG-8	8 S 2.9	F040/J2072 (161,46)
*AM 1318-432	13 18 20 −43 26.0	3.497 −0.315	308.62 18.86	160.2 −6.2	E w. I/A SPIRAL PKS 1318-43 NGC 5090/91 = 270-IG-2/4 = SS-15	1 S 1.1	F269/J1553 (39,260)
AM 1318-270	13 18 21 −27 04.0	3.302 −0.315	310.99 35.09	144.4 −1.6	DWARF	20 D 0.2	F508/J2347 (103,68)
AM 1318-473	13 18 37 −47 34.8	3.565 −0.315	308.16 14.74	164.2 −7.2	COMPACT 220-?-4	13 D 0.2	F270/J1313 (301,38)
AM 1318-353	13 18 42 −35 32.3	3.394 −0.314	309.76 26.68	152.6 −3.9	TIGHT GROUP of 10 I/A GALAXIES incl. 382-IG-49	18 D 4.5	F382/J1530 (106,149)
AM 1318-454	13 18 46 −45 40.8	3.534 −0.314	308.42 16.62	162.4 −6.7	CLOSE PAIR of SPIRALS Se 102/6 = 270-G-5/6 = Fa 322+469	23 S 2.9	F269/J1553 (41,139)
AM 1318-444	13 18 47 −44 48.2	3.520 −0.314	308.53 17.49	161.5 −6.5	DWARF	20 D 0.7	F269/J1553 (38,186)
*AM 1319-271	13 19 01 −27 10.2	3.305 −0.314	311.15 34.97	144.6 −1.4	SP. w. BR.INNER + FNT.OUTER ARMS NGC 5101 = 508-G-58	10 D 6.7	F444/J1484 (249,328)
AM 1319-362	13 19 07 −36 22.2	3.406 −0.314	309.73 25.85	153.4 −4.1	S0 w. LINEAR COMP'N. NGC 5102 = 382-G-50	8 S 5.8	F382/J1530 (103,104)
AM 1319-251	13 19 10 −25 14.0	3.286 −0.314	311.54 36.88	142.7 −0.8	IRR. w. TAIL 508-G-59	15,16 D 0.7	F508/J2347 (92,167)
AM 1319-370	13 19 16 −37 07.6	3.416 −0.314	309.66 25.10	154.1 −4.2	COMPACT w. JET 382-G-51	7,13 D 0.7	F324/J1381 (249,331)
AM 1319-392	13 19 43 −39 28.2	3.448 −0.314	309.42 22.76	156.4 −4.8	SMALL GALAXY w. JET	7 D 0.5	F324/J1381 (242,205)
AM 1320-330	13 20 00 −33 09.3	3.372 −0.314	310.42 29.01	150.4 −3.0	GROUP of 4 (incl. 3 PECULIARS) incl. 382-G-53 = Fa 470	12,18 D 1.8	F382/J1530 (89,277)
AM 1320-322	13 20 03 −32 24.6	3.363 −0.314	310.55 29.75	149.7 −2.7	GAL. w. PEC. ABS'N. in GROUP of 3 444-G-16 (in GROUP of 3)	2,14,15,24 S 1.2	F444/J1484 (234,47)
AM 1320-353	13 20 10 −35 30.7	3.400 −0.314	310.09 26.67	152.6 −3.6	SPIRAL w. 3 APP. COMP'NS. 382-G-54	8 S 1.6	F382/J1530 (90,150)
AM 1320-343	13 20 12 −34 36.9	3.389 −0.314	310.24 27.56	151.8 −3.4	ELONGATED GROUP of 10	18 D 9.9	F382/J1530 (89,198)
*AM 1320-265	13 20 19 −26 56.7	3.306 −0.314	311.54 35.15	144.4 −1.1	RING w. I/A COMP'N.	6,8 S 0.4	F444/J1484 (234,340)
AM 1320-230	13 20 46 −23 07.9	3.270 −0.313	312.40 38.90	140.8 0.1	DWARF	20 D 0.7	F508/J2347 (71,279)
AM 1320-273	13 20 53 −27 39.7	3.315 −0.313	311.56 34.42	145.2 −1.2	DWARF	20 D 0.3	F509/J0652 (333,32)
AM 1320-274	13 20 56 −27 42.6	3.316 −0.313	311.57 34.37	145.2 −1.2	DWARF	20 D 0.4	F509/J0652 (332,29)
AM 1320-274	13 20 57 −27 43.2	3.316 −0.313	311.57 34.36	145.2 −1.2	SPIRAL w. ASYMMETRIC ARMS 444-G-22	10 D 0.7	F509/J0652 (332,29)
AM 1321-241	13 21 04 −24 13.7	3.281 −0.313	312.26 37.81	141.9 −0.1	DWARF? (SMALL)	20 D 0.4	F509/J0652 (335,216)
AM 1321-270	13 21 11 −27 05.8	3.310 −0.313	311.75 34.97	144.6 −1.0	DWARF	20 D 0.2	F508/J2347 (69,66)
AM 1321-390	13 21 16 −39 00.2	3.449 −0.313	309.81 23.18	156.1 −4.4	CLOSE PAIR (E + COMPACT) 324-G-16	13,23 S 1.3	F324/J1381 (226,230)
AM 1321-242	13 21 20 −24 24.0	3.284 −0.313	312.31 37.63	142.1 −0.1	RESOLVED DWARF 508-G-66 = UKS 1321-244	20 D 1.3	F508/J2347 (65,211)
AM 1321-233	13 21 25 −23 36.7	3.276 −0.313	312.49 38.40	141.3 0.1	GALAXY + PROTUBERANCE 508-G-67	22 D 1.1	F509/J0652 (332,249)
AM 1321-304	13 21 48 −30 42.6	3.350 −0.313	311.26 31.38	148.1 −1.9	DWARF	20 D 0.6	F444/J1484 (215,138)
AM 1321-360	13 21 51 −36 02.2	3.413 −0.313	310.39 26.11	153.2 −3.4	LINE of 3	24 D 3.1	F383/J1449 (331,119)
AM 1321-372	13 21 51 −37 25.1	3.430 −0.313	310.18 24.74	154.6 −3.8	HI.S.BR. PEC.SP + FNT. DWARF [NE] NGC 5121 = 382-G-57	12,20,23 D 1.6	F324/J1381 (221,316)
AM 1321-375	13 21 56 −37 51.3	3.436 −0.313	310.13 24.30	155.0 −4.0	COMPACT (INTERNAL RING) 324-G-18	6,13 D 0.2	F383/J1449 (326,21)
AM 1322-422	13 22 04 −42 26.6	3.501 −0.313	309.48 19.75	159.4 −5.2	DWARF	20 D 0.4	F270/J1313 (277,315)
*AM 1322-300	13 22 20 −30 05.0	3.345 −0.313	311.50 31.98	147.6 −1.6	I/A QUARTET	4 D 2.9	F444/J1484 (209,172)
AM 1322-332	13 22 23 −33 23.7	3.383 −0.313	310.94 28.70	150.7 −2.6	SPIRAL w. APP. COMP'NS. MCG-5-32-11 = 382-G-58	8 S 8.1	F383/J1449 (330,261)

Identification	R.A. (1950.0) Declination	Precession R.A./Dec.	Gal. Long. Gal. Lat.	S.G.L. S.G.B.	Description Cross-Identification	Coding Size (Min.)	Field/Plate (x, y)
AM 1322-333	13 22 36 −33 32.1	3.385 −0.313	310.97 28.56	150.9 −2.6	IRR. + I/A COMPACT 382-G-60	1,13,16 S 0.5	F382/J1530 (60,256)
AM 1322-424	13 22 37 −42 45.7	3.509 −0.313	309.53 19.42	159.7 −5.2	RINGING, PEC. ABS'N. + FILAMENT Cen A = NGC 5128 = 270-IG-9 [PKS/MC]	7,14,16,22 D 9.0	F270/J1313 (272,298)
AM 1322-250	13 22 39 −25 07.3	3.294 −0.313	312.53 36.87	142.9 −0.1	I/A DBL.	2 S 0.5	F508/J2347 (50,172)
AM 1322-370	13 22 41 −37 07.0	3.430 −0.312	310.40 25.01	154.3 −3.6	DWARF w. STRUCTURE NGC 5121a = 382-G-61 = UKS 1322-371	20 D 1.6	F382/J1530 (65,63)
*AM 1322-372	13 22 46 −37 20.7	3.433 −0.312	310.39 24.78	154.6 −3.6	PLANETARY NEBULA (DBL. SHELL) Lo-8 = 382-PN-63 = pk 310+24-1	25 D 2.2	F324/J1381 (211,320)
AM 1322-242	13 22 48 −24 23.5	3.287 −0.312	312.73 37.58	142.2 0.2	SPIRAL w. DISTURBED DISK 508-G-75	12 D 0.8	F509/J0652 (314,208)
AM 1322-342	13 22 59 −34 22.7	3.397 −0.312	310.92 27.71	151.7 −2.7	GALAXY w. JETS	7 D 0.5	F382/J1530 (57,210)
AM 1323-365	13 23 03 −36 58.5	3.430 −0.312	310.51 25.14	154.2 −3.5	GROUP of 9	18 D 4.5	F382/J1530 (61,71)
AM 1323-262	13 23 14 −26 25.1	3.309 −0.312	312.43 35.57	144.1 −0.3	I/A DBL. IC 4245/6 = 509-IG-2	2 S 0.3	F508/J2347 (44,102)
AM 1323-222	13 23 36 −22 22.3	3.269 −0.312	313.39 39.54	140.3 1.0	SPIRAL w. PARTIAL RING + COMPACT MCG-4-32-14 = 576-G-57	6,8,12,13 D 1.1	F508/J2347 (36,320)
AM 1323-335	13 23 37 −33 51.8	3.393 −0.312	311.15 28.20	151.3 −2.5	DWARF 382-G-66	20 D 0.9	F383/J1449 (315,236)
AM 1323-295	13 23 44 −29 57.8	3.348 −0.312	311.88 32.05	147.6 −1.3	COMPACT	13 D 0.4	F444/J1484 (192,179)
AM 1323-271	13 23 58 −27 10.9	3.319 −0.312	312.48 34.79	144.9 −0.4	E w. 7 COMP'NS. in CHAIN MCG-4-32-16 = 509-G-8 = PKS 1323-271	8,17 D 3.4	F509/J0652 (297,58)
*AM 1323-362	13 23 58 −36 29.9	3.427 −0.312	310.78 25.59	153.8 −3.2	SINGLE w. RESOLVED STRUCTURE 382-G-67	22 D 0.7	F382/J1530 (50,96)
*AM 1323-293	13 23 59 −29 37.5	3.345 −0.312	312.00 32.38	147.2 −1.1	RING or DISTURBED SPIRAL IC 4248 = 444-G-35 = To 1323-296	6,16 D 1.3	F444/J1484 (190,197)
AM 1323-861	13 23 59 −86 16.7	0.174 −0.312	303.62 −23.71	202.9 −15.1	I/A DBL. in CLUSTER	2,19 S 0.3	F007/J2961 (108,98)
AM 1324-453	13 24 06 −45 31.3	3.561 −0.312	309.41 16.65	162.5 −5.8	SMALL I/A DBL.	2 S 1.3	F270/J1313 (253,150)
AM 1324-294	13 24 13 −29 48.9	3.348 −0.312	312.03 32.18	147.4 −1.1	DWARF/IRR. 444-G-37	20 D 2.2	F444/J1484 (187,187)
AM 1324-861	13 24 19 −86 16.8	0.174 −0.312	303.62 −23.71	202.9 −15.1	GROUP of 8	18 D 9.0	F008/J2963 (231,104)
AM 1324-450	13 24 31 −45 05.8	3.556 −0.312	309.55 17.06	162.1 −5.6	DWARF	20 D 0.3	F270/J1313 (250,173)
*AM 1324-375	13 24 35 −37 55.7	3.449 −0.312	310.69 24.15	155.2 −3.5	IRR. w. DUST + DIAMETRICS 324-G-23 (PLUME)	8,12,14,15,16 S 2.2	F382/J1530 (46,19)
AM 1324-250	13 24 38 −25 08.0	3.299 −0.311	313.08 36.78	143.0 0.4	GROUP of 9 incl. RING (Sa at CENTRE)	6,18 D 3.6	F508/J2347 (26,171)
*AM 1324-411	13 24 42 −41 13.5	3.496 −0.311	310.18 20.89	158.4 −4.4	RESOLVED DWARF Se 102/1 = 324-G-24 = UKS 1324-412	20 D 4.7	F324/J1381 (190,112)
AM 1324-564	13 24 44 −56 41.7	3.808 −0.311	307.92 5.58	173.3 −8.7	PLANETARY NEBULA	25	F173/J (85,89)
*AM 1324-431	13 24 56 −43 10.2	3.527 −0.311	309.92 18.96	160.3 −4.9	SHRED (ONE-ARMED SP., NO NUCLEUS) Se 102/5 = 270-G-12 = Fa 324	16,22 D 1.0	F270/J1313 (248,277)
AM 1324-265	13 24 59 −26 57.6	3.319 −0.311	312.79 34.97	144.8 −0.1	RICH CLUSTER (40 BRIGHT GALAXIES) IC 4254 = 509-G-17	19 D 17.9	F444/J1484 (178,340)
*AM 1325-405	13 25 00 −40 52.6	3.492 −0.311	310.29 21.23	158.1 −4.3	CIRCLE of 5 LINKED = KLEM 25 324-IG-25 + Fa 159 in DN 1325-408	18 D 0.6	F324/J1381 (187,130)
*AM 1325-292	13 25 03 −29 21.6	3.346 −0.311	312.33 32.60	147.1 −0.8	E + SP. (EDGE-ON RING?; GRP of 3) NGC 5152 = 444-G-44	2,24 S 0.4	F444/J1484 (177,211)
AM 1325-311	13 25 06 −31 14.0	3.367 −0.311	311.98 30.75	148.9 −1.4	RICH CLUSTER of 25 GALAXIES MCG-5-32-26 = 444-G-46	19 D 44.8	F444/J1484 (177,111)
*AM 1325-251	13 25 13 −25 16.3	3.302 −0.311	313.21 36.62	143.2 0.5	RING w. SMALL NUCLEUS + PLUME	6,15 D 0.3	F509/J0652 (283,161)
*AM 1325-274	13 25 16 −27 40.0	3.327 −0.311	312.72 34.27	145.5 −0.3	M51-TYPE	9 D 0.4	F508/J2347 (21,35)
AM 1325-265	13 25 16 −26 50.7	3.319 −0.311	312.89 35.07	144.7 0.0	I/A DBL. in LARGE CLUSTER 509-IG-21	2,18 S 0.3	F509/J0652 (281,77)
AM 1325-244	13 25 26 −24 42.6	3.297 −0.311	313.40 37.17	142.7 0.7	GROUP of 6 (PECULIAR SPIRAL + E's) MCG-4-32-21 = 509-G-23	18,12 D 5.8	F509/J0652 (281,191)
AM 1325-375	13 25 30 −37 54.6	3.452 −0.311	310.88 24.15	155.3 −3.3	STAR w. NEBULOSITY 324-G-26	21 D 0.3	F382/J1530 (37,20)
AM 1325-441	13 25 30 −44 13.3	3.547 −0.311	309.86 17.91	161.3 −5.2	SPIRAL w. DISTORTED OUTER RING 270-G-13	10 D 1.1	F270/J1313 (242,221)
AM 1325-271	13 25 35 −27 18.5	3.325 −0.311	312.88 34.60	145.1 −0.1	SPIRAL w. LO.S.BR. GALAXY(DWARF?) 509-G-25 = UKS 1325-273	20,23 S 0.6	F509/J0652 (277,52)

Identification	R.A. (1950.0) Declination	Precession R.A./Dec.	Gal. Long. Gal. Lat.	S.G.L. S.G.B.	Description Cross-Identification	Coding Size (Min.)	Field/Plate (x, y)
AM 1325-255	13 25 41 −25 58.3	3.311 −0.311	313.19 35.92	143.9 0.3	DWARF 509-G-27	20 D 0.4	F509/J0652 (277,124)
AM 1325-371	13 25 46 −37 19.1	3.445 −0.311	311.04 24.72	154.7 −3.1	CLOSE PAIR (incl. PEC. SPIRAL) 383-G-2	12,23 S 9.4	F324/J1381 (179,321)
AM 1325-371	13 25 56 −37 10.3	3.444 −0.311	311.10 24.86	154.6 −3.0	GALAXY w. DUST in NUCLEUS + TAILS 383-G-3	14,15 D 0.7	F382/J1530 (30,59)
AM 1326-475	13 26 01 −47 53.0	3.615 −0.311	309.39 14.27	164.9 −6.1	DWARF w. STRUCTURE 220-G-14	20 D 0.5	F270/J1313 (233,24)
AM 1326-340	13 26 33 −34 00.5	3.405 −0.310	311.81 27.96	151.6 −1.9	EDGE-ON w. MULTIPLE ABS'N. LANES 383-G-5 (+ APP. COMP'NS.)	8,14 D 2.7	F382/J1530 (17,229)
AM 1326-234	13 26 42 −23 42.3	3.290 −0.310	313.99 38.10	141.8 1.2	I/A DBL. (ONE w. TAIL)	2,15 S 0.3	F509/J0652 (267,246)
AM 1326-251	13 26 57 −25 15.2	3.306 −0.310	313.70 36.57	143.3 0.8	I/A DBL.	2 S 0.5	F509/J0652 (262,163)
AM 1327-291	13 27 05 −29 15.2	3.351 −0.310	312.87 32.63	147.1 −0.4	GALAXY + I/A COMP'N in GROUP of 3 444-G-55	1,24 D 1.7	F444/J1484 (153,217)
AM 1327-771	13 28 10 −77 23.3	5.278 −0.310	305.16 −14.87	193.8 −13.4	DWARF 40-G-10	20 D 1.3	F021/J2083 (188,323)
AM 1327-231	13 27 58 −23 14.9	3.288 −0.310	314.46 38.50	141.4 1.7	SPIRAL w. I/A COMP'N. 509-G-36	1 S 0.4	F509/J0652 (251,270)
AM 1328-215	13 28 20 −21 55.8	3.275 −0.309	314.90 39.77	140.2 2.1	SPIRAL w. COMP'N. (M51-TYPE?)	8,9 D 0.4	F509/J0652 (248,341)
AM 1328-250	13 28 27 −25 08.8	3.309 −0.309	314.14 36.61	143.3 1.2	DISTURBED GALAXY (CENTRE of 6) MCG-4-32-29 = 509-G-38	16,18 D 0.9	F509/J0652 (244,169)
AM 1328-772	13 28 31 −77 23.9	5.324 −0.309	305.22 −14.96	193.9 −13.3	PAIR of EDGE-ON SPIRALS	23 S 0.6	F021/J2083 (184,319)
*AM 1328-343	13 28 35 −34 32.2	3.420 −0.309	312.17 27.37	152.2 −1.7	SPIRAL w. DISTURBED LENS NGC 5188 = 383-G-9	12 D 2.9	F383/J1449 (259,201)
AM 1328-313	13 28 38 −31 33.5	3.383 −0.309	312.78 30.30	149.4 −0.8	CLUSTER of 20 GALAXIES	2,19 S 0.3	F444/J1484 (136,93)
AM 1328-325	13 28 59 −32 59.0	3.401 −0.309	312.57 28.89	150.8 −1.1	I/A DBL. in GROUP of 3 = SS-16 NGC 5193 = MCG-5-32-36 = 383-G-14	2,24 D 3.4	F444/J1484 (133,16)
AM 1329-334	13 29 07 −33 47.0	3.412 −0.309	312.45 28.09	151.5 −1.3	GALAXY w. TAIL	15 D 0.9	F383/J1449 (254,242)
AM 1329-324	13 29 19 −32 40.5	3.399 −0.309	312.72 29.18	150.5 −1.0	CLOSE PAIR	23 S 0.2	F383/J1449 (253,301)
AM 1329-601	13 29 34 −60 11.4	3.961 −0.309	308.03 2.03	176.9 −9.1	NEBULOSITY 132-EN-12	21	F132/J2097 (102,168)
AM 1329-223	13 29 45 −22 36.8	3.285 −0.309	315.14 39.04	141.0 2.2	I/A DBL. (E + SPIRAL)	2 S 0.9	F509/J0652 (230,305)
*AM 1329-325	13 29 48 −32 51.5	3.403 −0.309	312.79 28.98	150.7 −0.9	I/A DBL. PKS 1329-328?	2 S 0.4	F383/J1449 (247,291)
AM 1329-265	13 29 51 −26 54.8	3.332 −0.309	314.11 34.82	145.1 0.9	DISRUPTED GALAXY (in GROUP of 8) IC 4281 = 509-G-49	16,18 D 1.1	F509/J0652 (226,74)
AM 1329-231	13 29 55 −23 15.3	3.293 −0.309	315.02 38.40	141.6 2.1	GROUP of 4	18 D 2.2	F509/J0652 (227,270)
AM 1329-753	13 29 58 −75 31.1	5.052 −0.309	305.62 −13.11	192.1 −12.8	PLANETARY NEBULA 40-PN?-11 = PK 305-13-1	25 D 1.0	F096/J2072 (119,149)
AM 1329-654	13 29 59 −65 43.3	4.206 −0.309	307.20 −3.44	182.3 −10.5	'S' SHAPE PK 307-3-1 [PKS] NGC 5189 = IC 4274 = 96-PN-16	25	F096/J2079 (80,137)
AM 1330-235	13 30 07 −23 56.8	3.300 −0.309	314.90 37.72	142.3 1.9	SPIRAL + HEAVY ARM + COMP'NS. IC 4280 = 509-G-54	10,8 D 0.9	F509/J0652 (224,233)
AM 1330-222	13 30 21 −22 20.5	3.284 −0.308	315.38 39.28	140.7 2.5	E + 7 COMP'NS. MCG-4-32-37 = 577-G-9	8 D 0.4	F509/J0652 (222,319)
AM 1330-265	13 30 23 −26 53.9	3.333 −0.308	314.25 34.81	145.1 1.1	GALAXY w. SHARP JET (DEFECT?)	7 D 0.3	F509/J0652 (220,75)
*AM 1330-421	13 30 49 −42 18.2	3.542 −0.308	311.19 19.64	159.8 −3.7	TIGHT GROUP of 7 E's	18 D 4.0	F270/J1313 (191,324)
AM 1331-380	13 31 06 −38 05.1	3.478 −0.308	312.04 23.79	155.8 −2.3	CLOSE I/A DBL. 324-IG-37	2 S 0.6	F324/J1381 (123,280)
AM 1331-311	13 31 09 −31 13.9	3.387 −0.308	313.46 30.53	149.2 −0.1	TIGHT GROUP of 12	18 D 4.5	F444/J1484 (107,110)
AM 1331-451	13 31 39 −45 17.0	3.598 −0.308	310.81 16.68	162.6 −4.4	FOURCADE-FIGUEROA SHRED 270-G-17	16,20 D 12.1	F270/J1313 (183,164)
AM 1331-311	13 31 52 −31 11.6	3.389 −0.308	313.64 30.53	149.3 0.0	COMPACT	13 D 0.2	F444/J1484 (99,112)
AM 1331-231	13 31 55 −23 11.9	3.297 −0.307	315.61 38.36	141.7 2.5	I/A DBL.(CLOSE) 509-IG-66	2 S 0.4	F509/J0652 (203,273)
*AM 1332-290	13 32 11 −29 04.5	3.364 −0.307	314.21 32.60	147.3 0.8	I/A DBL. in CLOSE TRIPLE	2,24 D 1.3	F444/J1484 (94,226)
*AM 1332-290	13 32 15 −29 00.4	3.363 −0.307	314.24 32.66	147.2 0.8	M51-TYPE	9 S 0.4	F444/J1484 (93,229)

Identification	R.A. (1950.0) Declination	Precession R.A./Dec.	Gal. Long. Gal. Lat.	S.G.L. S.G.B.	Description Cross-Identification	Coding Size (Min.)	Field/Plate (x, y)
*AM 1332-331	13 32 17 −33 13.6	3.416 −0.307	313.30 28.52	151.2 −0.5	I/A DBL. in CHAIN of 5 NGC 5215 = 383-IG-28/29	2,17 S 0.7	F383/J1449 (219,272)
*AM 1332-262	13 32 22 −26 20.7	3.332 −0.307	314.92 35.26	144.7 1.7	I/A DBL. COMPACT w. PLUME	2,13,15 S 0.1	F509/J0652 (196,105)
AM 1332-321	13 32 27 −32 17.0	3.404 −0.307	313.54 29.44	150.3 −0.2	CLOSE I/A DBL.	2 S 0.2	F383/J1449 (217,323)
AM 1332-335 [1332-3357]	13 32 27 −33 55.6	3.426 −0.307	313.18 27.83	151.9 −0.7	I/A TRIPLE (#1) 383-G-32?	3 D 0.7	F383/J1449 (217,234)
AM 1332-335 [1332-3354]	13 32 30 −33 55.9	3.426 −0.307	313.20 27.82	151.9 −0.7	I/A TRIPLE (#2) 383-G-33	3 D 0.4	F383/J1449 (216,234)
AM 1332-335 [1332-3356]	13 32 30 −33 57.1	3.426 −0.307	313.19 27.80	151.9 −0.7	I/A TRIPLE (#3) 383-G-31	3 D 0.9	F383/J1449 (216,233)
AM 1332-224	13 32 38 −22 44.4	3.293 −0.307	315.94 38.78	141.3 2.8	DWARF? DISTURBED GALAXY?	16,20 D 0.7	F509/J0652 (194,298)
AM 1332-300	13 32 59 −30 08.0	3.379 −0.307	314.16 31.53	148.3 0.6	RING in CLOSE PAIR MCG-5-32-47 = 444-G-76	6,23 S 3.4	F444/J1484 (85,169)
AM 1333-254	13 33 07 −25 45.0	3.328 −0.307	315.27 35.81	144.2 2.0	E + COMPACT	13,23 S 0.2	F509/J0652 (188,136)
AM 1333-324	13 33 14 −32 45.4	3.413 −0.307	313.62 28.94	150.8 −0.2	SPIRAL w. APP. COMP'N MCG-5-32-48 = 383-G-37	8 S 0.9	F383/J1449 (208,297)
AM 1333-493	13 33 28 −49 33.4	3.695 −0.307	310.36 12.42	166.8 −5.4	GROUP of 3 (WIDER GROUP of 5)	18,24 D 5.6	F220/J1569 (149,203)
AM 1333-544	13 33 34 −54 48.6	3.825 −0.307	309.45 7.24	171.9 −7.0	CHAIN of 5 GALAXIES	17 D 26.0	F174/J1609 (265,186)
AM 1333-285	13 33 42 −28 58.8	3.367 −0.306	314.62 32.62	147.3 1.1	DWARF 444-G-78	20 D 0.9	F444/J1484 (76,231)
AM 1333-261	13 33 47 −26 18.0	3.336 −0.306	315.31 35.24	144.8 2.0	SPIRAL w. I/A COMP'N. IC 4298 = 509-G-80	1 D 1.6	F509/J0652 (179,107)
*AM 1333-334	13 33 48 −33 42.6	3.428 −0.306	313.54 27.99	151.8 −0.4	LARGE E w. APP. COMP'NS. IC 4296 = 383-G-39 = PKS 1333-33	8 D 2.7	F383/J1449 (202,246)
AM 1334-333	13 34 06 −33 30.0	3.426 −0.306	313.66 28.18	151.6 −0.3	SMALL I/A DBL.	2 S 0.5	F383/J1449 (198,257)
AM 1334-455	13 34 17 −45 57.8	3.626 −0.306	311.16 15.93	163.4 −4.2	IRR. w. FILAMENTS 270-G-21	16 D 1.0	F270/J1313 (158,127)
AM 1334-274	13 34 33 −27 47.6	3.356 −0.306	315.15 33.74	146.2 1.7	DWARF (STRUCTURE) 444-G-84 = UKS 1334-278	20 D 1.5	F509/J0652 (169,27)
AM 1334-423	13 34 42 −42 36.7	3.567 −0.306	311.88 19.21	160.3 −3.1	RECTANGULAR E GALAXY NGC 5237 270-IG-22 = Fa 160 = Ag-48	22 D 0.9	F270/J1313 (152,307)
AM 1334-221	13 34 43 −22 17.0	3.293 −0.306	316.67 39.12	141.0 3.4	I/A DBL.(COMPACT)	2,13 S 0.1	F509/J0652 (168,323)
AM 1334-420	13 34 51 −42 08.3	3.559 −0.306	312.00 19.67	159.8 −2.9	PAIR of CONTACT E's	2 S 0.1	F270/J1313 (151,333)
AM 1334-781	13 34 56 −78 16.3	5.663 −0.306	305.41 −15.87	194.9 −13.2	E + IRR.	23 S 0.3	F021/J2083 (166,272)
AM 1334-351	13 34 57 −35 14.7	3.453 −0.306	313.46 26.43	153.3 −0.7	SP. w. SHARP KNOT(STAR?) in ARM	12 D 0.5	F383/J1449 (189,164)
AM 1335-310	13 35 36 −31 00.6	3.398 −0.305	314.59 30.55	149.3 0.8	I/A DBL. To 1335-309 = 444-IG-89	2 S 0.2	F444/J1484 (56,121)
AM 1335-274	13 35 39 −27 42.1	3.357 −0.305	315.44 33.78	146.2 1.9	GROUP of 5	17 D 7.2	F509/J0652 (157,32)
AM 1336-480 [1337-4805]	13 36 42 −48 02.7	3.682 −0.305	311.18 13.80	165.7 −4.6	PAIR of LO.S.BR. SPIRALS (#1) 220-G-26 = UKS 1336-480	8,23 S 8.3	F220/J1569 (125,273)
AM 1336-253	13 36 09 −25 35.7	3.334 −0.305	316.13 35.81	144.3 2.7	GALAXY w. DUST LANE (?) on EDGE IC 4310 = 509-G-88	14 D 1.4	F509/J0652 (151,145)
AM 1336-504	13 36 36 −50 49.1	3.744 −0.305	310.63 11.09	168.2 −5.4	CLOSE TRIPLE	24 D 7.6	F220/J1569 (123,135)
*AM 1336-221	13 36 39 −22 15.1	3.297 −0.305	317.24 39.04	141.1 3.9	SP. w. CONDENS. in ARM (M51-TYPE)	9 D 0.8	F509/J0652 (144,324)
AM 1336-393	13 36 55 −39 36.1	3.527 −0.305	312.94 22.09	157.6 −1.7	CLOSE GROUP of 4 GALAXIES	18 D 2.7	F324/J1381 (64,197)
AM 1337-312	13 37 03 −31 23.5	3.408 −0.304	314.85 30.12	149.8 1.0	S0 w. FILAMENT to DWARF NGC 5253 = 445-G-4	7,16,20 D 4.5	F444/J1484 (40,100)
AM 1337-312	13 37 06 −31 22.5	3.408 −0.304	314.87 30.13	149.8 1.0	DWARF (LINEAR FEATURE)	20 D 0.9	F444/J1484 (39,101)
AM 1337-480 [1336-4802]	13 37 32 −48 03.8	3.685 −0.304	311.31 13.73	165.6 −4.4	PAIR of LO.S.BR. SPIRALS (#2) NGC 5266a = 220-G-30 = UKS 1337-480	23 S 6.9	F270/J1313 (133,14)
AM 1337-292	13 37 11 −29 29.9	3.384 −0.304	315.36 31.96	148.0 1.6	CHAIN of 5	17 D 5.2	F445/J0632 (302,204)
AM 1337-251	13 37 16 −25 13.3	3.332 −0.304	316.54 36.12	144.0 3.1	IRR. IC 4315 = 509-G-91	16 D 1.8	F509/J0652 (137,165)
AM 1337-283	13 37 29 −28 38.4	3.374 −0.304	315.66 32.78	147.2 2.0	DWARF (RESOLUTION into KNOTS?) IC 4316 = 445-G-6 = To 1337-286	20 D 1.6	F444/J1484 (31,248)

Identification	R.A. (1950.0) Declination	Precession R.A./Dec.	Gal. Long. Gal. Lat.	S.G.L. S.G.B.	Description Cross-Identification	Coding Size (Min.)	Field/Plate (x, y)
AM 1337-233	13 37 33 −23 36.1	3.314 −0.304	317.09 37.68	142.5 3.6	GALAXY + 9 COMP'NS. NGC 5260 = 509-G-92	8 D 1.6	F509/J0652 (133,252)
AM 1337-322	13 37 57 −32 24.6	3.425 −0.304	314.82 29.08	150.8 0.8	3-ARMED SPIRAL MCG-5-32-64 = 445-G-9	11 D 0.6	F445/J0632 (291,47)
AM 1338-232	13 38 10 −23 21.0	3.313 −0.304	317.34 37.89	142.3 3.9	I/A DBL. (RING + IRR.)	2,6,16 S 0.4	F509/J0652 (125,265)
AM 1338-472	13 38 13 −47 24.7	3.677 −0.304	311.56 14.38	165.0 −4.0	DWARF (STRUCTURE)	20 D 0.5	F270/J1313 (123,49)
AM 1338-225	13 38 17 −22 58.1	3.308 −0.304	317.49 38.26	141.9 4.0	I/A DWARFS	2,20,20 S 0.7	F509/J0652 (124,286)
AM 1338-293	13 38 46 −29 39.5	3.391 −0.303	315.72 31.72	148.3 1.9	IRR. (HALO + RESOLUTION?) NGC 5264 = 445-G-12 = UKS 1338-296	16,20 D 2.2	F444/J1484 (17,193)
AM 1338-365	13 38 58 −36 59.0	3.494 −0.303	313.95 24.57	155.2 −0.5	CLOSE PAIR	23 S 0.4	F324/J1381 (38,337)
AM 1339-445	13 39 04 −44 57.9	3.632 −0.303	312.22 16.75	162.8 −3.1	DWARF	20 D 0.6	F270/J1313 (112,180)
AM 1339-360	13 39 12 −36 06.0	3.482 −0.303	314.21 25.42	154.4 −0.1	SPIRAL w. CONNECTED COMP'N. 383-G-64	9 S 0.4	F383/J1449 (142,118)
AM 1339-471	13 39 56 −47 18.0	3.684 −0.303	311.88 14.44	165.0 −3.7	VERY SMALL DWARF	20 D 0.1	F270/J1313 (107,55)
*AM 1339-475	13 39 56 −47 54.8	3.698 −0.303	311.75 13.83	165.6 −3.9	'PROLATE' w. CURVED ABSORPTION NGC 5266 = 220-G-33	14 D 1.3	F270/J1313 (108,22)
AM 1340-443	13 40 11 −44 36.7	3.631 −0.303	312.49 17.06	162.5 −2.8	SMALL SP. w. VERY PECULIAR ARMS 270-G-26	10 D 0.8	F270/J1313 (101,199)
AM 1340-354	13 40 20 −35 44.1	3.481 −0.303	314.54 25.73	154.1 0.2	M51-TYPE	9 S 0.4	F383/J1449 (130,137)
AM 1340-254	13 40 34 −25 40.9	3.346 −0.302	317.30 35.50	144.7 3.6	SPIRAL w. DBL. RING ARMS 509-G-98	6,10 D 1.3	F509/J0652 (97,140)
AM 1340-603	13 40 36 −60 34.6	4.080 −0.302	309.30 1.40	177.7 −7.9	FAINT PLANETARY NEBULA? vdB-5 = 133-SNR-2 = PK 309+1-1	25 D 0.8	F133/J1615 (281,145)
AM 1340-453	13 40 44 −45 30.0	3.652 −0.302	312.40 16.17	163.4 −3.0	DWARF	20 D 0.3	F270/J1313 (97,151)
AM 1340-453	13 40 45 −45 39.6	3.655 −0.302	312.37 16.01	163.5 −3.0	DWARF (RESOLUTION)	20 D 0.4	F270/J1313 (97,143)
AM 1341-270	13 41 02 −27 06.7	3.365 −0.302	317.00 34.09	146.0 3.2	SPIRAL w. CONDENSATIONS in ARM MCG-4-32-52 = 509-G-10	10 D 1.5	F445/J0632 (259,333)
*AM 1341-265	13 41 16 −26 58.8	3.364 −0.302	317.10 34.20	145.9 3.3	E + LINE of ABSORPTION IC 4320 = 509-G-101	14 D 0.6	F509/J0652 (90,70)
AM 1341-465	13 41 28 −46 55.2	3.685 −0.302	312.23 14.75	164.7 −3.3	HI.S.BR. IRR. 270-G-28	16 D 0.7	F270/J1313 (93,75)
AM 1341-294	13 41 45 −29 41.2	3.400 −0.302	316.45 31.55	148.5 2.5	SPIRAL + I/A COMP'N.	1 S 0.2	F445/J0632 (249,194)
AM 1341-294	13 41 52 −29 48.4	3.402 −0.302	316.45 31.43	148.7 2.5	DWARF (SMALL) 445-G-23	20 D 0.6	F445/J0632 (247,188)
*AM 1342-413	13 42 00 −41 36.4	3.584 −0.301	313.50 19.92	159.8 −1.5	LARGE DWARF (RESOLUTION) 325-G?-11 = UKS 1342-416	20 D 4.0	F324/J1381 (16,87)
*AM 1342-355	13 42 04 −35 54.1	3.490 −0.301	314.89 25.49	154.4 0.5	SPIRAL w. PECULIAR ARMS 383-G-71	10 D 1.3	F383/J1449 (111,128)
AM 1342-295	13 42 37 −29 58.8	3.407 −0.301	316.58 31.22	148.9 2.6	M51-TYPE IC 4324 = 445-G-25	9 D 0.6	F445/J0632 (239,179)
AM 1343-452	13 43 13 −45 26.3	3.664 −0.301	312.86 16.14	163.5 −2.6	DWARF	20 D 0.3	F270/J1313 (74,154)
AM 1343-224	13 43 17 −22 46.2	3.317 −0.301	318.97 38.16	142.1 5.2	TIGHT GROUP of 4 in COMMON ENVEL.	4,18 D 0.4	F510/J2385 (336,298)
AM 1343-244	13 43 30 −24 41.6	3.341 −0.301	318.40 36.29	143.9 4.6	COMPACT	13 D 0.1	F510/J2385 (332,194)
AM 1344-215	13 44 12 −21 51.8	3.308 −0.300	319.55 38.97	141.3 5.7	COMPACT NUCLEUS w. IRR. RING	6,13 D 0.8	F509/J0652 (50,344)
AM 1344-315	13 44 21 −31 57.8	3.440 −0.300	316.44 29.20	150.9 2.3	COMPACT w. JETS + TAILS	7,13,15 D 0.4	F383/J1449 (82,339)
AM 1344-240	13 44 26 −24 07.3	3.336 −0.300	318.84 36.78	143.5 5.0	E w. JET PKS 1344-241 MCG-4-32-2 = 509-G-108	7 D 1.1	F509/J0652 (49,223)
AM 1344-301	13 44 31 −30 10.2	3.415 −0.300	316.99 30.93	149.2 2.9	I/A DBL + DEBRIS PKS 1344-302? NGC 5291 = 445-G-30	2,15 S 0.8	F445/J0632 (216,169)
*AM 1344-323	13 44 36 −32 36.9	3.450 −0.300	316.32 28.56	151.5 2.1	SPIRAL w. I/A COMP'N.in HALO of E MCG-5-33-2 = 383-G-76	1,9,23 S 2.9	F383/J1449 (80,304)
AM 1344-450	13 44 53 −45 08.5	3.666 −0.300	313.22 16.36	163.3 −2.2	CONTACT I/A DBL. in TRIPLE	2,24 S 0.2	F270/J1313 (57,169)
AM 1344-252	13 44 57 −25 22.0	3.353 −0.300	318.57 35.55	144.7 4.6	HI.S.BR. SPIRAL w. JET 509-G-109	7 D 0.5	F510/J2385 (313,158)
AM 1346-475	13 46 03 −47 57.2	3.734 −0.299	312.77 13.58	166.0 −3.0	DWARF	20 D 0.2	F270/J1313 (53,18)

Identification	R.A. (1950.0) Declination	Precession R.A./Dec.	Gal. Long. Gal. Lat.	S.G.L. S.G.B.	Description Cross-Identification	Coding Size (Min.)	Field/Plate (x, y)
AM 1346-274 [1346-2757]	13 46 14 −27 49.9	3.388 −0.299	318.13 33.10	147.1 4.1	WIDE GROUP of 3 SPIRALS (#1)	24 D 9.0	F510/J2385 (295,26)
AM 1346-300	13 46 14 −30 03.0	3.419 −0.299	317.45 30.96	149.2 3.3	E w. RINGS (FAINT) RICH FIELD IC 4329 = 445-G-46 = MC 1346-300	22 D 1.9	F445/J0632 (197,175)
AM 1346-274 [1346-2749]	13 46 18 −27 45.4	3.388 −0.299	318.17 33.17	147.1 4.1	WIDE GROUP of 3 SPIRALS (#2) 445-G-48	24 D 9.0	F510/J2385 (295,30)
*AM 1346-354	13 46 23 −35 49.0	3.505 −0.299	315.85 25.36	154.6 1.3	HI.S.BR. Sd-DWARF (RESOLUTION) 383-G-87 = UKS 1346-358	20 D 4.0	F384/J1559 (327,133)
AM 1346-275 [1346-2745]	13 46 31 −27 57.3	3.391 −0.299	318.16 32.96	147.3 4.1	WIDE GROUP of 3 SPIRALS (#3) 445-G-51	24 D 9.0	F510/J2385 (292,20)
AM 1346-223	13 46 37 −22 35.3	3.322 −0.299	319.98 38.12	142.2 5.9	NUCLEUS w. 2 SMALL KNOTTY ARMS	12 D 0.4	F509/J0652 (20,305)
AM 1346-483	13 46 37 −48 30.4	3.751 −0.299	312.74 13.02	166.5 −3.1	[N.S.] CHAIN of 13 incl. 221-IG-5	17 D 1.2	F221/J2204 (299,256)
AM 1346-324	13 46 46 −32 49.6	3.460 −0.298	316.76 28.24	151.9 2.5	SPIRAL at CENTRE of CHAIN of 6 incl. 383-G-88	17 D 20.2	F384/J1559 (329,293)
AM 1347-354	13 47 46 −35 40.1	3.507 −0.298	316.19 25.44	154.6 1.7	DISTURBED SINGLE GALAXY 383-G-92	16 D 0.5	F383/J1449 (49,139)
AM 1347-363	13 47 48 −36 30.3	3.521 −0.298	315.97 24.63	155.4 1.4	DISTURBED SINGLE 383-G-93 (V.CLOSE I/A DBL.?)	2,16 D 0.7	F384/J1559 (311,96)
AM 1347-505	13 47 57 −50 57.7	3.821 −0.298	312.38 10.58	168.9 −3.7	PLANETARY NEBULA NGC 5307 = PK312+10-1 = 221-PN-11	25	F221/J2204 (281,125)
AM 1348-524	13 48 10 −52 40.3	3.870 −0.297	312.02 8.90	170.5 −4.3	CLOSE PAIR of SPIRALS	23 S 0.1	F221/J2204 (275,34)
AM 1348-475	13 48 22 −47 50.2	3.745 −0.297	313.19 13.60	166.0 −2.6	SUB-GROUP of 3 in CHAIN 221-G-12/13/14 = Fa 325/472/473	17,24 D 0.4	F221/J2204 (284,293)
AM 1348-333	13 48 24 −33 34.1	3.477 −0.297	316.93 27.44	152.7 2.5	Sd/DWARF To 1348-335 = 384-G-2 = UKS1348-335	20 D 8.1	F384/J1559 (309,254)
AM 1348-372	13 48 25 −37 22.7	3.538 −0.297	315.86 23.75	156.2 1.2	GALAXY w. APP. COMP'NS. 384-G-3	8 S 2.2	F384/J1559 (303,50)
AM 1348-422	13 48 39 −42 24.9	3.630 −0.297	314.58 18.85	161.0 −0.6	EDGE-ON w. APP. COMP'N. 325-G-32	8 S 0.6	F270/J1313 (14,314)
AM 1349-433	13 49 23 −43 38.1	3.658 −0.297	314.41 17.64	162.1 −0.9	DWARF 271-G-3	20 D 0.8	F270/J1313 (10,248)
*AM 1349-273	13 49 31 −27 38.7	3.395 −0.297	319.02 33.08	147.2 4.8	SB(r) w. RING + 9 COMP'NS. MCG-5-33-26 = 445-G-64	8,12 D 2.8	F445/J0632 (157,304)
AM 1349-300	13 49 53 −30 03.5	3.430 −0.296	318.33 30.74	149.5 4.0	GROUP + 08 E's (LOOSE CHAIN)	17,18 D 3.7	F445/J0632 (154,174)
AM 1351-271	13 51 07 −27 12.7	3.393 −0.295	319.57 33.40	146.9 5.3	E + COMPACT COMP'N.	8,13 S 0.2	F510/J2385 (238,60)
AM 1351-365	13 51 09 −36 54.0	3.541 −0.295	316.57 24.07	156.0 1.8	SPIRAL w. PECULIAR ARMS	10 D 0.5	F384/J1559 (274,76)
*AM 1351-263	13 51 29 −26 38.3	3.386 −0.295	319.86 33.93	146.4 5.6	I/A DBL. MCG-4-33-8/9 = 510-G-9/10	2 S 1.0	F510/J2385 (234,91)
*AM 1351-261	13 51 45 −26 18.8	3.383 −0.295	320.05 34.22	146.1 5.8	GROUP of 3 (COMPACT E + SPIRAL) MCG-4-33-11 = 510-G-12	24 D 3.8	F510/J2385 (230,109)
*AM 1352-280	13 52 05 −28 05.8	3.408 −0.295	319.52 32.49	147.8 5.2	IRR. (TYPE II) AMORPHOUS (BAR) MCG-5-33-30 = 445-G-75	6,10,12,16 D 3.4	F445/J0632 (127,279)
*AM 1352-263	13 52 14 −26 32.3	3.387 −0.295	320.09 33.97	146.4 5.8	EDGE-ON Sa w. WARPED PLANE + ABSN MCG-4-33-13 = 510-G-13	12,14 D 1.8	F510/J2385 (225,97)
AM 1352-333	13 52 38 −33 39.4	3.493 −0.294	317.85 27.12	153.1 3.3	GALAXY w. PECULIAR ABSORPTION 384-G-12 = PKS 1352-336?	14 D 0.7	F384/J1559 (262,251)
*AM 1352-250	13 52 47 −25 04.6	3.368 −0.294	320.76 35.33	145.0 6.4	COMPACT	13 D 0.1	F510/J2385 (218,175)
AM 1353-233	13 53 21 −23 31.7	3.349 −0.294	321.50 36.77	143.6 7.1	I/A TRIPLE 510-G-19	3 D 1.3	F510/J2385 (212,258)
AM 1353-273	13 53 22 −27 30.3	3.403 −0.294	320.04 32.97	147.4 5.7	SP. w. PEC. THIN BRANCHING ARMS MCG-4-33-16 = 445-G-80	10 D 1.1	F510/J2385 (211,45)
AM 1353-255	13 53 24 −25 54.9	3.381 −0.294	320.62 34.49	145.9 6.2	GAL. w. TAIL + DIAMETRIC COMP'NS.	8,15 S 0.6	F510/J2385 (211,130)
*AM 1353-272	13 53 24 −27 21.2	3.401 −0.294	320.10 33.12	147.2 5.7	INTEGRAL SPIRAL + COMP'N. 510-G-20	8 S 1.0	F445/J0632 (111,319)
*AM 1353-265	13 53 31 −26 55.4	3.396 −0.294	320.29 33.52	146.8 5.9	STAR CLUSTER ?	22 D 1.3	F510/J2385 (209,76)
AM 1354-523	13 54 18 −52 31.6	3.907 −0.293	312.97 8.82	170.7 −3.4	SPIRAL w. APP. COMP'N. 174-G-3	8 D 0.1	F221/J2204 (225,43)
*AM 1354-243	13 54 23 −24 34.8	3.365 −0.293	321.37 35.70	144.7 6.9	GROUP of 7 in WIDER GROUP	18 D 4.0	F510/J2385 (199,202)
*AM 1354-250	13 54 25 −25 00.2	3.371 −0.293	321.22 35.29	145.1 6.8	SP.w. BRIGHT RING + I/A SP.COMP'N IC 4350 = 510-G-22	1,6 S 1.3	F510/J2385 (198,179)
AM 1354-265	13 54 32 −26 55.7	3.398 −0.293	320.54 33.45	146.9 6.1	DISRUPTED RING GALAXY MCG-4-33-20 = 510-G-24	6 D 0.5	F510/J2385 (197,76)

Identification	R.A. (1950.0) Declination	Precession R.A./Dec.	Gal. Long. Gal. Lat.	S.G.L. S.G.B.	Description Cross-Identification	Coding Size (Min.)	Field/Plate (x, y)
AM 1355-481	13 55 17 −48 15.3	3.795 −0.293	314.24 12.91	166.8 −1.6	CLOSE PAIR (E + SPIRAL) 221-G-20/21	23 S 0.1	F221/J2204 (222,273)
AM 1355-791	13 55 37 −79 11.9	6.458 −0.292	306.22 −17.00	196.1 −12.5	SMALL LINE of 3	24 D 0.1	F021/J2083 (115,219)
AM 1355-331 [1355-3314]	13 55 45 −33 13.3	3.496 −0.292	318.69 27.35	152.9 4.1	CLOSE PAIR (#1)	23 D 0.3	F384/J1559 (227,275)
AM 1355-241	13 55 48 −24 13.5	3.364 −0.292	321.89 35.93	144.5 7.4	E + CONTACT COMP'N. (GROUP of 6)	1,18 D 3.6	F510/J2385 (182,221)
*AM 1355-250	13 55 50 −25 09.2	3.377 −0.292	321.54 35.05	145.4 7.0	TIGHT CLUSTER of DISTANT GALAXIES	19 D 1.8	F510/J2385 (181,171)
AM 1355-331 [1355-3313]	13 55 56 −33 14.5	3.497 −0.292	318.73 27.32	152.9 4.1	CLOSE PAIR (#2)	23 D 0.5	F384/J1559 (225,273)
AM 1356-261	13 56 05 −26 18.3	3.393 −0.292	321.17 33.93	146.5 6.7	SPIRAL w. DISRUPTED ARM	10 D 0.5	F510/J2385 (178,109)
AM 1356-303	13 56 27 −30 34.1	3.457 −0.292	319.74 29.85	150.5 5.2	M51-TYPE	9 S 0.3	F445/J0632 (79,146)
AM 1356-251	13 56 38 −25 19.6	3.381 −0.292	321.68 34.83	145.6 7.1	IRR. DWARF 510-G-32	20 D 0.7	F510/J2385 (172,162)
*AM 1356-332	13 56 51 −33 26.3	3.504 −0.292	318.87 27.08	153.2 4.2	DISTURBED SPIRAL (ISOLATED?) 384-IG-25	16 D 2.0	F384/J1559 (215,263)
AM 1357-504	13 57 25 −50 47.7	3.876 −0.291	313.90 10.37	169.3 −2.3	DWARF? UKS 1357-507	20 D 0.1	F221/J2204 (201,137)
*AM 1357-253	13 57 27 −25 32.4	3.386 −0.291	321.81 34.57	145.9 7.2	M51-TYPE 510-G-35	9 S 0.5	F510/J2385 (162,150)
*AM 1358-221 [1358-2219]	13 58 20 −22 19.1	3.344 −0.290	323.35 37.55	142.9 8.6	PAIR (#1: SPIRAL) MCG-4-33-27 = 578-G-14	23 S 3.4	F510/J2385 (150,323)
*AM 1358-324	13 58 27 −32 49.3	3.499 −0.290	319.43 27.57	152.7 4.7	SPIRAL w. APP. COMP'NS. NGC 5398 = 384-G-32 = To 1358-328	8 D 3.6	F384/J1559 (197,296)
*AM 1358-221 [1358-2219]	13 58 37 −22 19.9	3.345 −0.290	323.42 37.51	142.9 8.6	PAIR (#2:RING w. OFF-SET NUCLEUS) MCG-4-33-27 = 578-G-15	6,23 S 3.4	F510/J2385 (146,322)
AM 1358-253	13 58 42 −25 34.2	3.389 −0.290	322.12 34.45	146.0 7.5	GROUP of 4 510-G-41	18 D 2.5	F510/J2385 (147,148)
AM 1358-265	13 58 43 −26 55.1	3.409 −0.290	321.61 33.17	147.2 7.0	SPIRAL w. PEC. ARMS (DEFECT?)	10 D 0.7	F446/J0661 (327,343)
AM 1358-225	13 58 46 −22 53.9	3.353 −0.290	323.22 36.97	143.5 8.5	LONG CHAIN of 5 MCG-4-33-29 = 510-G-43	17 D 17.9	F510/J2385 (145,292)
*AM 1359-330	13 59 37 −33 09.2	3.508 −0.290	319.59 27.18	153.1 4.8	SPIRAL w. GAP in SPIRAL STRUCTURE MCG-5-33-38 = 384-G-35	10,13 D 1.8	F384/J1559 (184,279)
AM 1359-465	13 59 52 −46 51.8	3.785 −0.289	315.39 14.04	165.8 −0.4	DWARF 271-G-14	20 D 0.3	F221/J2204 (181,348)
AM 1400-255	14 00 06 −25 56.3	3.398 −0.289	322.34 34.00	146.4 7.6	GAL. w. PEC. ABS'N. + 4 DIAMETRICS 510-G-48	8,14 S 1.3	F510/J2385 (130,129)
*AM 1400-221	14 00 34 −22 18.8	3.348 −0.289	323.96 37.38	143.1 9.1	SP. w. HEAVY ARM + DWARF COMP'N MCG-4-33-34 = 578-G-19	10,20 S 3.0	F510/J2385 (122,323)
AM 1400-374	14 00 39 −37 47.9	3.594 −0.289	318.28 22.68	157.5 3.3	GALAXY w. LUMPY NUCLEUS 325-G-48	16 D 0.9	F384/J1559 (172,29)
AM 1400-322	14 00 44 −32 28.8	3.501 −0.289	320.07 27.75	152.6 5.3	CLOSE DBL. (incl.EDGE-ON SPIRAL) 446-G-2	23 S 0.7	F384/J1559 (171,315)
AM 1400-314	14 00 56 −31 42.5	3.489 −0.289	320.39 28.47	151.9 5.7	E + 2 APP. COMP'NS. 446-G-3	8 S 2.2	F446/J0661 (294,86)
AM 1400-312	14 00 58 −31 20.9	3.483 −0.289	320.52 28.81	151.6 5.8	DEFORMED SPIRAL	10 D 0.7	F446/J0661 (294,105)
AM 1401-290	14 01 01 −29 06.5	3.448 −0.289	321.35 30.93	149.5 6.6	CLOSE PAIR (incl. COMPACT)	13,23 S 1.8	F446/J0661 (296,226)
*AM 1401-324	14 01 19 −32 47.2	3.508 −0.288	320.09 27.42	152.9 5.3	SPIRAL w. PEC. ARM + APP. COMP'N. 384-G-41	1,9,10 S 0.7	F446/J0661 (289,28)
AM 1401-222	14 01 23 −22 20.4	3.350 −0.288	324.17 37.28	143.2 9.2	COMPACT SPIRAL	12,13 D 0.4	F510/J2385 (112,322)
*AM 1401-252	14 01 26 −25 24.1	3.394 −0.288	322.90 34.40	146.1 8.1	SPIRAL + I/A COMP'N MCG-4-33-38 = 510-G-56	1 S 1.4	F510/J2385 (113,157)
AM 1401-265	14 01 32 −26 54.6	3.416 −0.288	322.32 32.97	147.5 7.6	SP. + I/A COMP'N.(BOTH DISTURBED) 510-G-57	1 S 1.1	F510/J2385 (113,76)
*AM 1401-243 [1401-2435]	14 01 48 −24 35.4	3.383 −0.288	323.33 35.14	145.3 8.5	CLOSE PAIR (#1: HI.S.BR. SPIRAL) MCG-4-33-39 = 510-G-58	12,23 S 1.8	F510/J2385 (108,201)
*AM 1401-243 [1401-2435]	14 01 57 −24 35.3	3.383 −0.288	323.36 35.13	145.3 8.5	CLOSE PAIR (#2: LO.S.BR. SPIRAL) MCG-4-33-40 = 510-G-59	20,23 S 1.8	F510/J2385 (107,201)
AM 1402-331	14 02 16 −33 13.3	3.518 −0.288	320.16 26.94	153.4 5.3	DISTURBED SINGLE GALAXY	16 D 0.4	F384/J1559 (154,275)
AM 1402-225	14 02 30 −22 55.5	3.361 −0.287	324.22 36.65	143.8 9.3	CLOSE PAIR of E's	23 S 0.4	F510/J2385 (99,290)
AM 1402-285	14 02 30 −28 53.5	3.449 −0.287	321.79 31.03	149.4 7.0	I/A DBL.	2 S 1.2	F446/J0661 (279,238)

Identification	R.A. (1950.0) Declination	Precession R.A./Dec.	Gal. Long. Gal. Lat.	S.G.L. S.G.B.	Description Cross-Identification	Coding Size (Min.)	Field/Plate (x, y)
*AM 1402-323	14 02 31 −32 35.8	3.508 −0.287	320.43 27.52	152.9 5.6	4 in a LINE (MIXED TYPES) 384-G-47	17 D 9.9	F446/J0661 (275,39)
AM 1403-550	14 03 18 −55 07.2	4.055 −0.287	313.55 5.97	173.6 −3.1	SPIRAL w. DISTURBED ARMS + COMP'N 175-G-1	8,10 S 5.8	F175/J1570 (292,170)
*AM 1403-300	14 03 23 −30 00.9	3.469 −0.287	321.57 29.91	150.5 6.8	ONE-ARMED SPIRAL 446-G-7	10 D 0.7	F446/J0661 (268,178)
*AM 1403-340	14 03 38 −34 04.3	3.537 −0.287	320.16 26.05	154.3 5.3	LARGE SPIRAL w. APP. COMP'NS. 'CONJUGAL GALAXY' = FRENK'S SYSTEM	8 S 2.9	F384/J1559 (139,229)
AM 1403-322	14 03 40 −32 20.2	3.508 −0.287	320.78 27.69	152.7 5.9	SPIRAL w. APP. COMP'N. 446-G-8	8 S 2.2	F384/J1559 (138,322)
AM 1403-305 [1404-3058]	14 03 46 −30 58.5	3.485 −0.286	321.31 28.97	151.5 6.5	PAIR of SPIRALS (#1) MCG-5-33-43 = 446-G-9	23 S 5.8	F446/J0661 (262,126)
AM 1404-294	14 04 10 −29 46.5	3.467 −0.286	321.85 30.08	150.4 7.0	DISTURBED GALAXY w. APP. COMP'NS NGC 5464 = 446-G-11 = To 1404-297	8,16 D 1.1	F446/J0661 (259,191)
AM 1404-305 [1403-3058]	14 04 11 −30 58.4	3.487 −0.286	321.40 28.94	151.5 6.6	PAIR of SPIRALS (#2) MCG-5-33-44 = 446-G-12	23 S 5.8	F446/J0661 (258,126)
AM 1404-233	14 04 19 −23 34.7	3.374 −0.286	324.42 35.89	144.6 9.4	CLOSE PAIR	23 S 0.4	F510/J2385 (77,255)
AM 1404-261 [1404-2619]	14 04 26 −26 18.8	3.414 −0.286	323.29 33.32	147.2 8.4	PAIR (#1)	23 S 4.3	F510/J2385 (78,108)
AM 1404-214	14 04 29 −21 48.2	3.349 −0.286	325.25 37.53	142.9 10.1	M51-TYPE MGC-4-33-45 = 578-G-24	9 S 0.4	F511/J2324 (345,347)
AM 1404-324	14 04 32 −32 45.3	3.517 −0.286	320.83 27.24	153.2 5.9	CLOSE PAIR	23 S 0.4	F384/J1559 (128,300)
AM 1404-264	14 04 39 −26 46.6	3.422 −0.286	323.15 32.87	147.6 8.3	E w. 7 APP. COMP'NS. IC 4374 = 510-G-69 = PKS 1404-267	8 D 11.2	F511/J2324 (337,80)
AM 1404-312	14 04 41 −31 22.7	3.495 −0.286	321.37 28.53	151.9 6.5	CLOSE TRIPLE 446-G-15	24 D 1.8	F446/J0661 (252,105)
AM 1404-261 [1404-2618]	14 04 49 −26 19.4	3.415 −0.286	323.38 33.28	147.2 8.5	PAIR (#2) MCG-4-33-48 = 510-G-70	23 S 4.3	F510/J2385 (73,107)
AM 1404-245	14 04 51 −24 52.8	3.394 −0.286	323.99 34.63	145.9 9.0	E w. DIAMETRC COMP'NS. 510-G-71	8 S 2.9	F510/J2385 (72,185)
AM 1405-263	14 05 04 −26 31.6	3.419 −0.285	323.36 33.07	147.4 8.5	GALAXY w. 6 APP. COMP'NS. MGC-4-33-49 = 510-G-72	8 S 5.4	F511/J2324 (332,93)
AM 1405-330	14 05 06 −33 04.9	3.525 −0.285	320.83 26.89	153.5 5.9	BRIGHT SPIRAL + APP. COMP'NS. NGC 5488 = IC 4375 = 384-G-58	8 D 4.0	F384/J1559 (122,282)
AM 1405-283	14 05 26 −28 32.6	3.451 −0.285	322.64 31.15	149.3 7.8	CLOSE QUARTET	18 D 1.6	F446/J0661 (245,257)
AM 1405-265 [1406-2656]	14 05 50 −26 56.1	3.427 −0.285	323.38 32.63	147.9 8.5	I/A DBL.(CONNECTION?) (#1)	2 D 0.3	F446/J0661 (241,344)
AM 1406-265 [1405-2656]	14 06 04 −26 56.8	3.428 −0.285	323.44 32.60	147.9 8.5	I/A DBL.(CONNECTION?) (#2)	2 D 0.2	F446/J0661 (239,343)
AM 1406-270	14 06 43 −27 08.6	3.433 −0.284	323.52 32.36	148.1 8.6	COMPACT w. SHARP ABS'N. NOTCH	13,14 D 0.3	F510/J2385 (51,62)
AM 1407-282	14 07 40 −28 22.7	3.455 −0.283	323.24 31.13	149.4 8.3	COMPACT in GROUP of 3	13,24 D 2.9	F446/J0661 (219,266)
AM 1408-231	14 08 03 −23 11.8	3.377 −0.283	325.57 35.94	144.5 10.3	GALAXY w. PUFF (in I/A DBL.) 511-G-5	2,15 S 5.8	F511/J2324 (299,273)
AM 1408-271	14 08 20 −27 19.6	3.440 −0.283	323.84 32.06	148.4 8.8	COMPACT	13 D 0.1	F446/J0661 (211,323)
AM 1408-292	14 08 31 −29 24.9	3.474 −0.283	323.02 30.09	150.4 8.0	I/A TRIPLE	3 D 2.7	F446/J0661 (208,210)
AM 1408-295	14 08 37 −29 56.6	3.483 −0.283	322.83 29.59	150.9 7.9	CLOSE TRIPLE	24 D 4.9	F446/J0661 (207,182)
*AM 1408-364	14 08 42 −36 46.8	3.605 −0.283	320.27 23.15	157.2 5.1	RING (w. OFF-SET, PEC. NUCLEUS) 384-G-63	6 D 0.6	F384/J1559 (85,83)
AM 1408-293	14 08 54 −29 39.0	3.479 −0.283	323.02 29.85	150.7 8.0	CLOSE DBL.	23 S 1.4	F446/J0661 (204,198)
AM 1409-340	14 09 12 −34 01.8	3.555 −0.282	321.38 25.71	154.7 6.3	CLOSE DBL. IC 4378/79 = 384-G-64/65	23 S 0.4	F384/J1559 (77,231)
AM 1409-265	14 09 31 −26 52.4	3.436 −0.282	324.32 32.39	148.1 9.2	SPIRAL + APP. COMP'NS. NGC 5495 = 511-G-10	8 D 1.8	F511/J2324 (278,76)
AM 1410-271	14 10 04 −27 11.6	3.442 −0.282	324.32 32.05	148.5 9.2	DISTURBED SINGLE (ARC of 3) 511-G-12	16,24 D 13.4	F446/J0661 (191,330)
*AM 1410-283	14 10 05 −28 32.2	3.464 −0.282	323.75 30.80	149.7 8.7	CHAIN of 4 (I/A SPIRALS + COMPACT) 446-G-28/29	2,17 D 7.6	F446/J0661 (190,258)
AM 1410-735	14 10 48 −73 58.8	5.587 −0.281	308.66 −12.27	191.4 −9.8	PLANETARY NEBULA? PK 308-12-1 = 41-PN-3	25 D 0.6	F041/J2067 (206,234)
AM 1411-275	14 11 05 −27 57.8	3.457 −0.281	324.24 31.25	149.3 9.1	GROUP of 3 446-G-34	24 D 3.6	F511/J2324 (259,17)
AM 1411-291	14 11 42 −29 19.6	3.481 −0.280	323.81 29.93	150.6 8.7	SPIRAL w. APP. COMP'N. 446-G-35	8 S 2.7	F446/J0661 (171,215)

Identification	R.A. (1950.0) Declination	Precession R.A./Dec.	Gal. Long. Gal. Lat.	S.G.L. S.G.B.	Description Cross-Identification	Coding Size (Min.)	Field/Plate (x, y)
AM 1411-434	14 11 57 −43 43.8	3.769 −0.280	318.47 16.38	163.8 2.8	OBSCURED SPIRAL + 2 I/A COMP'NS. IC 4386=271-IG-25=Se 111/1 [UKS?]	1 S 3.4	F272/J1656 (332,242)
AM 1412-241	14 12 07 −24 11.0	3.400 −0.280	326.17 34.67	145.8 10.8	WIDE I/A QUINTET 511-G-15	5 D 17.9	F511/J2324 (248,220)
AM 1412-365	14 12 32 −36 57.8	3.623 −0.280	320.99 22.72	157.7 5.8	CLOSE PAIR of SPIRALS 385-G-3 = PKS 1412-368?	23 S 0.7	F385/J1562 (302,71)
AM 1412-475	14 12 39 −47 53.1	3.881 −0.280	317.19 12.41	167.6 1.2	E w. DIAMETRIC COMP'NS NGC 5516 = 221-G-34	8 S 2.2	F272/J1656 (315,19)
*AM 1412-531	14 12 46 −53 10.9	4.050 −0.280	315.47 7.40	172.4 −1.1	PLANETARY NEBULA (DOUBLE JETS)	7,25 D 0.9	F175/J1570 (221,276)
AM 1413-313 [1413-3127]	14 13 07 −31 31.5	3.523 −0.279	323.22 27.77	152.8 8.1	CLOSE PAIR of SPIRALS (#1) IC 4388 = 446-G-38	23 D 0.8	F446/J0661 (155,97)
AM 1413-243	14 13 19 −24 36.7	3.409 −0.279	326.28 34.17	146.3 10.9	SPIRAL ARMS off RING(1 to COMP'N)	6,8 S 1.1	F511/J2324 (233,198)
AM 1413-774	14 13 25 −77 41.7	6.441 −0.279	307.58 −15.84	195.0 −11.1	TRIPLE	24 D 0.2	F021/J2083 (56,292)
AM 1413-312 [1413-3131]	14 13 30 −31 27.0	3.523 −0.279	323.34 27.81	152.7 8.2	CLOSE PAIR of SPIRALS (#2) IC 4391 = 446-G-41	23 D 0.8	F446/J0661 (151,101)
AM 1414-235	14 14 48 −23 57.2	3.402 −0.278	326.97 34.64	145.9 11.5	TRIANGULAR RING SPIRAL + 2 COMP'N MGC-4-34-2 = 511-G-18	6,8 D 4.9	F511/J2324 (216,233)
AM 1415-530	14 15 13 −53 08.3	4.064 −0.278	315.84 7.32	172.6 −0.7	CONDENS. in FAINT NEBULOSITY	22 D 2.2	F175/J1570 (202,279)
AM 1415-272	14 15 29 −27 25.6	3.459 −0.277	325.53 31.38	149.2 10.2	SPIRAL w. 2 APP. COMP'NS. 511-G-22	8 S 1.0	F446/J0661 (126,317)
AM 1415-271	14 15 57 −27 10.5	3.456 −0.277	325.76 31.58	149.0 10.4	LONG CLUSTER of BR. E's + SPIRALS MCG-4-34-7 = 511-G-26	17,19 D 17.9	F446/J0661 (120,331)
AM 1415-280 [1416-2806]	14 15 58 −28 08.2	3.472 −0.277	325.33 30.69	149.9 10.0	SMALL I/A DBL. (#1)	2 D 0.3	F446/J0661 (121,279)
AM 1415-450	14 15 59 −45 05.1	3.822 −0.277	318.70 14.86	165.4 2.9	DISTURBED SINGLE/DWARF? Se 111/2 = 272-G-04	16 D 1.1	F272/J1656 (290,171)
AM 1416-280 [1415-2808]	14 16 03 −28 06.7	3.472 −0.277	325.36 30.70	149.9 10.1	SMALL I/A DBL. (#2)	2 D 0.3	F446/J0661 (120,280)
AM 1416-611	14 16 10 −61 10.1	4.432 −0.277	313.31 −0.31	179.9 −4.1	DARK and BRIGHT CLOUD	22 D 7.8	F133/J1615 (49,112)
AM 1416-610	14 16 17 −61 01.7	4.425 −0.277	313.37 −0.18	179.7 −4.0	NEBULOSITY	21	F134/J2086 (294,120)
*AM 1416-262	14 16 30 −26 24.9	3.445 −0.276	326.23 32.23	148.3 10.9	THIN G.D. SPIRAL + I/A COMP'N. MGC-4-34-10 = 511-G-30	1 S 2.2	F511/J2324 (195,101)
AM 1416-270 [1416-2708]	14 16 30 −27 08.6	3.457 −0.276	325.90 31.56	149.0 10.6	I/A PAIR (#1: SB(r) + PLUME)	2,15 D 0.8	F511/J2324 (195,62)
AM 1416-270 [1416-2708]	14 16 36 −27 08.8	3.458 −0.276	325.92 31.55	149.0 10.6	I/A PAIR (#2: S0) MGC-4-34-11 = 511-G-32	2 D 0.6	F511/J2324 (193,62)
AM 1417-291	14 17 19 −29 16.4	3.495 −0.276	325.14 29.52	151.1 9.9	GALAXY w. TAIL	15 D 0.6	F446/J0661 (105,218)
*AM 1417-290	14 17 39 −29 00.7	3.492 −0.276	325.33 29.74	150.8 10.0	SPIRAL w. APP. COMP'NS.(I/A DBL.) NGC 5556 = 446-G-50	2,8 D 5.4	F446/J0661 (101,232)
AM 1418-270	14 18 20 −27 01.9	3.460 −0.275	326.39 31.50	149.1 11.0	ONE-ARMED SP. + COMP'N. (MERGER?) MGC-4-34-13 = 511-G-34	8,10 D 1.6	F511/J2324 (173,68)
*AM 1419-263	14 19 13 −26 37.9	3.455 −0.274	326.79 31.79	148.8 11.3	SINGLE DISTURBED + TAILS + COMP'N MGC-4-34-14 = 511-IG-35	8,15 D 3.1	F511/J2324 (162,89)
AM 1419-435	14 19 14 −43 55.3	3.807 −0.274	319.69 15.76	164.6 3.9	V.HI.S.BR. GALAXY (P.N.?) IC 4406 = PK 319+15.1 = 272-PN-6	16 D 2.0	F272/J1656 (261,234)
*AM 1419-340 [1420-3403]	14 19 26 −34 08.2	3.591 −0.274	323.53 24.85	155.7 8.2	CLOSE PAIR (incl. RING ?)(#1) 385-G-15	10,23 D 1.5	F385/J1562 (231,225)
AM 1419-330	14 19 34 −33 01.3	3.570 −0.274	324.03 25.88	154.7 8.7	GALAXY w. I/A COMP'N.(in PAIR)	1,23 S 0.3	F385/J1562 (230,285)
AM 1420-340 [1419-3408]	14 20 04 −34 04.1	3.591 −0.274	323.69 24.87	155.7 8.4	CLOSE PAIR (incl. RING ?)(#2) 385-G-17	10,23 D 0.5	F385/J1562 (224,229)
AM 1420-292	14 20 20 −29 28.2	3.507 −0.273	325.75 29.09	151.5 10.4	GALAXY w. JETS 446-G-57	7 D 0.9	F446/J0661 (70,207)
AM 1421-282	14 21 00 −28 27.5	3.491 −0.273	326.36 29.96	150.6 10.9	GROUP of 3 SPIRALS NGC 5592 = 446-G-58	24 D 11.6	F446/J0661 (61,261)
AM 1422-421	14 22 15 −42 18.3	3.780 −0.272	320.83 17.06	163.4 5.1	I/A SPIRALS (ARM to COMP'N.)	1,2 S 0.6	F272/J1656 (233,322)
AM 1422-351	14 22 38 −35 19.5	3.625 −0.271	323.70 23.51	157.1 8.3	CLOSE PAIR 385-G-22	23 S 1.1	F385/J1562 (195,162)
AM 1423-234 [1423-2342]	14 23 03 −23 45.4	3.417 −0.271	329.15 34.05	146.5 13.3	PAIR (#1: LO.S.BR. SPIRAL) MGC-4-34-15 = 511-G-40	20,23 D 0.7	F511/J2324 (114,243)
AM 1423-234 [1423-2345]	14 23 11 −23 42.9	3.416 −0.271	329.21 34.07	146.4 13.3	PAIR (#2: SPIRAL) 511-G-41	23 D 0.4	F511/J2324 (113,245)
*AM 1426-331	14 26 19 −33 13.7	3.595 −0.268	325.38 25.13	155.5 9.9	IRREGULAR-SHAPED E MCG-5-34-14 = 385-G-30	22 D 1.3	F385/J1562 (154,274)

Identification	R.A. (1950.0) Declination	Precession R.A./Dec.	Gal. Long. Gal. Lat.	S.G.L. S.G.B.	Description Cross-Identification	Coding Size (Min.)	Field/Plate (x, y)
AM 1426-224	14 26 32 −22 42.5	3.407 −0.268	330.58 34.64	145.8 14.4	DISTURBED SPIRAL + APP. COMP'NS. MGC-4-34-16 = 511-G-44	8,12 S 1.8	F511/J2324 (70,299)
*AM 1427-263	14 27 05 −26 35.3	3.473 −0.268	328.69 31.10	149.5 12.9	DISTURBED w. COMP'N. + ABS'N.	8,14,16 D 0.6	F511/J2324 (67,90)
AM 1427-432	14 27 06 −43 20.3	3.828 −0.268	321.29 15.77	164.7 5.5	SMALL M51-TYPE IC 4441 = 272-G-11	9 S 0.9	F272/J1656 (185,266)
AM 1427-340	14 27 14 −34 00.7	3.613 −0.267	325.22 24.34	156.3 9.7	SPIRAL w. APP. COMP'NS. A 1427-34 = 385-G-33	8 S 2.9	F385/J1562 (144,232)
AM 1428-241	14 28 02 −24 15.5	3.435 −0.267	330.12 33.10	147.4 14.1	DWARF (SOME STRUCTURE) 511-G-47	20 D 0.4	F512/J2317 (326,217)
AM 1428-215	14 28 05 −21 58.1	3.398 −0.267	331.37 35.14	145.3 15.1	CHAIN of 4 (COMMON ENVELOPE)	17 D 0.8	F511/J2324 (50,338)
AM 1428-444	14 28 21 −44 45.7	3.872 −0.266	320.93 14.37	166.1 5.0	I/A DBL. Se 114-2 = 272-IG-13	2 S 1.3	F272/J1656 (173,190)
AM 1428-431	14 28 26 −43 11.8	3.830 −0.266	321.57 15.81	164.7 5.7	HI.S.BR. SP.(BROKEN ARM) + DWARF IC 4444 = Se 114-1 = 272-G-14	10,12,20 D 1.8	F272/J1656 (172,274)
AM 1428-220	14 28 35 −22 08.8	3.401 −0.266	331.40 34.93	145.5 15.1	DISTURBED SPIRAL + APP. COMP'NS. MGC-4-34-18 = 579-G-25	8,12 S 2.5	F511/J2324 (45,329)
AM 1428-251	14 28 40 −25 10.1	3.452 −0.266	329.79 32.22	148.3 13.9	LUMPY COMPACT MGC-4-34-19 = 511-G-50	13 D 0.8	F511/J2324 (47,166)
AM 1429-435	14 29 26 −43 57.2	3.855 −0.265	321.44 15.04	165.4 5.6	MULTI-ARMED SP. w. NUCLEAR ABS'N. NGC 5643 = 272-G-16	10,14 D 5.4	F272/J1656 (162,233)
AM 1429-603	14 29 27 −60 36.4	4.512 −0.265	315.02 −0.35	180.1 −2.4	PLANETARY NEBULA PS 315-01 = 134-PN-7	25	F134/J2086 (209,146)
AM 1429-225	14 29 50 −22 58.3	3.417 −0.265	331.25 34.06	146.4 15.0	IRR. SPIRAL (LINEAR) 512-IG-2	16 D 0.8	F512/J2317 (306,287)
AM 1430-244	14 30 06 −24 43.8	3.448 −0.265	330.37 32.47	148.1 14.3	CLOSE PAIR incl. COMPACT	13,23 S 0.6	F512/J2317 (300,192)
AM 1430-333	14 30 20 −33 32.8	3.613 −0.265	326.07 24.50	156.2 10.5	GROUP of 4 385-G-41	18 D 4.5	F385/J1562 (109,257)
AM 1430-350	14 30 38 −35 03.5	3.646 −0.264	325.45 23.10	157.6 9.9	DWARF 385-G-44 ?	20 D 0.9	F385/J1562 (107,176)
AM 1430-381	14 30 55 −38 18.8	3.719 −0.264	324.07 20.11	160.5 8.4	PECULIAR GALAXY 326-G-30	16 D 1.0	F327/J0653 (314,266)
AM 1432-383	14 32 15 −38 39.1	3.732 −0.263	324.17 19.69	161.0 8.5	CHAIN of 5 (CONNECTED)	17 D 3.8	F327/J0653 (299,248)
AM 1432-223	14 32 45 −22 34.1	3.416 −0.263	332.20 34.11	146.3 15.8	SMALL I/A TRIPLE	3 D 0.9	F512/J2317 (270,309)
AM 1433-242	14 33 47 −24 28.1	3.451 −0.262	331.39 32.32	148.2 15.2	3-ARMED SPIRAL	11 D 0.4	F512/J2317 (255,207)
*AM 1434-783	14 34 22 −78 35.7	7.203 −0.261	308.26 −17.07	196.3 −10.5	HI.S.BR. RING (OFF-SET NUCLEUS) IC 4448 = 22-G-2 = Se 113/1	6,10 D 0.8	F022/J2142 (244,249)
AM 1434-231	14 34 29 −23 13.3	3.431 −0.261	332.25 33.35	147.1 15.9	CLOSE PAIR 512-G-8	23 S 1.2	F512/J2317 (248,274)
*AM 1435-220	14 35 36 −22 09.0	3.415 −0.260	333.14 34.16	146.2 16.6	SPIRAL w. I/A COMP'N. IC 4468 = 580-G-6	1 S 3.0	F512/J2317 (235,332)
AM 1436-444	14 36 15 −44 48.4	3.909 −0.259	322.24 13.77	166.8 6.2	SPIRAL w. APP. COMP'NS. NGC 5688 = 272-G-22	8 S 13.4	F272/J1656 (98,186)
AM 1436-622	14 36 31 −62 25.8	4.688 −0.259	315.09 −2.36	182.1 −2.5	PLANETARY NEBULA 134-SNR-11	25	F134/J2086 (163,49)
AM 1436-440	14 36 52 −44 06.1	3.892 −0.259	322.64 14.37	166.2 6.6	IRR. IC 4472 = 272-G-23	16 D 2.7	F272/J1656 (91,224)
AM 1437-225	14 37 15 −22 52.6	3.430 −0.258	333.12 33.34	147.0 16.6	DWARF 512-G-11	20 D 0.3	F512/J2317 (214,293)
AM 1437-253	14 37 17 −25 33.5	3.478 −0.258	331.61 30.98	149.6 15.4	EDGE-ON + COMP'N. + CHAIN of 4 MCG-4-35-3 = 512-G-12	8,17 S 4.5	F512/J2317 (212,149)
AM 1437-345 [1437-3455]	14 37 26 −34 54.4	3.664 −0.258	326.88 22.64	158.1 11.2	CLOSE PAIR (#1) 386-G-4	23 D 0.5	F385/J1562 (32,183)
AM 1437-345 [1437-3454]	14 37 42 −34 55.8	3.666 −0.258	326.93 22.59	158.2 11.2	CLOSE PAIR (#2) 386-G-6	23 D 0.7	F385/J1562 (29,181)
AM 1438-400	14 38 09 −40 06.9	3.790 −0.258	324.62 17.88	162.8 8.8	SPIRAL w. I/A COMP'N. 327-G-08	1 S 0.4	F327/J0653 (236,171)
AM 1438-330	14 38 35 −33 05.9	3.629 −0.257	327.99 24.15	156.6 12.3	SPIRAL w. PECULIAR ARM MCG-5-35-3 = 386-G-9	10 D 1.0	F447/J0633 (125,11)
AM 1439-402	14 39 41 −40 21.8	3.802 −0.256	324.78 17.53	163.2 8.9	SPIRAL w. DISTURBED EDGE 327-G-12	12 D 0.8	F327/J0653 (220,158)
AM 1440-231	14 40 36 −23 14.0	3.443 −0.255	333.71 32.66	147.7 17.2	HI.S.BR. IRR. MCG-4-35-5 = 512-G-17	16 D 0.6	F512/J2317 (172,274)
*AM 1440-241	14 40 40 −24 15.1	3.461 −0.255	333.14 31.76	148.7 16.7	VERY CLOSE I/A E + SPIRAL MCG-4-35-4 = 512-G-18	2 S 0.5	F512/J2317 (171,219)
AM 1440-365	14 40 41 −36 54.3	3.720 −0.255	326.57 20.56	160.2 10.8	RING w. OFF-SET NUCLEUS + CONDENS	6 D 0.4	F327/J0653 (212,343)

Identification	R.A. (1950.0) Declination	Precession R.A./Dec.	Gal. Long. Gal. Lat.	S.G.L. S.G.B.	Description Cross-Identification	Coding Size (Min.)	Field/Plate (x, y)
AM 1440-365	14 40 44	3.719	326.61	160.2	GALAXY w. DBL. NUCLEUS	22	F327/J0653
	−36 51.3	−0.255	20.60	10.8		D 0.3	(211,346)
AM 1442-233	14 42 04	3.452	333.86	148.2	SB(r) + COMP'NS.	8	F512/J2317
	−23 35.0	−0.254	32.19	17.3	MCG-4-35-8 = CG-16 = 512-G-20	S 2.2	(154,255)
*AM 1442-401	14 42 12	3.808	325.29	163.3	SMALL COMPACT NUCLEUS + JETS	7	F327/J0653
	−40 15.1	−0.254	17.42	9.4		D 0.8	(195,164)
AM 1442-501	14 42 54	4.118	320.98	172.0	PLANETARY NEBULA	25	F223/J1666
	−50 10.7	−0.253	8.43	4.5			(326,165)
AM 1443-273	14 43 53	3.529	332.03	152.0	EDGE-ON w. I/A COMP'N. (PEC.DUST)	1,14	F512/J2317
	−27 30.4	−0.252	28.57	15.9	MCG-5-35-5 = 447-G-36	S 1.3	(133,45)
AM 1444-221	14 44 11	3.433	335.11	147.3	2 SMALL S(r) GALAXIES	6,23	F512/J2317
	−22 19.6	−0.252	33.03	18.3		S 0.9	(128,323)
AM 1444-302	14 44 26	3.589	330.57	154.8	E w. CURVED ABSORPTION	14	F447/J0633
	−30 26.3	−0.252	25.94	14.6	MCG-5-35-6 = 447-G-37	D 0.8	(55,152)
AM 1444-220	14 44 36	3.431	335.31	147.1	LINE of 3 (PEC. + SB + IRR.)	24	F512/J2317
	−22 09.7	−0.251	33.12	18.5	580-G-25/26/27	S 3.0	(123,332)
AM 1444-650	14 44 37	4.963	314.81	184.8	DARK CLOUD, NEBULOSITY	21	F098/J1625
	−65 03.0	−0.251	−5.12	−3.1			(153,177)
AM 1444-392	14 44 56	3.799	326.14	162.9	I/ A TRIPLET	3	F327/J0653
	−39 29.1	−0.251	17.88	10.3	327-IG-19	D 1.8	(167,205)
AM 1445-275	14 45 38	3.541	332.20	152.6	SMALL SPIRAL w. STRONG KNOTS	12	F512/J2317
	−27 54.1	−0.250	28.03	16.0	448-G-1	D 0.5	(112,23)
AM 1445-274	14 45 56	3.537	332.39	152.4	DISTURBED + I/A COMP'N.	1,16	F512/J2317
	−27 40.2	−0.250	28.20	16.2	448-G-2	S 0.2	(108,36)
AM 1446-414	14 46 39	3.867	325.35	165.1	M51-TYPE	9	F327/J0653
	−41 47.3	−0.250	15.67	9.4		D 0.5	(150,81)
AM 1448-374	14 48 15	3.767	327.57	161.7	E + APP. COMP'N.	8	F327/J0653
	−37 46.7	−0.248	19.09	11.7	327-G-23	S 0.5	(131,296)
*AM 1448-262	14 48 18	3.518	333.62	151.5	SPIRAL w. HI.S.BR. I/A COMP'N.	1	F512/J2317
	−26 25.5	−0.248	29.02	17.3	MCG-4-35-12 = 512-G-23	S 0.7	(79,102)
AM 1448-365	14 48 35	3.748	328.03	161.1	DWARF	20	F327/J0653
	−36 59.4	−0.248	19.76	12.1		D 0.5	(127,338)
AM 1448-375	14 48 38	3.771	327.59	161.9	GROUP of 5	18	F327/J0653
	−37 53.3	−0.248	18.96	11.7		D 1.8	(127,290)
*AM 1449-420	14 49 40	3.888	325.71	165.7	CLOSE TRIPLET (in CHAIN of 6)	17,24	F327/J0653
	−42 07.1	−0.247	15.12	9.7	327-IG-26	D 0.6	(120,63)
AM 1449-223	14 49 54	3.448	336.29	148.1	GROUP of 6	18	F512/J2317
	−22 35.0	−0.246	32.12	19.4		D 0.8	(57,309)
AM 1451-402	14 51 08	3.846	326.79	164.4	CLOSE QUARTET (CHAIN of 8)	17	F327/J0653
	−40 25.6	−0.245	16.49	10.8		D 1.8	(103,153)
AM 1452-272	14 52 09	3.547	333.85	152.9	PAIR of SPIRALS + COMP'NS.	8,23	F512/J2317
	−27 28.5	−0.244	27.68	17.5	513-G-2	S 1.3	(34,45)
AM 1452-373	14 52 12	3.776	328.37	162.0	M51-TYPE	9	F327/J0653
	−37 38.1	−0.244	18.85	12.4		D 1.1	(89,303)
*AM 1452-234	14 52 15	3.475	336.09	149.5	RING w. OFF-SET NUCLEUS	6,10	F513/J2358
	−23 45.9	−0.244	30.83	19.3	MCG-4-35-15 = 513-G-3	D 0.8	(299,244)
AM 1452-380	14 52 25	3.788	328.19	162.4	DISTURBED GALAXY	16	F327/J0653
	−38 04.4	−0.244	18.44	12.2	327-G-31	D 2.7	(87,279)
AM 1452-252	14 52 42	3.508	335.18	151.1	SMALL RING w. OFF-SET NUCLEUS	6	F512/J2317
	−25 25.8	−0.244	29.36	18.6	513-G-7	D 0.5	(25,155)
AM 1452-372	14 52 56	3.775	328.58	162.0	CHAIN of 7 incl. I/A DBL.	2,17	F386/J2989
	−37 29.5	−0.243	18.90	12.6	386-G-33	D 12.1	(127,45)
*AM 1452-372	14 53 17	3.774	328.69	161.9	DISTURBED GALAXY in CHAIN of 13	2,16,17	F327/J0653
	−37 24.2	−0.243	18.95	12.7	386-IG-39	D 1.3	(77,315)
AM 1453-354	14 53 47	3.734	329.64	160.5	PAIR (#1: DISTURBED)	16,23	F386/J2989
[1453-3540]	−35 44.5	−0.242	20.35	13.7	386-G-40	D 0.5	(116,139)
AM 1453-354	14 53 51	3.733	329.69	160.5	PAIR (#2: E)	23	F386/J2989
[1453-3544]	−35 40.9	−0.242	20.40	13.7	386-G-41	D 0.6	(115,142)
*AM 1454-425	14 54 26	3.930	326.12	166.8	I/A DBL. w. STRAIGHT JET	2,7	F327/J0653
	−42 55.8	−0.242	13.99	10.0	IC 4518 = Ag-50 = 273-IG-12	S 0.7	(74,18)
AM 1454-381	14 54 27	3.801	328.44	162.9	DWARF	20	F327/J0653
	−38 18.2	−0.242	18.05	12.4	327-G-3	D 0.8	(66,266)
AM 1454-385	14 54 55	3.820	328.19	163.5	SPIRAL w. ASYMMETRIC OUTER ARM	10,12	F327/J0653
	−38 57.4	−0.241	17.43	12.2	327-G-35	D 0.7	(62,231)
AM 1455-473	14 55 03	4.081	324.01	170.8	PEC. Sm or REFL.NEBULA (OBSCURED)	22	F223/J1666
	−47 30.0	−0.241	9.92	7.6	273-EN?-14 = UKS 1455-475		(225,313)
AM 1455-375	14 55 19	3.793	328.81	162.6	PECULIAR SPIRAL	12	F327/J0653
	−37 52.9	−0.241	18.34	12.8		D 0.6	(56,289)
AM 1455-283	14 55 43	3.576	334.01	154.3	I/A DBL. + 2 APP. COMP'NS.	2,8	F448/J4971
	−28 30.3	−0.241	26.39	17.7	To 1455-284 = 448-G-10	S 0.4	(195,260)
AM 1457-222	14 57 12	3.459	338.03	148.8	GALAXY w. CLOSE I/A COMP'N.	1	F513/J2358
	−22 27.5	−0.239	31.32	20.9	581-IG-12	s 0.3	(239,315)

Identification	R.A. (1950.0) Declination	Precession R.A./Dec.	Gal. Long. Gal. Lat.	S.G.L. S.G.B.	Description Cross-Identification	Coding Size (Min.)	Field/Plate (x, y)
*AM 1457-370	14 57 15 −37 09.8	3.780 −0.239	329.54 18.77	162.2 13.5	PEC.SP.w. CONDENS. at BASE of ARM 386-G-44	10,12 D 1.1	F327/J0653 (34,327)
AM 1457-403	14 57 20 −40 36.5	3.873 −0.239	327.77 15.77	165.2 11.7	RING GALAXY 328-G?-1	6 D 0.7	F327/J0653 (41,142)
AM 1457-261	14 57 31 −26 15.1	3.534 −0.239	335.73 28.09	152.4 19.1	SMALL I/A DBL (V.HI.S.BR.) 513-IG-11	2 S 0.5	F513/J2358 (233,112)
AM 1458-364 [1458-3646]	14 58 00 −36 47.2	3.773 −0.238	329.88 19.02	161.9 13.8	GROUP of 3 (#1)	24 D 0.3	F387/J1319 (330,81)
AM 1458-364 [1458-3647]	14 58 06 −36 44.3	3.772 −0.238	329.92 19.05	161.9 13.9	GROUP of 3 (#2)	24 D 0.4	F387/J1319 (329,83)
AM 1458-374 [1458-3746]	14 58 20 −37 47.8	3.800 −0.238	329.40 18.11	162.8 13.4	CLOSE TRIPLE (#1: Sb) 328-G-5	24 D 0.4	F386/J2989 (69,28)
AM 1458-364 [1458-3644]	14 58 22 −36 46.8	3.774 −0.238	329.95 18.99	161.9 13.9	GROUP of 3 (#3)	24 D 0.4	F387/J1319 (326,81)
AM 1458-374 [1458-3747]	14 58 31 −37 45.0	3.800 −0.238	329.46 18.14	162.8 13.4	CLOSE TRIPLE (#2: Sd) 328-G-6	24 D 0.5	F386/J2989 (67,30)
AM 1458-374 [1458-3745]	14 58 50 −37 46.1	3.801 −0.237	329.51 18.09	162.9 13.5	CLOSE TRIPLE (#3: Sc) 328-G-7	24 D 0.8	F386/J2989 (64,29)
AM 1459-722	14 59 23 −72 25.0	6.046 −0.237	312.59 −12.27	191.7 −6.0	GALAXY w. PECULIAR DUST 67-G-5	14 D 0.7	F041/J2067 (13,304)
AM 1500-721	15 00 27 −72 14.4	6.028 −0.236	312.75 −12.16	191.6 −5.8	GALAXY w. PECULIAR DUST NGC 5799 = 67-G-6 = Fa 164	14 D 0.7	F041/J2067 (7,313)
AM 1502-610	15 02 01 −61 09.7	4.804 −0.234	318.35 −2.57	182.7 0.7	PLANETARY NEBULA PK 318-2-2 = 135-PN-3	25	F135/J2175 (244,116)
AM 1502-270	15 02 13 −27 08.8	3.562 −0.234	336.18 26.77	153.8 19.6	GROUP of 4 in WIDER GROUP	18 D 5.4	F513/J2358 (177,64)
AM 1503-312	15 03 50 −31 22.0	3.658 −0.232	333.99 23.03	157.8 17.7	I/A TRIPLE (COMP'N.?) 448-IG-12	3 D 0.9	F448/J4971 (101,106)
AM 1504-250	15 04 12 −25 07.3	3.524 −0.232	337.86 28.21	152.2 21.0	CLOSE TRIPLE	24 D 0.7	F513/J2358 (152,173)
AM 1504-754 [1505-7540]	15 04 32 −75 49.1	6.885 −0.231	311.14 −15.39	194.7 −7.7	CLOSE PAIR of SPIRALS (#1) 42-G-1	23 S 10.3	F042/J2132 (243,133)
AM 1504-613	15 04 39 −61 32.7	4.851 −0.231	318.44 −3.06	183.2 0.7	PLANETARY NEBULA 135-?-4 = PK 318-3-1	25	F135/J2175 (226,96)
AM 1504-331 [1504-3314]	15 04 56 −33 11.6	3.704 −0.231	333.15 21.36	159.6 17.0	CLOSE PAIR (#1)	23 D 0.3	F387/J1319 (259,276)
AM 1504-331 [1504-3311]	15 04 58 −33 14.1	3.705 −0.231	333.13 21.32	159.6 17.0	CLOSE PAIR (#2)	23 D 0.3	F387/J1319 (258,273)
AM 1505-523	15 05 04 −52 39.1	4.338 −0.231	322.91 4.63	176.0 6.0	PAIR of DARK GLOBULES	22 S 3.3	F176/J1571 (65,303)
*AM 1505-253	15 05 09 −25 35.0	3.535 −0.231	337.77 27.71	152.7 21.0	PEC.SP. w. PEC.ARMS (DBL.NUCLEUS) 513-G?-19 (or 3 STARS ?)	10,12 D 0.9	F513/J2358 (141,148)
AM 1505-754 [1504-7549]	15 05 42 −75 40.2	6.862 −0.230	311.29 −15.30	194.7 −7.5	CLOSE PAIR of SPIRALS (#2) IC 4522 = 42-G-2 = UKS 1505-756	23 S 10.3	F042/J2132 (240,141)
AM 1506-642	15 06 22 −64 29.2	5.107 −0.230	317.14 −5.70	185.6 −0.9	PLANETARY NEBULA NGC 5844 = PKS 317-5-1 = 99-PN-1	25	F098/J1625 (27,200)
AM 1507-374	15 07 59 −37 41.6	3.828 −0.228	331.17 17.22	163.8 15.0	I/A DBL. 328-IG-29	2 S 0.5	F387/J1319 (222,34)
AM 1508-552	15 08 09 −55 28.6	4.494 −0.228	321.88 1.96	178.5 4.7	PLANETARY NEBULA PK 321+1-1 = 177-PN-1	25	F177/J1628 (300,152)
AM 1508-223	15 08 27 −22 37.3	3.481 −0.227	340.41 29.71	150.3 23.1	COMPACT	13 D 0.1	F513/J2358 (99,307)
AM 1509-235	15 09 06 −23 51.3	3.507 −0.227	339.72 28.62	151.6 22.6	ONE-ARMED SPIRAL + COMPACT MCG-4-36-2 = 513-G-22	10,13,23 D 0.9	F513/J2358 (92,240)
AM 1510-232	15 10 49 −23 28.6	3.502 −0.225	340.34 28.69	151.4 23.1	SPIRAL w. I/A COMP'N. MCG-4-36-3 = 513-G-26	1 S 0.6	F513/J2358 (71,260)
AM 1510-623	15 10 57 −62 34.6	4.979 −0.225	318.55 −4.32	184.4 0.7	NEBULOSITY 99-EN-3	21	F135/J2175 (185,41)
AM 1511-225	15 11 59 −22 56.5	3.493 −0.224	340.95 28.97	151.1 23.6	DWARF w. KNOTS 513-G-29	20 D 0.5	F513/J2358 (56,289)
AM 1512-435	15 12 35 −43 51.0	4.029 −0.223	328.57 11.54	169.5 12.2	GROUP of 3 274-G-5/6	24 D 13.4	F273/J3984 (19,238)
AM 1514-361	15 14 22 −36 13.3	3.807 −0.221	333.14 17.77	163.3 16.9	COMPACT: 387-G-19	13 D 0.2	F387/J1319 (154,114)
AM 1514-220	15 14 23 −22 07.0	3.480 −0.221	342.02 29.30	150.6 24.5	DISTURBED SPIRAL w. FILAMENT MCG-4-36-5 = 582-G-1	10 D 1.6	F513/J2358 (25,333)
AM 1514-431 [1514-4319]	15 14 28 −43 19.1	4.018 −0.221	329.15 11.81	169.3 12.8	CLOSE PAIR (#1: S0) 274-G-8	8,23 S 0.6	F274/J1324 (273,268)
AM 1514-431 [1514-4319]	15 14 29 −43 19.4	4.018 −0.221	329.15 11.80	169.3 12.8	CLOSE PAIR (#2: SPIRAL) 274-G-9	8,23 S 0.6	F274/J1324 (273,268)
AM 1514-362	15 14 34 −36 24.1	3.812 −0.221	333.07 17.60	163.5 16.8	RINGED COMPACT + COMPACT COMP'N.	6,13,23 S 2.9	F387/J1319 (152,104)

Identification	R.A. (1950.0) Declination	Precession R.A./Dec.	Gal. Long. Gal. Lat.	S.G.L. S.G.B.	Description Cross-Identification	Coding Size (Min.)	Field/Plate (x, y)
AM 1514-331	15 14 59 −33 14.1	3.731 −0.220	335.01 20.18	160.8 18.7	SB(r) + 3 COMP'NS. 387-G-21	8 S 7.2	F387/J1319 (146,274)
AM 1515-235 [1515-2353]	15 15 17 −23 55.0	3.519 −0.220	340.98 27.74	152.4 23.8	PAIR (I/A COMP'N.) (#1: E) NGC 5898 = 514-G-2	1,8,18,23 S 5.4	F514/J2318 (286,236)
AM 1515-332	15 15 36 −33 24.8	3.737 −0.220	335.01 19.96	161.0 18.7	DWARF COMP'N.	20 D 0.7	F387/J1319 (139,264)
AM 1515-235 [1515-2355]	15 15 40 −23 53.3	3.519 −0.219	341.08 27.70	152.5 23.9	PAIR (I/A COMP'N) (#2: E) NGC 5903 = 514-G-4 (GROUP of 5)	1,8,18,23 S 5.4	F514/J2318 (282,238)
AM 1516-410	15 16 01 −41 03.0	3.951 −0.219	330.66 13.56	167.6 14.4	ONE-ARMED SPIRAL 328-G-43	10 D 2.2	F328/J0644 (118,123)
AM 1516-233	15 16 08 −23 38.5	3.515 −0.219	341.34 27.84	152.3 24.1	GALAXY w. STRONG ABS'N. MCG-4-36-9 = 514-G-5	14 D 1.4	F514/J2318 (276,251)
AM 1517-311	15 17 13 −31 14.8	3.688 −0.218	336.63 21.54	159.3 20.2	SPIRAL w. I/A COMP'N. MCG-5-36-1 = 449-G-5	1 S 7.2	F449/J1335 (209,110)
AM 1517-364	15 17 35 −36 44.9	3.830 −0.217	333.39 16.97	164.1 17.1	SPIRAL w. DIAMETRIC COMP'NS. 387-G-26	8 S 7.4	F387/J1319 (119,85)
*AM 1518-862	15 18 52 −86 22.8	0.320 −0.216	305.42 −24.49	203.7 −13.5	ONE-ARMED SPIRAL + COMP'N. 8-G-9	8,10 D 0.9	F007/J2961 (31,37)
AM 1520-321	15 20 05 −32 13.7	3.718 −0.215	336.56 20.39	160.6 20.2	SPIRAL + CHAIN of 6 COMP'NS. 449-G-9	8,17 D 0.8	F449/J1335 (176,58)
AM 1520-353	15 20 10 −35 34.0	3.805 −0.214	334.54 17.65	163.5 18.2	E + SB(r) + COMP'NS. 387-G-28	8,23 S 5.4	F387/J1319 (90,148)
AM 1521-682	15 21 21 −68 26.5	5.676 −0.213	316.30 −9.85	189.7 −2.1	CLOSE PAIR of SPIRALS 68-G-4	23 S 3.9	F068/J2139 (252,261)
AM 1523-300	15 23 08 −30 01.1	3.671 −0.211	338.52 21.79	159.0 21.9	GROUP of 3 SPIRALS incl. 449-G-10	24 D 9.0	F449/J1335 (141,176)
AM 1528-864	15 28 12 −86 47.0	0.364 −0.205	305.27 −24.90	204.1 −13.6	I/A DBL.	2 S 0.3	F007/J2961 (43,17)
AM 1529-272	15 29 04 −27 28.4	3.622 −0.204	341.31 23.02	157.5 24.5	SPIRAL w. I/A COMP'N. 514-G-23/24	1 S 1.6	F450/J2300 (344,310)
AM 1530-585	15 30 14 −58 59.2	4.838 −0.203	322.49 −2.60	183.2 4.8	PLANETARY NEBULA PK 322-2-1 = 135-PN-11	25 D 1.3	F135/J2175 (53,230)
AM 1531-383	15 31 47 −38 31.4	3.922 −0.201	334.70 13.90	167.4 18.3	GALAXY w. I/A COMP'N. 329-G-9	1 D 0.7	F329/J0635 (224,258)
AM 1531-664	15 31 50 −66 41.6	5.548 −0.201	318.15 −8.99	189.0 −0.2	SPIRAL + APP. COMP'N. NGC 5938 = 99-G-7	8 D 2.8	F099/J1657 (137,88)
AM 1535-680	15 35 39 −68 01.1	5.748 −0.197	317.65 −10.27	190.2 −0.9	VERY SMALL RING	6 D 0.2	F099/J1657 (120,16)
AM 1536-254	15 36 03 −25 48.8	3.596 −0.196	343.77 23.32	157.0 26.7	HI.S.BR. IRR. 515-G-4	16 D 0.8	F515/J2332 (299,133)
AM 1538-473	15 38 42 −47 39.0	4.269 −0.193	330.18 5.83	175.5 13.3	V.LO.S.BR. PLANETARY NEBULA PK 330+5-1	25 D 1.6	F275/J2976 (299,34)
AM 1539-383	15 39 19 −38 32.4	3.943 −0.192	335.90 12.99	168.4 19.4	GALAXY w. APP. COMP'NS. 329-G-15	8 S 2.0	F329/J0635 (145,258)
AM 1540-753 [1542-7531]	15 40 49 −75 37.9	7.356 −0.191	313.16 −16.52	196.0 −5.8	CLOSE PAIR of SPIRALS (#1) NGC 5967a = 42-G-9	23 S 8.4	F042/J2132 (123,144)
AM 1541-780	15 41 30 −78 03.4	8.271 −0.190	311.58 −18.42	197.8 −7.5	DISTURBED GAL. w. APP. COMP'NS. IC 4555? = 22-G-12	8 S 4.5	F023/J2166 (313,269)
AM 1542-753 [1540-7539]	15 42 09 −75 31.2	7.338 −0.189	313.30 −16.49	196.0 −5.7	CLOSE PAIR of SPIRALS (#2) NGC 5967 = 42-G-10	23 S 8.4	F042/J2132 (118,149)
AM 1543-564	15 43 13 −56 48.3	4.762 −0.188	325.18 −1.87	182.7 7.5	STAR w. NEBULOSITY	21	F178/J2206 (280,82)
AM 1543-283	15 43 39 −28 34.9	3.676 −0.187	343.22 20.12	160.6 26.4	SPIRAL w. I/A DBL. COMP'N. MCG-5-37-2 = 450-G-13	1 S 9.4	F450/J2300 (170,253)
AM 1544-304	15 44 27 −30 41.1	3.731 −0.186	341.91 18.41	162.6 25.2	CLOSE PAIR	23 S 1.1	F450/J2300 (161,140)
*AM 1546-284	15 46 15 −28 45.3	3.685 −0.184	343.55 19.61	161.2 26.7	GALAXY (2 LINEAR PARTS) + COMP'N. 450-G-18 + 450-G-17	16 S 2.2	F450/J2300 (139,243)
AM 1546-242	15 46 41 −24 25.9	3.580 −0.184	346.72 22.77	157.4 29.5	GALAXY w. TAIL	15 D 1.3	F515/J2332 (171,209)
AM 1546-394	15 46 44 −39 45.7	4.002 −0.183	336.27 11.14	170.4 19.7	CHAIN of 4	17 D 4.2	F330/J1542 (337,188)
AM 1547-744	15 47 10 −74 40.7	7.153 −0.183	314.12 −16.05	195.6 −4.9	WIDE TRIPLE IC 4578 = 42-G-13 + 42-G-12	24 D 13.4	F043/J2150 (330,182)
AM 1547-512	15 47 55 −51 22.6	4.475 −0.182	329.08 1.97	179.3 11.8	PLANETARY NEBULA PK 329+2-1 = 225-PN-2	25	F225/J1631 (280,103)
AM 1547-321	15 47 56 −32 15.5	3.780 −0.182	341.43 16.73	164.5 24.8	I/A DBL. (ONE COMPACT)	2,13 S 0.4	F389/J0730 (314,323)
AM 1549-270	15 49 57 −27 08.5	3.651 −0.180	345.35 20.27	160.3 28.4	SPIRAL w. APP. COMP'N. 515-G-13	8 S 4.2	F450/J2300 (95,330)
AM 1551-511	15 51 07 −51 13.7	4.480 −0.178	329.56 1.77	179.6 12.3	PLANETARY NEBULA vdB-7 = 225-PN-3 = PK 329+1-1	25 D 0.1	F225/J1631 (254,112)

Identification	R.A. (1950.0) Declination	Precession R.A./Dec.	Gal. Long. Gal. Lat.	S.G.L. S.G.B.	Description Cross-Identification	Coding Size (Min.)	Field/Plate (x, y)
AM 1551-675	15 51 19 −67 54.1	5.858 −0.178	318.89 −11.09	191.2 0.3	I/A DBL. 68-IG-14	2 S 0.2	F099/J1657 (41,16)
AM 1552-303 [1552-3030]	15 52 34 −30 31.2	3.742 −0.176	343.40 17.38	163.7 26.7	CLOSE PAIR (#1)	23 D 0.3	F450/J2300 (67,148)
AM 1552-303 [1552-3031]	15 52 39 −30 30.7	3.742 −0.176	343.42 17.38	163.7 26.7	CLOSE PAIR (#2)	23 D 0.5	F450/J2300 (66,148)
AM 1553-312	15 53 15 −31 25.3	3.768 −0.175	342.89 16.61	164.6 26.2	GROUP of 4 (SOME I/A, ABS'N.) 450-G-24	18 D 7.2	F450/J2300 (60,99)
AM 1553-311	15 53 19 −31 12.7	3.762 −0.175	343.05 16.76	164.4 26.4	GALAXY w. APP. COMP'N.	8 D 0.4	F451/J0702 (325,112)
AM 1554-281	15 54 43 −28 14.1	3.686 −0.174	345.39 18.76	162.1 28.5	SPIRAL w. APP. COMP'NS. 451-G-1	8 D 1.3	F450/J2300 (39,270)
AM 1554-302	15 54 47 −30 24.0	3.743 −0.174	343.86 17.15	163.9 27.1	SPIRAL w. APP. COMP'NS.	8 D 0.7	F450/J2300 (41,154)
AM 1555-275	15 55 24 −27 55.8	3.679 −0.173	345.72 18.88	161.9 28.8	GALAXY w. I/A COMP'N.	1 S 0.5	F516/J2349 (329,20)
AM 1555-661 [1555-6612]	15 55 28 −66 13.6	5.669 −0.173	320.31 −10.08	190.4 1.8	I/A PAIR of SPIRALS (#1) IC 4584 = 100-G-4	2 S 3.3	F100/J4339 (247,111)
AM 1555-661 [1555-6613]	15 55 34 −66 12.2	5.667 −0.173	320.34 −10.07	190.3 1.8	I/A PAIR of SPIRALS (#2) IC 4585 = 100-G-5	2 S 3.3	F100/J4339 (246,113)
AM 1556-263	15 56 03 −26 33.9	3.646 −0.172	346.82 19.77	160.8 29.8	LINE of 4	17 D 3.1	F516/J2349 (323,93)
AM 1556-325	15 56 10 −32 56.5	3.815 −0.172	342.31 15.07	166.3 25.7	I/A DBL. (TRIPLET?)	2,3 D 0.5	F389/J0730 (220,287)
AM 1556-235	15 56 40 −23 51.3	3.580 −0.171	348.94 21.63	158.4 31.6	I/A DBL. in GROUP of 4	2,18 S 1.0	F516/J2349 (318,239)
AM 1556-270	15 56 54 −27 06.0	3.660 −0.171	346.58 19.25	161.4 29.6	GALAXY w. PLUMES or I/A DBL. 515-PN?-19	2,15 D 0.7	F450/J2300 (11,330)
AM 1557-291	15 57 57 −29 12.3	3.716 −0.170	345.24 17.56	163.4 28.4	GALAXY w. APP. COMP'NS. 451-G-4	8 D 1.0	F451/J0702 (274,221)
AM 1558-342	15 58 08 −34 24.3	3.862 −0.169	341.61 13.71	167.8 25.0	PLANETARY NEBULA ? NGC 6026 = PK 341+13-1 = 389-PN-7	25 D 1.1	F389/J0730 (197,209)
AM 1558-635	15 58 40 −63 58.7	5.435 −0.169	322.07 −8.62	189.1 3.7	ONE-ARMED SPIRAL in I/A TRIPLE 100-G-8	3,10 D 1.1	F100/J4339 (235,233)
AM 1600-355	16 00 07 −35 52.7	3.910 −0.167	340.90 12.35	169.3 24.3	LO.S.BR. PLANETARY NEBULA Lo-11 = 389-PN-9 = PK 340+12-1	25 D 1.3	F389/J0730 (176,130)
AM 1601-413	16 01 19 −41 33.2	4.102 −0.165	337.21 7.98	173.8 20.5	CLOSE PAIR 330-G-10/11	23 S 3.6	F330/J1542 (186,96)
AM 1601-673	16 01 28 −67 35.3	5.889 −0.165	319.84 −11.49	191.7 1.1	SB(r)c w. COMP'N. in DISK 68-IG-16	1 S 1.1	F069/J2144 (315,301)
AM 1601-332 [1601-3322]	16 01 33 −33 21.9	3.838 −0.165	342.87 14.00	167.5 26.2	CLOSE PAIR (#1: E)	23 D 0.2	F389/J0730 (159,265)
AM 1601-332 [1601-3321]	16 01 47 −33 22.9	3.839 −0.165	342.89 13.95	167.5 26.2	CLOSE PAIR (#2: PECULIAR SP.) 389-G-12	12,23 D 0.3	F389/J0730 (157,264)
AM 1603-290	16 03 10 −29 08.8	3.723 −0.163	346.14 16.82	164.2 29.3	SPIRAL w. APP. COMP'N. 451-G-8	8 D 1.9	F451/J0702 (213,225)
AM 1603-332	16 03 29 −33 28.4	3.845 −0.163	343.10 13.64	167.9 26.4	PECULIAR SPIRAL ?	12,16 D 0.8	F389/J0730 (138,259)
AM 1604-324	16 04 32 −32 40.1	3.823 −0.161	343.83 14.08	167.4 27.2	M51-TYPE PK 343+14-1 ?	9 D 0.5	F389/J0730 (125,302)
AM 1606-285	16 06 22 −28 52.4	3.721 −0.159	346.86 16.53	164.5 30.0	CLOSE TRIPLE of ELONGATED GALS. 451-G-11	24 D 3.4	F451/J0702 (175,240)
AM 1606-304	16 06 36 −30 47.1	3.773 −0.159	345.51 15.13	166.2 28.8	PLANETARY NEBULA Lo-13 = 451-PN-13 = PK 345+15-1	25 D 1.1	F451/J0702 (172,137)
AM 1606-315	16 06 54 −31 57.0	3.807 −0.158	344.72 14.25	167.2 28.0	CHAIN of 4	17 D 11.2	F451/J0702 (169,75)
AM 1607-663	16 07 34 −66 31.4	5.786 −0.157	321.02 −11.12	191.5 2.3	COMPACT/PLANETARY NEBULA ? 100-G?-12 = PK 321-11-1	13,25 D 0.3	F100/J4339 (181,98)
AM 1609-502	16 09 28 −50 20.5	4.501 −0.155	332.33 0.53	181.1 14.9	DIFFUSE NEBULOSITY + PEC. ABS'N.	22	F225/J1631 (98,160)
AM 1609-360	16 09 41 −36 06.3	3.937 −0.155	342.18 10.86	171.0 25.5	HI.S.BR. PLANETARY NEBULA NGC 6072 = PK342+10-1 = 389-PN-15	25 D 1.8	F389/J0730 (72,116)
AM 1610-311	16 10 31 −31 14.0	3.793 −0.154	345.80 14.22	167.2 29.1	CLOSE PAIR 451-G-17	23 S 1.7	F451/J0702 (127,113)
AM 1610-544	16 10 34 −54 49.5	4.757 −0.153	329.38 −2.86	184.2 11.6	PLANETARY NEBULA PK 329-2-2 = 178-PN-15	25	F179/J1618 (345,184)
AM 1610-604	16 10 39 −60 47.0	5.195 −0.153	325.28 −7.20	188.1 7.0	CLUSTER (27 OBSCURED GALAXIES) MANY ESO GALAXIES PKS 1610-60	19 D 17.0	F137/J1588 (293,133)
AM 1612-283	16 12 52 −28 36.5	3.723 −0.150	348.11 15.70	165.4 31.3	LINE of 3	24 D 2.1	F451/J0702 (99,254)
AM 1613-822	16 13 20 −82 25.3	12.046 −0.150	309.38 −22.48	201.8 −9.8	GROUP of 3	24 D 2.5	F023/J2166 (208,48)

Identification	R.A. (1950.0) Declination	Precession R.A./Dec.	Gal. Long. Gal. Lat.	S.G.L. S.G.B.	Description Cross-Identification	Coding Size (Min.)	Field/Plate (x, y)
AM 1613-505	16 13 47 −50 58.7	4.548 −0.149	332.38 −0.40	182.1 14.9	SNR (VEIL NEBULOSITY) 225-SNR-10	22 D 0.2	F225/J1631 (63,125)
AM 1615-420	16 15 55 −42 08.5	4.159 −0.147	338.81 5.69	176.4 21.9	PLANETARY NEBULA IC 4599 = PK 338+5-1 = 331-PN-1	25 D 1.1	F331/J1535 (299,62)
AM 1620-313	16 20 14 −31 38.1	3.819 −0.141	347.00 12.46	169.2 30.2	PLANETARY NEBULA PK 346+12-1 = 452-PN-1	25 D 1.2	F451/J0702 (16,90)
AM 1622-651 [1623-6518]	16 22 48 −65 17.9	5.720 −0.137	323.02 −11.36	191.9 4.3	CLOSE PAIR (#1)	23 S 1.1	F100/J4339 (96,161)
AM 1623-651 [1622-6517]	16 23 00 −65 18.0	5.721 −0.137	323.04 −11.38	192.0 4.3	CLOSE PAIR (#2) 100-G-24	23 S 1.1	F100/J4339 (95,161)
AM 1623-535	16 23 56 −53 54.8	4.747 −0.136	331.41 −3.58	185.2 13.5	PLANETARY NEBULA PK 331-3-2 = 179-PN-8	25	F179/J1618 (243,239)
AM 1625-351	16 25 06 −35 17.1	3.939 −0.134	345.01 9.21	172.9 28.2	CLUSTER near RIFT	22	F390/J0728 (167,164)
AM 1627-824	16 27 12 −82 45.1	12.727 −0.132	309.41 −23.04	202.4 −9.8	RING w. OFF-SET NUCLEUS	6 D 0.6	F023/J2166 (184,32)
AM 1628-400	16 28 06 −40 08.6	4.110 −0.130	341.85 5.45	176.9 24.9	PLANETARY NEBULA NGC 6153 = PK 341+5-1 = 331-PN-6	25 D 0.6	F331/J1535 (178,172)
AM 1630-475	16 30 00 −47 58.5	4.443 −0.128	336.37 −0.17	182.3 18.9	BR. NEBULOSITY NGC 6164 = 226-EN-12 = PK 336-0+1	21,22 D 17.2	F277/J1552 (341,14)
AM 1633-682	16 33 01 −68 21.0	6.199 −0.124	321.39 −14.15	194.5 2.3	I/A DBL. w. TAILS 69-IG-6 = PKS 1632-68?	2,15 S 1.3	F069/J2144 (155,268)
AM 1635-524	16 35 01 −52 45.6	4.711 −0.121	333.39 −4.00	185.9 15.4	PLANETARY NEBULA	25	F179/J1618 (155,302)
AM 1635-524	16 35 40 −52 40.9	4.708 −0.120	333.51 −4.02	186.0 15.5	PLANETARY NEBULA	25	F226/J1486 (133,35)
AM 1639-812	16 39 10 −81 21.7	11.325 −0.115	310.89 −22.53	201.9 −8.4	I/A DBL. w. TAIL 23-IG-3	2 S 0.1	F023/J2166 (161,106)
AM 1639-772	16 39 37 −77 23.7	8.685 −0.115	314.33 −20.20	199.8 −5.0	DISTURBED SINGLE IC 4608 = 43-IG-4	16 D 0.7	F044/J2158 (343,30)
AM 1643-572	16 43 08 −57 20.1	5.041 −0.110	330.69 −7.88	189.5 12.2	CLOSE PAIR (HALO?) Se 125/1 = 179-IG-13 = UKS 1643-573	23 S 1.3	F137/J1588 (70,320)
AM 1644-814	16 44 44 −81 41.7	11.736 −0.108	310.72 −22.89	202.3 −8.6	GALAXY w. I/A COMP'NS.	1 S 0.9	F024/J2167 (322,64)
AM 1646-795	16 46 41 −79 59.6	10.261 −0.105	312.28 −21.99	201.5 −7.1	I/A DBL (incl. RING)	2,6 S 0.3	F023/J2166 (140,178)
AM 1648-590	16 48 23 −59 08.0	5.200 −0.103	329.73 −9.56	191.1 11.0	GALAXY w. LINEAR ARM + 4 COMP'NS. NGC 6221 = 138-G-3	8,10 D 3.4	F137/J1588 (39,221)
AM 1649-723	16 49 00 −72 30.3	7.110 −0.102	318.87 −17.77	197.8 −0.5	SPIRAL w. APP. COMP'N. NGC 6209 = 43-G-8	8 D 1.7	F043/J2150 (103,311)
AM 1650-765	16 50 36 −76 55.9	8.566 −0.100	315.11 −20.42	200.1 −4.3	DISTURBED ISOLATED SPIRAL	16 D 1.3	F023/J2166 (117,342)
AM 1654-782	16 54 26 −78 23.3	9.310 −0.094	313.94 −21.40	201.0 −5.5	COMPACT w. FAINT OUTER RING 23-G?-8 = PN 313-21-1	6,13 D 0.6	F023/J2166 (114,263)
AM 1656-774	16 56 46 −77 46.3	9.005 −0.091	314.56 −21.17	200.8 −4.9	LOOSE CHAIN of 8 (3 DISTURBED) 23-G-11	17 D 9.0	F023/J2166 (104,295)
AM 1701-704	17 01 28 −70 43.9	6.771 −0.084	320.98 −17.56	197.9 1.5	CLOSE PAIR 70-IG-5	23 S 1.5	F070/J1762 (261,137)
AM 1703-731	17 03 13 −73 15.2	7.377 −0.082	318.84 −19.06	199.1 −0.7	SPIRAL w. COMPACT off MAJOR AXIS 44-G-1	8,13 S 5.6	F044/J2158 (306,264)
AM 1705-740	17 05 14 −74 04.6	7.623 −0.079	318.18 −19.62	199.6 −1.4	SPIRAL w. I/A COMP'N. 44-IG-2	1 D 1.0	F043/J2150 (50,220)
AM 1705-773	17 05 39 −77 30.5	8.935 −0.078	315.07 −21.43	201.1 −4.4	OBSCURED SPIRAL (PAIR w. IC 4635) IC 4633 = 44-G-3 = UKS 1706-774	8,20,23 D 4.0	F024/J2167 (336,294)
AM 1706-604	17 06 18 −60 45.1	5.394 −0.078	329.79 −12.32	193.9 10.7	PECULIAR SINGLE GALAXY	16 D 0.6	F138/J1648 (177,139)
AM 1706-600	17 06 19 −60 00.3	5.324 −0.078	330.42 −11.89	193.5 11.4	SPIRAL w. APP. COMP'N.	8 S 0.6	F138/J1648 (177,179)
AM 1706-591	17 06 20 −59 15.6	5.258 −0.077	331.04 −11.47	193.2 12.0	DISTURBED SPIRAL w. APP. COMP'NS	8 D 0.1	F138/J1648 (177,219)
AM 1710-362	17 10 19 −36 21.8	4.034 −0.072	350.06 1.47	182.3 32.6	PLANETARY NEBULA ('BOW-TIE')	25	F392/J0680 (197,106)
AM 1710-730	17 10 23 −73 05.7	7.366 −0.072	319.27 −19.42	199.5 −0.3	CHAIN of 4 (SOME DISTURBED) 44-G-7/9	17 D 14.6	F044/J2158 (279,276)
AM 1710-333	17 10 52 −33 30.8	3.937 −0.071	352.43 3.06	180.7 35.2	DARK GLOBULES	22	F392/J0680 (192,259)
AM 1715-800	17 15 05 −80 07.4	10.599 −0.065	312.85 −23.11	202.6 −6.6	GROUP of 4 (SP. + S0 w. COMP'NS.) IC 4641 = 24-G-2 = PKS 1716-80?	8,18 D 9.0	F024/J2167 (282,162)
AM 1715-784	17 15 24 −78 40.0	9.613 −0.065	314.25 −22.44	202.1 −5.3	STAR w. NEBULOSITY 24-RN-3	21 D 4.0	F023/J2166 (60,240)
AM 1716-514	17 16 49 −51 42.1	4.735 −0.063	338.20 −8.37	191.2 19.5	PLANETARY NEBULA NGC 6326 = PK 338-8-1 = 228-PN-1	25	F228/J1756 (289,85)

Identification	R.A. (1950.0) Declination	Precession R.A./Dec.	Gal. Long. Gal. Lat.	S.G.L. S.G.B.	Description Cross-Identification	Coding Size (Min.)	Field/Plate (x, y)
AM 1718-645	17 18 44 −64 58.2	5.888 −0.060	326.96 −15.82	196.9 7.4	LO.S.BR. SPIRAL + APP.COMP'N. NGC 6328 = 102-G-3 = PKS 1718-649	8 S 3.4	F102/J1619 (275,175)
AM 1718-382	17 18 48 −38 26.4	4.116 −0.060	349.35 −1.11	185.2 31.6	PLANETARY NEBULA NGC 6337 = PK 349-1-1 = 333-PN-5	25 D 0.1	F333/J0801 (192,263)
AM 1718-702	17 18 54 −70 25.0	6.768 −0.060	322.07 −18.63	199.0 2.4	CLOSE PAIR 70-IG-9/10	23 S 1.4	F070/J1762 (184,157)
AM 1719-814	17 19 19 −81 40.7	12.064 −0.059	311.42 −23.96	203.4 −8.0	DWARF	20 D 0.6	F024/J2167 (258,82)
AM 1722-793	17 22 16 −79 35.0	10.242 −0.055	313.53 −23.15	202.7 −6.0	CLOSE PAIR	23 S 0.4	F024/J2167 (270,194)
AM 1722-595	17 22 18 −59 53.6	5.346 −0.055	331.64 −13.50	195.3 12.3	SPIRAL w. I/A COMP'N. 138-G-27	1 S 1.1	F139/J1589 (325,179)
AM 1722-605	17 22 37 −60 50.5	5.436 −0.054	330.84 −14.03	195.7 11.4	HI.S.BR. DISTURBED SINGLE IC 4653 = 138-G-28	13,16 D 0.9	F139/J1589 (319,128)
*AM 1723-663	17 23 34 −66 38.9	6.128 −0.053	325.72 −17.11	198.0 6.1	CHAIN of 8 GALAXIES	17 D 10.8	F102/J1619 (243,87)
*AM 1724-622	17 24 28 −62 23.9	5.598 −0.052	329.58 −15.03	196.5 10.0	"SACRED MUSHROOM" 138-IG-29/30	2,6 S 2.8	F102/J1619 (248,315)
AM 1726-234	17 26 18 −23 43.4	3.654 −0.049	2.43 5.85	178.1 45.3	PLANETARY NEBULA NGC 6369 = PK 2+5-1 = 520-PN-3	25 D 1.0	F520/J2326 (299,246)
AM 1729-605	17 29 46 −60 50.3	5.446 −0.044	331.30 −14.78	196.6 11.7	LINE of 3 139-G-5	24 D 17.9	F139/J1589 (272,132)
AM 1731-795	17 31 37 −79 59.7	10.588 −0.041	313.31 −23.70	203.2 −6.2	CLOSE PAIR	23 S 0.6	F024/J2167 (245,175)
AM 1732-400	17 32 11 −40 09.5	4.192 −0.040	349.36 −4.21	188.8 31.1	PLANETARY NEBULA Lo-16 = 333-PN-15 = PK 349-4-1	25	F334/J1543 (321,168)
AM 1733-595	17 33 09 −59 54.9	5.363 −0.039	332.35 −14.68	196.6 12.7	SP. w. APP.COMP'N.(LONG THIN ARM) 139-G-12	8,10 S 3.1	F139/J1589 (252,182)
*AM 1738-614	17 38 06 −61 40.3	5.540 −0.032	331.06 −16.06	197.8 11.3	S(r) w. DIAMETRIC COMP'NS NGC 6398 = 139-G-18	8 S 5.8	F139/J1589 (217,89)
AM 1739-375	17 39 33 −37 54.6	4.109 −0.030	352.03 −4.22	189.4 33.8	PLANETARY NEBULA	25	F393/J1503 (141,23)
AM 1740-790	17 40 28 −79 01.7	9.939 −0.028	314.46 −23.69	203.3 −5.2	SPIRAL w. PECULIAR ARMS 24-IG-9	10 D 0.6	F024/J2167 (229,229)
AM 1741-752	17 41 28 −75 26.2	8.199 −0.027	318.08 −22.36	202.3 −1.7	I/A DBL. 44-IG-20	2 S 1.5	F044/J2158 (161,156)
AM 1743-603	17 43 58 −60 33.6	5.434 −0.023	332.43 −16.17	198.2 12.6	WIDE GROUP of 9	18 D 17.9	F139/J1589 (180,149)
AM 1745-593	17 45 56 −59 32.8	5.341 −0.020	333.48 −15.92	198.1 13.6	SPIRAL w. APP. COMP'NS 139-G-34	8 S 2.7	F139/J1589 (166,204)
AM 1748-673	17 48 22 −67 37.1	6.313 −0.017	325.98 −19.69	200.6 5.9	LUMPY COMPACT	13,16 D 0.3	F102/J1619 (113,35)
AM 1748-460	17 48 24 −46 06.1	4.460 −0.017	345.78 −9.82	194.3 26.6	PLANETARY NEBULA	25	F279/J0814 (138,120)
AM 1749-634	17 49 35 −63 45.7	5.781 −0.015	329.73 −18.18	199.7 9.7	M51-TYPE ? 102-IG-18	9 S 0.5	F102/J1619 (96,241)
AM 1754-634	17 54 52 −63 41.3	5.774 −0.007	330.06 −18.68	200.3 9.9	E + SB(r) + N (ALIGNED) = SS-17 NGC 6483 = 102-G/GA-20 = Fa 486/8	13,24 S 5.6	F102/J1619 (64,243)
AM 1757-662	17 57 41 −66 25.7	6.135 −0.003	327.52 −20.05	201.2 7.3	SPIRAL w. APP. COMP'N. NGC 6492 = 102-G-22	8 S 8.5	F102/J1619 (60,95)
AM 1759-652	17 59 11 −65 27.1	5.998 −0.001	328.54 −19.81	201.1 8.3	E + 2 APP. COMP'NS. NGC 6502 = 103-G-2	8 S 1.3	F103/J2290 (295,150)
AM 1759-593 [1800-5936]	17 59 54 −59 37.4	5.352 0.000	334.21 −17.54	199.9 14.0	PAIR of SPIRALS (#1) 139-G-53 = Fa 491	23 S 2.5	F139/J1589 (72,196)
AM 1800-593 [1759-5937]	18 00 15 −59 36.0	5.350 0.000	334.26 −17.57	199.9 14.0	PAIR of SPIRALS (#2)	23 S 2.5	F139/J1589 (69,198)
AM 1803-510	18 03 19 −51 01.8	4.724 0.005	342.51 −14.30	198.3 22.5	V.HI.S.BR. RING/PLAN.NEB.? PK 342-14-1 = 229-PN-6	6,25 D 0.9	F229/J1596 (149,122)
AM 1805-452	18 05 22 −45 25.7	4.429 0.008	347.82 −12.11	197.3 28.0	POOR GLOBULAR CLUSTER 280-SC-6	22 D 1.5	F280/J0807 (242,156)
*AM 1806-852	18 06 15 −85 25.5	0.329 0.009	308.02 −26.45	205.8 −11.2	E + IRR. (ISOLATED I/A DBL.) NGC 6438 = 10-IG-2 = SS-18	2 S 0.7	F010/J2477 (249,146)
AM 1808-594 [1808-5944]	18 08 08 −59 45.2	5.363 0.012	334.53 −18.55	201.0 14.1	PAIR (#1) 140-G-8	23 S 1.9	F139/J1589 (16,185)
AM 1808-594 [1808-5945]	18 08 25 −59 44.3	5.361 0.012	334.56 −18.57	201.0 14.1	PAIR (#2)	23 S 1.9	F139/J1589 (14,186)
AM 1808-643	18 08 36 −64 31.6	5.875 0.013	329.85 −20.39	201.9 9.4	SPIRAL + APP. COMP'N. IC 4680 = 103-G-9	8 S 2.7	F103/J2290 (245,203)
*AM 1809-574	18 09 16 −57 44.6	5.188 0.014	336.53 −17.90	200.7 16.1	I/A DBL. (SPIRALS) Fa 44/45/46 IC 4687/89 = 140-IG-10/11	2 S 1.6	F140/J1649 (267,297)
AM 1809-600	18 09 46 −60 06.3	5.395 0.014	334.27 −18.87	201.2 13.8	DISTURBED GALAXY + APP. COMP'N. 140-G-12	8,16 D 2.2	F140/J1649 (258,170)

Identification	R.A. (1950.0) Declination	Precession R.A./Dec.	Gal. Long. Gal. Lat.	S.G.L. S.G.B.	Description Cross-Identification	Coding Size (Min.)	Field/Plate (x, y)
AM 1812-571	18 12 32 −57 14.9	5.147 0.018	337.19 −18.11	201.1 16.6	GALAXY w. PLUMES 182-G-7	15,18 D 1.3	F140/J1649 (245,324)
AM 1812-822	18 12 38 −82 25.5	13.104 0.018	311.32 −25.94	205.4 −8.2	M51-TYPE 24-G-13	9 S 0.7	F010/J2477 (284,304)
AM 1813-691	18 13 43 −69 11.5	6.582 0.020	325.32 −22.44	203.3 4.9	I/A(?) DBL. 71-G-7/8	2 S 3.8	F071/J0900 (169,220)
AM 1814-763	18 14 12 −76 36.2	8.672 0.021	317.59 −24.61	204.5 −2.4	CLOSE PAIR NGC 6557 = 45-G-1 ?	23 S 9.0	F044/J2158 (61,84)
AM 1814-424	18 14 23 −42 47.3	4.307 0.021	350.97 −12.38	198.5 30.9	GALAXY + COMP'N. 280-G-7	8 D 3.4	F280/J0807 (156,298)
AM 1814-774	18 14 44 −77 44.0	9.205 0.021	316.40 −24.92	204.7 −3.5	CLOSE DBL. E w. ABS'N. LANE 24-IG-15 = Fa 175/76	14,23 S 1.4	F024/J2167 (138,300)
AM 1814-740	18 14 50 −74 03.1	7.739 0.022	320.30 −23.96	204.1 0.1	DWARF 45-G-2	20 D 1.0	F044/J2158 (37,219)
AM 1815-710	18 15 01 −71 09.4	6.979 0.022	323.33 −23.14	203.7 3.0	DWARF	20 D 0.5	F071/J0900 (164,114)
AM 1815-791	18 15 07 −79 19.6	10.146 0.022	314.69 −25.33	205.0 −5.1	I/A TRIPLEGV180,4;ES,3	3 D 1.0	F024/J2167 (142,214)
AM 1815-644	18 15 23 −64 47.6	5.905 0.022	329.86 −21.16	202.7 9.2	SPIRAL + APP. COMP'N. IC 4696 = 103-G-13	8 S 5.8	F103/J2290 (206,190)
AM 1815-492	18 15 27 −49 27.9	4.632 0.023	344.83 −15.39	200.1 24.4	I/A DBL. in COMMON ENVELOPE	2,23 S 0.5	F230/J2530 (306,205)
AM 1816-421	18 16 02 −42 12.6	4.282 0.023	351.64 −12.40	198.8 31.5	GALAXY + 2 APP. COMP'NS. 335-G-8	8 D 0.6	F280/J0807 (140,329)
AM 1816-632	18 16 14 −63 24.5	5.735 0.024	331.29 −20.78	202.6 10.6	EDGE-ON + COMPACT COMP'N. IC 4698 = 103-G-15	8 S 3.1	F103/J2290 (202,265)
AM 1816-620	18 16 44 −62 04.8	5.587 0.024	332.64 −20.37	202.4 12.0	PAIR of I/A SPIRALS 140-G-17	2 S 0.7	F103/J2290 (200,336)
AM 1817-772	18 17 28 −77 24.9	9.040 0.025	316.77 −24.99	204.8 −3.2	SMALL RING/SPIRAL	6 S 0.3	F024/J2167 (129,316)
AM 1817-615	18 17 34 −61 53.8	5.567 0.026	332.86 −20.40	202.5 12.1	IRR. (JETS and TAILS) 140-G-18	15,16 D 1.3	F140/J1649 (205,76)
AM 1818-622	18 18 06 −62 20.3	5.614 0.026	332.44 −20.61	202.6 11.7	DWARF	20 D 0.9	F103/J2290 (191,322)
AM 1819-345	18 19 27 −34 59.8	4.005 0.028	358.53 −9.81	198.0 38.7	CLOSE PAIR	23 S 0.7	F395/J0664 (229,179)
AM 1819-430	18 19 27 −43 08.3	4.320 0.028	351.06 −13.37	199.7 30.7	GALAXY w. DIAMETRIC COMP'NS. 280-G-12/13	8 D 3.5	F280/J0807 (107,279)
AM 1820-765	18 20 27 −76 58.2	8.823 0.030	317.30 −25.04	204.9 −2.7	SPIRAL w. I/A DBL. COMP'NS. 45-G-5	2,8 S 3.7	F024/J2167 (118,340)
AM 1821-471	18 21 19 −47 15.0	4.512 0.031	347.34 −15.38	200.8 26.7	DISTURBED SINGLE 280-IG-14	16 D 0.6	F280/J0807 (94,58)
AM 1822-624	18 22 17 −62 47.8	5.660 0.032	332.15 −21.22	203.2 11.3	CLOSE PAIR	23 S 2.2	F103/J2290 (165,298)
*AM 1822-714	18 22 20 −71 43.9	7.101 0.033	322.90 −23.86	204.3 2.5	E + DISTURBED + DUST (GROUP of 4) IC 4704/5/12 = 71-G-11/2/3/4 = Fa 497	14,18 D 0.9	F071/J0900 (134,83)
AM 1823-791	18 23 06 −79 17.3	10.100 0.034	314.83 −25.68	205.3 −5.0	I/A DBL.	2 S 0.5	F024/J2167 (122,215)
AM 1823-670	18 23 30 −67 03.0	6.211 0.034	327.83 −22.65	203.9 7.1	KNOTTY Sd/DWARF + DWARF COMP'N. IC 4710 = Se 127/1 = 103-G-22	8,20,20 S 4.9	F103/J2290 (161,69)
*AM 1823-512	18 23 36 −51 29.5	4.743 0.034	343.42 −17.39	201.8 22.6	M51-TYPE in GROUP of 5 230-IG-4 = DN 1824-515	9,18 S 1.4	F230/J2530 (233,99)
AM 1824-821	18 24 02 −82 11.0	12.752 0.035	311.67 −26.27	205.7 −7.9	SMALL I/A DBL.	2 S 0.1	F024/J2167 (135,60)
AM 1824-363	18 24 21 −36 37.6	4.061 0.035	357.49 −11.42	199.6 37.3	PLANETARY NEBULA	25	F395/J0664 (175,92)
AM 1824-371	18 24 27 −37 17.8	4.085 0.036	356.88 −11.73	199.7 36.6	PLANETARY NEBULA (LO.S.BR.) Lo-17 = 395-PN-7 = PK 356-11-1	25 D 0.1	F335/J1407 (47,323)
AM 1825-795	18 25 19 −79 58.7	10.587 0.037	314.10 −25.92	205.5 −5.7	ONE-ARMED SPIRAL w. COMP'NS. 24-G-17	8,10 S 0.5	F024/J2167 (120,177)
AM 1826-424	18 26 09 −42 47.2	4.302 0.038	351.91 −14.34	201.0 31.2	SP. + SP. COMP'N + DBL. in CHAIN incl. 281-G-1/2	8,17,23 D 12.5	F336/J0787 (296,27)
AM 1826-345	18 26 28 −34 53.6	3.999 0.038	359.27 −11.05	199.8 39.1	CLOSE PAIR	23 S 0.7	F395/J0664 (152,185)
*AM 1827-625	18 27 15 −62 58.1	5.673 0.040	332.16 −21.81	203.8 11.2	RING?	6 D 0.9	F140/J1649 (145,18)
AM 1829-333	18 29 08 −33 33.0	3.952 0.042	0.75 −10.96	200.2 40.5	I/A DBL.	2 S 0.6	F395/J0664 (122,257)
AM 1829-392	18 29 08 −39 29.6	4.165 0.042	355.24 −13.49	201.2 34.6	LINE of 3 + I/A DBL. incl. 336-G-1/2	2,24 D 5.0	F336/J0787 (271,205)
AM 1829-413	18 29 23 −41 32.0	4.247 0.043	353.35 −14.37	201.5 32.6	I/A PAIR of SPIRALS Ag-52 = 336-G-3/4	2 S 4.3	F336/J0787 (266,96)

Identification	R.A. (1950.0) Declination	Precession R.A./Dec.	Gal. Long. Gal. Lat.	S.G.L. S.G.B.	Description Cross-Identification	Coding Size (Min.)	Field/Plate (x, y)
*AM 1830-583	18 30 00 −58 32.2	5.238 0.044	336.79 −20.73	203.6 15.7	SP. w. JET(?) + COMP'N. [DN] IC 4720/21 = Se 128/1 = 140-G-25/7	7,8,23 D 4.9	F140/J1649 (120,255)
*AM 1830-574	18 30 13 −57 49.9	5.179 0.044	337.51 −20.53	203.6 16.4	SPIRAL w. PECULIAR ARM IC 4722 = 140-G-28	10 D 1.3	F182/J1575 (25,19)
AM 1831-671	18 31 41 −67 18.0	6.236 0.046	327.81 −23.49	204.7 7.0	CLOSE TRIPLE (I/A ?) IC 4723 = 103-G-28/29/30	3 D 3.1	F071/J0900 (75,318)
AM 1831-625 [1832-6253]	18 31 48 −62 53.8	5.658 0.046	332.41 −22.28	204.3 11.4	CLOSE PAIR (#1) 103-G-31	23 D 0.3	F140/J1649 (117,21)
AM 1832-625 [1831-6253]	18 32 13 −62 53.6	5.657 0.047	332.42 −22.33	204.3 11.4	CLOSE PAIR (#2) IC 4726 = 103-G-32	23 D 0.4	F140/J1649 (114,21)
AM 1832-771	18 32 21 −77 11.0	8.887 0.047	317.22 −25.74	205.6 −2.9	E w. APP. COMP'N. 45-G-11 = PKS 1833-77	8 S 1.6	F024/J2167 (84,324)
AM 1832-652	18 32 47 −65 20.7	5.954 0.048	329.89 −23.07	204.6 8.9	PAIR of IRR. [E.W.] Fa 177	16,16,23 S 7.2	F103/J2290 (108,159)
AM 1833-414	18 33 41 −41 49.4	4.256 0.049	353.41 −15.22	202.5 32.4	I/A DBL.	2 S 1.0	F336/J0787 (222,81)
AM 1834-573	18 34 08 −57 32.1	5.150 0.050	337.99 −20.93	204.1 16.7	ARC of 5 (E + SPIRALS) IC 4734 = 140-G-33	17 D 17.9	F183/J4391 (236,42)
AM 1835-615 [1835-6156]	18 35 47 −61 57.3	5.550 0.052	333.52 −22.45	204.7 12.3	CLOSE PAIR (#1) IC 4738 = 140-G-35 = PKS 1835-619	23 D 0.5	F140/J1649 (90,70)
AM 1836-615 [1835-6157]	18 36 14 −61 56.8	5.549 0.053	333.55 −22.50	204.7 12.3	CLOSE PAIR (#2) IC 4739 = 140-G-36	23 D 0.6	F140/J1649 (87,70)
AM 1836-640 [1837-6354]	18 36 36 −64 03.6	5.785 0.053	331.36 −23.12	204.9 10.2	LINE of 3 (#1: S0) 103-G-46 = Fa 178	24 D 9.6	F104/J1590 (340,221)
AM 1836-635 [1836-6403]	18 36 54 −63 59.7	5.776 0.054	331.43 −23.13	205.0 10.3	LINE of 3 (#2: Sb) IC 4741 = 103-G-47 = Fa 179	24 D 9.6	F104/J1590 (339,225)
AM 1837-635 [1836-6359]	18 37 04 −63 54.5	5.766 0.054	331.53 −23.13	205.0 10.4	LINE of 3 (#3: SB0) IC 4742 = 103-G-48 = Fa 180	24 D 9.6	F104/J1590 (338,229)
AM 1837-631	18 37 14 −63 16.6	5.692 0.054	332.20 −22.98	204.9 11.0	IRR. IC 4744 = 103-G-50	16 D 0.6	F104/J1590 (341,263)
AM 1838-411	18 38 17 −41 13.0	4.227 0.056	354.34 −15.78	203.5 33.0	CLOSE PAIR	23 S 0.6	F336/J0787 (176,114)
AM 1838-494	18 38 20 −49 47.6	4.632 0.056	345.98 −18.95	204.2 24.5	SPIRAL w. I/A COMP'N. 230-G-6	1 S 0.1	F230/J2530 (107,189)
AM 1838-620 [1838-6209]	18 38 42 −62 09.6	5.567 0.056	333.41 −22.84	205.0 12.1	CLOSE PAIR (#1: S0) IC 4751 = 140-G-40 = Fa 183	23 S 1.7	F104/J1590 (338,324)
AM 1838-620 [1838-6209]	18 38 54 −62 09.0	5.565 0.056	333.43 −22.86	205.1 12.2	CLOSE PAIR (#2: E) IC 4753 = 140-G-41 = Fa 184	23 S 1.7	F104/J1590 (337,324)
*AM 1839-571	18 39 38 −57 13.1	5.117 0.057	338.56 −21.54	204.9 17.1	EDGE-ON GALAXY w. ABS'N. IC 4757 = 183-G-9	14 D 1.3	F183/J4391 (196,60)
AM 1839-652	18 39 41 −65 21.3	5.942 0.058	330.08 −23.77	205.3 9.0	COMPACT	13 D 0.4	F103/J2290 (69,156)
AM 1839-525	18 39 54 −52 54.1	4.813 0.058	342.96 −20.22	204.6 21.4	SB w. PECULIAR ARMS + 2 COMP'NS IC 4761 = 183-G-10	8,10 S 2.9	F183/J4391 (196,292)
AM 1840-564	18 40 07 −56 47.7	5.083 0.058	339.02 −21.47	204.9 17.5	GALAXY + STAR ? 183-G-11	22 D 0.6	F141/J0890 (322,344)
AM 1840-622 [1840-6220]	18 40 14 −62 25.1	5.591 0.058	333.19 −23.08	205.2 11.9	PAIR (#1: SPIRAL) 140-G-43 = Fa 51 = SS-19b	23 S 4.3	F104/J1590 (327,311)
AM 1840-505	18 40 20 −50 50.2	4.688 0.058	345.06 −19.60	204.6 23.5	I/A DBL. 230-IG-9	2 S 0.8	F230/J2530 (92,133)
AM 1840-622 [1840-6225]	18 40 27 −62 20.9	5.583 0.059	333.27 −23.09	205.3 12.0	PAIR (#2: E) = SS-19a NGC 6673 = 140-G-44 = Fa 186	23 S 4.3	F104/J1590 (326,315)
AM 1840-630	18 40 57 −63 08.4	5.669 0.059	332.46 −23.35	205.4 11.2	IRR. (I/A DBL?) IC 4759 = 103-IG-60	2,16 D 1.1	F104/J1590 (319,273)
AM 1841-385	18 41 34 −38 51.0	4.131 0.060	356.85 −15.44	204.0 35.4	CLOSE TRIPLE	24 D 5.6	F336/J0787 (142,241)
AM 1842-625	18 42 06 −62 59.2	5.649 0.061	332.66 −23.43	205.5 11.3	COMPACT	13 D 0.2	F104/J1590 (313,282)
*AM 1842-632	18 42 31 −63 25.2	5.697 0.062	332.21 −23.59	205.6 10.9	RICH CLUSTER (13 in CORE) incl. 104-G-1/2/4/5 = Fa 187/188	19 D 1.3	F103/J2290 (44,258)
AM 1842-412	18 42 39 −41 27.4	4.233 0.062	354.43 −16.63	204.5 32.9	CHAIN of 5 incl. 336-G-14	17 D 8.1	F336/J0787 (132,101)
AM 1842-574	18 42 39 −57 45.0	5.154 0.062	338.14 −22.08	205.3 16.6	E w. SMALL COMP'NS. both SIDES 141-G-03	8 D 0.4	F141/J0890 (300,294)
AM 1842-814	18 42 54 −81 41.9	12.069 0.062	312.31 −26.86	206.4 −7.4	SPIRAL w. I/A DBL. COMP'N.	1 S 2.1	F024/J2167 (96,81)
*AM 1842-632	18 42 55 −63 29.6	5.705 0.062	332.15 −23.65	205.6 10.8	"CRYSTAL FROG" (CROSSED JETS?) IC 4767 = 104-G-10	7,16 D 1.3	F103/J2290 (42,254)
AM 1843-631	18 43 00 −63 14.8	5.677 0.062	332.41 −23.60	205.6 11.1	SPIRAL w. PECULIAR STRAIGHT ARM IC 4769 = 104-G-11	10 D 2.5	F103/J2290 (41,267)
AM 1843-602	18 43 14 −60 24.3	5.384 0.063	335.41 −22.89	205.5 13.9	DISTURBED GALAXY	16 D 0.7	F141/J0890 (287,152)

Identification	R.A. (1950.0) Declination	Precession R.A./Dec.	Gal. Long. Gal. Lat.	S.G.L. S.G.B.	Description Cross-Identification	Coding Size (Min.)	Field/Plate (x, y)
AM 1843-393	18 43 20 −39 32.0	4.156 0.063	356.34 −16.02	204.5 34.8	I/A DBL. 336-IG-15	2 S 0.8	F336/J0787 (124,204)
AM 1843-651	18 43 59 −65 13.8	5.914 0.064	330.34 −24.18	205.8 9.1	HI.S.BR. SB(r) (MED.BR. ENVELOPE) NGC 6684 = GC1-102 = 104-G-16	12,22 D 4.0	F104/J1590 (291,163)
AM 1844-533	18 44 05 −53 35.6	4.851 0.064	342.47 −21.03	205.3 20.7	SPIRAL w. DIAMETRIC COMP'NS. 183-G-12 (ONE DWARF)	8,20 S 2.7	F183/J4391 (163,255)
AM 1845-785	18 45 58 −78 56.6	9.772 0.067	315.40 −26.69	206.4 −4.6	GALAXY w. MANY RINGS 25-G-2 = Fa 189 (LAST ONE BROKEN)	6 D 0.9	F024/J2167 (62,225)
AM 1846-564	18 46 19 −56 46.9	5.072 0.067	339.30 −22.28	205.8 17.6	CLOSE PAIR in COMMON ENVELOPE	2 S 0.4	F183/J4391 (147,83)
AM 1846-553	18 46 44 −55 33.5	4.981 0.068	340.58 −21.99	205.8 18.8	CLOSE I/A DBL. IC 4782 = 183-IG-17/18	2 S 1.0	F183/J4391 (143,149)
AM 1847-585	18 47 10 −58 52.4	5.239 0.068	337.15 −22.96	206.0 15.5	SPIRAL w. THICK ARM IC 4783 141-G-08	10 D 1.1	F141/J0890 (265,236)
AM 1847-523	18 47 23 −52 38.9	4.786 0.069	343.60 −21.22	205.8 21.7	SPIRAL w. 2 APP. COMP'NS 183-G-13	8 S 3.1	F183/J4391 (135,305)
*AM 1847-645	18 47 31 −64 53.7	5.863 0.069	330.78 −24.46	206.2 9.5	KNOTTY LO.S.BR. BARRED DWARF NGC 6684a = Se 129/3 = 104-G-19 [UKS]	20 D 2.9	F104/J1590 (273,182)
AM 1847-572	18 47 47 −57 22.7	5.115 0.069	338.74 −22.64	206.0 17.0	SPIRAL w. I/A DBL. COMP'NS. NGC 6699 = 183-G-21	2,8 S 7.2	F183/J4391 (137,51)
AM 1848-631 [1848-6320]	18 48 06 −63 19.3	5.673 0.070	332.48 −24.17	206.2 11.0	CLOSE PAIR (#1: E) IC 4784? = 104-G-20	23 S 2.2	F104/J1590 (275,266)
AM 1848-632 [1848-6319]	18 48 19 −63 20.8	5.676 0.070	332.46 −24.20	206.2 11.0	CLOSE PAIR (#2: S0)	23 S 2.2	F104/J1590 (274,265)
AM 1848-374	18 48 53 −37 48.1	4.086 0.071	358.43 −16.36	205.8 36.5	CLOSE PAIR	23 S 13.4	F336/J0787 (64,296)
AM 1849-795	18 49 25 −79 51.6	10.370 0.071	314.39 −26.94	206.6 −5.5	SMALL GALAXY w. JET 24-G-13	7 D 0.6	F024/J2167 (63,175)
AM 1849-422	18 49 33 −42 22.9	4.264 0.072	354.02 −18.16	206.0 32.0	COMPACT	13 D 0.5	F337/J0763 (322,48)
AM 1850-620	18 50 28 −62 06.6	5.536 0.073	333.84 −24.16	206.4 12.2	VERY SMALL I/A 'STARS'	2 S 0.1	F141/J0890 (236,63)
AM 1850-684 [1851-6837]	18 50 42 −68 44.7	6.424 0.073	326.69 −25.54	206.5 5.6	WIDE PAIR (#1: SPIRAL) IC 4787 = 72-G-4 = UKS 1850-687	23 S 6.0	F072/J4433 (245,244)
*AM 1850-645	18 50 50 −64 52.6	5.852 0.073	330.89 −24.80	206.5 9.5	LARGE LO.S.BR. DWARF 104-G-22 = UKS 1850-648	20 D 1.8	F104/J1590 (254,184)
AM 1851-683 [1850-6837]	18 51 00 −68 37.9	6.404 0.074	326.82 −25.55	206.6 5.7	WIDE PAIR (#2: SPIRAL w. COMP'NS) IC 4789 = 72-G-5	8,23 S 6.0	F072/J4433 (244,250)
AM 1851-535 [1851-5347]	18 51 20 −53 53.0	4.858 0.074	342.52 −22.14	206.5 20.5	PAIR (#1: SPIRAL) NGC 6707 = 183-G-25	23 S 0.4	F183/J4391 (105,238)
AM 1851-534 [1851-5353]	18 51 31 −53 47.3	4.852 0.074	342.63 −22.14	206.5 20.6	PAIR (#2: SPIRAL) NGC 6708 = 183-G-27	23 S 0.4	F183/J4391 (104,243)
AM 1852-414	18 52 02 −41 48.6	4.237 0.075	354.76 −18.39	206.6 32.5	DISTURBED SINGLE w. TAIL 337-IG-1	15,16 D 0.8	F336/J0787 (38,80)
AM 1852-695	18 52 18 −69 51.7	6.622 0.076	325.49 −25.87	206.7 4.5	DBL. NUCLEUS w. FAINT ENVELOPE 72-IG-6 = Fa 499 (+ APP. COMP'N.)	8,22 S 0.1	F072/J4433 (233,184)
AM 1852-542	18 52 26 −54 22.3	4.889 0.076	342.06 −22.44	206.7 20.0	LINE of 5 DN 1852-542 = SS-20 IC 4796/97 = Se130/3 = 183-G-28/29	17 D 4.5	F183/J4391 (98,212)
AM 1854-631	18 54 02 −63 12.5	5.646 0.078	332.76 −24.80	206.9 11.1	PEC. SPIRAL w. CONDENS. at ENDS IC 4800 = 104-G-25 (in PAIR)	12,23 D 1.6	F104/J1590 (240,274)
*AM 1854-640	18 54 10 −64 00.1	5.736 0.078	331.91 −24.98	206.9 10.4	SB w. MULTIPLE ARMS ONE SIDE IC 4799 = 104-G-27 = fA 335	10,11 D 1.6	F104/J1590 (237,232)
*AM 1854-490	18 54 17 −49 04.6	4.571 0.078	347.61 −21.16	207.0 25.3	RING w. OFF-SET NUCLEUS + COMP'N. 231-G-12 = Fa 500 + Fa 501?	6,8 S 0.3	F231/J1597 (227,226)
AM 1856-620	18 56 12 −62 08.3	5.525 0.081	333.97 −24.82	207.1 12.2	I/A TRIPLET IC 4803 = 141-IG-17	3 S 0.4	F141/J0890 (200,62)
*AM 1856-570	18 56 36 −57 04.8	5.074 0.082	339.39 −23.71	207.3 17.3	SPIRAL w. CONDENS. in BROKEN ARM	10 D 0.4	F184/J1576 (311,61)
AM 1856-491	18 56 48 −49 10.6	4.572 0.082	347.64 −21.58	207.5 25.2	CLOSE PAIR in CLUSTER 231-G-14	19,23 S 0.8	F231/J1597 (205,221)
AM 1856-540	18 56 59 −54 02.5	4.858 0.082	342.60 −22.99	207.4 20.3	CONTACT PAIR 183-G-35	2 S 0.3	F183/J4391 (61,228)
AM 1857-573	18 57 18 −57 36.1	5.113 0.083	338.86 −23.93	207.4 16.7	E + SPIRAL w. PECULIAR NUCLEUS IC 4806 = 141-G-20 = Fa 503	12,23 S 11.6	F183/J4391 (69,37)
*AM 1857-535	18 57 53 −53 56.0	4.849 0.083	342.75 −23.09	207.5 20.4	S0 + APP. I/A DBL. (in CHAIN) NGC 6725 = 183-G-36	2,8,17 D 2.0	F184/J1576 (312,230)
AM 1858-565	18 58 03 −56 59.8	5.064 0.084	339.53 −23.88	207.5 17.4	GALAXY w. 3rd. ARM w. KNOT on END IC 4807 = Se 131/1 = 183-G-37	11,12,13 D 0.7	F141/J0890 (190,338)
AM 1858-542	18 58 27 −54 24.7	4.880 0.084	342.27 −23.29	207.6 19.9	I/A DBL.	2 S 0.3	F184/J1576 (306,205)
AM 1859-380	18 59 12 −38 03.1	4.084 0.085	358.97 −18.34	208.3 36.3	GROUP of 4 incl. 337-G-8	18 D 0.3	F337/J0763 (230,283)

Identification	R.A. (1950.0) Declination	Precession R.A./Dec.	Gal. Long. Gal. Lat.	S.G.L. S.G.B.	Description Cross-Identification	Coding Size (Min.)	Field/Plate (x, y)
AM 1859-531	18 59 35 −53 14.9	4.802 0.086	343.54 −23.15	207.8 21.1	E + SPIRAL 184-G-4 = Fa 506	23 S 2.8	F183/J4391 (38,269)
AM 1900-411	19 00 13 −41 17.2	4.206 0.087	355.84 −19.64	208.4 33.0	I/A DBL.	2 S 0.9	F337/J0763 (217,110)
AM 1900-555	19 00 33 −55 55.9	4.980 0.087	340.75 −23.96	207.9 18.4	DISTURBED SINGLE GALAXY 184-IG-7	16 D 0.9	F184/J1576 (286,124)
AM 1901-385	19 01 32 −38 57.6	4.115 0.089	358.24 −19.09	208.8 35.4	PAIR of CONTACT E's	2 S 0.4	F337/J0763 (205,235)
AM 1901-865	19 01 56 −86 55.9	0.451 0.089	306.46 −27.52	206.8 −12.6	LINEAR NEBULOSITY	22 D 11.2	F011/J2562 (281,39)
AM 1902-561	19 02 07 −56 14.5	4.999 0.089	340.48 −24.25	208.1 18.1	PAIR of I/A SPIRALS IC 4817 = 184-G-10 (w. 184-G-11)	2 S 1.5	F184/J1576 (273,108)
AM 1902-653 [1902-6530]	19 02 21 −65 32.3	5.902 0.090	330.41 −26.11	207.7 8.8	PAIR in GROUP of 4 (#1: S0:) [DN] NGC 6734 = Se 132/2 = 104-G-36 [Fa]	18,23 D 4.9	F104/J1590 (188,151)
AM 1902-422	19 02 27 −42 26.7	4.250 0.090	354.82 −20.42	208.8 31.9	E w. I/A COMP'N 337-G-9/10 = SS-21	1 S 0.1	F337/J0763 (194,48)
AM 1902-653 [1902-6532]	19 02 38 −65 30.5	5.897 0.090	330.45 −26.14	207.8 8.8	PAIR in GROUP of 4 (#2: E) [DN] NGC 6736 = Se 132/2 = 104-G-37	18,23 D 4.9	F104/J1590 (187,152)
AM 1903-335	19 03 02 −33 59.5	3.940 0.091	3.25 −17.59	209.5 40.3	CLOSE PAIR	23 S 0.7	F397/J0704 (279,232)
AM 1903-552	19 03 42 −55 22.8	4.934 0.092	341.45 −24.26	208.4 18.9	CLOSE PAIR 184-IG-15	23 S 2.0	F184/J1576 (263,155)
AM 1904-634	19 04 01 −63 48.4	5.684 0.092	332.35 −26.00	208.0 10.5	DWARF in DISK of LARGE SPIRAL 104-G-38	20 D 1.6	F104/J1590 (179,244)
AM 1904-633	19 04 31 −63 32.7	5.652 0.093	332.64 −26.02	208.1 10.8	DISTURBED DWARF SPIRAL IC 4820 = 104-IG-39	20 D 1.3	F104/J1590 (176,258)
AM 1904-844	19 04 38 −84 47.2	17.135 0.093	308.87 −27.62	207.0 −10.4	COMPACT	13 D 0.1	F010/J2477 (189,191)
AM 1905-635	19 05 03 −63 56.5	5.696 0.094	332.22 −26.14	208.1 10.4	SPIRAL w. COMP'N. NORMAL to ARM NGC 6744 = 104-G-42	1 D 20.2	F104/J1590 (173,236)
AM 1905-554	19 05 40 −55 40.3	4.950 0.094	341.21 −24.60	208.6 18.6	PLANETARY NEBULA? 184-G-18 = PK 341-24-1	25 D 0.8	F184/J1576 (248,140)
AM 1905-621	19 05 52 −62 18.1	5.513 0.095	334.03 −25.95	208.3 12.0	DWARF	20 D 0.4	F104/J1590 (168,324)
AM 1905-845	19 05 54 −84 53.0	17.383 0.095	308.76 −27.65	207.0 −10.5	HI.S.BR. SB(r) + FAINT OUTER ARMS 10-G-3 = Fa 336	10 D 0.5	F010/J2477 (187,185)
AM 1906-641	19 06 39 −64 18.4	5.733 0.096	331.85 −26.37	208.2 10.0	DWARF Se 129/2 = 104-G-44 = UKS1906-643	20 D 1.8	F104/J1590 (164,217)
AM 1906-621	19 06 56 −62 15.5	5.506 0.096	334.11 −26.07	208.4 12.0	DWARF	20 D 0.4	F104/J1590 (161,327)
AM 1907-570	19 07 08 −57 08.1	5.053 0.096	339.69 −25.12	208.8 17.2	BR. SPIRAL + 6 APP. COMP'NS. NGC 6753 = 184-G-22 (HALF RING)	8,12 S 6.3	F184/J1576 (235,62)
AM 1907-525	19 07 27 −52 57.0	4.767 0.097	344.18 −24.22	209.1 21.3	M51-TYPE Se 130/2 = 184-IG-23	9 D 1.0	F184/J1576 (238,286)
AM 1907-621	19 07 30 −62 18.0	5.508 0.097	334.07 −26.14	208.5 12.0	DWARF?	20 D 1.8	F104/J1590 (158,324)
AM 1907-504	19 07 31 −50 45.2	4.638 0.097	346.52 −23.68	209.3 23.5	SPIRAL + I/A COMP'N. NGC 6754 = 231-G-25	1 S 1.3	F231/J1597 (113,136)
AM 1907-640	19 07 32 −64 03.8	5.701 0.097	332.13 −26.43	208.4 10.2	E w. JET + CONDENSATION IC 4823 = 104-IG-25	7 D 1.1	F104/J1590 (159,230)
AM 1908-313	19 08 18 −31 34.5	3.858 0.098	6.03 −17.73	211.2 42.6	SPIRAL + APP. COMP'N.	8	F397/J0704 (222,363)
*AM 1908-535	19 08 39 −53 57.3	4.827 0.099	343.16 −24.63	209.2 20.3	DISTURBED SINGLE GALAXY 184-IG-32	16 D 1.1	F184/J1576 (227,233)
AM 1908-621	19 08 45 −62 10.6	5.491 0.099	334.23 −26.26	208.6 12.1	DWARF IC 4824 = 141-G-33 = UKS 1908-621	20 D 1.8	F104/J1590 (150,331)
AM 1909-621 [1911-6227]	19 09 06 −62 10.0	5.489 0.099	334.25 −26.30	208.7 12.1	GROUP of 4 (#1: SPIRAL) IC 4828 = 141-G-36	8 S 12.5	F104/J1590 (148,331)
AM 1909-460	19 09 08 −46 05.7	4.398 0.099	351.47 −22.67	209.9 28.2	GALAXY w. BROKEN ARMS + COMP'NS. 282-G-21	8,10 D 1.7	F282/J0808 (169,121)
AM 1909-615	19 09 21 −61 59.9	5.471 0.100	334.44 −26.30	208.7 12.3	DWARF	20 D 0.3	F104/J1590 (146,340)
*AM 1909-562	19 09 41 −56 24.0	4.992 0.100	340.56 −25.30	209.2 17.9	CHAIN of 6 (PEC. ABS'N. in ONE) NGC 6758 = 184-G-37	14,17 D 26.9	F184/J1576 (217,102)
*AM 1910-543	19 10 10 −54 39.6	4.869 0.101	342.46 −25.00	209.4 19.6	CHAIN of 4 DN 1910-547	17 D 9.0	F184/J1576 (215,195)
*AM 1910-622 [1909-6210]	19 10 11 −62 21.6	5.506 0.101	334.06 −26.46	208.8 11.9	GROUP of 4 (#2: EDGE-ON SPIRAL) IC 4831 = 141-G-38 = Se 132/1	8 S 12.5	F104/J1590 (141,321)
AM 1910-464	19 10 53 −46 40.8	4.422 0.102	350.96 −23.13	210.2 27.5	LONG CHAIN of 13 282-G-28 = PKS 1910-466	17 D 7.8	F282/J0808 (152,89)
AM 1911-622 [1910-6221]	19 11 06 −62 24.8	5.508 0.102	334.02 −26.57	208.9 11.9	GROUP of 4 (#3: IRR./LOOP) IC 4833 = 141-G-40	8,13 S 12.5	F104/J1590 (135,318)

Identification	R.A. (1950.0) Declination	Precession R.A./Dec.	Gal. Long. Gal. Lat.	S.G.L. S.G.B.	Description Cross-Identification	Coding Size (Min.)	Field/Plate (x, y)
*AM 1911-544 [1911-5442]	19 11 10 −54 45.0	4.873 0.102	342.40 −25.16	209.5 19.5	CLOSE PAIR (#1 + I/A COMP'N.) IC 4837 = 184-IG-46	1,23 D 1.5	F184/J1576 (207,190)
AM 1911-541	19 11 15 −54 12.8	4.837 0.102	342.98 −25.06	209.6 20.0	SPIRAL w. APP. COMP'N. IC 4837a = 184-G-47	8 S 11.2	F184/J1576 (207,219)
AM 1911-571	19 11 21 −57 11.1	5.045 0.102	339.76 −25.69	209.4 17.1	DWARF	20 D 0.5	F184/J1576 (204,60)
*AM 1911-544 [1911-5445]	19 11 29 −54 42.7	4.870 0.102	342.45 −25.20	209.6 19.5	CLOSE PAIR (#2) IC 4839 = 184-IG-46	23 D 2.0	F184/J1576 (204,193)
AM 1911-622	19 11 30 −62 29.1	5.514 0.103	333.95 −26.63	208.9 11.8	I/A DBL. in CHAIN	2,17 D 0.6	F142/J0906 (341,36)
AM 1911-622 [1911-6224]	19 11 36 −62 27.5	5.511 0.103	333.98 −26.64	208.9 11.8	GROUP of 4 (#4: LINEAR DWARF) 141-G-42 = UKS 1911-624	8,20 S 12.5	F104/J1590 (132,315)
AM 1912-601	19 12 13 −60 10.5	5.289 0.104	336.51 −26.34	209.2 14.1	COMPACT	13 D 0.7	F141/J0890 (95,166)
AM 1912-572	19 12 14 −57 25.5	5.061 0.104	339.53 −25.85	209.5 16.8	COMPACT	13 D 0.8	F141/J0890 (87,313)
AM 1912-573	19 12 18 −57 33.7	5.071 0.104	339.38 −25.89	209.5 16.7	SMALL COMPACT	13 D 0.1	F184/J1576 (197,40)
AM 1912-614	19 12 19 −61 42.5	5.433 0.104	334.83 −26.60	209.1 12.5	GALAXY w. I/A COMP'N. IC 4838 = 141-IG-45	1 D 1.0	F141/J0890 (98,83)
AM 1912-860	19 12 46 −86 06.8	0.363 0.104	307.37 −27.72	207.1 −11.8	SPIRAL w. COMPACT COMP'N. [E.] 10-G-4	8 S 2.7	F011/J2562 (301,80)
AM 1912-401	19 12 56 −40 12.9	4.146 0.105	357.75 −21.57	211.4 33.9	GROUP of 4 CONTACT E's incl. 337-G-16 + SP. w. PEC. ARMS	2,10,18 D 0.2	F337/J0763 (88,167)
AM 1913-540 [1913-5409]	19 13 01 −54 07.1	4.827 0.105	343.14 −25.29	209.9 20.1	PAIR of COMPACTS (#1)	13,23 D 0.2	F184/J1576 (193,224)
AM 1913-540 [1930-5407]	19 13 08 −54 09.4	4.829 0.105	343.10 −25.31	209.9 20.1	PAIR of COMPACTS (#2)	13,23 D 0.2	F184/J1576 (192,222)
*AM 1913-673	19 13 46 −67 31.7	6.137 0.106	328.37 −27.50	208.7 6.7	THIN-ARMED SPIRAL w. DIAMETRICS 72-G-13	8,10 S 1.3	F072/J4433 (132,312)
AM 1913-605	19 13 49 −60 56.7	5.354 0.106	335.70 −26.66	209.3 13.3	CLOSE I/A DBL. (near BR. GAL.)	2,8 D 0.3	F141/J0890 (86,124)
*AM 1914-603	19 14 11 −60 35.3	5.320 0.106	336.10 −26.65	209.4 13.6	I/A TRIPLE (6 COMP'NS.) NGC 6769/70/71 = 141-IG-48/49/50	3,8 D 5.4	F141/J0890 (83,143)
*AM 1916-624	19 16 12 −62 44.8	5.525 0.109	333.75 −27.20	209.5 11.5	COMPACT	13 D 0.3	F142/J0906 (311,24)
AM 1916-602	19 16 48 −60 25.0	5.295 0.110	336.35 −26.94	209.8 13.8	COMPACT w. PLUME (SPIRAL COMP'N.)	8,13,15 D 0.3	F141/J0890 (65,151)
AM 1916-584	19 16 58 −58 45.5	5.152 0.110	338.19 −26.70	210.0 15.4	SPIRAL w. DEFORMED ARM 141-G-55	10 D 0.9	F141/J0890 (58,240)
AM 1918-740	19 18 30 −74 04.0	7.481 0.112	321.01 −28.27	208.5 0.2	PAIR of I/A DBL. Fa 513	2 S 1.8	F046/J2588 (288,222)
AM 1918-565 [1918-5653]	19 18 31 −56 54.4	5.004 0.112	340.27 −26.59	210.4 17.2	I/A DBL. (#1: COMPACT)	2,13 D 0.2	F184/J1576 (151,75)
AM 1918-555	19 18 43 −55 52.4	4.929 0.112	341.41 −26.44	210.5 18.3	SPIRAL w. APP. COMP'N. NGC 6780 = 184-G-62	8 S 3.8	F184/J1576 (149,130)
AM 1918-565 [1918-5654]	19 18 45 −56 53.2	5.002 0.113	340.30 −26.62	210.4 17.3	I/A DBL. (#2: PECULIAR SPIRAL) IC 4848 = 184-G-61	2,12 D 0.5	F184/J1576 (150,76)
AM 1918-492	19 18 53 −49 29.2	4.545 0.113	348.35 −25.14	211.4 24.6	I/A DBL. or GALAXY w. JET 232-IG-3	2,7 S 0.5	F232/J1465 (276,205)
AM 1919-504	19 19 24 −50 41.8	4.608 0.113	347.07 −25.49	211.3 23.4	GALAXY w. JETS	7	F232/J1465 (269,140)
AM 1919-550	19 19 36 −55 09.9	4.878 0.114	342.21 −26.43	210.8 19.0	ISOLATED SPIRAL w. WARPED PLANE 184-G-63 = PKS 1919-551	8,12 D 2.9	F184/J1576 (142,168)
AM 1920-630	19 20 24 −63 03.7	5.542 0.115	333.46 −27.71	209.9 11.1	CONTACT PAIR + COMP'NS. 104-IG-51	2,8 S 0.5	F104/J1590 (80,281)
AM 1920-635	19 20 36 −63 57.9	5.641 0.115	332.45 −27.83	209.8 10.2	S0 w. COMP'N. + TAIL [S] SS-23 NGC 6776 = 104-G-53 = MC 1920-639	8,15 S 2.6	F104/J1590 (82,232)
AM 1921-693	19 21 03 −69 37.7	6.448 0.116	326.05 −28.32	209.2 4.6	COMPACT (LUMPY)	13 D 4.0	F073/J0916 (339,186)
*AM 1921-534	19 21 27 −53 42.8	4.779 0.116	343.86 −26.42	211.2 20.4	PEC. SPIRAL w. DIAMETRIC COMP'NS. 184-G-64 = Fa 514/5	8 S 4.0	F184/J1576 (126,246)
*AM 1921-530	19 21 37 −53 05.7	4.740 0.116	344.54 −26.33	211.4 21.0	INTEGRAL SPIRAL w. DBL. NUCLEUS 184-IG-65	12 D 0.9	F184/J1576 (124,279)
AM 1921-654	19 21 39 −65 43.5	5.849 0.116	330.47 −28.11	209.7 8.4	I/A DBL. NGC 6784 = 104-IG-55	2 S 0.8	F105/J0891 (322,131)
*AM 1922-602	19 22 04 −60 26.1	5.279 0.117	336.43 −27.59	210.4 13.7	SPIRAL w. BROKEN ARM IC 4852 = 142-G-6	10 D 1.8	F141/J0890 (30,147)
AM 1922-630	19 22 24 −63 03.1	5.533 0.117	333.50 −27.93	210.1 11.1	CONTACT PAIR (RING)	2 S 0.1	F104/J1590 (68,280)
*AM 1922-550	19 22 53 −55 02.8	4.861 0.118	342.44 −26.87	211.3 19.0	GALAXY w. I/A COMP'N. NGC 6788 = 184-G-67	1 S 6.3	F184/J1576 (117,174)

Identification	R.A. (1950.0) Declination	Precession R.A./Dec.	Gal. Long. Gal. Lat.	S.G.L. S.G.B.	Description Cross-Identification	Coding Size (Min.)	Field/Plate (x, y)
AM 1923-592	19 23 07 −59 24.7	5.186 0.118	337.59 −27.58	210.7 14.7	SP. w. CONDENSATIONS in DISK IC 4854/55 = 142-G-7 = Fa 58	12 D 0.9	F141/J0890 (18,201)
AM 1923-585	19 23 27 −58 54.9	5.144 0.119	338.16 −27.56	210.8 15.2	COMPACT 142-IG-8	13 D 0.4	F142/J0906 (277,233)
AM 1923-342	19 23 56 −34 25.1	3.928 0.120	4.42 −21.79	215.0 39.4	CLOSE PAIR (E + SPIRAL)	23	F397/J0704 (47,209)
AM 1924-414 [1924-4140]	19 24 02 −41 45.0	4.186 0.120	356.84 −24.03	213.6 32.2	PAIR of IRR. (#1)	16,23 D 0.4	F338/J1515 (239,85)
*AM 1924-585	19 24 21 −58 51.9	5.136 0.120	338.23 −27.66	210.9 15.2	SPIRAL w. I/A COMP'N. Se 131/2 = IC 4857/8 = 142-G-12	1 S 1.8	F142/J0906 (271,236)
AM 1924-414 [1924-4145]	19 24 28 −41 40.7	4.183 0.120	356.93 −24.09	213.7 32.2	PAIR of IRR. (#2) 338-IG-4	16,23 D 0.5	F338/J1515 (235,89)
AM 1924-612	19 24 30 −61 25.0	5.360 0.120	335.37 −28.00	210.6 12.7	DISTURBED SPIRAL (HI.S.BR. BAR) 142-IG-13	12 D 0.9	F142/J0906 (263,99)
AM 1924-364	19 24 31 −36 40.6	4.001 0.120	2.15 −22.63	214.7 37.1	COMPACT w. JET 338-IG-4	7,13 D 0.6	F388/J0795 (236,87)
*AM 1925-724	19 25 00 −72 45.4	7.085 0.121	322.50 −28.72	209.1 1.4	I/A DBL. w. VERY LONG TAILS	2,15 D 0.4	F046/J2588 (271,295)
AM 1925-343	19 25 10 −34 32.0	3.930 0.121	4.39 −22.07	215.3 39.2	CLOSE PAIR	23	F398/J0665 (310,202)
AM 1925-345	19 25 23 −34 53.5	3.941 0.122	4.04 −22.23	215.3 38.9	I/A DBL. 338-IG-6	2 S 0.4	F388/J0795 (227,183)
*AM 1925-724	19 25 24 −72 45.2	7.081 0.122	322.51 −28.75	209.1 1.4	VERY CLOSE I/A DBL. (2 STRAIGHT TAILS or JETS?)	2,7,12 S 0.1	F072/J4433 (91,30)
AM 1926-672 [1926-6725]	19 26 25 −67 28.6	6.068 0.123	328.53 −28.70	210.0 6.7	CLOSE PAIR of SPIRALS (#1) IC 4860 = 105-G-1	23 S 3.1	F104/J1590 (64,42)
AM 1926-672 [1926-6728]	19 26 37 −67 25.9	6.060 0.123	328.58 −28.72	210.0 6.7	CLOSE PAIR of SPIRALS (#2) IC 4862 = 105-G-3	23 S 3.1	F104/J1590 (62,44)
*AM 1926-393	19 26 49 −39 30.8	4.097 0.124	359.35 −23.90	214.7 34.3	HI.S.BR. I/A DBL. (TORN UP !) 338-IG-8	2 S 0.5	F338/J1515 (212,205)
AM 1926-343	19 26 51 −34 31.0	3.927 0.124	4.53 −22.39	215.8 39.2	I/A TRIPLE 338-IG-8	3 D 0.9	F388/J0795 (211,203)
AM 1927-533 [1927-5339]	19 27 00 −53 38.7	4.759 0.124	344.10 −27.22	212.1 20.3	PAIR of SPIRALS (#1) 184-G-75 (in CHAIN of 4)	17,23 D 0.7	F185/J2370 (343,244)
AM 1927-533 [1927-5338]	19 27 37 −53 39.7	4.758 0.125	344.11 −27.31	212.2 20.3	PAIR of SPIRALS (#2) 184-G-77 (in CHAIN of 4)	23 D 0.8	F185/J2370 (338,243)
AM 1927-402	19 27 57 −40 26.3	4.129 0.125	358.44 −24.37	214.7 33.3	CLOSE PAIR of E's	23 S 1.1	F338/J1515 (200,156)
AM 1929-451	19 29 17 −45 19.9	4.323 0.127	353.29 −25.87	214.0 28.5	GALAXY w. APP. COMP'NS. 283-G-4	8 D 1.5	F283/J 803 (242,161)
AM 1929-394	19 29 59 −39 46.5	4.101 0.128	359.26 −24.56	215.3 33.9	ISOLATED PAIR of ENVELOPED E's 338-IG-11 = PKS 1929-397	2 S 0.4	F338/J1515 (179,191)
AM 1930-344	19 30 00 −34 46.7	3.930 0.128	4.48 −23.09	216.5 38.8	PAIR of I/A E's 338-IG-11	2 S 0.4	F388/J0795 (176,189)
AM 1931-525	19 31 08 −52 57.1	4.705 0.129	344.99 −27.71	212.9 20.9	CHAIN of 4 184-G-82 = DN 1931-529	17 D 5.8	F184/J1576 (47,283)
AM 1931-773 [1932-7740]	19 31 16 −77 38.9	8.697 0.130	316.91 −28.96	208.8 −3.5	PAIR of SPIRALS (#1) 25-G-12	1,23 S 4.7	F046/J2588 (227,35)
AM 1931-404	19 31 36 −40 49.6	4.136 0.130	358.24 −25.15	215.5 32.8	E w. 5 CLOSE COMP'NS.	8 S 0.7	F338/J1515 (163,135)
*AM 1931-610	19 31 36 −61 08.4	5.306 0.130	335.78 −28.82	211.5 12.8	DWARF w. APP. COMP'N. IC 4869 = 142-G-25 = UKS 1931-611	8,20 S 3.4	F142/J0906 (218,116)
AM 1932-655	19 32 46 −65 55.4	5.821 0.132	330.33 −29.26	210.8 8.1	DISTURBED SINGLE GALAXY IC 4870 = 105-IG-11	16 D 0.9	F105/J0891 (260,125)
*AM 1932-532	19 32 50 −53 25.9	4.728 0.132	344.50 −28.04	213.1 20.4	CHAIN of 5 incl. COMPACT	13,17 D 5.4	F184/J1576 (35,257)
AM 1932-774 [1931-7738]	19 32 55 −77 40.2	8.690 0.132	316.87 −29.05	208.9 −3.5	PAIR of SPIRALS (#2) IC 4864 = 25-G-13	1,23 S 4.7	F046/J2588 (222,34)
AM 1933-372	19 33 34 −37 24.3	4.011 0.133	1.98 −24.57	216.7 36.1	SPIRAL w. 2 I/A COMP'NS. NGC 6806 = 338-G-15	1 S 3.7	F388/J0795 (139,48)
AM 1933-422	19 33 35 −42 24.4	4.193 0.133	356.65 −25.91	215.5 31.2	DISTURBED SPIRAL w. 2 I/A COMP'NS NGC 6806 = Ag-58 = 338-G-15	1,12 S 2.9	F338/J1515 (144,50)
AM 1933-525	19 33 47 −52 57.0	4.697 0.133	345.07 −28.10	213.3 20.8	SP. + APP.COMP'N + I/A DBL.[S.W.] IC 4876 = 185-G-4	1,2 S 6.7	F184/J1576 (25,282)
*AM 1934-563	19 34 34 −56 34.1	4.927 0.134	341.01 −28.72	212.7 17.3	I/A TRIPLE (SPIKED CENTRAL)	3 S 3.4	F184/J1576 (33,88)
AM 1934-562	19 34 47 −56 24.3	4.915 0.134	341.20 −28.73	212.8 17.4	LOOSE CHAIN of 4	17 D 5.4	F184/J1576 (30,96)
AM 1934-390	19 34 53 −39 00.4	4.064 0.134	0.36 −25.26	216.6 34.5	ONE-ARMED SPIRAL + APP. COMP'N. 338-G-17	8,10 D 9.0	F338/J1515 (128,232)
AM 1935-703 [1935-7032]	19 35 11 −70 33.9	6.537 0.135	325.01 −29.53	210.2 3.5	I/A DBL (#1) 73-IG-1b = Fa 517	2 S 0.9	F072/J4433 (40,143)

Identification	R.A. (1950.0) Declination	Precession R.A./Dec.	Gal. Long. Gal. Lat.	S.G.L. S.G.B.	Description Cross-Identification	Coding Size (Min.)	Field/Plate (x, y)
AM 1935-703 [1935-7033]	19 35 12 −70 32.8	6.534 0.135	325.03 −29.54	210.2 3.5	I/A DBL.(#2: TAIL or PEC. ARM) 73-IG-1a	2,15 S 0.9	F072/J4433 (39,144)
AM 1935-522	19 35 42 −52 29.1	4.664 0.135	345.64 −28.31	213.7 21.2	DISTURBED SINGLE IC 4879 = 232-G-16	16	F232/J1465 (132,46)
AM 1936-612	19 36 17 −61 23.3	5.309 0.136	335.55 −29.40	212.0 12.5	GALAXY w. JET(?) and TAILS 142-G-31	7,15 D 0.7	F142/J0906 (188,103)
*AM 1939-515	19 39 22 −51 55.6	4.621 0.140	346.37 −28.78	214.4 21.7	GAL. (DISTURBED ARMS + 6 COMP'NS) IC 4886 = 232-IG-18 = DN 1939-519	1,6,8,10 S 1.8	F185/J2370 (248,341)
AM 1939-273	19 39 56 −27 32.7	3.705 0.141	12.65 −22.72	221.7 45.3	GAL. w. PEC. ABS'N. (TILTED SP.?) 460-G-26	14 D 0.4	F526/J0753 (275,42)
AM 1940-520	19 40 02 −52 01.7	4.624 0.141	346.27 −28.90	214.5 21.6	I/A DBL. in GROUP of 4 232-IG-20	2,18 D 1.1	F185/J2370 (242,336)
AM 1940-831	19 40 51 −83 13.3	13.243 0.142	310.53 −28.70	208.1 −9.0	GALAXY w. PLUME/JETS 10-G-7	7,15 D 0.7	F010/J2477 (135,272)
AM 1941-515	19 41 06 −51 53.7	4.613 0.143	346.45 −29.04	214.7 21.6	GROUP of 3 in WIDER GROUP incl. IC 4886, 232-G-22/21	24 D 1.3	F233/J1488 (336,72)
AM 1941-542	19 41 14 −54 27.9	4.764 0.143	343.53 −29.40	214.1 19.1	E + 5 APP. COMP'NS. IC 4889 = 185-G-14 = SS-24a	8 S 6.3	F185/J2370 (229,205)
AM 1941-552	19 41 18 −55 28.2	4.828 0.143	342.39 −29.53	213.9 18.1	I/A TRIPLE (E GALAXIES) NGC 6812 = 185-G-15 = PKS 1941-554	3 D 2.2	F185/J2370 (227,151)
AM 1941-564	19 41 27 −56 40.4	4.909 0.143	341.01 −29.67	213.7 17.0	SPIRAL + I/A COMP'N. IC4890 = 185-G-16	1 S 0.9	F185/J2370 (225,87)
AM 1941-812	19 41 35 −81 27.0	11.101 0.143	312.53 −29.01	208.5 −7.3	STAR w. NEBULOSITY	21 D 3.1	F026/J1818 (331,73)
AM 1941-502 [1942-5025]	19 41 41 −50 27.7	4.535 0.143	348.08 −28.92	215.1 23.0	PAIR of LO.S.BR. SPIRALS (#1)	23 S 3.8	F233/J1488 (336,150)
AM 1942-502 [1941-5027]	19 42 06 −50 25.5	4.532 0.144	348.14 −28.98	215.2 23.0	PAIR of LO.S.BR. SPIRALS (#2) 232-G-23	23 S 3.8	F233/J1488 (332,152)
AM 1942-402 [1943-4028]	19 42 46 −40 20.9	4.096 0.145	359.35 −27.07	218.0 32.8	WIDE TRIPLE (#1: COMPACT)	13,24 D 0.3	F338/J1515 (49,157)
AM 1943-403 [1942-4020]	19 43 09 −40 33.5	4.103 0.145	359.14 −27.20	218.1 32.6	WIDE TRIPLE (#2)	24 D 0.4	F338/J1515 (45,146)
AM 1943-515	19 43 09 −51 58.3	4.611 0.145	346.41 −29.37	215.0 21.5	KNOTTY GALAXY w. I/A COMP'N. IC 4894 = 232-IG-25	2	F232/J1465 (70,71)
AM 1943-621	19 43 10 −62 14.6	5.359 0.145	334.61 −30.26	212.7 11.5	STAR w. APP. NEBULOSITY	21 D 0.3	F105/J0891 (207,325)
AM 1943-352	19 43 21 −35 28.6	3.930 0.146	4.64 −25.90	219.7 37.5	CLOSE PAIR (E + SPIRAL) 339-G-1	23 S 5.4	F388/J0795 (31,149)
AM 1943-402 [1943-4033]	19 43 21 −40 28.8	4.099 0.146	359.24 −27.21	218.1 32.7	WIDE TRIPLE (#3: SPIRAL) 339-G-1	24 D 0.6	F338/J1515 (43,150)
AM 1943-231	19 43 34 −23 15.6	3.590 0.146	17.30 −21.97	224.7 49.2	PLANETARY NEBULA PK 17-21-1 = 526-PN-3	25 D 2.0	F526/J0753 (234,272)
AM 1943-345	19 43 42 −34 54.7	3.912 0.146	5.26 −25.81	220.0 38.0	I/A DBL.	2 S 0.3	F388/J0795 (26,179)
AM 1944-480	19 44 01 −48 03.0	4.409 0.146	350.87 −28.90	216.2 25.3	OVAL COMPACT	13 D 0.2	F283/J 803 (107,15)
AM 1945-541	19 45 24 −54 19.2	4.740 0.148	343.78 −29.99	214.8 19.1	CLOSE PAIR 185-G-21	23 S 1.6	F185/J2370 (197,214)
AM 1946-614	19 46 36 −61 42.6	5.291 0.150	335.25 −30.65	213.2 11.9	3 ALIGNED (COMPACTS?)	13,24 D 0.7	F142/J0906 (122,85)
AM 1946-512	19 46 41 −51 29.4	4.573 0.150	347.05 −29.84	215.7 21.8	CHAIN of 4 incl. IC 4900	17 D 7.2	F233/J1488 (290,97)
AM 1947-445	19 47 51 −44 59.3	4.263 0.151	354.45 −28.99	217.7 28.1	I/A DBL. (V.FNT. RING) + 2 GAL. 283-G-19/20	2,6,18 S 2.0	F284/J5235 (333,176)
*AM 1950-585	19 50 12 −58 50.7	5.032 0.154	338.60 −30.99	214.3 14.6	GALAXY w. M51-TYPE COMP'N. IC 4901 = 140-G-50	8,9 D 4.5	F142/J0906 (92,237)
AM 1950-620	19 50 34 −62 03.9	5.305 0.155	334.84 −31.12	213.6 11.5	DISTURBED GAL.w. CONDENS. in DISK 142-G-52	16 D 0.5	F105/J0891 (160,335)
AM 1950-380	19 50 40 −38 06.9	4.001 0.155	2.22 −28.01	220.5 34.5	INTEGRAL-SIGN SPIRAL 339-IG-4 (I/A w. E)	2 S 1.5	F399/J0734 (275,10)
AM 1950-471	19 50 48 −47 10.4	4.349 0.155	352.08 −29.88	217.6 25.8	COMPACT	13 D 0.2	F233/J1488 (263,330)
AM 1951-573	19 51 23 −57 38.2	4.937 0.156	340.02 −31.09	214.8 15.7	GALAXY w. CONDENSATION in DISK 142-G-53	12 D 0.8	F142/J0906 (80,302)
AM 1951-605	19 51 39 −60 52.1	5.191 0.156	336.24 −31.23	214.0 12.6	COMPACT 142-G-54	13 D 0.3	F142/J0906 (87,128)
AM 1951-374	19 51 50 −37 42.4	3.985 0.157	2.73 −28.14	221.0 34.9	GALAXY w. SMALL FAN	15 D 0.4	F339/J0809 (220,300)
AM 1952-232 [1952-2324	19 52 03 −23 25.1	3.584 0.157	17.90 −23.84	227.4 48.3	CLOSE PAIR (#1) MCG-4-47-1 = 526-G-15	23 D 0.8	F526/J0753 (130,264)
AM 1952-570	19 52 08 −57 02.7	4.892 0.157	340.71 −31.16	215.1 16.3	E at CENTRE of CHAIN of 4 185-G-27 = DN 1952-570	17 D 13.4	F185/J2370 (146,67)

Identification	R.A. (1950.0) Declination	Precession R.A./Dec.	Gal. Long. Gal. Lat.	S.G.L. S.G.B.	Description Cross-Identification	Coding Size (Min.)	Field/Plate (x, y)
AM 1952-841	19 52 09 −84 19.5	14.937 0.157	309.23 −28.79	208.2 −10.2	I/A DBL.	2 S 0.2	F011/J2562 (313,184)
AM 1952-232 [1952-2325]	19 52 13 −23 24.0	3.583 0.157	17.94 −23.87	227.5 48.3	CLOSE PAIR (#2) MCG-4-47-2 = 526-G-16	23 D 1.1	F526/J0753 (127,265)
AM 1952-570	19 52 15 −57 04.0	4.893 0.157	340.69 −31.17	215.1 16.2	COMPACT	13 D 0.3	F142/J0906 (72,332)
AM 1952-831	19 52 22 −83 13.8	13.003 0.157	310.45 −29.04	208.5 −9.1	I/A TRIPLE	3 D 0.6	F010/J2477 (117,269)
AM 1952-603	19 52 30 −60 36.1	5.164 0.157	336.56 −31.33	214.2 12.8	PAIR of COMPACTS IC 4906 = 142-G-57 = Fa 64?	13,23 S 3.9	F143/J0917 (334,127)
AM 1952-481	19 52 59 −48 13.4	4.390 0.158	350.95 −30.40	217.7 24.7	COMPACT	13 D 0.2	F233/J1488 (242,274)
AM 1952-834	19 52 59 −83 49.9	13.964 0.158	309.78 −28.92	208.3 −9.7	TIGHT CHAIN of 4	17 D 4.7	F010/J2477 (121,237)
AM 1953-553	19 53 09 −55 31.8	4.787 0.158	342.49 −31.20	215.6 17.7	E + SPIRAL	23 S 1.0	F185/J2370 (137,148)
*AM 1953-260	19 53 27 −26 02.9	3.648 0.159	15.31 −25.05	226.4 45.8	RING GALAXY 526-G-18	6 S 0.4	F526/J0753 (114,123)
AM 1954-553	19 54 11 −55 36.6	4.788 0.160	342.41 −31.35	215.7 17.6	GROUP of 3 S0's 185-G-37	24 D 2.5	F185/J2370 (130,144)
AM 1954-524	19 54 42 −52 46.7	4.616 0.160	345.72 −31.21	216.6 20.3	CHAIN of 7 E's IC 4915 = 185-G-38	17 D 9.4	F185/J2370 (122,296)
AM 1955-522	19 55 03 −52 24.8	4.595 0.161	346.15 −31.23	216.8 20.6	PAIR in GROUP of 5 (+ COMPACT) IC 4917 = 233-G-15	13,18,23 S 0.9	F233/J1488 (220,49)
*AM 1955-570	19 55 06 −57 07.7	4.885 0.161	340.64 −31.56	215.5 16.1	M51-TYPE 185-IG-40 = B 1955-571	9 S 0.7	F142/J0906 (52,328)
AM 1955-832	19 55 25 −83 28.7	13.310 0.161	310.15 −29.07	208.5 −9.4	CHAIN of 4	17 D 2.4	F011/J2562 (328,227)
AM 1956-553	19 56 09 −55 31.1	4.774 0.162	342.53 −31.63	216.1 17.6	SPIRALS + I/A COMP'N. IC 4919 = 185-G-43	1 S 1.7	F185/J2370 (115,148)
AM 1956-392	19 56 49 −39 23.2	4.030 0.163	1.12 −29.47	221.4 33.0	STAR w. NEBULOSITY	21 D 0.3	F339/J0809 (167,210)
AM 1956-405	19 56 57 −40 51.9	4.082 0.163	359.47 −29.81	220.9 31.6	EDGE-ON SPIRAL (BOX NUCLEUS) 339-G-16	12 D 1.1	F339/J0809 (166,131)
AM 1956-583	19 56 57 −58 39.9	4.988 0.163	338.84 −31.86	215.3 14.5	CLOSE PAIR incl. I/A DBL.	2,23 S 6.5	F143/J0917 (312,233)
AM 1957-345	19 57 22 −34 55.0	3.886 0.164	6.12 −28.51	223.4 37.2	PEC. SPIRAL (CONNECTED COMP'N.?) 399-G-14 = UKS 1957-349	8,12 D 1.8	F399/J0734 (204,183)
*AM 1957-394	19 57 23 −39 47.9	4.043 0.164	0.69 −29.67	221.4 32.5	RING w. OFF-CENTRE NUCLEUS 339-G-20	6 D 0.9	F339/J0809 (162,188)
*AM 1957-471	19 57 23 −47 12.4	4.331 0.164	352.24 −30.99	218.8 25.5	I/A QUARTET (KLEM 30 = DN 1957-472) NGC 6845 = Se135/3 = 284-IG-8 [B]	4 D 2.7	F233/J1488 (203,329)
AM 1957-384	19 57 36 −38 43.3	4.006 0.164	1.91 −29.47	221.9 33.6	CLUSTER of GALAXIES (MIXED TYPES)	18 D 7.8	F339/J0809 (159,246)
AM 1957-420	19 57 39 −42 06.2	4.125 0.164	358.10 −30.19	220.6 30.3	GAL.w. I/A COMP'N. + SHARP NUCLEUS	1 D 0.9	F339/J0809 (160,64)
AM 1958-383	19 58 19 −38 33.5	3.999 0.165	2.13 −29.57	222.1 33.7	SMALL HI.S.BR. SPIRAL. w. KNOTS	12 D 0.7	F339/J0809 (151,254)
AM 1958-374	19 58 44 −37 43.4	3.971 0.165	3.08 −29.46	222.5 34.4	BRIGHT-RIMMED SP. + APP. COMP'N. 339-G-26	8,10 D 1.3	F399/J0734 (190,32)
AM 1958-561	19 58 47 −56 13.8	4.809 0.165	341.72 −32.03	216.2 16.8	SPIRAL (BULGE w. ABS'N. + COMP'N.) NGC 6848 = 185-G-52 (in GROUP)	8,12,18 S 3.6	F186/J1591 (345,106)
*AM 1959-560	19 59 27 −56 05.4	4.797 0.166	341.89 −32.12	216.4 16.9	E (BOX NUCLEUS) + 7 COMP'NS. 185-G-54	8,16,23 S 0.4	F186/J1591 (340,114)
AM 1959-550 [1959-5459]	19 59 32 −55 07.4	4.735 0.166	343.03 −32.09	216.7 17.8	PAIR + COMP'NS. (#1:SB(r)) IC 4933 = 185-G-55 = Fa 338 [Se]	8,23 S 7.2	F186/J1591 (344,166)
AM 1959-545 [1959-5507]	19 59 33 −54 59.4	4.727 0.166	343.19 −32.08	216.7 18.0	PAIR + COMP'NS. (#2: IRR.) NGC 6850 = 185-G-56 (Se 138/8)	8,23 S 7.2	F186/J1591 (344,173)
AM 1959-565	19 59 52 −56 50.8	4.845 0.167	341.00 −32.20	216.2 16.2	CLOSE TRIPLE	24 D 2.1	F185/J2370 (89,76)
AM 2000-602	20 00 22 −60 26.2	5.111 0.167	336.75 −32.30	215.2 12.7	COMPACT	13 D 0.2	F142/J0906 (29,148)
AM 2000-574 [2000-5749]	20 00 28 −57 44.5	4.904 0.168	339.94 −32.30	216.0 15.3	CLOSE PAIR (#1: SPIRAL) IC 4935 = 143-G-3	23 S 5.4	F186/J1591 (326,26)
AM 2000-600	20 00 37 −60 09.4	5.087 0.168	337.08 −32.33	215.3 13.0	I/A DBL. + I/A DBL.	2,2 S 1.3	F143/J0917 (282,155)
AM 2000-574 [2000-5744]	20 00 42 −57 49.6	4.909 0.168	339.84 −32.34	216.0 15.2	CLOSE PAIR (#2: ONE-ARMED SPIRAL) 143-G-4 = Fa 65 = B 2000-574	10,23 S 5.4	F186/J1591 (324,22)
AM 2000-263 [2001-2634]	20 00 52 −26 36.6	3.651 0.168	15.34 −26.81	228.3 44.6	GROUP of 3 (incl. I/A DBL.) (#1) MCG-4-47-6 = 527-G-5	2,242 D 0.5	F526/J0753 (24,91)
AM 2000-601	20 00 57 −60 17.4	5.096 0.168	336.93 −32.37	215.3 12.8	I/A TRIPLE w. BRIDGE 143-IG-5 = B 2000-603	3 D 1.3	F142/J0906 (24,155)

Identification	R.A. (1950.0) Declination	Precession R.A./Dec.	Gal. Long. Gal. Lat.	S.G.L. S.G.B.	Description Cross-Identification	Coding Size (Min.)	Field/Plate (x, y)
AM 2000-571	20 00 58 −57 10.6	4.862 0.168	340.61 −32.36	216.2 15.8	COMPACT	13 D 0.3	F143/J0917 (289,315)
AM 2001-263 [2000-2636]	20 01 10 −26 31.9	3.649 0.168	15.44 −26.84	228.4 44.7	GROUP of 3 (incl. I/A DBL.) (#2) MCG-4-47-7 = 527-G-6	24 D 0.5	F526/J0753 (21,95)
AM 2001-391	20 01 17 −39 12.5	4.014 0.169	1.54 −30.28	222.5 32.9	E + APP. COMP'N.	8 S 0.3	F339/J0809 (121,219)
AM 2001-263 [2001-2631]	20 01 30 −26 34.4	3.650 0.169	15.43 −26.92	228.5 44.6	GROUP of 3 (incl. I/A DBL.) (#3) 527-IG-7	2,13,24 D 0.4	F526/J0753 (17,93)
AM 2001-613	20 01 32 −61 34.5	5.202 0.169	335.40 −32.42	215.0 11.6	DISTURBED GALAXY (PARALLEL JETS) IC 4936 = 143-G-6	7,16 D 2.0	F143/J0917 (271,80)
AM 2001-543	20 01 44 −54 31.3	4.689 0.169	343.76 −32.37	217.2 18.3	E + APP. COMP'N. NGC 6854 = 185-G-61	8 S 2.1	F185/J2370 (69,200)
*AM 2001-602	20 01 59 −60 21.4	5.096 0.170	336.84 −32.50	215.4 12.7	SPIRAL w. BRIGHT ARCS + RING IC 4938 = 143-G-7 = Fa 66	6,12 D 2.7	F143/J0917 (272,145)
*AM 2002-480	20 02 12 −48 07.4	4.356 0.170	351.30 −31.91	219.4 24.4	GALAXY w. LINE of BRIGHT KNOTS NGC 6851b = 233-G-25	12 S 1.2	F233/J1488 (160,280)
AM 2002-565	20 02 18 −56 57.5	4.841 0.170	340.88 −32.54	216.5 16.0	DWARF	20 D 0.5	F186/J1591 (316,69)
AM 2002-524	20 02 37 −52 43.4	4.583 0.170	345.89 −32.40	217.9 20.0	CHAIN of 4 IC 4941/42/44 = 185-G-65/66/67	17 D 10.8	F185/J2370 (57,296)
*AM 2002-563	20 02 51 −56 32.0	4.811 0.171	341.38 −32.60	216.7 16.3	GALAXY w. PARTIAL RING + COMP'N. NGC 6855 = 185-G-63	6,8 S 2.5	F186/J1591 (313,92)
AM 2002-515	20 02 54 −51 52.5	4.536 0.171	346.90 −32.39	218.2 20.8	I/A DBL. 233-G-29/30	2 S 0.5	F185/J2370 (52,342)
AM 2003-415	20 03 17 −41 52.9	4.102 0.171	358.57 −31.17	221.8 30.2	CHAIN of 9	17 S 67.2	F339/J0809 (104,75)
AM 2003-344	20 03 44 −34 41.7	3.866 0.172	6.74 −29.72	225.0 36.9	I/A DBL. (SMALL) 399-IG-20	2 S 0.3	F399/J0734 (134,194)
*AM 2004-611	20 04 36 −61 14.9	5.157 0.173	335.78 −32.80	215.4 11.8	V.LO.S.BR. DWARF + HI.S.BR.GALAXY NGC 6860 = 143-G-9	20,23 S 3.8	F143/J0917 (253,98)
*AM 2004-295	20 04 41 −29 57.5	3.732 0.173	12.03 −28.63	227.5 41.2	SPIRAL w. (3?) BRANCHED ARMS MCG-5-47-23 = 461-G-43	10,11 D 1.0	F461/J0744 (75,178)
AM 2004-662	20 04 42 −66 21.7	5.685 0.173	329.72 −32.47	213.9 6.9	ELONGATED GALAXY + I/A COMP'N 105-G-26 = B 2004-663 (SHREAD?)	1,16,22 D 1.1	F106/J0902 (323,97)
AM 2005-543	20 05 10 −54 31.9	4.675 0.173	343.77 −32.87	217.7 18.1	CHAIN of 4 (incl.I/A DBL.)	2,17 D 2.8	F185/J2370 (43,198)
*AM 2005-282	20 05 33 −28 25.0	3.690 0.174	13.77 −28.36	228.6 42.5	RING 461-G-45	6 D 1.0	F462/J0682 (335,260)
AM 2005-555	20 05 35 −55 50.4	4.754 0.174	342.22 −32.97	217.3 16.9	CLOSE SP. in WIDE GROUP of 9 IC 4957? = 186-G-3 = DN 2005-545	18,23 S 2.9	F186/J1591 (295,131)
AM 2005-424	20 05 49 −42 45.8	4.127 0.174	357.65 −31.79	222.0 29.2	SPIRAL w. CONNECTED COMP'N.	1 S 0.7	F339/J0809 (80,27)
AM 2005-492 [2006-4928]	20 05 55 −49 22.1	4.401 0.174	349.91 −32.65	219.5 23.0	PAIR (#1) 233-G-36 = Fa 531	23 S 5.6	F233/J1488 (128,213)
AM 2006-492 [2005-4922]	20 06 00 −49 28.7	4.406 0.175	349.78 −32.68	219.5 22.9	PAIR (#2) 233-G-37 = Fa 532	23 S 5.6	F233/J1488 (127,207)
AM 2006-483	20 06 16 −48 31.6	4.361 0.175	350.91 −32.63	219.9 23.8	E w. APP. COMP'N in GROUP of 9 NGC 6868 (IC 4949, NGC 6861/70)	8,18 S 1.3	F233/J1488 (123,258)
AM 2006-740	20 06 18 −74 09.9	7.086 0.175	320.58 −31.52	211.7 −0.6	GALAXY w. FAINT JET	7 D 0.4	F047/J1797 (353,206)
AM 2006-545	20 06 34 −54 56.2	4.694 0.175	343.30 −33.09	217.7 17.7	DISTURBED SPIRAL + APP. COMP'NS. NGC 6867 = 186-G-6	8,16 S 1.9	F185/J2370 (33,175)
AM 2006-502	20 06 47 −50 26.2	4.449 0.175	348.66 −32.88	219.3 21.9	COMPACT	13 D 0.1	F233/J1488 (121,155)
*AM 2006-564	20 06 51 −56 49.5	4.811 0.176	341.04 −33.16	217.1 15.9	RINGED E at CENTRE of CLUSTER PKS 2006-56	6,8,19 D 4.9	F186/J1591 (283,78)
AM 2008-562	20 08 04 −56 22.8	4.777 0.177	341.58 −33.32	217.4 16.3	I/A DBL. 186-IG-9	2 S 0.5	F185/J2370 (28,97)
*AM 2008-565	20 08 27 −56 58.5	4.814 0.178	340.86 −33.37	217.3 15.7	E w. 9 COMP'NS. (ELONGATED) IC 4965 = 186-G-11 = DN 2008-569	8 D 5.4	F186/J1591 (271,71)
AM 2008-543	20 08 29 −54 30.6	4.660 0.178	343.82 −33.35	218.1 18.0	SB(r) + FAINT OUTER RING 186-G-12 = Fa 340	6 D 1.3	F186/J1591 (277,203)
AM 2008-290	20 08 38 −29 00.8	3.700 0.178	13.35 −29.18	229.1 41.7	CHAIN of 5 (I/A DBL. at CENTRE) 462-G-2	2,17 S 0.9	F462/J0682 (298,229)
AM 2008-314 [2008-3142]	20 08 55 −31 40.7	3.770 0.178	10.42 −29.98	227.7 39.3	CLOSE PAIR (S0 + Sd) #1 462-G-3	23 D 0.7	F462/J0682 (292,86)
AM 2008-314 [2008-3140]	20 08 55 −31 42.8	3.771 0.178	10.39 −29.99	227.7 39.2	CLOSE PAIR (S0 + Sd) #2	23 D 0.5	F462/J0682 (292,84)
AM 2009-461	20 09 40 −46 18.7	4.254 0.179	353.60 −32.96	221.3 25.7	S0 + CONN. COMP'NS (+ I/A DBL.) NGC 6875 = 284-G-28 = B 2009-463	1,2,23 S 1.3	F284/J5235 (127,109)
AM 2010-371	20 10 10 −37 20.4	3.932 0.180	4.12 −31.58	225.2 34.0	'E' w. FAINT SPIRAL ARMS 399-G-25	12 D 3.0	F339/J0809 (25,318)

Identification	R.A. (1950.0) Declination	Precession R.A./Dec.	Gal. Long. Gal. Lat.	S.G.L. S.G.B.	Description Cross-Identification	Coding Size (Min.)	Field/Plate (x, y)
AM 2010-581	20 10 18 −58 17.4	4.895 0.180	339.28 −33.60	217.1 14.4	ISOLATED PECULIAR GALAXY	12,16 D 0.7	F143/J0917 (220,258)
AM 2010-215	20 10 26 −21 57.7	3.527 0.180	21.07 −27.29	234.1 47.8	I/A TRIPLE near LARGE GALAXY 596-IG-5	3 D 0.8	F527/J0796 (172,339)
AM 2010-665	20 10 54 −66 58.5	5.718 0.181	328.91 −33.01	214.3 6.1	ELONGATED COMPACT	13 D 0.2	F106/J0902 (287,67)
AM 2010-570	20 10 56 −57 05.9	4.810 0.181	340.71 −33.71	217.6 15.5	CLOSE PAIR	23 S 0.9	F143/J0917 (217,322)
AM 2011-564	20 11 05 −56 46.0	4.788 0.181	341.11 −33.74	217.7 15.8	VERY SMALL I/A TRIPLE 186-IG-20	3 D 0.6	F186/J1591 (252,83)
AM 2011-620	20 11 10 −62 00.8	5.187 0.181	334.81 −33.53	215.9 10.8	I/A DBL. in CHAIN of 4 IC 4947 = 143-IG-16 = Fa 68/69/70	2,17 D 3.6	F143/J0917 (209,58)
AM 2011-580	20 11 27 −58 04.5	4.874 0.181	339.54 −33.76	217.3 14.5	ISOLATED GAL. w. PECULIAR ABS'N. IC 4980 = 143-G-18	14 D 1.8	F143/J0917 (212,270)
*AM 2011-740	20 11 37 −74 02.6	6.996 0.181	320.64 −31.90	212.0 −0.6	SPIRAL w. 3 APP. COMP'NS IC 4964 = 46-G-8	8 S 1.3	F046/J2588 (92,227)
*AM 2011-705	20 11 42 −70 55.1	6.315 0.182	324.26 −32.50	213.1 2.4	INTEGRAL-SIGN SPIRAL + I/A COMP'N NGC 6872 + IC 4970 = 73-IG-32/33 [B]	1,12,17 D 7.1	F073/J0916 (107,125)
AM 2012-511	20 12 03 −51 10.5	4.465 0.182	347.84 −33.76	219.9 21.0	I/A DBL. B 2012-511	2 S 0.8	F233/J1488 (78,114)
*AM 2012-282	20 12 09 −28 27.6	3.680 0.182	14.21 −29.75	230.4 41.9	DISK w. OFF-SET NUCLEUS (in GROUP of 3)	6,12,13,24 D 0.5	F462/J0682 (257,259)
*AM 2012-521	20 12 18 −52 14.4	4.518 0.182	346.56 −33.85	219.5 20.0	DWARF SPIRAL + Sd COMP'N. IC 4983 = 233-G-48	8,20 S 1.3	F234/J2363 (322,53)
AM 2012-475	20 12 43 −47 55.3	4.311 0.183	351.75 −33.64	221.2 24.0	CONTACT DBL. in CHAIN of 4	2,17 S 0.2	F233/J1488 (65,289)
*AM 2013-470 [2013-4701]	20 13 02 −47 07.3	4.276 0.183	352.71 −33.62	221.6 24.7	PAIR (#1: E w. ABS'N.) 284-G-44 = B 2013-471	14,23 S 4.9	F233/J1488 (60,331)
AM 2013-710	20 13 15 −71 01.1	6.319 0.183	324.12 −32.60	213.2 2.2	E + I/A COMP'N. (in a GROUP) NGC 6880 + IC 4981 = 73-IG-37/38	1,8,18 S 1.6	F073/J0916 (101,120)
*AM 2013-470 [2013-4707]	20 13 17 −47 01.8	4.271 0.183	352.83 −33.65	221.7 24.8	PAIR (#2: RING)	14,23 S 4.9	F233/J1488 (58,336)
AM 2013-525	20 13 30 −52 57.0	4.551 0.184	345.71 −34.05	219.4 19.2	SP. w. STRAIGHT ARM to COMP'NS. NGC 6887 = 186-G-27	1,8 S 2.2	F234/J2363 (310,15)
AM 2013-522	20 13 33 −52 25.8	4.523 0.184	346.34 −34.05	219.6 19.7	SPIRAL w. COMPACT COMP'N. IC 4987 = 233-G-52	8,13 S 0.9	F233/J1488 (68,46)
AM 2013-370	20 13 36 −37 08.5	3.918 0.184	4.49 −32.22	226.1 33.9	SPIRAL + PECULIAR ARM 400-G-5	10 D 1.8	F340/J0804 (264,331)
AM 2014-710	20 14 15 −71 01.0	6.309 0.185	324.10 −32.68	213.2 2.2	GAL. w. I/A COMP'N. in DN 2014-711 NGC 6876/7 = Se 137/2 = 73-IG-35/6	1,8 S 1.1	F073/J0916 (96,119)
AM 2014-321	20 14 21 −32 12.9	3.774 0.185	10.17 −31.24	228.8 38.3	CLOSE GROUP of 5	18 D 4.0	F462/J0682 (229,58)
AM 2014-453 [2014-4540]	20 14 32 −45 38.9	4.211 0.185	354.51 −33.73	222.5 26.0	PAIR (#1) in GROUP ? 284-G-50	23 S 1.2	F285/J1581 (343,140)
AM 2014-454 [2014-4538]	20 14 32 −45 40.2	4.212 0.185	354.48 −33.73	222.5 26.0	PAIR (#2) in GROUP ? 284-G-51	23 S 1.2	F285/J1581 (343,139)
AM 2015-525	20 15 00 −52 50.8	4.539 0.186	345.85 −34.28	219.7 19.2	CHAIN of 7 DN 2015-529	17 D 9.0	F186/J1591 (228,294)
AM 2015-411	20 15 01 −41 12.6	4.046 0.186	359.79 −33.22	224.5 30.0	D GALAXY + PECULIAR COMP'NS. IC 4991 = 340-G-11	8 D 1.6	F340/J0804 (246,113)
*AM 2015-273	20 15 19 −27 36.5	3.653 0.186	15.39 −30.17	231.7 42.3	EDGE-ON w. JET 462-G-7	7 D 1.3	F462/J0682 (220,305)
AM 2015-451	20 15 27 −45 14.1	4.192 0.186	355.02 −33.84	222.8 26.3	EDGE-ON w. COMP'N. 284-G-55	1 S 0.5	F285/J1581 (336,163)
AM 2016-660	20 16 19 −66 04.7	5.567 0.187	329.89 −33.67	215.1 6.8	SMALL SPIRAL w. EXTENDED MATERIAL 106-G-2	12,15 D 1.1	F106/J0902 (262,117)
AM 2016-392	20 16 21 −39 26.7	3.983 0.187	1.93 −33.18	225.6 31.6	DISTURBED SP. + I/A DBL. + COMP'NS Ag-62 = 340-IG-17	2,8,12 D 2.7	F340/J0804 (233,208)
AM 2016-330	20 16 26 −33 00.3	3.791 0.187	9.40 −31.86	228.9 37.4	I/A DBL. (E + PECULIAR) 400-IG-7	2 S 0.4	F462/J0682 (206,16)
AM 2016-670	20 16 42 −67 03.2	5.684 0.188	328.72 −33.56	214.8 5.9	SPIRAL w. PLUME IC 4990 = 105-G-3	12,15 D 0.9	F106/J0902 (257,65)
AM 2017-491	20 17 06 −49 17.0	4.356 0.188	350.17 −34.46	221.4 22.4	DWARF 234-G-4	20 D 1.8	F234/J2363 (289,214)
AM 2017-525	20 17 06 −52 50.6	4.530 0.188	345.86 −34.59	220.0 19.1	I/A CONTACT DBL. in COMMON HALO 186-IG-35	2 S 0.2	F234/J2363 (281,22)
*AM 2017-482	20 17 13 −48 24.0	4.316 0.188	351.25 −34.43	221.8 23.2	SPIRAL w. PECULIAR ABS'N. + HALO NGC 6893 = 234-G-6 = Fa 537	12,14,16 D 2.9	F234/J2363 (290,261)
AM 2017-651	20 17 41 −65 14.0	5.462 0.189	330.87 −33.92	215.5 7.5	COMPACT (SOME STRUCTURE)	13 D 0.2	F105/J0891 (10,155)
*AM 2018-495	20 18 11 −49 51.7	4.378 0.189	349.49 −34.67	221.4 21.8	DISTURBED E w. PECULIAR ABS'N. (in CHAIN of 6)	14,17 D 1.8	F234/J2363 (279,183)

Identification	R.A. (1950.0) Declination	Precession R.A./Dec.	Gal. Long. Gal. Lat.	S.G.L. S.G.B.	Description Cross-Identification	Coding Size (Min.)	Field/Plate (x, y)
AM 2018-481	20 18 14 −48 10.1	4.302 0.189	351.54 −34.58	222.1 23.4	CLOSE TRIPLE 234-G-9/11 = Fa 538/9	24 D 3.6	F234/J2363 (282,274)
AM 2018-472	20 18 39 −47 23.1	4.267 0.190	352.50 −34.60	222.5 24.1	SPIRAL w. APP. COMP'N.	8 S 1.6	F234/J2363 (279,316)
AM 2018-613	20 18 39 −61 37.3	5.107 0.190	335.20 −34.45	216.9 10.9	SMALL Sd w. FILAMENT to 'STAR' 143-G-22	12,15,20 D 1.1	F143/J0917 (162,80)
AM 2018-463	20 18 56 −46 33.8	4.233 0.190	353.49 −34.58	222.9 24.8	3-ARMED SPIRAL (HI.S.BR.)	11 D 0.4	F285/J1581 (300,93)
AM 2019-423 [2019-4235]	20 19 02 −42 32.9	4.080 0.190	358.33 −34.16	224.7 28.5	I/A DBL. (#1)	2 S 3.4	F285/J1581 (307,308)
AM 2019-465	20 19 08 −46 59.5	4.249 0.191	352.98 −34.65	222.7 24.4	DWARF Se 136/9 = 285-G-3	20 D 0.5	F285/J1581 (297,70)
AM 2019-300	20 19 09 −30 00.9	3.707 0.191	12.97 −31.66	231.2 39.8	SPIRAL w. 6 APP. COMP'NS. MCG-5-48-4 = 462-G-10	8 S 5.8	F462/J0682 (175,176)
AM 2019-423 [2019-4232]	20 19 18 −42 35.7	4.081 0.191	358.29 −34.21	224.7 28.5	I/A DBL. (#2)	2 S 3.4	F285/J1581 (305,306)
*AM 2019-442	20 19 34 −44 25.7	4.147 0.191	356.09 −34.48	223.9 26.8	COMPACT NUCLEUS w. LO.S.BR.MATR'L NGC 6902a = Se 136/4 = 285-G-4 [B]	13,15 D 0.9	F285/J1581 (298,208)
AM 2019-623	20 19 55 −62 30.0	5.176 0.191	334.12 −34.50	216.7 10.0	DWARF w. RESOLUTION to 6 STARS? 142-G-26	20 D 1.3	F143/J0917 (155,33)
AM 2019-314	20 19 56 −31 40.6	3.748 0.192	11.13 −32.25	230.5 38.3	3 in a LINE MCG-5-48-7 = 462-G-14	24 D 9.9	F462/J0682 (166,87)
*AM 2020-504	20 20 13 −50 48.7	4.414 0.192	348.34 −35.03	221.3 20.8	E + RING on MI.AX. (CHAIN of 5) Fa 342? = 234-G-16?	6,17,22 D 0.6	F234/J2363 (260,133)
*AM 2020-483	20 20 15 −48 31.2	4.310 0.192	351.14 −34.94	222.3 22.9	SPIRAL w. I/A COMP'N. 234-G-19	1 S 1.6	F234/J2363 (263,256)
AM 2020-254	20 20 22 −25 44.9	3.600 0.192	17.84 −30.68	234.3 43.4	I/A DBL. (SMALL)	2 S 0.3	F527/J0796 (52,135)
AM 2020-370 [2020-3705]	20 20 22 −37 02.8	3.898 0.192	4.92 −33.52	227.6 33.4	CHAIN of 3 + (#1: EDGE-ON) [PKS?] 400-G-11 = KLEM 31 = DN 2021-371	8,24 S 154.6	F400/J0627 (215,68)
AM 2020-370 [2020-3702]	20 20 29 −37 04.8	3.899 0.192	4.89 −33.55	227.6 33.4	CHAIN of 3 + (#2: DISTURBED)[PKS?] 400-G-12 = KLEM 31 = DN 2021-371	8,16,24 S 154.6	F400/J0627 (214,66)
AM 2020-841	20 20 35 −84 13.4	13.870 0.192	309.12 −29.50	208.9 −10.3	GALAXY w. TAILS + APP.COMP'N.	8,15 S 0.2	F010/J2477 (89,208)
AM 2020-513	20 20 39 −51 38.5	4.453 0.192	347.33 −35.12	221.0 20.0	HI.S.BR. IRR.	16 D 0.5	F234/J2363 (254,88)
AM 2020-370 [2020-3704]	20 20 47 −37 05.8	3.899 0.193	4.88 −33.61	227.7 33.3	CHAIN of 3 + (#3: DUST?) [PKS?] 400-G-13 = KLEM 31 = DN 2021-371	8,14,24 S 154.6	F400/J0627 (211,65)
AM 2020-261 [2022-2559]	20 20 54 −26 10.7	3.609 0.193	17.41 −30.93	234.1 43.0	WIDE PAIR (PEC.ARMS) (#1: SPIRAL) IC 4999 = 527-G-21	12,23 D 1.6	F528/J0649 (313,115)
*AM 2021-434	20 21 02 −43 49.0	4.120 0.193	356.86 −34.68	224.5 27.2	MULTI-ARMED SPIRAL + APP. COMP'N. NGC 6902 = Se 136/2 = 285-G-8	8,10 S 4.9	F285/J1581 (285,241)
AM 2021-233	20 21 11 −23 37.6	3.550 0.193	20.24 −30.19	236.0 45.2	I/A DBL. (SPIRAL + S0) MCG-4-48-5 = 528-G-1	2 S 1.0	F528/J0649 (312,252)
*AM 2021-724	20 21 16 −72 46.7	6.590 0.193	321.92 −32.85	213.1 0.4	RING GALAXY. w. OFF-SET NUCLEUS 46-IG-10 + APP. COMP'N.	6,8 D 2.7	F073/J0916 (76,22)
AM 2021-231	20 21 43 −23 16.5	3.541 0.194	20.67 −30.19	236.4 45.4	I/A SPIRALS (ONE V.LONG ARM) 528-IG-2	2,10 S 0.5	F528/J0649 (306,271)
AM 2021-410	20 21 50 −41 04.0	4.021 0.194	0.20 −34.47	225.9 29.6	SPIRAL w. PECULIAR ARM	10 D 0.8	F340/J0804 (177,121)
AM 2021-842	20 21 50 −84 22.3	14.114 0.194	308.94 −29.48	208.9 −10.4	SMALL I/A DBL. in CHAIN of 5	2,17 D 4.9	F011/J2562 (275,197)
AM 2021-274	20 21 58 −27 44.4	3.645 0.194	15.74 −31.62	233.4 41.5	GROUP of 5 (CLOSE)	18 D 1.8	F462/J0682 (141,298)
AM 2022-311	20 22 11 −31 15.9	3.733 0.194	11.75 −32.61	231.2 38.4	CLOSE PAIR (SPIRALS)	23 S 1.3	F462/J0682 (140,109)
AM 2022-255 [2020-2610]	20 22 20 −25 59.6	3.603 0.194	17.73 −31.18	234.6 43.0	WIDE PAIR (PEC.ARMS)(#2: SPIRAL) IC 5005 = 528-G-4	12,23 D 1.4	F528/J0649 (296,125)
AM 2022-334 [2022-3342]	20 22 30 −33 42.6	3.798 0.195	8.94 −33.25	229.9 36.2	PAIR (#1: INTEGRAL SIGN) 400-G-17	2,12 S 56.0	F400/J0627 (192,247)
AM 2022-545 [2022-5457]	20 22 31 −54 56.2	4.620 0.195	343.29 −35.38	220.0 16.9	PAIR (#1: DISTURBED SPIRAL) IC 5001 = 186-G-42	16,23 S 3.6	F186/J1591 (167,183)
AM 2022-334 [2022-3342]	20 22 40 −33 42.6	3.797 0.195	8.95 −33.29	229.9 36.2	PAIR (#2: DISTURBED S0/E)	2,16 S 56.0	F400/J0627 (191,247)
AM 2022-545 [2022-5456]	20 22 51 −54 57.8	4.620 0.195	343.26 −35.42	220.0 16.8	PAIR (#2: HI.S.BR. NUCLEUS) IC 5002 = 186-G-44	23 S 3.6	F186/J1591 (165,181)
AM 2023-490	20 23 18 −49 03.4	4.322 0.196	350.51 −35.47	222.5 22.3	DWARF 234-G-27	20 D 1.3	F234/J2363 (235,227)
AM 2023-231	20 23 22 −23 16.9	3.539 0.196	20.81 −30.55	236.9 45.2	GALAXY w. PECULIAR ABS'N. MCG-4-48-8 = 528-G-5	14 D 1.1	F528/J0649 (286,271)
*AM 2023-325	20 23 22 −32 53.7	3.773 0.196	9.94 −33.25	230.6 36.8	CLOSE PAIR (ONE w. JET)	7,23 S 2.5	F462/J0682 (127,21)

Identification	R.A. (1950.0) Declination	Precession R.A./Dec.	Gal. Long. Gal. Lat.	S.G.L. S.G.B.	Description Cross-Identification	Coding Size (Min.)	Field/Plate (x, y)
AM 2023-420	20 23 27 −42 02.9	4.049 0.196	359.06 −34.91	225.8 28.6	I/A DBL. B 2023-420	2 S 0.4	F285/J1581 (264,337)
AM 2023-363	20 23 54 −36 30.5	3.873 0.196	5.72 −34.11	228.7 33.6	DWARF	20 D 24.6	F400/J0627 (178,97)
AM 2024-523	20 24 01 −52 33.1	4.484 0.196	346.22 −35.64	221.2 19.0	SP. + I/A DBL. COMP'N + COMPACT 186-G-47 = Fa 343	1,13 S 3.1	F234/J2363 (226,40)
AM 2024-544	20 24 04 −54 47.7	4.605 0.196	343.46 −35.60	220.2 16.9	RING + RING, CONN. to CONDENS. 186-G-48 = Fa 540	6,6 S 0.5	F186/J1591 (155,190)
AM 2024-305	20 24 05 −30 59.1	3.722 0.196	12.19 −32.93	231.9 38.5	GALAXY w. TAIL 462-G-21	15 D 1.3	F462/J0682 (118,124)
*AM 2024-471	20 24 06 −47 11.6	4.239 0.196	352.81 −35.50	223.5 23.9	LARGE E w. DIAMETRIC COMP'NS. NGC 6909 = 285-G-12	8 S 2.7	F285/J1581 (251,61)
AM 2024-620	20 24 09 −62 08.4	5.117 0.197	334.48 −35.03	217.3 10.2	COMPACT	13 D 0.1	F106/J0902 (225,331)
AM 2024-515	20 24 25 −51 51.6	4.448 0.197	347.07 −35.70	221.5 19.6	WIDE PAIR of I/A SPIRALS 234-G-32	2 S 6.7	F234/J2363 (223,77)
AM 2024-620	20 24 26 −62 07.5	5.114 0.197	334.50 −35.07	217.4 10.2	IRR.	16 D 0.3	F106/J0902 (223,331)
AM 2024-430	20 24 30 −43 05.2	4.082 0.197	357.83 −35.22	225.5 27.6	I/A DBL. in GROUP of 3 285-IG-15 = B 2024-430	2,24 S 0.4	F285/J1581 (253,281)
*AM 2024-444	20 24 30 −44 46.6	4.144 0.197	355.77 −35.39	224.7 26.1	TRIPLE NUCLEUS w.TAIL 285-IG-14 = B 2024-447	15,16 D 0.9	F285/J1581 (251,190)
AM 2024-525	20 24 31 −52 51.0	4.497 0.197	345.85 −35.71	221.1 18.7	GAL. w. TAIL + 3 COMP'NS. (2 I/A) Se 138/3 = 186-IG-51 = B02024-528	1,15 D 1.8	F234/J2363 (221,24)
AM 2024-612	20 24 35 −61 28.7	5.058 0.197	335.28 −35.16	217.6 10.7	DISTURBED GALAXY w. LOOPS	15,16 D 0.7	F143/J0917 (124,87)
AM 2024-315	20 24 36 −31 50.5	3.743 0.197	11.23 −33.25	231.5 37.7	CLOSE PAIR	23 S 0.8	F462/J0682 (113,78)
AM 2024-355	20 24 56 −35 58.8	3.855 0.197	6.39 −34.22	229.2 34.0	PAIR (#1: IRR.)	16,23 D 39.2	F400/J0627 (166,125)
AM 2024-473	20 24 59 −47 39.0	4.255 0.197	352.26 −35.68	223.4 23.4	GALAXY w. FILAMENTS 234-G-34	15 D 0.6	F234/J2363 (222,303)
AM 2025-360	20 25 05 −36 00.9	3.856 0.198	6.36 −34.25	229.2 33.9	PAIR (#2: SPIRAL) 400-G-27	23 D 25.8	F400/J0627 (165,123)
AM 2025-484	20 25 12 −48 40.2	4.297 0.198	351.00 −35.76	223.0 22.5	LONG CHAIN of 9 234-G-35	17 D 17.9	F234/J2363 (219,248)
AM 2025-494	20 25 13 −49 48.6	4.348 0.198	349.60 −35.80	222.5 21.4	PECULIAR DWARF (NO NUCLEUS) 234-G-36	20 D 1.1	F234/J2363 (218,187)
AM 2025-361	20 25 20 −36 11.7	3.861 0.198	6.16 −34.34	229.1 33.7	S0 + CONNECTED COMP'N. SS-25 400-G-29 = IC 5011/13 + 400-G-30	1 S 28.0	F400/J0627 (162,114)
AM 2025-483	20 25 47 −48 39.2	4.294 0.198	351.03 −35.86	223.1 22.5	LO.S.BR. DWARF 234-G-38	20 D 1.1	F234/J2363 (214,249)
AM 2025-370	20 25 56 −37 08.1	3.886 0.199	5.06 −34.63	228.8 32.8	M51-TYPE (INTEGRAL-SIGN) 400-IG-31	9 D 14.0	F400/J0627 (156,63)
AM 2025-315	20 25 59 −31 51.0	3.740 0.199	11.31 −33.54	231.8 37.5	DWARF 462-G-24	20 D 1.1	F462/J0682 (97,77)
*AM 2026-424 ?	20 26 10 −42 40.2	4.062 0.199	358.38 −35.48	226.0 27.9	RING w. COMP'N. in RING [B] 285-G-19 (I/A w. NUCLEUS)	1,6 D 0.5	F285/J1581 (237,304)
AM 2026-515	20 26 12 −51 53.4	4.441 0.199	347.03 −35.98	221.8 19.5	KNOTTY SINGLE GALAXY Se 138/1 = 234-IG-39	16 D 0.5	F234/J2363 (208,76)
*AM 2026-225	20 26 16 −22 50.2	3.525 0.199	21.55 −31.03	238.0 45.2	DBL. M51-TYPE (GROUP of 5) MCG-4-48-9 = 528-G-8	9,18 D 1.8	F528/J0649 (250,295)
AM 2027-443 [2027-4433]	20 27 21 −44 34.5	4.126 0.200	356.07 −35.88	225.3 26.0	PAIR (#1: VERY DISTURBED) 285-G-24	2 S 1.1	F285/J1581 (224,202)
AM 2027-443 [2027-4334]	20 27 22 −44 33.1	4.126 0.200	356.10 −35.88	225.3 26.1	PAIR (#2: ONE-ARMED SPIRAL) 285-G-23	2,10 S 1.1	F285/J1581 (224,203)
AM 2027-333 [2028-3338]	20 27 29 −33 39.6	3.785 0.200	9.27 −34.25	231.1 35.8	PAIR (#1: Sb) IC 5020 = 400-G-35	23 D 72.8	F400/J0627 (137,249)
AM 2027-253	20 27 52 −25 32.1	3.583 0.201	18.68 −32.23	236.4 42.8	I/A DBL. (incl. PECULIAR SPIRAL) MCG-4-48-11	2,12 S 0.1	F528/J0649 (229,150)
AM 2028-333 [2027-3339]	20 28 06 −33 38.8	3.783 0.201	9.32 −34.38	231.2 35.7	PAIR (#2: Sc + COMP'N.) 400-G-37	23 D 50.4	F400/J0627 (130,250)
AM 2028-830	20 28 08 −83 07.0	11.907 0.201	310.22 −30.09	209.5 −9.3	DWARF (3 KNOTS) + OBSCURED SPIRAL 11-G-2	20,23 S 7.2	F011/J2562 (287,264)
AM 2028-592	20 28 15 −59 24.8	4.877 0.201	337.74 −35.85	218.9 12.5	SP. w. 2 HI.S.BR. COMP'NS. 143-G-30	8 S 4.5	F143/J0917 (96,197)
*AM 2028-535	20 28 18 −53 54.7	4.535 0.201	344.52 −36.25	221.2 17.5	SPIRAL w. DISTURBED COMP'NS. [B] 186-IG-59 = Se 138/7 = DN 2028-539	8 S 5.4	F186/J1591 (122,237)
*AM 2028-310	20 28 34 −31 00.1	3.713 0.202	12.46 −33.87	232.9 38.0	SPIRAL + I/A + APP. COMP'NS. NGC 6923 = 462-G-29 [A82]	1,8 S 3.1	F463/J0666 (334,123)
AM 2028-422 [2029-4223]	20 28 51 −42 21.9	4.043 0.202	358.82 −35.94	226.6 27.9	PAIR (#1)	23 S 2.1	F285/J1581 (210,321)

Identification	R.A. (1950.0) Declination	Precession R.A./Dec.	Gal. Long. Gal. Lat.	S.G.L. S.G.B.	Description Cross-Identification	Coding Size (Min.)	Field/Plate (x, y)
AM 2028-662	20 28 54	5.507	329.26	216.2	SMALL I/A DBL.	2	F106/J0902
	−66 23.4	0.202	−34.87	6.0		S 0.7	(193,103)
AM 2029-422	20 29 03	4.043	358.80	226.6	PAIR (#2)	23	F285/J1581
[2028-4221]	−42 23.2	0.202	−35.97	27.9	340-G-41	S 2.1	(208,319)
AM 2029-235	20 29 28	3.543	20.70	238.1	WIDE I/A DBL. (#1)	2	F528/J0649
[2029-2352]	−23 51.6	0.203	−32.06	44.0		D 0.3	(210,240)
AM 2029-573	20 29 29	4.750	339.87	219.8	CONNECTED I/A DBL. near BR.SPIRAL	2,8	F143/J0917
	−57 39.9	0.203	−36.18	14.0	143-IG-31 = B 2029-576	S 0.4	(83,290)
AM 2029-291	20 29 33	3.667	14.56	234.3	DISTURBED HI.S.BR. GALAXY	16	F463/J0666
	−29 14.8	0.203	−33.63	39.4	(I/A TRIPLE ?)	D 0.4	(325,218)
AM 2029-544	20 29 36	4.572	343.53	221.0	PAIR (#1: LO.S.BR. SPIRAL)	23	F186/J1591
[2029-5441]	−54 42.3	0.203	−36.41	16.7	186-G-60	S 1.8	(113,194)
*AM 2029-282	20 29 37	3.648	15.49	234.8	DISTURBED GALAXY w.TAIL	15,16	F462/J0682
	−28 27.2	0.203	−33.44	40.1	462-IG-34	D 1.8	(51,258)
AM 2029-235	20 29 39	3.543	20.69	238.1	WIDE I/A DBL. (#2)	2	F528/J0649
[2029-2351]	−23 52.7	0.203	−32.10	43.9		D 0.4	(208,239)
AM 2029-270	20 29 40	3.615	17.07	235.8	I/A DBL.	2	F463/J0666
	−27 04.4	0.203	−33.06	41.2		S 0.2	(327,334)
AM 2029-733	20 29 43	6.684	320.75	213.4	LARGE COMPACT	13	F047/J1797
	−73 37.6	0.203	−33.22	−0.6	IC 5014 = 047-IG-02	D 0.4	(271,248)
AM 2029-544	20 29 48	4.571	343.54	221.1	PAIR (#2: MAGELLANIC IRR.)	23	F186/J1591
[2029-5442]	−54 41.8	0.203	−36.44	16.7	IC 5021 = 186-IG-61	S 1.8	(111,195)
AM 2029-662	20 29 51	5.509	329.14	216.2	GALAXY w. JET(?)	7	F106/J0902
	−66 28.3	0.203	−34.94	5.9		D 0.2	(188,99)
AM 2029-493	20 29 56	4.315	349.97	223.4	CLOSE GROUP of 3	24	F234/J2363
	−49 31.5	0.203	−36.56	21.4		D 1.6	(177,203)
AM 2030-272	20 30 03	3.622	16.75	235.7	I/A DBL. incl. LUMPY COMPACT	2,13	F463/J0666
	−27 22.9	0.203	−33.23	40.9		S 0.2	(322,318)
AM 2030-312	20 30 04	3.722	12.00	233.0	I/A DBL. (#1)	2	F462/J0682
[2030-3126]	−31 28.3	0.203	−34.29	37.4	MCG-5-48-20 = 463-G-1	D 0.4	(50,96)
AM 2030-312	20 30 11	3.721	12.03	233.0	I/A DBL. (#2)	2	F462/J0682
[2030-3128]	−31 26.8	0.204	−34.31	37.4		D 0.4	(49,97)
AM 2030-253	20 30 20	3.581	18.76	237.0	E w. COMPACT COMP'N.	8	F528/J0649
	−25 38.7	0.204	−32.79	42.4	NGC 6924 = 528-G-16	S 1.2	(199,145)
AM 2030-244	20 30 32	3.562	19.73	237.7	3 in a LINE (#1)	24	F528/J0649
[2031-2450]	−24 47.9	0.204	−32.58	43.1	MCG-4-48-15 = 528-G-18	D 0.8	(197,190)
AM 2030-442	20 30 35	4.111	356.28	225.9	I/A PAIR of SPIRALS (#1)	2	F285/J1581
[2030-4427]	−44 27.0	0.204	−36.44	25.9	285-G-31	S 2.7	(193,209)
AM 2030-731	20 30 35	6.595	321.13	213.6	DWARF (RESOLUTION and NUCLEUS)	20	F047/J1797
	−73 17.1	0.204	−33.37	−0.4	47-G-04	D 0.9	(270,267)
AM 2030-244	20 30 51	3.561	19.75	237.7	3 in a LINE (#2)	24	F528/J0649
[2030-2447]	−24 48.4	0.204	−32.65	43.0		D 0.4	(193,190)
AM 2030-442	20 30 51	4.111	356.26	225.9	I/A PAIR of SPIRALS (#2)	2	F285/J1581
[2030-4428]	−44 28.5	0.204	−36.49	25.9	285-G-32	S 2.7	(190,207)
*AM 2030-303	20 30 53	3.697	13.14	233.8	SPIRAL w. I/A (DBL.) COMP'N.	1	F462/J0682
	−30 32.6	0.204	−34.24	38.1	463-IG-3	S 0.9	(39,146)
AM 2031-475	20 31 01	4.239	352.08	224.3	I/A QUARTET in GROUP of 9	4,18	F285/J1581
	−47 50.2	0.205	−36.70	22.8	234-IG-48 incl. Fa 546 in GROUP?	D 0.7	(188,27)
AM 2031-301	20 31 04	3.690	13.46	234.0	COMPACT	13	F462/J0682
	−30 17.0	0.205	−34.21	38.3		D 0.2	(37,160)
AM 2031-310	20 31 05	3.708	12.58	233.5	CHAIN of 6(E + HI.S.BR CENTRAL)	2,17	F463/J0666
	−31 01.8	0.205	−34.40	37.7		D 22.4	(305,122)
AM 2031-320	20 31 14	3.736	11.27	232.8	SP. w. DISTURBED OUTER REGIONS	8,12	F463/J0666
	−32 09.1	0.205	−34.69	36.7	NGC 6925 = 463-G-4	D 4.0	(302,62)
*AM 2031-440	20 31 14	4.093	356.84	226.2	PLUMES + MATR'L out of CONDENS.	15	F285/J1581
	−44 00.7	0.205	−36.52	26.2	285-IG-35 = B 2031-440	D 1.1	(187,232)
AM 2031-322	20 31 16	3.744	10.93	232.6	DISTURBED SINGLE GALAXY	16	F462/J0682
	−32 26.4	0.205	−34.76	36.4	463-G-5	D 0.9	(38,44)
AM 2031-245	20 31 17	3.562	19.74	237.8	3 in a LINE (#3)	24	F528/J0649
[2030-2448]	−24 50.9	0.205	−32.76	42.9		D 0.4	(188,187)
AM 2031-264	20 31 20	3.604	17.60	236.5	CLOSE TRIPLE	24	F528/J0649
	−26 43.6	0.205	−33.32	41.3		D 1.8	(187,87)
AM 2031-500	20 31 41	4.330	349.34	223.4	DISTURBED SP. w. I/A COMP'N.	1,12	F234/J2363
	−50 02.0	0.205	−36.84	20.8	234-IG-49 = B 2031-500	S 1.4	(162,175)
AM 2031-305	20 31 44	3.705	12.73	233.7	ONE-ARMED SP. w. CONDENS. in ARM	12	F463/J0666
	−30 56.4	0.205	−34.51	37.7	(in CHAIN of 6)	D 0.7	(297,127)
AM 2031-575	20 31 47	4.753	339.52	220.0	I/A DBL.	2	F144/J0892
	−57 54.9	0.205	−36.46	13.7	143-IG-32	S 0.2	(335,283)
AM 2031-302	20 31 53	3.689	13.45	234.1	HI. and LO.S.BR. PAIR of SP.	23	F462/J0682
	−30 20.0	0.205	−34.40	38.2		S 1.2	(27,157)
AM 2032-253	20 32 02	3.576	19.01	237.5	I/A DBL. SPIRAL (#1)	2	F528/J0649
[2032-2530]	−25 32.5	0.206	−33.13	42.2	528-IG-20	D 0.7	(179,150)

Identification	R.A. (1950.0) Declination	Precession R.A./Dec.	Gal. Long. Gal. Lat.	S.G.L. S.G.B.	Description Cross-Identification	Coding Size (Min.)	Field/Plate (x, y)
AM 2032-253 [2032-2532]	20 32 06 −25 30.9	3.575 0.206	19.04 −33.14	237.5 42.3	I/A DBL. SPIRAL (#2) MCG-4-48-20 = 528-G-21	2 D 1.4	F528/J0649 (178,152)
AM 2032-433	20 32 06 −43 35.3	4.075 0.206	357.38 −36.64	226.6 26.6	CLOSE GROUP of 5 (PECULIARS)	18 D 2.9	F285/J1581 (178,255)
AM 2032-494	20 32 23 −49 46.2	4.316 0.206	349.67 −36.96	223.7 21.0	GALAXY w. COMP'NS. (ONE I/A) 234-G-51 = Fa 547	1 S 3.6	F234/J2363 (156,190)
AM 2032-451	20 32 25 −45 16.7	4.135 0.206	355.28 −36.82	225.8 25.0	SMALL I/A DBL.	2 S 0.5	F285/J1581 (175,164)
AM 2032-533	20 32 29 −53 36.3	4.499 0.206	344.87 −36.88	221.9 17.5	DISTURBED SINGLE + 2 COMP'NS. 186-IG-66 = B 2032-536	8,12 D 0.5	F186/J1591 (88,252)
AM 2032-492	20 32 48 −49 26.2	4.299 0.207	350.08 −37.02	223.9 21.3	EDGE-ON w. APP.COMP'NS. 234-G-53	8 S 7.0	F234/J2363 (152,207)
AM 2033-215	20 33 03 −21 59.9	3.497 0.207	23.09 −32.24	240.5 45.0	SP. w. PECULIAR ARM + I/A COMP'N.	1,10 S 2.0	F528/J0649 (166,341)
AM 2033-620	20 33 05 −62 00.3	5.046 0.207	334.46 −36.09	218.4 9.9	COMPACT DBL.	2 D 0.2	F106/J0902 (169,338)
*AM 2033-512	20 33 21 −51 20.6	4.383 0.207	347.69 −37.10	223.1 19.5	PECULIAR SPIRAL (ONE-ARMED?) 234-G-55 = Fa 347	10,12 D 0.9	F234/J2363 (149,105)
AM 2033-260	20 33 24 −26 07.6	3.587 0.207	18.45 −33.59	237.4 41.6	CLOSE PAIR incl. RING	6,23 S 1.0	F528/J0649 (162,119)
AM 2033-545	20 33 25 −54 50.0	4.560 0.207	343.33 −36.95	221.5 16.3	I/A E's + MATERIAL 186-IG-69 = B 2033-548	2 S 0.7	F186/J1591 (83,186)
*AM 2033-501	20 33 26 −50 16.3	4.334 0.207	349.04 −37.12	223.6 20.5	V.CLOSE LINE of 3 (HI.S.BR.) 234-IG-56 = Fa 348 [A82]	24 D 0.5	F234/J2363 (147,162)
AM 2033-692	20 33 27 −69 21.7	5.854 0.207	325.62 −34.63	215.4 3.1	DWARF (VERY LO.S.BR.) Se 137/1 = 74-G-2	20 D 0.5	F074/J0907 (246,209)
AM 2033-620	20 33 45 −62 04.9	5.048 0.208	334.36 −36.15	218.4 9.8	GALAXY w. MATERIAL + PROJECTIONS 143-G-38	15 D 0.7	F106/J0902 (165,334)
AM 2033-314 [2033-3148]	20 33 48 −31 48.4	3.722 0.208	11.83 −35.14	233.6 36.7	I/A TRIPLE (#1)	3 D 0.6	F462/J0682 (8,77)
AM 2033-314 [2033-3148]	20 33 49 −31 49.5	3.722 0.208	11.80 −35.15	233.6 36.7	I/A TRIPLE (#2)	3 D 0.3	F462/J0682 (7,76)
AM 2033-314 [2033-3149]	20 33 58 −31 48.6	3.722 0.208	11.83 −35.18	233.7 36.7	I/A TRIPLE (#3: COMPACT)	3 D 0.2	F462/J0682 (6,77)
*AM 2034-274	20 34 32 −27 45.1	3.622 0.209	16.65 −34.29	236.5 40.1	I/A SPIRALS MCG-5-48-24 = 463-G-8	2 S 1.2	F463/J0666 (268,299)
*AM 2034-483	20 34 34 −48 34.0	4.255 0.209	351.18 −37.31	224.6 21.9	PECULIAR RING w. COMP'N. 234-IG-58	2,6 S 0.4	F234/J2363 (136,254)
AM 2034-521 [2035-5219]	20 34 40 −52 17.1	4.423 0.209	346.49 −37.27	222.8 18.6	PAIR (#1: Sa) NGC 6935 = 234-G-59 = Lu 2034-52	23 S 4.9	F234/J2363 (138,53)
AM 2035-441	20 35 04 −44 13.8	4.087 0.209	356.63 −37.23	226.8 25.7	GALAXY w. JET + COMP'NS.	7,8 S 2.0	F285/J1581 (150,220)
*AM 2035-521 [2034-5217]	20 35 05 −52 19.0	4.421 0.209	346.46 −37.33	222.9 18.5	PAIR (#2: 3-ARMED SPIRAL) NGC 6937 = 234-G-60 = Fa 349	11,23 S 4.9	F186/J1591 (64,321)
AM 2035-252 [2035-2530]	20 35 09 −25 28.0	3.569 0.209	19.34 −33.78	238.3 41.9	CLOSE TRIPLE (#1 and #2) 528-G-28	24 D 0.9	F528/J0649 (141,154)
AM 2035-253 [2035-2528]	20 35 12 −25 30.6	3.570 0.209	19.30 −33.80	238.3 41.9	CLOSE TRIPLE (#3) 528-G-29	24 D 0.7	F528/J0649 (140,152)
AM 2035-320	20 35 17 −32 05.8	3.726 0.209	11.56 −35.51	233.8 36.3	PECULIAR SP. + APP. COMP'NS. MCG-5-48-26 = 463-G-10	8,12 S 1.9	F463/J0666 (256,66)
*AM 2035-251 [2035-2517]	20 35 22 −25 17.6	3.565 0.209	19.56 −33.77	238.5 42.0	CLOSE PAIR(E + SPIRAL) (#1)	23 D 0.4	F528/J0649 (138,163)
*AM 2035-411 [2035-4121]	20 35 34 −41 17.5	3.986 0.210	0.31 −37.06	228.4 28.3	PAIR of SPIRALS + COMPACT (#1) 341-G-1	13,24 S 3.8	F341/J2375 (305,108)
*AM 2035-251 [2035-2517]	20 35 41 −25 17.9	3.564 0.210	19.58 −33.85	238.6 42.0	CLOSE PAIR(E + SPIRAL) (#2)	23 D 0.5	F528/J0649 (134,163)
AM 2035-412 [2035-4117]	20 35 44 −41 21.6	3.988 0.210	0.23 −37.10	228.4 28.2	PAIR of SPIRALS + COMPACT (#2) 341-G-2	13,24 S 3.8	F341/J2375 (303,105)
AM 2036-304 [2036-3047]	20 36 01 −30 42.7	3.690 0.210	13.26 −35.35	234.8 37.4	PAIR OF SPIRALS (#1) 463-G-11	23 D 0.5	F463/J0666 (248,141)
AM 2036-304 [2036-3042]	20 36 21 −30 47.8	3.691 0.211	13.18 −35.44	234.9 37.3	PAIR OF SPIRALS (#2: w. COMP'N.) 463-G-12	8,23 D 0.9	F463/J0666 (244,136)
AM 2036-801	20 36 32 −80 11.0	9.062 0.211	313.27 −31.45	211.0 −6.8	CLOSE PAIR NGC 6920 = 26-G-4	23 S 9.0	F026/J1818 (230,165)
AM 2036-510	20 36 35 −51 08.1	4.359 0.211	347.94 −37.61	223.7 19.5	INTEGRAL E + I/A COMP'N. (PAIR 234-IG-2	1,23 S 2.2	F234/J2363 (121,116)
AM 2036-513	20 36 36 −51 36.6	4.381 0.211	347.34 −37.59	223.4 19.0	GALAXY w. LOOP and TAIL 234-IG-63	15 D 0.9	F234/J2363 (122,90)
AM 2036-734	20 36 39 −73 47.8	6.638 0.211	320.39 −33.63	213.7 −1.0	SB0 w. FAINT OUTER RING NGC 6932 = 47-G-08	6 D 2.2	F047/J1797 (244,241)
*AM 2036-542	20 36 52 −54 28.7	4.523 0.211	343.73 −37.47	222.1 16.5	SB(r) + COMP'NS. (I/A DBL.) NGC 6942 = 186-G-73 = DN 2036-532	2,8 S 5.8	F186/J1591 (56,204)

Identification	R.A. (1950.0) Declination	Precession R.A./Dec.	Gal. Long. Gal. Lat.	S.G.L. S.G.B.	Description Cross-Identification	Coding Size (Min.)	Field/Plate (x, y)
*AM 2037-445	20 37 29 −44 59.6	4.106 0.212	355.70 −37.70	226.8 24.9	FILAMENTS w. CONDENS. + COMPACT 285-G-44	13,22 S 2.0	F285/J1581 (127,179)
AM 2037-550	20 37 41 −55 09.5	4.556 0.212	342.86 −37.53	221.9 15.8	CURVED FILAMENT w. CONDENSATION 187-IG-1	7,16,22 D 0.8	F187/J1577 (301,165)
AM 2038-382	20 38 00 −38 22.6	3.889 0.212	4.01 −37.18	230.5 30.6	DISTURBED w. LOOPS and MATERIAL MCG-6-45-9 = 341-IG-4 = B2038-383	15,16 D 1.3	F341/J2375 (284,266)
*AM 2038-323	20 38 08 −32 39.7	3.733 0.213	11.04 −36.22	234.0 35.5	SB w. BROKEN RING + COMP'N. NGC 6947 = 401-G-3 [A82]	1,6,8 S 2.1	F401/J3557 (291,304)
AM 2038-502	20 38 41 −50 24.6	4.316 0.213	348.84 −37.96	224.3 20.0	I/A DBL. Se 143/8 = 234-IG-65	2 S 0.4	F234/J2363 (102,154)
AM 2038-654	20 38 48 −65 49.6	5.363 0.213	329.68 −35.96	217.3 6.1	RESOLVED DWARF (V.LO.S.BR.) IC 5028 = 106-G-10	20 D 0.8	F106/J0902 (139,133)
AM 2039-242 [2039-2419]	20 39 13 −24 23.5	3.538 0.214	20.92 −34.34	240.2 42.3	I/A DBL. w. TAILS in PAIR (#1) MCG-4-48-29 = 528-IG-33	2,15,23 D 1.0	F528/J0649 (91,211)
AM 2039-295	20 39 20 −29 57.3	3.664 0.214	14.37 −35.87	236.1 37.7	I/A DBL. 463-IG-19	2 S 0.2	F463/J0666 (210,182)
AM 2039-613	20 39 29 −61 35.3	4.969 0.214	334.82 −36.90	219.2 9.9	SP. w. PEC ARMS + 7 COMP'NS.	8,10 S 3.1	F143/J0917 (29,75)
*AM 2039-241 [2039-2423]	20 39 32 −24 19.4	3.536 0.214	21.02 −34.39	240.3 42.3	I/A DBL. w. TAILS in PAIR (#2) MCG-4-48-30 = 528-G-34	2,15,23 D 1.0	F528/J0649 (87,215)
AM 2039-533	20 39 48 −53 32.5	4.459 0.214	344.87 −37.97	223.0 17.1	PECULIAR SP. w. APP. COMP'N. NGC 6948 = 187-G-9 = B 2039-535	8,12 S 4.5	F186/J1591 (29,253)
AM 2040-620	20 40 06 −62 09.2	5.010 0.215	334.11 −36.87	219.1 9.4	GALAXY w. PROJECTIONS Fa 75	15 D 0.7	F106/J0902 (125,329)
AM 2040-300 [2040-2953]	20 40 11 −30 01.7	3.664 0.215	14.34 −36.07	236.3 37.5	WIDE PAIR BR. SP. + COMP'NS. (#1) 463-G-20	8,23 D 1.9	F463/J0666 (200,178)
AM 2040-245	20 40 19 −24 53.7	3.547 0.215	20.42 −34.73	240.1 41.7	3 in a LINE incl. COMPACT	13,24 D 3.1	F528/J0649 (78,184)
AM 2040-295 [2040-3001]	20 40 31 −29 53.2	3.660 0.215	14.53 −36.10	236.4 37.6	WIDE PAIR BR. SP. + COMP'NS. (#2) IC 5041 = IC 5047 = 463-G-21	8,23 D 1.5	F463/J0666 (196,185)
AM 2040-434	20 40 31 −43 41.9	4.049 0.215	357.38 −38.17	228.0 25.7	I/A DBL. (CONTACT)	2 S 0.1	F285/J1581 (97,248)
AM 2040-263	20 40 43 −26 36.0	3.584 0.216	18.46 −35.30	238.9 40.3	SP. w. I/A COMP'N. in a GROUP	1,18 S 0.2	F529/J0745 (348,83)
AM 2040-674	20 40 45 −67 43.8	5.565 0.216	327.34 −35.69	216.7 4.3	I/A DBL. B 2040-677 IC 5031/32 = Se 141/2 = 74-IG-8	2 D 0.9	F106/J0902 (132,30)
AM 2040-264	20 40 47 −26 43.7	3.587 0.216	18.31 −35.35	238.8 40.2	GROUP of 4 528-G-36 = PKS 2040-26	18 D 4.0	F528/J0649 (74,86)
AM 2040-435	20 40 52 −43 53.2	4.054 0.216	357.14 −38.25	228.0 25.5	COMPACT	13 D 0.2	F285/J1581 (93,238)
AM 2040-344	20 40 54 −34 41.5	3.779 0.216	8.70 −37.18	233.3 33.5	I/A DBL.	2 S 1.1	F401/J3557 (257,195)
AM 2041-250	20 41 03 −25 00.4	3.549 0.216	20.35 −34.92	240.2 41.5	CLOSE PAIR (COMPACT w. TAIL) 528-IG-37 (I/A TRIPLE?)	3,13,15,23 S 0.2	F528/J0649 (69,178)
AM 2041-530	20 41 07 −53 00.2	4.426 0.216	345.53 −38.21	223.4 17.5	SPIRAL w. I/A COMP'N. 187-IG-13 = B 2041-530	1,9 S 0.6	F235/J3389 (332,13)
AM 2041-335	20 41 09 −33 50.4	3.756 0.216	9.76 −37.08	233.9 34.2	CLOSE PAIR (ONE SPIRAL) 401-G-5	23 S 1.7	F401/J3557 (255,241)
AM 2041-381	20 41 14 −38 10.4	3.874 0.216	4.37 −37.78	231.3 30.5	S0 w. MANY COMP'NS.	8 D 1.8	F341/J2375 (250,277)
AM 2041-361	20 41 21 −36 11.0	3.818 0.216	6.86 −37.52	232.5 32.2	CLOSE PAIR of E's 401-G-6	23 S 0.5	F401/J3557 (251,115)
AM 2041-505	20 41 43 −50 54.9	4.325 0.217	348.18 −38.42	224.5 19.3	DISTORTED DWARF 234-IG-67	16,20 D 1.0	F234/J2363 (77,126)
AM 2042-434	20 42 06 −43 42.4	4.044 0.217	357.39 −38.46	228.3 25.6	I/A DBL. 285-G-50 = B 2042-437	2 S 0.6	F285/J1581 (81,247)
*AM 2042-382	20 42 15 −38 21.4	3.876 0.217	4.17 −38.00	231.4 30.2	SPIRAL w. I/A COMP'N + 7 COMP'NS 341-G-11 = MCG-6-45-12= B2042-383	1,8 S 2.0	F341/J2375 (239,267)
*AM 2042-382	20 42 23 −38 24.8	3.877 0.217	4.11 −38.04	231.4 30.1	DIFFRACTION SPIKE on NUCLEUS 341-IG-12	22 D 0.7	F341/J2375 (238,264)
AM 2042-651	20 42 28 −65 12.2	5.268 0.217	330.32 −36.48	217.9 6.5	SPIRAL w. PECULIAR ARMS IC 5038 = 106-G-12	10 D 0.9	F106/J0902 (117,165)
AM 2042-442 [2043-4423]	20 42 54 −44 27.4	4.066 0.218	356.44 −38.63	228.1 24.9	GROUP of FOUR (#1)	18 D 9.9	F286/J2405 (340,203)
AM 2043-443 [2042-4427]	20 43 02 −44 30.8	4.068 0.218	356.37 −38.66	228.0 24.8	GROUP of FOUR (#2)	18 D 9.9	F286/J2405 (339,200)
AM 2043-442 [2043-4430]	20 43 19 −44 27.6	4.065 0.218	356.44 −38.71	228.1 24.8	GROUP of FOUR (#3)	18 D 9.9	F286/J2405 (336,203)
AM 2043-442 [2043-4427]	20 43 30 −44 23.6	4.062 0.219	356.52 −38.74	228.2 24.9	GROUP of FOUR (#4)	18 D 9.9	F286/J2405 (334,206)
AM 2043-274	20 43 48 −27 48.2	3.605 0.219	17.25 −36.28	238.7 38.9	SP. w. LINEAR CONDENS. in ARM 463-G-24 + 7 COMP'NS	8,10 D 0.8	F529/J0745 (310,19)

Identification	R.A. (1950.0) Declination	Precession R.A./Dec.	Gal. Long. Gal. Lat.	S.G.L. S.G.B.	Description Cross-Identification	Coding Size (Min.)	Field/Plate (x, y)
AM 2044-691	20 44 08 −69 16.9	5.737 0.219	325.40 −35.56	216.2 2.8	RING w. OFF-SET NUCLEUS 74-IG-9 = Se 141/3 = Fa 563	6 D 0.5	F074/J0907 (195,216)
AM 2044-383	20 44 09 −38 36.1	3.877 0.219	3.92 −38.40	231.6 29.8	CONTACT E's IC 5049 = 341-IG-14	2 S 0.3	F341/J2375 (219,254)
AM 2044-224 [2044-2240]	20 44 27 −22 43.3	3.494 0.220	23.30 −34.96	242.9 42.9	CLOSE PAIR (#1)	23 D 0.4	F528/J0649 (25,300)
AM 2044-224 [2044-2243]	20 44 39 −22 40.9	3.493 0.220	23.36 −34.99	242.9 42.9	CLOSE PAIR (#2) MCG-4-49-3 = 529-G-2	23 D 0.5	F528/J0649 (22,302)
AM 2044-453	20 44 48 −45 37.8	4.101 0.220	354.95 −39.00	227.7 23.7	COMPACT	13 D 0.2	F285/J1581 (59,143)
*AM 2045-432	20 45 13 −43 27.9	4.024 0.220	357.73 −39.01	229.0 25.5	GALAXY w. LOOPS and TAILS 285-G-55	15 D 0.9	F285/J1581 (50,259)
AM 2045-393	20 45 14 −39 35.0	3.903 0.220	2.70 −38.72	231.2 28.9	CHAIN of 5	17 D 9.0	F341/J2375 (208,202)
*AM 2045-470	20 45 17 −47 00.8	4.149 0.221	353.16 −39.10	227.1 22.4	3-ARMED SPIRAL w. I/A COMP'N.	1,11 S 1.3	F286/J2405 (311,66)
AM 2045-720 [2045-7205]	20 45 34 −72 05.1	6.173 0.221	322.10 −34.80	215.1 0.3	CLOSE PAIR (#1: E) IC 5044 = 74-G-10	23 S 0.9	F047/J1797 (214,335)
AM 2045-720 [2045-7205]	20 45 43 −72 05.8	6.174 0.221	322.08 −34.81	215.1 0.2	CLOSE PAIR (#2: SPIRAL) IC 5045 = 74-G-11	23 S 0.9	F047/J1797 (214,335)
AM 2045-521 [2046-5205]	20 45 44 −52 11.0	4.364 0.221	346.51 −38.97	224.5 17.9	GROUP of 5 (#1: E)	18 S 9.4	F235/J3389 (297,59)
AM 2045-472	20 45 48 −47 23.5	4.162 0.221	352.68 −39.18	227.0 22.1	CLOSE PAIR	23 S 0.7	F235/J3389 (308,316)
AM 2045-442	20 45 51 −44 23.5	4.053 0.221	356.55 −39.16	228.6 24.7	SMALL ONE-ARMED SPIRAL 285-G-57	10 D 0.8	F285/J1581 (46,209)
AM 2045-392	20 45 58 −39 24.3	3.895 0.221	2.95 −38.84	231.5 28.9	SPIRAL w. I/A COMP'N. IC 5056? = 341-IG-16 = PKS 2045-395	1 S 0.4	F341/J2375 (200,211)
AM 2046-295	20 46 03 −29 52.4	3.648 0.221	14.90 −37.26	237.7 36.9	I/A DBL. w. TAIL 463-IG-27	2,15 S 0.4	F463/J0666 (132,186)
AM 2046-390	20 46 04 −39 01.4	3.884 0.221	3.44 −38.82	231.7 29.3	I/A DBL. 341-G-17	2 S 0.4	F341/J2375 (199,232)
AM 2046-520 [2045-5211]	20 46 07 −52 08.2	4.359 0.221	346.56 −39.03	224.6 17.9	GROUP of 5 (#2: S0)	18 S 9.4	F235/J3389 (294,62)
AM 2046-530	20 46 09 −53 03.9	4.403 0.221	345.37 −38.96	224.1 17.1	LUMPY COMPACT Fa 567?	13 D 0.2	F234/J2363 (46,9)
AM 2046-521 [2046-5208]	20 46 10 −52 10.3	4.361 0.222	346.52 −39.03	224.5 17.9	GROUP of 5 (#3: PECULIAR SPIRAL) 235-IG-3	18 S 9.4	F235/J3389 (293,60)
AM 2046-521 [2046-5210]	20 46 12 −52 16.5	4.365 0.222	346.39 −39.03	224.5 17.8	GROUP of 5 (#4: E)	18 S 9.4	F235/J3389 (293,55)
AM 2046-470	20 46 19 −47 01.1	4.145 0.222	353.16 −39.28	227.2 22.3	SPIRAL w. APP. COMP'NS. 286-G-3	8 S 1.2	F234/J2363 (27,333)
AM 2046-501	20 46 25 −50 19.3	4.277 0.222	348.89 −39.19	225.5 19.5	LO.S.BR. SP. w. PECULIAR ARMS Se 143/6? = 235-IG-4 = B 2046-503	10,14,20 D 1.8	F234/J2363 (36,156)
AM 2046-520 [2046-5216]	20 46 28 −52 05.3	4.355 0.222	346.62 −39.09	224.6 17.9	GROUP of 5 (#5: E)	18 S 9.4	F235/J3389 (291,65)
AM 2046-513	20 46 36 −51 30.4	4.328 0.222	347.37 −39.15	224.9 18.4	I/A DBL. (incl.DWARF) 235-IG-5 = B 2046-514	2,20 S 1.0	F234/J2363 (38,92)
AM 2046-473	20 46 43 −47 39.2	4.168 0.222	352.34 −39.34	227.0 21.8	I/A DBL. 235-IG-6	2 S 1.0	F234/J2363 (25,299)
AM 2046-255	20 46 49 −25 53.3	3.557 0.222	19.77 −36.42	240.9 40.0	RING GALAXY MCG-4-49-4 = 529-G-5	6 D 1.0	F529/J0745 (275,123)
AM 2046-680	20 46 49 −68 07.1	5.556 0.222	326.69 −36.13	217.0 3.7	M51-TYPE in CHAIN of 5 74-IG-13 = Se 141/4	1,9,17 S 0.5	F074/J0907 (183,278)
*AM 2047-330	20 47 14 −33 00.1	3.720 0.223	11.10 −38.17	235.8 34.2	I/A DBL. (HI.S.BR.) Lu 2047-33 = 401-G-14/15	2 S 0.4	F401/J3557 (188,287)
AM 2047-692	20 47 22 −69 23.2	5.719 0.223	325.18 −35.80	216.4 2.6	DWARF (EDGE-ON) IC 5052 = 74-G-15	20 D 4.9	F074/J0907 (180,210)
AM 2047-305	20 47 45 −30 50.5	3.666 0.223	13.81 −37.84	237.4 35.9	EDGE-ON SP. + 7 APP. COMP'NS. 463-G-28	8 S 2.0	F463/J0666 (113,134)
AM 2047-250	20 47 49 −25 05.1	3.538 0.223	20.81 −36.41	241.7 40.6	PEC. SP. (MATR'L at ONE END) 529-G-7	8,12 D 0.8	F529/J0745 (264,166)
AM 2047-222	20 47 53 −22 27.2	3.483 0.223	23.91 −35.63	243.9 42.6	I/A CLOSE COMPACT	2,13 D 0.3	F529/J0745 (265,307)
AM 2048-450	20 48 05 −45 04.4	4.068 0.224	355.68 −39.57	228.6 23.9	COMPACT I/A DBL. 286-IG-4	2,13 S 0.4	F285/J1581 (27,171)
AM 2048-362	20 48 10 −36 29.9	3.807 0.224	6.73 −38.92	233.7 31.2	I/A DBL. 401-G-16	2 S 0.4	F401/J3557 (177,99)
*AM 2048-571	20 48 10 −57 15.4	4.616 0.224	339.99 −38.72	222.3 13.3	GAL. w. PEC. ABS'N + APP.COMP'NS. IC 5063 = 187-G-23 PKS-2048-57	8,14 D 2.0	F187/J1577 (219,56)
AM 2048-533	20 48 17 −53 39.9	4.421 0.224	344.56 −39.21	224.1 16.4	E + I/A COMP'N. 187-G-27	1 S 1.0	F187/J1577 (222,249)

Identification	R.A. (1950.0) Declination	Precession R.A./Dec.	Gal. Long. Gal. Lat.	S.G.L. S.G.B.	Description Cross-Identification	Coding Size (Min.)	Field/Plate (x, y)
AM 2048-525	20 48 19	4.388	345.43	224.4	CLUSTER of E's + SP's = NEST of 8	18,19	F187/J1577
	−52 59.4	0.224	−39.29	17.0	187-IG-25 incl. Fa 569/70/2/3/4	D 9.0	(222,285)
AM 2048-323	20 48 27	3.707	11.64	236.3	SMALL I/A DBL.	2	F463/J0666
	−32 36.9	0.224	−38.34	34.4	401-IG-18	S 0.2	(106,38)
AM 2048-440	20 48 29	4.033	356.95	229.2	RING w. BAR + PEC. ABS'N.	6,14	F286/J2405
	−44 05.6	0.224	−39.62	24.7	286-IG-4	D 1.1	(287,224)
AM 2048-485	20 48 39	4.211	350.62	226.6	M51-TYPE + I/A COMP'N. + 4 APP.	1,8,9	F234/J2363
	−48 58.0	0.224	−39.62	20.5	NGC 6970 = 235-G-8	S 2.2	(12,227)
*AM 2048-711	20 48 42	6.003	322.87	215.6	WIDE I/A DBL. (#1: EDGE-ON)	2	F074/J0907
[2048-7112]	−71 19.9	0.224	−35.28	0.8	IC 5053 = 74-G-18	D 0.8	(174,106)
AM 2048-300	20 48 45	3.645	14.87	238.2	GALAXY w. MATERIAL + COMP'N.	8	F463/J0666
	−30 01.9	0.224	−37.87	36.5	IC 5065 = 463-G-30	D 1.3	(101,177)
*AM 2048-711	20 48 46	5.982	323.00	215.7	WIDE I/A DBL. (#2: RING ?)	2,6	F074/J0907
[2048-7120]	−71 12.6	0.224	−35.33	0.9	IC 5054 = 74-G-19	D 0.8	(174,112)
AM 2048-515	20 48 49	4.339	346.71	225.0	SPIRAL w. APP. COMP'N.	8	F187/J1577
	−51 59.3	0.224	−39.45	17.8	235-G-10	S 0.9	(219,339)
AM 2049-635	20 49 03	5.085	331.77	219.2	COMPACT	13	F107/J0908
	−63 50.1	0.225	−37.50	7.4		D 0.2	(336,231)
*AM 2049-522	20 49 05	4.358	346.15	224.8	CLOSE PAIR incl. RING (#1)	6,23	F234/J2363
[2049-5221]	−52 25.1	0.225	−39.46	17.4		D 0.6	(21,42)
AM 2049-532	20 49 07	4.406	344.83	224.3	I/A DBL. (E + SPIRAL)	2	F187/J1577
	−53 26.5	0.225	−39.36	16.5		S 1.2	(215,261)
*AM 2049-522	20 49 09	4.354	346.23	224.9	CLOSE PAIR (#2)	23	F234/J2363
[2049-5225]	−52 21.2	0.225	−39.47	17.5	235-G-9	D 0.7	(20,45)
AM 2049-691	20 49 25	5.676	325.29	216.7	I/A DBL. (BAR = MERGER!?)	2	F074/J0907
	−69 13.7	0.225	−36.02	2.6	74-IG-20 = Se 141/6? = B 2049-692	S 0.5	(170,219)
AM 2049-390	20 49 43	3.875	3.42	232.3	DWARF/PROBABLE DEFECT	20	F341/J2375
	−39 07.0	0.225	−39.54	28.8		D 0.3	(161,227)
*AM 2050-441	20 50 59	4.029	356.76	229.5	SP.w.THIN DIST'BD CIRC. ARMS	8,10	F286/J2405
	−44 15.2	0.227	−40.07	24.3	286-G-10	D 3.6	(263,216)
AM 2051-281	20 51 11	3.601	17.15	240.0	CLOSE PAIR	23	F463/J0666
	−28 18.8	0.227	−37.99	37.5		S 0.6	(71,269)
AM 2051-410	20 51 14	3.925	1.00	231.5	SPIRAL w. 3 APP. COMP'NS	8	F341/J2375
	−41 00.2	0.227	−39.98	27.1	Ag-63 = 341-G-21	S 1.1	(147,125)
*AM 2052-301	20 52 03	3.644	14.72	238.7	I/A DBL. or M51-TYPE w. LOOP	2,9,15	F463/J0666
	−30 19.1	0.228	−38.63	35.8		D 0.5	(63,161)
AM 2052-311	20 52 04	3.667	13.50	238.0	DISTURBED + I/A + APP.COMP'NS.	1,8	F463/J0666
	−31 17.0	0.228	−38.83	35.0	MCG-5-49-6 = 463-IG-34	S 2.0	(64,109)
AM 2052-270	20 52 10	3.572	18.75	241.2	LONG,THIN ARMS + COMPACT COMP.	10,13	F529/J0745
	−27 03.8	0.228	−37.89	38.4		D 0.7	(210,60)
AM 2052-434	20 52 10	4.009	357.37	230.0	I/A DBL. in LOOSE CHAIN	2,17	F286/J2405
	−43 47.5	0.228	−40.28	24.6	286-IG-11 = DN 2052-438	S 0.2	(252,241)
AM 2052-260	20 52 22	3.551	19.96	242.0	I/A TRIPLE (FAINT RING +	3,6	F529/J0745
	−26 05.2	0.228	−37.67	39.1	+ CONNECTED COM'N.)	S 1.0	(208,113)
AM 2052-440	20 52 36	4.015	357.08	229.9	RING	6	F286/J2405
	−44 00.5	0.228	−40.36	24.4		D 0.6	(248,230)
AM 2052-252	20 52 39	3.536	20.80	242.6	PECULIAR SP.w. STELLAR NUCLEUS?	12,21	F529/J0745
	−25 24.4	0.228	−37.55	39.6	529-G-12	D 0.9	(205,149)
AM 2052-594	20 52 43	4.742	336.72	221.6	CLOSE PAIR	23	F144/J0892
	−59 43.9	0.229	−38.85	10.8	Fa 79/80	S 1.3	(185,192)
*AM 2052-221	20 52 46	3.472	24.59	245.3	M51-TYPE	9	F529/J0745
	−22 15.3	0.229	−36.64	42.0	MCG-4-49-8 = 593-G-3 [A82]	S 0.8	(205,318)
AM 2053-234	20 53 28	3.501	22.84	244.1	COMPACT	13	F529/J0745
	−23 46.5	0.229	−37.26	40.8		D 0.1	(196,237)
AM 2053-664	20 53 30	5.326	328.16	218.2	I/A DBL. (E + SPIRAL)	2	F107/J0908
	−66 40.9	0.229	−37.16	4.7	106-IG-15 = PKS 2053-666	S 0.3	(296,81)
AM 2053-232	20 53 53	3.494	23.25	244.5	I/A DBL.	2	F529/J0745
	−23 27.5	0.230	−37.25	41.0		S 0.4	(191,254)
*AM 2054-520	20 54 18	4.315	346.50	225.7	SP. w. HEAVY ARM + 4 I/A COMPS.	1,10	F235/J3389
	−52 04.1	0.230	−40.29	17.3	NGC 6984 = 235-G-20	S 17.9	(226,68)
AM 2054-505	20 54 19	4.262	348.07	226.4	CHAIN of 4 in GROUP of 5	17,18	F235/J3389
	−50 51.8	0.230	−40.41	18.4	Se 143/7 = DN 2054-509	D 2.7	(227,133)
AM 2054-245	20 54 26	3.523	21.51	243.4	PAIR (#1: HI.S.BR. ONE-ARMED SP.)	10,12	F529/J0745
	−24 56.5	0.230	−37.80	39.7		D 0.7	(184,174)
*AM 2054-433	20 54 30	3.993	357.69	230.5	I/A DBL. (#1: E + SPIRAL)	2	F286/J2405
[2054-4334]	−43 33.5	0.230	−40.69	24.6	Ag-64 = 286-G-17	S 1.6	(230,254)
*AM 2054-433	20 54 31	3.993	357.66	230.5	I/A DBL. (#2: E + SPIRAL)	2	F286/J2405
[2054-4333]	−43 34.8	0.230	−40.70	24.6	Ag-64 = 286-G-18	S 1.6	(229,253)
AM 2054-335	20 54 41	3.722	10.32	236.7	CLOSE I/A DBL. (CHAIN of 6)	2,17	F401/J3557
	−33 53.6	0.231	−39.85	32.6		S 0.2	(105,238)
AM 2054-244	20 54 51	3.520	21.69	243.6	PAIR (#2: SP. w. OFF-SET NUCLEUS)	12,23	F529/J0745
	−24 49.2	0.231	−37.86	39.8	MCG-4-49-9 = 529-G-13	D 0.9	(179,181)

Identification	R.A. (1950.0) Declination	Precession R.A./Dec.	Gal. Long. Gal. Lat.	S.G.L. S.G.B.	Description Cross-Identification	Coding Size (Min.)	Field/Plate (x, y)
AM 2054-232	20 54 52 −23 28.9	3.492 0.231	23.31 −37.48	244.7 40.8	I/A DBL. in GROUP of 5 near 529-G-14	2,18 S 0.4	F529/J0745 (178,253)
*AM 2054-322 [2055-3223]	20 54 57 −32 29.2	3.688 0.231	12.13 −39.66	237.7 33.7	THREE in a LINE (#1) 464-G-3	24 D 0.8	F463/J0666 (32,44)
*AM 2055-492	20 55 04 −49 28.8	4.202 0.231	349.88 −40.64	227.2 19.5	I/A DBL. (MERGER ?) B 2055-494 Se 143/5 = 235-IG-23 = DN 2055-495	2 S 1.2	F235/J3389 (222,207)
AM 2055-321	20 55 08 −32 14.1	3.681 0.231	12.46 −39.66	238.0 33.9	WIDE I/A QUARTET MCG-5-49-9? = 464-G-4	4 D 5.8	F401/J3557 (98,327)
AM 2055-322 [2054-3229]	20 55 09 −32 27.5	3.686 0.231	12.18 −39.70	237.8 33.7	THREE in a LINE (#2)	24 D 0.2	F463/J0666 (30,45)
*AM 2055-425	20 55 09 −42 50.6	3.967 0.231	358.63 −40.79	231.1 25.1	MULTIPLE NUCLEUS + STRONG ARM 286-IG-19 = B 2055-428	10,12 D 0.7	F341/J2375 (109,26)
AM 2055-483	20 55 10 −48 30.0	4.163 0.231	351.16 −40.72	227.8 20.3	SMALL I/A DBL. 235-IG-24	2 S 0.6	F235/J3389 (222,260)
AM 2055-541	20 55 10 −54 19.8	4.417 0.231	343.54 −40.13	224.6 15.3	DISTURBED SPIRAL in PAIR 187-G-38/40 = Fa 351/578	2,10 S 3.8	F187/J1577 (167,213)
AM 2055-321	20 55 11 −32 14.5	3.681 0.231	12.46 −39.67	238.0 33.9	GROUP of 4 MCG-5-49-9 = 464-G-4	18 D 5.8	F463/J0666 (29,57)
*AM 2055-425	20 55 11 −42 51.1	3.968 0.231	358.62 −40.80	231.1 25.1	GALAXY w. TAIL (LUMPY NUCLEUS) 286-IG-19	12,15 D 1.1	F286/J2405 (223,292)
AM 2055-625	20 55 12 −62 58.4	4.963 0.231	332.60 −38.38	220.2 7.8	CLOSE TRIPLE	24 D 1.8	F144/J0892 (170,17)
AM 2055-314 [2055-3137]	20 55 22 −31 40.1	3.667 0.231	13.20 −39.60	238.4 34.3	PAIR of I/A SPIRALS (#1) MCG-5-49-10 = 464-G-5	2 D 0.4	F463/J0666 (26,88)
AM 2055-313 [2055-3140]	20 55 23 −31 37.8	3.666 0.231	13.25 −39.59	238.4 34.3	PAIR of I/A SPIRALS (#2) 464-G-6	2 D 0.4	F463/J0666 (26,90)
AM 2055-322 [2055-3227]	20 55 29 −32 23.3	3.684 0.231	12.28 −39.76	237.9 33.7	THREE in a LINE (#3)	24 D 0.5	F463/J0666 (26,49)
AM 2055-230	20 55 35 −23 01.9	3.482 0.232	23.92 −37.50	245.3 41.0	LOOSE [NNE] CHAIN of 4	17 D 7.8	F529/J0745 (170,277)
AM 2055-521	20 55 36 −52 12.1	4.314 0.232	346.30 −40.47	225.8 17.1	I/A DBL. 235-IG-26	2 D 0.3	F187/J1577 (163,327)
AM 2055-273 [2056-2734]	20 55 43 −27 38.2	3.577 0.232	18.29 −38.79	241.5 37.5	LINE of 3 (#1)	24 D 0.4	F464/J0640 (290,305)
AM 2055-273 [2055-2738]	20 55 57 −27 37.0	3.576 0.232	18.34 −38.84	241.6 37.4	LINE of 3 (#2: PECULIAR)	24 D 0.4	F464/J0640 (287,306)
AM 2056-825	20 56 00 −82 55.6	10.818 0.232	310.01 −30.93	210.4 −9.5	2 E's in COMMON ENVELOPE	2 S 0.4	F011/J2562 (246,285)
AM 2056-425	20 56 10 −42 58.1	3.968 0.232	358.48 −40.98	231.2 24.9	I/A DBL. 286-IG-20 = B 2056-429	2 S 0.8	F286/J2405 (214,286)
AM 2056-725	20 56 11 −72 50.6	6.183 0.232	320.89 −35.28	215.4 −0.8	SPIRAL + SPIRAL COMP'N. IC 5071 = 047-G-19	8 S 9.2	F047/J1797 (171,295)
AM 2056-273 [2055-2737]	20 56 13 −27 34.8	3.575 0.232	18.40 −38.89	241.7 37.4	LINE of 3 (#3)	24 D 1.0	F464/J0640 (284,308)
AM 2056-272	20 56 16 −27 25.5	3.571 0.232	18.60 −38.86	241.8 37.5	CLOSE I/A DBL. E's	2 S 0.2	F529/J0745 (161,41)
AM 2056-714	20 56 41 −71 47.8	5.987 0.233	322.06 −35.71	216.0 0.1	ONE-ARMED SPIRAL	10 D 0.6	F074/J0907 (141,80)
AM 2056-392	20 56 55 −39 29.7	3.862 0.233	3.08 −40.95	233.4 27.7	RING in LINE of 3	6,13,24 D 4.9	F341/J2375 (87,205)
AM 2057-352	20 57 01 −35 23.6	3.753 0.233	8.47 −40.55	236.1 31.1	CLOSE PAIR	23 S 1.0	F401/J3557 (80,157)
*AM 2057-650	20 57 06 −65 09.7	5.140 0.233	329.86 −37.97	219.3 5.8	ONE-ARMED SPIRAL w. I/A COMPACT 107-G-2 in DN 2057-651?	1,9,10 D 1.1	F106/J0902 (35,161)
AM 2057-572	20 57 11 −57 25.5	4.570 0.233	339.50 −39.89	223.3 12.5	SP. w. CONDENSATION in GROUP of 3 187-G-45 = B 2057-574	12,24 D 0.7	F144/J0892 (153,315)
AM 2058-241	20 58 21 −24 16.5	3.502 0.234	22.64 −38.47	244.8 39.7	PECULIAR SPIRAL, COMPACT	12,13 D 0.4	F529/J0745 (136,210)
*AM 2058-381	20 58 28 −38 16.6	3.823 0.235	4.72 −41.16	234.5 28.6	M51-TYPE 341-G-30 = B 2058-382	9 S 0.7	F342/J2611 (343,268)
AM 2058-413	20 58 34 −41 37.3	3.918 0.235	0.28 −41.39	232.4 25.8	GROUP of 5 (MOSTLY E's)	18 D 2.7	F341/J2375 (73,90)
AM 2058-325	20 58 44 −32 59.2	3.690 0.235	11.67 −40.53	238.1 32.8	I/A QUARTET in a LINE incl. 401-G-28	4,17 D 2.2	F464/J0640 (249,18)
AM 2058-275	20 58 51 −27 55.0	3.576 0.235	18.16 −39.53	242.0 36.8	LOOSE CHAIN of 12	17,23 D 9.0	F464/J0640 (252,291)
*AM 2058-235	20 58 56 −23 50.3	3.492 0.235	23.23 −38.47	245.3 39.9	E w. PLUME + 2 CONDENSATIONS	15 D 1.0	F529/J0745 (129,233)
AM 2059-244	20 59 00 −24 47.2	3.511 0.235	22.07 −38.76	244.5 39.2	SB w.CONDENS.in ARM + E in PAIRGV180,4;ES8 529-G-16	3,12 D 1.0	F529/J0745 (128,182)
AM 2059-730	20 59 20 −73 00.4	6.173 0.235	320.59 −35.43	215.5 −1.0	CONTACT E's in GROUP of 11 incl. IC 5073 = 47-G-21 = Fa 579	2,18 D 7.2	F047/J1797 (158,286)

Identification	R.A. (1950.0) Declination	Precession R.A./Dec.	Gal. Long. Gal. Lat.	S.G.L. S.G.B.	Description Cross-Identification	Coding Size (Min.)	Field/Plate (x, y)
AM 2100-352 [2100-3524]	21 00 11 −35 26.3	3.745 0.236	8.52 −41.19	236.7 30.7	4 in CHAIN of 6 (#1: S0) 402-G-2	17 D 0.8	F402/J0735 (304,154)
AM 2100-352 [2100-3526]	21 00 12 −35 20.4	3.742 0.236	8.65 −41.18	236.8 30.7	4 in CHAIN of 6 (#2: SPIRAL) 402-G-1	17 D 0.8	F402/J0735 (304,159)
AM 2100-352 [2100-3520]	21 00 15 −35 22.5	3.743 0.236	8.60 −41.20	236.8 30.7	4 in CHAIN of 6 (#3: E)	17 D 0.3	F402/J0735 (304,157)
AM 2100-583	21 00 15 −58 34.5	4.617 0.236	337.92 −40.05	223.0 11.3	I/A DBL. Fa-87	2 S 0.5	F144/J0892 (133,253)
AM 2100-503	21 00 18 −50 34.3	4.220 0.236	348.34 −41.38	227.4 18.1	GALAXY w. 4 COMP'NS. in CHAIN 235-G-45	1,17 S 1.3	F235/J3389 (177,149)
AM 2100-352 [2100-3522]	21 00 20 −35 24.2	3.743 0.237	8.57 −41.22	236.8 30.7	4 in CHAIN of 6 (#4: SPIRAL) 402-G-3	17 D 1.1	F402/J0735 (303,156)
*AM 2100-393	21 00 21 −39 38.7	3.855 0.237	2.94 −41.63	234.0 27.2	IRR./Sd I (!) 341-G-32	16,224 D 1.6	F341/J2375 (51,196)
*AM 2100-280	21 00 26 −28 03.3	3.575 0.237	18.09 −39.91	242.2 36.5	SP. w. PEC. FILAMENTARY ARMS [E] MCG-5-49-15 = 464-G-19 [PKS?]	10 D 1.3	F529/J0745 (112,7)
AM 2100-725	21 00 33 −72 59.0	6.152 0.237	320.57 −35.52	215.6 −1.1	RING + I/A DBL.	2,6 S 1.3	F074/J0907 (128,16)
AM 2100-262	21 00 42 −26 23.8	3.540 0.237	20.20 −39.56	243.6 37.7	LUMPY COMPACT in GROUP of 9	13,18 D 0.2	F529/J0745 (108,96)
*AM 2100-381	21 00 45 −38 16.6	3.816 0.237	4.77 −41.60	234.9 28.3	DBL. M51-TYPE in PAIR	9,23 S 1.0	F341/J2375 (44,269)
AM 2100-434	21 00 51 −43 45.0	3.974 0.237	357.44 −41.84	231.5 23.8	I/A DBL. 286-IG-33	2 S 0.2	F286/J2405 (168,244)
AM 2101-271	21 01 04 −27 13.1	3.557 0.237	19.19 −39.84	243.0 37.0	INTEGRAL SP. w. (CONN?) COMP'NS 529-G-18	1 S 0.5	F529/J0745 (104,52)
AM 2101-353	21 01 08 −35 34.4	3.745 0.237	8.37 −41.40	236.8 30.4	GALAXY + I/A COMP'N.	1 S 0.4	F402/J0735 (294,147)
*AM 2101-480	21 01 09 −48 02.4	4.118 0.237	351.69 −41.74	229.0 20.2	SMALL I/A TRIPLE 235-IG-48	3 D 1.3	F286/J2405 (166,14)
AM 2101-221	21 01 17 −22 16.5	3.458 0.238	25.34 −38.52	247.2 40.7	PECULIAR SPIRAL	12 D 0.5	F529/J0745 (99,317)
*AM 2101-215	21 01 32 −21 58.8	3.452 0.238	25.72 −38.48	247.5 40.9	SP. w. PEC. ARMS (COMP'NS[E]+[N]) MCG-4-49-10 = 598-G-20	1,8,10 D 1.5	F529/J0745 (96,333)
AM 2101-475	21 01 45 −47 59.3	4.114 0.238	351.75 −41.84	229.1 20.2	EDGE-ON w. 10 COMP'NS. 235-G-53 (PECULIAR NUCLEUS)	8 S 7.6	F235/J3389 (164,287)
AM 2102-482	21 02 30 −48 24.5	4.126 0.239	351.17 −41.94	228.9 19.7	FACE-ON SP.+ PEC. (in CHAIN of 6) 235-G-55	17 D 26.9	F235/J3389 (157,265)
AM 2102-285 [2102-2855]	21 02 38 −28 53.5	3.588 0.239	17.17 −40.56	242.0 35.5	LINE of THREE (#1)	24 D 0.3	F464/J0640 (207,239)
AM 2102-285 [2102-2853]	21 02 44 −28 54.1	3.588 0.239	17.16 −40.59	242.0 35.5	LINE OF THREE (#2: DISTURBED) 464-IG-22	24 D 0.4	F464/J0640 (206,238)
AM 2102-453	21 02 44 −45 38.5	4.027 0.239	354.89 −42.14	230.6 22.0	DISTURBED w. I/A COMP'N. 286-G-46	1 S 2.0	F286/J2405 (151,143)
AM 2102-234	21 02 46 −23 45.2	3.484 0.239	23.66 −39.29	246.3 39.4	SP. w. I/A COMP'NS. in PAIR	1,3,8 S 0.7	F529/J0745 (81,237)
*AM 2102-252	21 02 48 −25 26.5	3.517 0.239	21.56 −39.76	244.8 38.1	I/A TRIPLE (RING,SP.w.I/A COMP'N) 529-G-20	1,3,6 D 1.7	F529/J0745 (82,147)
AM 2102-285 [2102-2854]	21 02 50 −28 55.6	3.589 0.239	17.14 −40.61	242.0 35.5	LINE of THREE (#3)	24 D 0.4	F464/J0640 (205,237)
AM 2102-395	21 02 53 −39 51.4	3.852 0.239	2.69 −42.12	234.3 26.8	GROUP of 6 E's	18 D 3.6	F341/J2375 (26,183)
AM 2102-253	21 02 54 −25 33.5	3.519 0.239	21.42 −39.82	244.7 38.0	HI.S.BR. ARMED SP. + 2 COMP'NS.GV180,4;ES 529-G-21	10,12 D 0.9	F529/J0745 (81,141)
AM 2103-434	21 03 04 −43 42.5	3.964 0.239	357.50 −42.24	231.9 23.6	GAL.w.COMPACT in LONG PLUME 286-G-47 in DN 2102-436? [PKS?]	1,13,15 S 4.0	F286/J2405 (146,246)
AM 2103-243 [2103-2436]	21 03 12 −24 32.0	3.498 0.239	22.73 −39.60	245.7 38.7	GROUP of 4 (#1: RS)	18 D 0.5	F529/J0745 (77,196)
AM 2103-243 [2103-2432]	21 03 15 −24 33.0	3.498 0.240	22.71 −39.62	245.7 38.7	GROUP of 4 (#2: Sa)	18 D 0.5	F529/J0745 (76,195)
AM 2103-450	21 03 27 −45 01.5	4.004 0.240	355.72 −42.29	231.1 22.5	I/A DBL. + COMP'NS. (GROUP of 17) NGC 7012 = 286-G-51 = DN 2103-450	2,18 S 1.6	F286/J2405 (144,176)
AM 2103-243 [2103-2433]	21 03 28 −24 34.1	3.498 0.240	22.71 −39.67	245.7 38.7	GROUP of 4 (#3: LO.S.BR.SP.) 529-G-23	18 D 0.5	F529/J0745 (73,194)
AM 2103-243 [2103-2434]	21 03 31 −24 36.8	3.499 0.240	22.65 −39.69	245.7 38.6	GROUP of 4 (#4: HI.S.BR.SP.) NGC 7019 = 529-G-22	18 D 0.7	F529/J0745 (73,191)
AM 2103-724	21 03 42 −72 47.3	6.073 0.240	320.68 −35.81	215.9 −1.0	COMPACT	13 D 0.1	F047/J1797 (141,297)
AM 2103-474	21 03 50 −47 45.4	4.096 0.240	352.03 −42.21	229.5 20.2	M51-TYPE 235-IG-61 = Lu 2103-47	9 S 0.4	F235/J3389 (145,300)
AM 2103-301	21 03 54 −30 17.9	3.616 0.240	15.43 −41.13	241.2 34.3	DWARF 464-G-23	20 D 0.5	F464/J0640 (192,163)

Identification	R.A. (1950.0) Declination	Precession R.A./Dec.	Gal. Long. Gal. Lat.	S.G.L. S.G.B.	Description Cross-Identification	Coding Size (Min.)	Field/Plate (x, y)
AM 2103-550	21 03 54 −55 09.3	4.407 0.240	342.19 −41.24	225.3 14.0	DWARF (RESOLUTION ?) 187-G-51	20 D 2.0	F187/J1577 (100,167)
AM 2103-583	21 03 58 −58 30.0	4.588 0.240	337.87 −40.54	223.5 11.1	CLOSE TRIPLE	24 D 5.4	F144/J0892 (107,256)
AM 2104-241	21 04 14 −24 14.8	3.491 0.241	23.17 −39.75	246.1 38.8	RING w. 3 APP. COMP'NS. 529-G-24	6,8 S 1.1	F530/J0797 (323,218)
AM 2104-810	21 04 16 −81 02.8	8.956 0.241	311.78 −32.09	211.6 −8.1	I/A DBL. + COMP'N.	2 S 0.6	F026/J1818 (168,121)
AM 2104-254 [2104-2537]	21 04 21 −25 40.1	3.518 0.241	21.40 −40.16	245.0 37.7	4 in GROUP of 15 (#1: E) NGC 7016 = 529-G-25 = PKS 2104-25	18 D 0.6	F529/J0745 (64,134)
AM 2104-254 [2104-2540]	21 04 24 −25 41.2	3.519 0.241	21.38 −40.17	245.0 37.7	4 in GROUP of 15 (#2: MERGER?) NGC 7017 = 529-IG-26 [PKS]	18,22 D 0.7	F529/J0745 (63,133)
AM 2104-253 [2104-2541]	21 04 29 −25 37.8	3.517 0.241	21.46 −40.18	245.0 37.7	4 in GROUP of 15 (#3: MERGER?) NGC 7018 = 529-IG-27 [PKS]	18 D 0.6	F529/J0745 (62,136)
*AM 2104-574	21 04 34 −57 42.1	4.537 0.241	338.87 −40.80	224.0 11.8	DISTURBED SP. + I/A COMP'N. 144-IG-18 = B 2104-577	1 S 0.5	F188/J1592 (340,27)
AM 2104-253 [2104-2537]	21 04 43 −25 37.1	3.517 0.241	21.49 −40.23	245.1 37.7	4 in GROUP of 15 (#4: Sa)	18 D 0.6	F529/J0745 (59,137)
AM 2105-855	21 05 02 −85 58.0	0.270 0.241	306.84 −29.59	209.0 −12.3	COMPACT E	13 D 0.3	F010/J2477 (77,102)
*AM 2105-332	21 05 03 −33 26.6	3.683 0.241	11.34 −41.91	239.1 31.6	M51-TYPE, SMALL COMP'N. OPPOSITE 402-IG-10(+9) [A82]	8,9 S 3.7	F402/J0735 (252,262)
AM 2105-614	21 05 06 −61 43.3	4.789 0.241	333.75 −39.83	221.9 8.3	CLOSE PAIR	23 S 0.5	F144/J0892 (106,83)
AM 2105-632	21 05 07 −63 29.2	4.924 0.241	331.56 −39.30	220.9 6.8	GALAXY w. PECULIAR ABSORPTION IC 5084 - 107-G-7	14 D 1.7	F107/J0908 (242,257)
AM 2105-434 [2105-4342]	21 05 13 −43 42.3	3.955 0.241	357.50 −42.63	232.2 23.4	GROUP in LONG CHAIN (#1)	17,18 D 6.3	F286/J2405 (126,246)
AM 2105-273 [2105-2730]	21 05 15 −27 30.0	3.553 0.242	19.13 −40.81	243.6 36.2	CLOSE PAIR of E's (#1)	23 D 0.4	F464/J0640 (177,314)
AM 2105-433 [2105-4342]	21 05 23 −43 38.1	3.952 0.242	357.59 −42.66	232.3 23.4	GROUP in LONG CHAIN (#2)	17,18 D 6.3	F286/J2405 (124,250)
AM 2105-434 [2105-4338]	21 05 23 −43 43.3	3.955 0.242	357.47 −42.66	232.2 23.3	GROUP in LONG CHAIN (#3) 286-G-59 ?	17,18 D 6.3	F286/J2405 (124,245)
AM 2105-434 [2105-4343]	21 05 25 −43 43.5	3.955 0.242	357.47 −42.67	232.2 23.3	GROUP in LONG CHAIN (#4)	17,18 D 6.3	F286/J2405 (124,245)
AM 2105-400 [2105-4003]	21 05 29 −40 04.5	3.848 0.242	2.43 −42.63	234.6 26.3	2 E's in ENVEL.in GROUP of 4 (#1) MCG-7-43-25 = 342-G-8 = B2105-400	2,18 D 0.4	F342/J2611 (266,174)
AM 2105-295	21 05 31 −29 58.3	3.605 0.242	15.94 −41.40	241.7 34.3	LARGE E w. 3 APP. COMP'NS. IC 5086 = MCG-5-50-2 = 464-G-25	8 S 11.2	F464/J0640 (173,181)
AM 2105-400 [2105-4004]	21 05 36 −40 03.4	3.848 0.242	2.46 −42.65	234.6 26.3	2 E's in ENVEL.in GROUP of 4 (#2) MCG-7-43-26 = 342-IG-10 = B2105-400	2,18 D 0.5	F342/J2611 (265,175)
AM 2105-240	21 05 39 −24 09.8	3.486 0.242	23.39 −40.03	246.5 38.7	PECULIAR SPIRAL (DBL. NUCLEUS ?) 530-G-3	12 D 0.8	F529/J0745 (46,215)
AM 2105-273 [2105-2730]	21 05 41 −27 30.5	3.553 0.242	19.15 −40.91	243.7 36.1	CLOSE PAIR of E's (#2)	23 D 0.5	F464/J0640 (171,313)
AM 2105-434 [2105-4343]	21 05 41 −43 42.5	3.953 0.242	357.49 −42.72	232.3 23.3	GROUP in LONG CHAIN (#5)	17,18 D 6.3	F286/J2405 (121,246)
*AM 2105-530	21 05 55 −53 08.0	4.300 0.242	344.78 −41.89	226.7 15.5	E + I/A SPIRAL in GROUP of 3 187-G-54	2,24 S 1.8	F187/J1577 (80,275)
AM 2105-243	21 05 57 −24 35.3	3.494 0.242	22.89 −40.22	246.2 38.3	HI.S.BR.SP.(CONDENS.in OUTER RIM) 530-G-5	1,12 D 0.6	F529/J0745 (43,192)
*AM 2106-624	21 06 05 −62 48.2	4.862 0.242	332.37 −39.61	221.4 7.4	INTEGRAL SPIRAL + I/A COMP'NS. 107-IG-9	1,8,12 D 1.0	F107/J0908 (238,294)
AM 2106-270	21 06 09 −27 06.9	3.544 0.242	19.69 −40.91	244.1 36.4	ONE-ARMED SPIRAL MCG-5-50-3 = 530-G-6	10 D 0.6	F529/J0745 (43,56)
*AM 2106-641	21 06 13 −64 14.8	4.977 0.243	330.58 −39.18	220.6 6.1	COMPACT + RING (near BR. GALAXY) near NGC 7020 [A82]	6,8,13 S 2.0	F107/J0908 (234,216)
AM 2106-382	21 06 19 −38 25.2	3.801 0.243	4.69 −42.70	235.8 27.5	DISTURBED SINGLE	16 D 0.4	F342/J2611 (260,263)
*AM 2106-272	21 06 21 −27 23.6	3.549 0.243	19.34 −41.02	244.0 36.1	I/A E + SP. w. ONE VERY LONG ARM 530-G-8	2 S 0.7	F529/J0745 (41,41)
AM 2106-453	21 06 31 −45 38.6	4.011 0.243	354.85 −42.80	231.2 21.6	I/A DBL. 286-IG-62	2 S 0.3	F286/J2405 (115,142)
AM 2106-454	21 06 32 −45 44.3	4.014 0.243	354.72 −42.80	231.2 21.6	HI.S.BR. GALAXY w. KNOTS 286-G-63	12 D 1.1	F286/J2405 (115,137)
AM 2106-230	21 06 34 −23 04.9	3.464 0.243	24.83 −39.92	247.7 39.3	Sa w. AMORPHOUS COMPONENT IC 5088 = MCG-5-50-1 = 530-G-10	10,22 D 2.1	F529/J0745 (34,273)
AM 2106-500	21 06 35 −50 03.5	4.168 0.243	348.87 −42.43	228.6 18.0	SMALL I/A DBL. 235-IG-68	2 S 0.3	F235/J3389 (123,176)
AM 2106-374	21 06 52 −37 42.3	3.781 0.243	5.67 −42.76	236.4 28.0	PEC. SP. at END of CHAIN of 4 MCG-6-46-10 = 342-IG-13 = B2106-377	12,17 D 11.2	F402/J0735 (231,33)

Identification	R.A. (1950.0) Declination	Precession R.A./Dec.	Gal. Long. Gal. Lat.	S.G.L. S.G.B.	Description Cross-Identification	Coding Size (Min.)	Field/Plate (x, y)
AM 2107-251	21 07 01 −25 19.3	3.506 0.243	22.04 −40.65	245.8 37.6	CLOSE PAIR	23 S 1.1	F530/J0797 (288,161)
AM 2107-474	21 07 09 −47 42.8	4.079 0.243	352.02 −42.77	230.0 19.9	HI.S.BR. RING w. LARGE NUCLEUS	6,12 D 0.7	F235/J3389 (115,301)
AM 2107-242	21 07 24 −24 25.7	3.488 0.244	23.21 −40.49	246.7 38.2	DBL. ARM + THIN FAINT OUTER ARM MCG-4-50-2 = 530-G-13	10,12 D 1.2	F529/J0745 (25,200)
AM 2107-425	21 07 37 −42 50.7	3.920 0.244	358.66 −43.08	233.1 23.8	I/A TRIPLE 286-IG-69	3 D 0.2	F342/J2611 (241,26)
AM 2107-461	21 07 38 −46 16.6	4.027 0.244	353.97 −42.96	231.0 21.0	EDGE-ON w. I/A DBL. COMP'N. 286-IG-67	2,8 S 2.7	F286/J2405 (106,108)
*AM 2108-452	21 08 12 −45 28.5	3.998 0.244	355.06 −43.10	231.6 21.6	HORSE-SHOE w. I/A COMP'N. 286-IG-73	1 S 0.6	F286/J2405 (99,151)
AM 2108-492	21 08 27 −49 29.3	4.137 0.245	349.59 −42.80	229.2 18.3	BOX FORM (MERGER?)	17,22 D 1.9	F235/J3389 (106,206)
AM 2108-232	21 08 47 −23 22.4	3.465 0.245	24.65 −40.49	247.9 38.7	ELONGATED CLUSTER MCG-4-50-6/7/8 = 530-G-20	19 D 10.3	F530/J0797 (268,266)
AM 2108-672	21 08 54 −67 29.8	5.263 0.245	326.55 −38.30	219.1 3.2	COMPACT GALAXY	13 D 0.3	F107/J0908 (214,42)
AM 2109-220	21 09 08 −22 04.7	3.441 0.245	26.31 −40.19	249.1 39.6	RING	6 D 0.5	F530/J0797 (265,336)
AM 2109-585 [2109-5855]	21 09 17 −58 56.2	4.576 0.246	337.08 −41.10	223.8 10.4	CLOSE PAIR (#1: PEC. ABS'N.)	14,23 D 0.4	F145/J1578 (334,228)
AM 2109-391	21 09 22 −39 19.8	3.815 0.246	3.49 −43.35	235.7 26.5	SPIRAL + APP. COMP'N. 342-G-20	8 S 0.4	F342/J2611 (227,215)
AM 2109-583	21 09 23 −58 38.2	4.558 0.246	337.46 −41.19	224.0 10.6	SPIRAL w. I/A COMP'N. 144-G-19	1 D 1.3	F145/J1578 (334,244)
AM 2109-585 [2109-5856]	21 09 36 −58 55.3	4.573 0.246	337.08 −41.14	223.9 10.4	CLOSE PAIR (#2: PEC. ABS'N.)	14,23 D 0.5	F144/J0892 (68,232)
AM 2109-633	21 09 38 −63 32.5	4.890 0.246	331.29 −39.77	221.3 6.5	GAL. w. WARPED PLANE (or PAIR?) 107-G-15	16 D 1.8	F107/J0908 (215,255)
AM 2109-380 [2109-3805]	21 09 47 −38 04.4	3.781 0.246	5.22 −43.36	236.7 27.4	CLOSE PAIR (#1: SPIRAL) 342-G-22	23 D 0.7	F402/J0735 (200,13)
AM 2109-380 [2109-3804]	21 09 53 −38 05.7	3.781 0.246	5.20 −43.38	236.7 27.4	CLOSE PAIR (#2: SPIRAL) 342-G-23	23 D 0.4	F402/J0735 (199,12)
*AM 2109-613	21 09 56 −61 30.0	4.736 0.246	333.81 −40.44	222.5 8.2	GALAXY w. JETS (EDGE-ON) 144-G-20	7 D 1.6	F145/J1578 (318,91)
AM 2110-432	21 10 08 −43 24.6	3.926 0.246	357.87 −43.53	233.2 23.1	I/A DBL.	2 S 0.7	F286/J2405 (77,261)
*AM 2110-810	21 10 35 −81 09.5	8.859 0.247	311.52 −32.25	211.8 −8.3	COMPACT (STAR?) w. JET	7,13 D 0.3	F027/J2485 (340,89)
AM 2111-554 [2112-5543]	21 11 35 −55 44.1	4.388 0.248	341.13 −42.18	225.9 12.9	CHAIN of 4 (#1)	17 S 10.3	F188/J1592 (296,136)
*AM 2111-472	21 11 46 −47 25.8	4.047 0.248	352.31 −43.57	230.9 19.7	SP. w. PEC. ARM + APP. COMP'NS. NGC 7038 = 286-G-79	8,10 S 20.2	F235/J3389 (72,315)
AM 2111-274	21 11 50 −27 41.7	3.543 0.248	19.33 −42.28	244.8 35.1	COMPACT	13 D 0.2	F530/J0797 (228,35)
AM 2112-592	21 12 05 −59 26.5	4.586 0.248	336.30 −41.30	223.9 9.8	CLOSE PAIR 144-IG-21 = Fa 88	23 S 0.8	F145/J1578 (312,202)
AM 2112-554 [2111-5544]	21 12 07 −55 43.2	4.384 0.248	341.13 −42.26	226.0 12.8	CHAIN of 4 (#2)	17 S 10.3	F188/J1592 (292,137)
*AM 2112-470	21 12 11 −47 00.1	4.031 0.248	352.89 −43.68	231.2 20.0	SPIRAL w. PEC. THIN OUTER ARM 286-G-80 in DN 2112-469	10 D 1.6	F286/J2405 (65,67)
AM 2112-554 [2112-5543]	21 12 28 −55 43.3	4.382 0.249	341.11 −42.31	226.0 12.8	CHAIN of 4 (#3)	17 S 10.3	F188/J1592 (289,137)
AM 2112-554 [2112-5543]	21 12 39 −55 43.2	4.380 0.249	341.10 −42.33	226.1 12.8	CHAIN of 4 (#4: SPIRAL) 188-G-1	17 D 10.3	F188/J1592 (288,138)
AM 2112-483	21 12 52 −48 38.0	4.084 0.249	350.63 −43.62	230.3 18.6	DWARF NGC 7029 = 235-G-72	20 D 0.2	F235/J3389 (65,250)
AM 2112-705	21 12 59 −70 51.6	5.636 0.249	322.47 −37.28	217.6 0.2	RING in GROUP of 7 IC 5091 = 74-G-27 in DN 2113-708	6,17,18 D 10.3	F075/J0953 (295,124)
AM 2113-483 [2114-4837]	21 13 08 −48 34.6	4.081 0.249	350.71 −43.67	230.4 18.6	PAIR (#1: S0) NGC 7041 = 235-G-82	23 D 2.7	F235/J3389 (63,253)
AM 2113-593	21 13 08 −59 32.6	4.585 0.249	336.12 −41.40	223.9 9.6	DISTURBED GALAXY w. PEC. ABS'N 145-IG-1 = B2113-595 = Fa 96 [DN]	14,16 D 0.9	F/J1578: (305,197)
AM 2113-222	21 13 22 −22 26.6	3.440 0.250	26.24 −41.23	249.7 38.7	GROUP of 5 599-G-4	18 D 8.5	F530/J0797 (212,317)
AM 2113-483	21 13 25 −48 37.4	4.081 0.250	350.63 −43.71	230.4 18.5	DWARF (NUCLEATED/SPIRAL?)	20 D 0.2	F235/J3389 (60,250)
*AM 2113-422 [2113-4228]	21 13 30 −42 28.2	3.885 0.250	359.16 −44.17	234.3 23.5	I/A DBL.(#1): SS-26a = B 2113-424 MCG-7-43-33 = Ag-66 = 342-G-26	2 D 1.0	F342/J2611 (183,47)
AM 2113-560	21 13 33 −56 09.0	4.396 0.250	340.50 −42.35	225.9 12.4	SMALL I/A DBL. 188-IG-2	2 S 0.2	F188/J1592 (280,115)

Identification	R.A. (1950.0) Declination	Precession R.A./Dec.	Gal. Long. Gal. Lat.	S.G.L. S.G.B.	Description Cross-Identification	Coding Size (Min.)	Field/Plate (x, y)
*AM 2113-422 [2113-4228]	21 13 41 −42 28.3	3.885 0.250	359.16 −44.20	234.4 23.5	I/A DBL.(#2): SS-26b = B 2113-424 MCG-7-43-34 = Ag-66 = 342-G-27	2 D 1.0	F342/J2611 (182,47)
AM 2113-590	21 13 55 −59 04.7	4.551 0.250	336.68 −41.63	224.3 9.9	GALAXY w. CURVED FILAMENT 145-IG-3 = B 2113-590	7,15 D 1.7	F145/J1578 (301,222)
AM 2113-341	21 13 59 −34 12.0	3.675 0.250	10.66 −43.85	240.2 29.9	DISTURBED GALAXY + ABSORPTION 402-G-21 (CENTRE of GROUP)	14,18 D 1.6	F402/J0735 (153,221)
AM 2114-412	21 14 18 −41 27.6	3.854 0.250	0.57 −44.33	235.1 24.2	GALAXY + MATERIAL 342-IG-32	15 D 0.7	F342/J2611 (176,101)
AM 2114-483 [2113-4834]	21 14 28 −48 36.3	4.075 0.251	350.63 −43.89	230.6 18.5	PAIR (#1: HI.S.BR. GALAXY)	13,23 S 1.7	F235/J3389 (51,251)
AM 2114-484	21 14 35 −48 44.9	4.079 0.251	350.43 −43.89	230.5 18.3	DWARF (NUCLEUS)	20 D 0.3	F235/J3389 (50,243)
AM 2114-483 [2115-4846]	21 14 36 −48 36.9	4.075 0.251	350.61 −43.91	230.6 18.4	PAIR (#2: LO.S.BR. SPIRAL) 235-G-84 = UKS 2114-486?	20,23 S 1.7	F235/J3389 (50,250)
AM 2114-440 [2114-4407]	21 14 38 −44 06.4	3.928 0.251	356.87 −44.31	233.4 22.1	I/A DBL.(#1)	2 S 1.7	F286/J2405 (35,222)
AM 2114-440 [2114-4406]	21 14 38 −44 07.9	3.929 0.251	356.83 −44.31	233.4 22.0	I/A DBL.(#2)	2 S 1.7	F286/J2405 (35,220)
AM 2114-463	21 14 42 −46 30.9	4.003 0.251	353.51 −44.15	231.9 20.1	ISOLATED SP. w. BIFURCATED ARM 287-G-4	10 D 1.6	F286/J2405 (41,92)
AM 2114-230	21 14 56 −23 00.9	3.447 0.251	25.65 −41.74	249.5 38.0	GROUP of 6 MCG-5-50-14/15 = 530-G-31/32	18 D 6.3	F530/J0797 (193,286)
AM 2115-272	21 15 07 −27 28.8	3.531 0.251	19.83 −42.93	245.7 34.8	GROUP of 3	24 D 1.3	F464/J0640 (59,314)
*AM 2115-574	21 15 14 −57 43.2	4.466 0.251	338.37 −42.17	225.2 11.0	I/A QUARTET [N.] of BR. GALAXY	4,8 D 0.5	F145/J1578 (297,296)
AM 2115-602	21 15 21 −60 20.3	4.617 0.251	335.00 −41.43	223.7 8.8	CLOSE TRIPLE	24 D 1.8	F145/J1578 (287,155)
AM 2115-273	21 15 25 −27 33.7	3.532 0.252	19.74 −43.02	245.7 34.7	I/A DBL. w. MATERIAL MCG-5-50-9 = 464-G-31	2,15 D 1.0	F530/J0797 (186,42)
*AM 2115-484 [2114-4836]	21 15 38 −48 46.5	4.075 0.252	350.36 −44.06	230.6 18.2	PAIR (#2: S0 w. ABSORPTION) NGC 7049 = 236-G-1	14,23 D 1.7	F235/J3389 (41,242)
AM 2115-743	21 15 42 −74 32.3	6.246 0.252	318.28 −35.78	215.7 −3.0	DWARF 47-G-28	20 D 0.7	F048/J1819 (334,188)
AM 2115-685	21 15 56 −68 56.5	5.350 0.252	324.53 −38.32	218.9 1.7	COMPACT in GROUP of 4	13,18 D 0.2	F074/J0907 (43,226)
*AM 2116-725	21 16 08 −72 58.7	5.934 0.252	319.97 −36.55	216.6 −1.7	I/A SPIRALS (ARM into NUCLEUS) 47-IG-29 = B 2116-729	2,10 S 0.6	F047/J1797 (93,284)
AM 2116-451 [2116-4520]	21 16 25 −45 19.8	3.957 0.252	355.15 −44.55	232.9 20.9	CLOSE GROUP of 3 (#1) 287-G-07	24 D 1.6	F286/J2405 (21,156)
AM 2116-452 [2116-4519]	21 16 30 −45 20.8	3.958 0.253	355.10 −44.56	232.9 20.9	CLOSE GROUP of 3 (#2)	24 D 1.6	F286/J2405 (21,154)
AM 2116-851	21 16 30 −85 13.6	0.226 0.253	307.42 −30.18	209.6 −11.8	LOOSE CHAIN of 4 Fa 583	17 D 3.4	F011/J2562 (201,166)
AM 2116-452 [2116-4520]	21 16 32 −45 20.3	3.958 0.253	355.11 −44.57	232.9 20.9	CLOSE GROUP of 3 (#3)	24 D 1.6	F286/J2405 (20,155)
AM 2116-511	21 16 32 −51 16.2	4.163 0.253	346.91 −43.81	229.2 16.1	SPIRAL + APP. COMP'N. 236-G-3	8 S 0.9	F236/J1823 (290,106)
AM 2116-851	21 16 49 −85 13.4	0.225 0.253	307.42 −30.19	209.6 −11.8	GROUP of 4	18 D 3.4	F002/J2563 (347,105)
AM 2116-220	21 16 55 −22 02.4	3.426 0.253	27.09 −41.89	250.8 38.4	I/A DBL.	2 S 0.6	F530/J0797 (168,339)
AM 2116-395	21 16 57 −39 58.6	3.804 0.253	2.65 −44.82	236.6 25.0	SPIRAL(!) w.LO.S.BR. RING COMP'N 342-G-35	6,8 S 2.8	F342/J2611 (149,180)
AM 2117-371	21 17 21 −37 12.7	3.734 0.253	6.55 −44.80	238.6 27.1	GALAXY w. I/A COMP'N.	1 S 0.3	F342/J2611 (144,329)
AM 2117-472	21 17 24 −47 24.8	4.020 0.253	352.19 −44.52	231.7 19.1	I/A DBL. (incl. N GALAXY) 287-G-8	2 S 1.2	F286/J2405 (19,43)
AM 2117-672	21 17 27 −67 26.2	5.167 0.253	326.21 −39.08	219.8 2.8	I/A DBL. (in GROUP of 3) 107-IG-23 = B 2117-674	2,15,24 S 0.2	F107/J0908 (170,46)
AM 2117-682	21 17 27 −68 22.9	5.268 0.253	325.11 −38.69	219.3 2.0	GALAXY w. I/A COMP'N.	1 S 0.2	F075/J0953 (288,259)
AM 2117-673 [2118-6734]	21 17 30 −67 33.5	5.179 0.254	326.07 −39.03	219.8 2.7	I/A TRIPLE (#1)	3 D 0.2	F075/J0953 (292,303)
AM 2117-515	21 17 44 −51 56.8	4.183 0.254	345.93 −43.87	228.9 15.4	SPIRAL w. I/A COMP'N. 236-IG-5	1 S 1.3	F188/J1592 (256,342)
AM 2117-304	21 17 51 −30 44.3	3.589 0.254	15.59 −44.16	243.6 32.0	CHAIN of 7	17 D 13.4	F464/J0640 (31,138)
AM 2117-673 [2117-6733]	21 17 57 −67 35.1	5.177 0.254	326.01 −39.06	219.8 2.7	I/A TRIPLE (#2)	3 D 0.4	F075/J0953 (289,302)
AM 2118-464	21 18 02 −46 48.6	3.997 0.254	353.02 −44.69	232.2 19.5	CONTACT E's + ENVELOPE (in 3) 287-IG-10	2,24 S 0.2	F287/J3653 (271,81)

Identification	R.A. (1950.0) Declination	Precession R.A./Dec.	Gal. Long. Gal. Lat.	S.G.L. S.G.B.	Description Cross-Identification	Coding Size (Min.)	Field/Plate (x, y)
AM 2118-673 [2117-6735]	21 18 27 −67 34.6	5.171 0.254	326.00 −39.11	219.8 2.7	I/A TRIPLE (#3) 75-IG-04	3 D 0.3	F075/J0953 (287,302)
AM 2118-695	21 18 55 −69 53.9	5.433 0.255	323.29 −38.15	218.5 0.7	SB(r) + DISTURBED ARM + COMP'N. 75-G-6 = Fa 584	2,12 S 1.5	F075/J0953 (274,178)
AM 2118-410 [2118-4100]	21 18 58 −41 03.2	3.825 0.255	1.13 −45.21	236.2 24.0	CLOSE PAIR (#1: E)	23 D 0.6	F342/J2611 (129,123)
AM 2118-410 [2118-4103]	21 18 59 −41 00.8	3.824 0.255	1.18 −45.21	236.2 24.0	CLOSE PAIR (#2: SPIRAL)	23 D 0.7	F342/J2611 (128,125)
AM 2119-503	21 19 12 −50 39.5	4.125 0.255	347.64 −44.33	229.9 16.3	SP. + COMP'NS (ONE I/A DBL.) B 2119-506	1,2 S 3.4	F236/J1823 (269,139)
AM 2119-695	21 19 15 −69 50.2	5.421 0.255	323.35 −38.20	218.6 0.8	I/A DBL. across SB(r)	2,8 S 1.4	F074/J0907 (33,177)
AM 2119-365	21 19 26 −36 53.6	3.720 0.255	7.03 −45.19	239.2 27.1	GALAXY w. I/A COMP'N. 402-G-26	1 S 3.6	F342/J2611 (121,346)
AM 2119-272	21 19 36 −27 23.6	3.519 0.255	20.25 −43.88	246.6 34.1	DISTURBED. SP. w. "ANTENNAE" 530-IG-38	15 D 0.9	F465/J0765 (275,317)
AM 2119-595 [2119-5953]	21 19 39 −59 51.8	4.555 0.256	335.38 −42.08	224.4 8.9	CLOSE PAIR (SPIRALS) (#1)	23 D 0.4	F145/J1578 (260,182)
AM 2119-475	21 19 44 −47 51.9	4.024 0.256	351.50 −44.85	231.8 18.5	I/A DBL.	2 S 1.2	F236/J1823 (270,289)
AM 2119-595 [2119-5952]	21 19 56 −59 53.4	4.554 0.256	335.33 −42.11	224.4 8.8	CLOSE PAIR (SPIRALS) (#2)	23 D 0.4	F145/J1578 (258,181)
AM 2120-362	21 20 36 −36 20.3	3.703 0.256	7.84 −45.39	239.8 27.4	COMPACT	13 D 0.3	F403/J3544 (346,104)
AM 2121-322	21 21 42 −32 23.3	3.613 0.257	13.47 −45.22	243.0 30.2	STAR + WISP of NEBULOSITY	21 D 1.0	F402/J0735 (64,316)
AM 2121-424 [2122-4237]	21 21 45 −42 40.4	3.857 0.258	358.80 −45.68	235.5 22.4	CLOSE TRIPLE (#1) NGC 7057 = 287-G-17	24 D 0.9	F342/J2611 (102,35)
AM 2122-394	21 22 18 −39 44.4	3.778 0.258	3.00 −45.85	237.6 24.6	DWARF	20 D 0.3	F342/J2611 (93,193)
*AM 2122-402	21 22 22 −40 29.3	3.797 0.258	1.92 −45.86	237.1 24.0	SPIRAL + I/A COMP'N. + APP. COMP'N IC 5105a = 342-G-43	1,8 S 1.1	F342/J2611 (94,152)
AM 2122-424 [2121-4240]	21 22 26 −42 42.1	3.855 0.258	358.75 −45.80	235.6 22.3	CLOSE TRIPLE (#2)	24 D 0.5	F342/J2611 (95,33)
AM 2122-423 [2121-4242]	21 22 42 −42 37.6	3.852 0.258	358.85 −45.85	235.7 22.3	CLOSE TRIPLE (#3) NGC 7060 = 287-G-22	24 D 1.3	F342/J2611 (93,37)
AM 2122-452	21 22 44 −45 21.8	3.930 0.258	354.94 −45.65	233.8 20.2	I/A DBL. (SPIRAL + SPIRAL ?) 287-IG-23	2 S 0.4	F287/J3653 (229,159)
AM 2122-345	21 22 50 −34 56.2	3.664 0.259	9.89 −45.74	241.2 28.1	TIGHT CLUSTER	19 D 5.4	F402/J0735 (56,179)
AM 2122-410	21 22 51 −41 02.9	3.810 0.259	1.12 −45.94	236.8 23.5	DISTURBED HI.S.BR. GALAXY IC 5105b = 342-G-46	16 D 1.0	F342/J2611 (89,122)
AM 2123-481	21 23 04 −48 16.6	4.021 0.259	350.81 −45.35	232.0 17.9	V.FAINT, DISTANT RICH CLUSTER	19 D 9.0	F236/J1823 (239,268)
AM 2123-225	21 23 05 −22 57.6	3.431 0.259	26.46 −43.52	251.2 36.7	CLOSE PAIR of SPIRALS w. COMP'NS MCG-4-50-23 = 530-G-43/46	8,23 S 1.7	F530/J0797 (92,289)
AM 2123-610 [2124-6105]	21 23 24 −61 07.0	4.602 0.259	333.58 −42.10	224.0 7.6	GROUP of 3 + COMP'NS.(#1)	8,24 D 0.6	F145/J1578 (232,116)
AM 2123-491	21 23 40 −49 19.7	4.053 0.259	349.31 −45.27	231.4 17.0	GROUP of 8 GALAXIES NGC 7061 = 236-G-13	18 D 17.9	F236/J1823 (233,212)
AM 2123-610 [2123-6107]	21 23 44 −61 02.5	4.595 0.259	333.65 −42.16	224.1 7.6	GROUP of 3 + COMP'NS.(#2) 145-G-6	8,24 D 0.7	F145/J1578 (230,120)
AM 2123-513	21 23 49 −51 33.4	4.133 0.259	346.20 −44.87	230.0 15.2	CLOSE I/A DBL.	2 S 0.3	F236/J1823 (229,92)
AM 2123-215	21 23 55 −21 58.8	3.412 0.259	27.83 −43.42	252.2 37.2	GROUP of 4 599-G-13	18 D 6.3	F530/J0797 (81,341)
AM 2124-610 [2123-6102]	21 24 12 −61 05.1	4.593 0.260	333.57 −42.20	224.1 7.6	GROUP of 3 + COMP'NS.(#3)	8,24 D 0.8	F145/J1578 (227,118)
AM 2124-393	21 24 14 −39 39.6	3.769 0.260	3.11 −46.22	238.0 24.4	DWARF	20 D 0.4	F342/J2611 (73,196)
AM 2124-270	21 24 20 −27 00.3	3.501 0.260	21.11 −44.83	247.8 33.7	SPIRAL w. PECULIAR ARM 530-G-49	10 D 0.7	F531/J0650 (340,68)
AM 2124-841	21 24 23 −84 11.2	11.314 0.260	308.28 −30.93	210.4 −11.0	CONTACT I/A DBL. (WIDE GROUP)	2 S 0.1	F011/J2562 (195,223)
*AM 2125-532	21 25 06 −53 20.9	4.196 0.261	343.68 −44.65	229.0 13.7	IRR. SINGLE/DWARF 188-IG-8	16,20 D 0.9	F188/J1592 (194,268)
*AM 2125-380	21 25 07 −38 04.7	3.728 0.261	5.41 −46.37	239.3 25.5	DISTURBED SP. w. COMPACT COMP'N. A 2125-38 = 342-G-50	1,10,13 D 1.8	F342/J2611 (62,281)
*AM 2125-533	21 25 44 −53 32.4	4.200 0.261	343.39 −44.70	228.9 13.4	V.SMALL LO.S.BR. GAL. w. KNOTS	20 D 0.3	F188/J1592 (189,257)
AM 2125-280	21 25 56 −28 05.1	3.517 0.261	19.72 −45.40	247.2 32.7	E + SPIRAL (ARM BENT from E)	2 S 0.7	F531/J0650 (320,10)

Identification	R.A. (1950.0) Declination	Precession R.A./Dec.	Gal. Long. Gal. Lat.	S.G.L. S.G.B.	Description Cross-Identification	Coding Size (Min.)	Field/Plate (x, y)
AM 2126-764	21 26 18 −76 42.0	6.585 0.262	315.59 −35.25	215.0 −5.1	S0 in PAIR w. EDGE-ON 47-G-34	23 D 1.8	F047/J1797 (80,81)
AM 2126-565	21 26 24 −56 58.3	4.349 0.262	338.75 −43.83	226.9 10.7	I/A DBL. B 2126-569	2 S 0.5	F145/J1578 (217,339)
AM 2126-395	21 26 34 −39 53.5	3.766 0.262	2.77 −46.67	238.2 23.9	I/A DBL. in GROUP of 4 342-IG-55	2,18 D 1.4	F342/J2611 (50,183)
AM 2126-601	21 26 57 −60 13.4	4.519 0.262	334.50 −42.82	224.9 8.0	SPIRAL w. PECULIAR ARM IC 5110 = 145-IG-7 = B 2126-602	10 D 1.3	F145/J1578 (210,165)
AM 2126-244	21 26 59 −24 45.9	3.455 0.262	24.38 −44.87	250.3 34.8	SMALL GROUP of 3 530-G-54	24 D 0.9	F530/J0797 (45,191)
AM 2127-585	21 27 06 −58 53.4	4.443 0.262	336.20 −43.30	225.8 9.1	CLOSE PAIR	23 S 1.0	F145/J1578 (210,236)
*AM 2127-431	21 27 12 −43 18.4	3.852 0.262	357.79 −46.63	235.9 21.3	SPIRAL w. 3 HI.S.BR. COMP'NS. NGC 7070 = 287-G-28	8 S 11.2	F287/J3653 (187,270)
AM 2127-403	21 27 14 −40 39.4	3.782 0.263	1.65 −46.78	237.7 23.3	RING-SPIRAL + APP. COMP'N. 343-G-1	6,8 S 1.8	F342/J2611 (44,142)
AM 2127-443	21 27 21 −44 39.2	3.889 0.263	355.84 −46.54	235.0 20.2	GALAXY w. TAIL 287-IG-30	15 D 0.7	F287/J3653 (186,198)
AM 2127-830 [2133-8301]	21 27 28 −83 07.3	9.912 0.263	309.22 −31.63	211.1 −10.2	PAIR (#1: E) 11-G-3	23 D 0.4	F011/J2562 (194,280)
AM 2127-751	21 27 46 −75 19.3	6.217 0.263	316.96 −36.06	215.9 −4.0	DWARF 47-G-36	20 D 0.7	F048/J1819 (286,153)
*AM 2127-221	21 27 47 −22 15.3	3.410 0.263	27.84 −44.36	252.7 36.4	GALAXY w. CONDENSATION in PLUME 599-IG-17	15 D 0.9	F531/J0650 (303,324)
AM 2128-291	21 28 08 −29 15.6	3.533 0.263	18.20 −46.10	246.6 31.5	SPIRAL + CONNECTED E 465-G-3	2 S 0.3	F465/J0765 (174,217)
AM 2128-385 [2128-3850]	21 28 08 −38 50.1	3.735 0.263	4.32 −46.97	239.2 24.5	CLOSE PAIR (#1) 343-G-3	23 D 0.7	F342/J2611 (32,239)
AM 2128-302	21 28 11 −30 29.5	3.557 0.263	16.44 −46.31	245.6 30.6	E + RING 465-G-4	6,23 D 1.1	F465/J0765 (174,151)
AM 2128-470	21 28 17 −47 06.7	3.957 0.263	352.27 −46.39	233.4 18.2	SPIRAL + LONG ARM + COMP'NS. 287-G-33	8,10 S 2.7	F236/J1823 (193,332)
AM 2128-371	21 28 25 −37 16.5	3.698 0.264	6.60 −47.00	240.4 25.6	I/A DBL (+ STAR ?)	2 S 0.5	F343/J0828 (304,324)
AM 2128-385 [2128-3850]	21 28 28 −38 50.0	3.733 0.264	4.32 −47.04	239.3 24.5	CLOSE PAIR (#2) NGC 7075 = 343-G-4 = PKS 2128-388	23 D 0.7	F342/J2611 (28,239)
AM 2128-380 [2128-3802]	21 28 35 −38 01.0	3.714 0.264	5.52 −47.05	239.9 25.1	CLOSE PAIR (M51-TYPE ?) (#1)	9,23 D 0.5	F343/J0828 (301,284)
*AM 2128-430	21 28 35 −43 04.2	3.839 0.264	358.11 −46.90	236.3 21.3	GAL.w. FNT. CIRC. ARMS+ PEC. ABSN Se 145/2 = NGC 7070a? = 287-G-34	10,14,22 D 2.7	F287/J3653 (174,283)
AM 2128-461	21 28 36 −46 16.7	3.930 0.264	353.46 −46.56	234.1 18.8	I/A DBL. + 3 COMP'NS.(#1: SP.) 287-G-35	2,8 S 2.5	F287/J3653 (174,111)
AM 2128-380 [2128-3801]	21 28 43 −38 02.4	3.714 0.264	5.48 −47.08	239.9 25.0	CLOSE PAIR (M51-TYPE ?) (#2)	9,23 D 0.4	F343/J0828 (299,283)
AM 2128-461	21 28 49 −46 15.9	3.929 0.264	353.47 −46.60	234.1 18.8	I/A DBL. + 3 COMP'NS. (#2: E:)	2,8 S 2.5	F287/J3653 (172,112)
AM 2128-501	21 28 53 −50 14.1	4.057 0.264	347.81 −45.93	231.5 15.7	I/A DBL.	2 S 0.6	F236/J1823 (187,164)
AM 2129-474	21 29 17 −47 42.4	3.971 0.264	351.38 −46.46	233.2 17.6	I/A DBL. 236-G-20	2 S 2.1	F236/J1823 (184,300)
AM 2129-343 [2129-3437]	21 29 22 −34 35.9	3.636 0.264	10.54 −47.05	242.6 27.5	CLOSE PAIR (#1) 403-G-6	23 D 0.3	F403/J3544 (252,200)
AM 2129-441	21 29 22 −44 17.5	3.869 0.264	356.31 −46.93	235.5 20.3	GALAXY w. BROAD FUZZY ARMS NGC 7079 = 287-G-36	10,12 D 1.8	F287/J3653 (166,217)
AM 2129-343 [2129-3435]	21 29 30 −34 37.8	3.636 0.265	10.50 −47.08	242.6 27.4	CLOSE PAIR (#2: COMPACT)	13,23 D 0.2	F403/J3544 (251,199)
AM 2129-253	21 29 58 −25 33.7	3.462 0.265	23.52 −45.72	250.1 33.8	I/A DBL. 531-IG-3	2 S 0.4	F531/J0650 (273,147)
AM 2130-294	21 30 32 −29 43.9	3.536 0.265	17.65 −46.69	246.7 30.8	GALAXY w. PLUME + RING MCG-5-50-17 = 465-G-7	6,15 D 0.8	F465/J0765 (146,192)
AM 2130-763	21 30 44 −76 34.2	6.465 0.266	315.55 −35.53	215.2 −5.1	EDGE-ON SPIRAL w. 7 COMP'NS. 48-G-2	8 S 13.4	F047/J1797 (65,86)
AM 2130-425	21 30 50 −42 52.2	3.824 0.266	358.36 −47.33	236.7 21.2	RICH CLUSTER	19 D 3.6	F343/J0828 (269,24)
AM 2130-621	21 30 50 −62 17.2	4.614 0.266	331.66 −42.47	224.0 6.1	GROUP of E GALAXIES Fa 106?	18 D 4.5	F107/J0908 (84,320)
AM 2130-413	21 30 56 −41 35.7	3.791 0.266	0.23 −47.43	237.6 22.1	M51-TYPE	9 S 0.4	F343/J0828 (270,92)
AM 2130-430	21 30 56 −43 01.1	3.828 0.266	358.14 −47.33	236.6 21.0	DISTANT RICH CLUSTER of GALAXIES	19 D 2.5	F342/J2611 (12,14)
AM 2130-260	21 30 59 −26 02.4	3.468 0.266	22.93 −46.06	249.9 33.3	I/A DBL. + I/A DBL.	2,2 D 0.4	F531/J0650 (261,121)

Identification	R.A. (1950.0) Declination	Precession R.A./Dec.	Gal. Long. Gal. Lat.	S.G.L. S.G.B.	Description Cross-Identification	Coding Size (Min.)	Field/Plate (x, y)
AM 2131-700	21 31 08 −70 03.0	5.299 0.266	322.49 −39.00	219.3 0.0	I/A DBL. 75-IG-15	2 S 0.2	F075/J0953 (217,174)
AM 2131-410 [2131-4102]	21 31 09 −41 02.8	3.777 0.266	1.04 −47.50	238.1 22.5	CLOSE PAIR (#1) 343-G-7	23 D 0.7	F343/J0828 (269,122)
AM 2131-833	21 31 10 −83 33.6	10.230 0.266	308.73 −31.46	210.9 −10.6	GALAXY w. PLUMES + I/A COMP'NS. 11-G-4	1,15 S 1.3	F011/J2562 (187,257)
*AM 2131-495	21 31 17 −49 56.0	4.033 0.266	348.12 −46.37	232.0 15.7	SMALL RING (NUCLEUS thru RIM)	6 D 0.7	F236/J1823 (166,180)
AM 2131-323	21 31 22 −32 31.4	3.588 0.266	13.64 −47.26	244.6 28.7	CLOSE PAIR in GROUP of 3	23,24 S 1.2	F403/J3544 (232,312)
AM 2131-333 [2131-3336]	21 31 23 −33 38.5	3.610 0.266	12.00 −47.38	243.7 27.9	I/A DBL. (#1)	2 D 0.2	F403/J3544 (231,252)
AM 2131-333 [2131-3338]	21 31 24 −33 36.5	3.609 0.266	12.05 −47.38	243.7 27.9	I/A DBL. (#2) 403-G-9	2 D 0.9	F403/J3544 (231,254)
AM 2131-410 [2131-4102]	21 31 25 −41 02.5	3.775 0.266	1.04 −47.55	238.1 22.5	CLOSE PAIR (#2) in DN 2131-411 NGC 7087 = 343-G-8 in Se 145/1	23 D 0.9	F343/J0828 (266,122)
AM 2131-254	21 31 39 −25 42.2	3.461 0.266	23.45 −46.12	250.3 33.4	RING GALAXY w. I/A COMP'N.	1,6 S 0.6	F531/J0650 (253,139)
AM 2131-372	21 31 42 −37 21.1	3.688 0.266	6.52 −47.66	240.9 25.1	GALAXY w. I/A COMP'N. 403-G-11	1 S 0.6	F343/J0828 (269,321)
AM 2131-572	21 31 43 −57 21.7	4.330 0.266	337.92 −44.39	227.2 9.9	GALAXY w. JET	7 D 0.3	F145/J1578 (179,319)
*AM 2131-562	21 31 47 −56 27.8	4.288 0.267	339.10 −44.69	227.8 10.6	ONE-ARMED SP. + I/A DBL. COMP'N. Fa 585	2,8,10 S 1.0	F188/J1592 (144,100)
AM 2132-500	21 32 00 −50 09.6	4.037 0.267	347.76 −46.43	231.9 15.5	SMALL I/A TRIPLE (QUINTET?) 236-IG-23 = B 2132-501	3,5 D 2.7	F236/J1823 (160,168)
AM 2132-513	21 32 13 −51 32.3	4.084 0.267	345.81 −46.15	231.0 14.4	LUMPY COMPACT	13 D 0.3	F236/J1823 (159,94)
AM 2132-664	21 32 21 −66 42.3	4.937 0.267	326.24 −40.70	221.4 2.5	GAL. w. I/A COMP'N (GROUP of 3) 107-G-38	1,24 S 1.1	F107/J0908 (90,83)
AM 2132-390	21 32 25 −39 04.8	3.724 0.267	3.95 −47.81	239.7 23.8	GALAXY w. I/A COMP'N.	1 S 0.2	F343/J0828 (259,228)
AM 2132-643	21 32 29 −64 39.1	4.765 0.267	328.67 −41.64	222.7 4.2	COMPACT Fa 108	13 D 0.3	F107/J0908 (82,193)
AM 2132-244	21 32 30 −24 42.1	3.442 0.267	24.92 −46.07	251.4 33.9	2 E's (COMMON ENV.) + COMPACTS	2,13 D 0.7	F531/J0650 (243,193)
AM 2132-265	21 32 31 −26 53.1	3.479 0.267	21.84 −46.58	249.5 32.4	SPIRAL w. I/A COMP'N. MGC-5-51-2 = 531-G-8	1 S 0.4	F531/J0650 (242,76)
AM 2132-535	21 32 43 −53 54.9	4.172 0.267	342.49 −45.59	229.5 12.5	FAINT RING GALAXY Fa 586	6 D 0.5	F188/J1592 (134,237)
AM 2132-722	21 32 47 −72 21.1	5.589 0.267	319.88 −37.94	218.0 −1.9	I/A DBL. (CLOSE)	2 S 0.2	F048/J1819 (287,313)
AM 2133-544	21 33 01 −54 46.9	4.205 0.268	341.28 −45.38	229.0 11.8	DUSTY IRR. w. ASYMMETRIC TAIL NGC 7090 = 188-G-12	15,20 D 8.1	F188/J1592 (133,190)
*AM 2133-384	21 33 05 −38 46.0	3.715 0.268	4.41 −47.94	240.0 23.9	SINGLE V. DISTURBED GALAXY(TAILS) VV 714 = 343-IG-13 = B 2133-387	15,16 D 1.1	F343/J0828 (252,245)
AM 2133-830 [2127-8307]	21 33 08 −83 01.5	9.602 0.268	309.19 −31.83	211.3 −10.3	PAIR (#2: EDGE-ON + COMP'NS.) 11-G-5	8,23 S 3.4	F011/J2562 (185,285)
*AM 2133-451	21 33 20 −45 17.3	3.879 0.268	354.73 −47.51	235.4 19.0	LUMPY IRR. + COMPACT COMP'N. 287-IG-39 = B 2133-452	13,16 D 0.7	F287/J3653 (129,164)
*AM 2134-471	21 34 10 −47 15.6	3.932 0.269	351.82 −47.35	234.2 17.5	SPIRAL w. OFF-SET NUCLEUS 287-G-40 = B 2134-462 (4 COMP'NS)	6,8,10,17 S 2.0	F236/J1823 (140,323)
AM 2134-322	21 34 11 −32 27.7	3.578 0.269	13.83 −47.84	245.1 28.3	I/A DBL. (COMPACT + LO.S.BR. SPIRAL)	2,13 S 0.5	F403/J3544 (200,316)
AM 2134-562	21 34 16 −56 27.2	4.270 0.269	338.95 −45.02	228.1 10.4	SP. + COMP'N. SP. (STAR BETWEEN) 188-GA-12 = Lu 2134-56	23 S 3.6	F188/J1592 (125,100)
AM 2134-481	21 34 32 −48 10.3	3.958 0.269	350.48 −47.24	233.6 16.7	I/A TRIPLE	3 D 2.2	F236/J1823 (137,274)
AM 2134-520 [2135-5202]	21 34 39 −52 02.1	4.088 0.269	344.97 −46.38	231.0 13.8	CHAIN of 3 (#1)	24 S 4.0	F188/J1592 (116,338)
AM 2134-425	21 34 47 −42 56.6	3.809 0.269	358.15 −48.04	237.3 20.6	GALAXY w. I/A COMP'N. NGC 7095? = 287-IG-41/42	1 S 0.9	F343/J0828 (230,21)
AM 2134-520 [2134-5202]	21 34 51 −52 02.3	4.086 0.269	344.96 −46.41	231.0 13.7	CHAIN of 3 (#2: HI.S.BR. SP.)	24 S 4.0	F188/J1592 (114,337)
AM 2134-425	21 34 58 −42 59.2	3.809 0.269	358.08 −48.07	237.3 20.6	GALAXY w. I/A COMP'N. 287-IG-42 = B2134-429 = NGC 7095?	1 S 1.2	F287/J3653 (111,287)
AM 2135-520 [2134-5202]	21 35 01 −52 02.3	4.085 0.269	344.95 −46.44	231.0 13.7	CHAIN of 3 (#3)	24 S 4.0	F188/J1592 (113,337)
AM 2135-513	21 35 12 −51 37.7	4.069 0.270	345.51 −46.57	231.3 14.0	CLOSE TRIPLE (SMALL)	24 D 0.3	F236/J1823 (134,89)
AM 2135-592	21 35 50 −59 28.9	4.406 0.270	334.88 −44.13	226.3 7.9	GALAXY w. TAIL (both HI.S.BR.) 145-IG-9	15 D 0.5	F145/J1578 (150,205)

Identification	R.A. (1950.0) Declination	Precession R.A./Dec.	Gal. Long. Gal. Lat.	S.G.L. S.G.B.	Description Cross-Identification	Coding Size (Min.)	Field/Plate (x, y)
AM 2136-492	21 36 07 −49 23.3	3.988 0.270	348.65 −47.25	233.0 15.6	GROUP of 3 (incl. DBL.RINGED SP.) 236-G-29	6,24 D 7.6	F236/J1823 (124,209)
AM 2136-640	21 36 08 −64 07.9	4.691 0.270	329.06 −42.23	223.3 4.3	DWARF 107-G-44 = PKS 2136-642?	20 D 0.5	F108/J0946 (316,218)
AM 2136-640	21 36 35 −64 02.3	4.680 0.271	329.14 −42.31	223.4 4.4	DWARF 107-G-45	20 D 0.6	F108/J0946 (313,223)
AM 2136-695	21 36 54 −69 51.7	5.203 0.271	322.38 −39.52	219.8 −0.2	CLOSE PAIR	23 S 0.6	F075/J0953 (191,184)
*AM 2137-424	21 37 03 −42 46.1	3.795 0.271	358.35 −48.47	237.7 20.5	E w. DIAMETRIC COMP'NS. NGC 7097 = 287-G-48 [A82]	8 S 3.4	F343/J0828 (208,31)
AM 2137-483	21 37 05 −48 37.6	3.959 0.271	349.70 −47.57	233.6 16.1	LINE of 3 (ONE DISTURBED)	24 D 2.7	F236/J1823 (115,250)
AM 2137-225	21 37 09 −22 55.1	3.403 0.271	27.82 −46.62	253.8 34.3	CHAIN of 5 (BR.GAL. + FAINTER) NGC 7103/4 + IC 5122/24 + IC 1393	17 D 10.1	F531/J0650 (186,289)
AM 2137-565	21 37 15 −56 58.8	4.272 0.271	338.06 −45.23	228.0 9.7	I/A DBL.	2 S 0.4	F145/J1578 (138,339)
AM 2137-640	21 37 26 −64 07.9	4.678 0.271	328.98 −42.35	223.5 4.2	TRIPLE off MI.AXIS of BR.GAL. NGC 7096 = IC-5121 = 107-G-46	8,24 D 7.8	F108/J0946 (308,218)
AM 2137-695	21 37 32 −69 51.2	5.194 0.272	322.35 −39.57	219.8 −0.2	CURVED FILAMENT w. NUCLEUS	22 D 0.4	F075/J0953 (188,185)
*AM 2137-515	21 37 55 −51 59.4	4.066 0.272	344.84 −46.88	231.4 13.5	SPIRAL w. I/A COMP'N. (JET?) 236-G-31	1,7 S 1.1	F188/J1592 (89,339)
AM 2137-490	21 37 56 −49 07.7	3.970 0.272	348.92 −47.59	233.4 15.6	COMPACT	13 D 0.3	F236/J1823 (108,222)
AM 2137-610	21 37 58 −61 07.1	4.480 0.272	332.65 −43.72	225.4 6.5	I/A DBL. Fa 113 = 145-IG-11	2 S 0.3	F145/J1578 (138,116)
AM 2138-545	21 38 06 −54 55.1	4.177 0.272	340.77 −46.03	229.5 11.2	SMALL PECULIAR SPIRAL 188-IG-15	12 D 0.4	F188/J1592 (94,182)
*AM 2138-461	21 38 43 −46 14.2	3.879 0.273	353.12 −48.30	235.5 17.7	SMALL GALAXY w. JETS 287-IG-50	7 D 0.4	F288/J1633 (339,108)
AM 2138-643	21 38 50 −64 39.8	4.703 0.273	328.24 −42.24	223.2 3.7	DISTURBED GALAXY 108-IG-1	16 D 0.6	F107/J0908 (46,189)
AM 2139-641	21 39 05 −64 18.2	4.675 0.273	328.66 −42.43	223.5 4.0	SPIRAL w. ARM through COMP'N.	1 S 0.4	F108/J0946 (298,210)
*AM 2139-525	21 39 12 −52 55.7	4.092 0.273	343.45 −46.80	230.9 12.6	SP.(I/A DBL. + PEC. SP. DIAMETRIC) NGC 7106 = 188-G-17 (+ B2139-529)	2,8,12 D 11.2	F188/J1592 (81,288)
AM 2139-450	21 39 14 −45 01.4	3.844 0.273	354.90 −48.58	236.4 18.5	SBd/DWARF NGC 7107 = 287-G-52 = Lu 2139-45	20 D 1.6	F287/J3653 (73,176)
AM 2139-643	21 39 24 −64 31.6	4.687 0.273	328.37 −42.36	223.4 3.8	GALAXY w. I/A COMP'N. (BRIDGE?)	1,15 S 0.4	F107/J0908 (42,196)
*AM 2139-481	21 39 27 −48 13.2	3.934 0.273	350.16 −48.04	234.2 16.1	SPIRAL + I/A + APP. COMP'NS. 236-G-35	1,8 S 9.0	F236/J1823 (93,271)
AM 2139-704	21 39 57 −70 40.3	5.258 0.274	321.31 −39.31	219.5 −1.0	I/A DBL. w. TAIL	2,15 S 1.3	F075/J0953 (177,141)
AM 2140-492	21 40 04 −49 28.6	3.969 0.274	348.30 −47.85	233.4 15.1	V.COMPACT	13 D 0.3	F236/J1823 (90,203)
AM 2140-652	21 40 17 −65 20.5	4.739 0.274	327.34 −42.06	222.9 3.1	GALAXY w. JETS (SMALL)	7 D 0.4	F108/J0946 (287,155)
AM 2140-493	21 40 27 −49 33.4	3.969 0.274	348.16 −47.89	233.4 15.0	SMALL CLOSE DBL.	2 S 0.4	F236/J1823 (86,199)
AM 2141-342	21 41 34 −34 20.2	3.591 0.275	11.19 −49.54	244.8 25.9	I/A SPIRALS (TAILS) 403-G-19/20	2,15 S 0.9	F403/J3544 (117,215)
*AM 2141-491 [2141-4914]	21 41 37 −49 14.6	3.953 0.275	348.55 −48.15	233.8 15.1	I/A TRIPLE (#1 and #2)	3 S 0.5	F236/J1823 (76,215)
AM 2141-361	21 41 38 −36 18.5	3.630 0.275	8.15 −49.62	243.2 24.5	CLOSE PAIR	23 S 1.4	F403/J3544 (118,109)
AM 2141-340	21 41 50 −34 04.3	3.585 0.275	11.60 −49.58	245.0 26.0	CLOSE I/A DBL. 403-IG-21	2 S 0.2	F403/J3544 (114,229)
AM 2141-515	21 41 54 −51 55.6	4.040 0.275	344.68 −47.49	231.9 13.1	RING w. COMMA-SHAPED NUCLEUS 236-IG-38	6 D 0.6	F236/J1823 (79,71)
*AM 2141-491 [2141-4914]	21 41 59 −49 14.2	3.951 0.275	348.54 −48.21	233.8 15.1	I/A TRIPLE (#3) 236-G-39	3 D 0.8	F236/J1823 (72,216)
AM 2142-725	21 42 11 −72 57.9	5.540 0.275	318.74 −38.18	218.1 −2.8	GALAXY w. JET (LO.S.BR.)	7 D 0.4	F048/J1819 (246,284)
AM 2142-483 [2142-4834]	21 42 32 −48 39.4	3.930 0.276	349.35 −48.44	234.3 15.5	PAIR of E's (+ COMP'N.) #1 NGC 7117 = SS-27a	8,23 D 0.7	F236/J1823 (66,246)
AM 2142-490	21 42 34 −49 02.5	3.942 0.276	348.79 −48.35	234.0 15.2	IRR. (near I/A TRIPLE) 236-G-41	16 D 0.7	F237/J3558 (333,227)
AM 2142-523	21 42 48 −52 30.0	4.054 0.276	343.81 −47.45	231.6 12.6	I/A DBL. 236-IG-44	2 S 1.5	F236/J1823 (73,40)
AM 2142-573	21 42 49 −57 31.2	4.255 0.276	336.94 −45.73	228.3 8.8	CENTRAL E in ELONGATED CLUSTER Fa 116	19 D 40.3	F145/J1578 (98,308)

Identification	R.A. (1950.0) Declination	Precession R.A./Dec.	Gal. Long. Gal. Lat.	S.G.L. S.G.B.	Description Cross-Identification	Coding Size (Min.)	Field/Plate (x, y)
AM 2142-483 [2142-4835]	21 42 54 −48 34.8	3.926 0.276	349.45 −48.52	234.4 15.5	PAIR of E's (+COMP'N.) #2 NGC 7118 = 236-G-45 = SS-27b	8,23 D 0.8	F236/J1823 (63,250)
AM 2143-481	21 43 01 −48 12.0	3.914 0.276	350.00 −48.62	234.7 15.8	GALAXY + I/A COMP'N. 236-IG-46	1 S 0.6	F236/J1823 (61,271)
*AM 2143-464	21 43 02 −46 45.0	3.872 0.276	352.15 −48.93	235.7 16.8	DISTURBED SP. w. I/A COMP'NS. NGC 7119b = 288-IG-1 = B 2143-467	1 S 1.1	F288/J1633 (298,83)
*AM 2143-464	21 43 04 −46 45.0	3.872 0.276	352.15 −48.94	235.7 16.8	GALAXY w. I/A + 3 APP. COMP'NS. NGC 7119a = 288-IG-2 = B 2143-467	1,8 S 4.9	F287/J3653 (41,82)
AM 2143-492	21 43 44 −49 25.5	3.947 0.277	348.16 −48.44	233.9 14.8	DWARF	20 D 0.4	F237/J3558 (321,207)
AM 2143-872	21 43 45 −87 21.0	0.321 0.277	305.19 −29.20	208.6 −13.7	ISOLATED IRR.	16 D 0.7	F010/J2477 (92,23)
AM 2143-835	21 43 47 −83 57.4	10.139 0.277	308.12 −31.47	210.9 −11.2	LINE of 3	24 D 2.8	F011/J2562 (169,235)
AM 2144-360	21 44 08 −36 01.4	3.616 0.277	8.61 −50.13	243.9 24.4	GALAXY w. JET + I/A COMP'N. 403-IG-26 (in PAIR)	7,9,23 S 0.6	F403/J3544 (91,124)
AM 2144-350 [2145-3511]	21 44 25 −35 07.2	3.597 0.277	10.02 −50.16	244.6 24.9	PAIR (#1) IC 5131 = 403-G-27	1,23 D 0.3	F403/J3544 (87,172)
*AM 2144-504	21 44 45 −50 47.8	3.984 0.278	346.10 −48.24	233.1 13.7	PECULIAR ARMED SPIRAL + COMP'NS. NGC 7124 = 236-G-49	2,8,10 D 3.6	F236/J1823 (52,131)
AM 2144-551 [2144-5512]	21 44 51 −55 13.1	4.143 0.278	339.89 −46.84	230.0 10.4	PAIR (#1: G.D. SPIRAL) Fa 591	23 S 1.7	F188/J1592 (43,163)
AM 2144-590	21 44 56 −59 06.9	4.315 0.278	334.70 −45.34	227.4 7.5	I/A (CONTACT) SPIRALS 145-IG-15	2 S 0.2	F145/J1578 (88,222)
AM 2144-705	21 44 57 −70 56.3	5.222 0.278	320.73 −39.51	219.6 −1.4	SPIRAL w. I/A COMP'N. 75-IG-25	1 S 0.5	F075/J0953 (155,126)
AM 2144-551 [2144-5513]	21 44 58 −55 12.1	4.141 0.278	339.90 −46.86	230.1 10.4	PAIR (#2: E)	23 S 1.7	F188/J1592 (42,164)
AM 2145-591	21 45 01 −59 15.3	4.321 0.278	334.51 −45.29	227.3 7.3	I/A DBL. (2 E's in GROUP of 10) DN 2145-593	2,18 S 0.1	F145/J1578 (87,215)
AM 2145-483	21 45 03 −48 37.1	3.915 0.278	349.26 −48.85	234.6 15.2	DWARF (NUCLEUS)	20 D 0.4	F237/J3558 (312,250)
AM 2145-351	21 45 19 −35 11.2	3.595 0.278	9.93 −50.35	244.7 24.8	I/A DBL + COMPN'S. NGC 7130 = IC 5135 = 403-G-32	2,8 D 1.1	F404/J0736 (337,165)
AM 2145-642	21 45 19 −64 24.6	4.619 0.278	328.09 −42.98	223.9 3.5	COMPACT	13 D 0.2	F108/J0946 (261,207)
AM 2145-351 [2144-3507]	21 45 20 −35 11.0	3.595 0.278	9.93 −50.35	244.7 24.8	PAIR(#2:DISTURBED SP + I/A COMP'N.) NGC 7130 = IC 5135 = 403-G-32	1,23 D 0.4	F403/J3544 (77,169)
AM 2145-322 [2145-3224]	21 45 22 −32 22.4	3.542 0.278	14.32 −50.18	247.0 26.7	CLOSE PAIR (#1: COMPACT)	13,23 D 0.3	F403/J3544 (73,319)
AM 2145-605 [2145-6050]	21 45 34 −60 56.8	4.404 0.278	332.32 −44.62	226.3 6.0	PAIR of SPIRALS (#1) + COMP'N NGC 7125 = 145-G-17	8,23 S 5.4	F146/J1873 (338,121)
AM 2145-322 [2145-3222]	21 45 37 −32 24.5	3.542 0.278	14.27 −50.24	247.0 26.6	CLOSE PAIR (#2: SPIRAL)	23 D 0.9	F403/J3544 (70,318)
AM 2145-565 [2146-5657]	21 45 43 −56 52.9	4.205 0.278	337.58 −46.33	229.0 9.1	CHAIN of 4 (#1) DN 2146-569	17 S 6.7	F188/J1592 (42,74)
AM 2145-605 [2145-6056]	21 45 43 −60 50.8	4.397 0.278	332.43 −44.68	226.3 6.1	PAIR of SPIRALS (#2) + COMP'N NGC 7126 = 145-G-18	8,23 S 5.4	F146/J1873 (338,126)
*AM 2145-543	21 45 44 −54 32.3	4.110 0.278	340.76 −47.20	230.6 10.8	RING w. OFF-SET NUCLEUS + COMP'N. 189-IG-2	1,6 S 0.4	F188/J1592 (33,199)
AM 2145-223	21 45 47 −22 38.1	3.381 0.278	29.04 −48.45	255.6 32.9	COMPACT	13 D 0.3	F531/J0650 (79,303)
AM 2145-814 [2147-8153]	21 45 47 −81 46.0	8.176 0.278	310.03 −32.95	212.4 −9.6	SBc w. I/A COMP'N. (#1: SBc) 27-G-1 = UKS 2145-817 = Fa 353	1 S 9.0	F027/J2485 (264,76)
AM 2145-565 [2145-5652]	21 45 49 −56 54.2	4.206 0.278	337.54 −46.34	229.0 9.0	CHAIN of 4 (#2) DN 2146-569	17 S 6.7	F188/J1592 (41,72)
AM 2146-565 [2145-5654]	21 46 10 −56 55.4	4.204 0.279	337.49 −46.37	229.0 9.0	CHAIN of 4 (#3) DN 2146-569	17 S 6.7	F188/J1592 (39,71)
AM 2146-565 [2145-5655]	21 46 16 −56 57.2	4.204 0.279	337.44 −46.38	229.0 9.0	CHAIN of 4 (#4) DN 2146-569	17 S 6.7	F188/J1592 (38,69)
AM 2146-452	21 46 20 −45 25.5	3.820 0.279	353.98 −49.75	237.1 17.4	SPIRAL w. DIAMETRIC COMP'NS. 288-G-10	8 D 4.0	F288/J1633 (270,155)
*AM 2146-504	21 46 22 −50 49.9	3.976 0.279	345.95 −48.47	233.2 13.5	DWARF (SHARP NUCLEUS)	20 D 0.5	F237/J3558 (295,132)
AM 2146-525	21 46 33 −52 55.2	4.045 0.279	342.95 −47.86	231.8 11.9	E w. PECULIAR COMP'N. (TAIL) 189-G-3 = B 2146-529	8,15 S 2.2	F237/J3558 (288,20)
AM 2146-593	21 46 33 −59 34.5	4.324 0.279	333.98 −45.33	227.3 7.0	LINE of 3	24 D 4.5	F146/J1873 (339,195)
AM 2146-713	21 46 37 −71 38.9	5.287 0.279	319.88 −39.21	219.2 −2.1	E + APP. COMP'NS.	8 D 0.5	F075/J0953 (149,88)
*AM 2146-504	21 46 43 −50 41.5	3.969 0.279	346.12 −48.57	233.4 13.5	CLOSE DBL. in GROUP of 6	18,23 S 1.1	F236/J1823 (35,136)

Identification	R.A. (1950.0) Declination	Precession R.A./Dec.	Gal. Long. Gal. Lat.	S.G.L. S.G.B.	Description Cross-Identification	Coding Size (Min.)	Field/Plate (x, y)
*AM 2146-350	21 46 47 −35 06.5	3.589 0.279	10.06 −50.64	245.0 24.6	DISTURBED GAL. w. TAIL (IRR near) NGC 7135 = 403-IG-35	1,7,15 D 3.1	F404/J0736 (321,170)
AM 2147-751	21 47 01 −75 13.2	5.849 0.279	316.19 −37.13	216.9 −4.7	LOOSE GROUP of 8	18 D 9.0	F048/J1819 (221,164)
AM 2147-815 [2145-8146]	21 47 05 −81 53.5	8.211 0.279	309.88 −32.90	212.4 −9.7	SBc w. I/A COMP'N.(#2: DWARF) 27-G-3	1,20 D 0.9	F027/J2485 (260,70)
AM 2147-555	21 47 12 −55 59.3	4.157 0.279	338.66 −46.87	229.8 9.6	SMALL I/A TRIPLE 189-IG-4	3 D 1.0	F189/J3474 (275,124)
AM 2147-502	21 47 13 −50 24.9	3.958 0.279	346.49 −48.72	233.6 13.7	GROUP of 4 DN 2147-560	18 D 4.5	F236/J1823 (30,150)
AM 2147-351	21 47 26 −35 10.9	3.588 0.280	9.95 −50.78	245.0 24.4	GROUP of 4	18 D 3.6	F403/J3544 (54,168)
AM 2147-612 [2147-6123]	21 47 30 −61 26.0	4.414 0.280	331.56 −44.61	226.1 5.5	I/A DBL. end of CHAIN (#1) 145-IG-21 = B 2147-614	2,17 D 0.5	F145/J1578 (77,97)
AM 2147-612 [2147-6126]	21 47 33 −61 23.9	4.411 0.280	331.60 −44.63	226.1 5.5	I/A DBL. end of CHAIN (#2)	2,17 D 0.3	F145/J1578 (77,99)
AM 2147-544	21 47 36 −54 47.1	4.107 0.280	340.28 −47.36	230.6 10.4	G.D. SPIRAL w. COMP'N. in ARM 189-G-5	1 S 0.7	F188/J1592 (20,185)
AM 2147-611	21 47 45 −61 12.8	4.400 0.280	331.81 −44.74	226.3 5.7	I/A TRIPLE	3 D 0.8	F145/J1578 (75,109)
AM 2147-724	21 47 47 −72 42.8	5.414 0.280	318.70 −38.67	218.6 −2.9	CLOSE PAIR 48-G-10	23 S 1.3	F075/J0953 (146,31)
*AM 2147-490	21 47 51 −49 02.6	3.912 0.280	348.46 −49.19	234.7 14.6	GAL. w. I/A COMP'N. (GROUP of 4) 237-IG-6 = B 2147-490 (RING?)	1,6,18 D 0.7	F237/J3558 (286,229)
AM 2148-805	21 48 15 −80 52.2	7.592 0.280	310.77 −33.61	213.1 −9.0	DWARF	20 D 0.9	F027/J2485 (268,125)
AM 2148-725	21 48 28 −72 55.5	5.434 0.280	318.44 −38.58	218.5 −3.1	GROUP of 4 (incl. 3 COMPACTS)	13,18 D 1.3	F075/J0953 (143,19)
AM 2148-432	21 48 38 −43 21.8	3.757 0.281	357.04 −50.50	238.9 18.6	Sm/DWARF 288-G-13 = Lu 2148-43	20 D 0.8	F288/J1633 (251,266)
AM 2148-554	21 48 51 −55 48.1	4.138 0.281	338.79 −47.15	230.1 9.6	SB w. CHAIN of 7 to [N.E.] NGC 7141 + 40 = 189-G-7 = Fa 354	17 D 3.1	F189/J3474 (263,135)
AM 2148-634	21 48 56 −63 48.1	4.542 0.281	328.55 −43.62	224.6 3.6	I/A DBL. 108-IG-5	2 S 0.2	F108/J0946 (241,240)
AM 2149-483	21 49 08 −48 38.1	3.893 0.281	348.98 −49.50	235.1 14.7	DWARF(RESOLUTION) (CLUSTER near) 237-G-10	20 D 0.8	F237/J3558 (276,251)
AM 2149-613	21 49 29 −61 35.0	4.404 0.281	331.22 −44.75	226.2 5.2	CLOSE PAIR	23 S 0.1	F145/J1578 (65,88)
AM 2149-594	21 49 43 −59 44.0	4.305 0.281	333.53 −45.62	227.5 6.6	SPIRAL w. 8 COMP'NS. IC 5141 = 145-G-22	8 S 5.4	F146/J1873 (316,188)
AM 2149-620	21 49 44 −62 08.0	4.433 0.281	330.52 −44.51	225.8 4.8	I/A DBL. (E + SPIRAL) 145-IG-23	2 S 0.6	F108/J0946 (240,330)
AM 2149-574	21 49 57 −57 41.0	4.208 0.282	336.18 −46.54	228.9 8.1	CLOSE PAIR	23 S 0.8	F145/J1578 (48,297)
AM 2149-520 [2150-5205]	21 49 58 −52 07.7	3.996 0.282	343.82 −48.61	232.7 12.1	CLOSE PAIR (#1) 237-G-12 = PKS 2150-52	23 D 0.3	F236/J1823 (12,57)
AM 2150-520 [2149-5207]	21 50 13 −52 05.9	3.994 0.282	343.85 −48.65	232.8 12.1	CLOSE PAIR (#2) PKS 2150-52	23 D 0.3	F236/J1823 (10,58)
AM 2150-673	21 50 33 −67 33.6	4.805 0.282	324.07 −41.79	222.2 0.8	I/A DBL. (near BRIGHT GAL.) NEAR IC 5140 = 75-G-37	2,8 S 0.2	F075/J0953 (122,307)
AM 2150-575	21 50 35 −57 51.1	4.211 0.282	335.90 −46.54	228.8 7.9	LO.S.BR. DWARF (SOME RESOLUTION) Se 149/6 = 145-G-25 = UKS2150-578	20 D 2.0	F145/J1578 (44,288)
AM 2150-241	21 50 36 −24 15.4	3.395 0.282	27.10 −49.94	254.9 31.1	GALAXY w. JET (SMALL)	7 D 0.7	F532/J0817 (291,217)
AM 2150-580	21 50 42 −58 05.1	4.220 0.282	335.59 −46.46	228.7 7.7	CLUSTER of GALAXIES DN 2150-581	19 D 13.4	F146/J1873 (316,277)
AM 2150-451	21 50 43 −45 19.1	3.795 0.282	353.91 −50.52	237.7 16.9	CLOSE I/A DBL. w. TAILS 288-IG-18 = B 2150-453	2,15 S 0.2	F288/J1633 (229,162)
AM 2151-702	21 51 58 −70 22.3	5.058 0.283	320.90 −40.31	220.4 −1.4	I/A DBL. 75-IG-40	2 S 0.3	F075/J0953 (123,156)
AM 2152-562	21 52 18 −56 20.7	4.134 0.283	337.77 −47.38	230.0 8.8	I/A DBL. in CHAIN of 4 Se 149/3 = 189-IG-13 = DN 2152-564	2,17 D 2.2	F189/J3474 (236,106)
AM 2152-432 [2152-4332]	21 52 20 −43 29.5	3.743 0.284	356.67 −51.14	239.3 18.0	GROUP of 3 (#1: DISTURBED) 288-G-20	24 D 1.1	F288/J1633 (215,260)
AM 2152-612	21 52 36 −61 29.5	4.371 0.284	331.08 −45.12	226.5 5.1	I/A DBL.	2 S 0.5	F146/J1873 (291,95)
AM 2152-371	21 52 38 −37 13.7	3.609 0.284	6.67 −51.82	244.2 22.3	LINE of 3 (incl. I/A DBL.)	2,24 D 1.2	F343/J0828 (45,327)
AM 2152-432 [2152-4329]	21 52 40 −43 27.5	3.741 0.284	356.71 −51.20	239.4 18.0	GROUP of 3 (#2: Sc) 288-G-21	24 D 1.5	F288/J1633 (212,262)
*AM 2152-592	21 52 48 −59 21.3	4.261 0.284	333.76 −46.14	228.0 6.6	RING GALAXY + I/A TRIPLET near 146-G-1	3,6 D 0.7	F145/J1578 (34,206)

Identification	R.A. (1950.0) Declination	Precession R.A./Dec.	Gal. Long. Gal. Lat.	S.G.L. S.G.B.	Description Cross-Identification	Coding Size (Min.)	Field/Plate (x, y)
AM 2152-433 [2152-4327]	21 52 58 −43 32.2	3.741 0.284	356.57 −51.25	239.3 17.9	GROUP of 3 (#3: SB(r))	24 D 0.6	F288/J1633 (209,258)
AM 2153-702 [2153-7024]	21 53 06 −70 24.8	5.047 0.284	320.79 −40.36	220.5 −1.5	CLOSE PAIR (E + SPIRAL) (#1) 75-G-42	23 D 0.5	F075/J0953 (118,153)
AM 2153-554	21 53 10 −55 47.7	4.106 0.284	338.44 −47.71	230.5 9.2	CLOSE PAIR Fa 595/7	23 S 1.0	F189/J3474 (231,136)
AM 2153-702 [2153-7024]	21 53 23 −70 24.6	5.043 0.284	320.77 −40.38	220.5 −1.5	CLOSE PAIR (E + SPIRAL) (#2) 75-G-43	23 D 0.5	F075/J0953 (116,153)
AM 2153-540	21 53 25 −54 06.3	4.041 0.284	340.75 −48.40	231.7 10.3	CLOSE PAIR 189-G-14	23 S 2.0	F189/J3474 (231,227)
AM 2153-462	21 53 48 −46 24.1	3.807 0.285	352.06 −50.81	237.3 15.8	I/A TRIPLE	3 D 6.7	F288/J1633 (199,104)
AM 2154-371	21 54 03 −37 13.9	3.603 0.285	6.65 −52.10	244.4 22.1	GROUP of 5 E's (+EARLY-TYPES)	18 D 2.7	F404/J0736 (240,57)
*AM 2154-512	21 54 03 −51 27.4	3.949 0.285	344.47 −49.42	233.7 12.2	DWARF w. JET 237-G-19	7,20 D 0.8	F237/J3558 (229,101)
AM 2154-382	21 54 05 −38 28.8	3.627 0.285	4.62 −52.05	243.4 21.2	CLOSE PAIR (I/A ?) 344-G-1	2,23 S 1.0	F343/J0828 (32,259)
AM 2154-253	21 54 06 −25 35.4	3.407 0.285	25.35 −51.02	254.3 29.6	SPIRAL + I/A COMP'N. NGC 7157 = MCG-4-51-15 = 532-G-3	1 S 0.8	F532/J0817 (248,146)
*AM 2154-592	21 54 06 −59 29.4	4.257 0.285	333.48 −46.23	228.0 6.4	3-ARMED SPIRAL (ONE to a 'STAR')	11 D 0.8	F145/J1578 (26,198)
AM 2154-371	21 54 07 −37 14.2	3.603 0.285	6.65 −52.12	244.4 22.1	E + 5 APP. COMP'NS. (2 I/A)	2,8 D 1.8	F344/J1508 (307,325)
AM 2154-662	21 54 17 −66 21.5	4.664 0.285	325.16 −42.76	223.3 1.4	CHAIN of 4 DN 2154-664	17 D 1.1	F108/J0946 (207,104)
AM 2154-603	21 54 23 −60 32.9	4.306 0.285	332.11 −45.76	227.3 5.6	DWARF 146-GA-2 = UKS 2154-605	20 D 0.7	F146/J1873 (282,147)
*AM 2154-572	21 54 41 −57 22.8	4.158 0.285	336.18 −47.24	229.6 7.9	SPIRAL w. DIAMETRIC COMP'NS. 189-G-16 = Fa 598	8 S 2.9	F146/J1873 (290,317)
AM 2154-382	21 54 43 −38 23.9	3.623 0.285	4.75 −52.18	243.6 21.2	CLOSE PAIR (LATE TYPE) (I/A?) 344-G-3/4	2,23 S 1.1	F343/J0828 (25,263)
AM 2154-521	21 54 48 −52 19.1	3.972 0.285	343.17 −49.24	233.1 11.5	SPIRAL w. CLOSE I/A COMP'N. 237-G-20 = B 2154-523	1 S 0.3	F189/J3474 (222,323)
AM 2154-584	21 54 55 −58 48.7	4.219 0.285	334.28 −46.63	228.6 6.8	I/A DBL. in TRIPLET 146-IG-3	2,24 S 0.2	F145/J1578 (17,234)
AM 2155-670	21 55 05 −67 01.3	4.707 0.286	324.35 −42.46	222.9 0.8	I/A DBL. 108-IG-7	2 S 0.4	F075/J0953 (97,335)
AM 2155-694	21 55 06 −69 43.4	4.948 0.286	321.39 −40.90	221.1 −1.1	GALAXY w. JET 75-G-45?	7 D 0.7	F075/J0953 (106,190)
AM 2155-451	21 55 07 −45 15.7	3.772 0.286	353.74 −51.29	238.3 16.4	N GALAXY DN 2155-455	12 D 0.4	F288/J1633 (187,165)
AM 2155-273	21 55 17 −27 37.8	3.435 0.286	22.23 −51.69	252.6 28.1	SPIRAL w. I/A COMP'NS. (PAIR) IC 5149 = MCG-5-51-29 = 466-G-24	1,23 S 1.3	F532/J0817 (233,36)
AM 2155-665	21 55 18 −66 59.7	4.702 0.286	324.36 −42.49	222.9 0.9	I/A DBL. 108-IG-7	2 S 0.4	F108/J0946 (201,70)
AM 2155-343	21 55 22 −34 32.8	3.549 0.286	11.04 −52.39	246.7 23.7	COMPACT + IRR. 404-IG-13	13,16,23 S 0.6	F404/J0736 (227,202)
*AM 2155-263	21 55 46 −26 37.2	3.418 0.286	23.87 −51.60	253.6 28.7	SMALL RING + COMP'N.	6,8 S 0.4	F532/J0817 (227,90)
AM 2156-440 [2157-4322]	21 56 10 −44 06.6	3.739 0.286	355.49 −51.71	239.3 17.1	GROUP of 4 (#1: EDGE-ON) 288-G-25	18 D 2.0	F288/J1633 (178,227)
AM 2156-441	21 56 27 −44 12.3	3.740 0.287	355.33 −51.74	239.3 17.0	LUMPY COMPACT	13 D 0.2	F288/J1633 (175,222)
AM 2156-232	21 56 32 −23 29.6	3.371 0.287	28.80 −51.06	256.6 30.4	I/A DBL.	2 S 0.4	F532/J0817 (219,258)
AM 2156-393	21 56 33 −39 37.6	3.640 0.287	2.70 −52.43	242.8 20.1	BRIGHT PLANETARY NEBULA IC 5148/50 = 344-PN-5	25 D 2.3	F344/J1508 (277,198)
AM 2156-480	21 56 34 −48 05.4	3.836 0.287	349.28 −50.84	236.4 14.3	GALAXY w. TAIL 237-IG-23 = B 2156-480	1,15 D 0.5	F237/J3558 (210,282)
AM 2156-433 [2156-4407]	21 56 36 −43 32.8	3.724 0.287	356.36 −51.89	239.8 17.4	GROUP of 4 (#2: Sc w. KNOT) NGC 7162 = 288-G-26	18 D 2.7	F288/J1633 (174,258)
AM 2157-581	21 57 07 −58 14.6	4.176 0.287	334.83 −47.15	229.2 7.0	I/A DBL.	2 S 0.5	F146/J1873 (270,271)
AM 2157-433 [2156-4332]	21 57 28 −43 37.8	3.722 0.287	356.18 −52.03	239.9 17.2	GROUP of 4 (#3: S0 + COMP'NS) NGC 7166 = 288-G-27	18 D 1.3	F288/J1633 (165,253)
AM 2157-432 [2157-4337]	21 57 31 −43 22.9	3.716 0.287	356.58 −52.09	240.1 17.4	GROUP of 4 (#4: SBd) NGC 7162a = 288-G-28	18 S 2.5	F288/J1633 (165,266)
AM 2157-245	21 57 40 −24 52.7	3.388 0.288	26.75 −51.65	255.5 29.4	SPIRAL w. COMP'N. in DISK NGC 7167 = 532-G-9	8,12 D 2.2	F532/J0817 (205,184)
AM 2158-380 [2158-3801]	21 58 05 −38 04.3	3.604 0.288	5.21 −52.86	244.3 20.9	PAIR (#1) TWO COMPACTS NEAR-BY	23 D 0.6	F404/J0736 (197,12)

Identification	R.A. (1950.0) Declination	Precession R.A./Dec.	Gal. Long. Gal. Lat.	S.G.L. S.G.B.	Description Cross-Identification	Coding Size (Min.)	Field/Plate (x, y)
AM 2158-380 [2158-3804]	21 58 17 −38 01.2	3.602 0.288	5.29 −52.90	244.4 20.9	PAIR (#2) PKS 2158-380	23 D 0.4	F404/J0736 (195,15)
AM 2158-601	21 58 23 −60 11.1	4.253 0.288	332.23 −46.37	227.9 5.5	RICH, SYMMETRIC DISTANT CLUSTER 146-IG-5	19 D 9.0	F146/J1873 (257,168)
AM 2158-424	21 58 30 −42 41.4	3.696 0.288	357.64 −52.39	240.7 17.8	PAIR of I/A SPIRALS MCG-7-45-6/7 = 288-G-30/2 = B2158-426	2 S 1.0	F288/J1633 (155,304)
AM 2158-433	21 58 35 −43 33.3	3.715 0.288	356.24 −52.24	240.1 17.2	DWARF (NUCLEATED ?) in DN 2158-434?	20 D 0.5	F288/J1633 (154,257)
AM 2158-261	21 58 42 −26 12.8	3.405 0.288	24.72 −52.17	254.4 28.4	I/A DBL. (EDGE-ON GALAXY ?)	2 S 0.5	F532/J0817 (192,112)
AM 2158-374 [2158-3740]	21 58 43 −37 40.7	3.594 0.288	5.85 −53.01	244.7 21.1	CLOSE PAIR (#1) 344-G-6	23 D 0.5	F344/J1508 (257,303)
*AM 2158-515	21 58 49 −51 59.2	3.935 0.288	343.31 −49.93	233.8 11.3	E w. PECULIAR COMP'N.(JETS?) NGC 7168 = 237-G-26 = SS-29	7,8 S 2.7	F237/J3558 (189,73)
AM 2158-374 [2158-3740]	21 58 52 −37 40.2	3.593 0.288	5.86 −53.04	244.7 21.1	CLOSE PAIR (#2: COMPACT)	13,23 D 0.2	F344/J1508 (255,303)
AM 2158-363	21 58 57 −36 30.9	3.571 0.288	7.78 −53.11	245.7 21.8	SMALL COMETARY NEBULA	21 D 0.9	F404/J0736 (188,96)
AM 2159-224	21 59 05 −22 40.5	3.354 0.289	30.31 −51.41	257.7 30.4	CLOSE GROUP of 4	18 D 2.2	F532/J0817 (187,302)
*AM 2159-320 [2159-3214]	21 59 05 −32 06.4	3.495 0.289	15.12 −53.05	249.3 24.7	TRIPLE (#1: DUST LANE) KLEM 4 NGC 7172 = 466-G-38 in DN 2159-322	3,14 D 1.8	F404/J0736 (186,333)
AM 2159-321 [2159-3206]	21 59 08 −32 12.7	3.497 0.289	14.95 −53.07	249.3 24.6	TRIPLE (#2: E) KLEM 4 NGC 7173 = 466-G-39 in DN 2159-322	3 D 0.6	F404/J0736 (186,327)
AM 2159-321 [2159-3212]	21 59 13 −32 14.2	3.497 0.289	14.91 −53.09	249.2 24.6	TRIPLE (#3: GAL. w. TAIL) KLEM 4 NGC 7174 = 466-G-40 in DN 2159-322	3,15 D 2.0	F404/J0736 (185,326)
AM 2159-412	21 59 14 −41 20.1	3.664 0.289	359.80 −52.73	241.9 18.6	SPIRAL w. APP. COMP'NS + COMPACT 344-G-7	8,13 S 6.5	F344/J1508 (247,107)
AM 2159-541	21 59 18 −54 19.1	4.008 0.289	339.95 −49.11	232.2 9.6	RESOLVED DWARF	20 D 2.2	F189/J3474 (185,216)
AM 2159-513	21 59 26 −51 32.4	3.918 0.289	343.91 −50.18	234.2 11.5	HIGHLY RESOLVED Sm IC 5152 = 237-G-27	20 D 4.5	F237/J3558 (184,97)
*AM 2159-472	21 59 41 −47 26.4	3.802 0.289	350.05 −51.52	237.3 14.3	I/A TRIPLE (LONG TAIL !) 288-IG-35 = B 2159-474	2,3,8,10,15 D 3.1	F288/J1633 (145,48)
AM 2159-434	21 59 43 −43 41.2	3.712 0.289	355.97 −52.42	240.1 16.9	DWARF	20 D 0.7	F288/J1633 (143,250)
AM 2159-330	21 59 48 −33 02.7	3.508 0.289	13.57 −53.26	248.6 23.9	PAIR incl. BULL'S-EYE NGC 7187 = 404-G-24	6,23 D 1.7	F404/J0736 (178,282)
AM 2200-215	22 00 27 −21 57.3	3.341 0.290	31.56 −51.50	258.6 30.6	CLOSE PAIR	8,23 S 1.0	F532/J0817 (170,341)
AM 2200-491	22 00 30 −49 10.0	3.843 0.290	347.34 −51.14	236.1 13.1	CLOSE PAIR 237-G-31 = PKS 2200-492?	23 S 1.3	F237/J3558 (175,224)
*AM 2200-662	22 00 38 −66 21.7	4.592 0.290	324.67 −43.28	223.8 0.9	DISTURBED COMPACT IC 5154 = Se 150/9 = 108-G-10	13,16 D 0.7	F108/J0946 (173,104)
*AM 2200-715	22 00 47 −71 56.7	5.111 0.290	318.70 −39.91	219.8 −3.0	RING in GROUP of 8 Se 149/7 = DN 2201-577	6,17 D 13.4	F075/J0953 (90,69)
AM 2201-575	22 01 08 −57 51.0	4.127 0.290	334.99 −47.80	229.8 6.9	CHAIN of 5 in GROUP of 8	17,18 D 4.4	F146/J1873 (243,294)
AM 2201-653	22 01 16 −65 39.5	4.535 0.290	325.41 −43.74	224.3 1.4	I/A DBL. (SMALL)	2 S 0.2	F108/J0946 (169,142)
AM 2201-230	22 01 32 −23 2.4	3.354 0.290	29.99 −52.05	257.8 29.8	SMALL I/A DBL. (RING?)	2,6 S 0.3	F532/J0817 (157,283)
AM 2201-411	22 01 38 −41 16.0	3.652 0.290	359.80 −53.19	242.2 18.3	GAL. w. FNT. PLUME + I/A COMP'N.	1,15 S 1.3	F344/J1508 (223,111)
*AM 2201-583	22 01 40 −58 31.9	4.151 0.290	334.04 −47.54	229.4 6.4	CHAIN of 4 (incl. V.KNOTTY PEC.) Se 149/11 = DN 2201-585 + Fa 126?	16,17 D 3.8	F146/J1873 (238,257)
AM 2202-502	22 02 44 −50 21.8	3.863 0.291	345.36 −51.08	235.5 12.0	E + 5 APP. COMP'NS. NGC 7196 = 237-G-36	8 S 1.1	F237/J3558 (156,160)
AM 2202-701	22 02 51 −70 10.8	4.886 0.291	320.37 −41.16	221.2 −1.9	2 E's in COMMON ENVELOPE 75-IG-56	2 S 0.4	F076/J1792 (310,159)
AM 2202-754	22 02 51 −75 40.9	5.634 0.291	314.94 −37.60	217.3 −5.8	DWARF (V.LO.S.BR.)	20 D 0.4	F048/J1819 (167,141)
AM 2203-270	22 03 17 −27 09.1	3.407 0.292	23.52 −53.35	254.3 27.0	CLOSE PAIR	23 S 0.8	F532/J0817 (137,62)
*AM 2203-281	22 03 21 −28 12.2	3.422 0.292	21.78 −53.54	253.4 26.4	E w. I/A SPIRAL w. I/A COMP'N. MCG-5-52-23/24 = 467-G-3	1,23 S 0.7	F466/J0798 (29,271)
AM 2203-440	22 03 21 −44 04.9	3.703 0.292	355.10 −52.98	240.3 16.2	ONE-ARMED SPIRAL	10 D 0.6	F288/J1633 (108,228)
AM 2203-430	22 03 31 −43 06.3	3.681 0.292	356.68 −53.21	241.1 16.8	DWARF 288-G-40	20 D 1.1	F288/J1633 (106,281)
AM 2204-311	22 04 00 −31 17.8	3.467 0.292	16.61 −54.04	250.7 24.4	I/A DBL. in CHAIN of 3 = KLEM 36 NGC 7204 = 467-IG-8 = DN 2204-314	2,24 D 1.1	F467/J0691 (292,105)

Identification	R.A. (1950.0) Declination	Precession R.A./Dec.	Gal. Long. Gal. Lat.	S.G.L. S.G.B.	Description Cross-Identification	Coding Size (Min.)	Field/Plate (x, y)
AM 2204-463	22 04 00 −46 39.8	3.759 0.292	350.93 −52.44	238.4 14.3	DWARF	20 D 0.2	F288/J1633 (105,90)
AM 2204-620	22 04 06 −62 05.6	4.295 0.292	329.35 −45.98	227.1 3.7	GALAXY w. PEC. ARM + APP. COMP'N.	8,10 D 0.4	F108/J0946 (150,333)
AM 2204-405	22 04 10 −40 53.9	3.633 0.292	0.30 −53.72	242.9 18.2	3 in a LINE	24 D 2.3	F344/J1508 (197,131)
AM 2204-525	22 04 48 −52 57.6	3.926 0.293	341.36 −50.40	233.8 9.9	POOR I/A GROUP of 4 = Se 148/4 IC 5162 = 189-IG-25 = DN 2205-526	4 D 1.6	F189/J3474 (141,289)
AM 2205-575	22 05 06 −57 50.3	4.094 0.293	334.63 −48.27	230.2 6.6	DWARF	20 D 0.6	F146/J1873 (214,295)
AM 2205-564	22 05 11 −56 40.8	4.049 0.293	336.16 −48.83	231.1 7.3	I/A DBL.	2 S 1.1	F189/J3474 (141,89)
AM 2205-574	22 05 11 −57 41.1	4.087 0.293	334.82 −48.35	230.3 6.7	LARGE SPIRAL + APP. COMP'NS. NGC 7205 = 146-G-9	8 S 7.6	F146/J1873 (214,303)
AM 2205-291	22 05 33 −29 17.8	3.432 0.293	20.06 −54.17	252.7 25.3	COMPACT NGC 7208 = MCG-5-52-32 = 467-G-10	13 D 0.6	F466/J0798 (5,212)
AM 2206-352	22 06 02 −35 21.5	3.525 0.294	9.67 −54.57	247.6 21.5	TIGHT GROUP(15) E's + SPIRALS	18 D 11.2	F404/J0736 (110,157)
AM 2206-272	22 06 05 −27 28.0	3.404 0.294	23.18 −54.02	254.4 26.3	COMPACT in LINE	13,24 D 0.4	F532/J0817 (104,44)
AM 2206-472	22 06 10 −47 25.0	3.766 0.294	349.57 −52.58	238.1 13.6	S0 w. PEC. DUST + COMP'NS.[PKS?] NGC 7213 = 288-G-43 = DN 2206-474	8,14 D 5.6	F288/J1633 (86,49)
*AM 2206-280	22 06 15 −28 03.2	3.412 0.294	22.20 −54.15	253.9 26.0	SPIRAL + I/A COMP'N. NGC 7214 = MCG-5-52-36 = 467-G-12	1,17 D 2.9	F467/J0691 (270,280)
AM 2206-501	22 06 17 −50 15.2	3.838 0.294	345.20 −51.65	235.9 11.6	DWARF	20 D 0.5	F237/J3558 (125,165)
AM 2206-272	22 06 22 −27 24.9	3.403 0.294	23.28 −54.07	254.5 26.3	DISTURBED OUTER ARMS (KNOTTY) MCG-5-52-38 = 532-G-22	12,18 D 1.1	F532/J0817 (101,47)
AM 2206-634	22 06 33 −63 43.3	4.358 0.294	327.19 −45.31	226.1 2.3	DISTURBED IRR.(in GROUP of 4)	16,18 D 0.4	F108/J0946 (137,245)
AM 2206-253	22 06 34 −25 39.8	3.378 0.294	26.22 −53.79	256.1 27.3	COMPACT w. CROOKED ABS'N. PKS 2206-251?	13,14 D 0.5	F532/J0817 (97,141)
AM 2206-500	22 06 37 −50 02.9	3.830 0.294	345.48 −51.77	236.1 11.7	HI.S.BR. ONE-ARMED SP. + I/A COMP'N Se 148/3 = 237-IG-42	2,8,10 D 0.7	F237/J3558 (122,176)
*AM 2206-380	22 06 54 −38 09.7	3.570 0.294	4.82 −54.58	245.4 19.6	BRIGHT-ARMED SPIRAL 344-G-10	10,12 D 0.8	F404/J0736 (104,7)
*AM 2207-670	22 07 00 −67 06.7	4.571 0.294	323.32 −43.35	223.7 0.0	GALAXY + JET w. I/A COMP'N.	1,7 D 0.8	F076/J1792 (308,325)
AM 2207-380	22 07 38 −38 03.1	3.565 0.295	4.99 −54.73	245.6 19.5	DWARF 344-G-11	20 D 0.7	F344/J1508 (162,284)
AM 2207-515	22 07 44 −51 57.5	3.876 0.295	342.53 −51.21	234.8 10.3	SMALL I/A DBL.	2 S 0.1	F189/J3474 (116,342)
*AM 2207-225	22 07 48 −22 54.1	3.338 0.295	30.85 −53.40	258.9 28.7	SB w. THIRD ARM + I/A COMP'N. MCG-4-52-18 = 532-G-25	1,11 S 6.0	F532/J0817 (79,289)
AM 2207-573	22 07 49 −57 34.6	4.062 0.295	334.70 −48.71	230.7 6.5	DWARF 146-G-13	20 D 0.7	F146/J1873 (195,309)
AM 2207-535	22 07 57 −53 51.5	3.932 0.295	339.77 −50.45	233.4 9.0	DISTURBED SPIRAL w. PEC. NUCLEUS 189-IG-27 = B 2207-538	16 D 0.9	F189/J3474 (116,240)
*AM 2208-251	22 08 06 −25 19.3	3.370 0.295	26.91 −54.05	256.7 27.2	EDGE-ON + 5 COMP'NS.(I/A) MCG-4-52-19 = 532-G-26	1,2,8 S 6.0	F532/J0817 (78,159)
*AM 2208-304	22 08 21 −30 48.9	3.446 0.295	17.56 −54.93	251.8 23.9	SPIRAL w. INTERSECTING ARM NGC 7221 = 467-G-18	10 D 2.7	F467/J0691 (243,132)
AM 2208-460	22 08 31 −46 09.5	3.723 0.295	351.36 −53.33	239.3 14.1	DWARF	20 D 0.2	F288/J1633 (63,115)
AM 2208-455	22 08 44 −45 50.4	3.715 0.295	351.86 −53.46	239.6 14.3	Sc + COMP'N. (in WIDE TRIPLE) Se 151-1 = 288-G-49	8,24 S 2.0	F288/J1633 (60,132)
AM 2208-340	22 08 47 −34 07.8	3.495 0.295	11.79 −55.16	249.0 21.8	GALAXY w. I/A COMP'N. 404-G-39	1 S 1.0	F404/J0736 (79,223)
AM 2208-625	22 08 47 −62 58.3	4.295 0.295	327.87 −45.94	226.8 2.7	SMALL I/A E + SPIRAL Se 150/1 = 108-IG-18 = B 2209-629	2 S 0.6	F146/J1873 (186,20)
AM 2209-715	22 09 02 −71 59.9	4.987 0.296	318.11 −40.37	220.2 −3.5	I/A DBL. w. RING + MATERIAL 76-IG-5 = B 2209-719	2,6 S 0.9	F075/J0953 (56,63)
AM 2209-543	22 09 03 −54 32.6	3.946 0.296	338.70 −50.30	233.0 8.4	3-ARMED SPIRAL + I/A COMP'NS. 189-G-28 = B 2209-545	1,11 S 2.0	F189/J3474 (109,203)
AM 2209-432	22 09 04 −43 25.4	3.661 0.296	355.79 −54.13	241.5 15.8	DWARF (KNOTS)	20 D 0.5	F288/J1633 (52,262)
AM 2209-280 [2209-2806]	22 09 20 −28 04.6	3.404 0.296	22.34 −54.83	254.3 25.4	I/A DBL. (#1)	2 D 0.3	F532/J0817 (66,11)
AM 2209-472	22 09 22 −47 28.2	3.749 0.296	349.20 −53.07	238.4 13.1	HI.S.BR. GALAXY w. PEC. ABS'N. IC 5170 = 237-G-46	14 D 1.6	F237/J3558 (94,314)
AM 2209-280 [2209-2804]	22 09 23 −28 06.7	3.404 0.296	22.28 −54.84	254.3 25.4	I/A DBL. (#2) MCG-5-52-47 = 467-G-21	2 D 0.6	F532/J0817 (65,9)

Identification	R.A. (1950.0) Declination	Precession R.A./Dec.	Gal. Long. Gal. Lat.	S.G.L. S.G.B.	Description Cross-Identification	Coding Size (Min.)	Field/Plate (x, y)
AM 2209-621	22 09 28 −62 18.7	4.254 0.296	328.59 −46.38	227.3 3.1	PECULIAR LINEAR GALAXY + COMP'N. 146-G-14	8,22 S 1.6	F146/J1873 (182,55)
AM 2209-361 [2209-3612]	22 09 38 −36 11.3	3.526 0.296	8.17 −55.27	247.4 20.4	PAIR (#1: S0) 404-G-40	23 D 1.0	F404/J0736 (72,112)
AM 2209-361 [2209-3611]	22 09 41 −36 12.5	3.526 0.296	8.14 −55.28	247.4 20.4	PAIR (#2: DWARF SPIRAL) 404-G-41	20,23 D 0.7	F404/J0736 (72,111)
AM 2209-532	22 09 42 −53 28.8	3.908 0.296	340.13 −50.85	233.9 9.1	GROUP of 3 (incl. I/A COMP'N.) 189-G-29	1,24 S 1.8	F189/J3474 (102,260)
*AM 2209-382 [2209-3822]	22 09 47 −38 25.3	3.563 0.296	4.27 −55.11	245.6 19.0	I/A DBL.(#1: SP. w. LONG ARMS) IC 5174 = MCG-6-48-28 = 344-G-13	2,10 D 0.5	F344/J1508 (140,264)
*AM 2209-382 [2209-3825]	22 09 47 −38 25.3	3.563 0.296	4.27 −55.11	245.6 19.0	I/A DBL.(#2) B 2209-384 IC 5175 = MCG-6-48-29 = 344-G-14	2 D 0.5	F344/J1508 (140,264)
AM 2209-711	22 09 59 −71 18.2	4.896 0.296	318.73 −40.89	220.8 −3.1	COMPACT (w. HALO)	13 D 0.3	F076/J1792 (272,102)
*AM 2210-490	22 10 03 −49 05.6	3.785 0.296	346.61 −52.64	237.2 12.0	3-ARMED SPIRAL 237-G-47	11 D 0.9	F237/J3558 (91,227)
AM 2210-814	22 10 08 −81 48.2	7.351 0.296	309.28 −33.55	213.0 −10.2	COMPACT w. LO.S.BR. HALO	13,22 D 0.9	F027/J2485 (217,81)
AM 2210-284	22 10 13 −28 46.3	3.411 0.296	21.19 −55.11	253.9 24.8	I/A DBL. + TAIL 467-IG-22	2,15 S 0.4	F467/J0691 (222,242)
AM 2210-461	22 10 15 −46 15.7	3.716 0.296	351.05 −53.59	239.4 13.8	CHAIN of 4 (+ APP. COMP'NS.) IC 5181 = 289-G-1	8,17 D 12.1	F288/J1633 (47,109)
AM 2210-462	22 10 16 −46 22.1	3.719 0.296	350.88 −53.56	239.3 13.7	DWARF (in WIDE GROUP)	20 D 0.1	F288/J1633 (47,103)
*AM 2210-262	22 10 19 −26 23.8	3.378 0.296	25.27 −54.76	256.0 26.2	GAL. w. V.PEC. NARROW DUST LANES NGC 7225 = 532-G-33	14 D 1.8	F533/J4469 (319,102)
AM 2210-303	22 10 19 −30 38.2	3.437 0.296	17.94 −55.33	252.2 23.7	CONNECTED E's	2 S 0.6	F467/J0691 (220,142)
AM 2210-371	22 10 19 −37 15.8	3.541 0.296	6.27 −55.33	246.6 19.6	I/A DBL. 404-IG-42	2 S 0.2	F344/J1508 (134,326)
AM 2210-222	22 10 40 −22 20.5	3.325 0.297	32.07 −53.88	259.8 28.4	SPIRAL w. 4 APP. COMP'NS. IC 1435 = 601-G-30	8 S 8.1	F532/J0817 (43,319)
AM 2210-274	22 10 41 −27 48.3	3.396 0.297	22.88 −55.08	254.8 25.3	SPIRAL w. I/A COMP'N. MCG-5-52-49 = 467-G-23	1 D 1.3	F532/J0817 (50,25)
*AM 2210-693	22 10 49 −69 36.6	4.721 0.297	320.38 −42.05	222.1 −2.0	I/A DBL. ('THE SCORPION') IC 5173 = 76-IG-8	2 D 2.0	F076/J1792 (276,193)
AM 2210-215	22 10 54 −21 58.8	3.320 0.297	32.68 −53.83	260.2 28.6	IRR. SINGLE MCG-4-52-26 = 601-G-31	16 D 1.3	F532/J0817 (40,338)
AM 2210-620	22 10 58 −62 04.1	4.227 0.297	328.73 −46.66	227.6 3.1	M51-TYPE	9 D 0.4	F108/J0946 (107,333)
AM 2210-243	22 10 59 −24 35.0	3.353 0.297	28.39 −54.52	257.8 27.1	CLOSE I/A DBL. 533-IG-2	2 S 0.2	F533/J4469 (314,200)
AM 2211-542	22 11 03 −54 22.0	3.926 0.297	338.74 −50.63	233.3 8.3	3-ARMED SPIRAL + APP. COMP'N. 189-G-31	8,11 D 0.9	F189/J3474 (93,212)
AM 2211-365	22 11 08 −36 55.7	3.532 0.297	6.84 −55.52	247.0 19.7	I/A DBL in GROUP of 4 = KLEM 41 incl. 404-G-44 = DN 2211-369	2,18 S 0.4	F344/J1508 (125,344)
AM 2211-670	22 11 12 −67 05.8	4.518 0.297	322.99 −43.68	224.0 −0.3	GALAXY w. PECULIAR APP. COMP'N. IC 5176 = 108-G-20	1 S 4.7	F075/J0953 (14,323)
AM 2211-405	22 11 29 −40 54.3	3.601 0.297	359.89 −55.08	243.8 17.1	I/A DBL. + CHAIN COMP'NS. 344-G-15 = PKS 2211-408?	2,17 S 1.2	F344/J1508 (123,131)
AM 2211-654	22 11 37 −65 48.1	4.427 0.297	324.37 −44.51	224.9 0.5	PECULIAR SPIRAL 108-IG-21 = B 2211-657	12 D 1.2	F108/J0946 (113,132)
AM 2211-390	22 11 44 −39 03.2	3.566 0.297	3.09 −55.41	245.3 18.3	SP. w. APP. COMP'NS. (I/A DBL.) 344-G-16	2,8 S 4.0	F344/J1508 (120,230)
AM 2211-292	22 11 45 −29 24.7	3.416 0.297	20.14 −55.52	253.5 24.2	E at CENTRE of CHAIN of 4 467-G-26	17 D 0.8	F467/J0691 (204,208)
*AM 2211-274	22 11 51 −27 43.0	3.392 0.297	23.10 −55.32	255.0 25.2	THICK STRAIGHT + THIN CIRC. ARMS MCG-5-52-54? = 467-G-27	10 D 1.6	F532/J0817 (36,29)
AM 2211-510	22 11 56 −51 00.0	3.822 0.298	343.53 −52.20	236.0 10.5	I/A DBL.	2 S 0.5	F238/J0961 (331,120)
AM 2212-463	22 12 02 −46 31.9	3.713 0.298	350.46 −53.80	239.4 13.4	DWARF/IRR. (in WIDE GROUP) 289-G-5	18,20 D 0.9	F288/J1633 (31,94)
AM 2212-365	22 12 32 −36 56.7	3.527 0.298	6.77 −55.80	247.2 19.5	I/A DBL. off MI.AX. of IC 5179 405-IG-4	2,8 S 0.4	F344/J1508 (110,343)
*AM 2212-460	22 12 32 −46 06.1	3.700 0.298	351.11 −54.01	239.8 13.6	CLOSE TRIPLE (LARGE) NGC 7232/33 = 289-IG-7/8	24 D 4.9	F289/J1593 (286,118)
AM 2212-503	22 12 45 −50 30.3	3.804 0.298	344.19 −52.51	236.4 10.7	SPIRAL w. PECULIAR ARM	10 D 0.5	F238/J0961 (326,147)
AM 2212-550	22 12 46 −55 04.0	3.936 0.298	337.58 −50.53	233.0 7.7	GALAXY w. COMP'N. on END of JET	1,7 S 0.7	F190/J1765 (335,170)
AM 2213-802	22 13 12 −80 20.5	6.600 0.298	310.37 −34.70	214.2 −9.3	VERY CLOSE PAIR	23 S 0.1	F027/J2485 (217,160)

Identification	R.A. (1950.0) Declination	Precession R.A./Dec.	Gal. Long. Gal. Lat.	S.G.L. S.G.B.	Description Cross-Identification	Coding Size (Min.)	Field/Plate (x, y)
AM 2213-370	22 13 14 −37 05.6	3.527 0.298	6.48 −55.93	247.1 19.3	BR. GALAXY between 2 GROUPS IC 5179 = IC5183/4 = 405-G-5	8,16,22 D 2.2	F404/J0736 (35,62)
AM 2213-265	22 13 18 −26 56.2	3.378 0.298	24.55 −55.51	255.9 25.3	M51-TYPE ? MCG-5-52-59 = 533-G-6	9 D 1.3	F532/J0817 (17,71)
AM 2213-703	22 13 48 −70 37.7	4.771 0.299	319.13 −41.57	221.5 −2.9	GALAXY w. FAN-LIKE JETS	7 D 0.5	F076/J1792 (258,139)
AM 2213-363	22 13 53 −36 39.1	3.517 0.299	7.25 −56.09	247.6 19.4	DWARF SPIRAL 405-G-6	20 D 1.5	F404/J0736 (27,86)
AM 2213-425	22 13 53 −42 59.7	3.629 0.299	356.16 −55.08	242.4 15.4	DWARF	20 D 0.5	F344/J1508 (101,18)
AM 2214-620	22 14 06 −62 06.7	4.198 0.299	328.37 −46.94	227.9 2.8	COMPACT	13 D 0.2	F108/J0946 (87,330)
AM 2214-484	22 14 07 −48 43.5	3.751 0.299	346.78 −53.39	238.0 11.7	GROUP of 6 237-G-50	18 D 2.7	F237/J3558 (54,245)
AM 2214-371	22 14 09 −37 11.9	3.525 0.299	6.27 −56.10	247.2 19.0	GROUP of 5	18 D 5.4	F405/J0766 (282,58)
AM 2214-481	22 14 09 −48 18.8	3.741 0.299	347.42 −53.55	238.3 12.0	DWARF	20 D 0.3	F238/J0961 (321,266)
AM 2214-451	22 14 12 −45 19.0	3.675 0.299	352.25 −54.53	240.6 13.9	DWARF 289-G-11	20 D 1.3	F289/J1593 (272,161)
*AM 2214-524	22 14 32 −52 46.3	3.854 0.299	340.65 −51.81	234.9 9.0	M51-TYPE Fa 355	9 S 0.5	F237/J3558 (61,28)
AM 2214-530	22 14 39 −53 05.3	3.862 0.299	340.17 −51.69	234.7 8.8	SINGLE, DISTURBED GALAXY w. LOOP	15,16,18 D 0.5	F189/J3474 (61,280)
AM 2214-470	22 14 47 −47 09.8	3.711 0.299	349.20 −54.04	239.3 12.6	I/A TRIPLE 289-IG-14 = B 2214-471	3 D 0.7	F289/J1593 (263,62)
AM 2214-355	22 14 50 −35 58.8	3.503 0.299	8.43 −56.33	248.3 19.7	CLUSTER of GALAXIES	19 D 4.5	F405/J0766 (276,123)
AM 2214-520	22 14 50 −52 01.0	3.831 0.299	341.71 −52.19	235.5 9.5	COMPACT	13 D 0.1	F190/J1765 (329,335)
AM 2214-604	22 14 55 −60 49.3	4.132 0.300	329.85 −47.76	228.9 3.6	PAIR of E's	23 S 0.3	F146/J1873 (147,135)
AM 2215-475	22 15 01 −47 52.7	3.726 0.300	348.02 −53.83	238.7 12.1	IRR. w. KNOTS	16 D 0.4	F289/J1593 (260,23)
AM 2215-595	22 15 03 −59 53.5	4.091 0.300	330.99 −48.30	229.6 4.2	SP. w. 6 COMP'NS. (COMPACT,PEC.) IC 5188 = 146-IG-17 (DIAMETRIC)	8,13 S 4.9	F146/J1873 (145,185)
AM 2215-425	22 15 07 −42 55.9	3.622 0.300	356.17 −55.32	242.6 15.3	CHAIN of 7 (I/A DBL. + E) incl. 289-G-15	2,8,17 D 9.9	F344/J1508 (89,21)
AM 2215-483	22 15 16 −48 30.1	3.739 0.300	347.01 −53.65	238.3 11.7	DISTURBED SINGLE GALAXY	16 D 0.5	F238/J0961 (310,256)
AM 2215-431	22 15 28 −43 16.0	3.627 0.300	355.57 −55.30	242.4 15.0	I/A DBL.	2 S 0.5	F289/J1593 (262,271)
*AM 2215-390	22 15 41 −39 08.7	3.551 0.300	2.73 −56.16	245.8 17.6	DBL. cD in ELONGATED CLUSTER 344-G-19 = PKS 2215-391	19 D 17.9	F344/J1508 (79,224)
AM 2216-802	22 16 53 −80 29.2	6.540 0.301	310.11 −34.70	214.2 −9.5	I/A DBL. in GROUP of 9 27-IG-7	2,18 S 0.2	F027/J2485 (208,152)
AM 2217-355	22 17 12 −35 51.0	3.492 0.301	8.61 −56.82	248.7 19.4	COMPACT (POSSIBLE GROUP)	13 D 0.1	F405/J0766 (250,131)
AM 2217-474	22 17 31 −47 41.8	3.708 0.301	348.06 −54.29	239.1 11.9	LUMPY COMPACT	13 D 0.4	F238/J0961 (292,300)
AM 2217-325	22 17 33 −32 56.3	3.447 0.301	13.94 −56.98	251.2 21.1	I/A TRIPLET 405-G-9	3 D 1.6	F405/J0766 (249,287)
AM 2217-383	22 17 38 −38 35.5	3.534 0.301	3.62 −56.62	246.5 17.6	SPIRAL-ARC + HI.S.BR. INNER SP.	10 D 1.1	F345/J0844 (327,250)
AM 2217-490	22 17 44 −49 01.5	3.737 0.301	345.94 −53.83	238.1 11.0	SP. w. SQUARE ARMS + I/A COMP'N. 238-G-2 = Fa 356 in CHAIN	1,10 S 1.2	F238/J0961 (287,229)
AM 2217-801	22 17 44 −80 15.1	6.429 0.301	310.27 −34.90	214.4 −9.4	Sd w. 2 COMP'NS. 27-G-8	8 D 3.1	F027/J2485 (207,165)
AM 2217-490	22 17 47 −49 01.9	3.737 0.301	345.92 −53.83	238.1 11.0	CHAIN of 4 (3 DISTURBED) incl. 238-G-2	8,17 D 7.6	F237/J3558 (23,227)
*AM 2217-245	22 17 58 −24 55.9	3.340 0.301	28.42 −56.15	258.4 25.6	"4-PETAL DAISY" (FILAMENTS) NGC 7252 = ARP 226 = 533-IG-15	15 D 3.6	F533/J4469 (228,183)
AM 2218-440	22 18 06 −44 07.9	3.630 0.302	353.88 −55.53	242.0 14.1	COMPACT 289-G-19 = PK 353-55-1	13 D 0.3	F289/J1593 (236,225)
AM 2218-351	22 18 09 −35 14.1	3.479 0.302	9.71 −57.05	249.3 19.6	COMPACT	13 D 0.1	F405/J0766 (240,164)
AM 2218-340	22 18 24 −34 04.3	3.461 0.302	11.85 −57.15	250.4 20.3	IRR. GALAXY	13,16 D 0.4	F405/J0766 (238,227)
AM 2218-840	22 18 34 −84 04.2	8.585 0.302	307.26 −32.02	211.5 −11.9	M51-TYPE in CHAIN of 5 11-IG-7 = B 2218-840	9,17 S 0.4	F011/J2562 (121,224)
AM 2219-384	22 19 00 −38 48.1	3.531 0.302	3.17 −56.85	246.5 17.3	I/A DBL.	2 S 0.7	F345/J0844 (312,239)

Identification	R.A. (1950.0) Declination	Precession R.A./Dec.	Gal. Long. Gal. Lat.	S.G.L. S.G.B.	Description Cross-Identification	Coding Size (Min.)	Field/Plate (x, y)
*AM 2219-255	22 19 05 −25 50.1	3.349 0.302	26.92 −56.58	257.8 24.9	I/A DBL. w. CONNECTED FILAMENT	2,9,15 S 0.8	F533/J4469 (214,134)
AM 2219-414	22 19 09 −41 40.1	3.579 0.302	358.05 −56.34	244.1 15.5	GROUP of 5	18 D 5.4	F344/J1508 (47,88)
AM 2219-600	22 19 13 −60 01.5	4.059 0.302	330.37 −48.65	229.8 3.7	ROUGH CHAIN of DISTURBED + PEC. IC 5203 = 146-IG-20	17 D 8.3	F146/J1873 (117,177)
AM 2219-483	22 19 23 −48 39.5	3.718 0.302	346.34 −54.22	238.6 11.1	DWARF (RESOLUTION ?) 238-G-5	20 D 0.5	F238/J0961 (273,249)
AM 2219-235	22 19 25 −23 51.3	3.324 0.302	30.44 −56.22	259.6 25.9	GALAXY w. PECULIAR DUST LANE	14 D 0.3	F533/J4469 (211,241)
*AM 2219-432	22 19 30 −43 26.8	3.610 0.302	354.93 −55.96	242.7 14.4	SMALL M51-TYPE + SPIRAL 289-IG-23 = B 2219-434	9,23 D 1.1	F289/J1593 (223,262)
AM 2219-265	22 19 44 −26 59.9	3.361 0.303	24.87 −56.93	256.8 24.1	SPIRAL + COMP'N. on THIRD ARM	11 S 0.7	F467/J0691 (109,337)
*AM 2219-234	22 19 47 −23 45.9	3.322 0.303	30.63 −56.28	259.8 25.9	I/A DBL. w. TAILS MCG-4-52-41 = 533-G-21	2,15 S 0.5	F533/J4469 (207,245)
*AM 2219-571	22 19 50 −57 14.7	3.952 0.303	333.85 −50.27	232.0 5.5	DBL. M51-TYPE 190-IG-3	9 S 0.4	F146/J1873 (108,326)
AM 2220-570	22 20 00 −57 02.2	3.944 0.303	334.11 −50.40	232.2 5.6	GROUP of 4	18 D 5.6	F146/J1873 (106,337)
AM 2220-660	22 20 01 −66 05.7	4.347 0.303	323.29 −44.99	225.3 −0.3	CHAIN of FOUR	17 D 13.4	F109/J0954 (307,111)
AM 2220-421	22 20 05 −42 11.2	3.584 0.303	357.07 −56.39	243.8 15.1	DISTURBED in CHAIN of 4	16,17 D 7.6	F344/J1508 (38,60)
*AM 2220-493	22 20 06 −49 32.6	3.734 0.303	344.87 −53.98	238.0 10.4	ONE-ARMED SP.(RING?) + I/A COMP'N 238-IG-6 = B 2220-495	1,6,10 S 0.4	F238/J0961 (265,202)
AM 2220-291	22 20 17 −29 12.4	3.388 0.303	20.87 −57.35	254.9 22.8	M51-TYPE NGC 7259 = 467-G-50	9 D 1.1	F467/J0691 (104,218)
AM 2220-612	22 20 25 −61 27.0	4.107 0.303	328.51 −47.93	228.8 2.7	E w. DIAMETRIC GROUPS (3 and 5 COMPONENTS EACH)	8 D 7.2	F146/J1873 (112,100)
AM 2220-381	22 20 28 −38 15.1	3.516 0.303	4.10 −57.21	247.1 17.4	CLOSE PAIR (LATE-TYPE SPIRAL) IC 5209/12 = 345-G-4/6	23 S 5.4	F344/J1508 (27,271)
AM 2220-632	22 20 33 −63 26.7	4.197 0.303	326.17 −46.72	227.3 1.4	SP. w. LONG ARM (in GROUP of 7) Se 150/2 = DN 2220-634(B2220-633?)	10,17 D 0.9	F109/J0954 (317,253)
AM 2220-423	22 20 35 −42 31.9	3.588 0.303	356.42 −56.39	243.6 14.8	Sd-DWARF UKS 2220-425	20 D 2.7	F344/J1508 (34,41)
*AM 2220-661	22 20 40 −66 17.6	4.351 0.303	323.02 −44.91	225.2 −0.5	SPIRAL w. TAILS	15 D 0.6	F108/J0946 (65,103)
AM 2220-460	22 20 42 −46 09.3	3.657 0.303	350.24 −55.34	240.7 12.5	DWARF (CONDENSATIONS)	20 D 0.5	F289/J1593 (210,117)
AM 2220-374	22 20 52 −37 41.1	3.506 0.303	5.11 −57.37	247.6 17.7	COMPACT	13 D 0.2	F345/J0844 (294,300)
AM 2220-382	22 20 59 −38 28.5	3.518 0.303	3.66 −57.28	247.0 17.2	ISOLATED GALAXY w. TAIL	15 D 0.9	F345/J0844 (291,258)
AM 2221-545	22 21 11 −54 54.5	3.868 0.303	336.85 −51.66	233.9 6.9	I/A SPIRALS w. COMMON ARM 190-IG-4	2,10 S 0.2	F190/J1765 (270,182)
AM 2221-413	22 21 18 −41 36.3	3.568 0.304	358.00 −56.74	244.4 15.2	MULTIPLE NUCLEI + LOOP + SPIKES	22 D 0.8	F345/J0844 (284,90)
AM 2221-625	22 21 30 −62 54.0	4.161 0.304	326.70 −47.14	227.8 1.7	ONE-ARMED SPIRAL + JET? (SMALL) 108-IG-30	7,12 D 0.4	F108/J0946 (44,284)
AM 2222-420	22 22 01 −42 04.7	3.573 0.304	357.10 −56.76	244.1 14.8	CLOSE I/A DBL. 345-IG-8	2 S 0.5	F344/J1508 (19,65)
AM 2222-572	22 22 03 −57 20.6	3.937 0.304	333.47 −50.47	232.1 5.2	I/A DBL.	2 S 0.2	F146/J1873 (92,320)
AM 2222-313	22 22 31 −31 36.6	3.412 0.304	16.45 −58.00	253.0 21.0	I/A DBL. MCG-5-52-73 = 467-IG-55	2 S 0.3	F468/J0659 (345,90)
AM 2222-863	22 22 36 −86 30.6	0.202 0.304	305.41 −30.17	209.6 −13.5	COMPACT w. PLUME (in PAIR)	7,13,15,23 S 1.3	F011/J2562 (141,95)
*AM 2222-275	22 22 40 −27 56.9	3.365 0.304	23.32 −57.72	256.3 23.1	I/A DBL. + M51-TYPE 467-IG-56	1,2,9,24 D 1.3	F533/J4469 (170,21)
AM 2222-572	22 22 49 −57 29.8	3.936 0.304	333.18 −50.47	232.1 5.0	DISTANT CLUSTER of GALAXIES R 75	19 D 6.7	F146/J1873 (87,312)
AM 2222-312	22 22 51 −31 27.3	3.409 0.304	16.75 −58.06	253.2 21.0	E + SPIRAL (CENTRE of GROUP) NGC 7268 = MCG-5-53-1 = 467-IG-57	18,23 D 0.8	F467/J0691 (76,97)
AM 2222-641 [2223-6417]	22 22 56 −64 15.8	4.212 0.304	325.01 −46.41	226.9 0.7	I/A DBL. (#1)	2 D 0.2	F109/J0954 (299,211)
AM 2223-635	22 23 11 −63 57.3	4.194 0.305	325.32 −46.63	227.1 0.8	CLOSE TRIPLE (incl. RING)	6,24 D 2.5	F108/J0946 (40,227)
AM 2223-310	22 23 18 −31 07.6	3.403 0.305	17.38 −58.14	253.6 21.2	CLOSE PAIR (incl. PEC. ABS'N.) incl. MCG-5-53-3 = 467-G-58	14,23 S 2.5	F468/J0659 (337,117)
AM 2223-345	22 23 19 −34 58.6	3.456 0.305	10.08 −58.12	250.2 18.9	GALAXY + I/A COMP'N.	1 D 4.5	F405/J0766 (184,178)

Identification	R.A. (1950.0) Declination	Precession R.A./Dec.	Gal. Long. Gal. Lat.	S.G.L. S.G.B.	Description Cross-Identification	Coding Size (Min.)	Field/Plate (x, y)
AM 2223-641 [2222-6415]	22 23 21 −64 17.9	4.209 0.305	324.93 −46.42	226.9 0.6	I/A DBL. (#2)	2 D 0.2	F109/J0954 (297,209)
AM 2223-373	22 23 22 −37 37.8	3.495 0.305	5.10 −57.87	248.0 17.3	GALAXY w. DUST LANE + MATERIAL 345-G-11	14 D 0.9	F345/J0844 (267,304)
AM 2223-284	22 23 29 −28 48.5	3.373 0.305	21.75 −58.00	255.7 22.4	COMPACT	13 D 0.3	F468/J0659 (338,241)
AM 2223-645	22 23 29 −64 59.6	4.244 0.305	324.15 −45.98	226.4 0.1	GROUP of 4	18 D 6.3	F109/J0954 (293,172)
AM 2223-433	22 23 33 −43 31.8	3.591 0.305	354.42 −56.64	243.1 13.7	E w. MANY COMP'NS.	8 S 3.1	F289/J1593 (183,258)
AM 2223-523	22 23 54 −52 30.8	3.782 0.305	339.94 −53.19	236.0 8.1	SBd/DWARF 190-G-11	20 D 1.1	F238/J0961 (228,44)
*AM 2224-310	22 24 04 −31 08.5	3.401 0.305	17.36 −58.31	253.7 21.0	GALAXY w. I/A COMP'N. MCG-5-53-5 = 467-G-62	1 S 1.4	F468/J0659 (328,116)
AM 2224-581	22 24 08 −58 11.3	3.948 0.305	332.13 −50.22	231.6 4.4	I/A DBL. in GROUP of 3 146-IG-24	2,24 S 0.4	F146/J1873 (79,274)
AM 2224-405	22 24 22 −40 52.9	3.542 0.305	359.05 −57.47	245.4 15.2	DWARF (RESOLVED) 345-G-13	20 D 0.6	F345/J0844 (254,129)
AM 2224-490	22 24 22 −49 09.9	3.699 0.305	344.97 −54.77	238.7 10.1	2 E's in ENVELOPE (FILAMENT?) 238-IG-11 + COMP'NS.	2,8 S 0.2	F238/J0961 (229,224)
AM 2224-431	22 24 26 −43 16.3	3.582 0.305	354.78 −56.87	243.4 13.8	CLOSE PAIR of SPIRALS 289-G-28 = DN 2224-435?	23 S 2.7	F289/J1593 (175,272)
AM 2224-405	22 24 36 −40 55.0	3.541 0.305	358.97 −57.50	245.4 15.2	COMPACT w. 2 CLOSE COMP'NS.	8,13 D 0.8	F345/J0844 (251,127)
AM 2224-545	22 24 37 −54 59.9	3.844 0.305	336.32 −52.03	234.2 6.4	SPIRAL w. 2 APP. COMP'NS 190-G-12 (in GROUP of 6)	8,18 S 1.5	F190/J1765 (244,179)
AM 2224-352	22 24 44 −35 24.7	3.457 0.306	9.22 −58.38	250.0 18.4	COMPACT	13 D 0.1	F405/J0766 (168,155)
AM 2225-635	22 25 09 −63 51.7	4.168 0.306	325.23 −46.85	227.3 0.7	I/A DBL. (ONE M51-TYPE)	2,9 S 0.8	F109/J0954 (288,233)
AM 2225-660	22 25 17 −66 09.3	4.287 0.306	322.73 −45.35	225.6 −0.8	CLOSE DBL. IC 5219/21 = 109-G-4/5	23 S 1.7	F109/J0954 (278,111)
AM 2225-795	22 25 17 −79 57.2	6.102 0.306	310.21 −35.34	214.9 −9.5	I/A DBL. (DWARFS) 27-IG-12	2,20 S 0.7	F027/J2485 (190,182)
AM 2225-610	22 25 31 −61 08.2	4.044 0.306	328.33 −48.61	229.5 2.4	E + APP. COMP'NS. 146-G-28	8 S 2.7	F146/J1873 (79,115)
AM 2225-295	22 25 45 −29 59.8	3.381 0.306	19.59 −58.60	254.9 21.4	DWARF 468-G-2	20 D 0.4	F468/J0659 (310,178)
*AM 2225-250	22 25 50 −25 06.1	3.323 0.306	28.81 −57.93	259.3 24.0	I/A DBL. w. EXTENDED PLUME NGC7284/5 = VV74a/b = 533-IG-31/2	2,9,15 S 0.5	F533/J4469 (132,174)
AM 2225-801	22 25 51 −80 17.7	6.192 0.306	309.92 −35.09	214.6 −9.7	CHAIN of 5	17 D 8.5	F027/J2485 (189,163)
AM 2226-584	22 26 02 −58 40.1	3.948 0.306	331.29 −50.15	231.4 3.9	GALAXY w. I/A COMP'N. and TAILS 148-IG-29	1,15 S 0.1	F146/J1873 (67,248)
AM 2226-655	22 26 23 −65 55.0	4.260 0.307	322.87 −45.59	225.8 −0.7	SB w. CONDENS. at END of BAR IC 5222 = 109-G-6 = fA 193	12 D 2.0	F109/J0954 (273,124)
AM 2226-502	22 26 29 −50 25.7	3.714 0.307	342.75 −54.52	237.9 9.1	GALAXY w. ABS'N.	14 D 0.4	F238/J0961 (209,156)
AM 2226-570	22 26 31 −57 04.2	3.891 0.307	333.29 −51.12	232.7 4.9	E w. 2 PECULIAR COMP'NS.	8 S 1.9	F190/J1765 (226,68)
AM 2226-652	22 26 32 −65 26.0	4.232 0.307	323.37 −45.93	226.2 −0.4	I/A DBL (w. POSSIBLE JETS)	2,7 D 0.5	F109/J0954 (274,150)
AM 2226-610	22 26 34 −61 05.4	4.032 0.307	328.26 −48.74	229.6 2.4	E w. 6 APP. COMP'NS.	8 S 9.0	F147/J1579 (316,113)
AM 2226-613 [2227-6137]	22 26 58 −61 34.5	4.048 0.307	327.64 −48.47	229.2 2.0	WIDE I/A DBL. in GROUP of 6 (#1) DN 2226-611	18,23 S 2.7	F146/J1873 (71,91)
AM 2227-304	22 27 01 −30 46.7	3.387 0.307	18.12 −58.92	254.4 20.7	GALAXY w. JET or I/A COMP'N. MCG-5-53-9 = 468-G-3	7 S 0.7	F468/J0659 (294,136)
AM 2227-613 [2226-6134]	22 27 15 −61 37.0	4.046 0.307	327.56 −48.47	229.2 2.0	WIDE I/A DBL. in GROUP of 6 (#2) 146-GA-29 = B 2227-612 in [DN]	18,23 S 2.7	F146/J1873 (69,89)
AM 2227-413	22 27 23 −41 30.2	3.538 0.307	357.67 −57.87	245.2 14.4	COMPACT + COMP'NS.(?)	8,13 D 0.2	F345/J0844 (223,96)
AM 2227-540	22 27 51 −54 01.3	3.793 0.307	337.27 −52.94	235.2 6.7	CLOSE PAIR	23 S 0.4	F190/J1765 (220,232)
AM 2227-271	22 27 52 −27 11.3	3.341 0.307	25.06 −58.75	257.7 22.5	GALAXY + JET (DEFECT?) MCG-5-53-11 = 533-G-37	7 D 1.1	F467/J0691 (12,325)
AM 2228-380 [2230-3744]	22 28 36 −38 04.0	3.479 0.308	3.98 −58.82	248.3 16.2	ARC (#1: PAIR of GALAXIES) NGC 7299 = 345-G-19	17,23	F405/J0766 (128,12)
*AM 2228-423	22 28 41 −42 30.9	3.548 0.308	355.71 −57.83	244.6 13.6	2 I/A DBLS. I/A 289-IG-34	2,2 D 0.4	F345/J0844 (209,42)
AM 2228-640	22 28 43 −64 07.1	4.141 0.308	324.57 −46.98	227.4 0.3	GAL. w. VERY EXTENDED MATERIAL MC 2228-641 + COMP'N.	8,15 D 0.7	F109/J0954 (266,221)

Identification	R.A. (1950.0) Declination	Precession R.A./Dec.	Gal. Long. Gal. Lat.	S.G.L. S.G.B.	Description Cross-Identification	Coding Size (Min.)	Field/Plate (x, y)
AM 2229-281	22 29 13 −28 16.8	3.350 0.308	23.04 −59.19	256.9 21.7	DWARF 468-G-8	20 D 0.7	F468/J0659 (271,271)
AM 2229-253	22 29 22 −25 39.4	3.320 0.308	28.08 −58.82	259.3 23.0	S0 + 18 COMP'NS PAIRED w. SPIRAL NGC 7294 = 533-G-44 = MC 2229-256	8,23 S 3.8	F533/J4469 (90,144)
AM 2229-524	22 29 30 −52 48.6	3.750 0.308	338.80 −53.78	236.3 7.2	I/A SPIRALS 190-IG-15	2 S 0.3	F238/J0961 (183,29)
AM 2229-372	22 29 33 −37 26.7	3.466 0.308	5.11 −59.11	248.9 16.5	M51-TYPE (both ARMS w. COMP'NS)	9 D 0.9	F345/J0844 (202,314)
AM 2229-295	22 29 42 −29 50.2	3.367 0.308	20.02 −59.45	255.5 20.7	DWARF 468-G-9	20 D 0.5	F468/J0659 (264,188)
*AM 2229-735	22 29 49 −73 56.6	4.853 0.308	314.90 −40.07	219.8 −5.9	GALAXY w. NUCLEATED ARC 48-IG-26 = B 2229-739	2 S 0.4	F049/J3647 (313,225)
AM 2229-242	22 29 57 −24 26.6	3.305 0.309	30.42 −58.70	260.5 23.6	CLOSE PAIR	23 S 0.6	F533/J4469 (82,209)
AM 2230-373 [2230-3737]	22 30 25 −37 34.6	3.465 0.309	4.80 −59.26	248.9 16.2	ARC (#2: GAL. w. BR. SPOT in RIM)	10,17 D 0.6	F405/J0766 (108,38)
*AM 2230-273 [2230-2730]	22 30 28 −27 30.4	3.338 0.309	24.61 −59.37	257.7 21.9	CLOSE PAIR (#1: SP. + I/A COMP'N.) MCG-5-53-14 = 468-G-11	1,8,10,23 D 0.9	F468/J0659 (257,313)
AM 2230-284	22 30 29 −28 44.3	3.352 0.309	22.20 −59.52	256.6 21.2	COMPACT	13 D 0.1	F468/J0659 (256,247)
AM 2230-373 [2228-3804]	22 30 29 −37 37.2	3.465 0.309	4.72 −59.26	248.9 16.2	ARC (#3:INTEGRAL SP. + I/A COMP'N.) 345-IG-22 = B 2230-376	1,17 S 0.6	F405/J0766 (108,36)
AM 2230-380	22 30 34 −38 01.9	3.470 0.309	3.92 −59.21	248.5 15.9	COMPACT	13,17 D 0.2	F405/J0766 (107,14)
*AM 2230-645	22 30 39 −64 57.3	4.160 0.309	323.46 −46.57	226.9 −0.4	SPIRAL + I/A DBL.COMP'N. IC 5227 = 109-G-8 [A82]	1 S 2.9	F109/J0954 (253,177)
*AM 2230-481	22 30 42 −48 16.9	3.642 0.309	345.60 −56.09	240.1 9.8	RING GALAXY w. I/A COMP'N. 238-G-16 = B 2230-482	1,6 S 0.4	F238/J0961 (173,272)
AM 2230-374 [2230-3734]	22 30 44 −37 44.2	3.466 0.309	4.47 −59.29	248.8 16.1	ARC (#4: GAL. + I/A + APP.COMP'NS) 345-IG-25 = B 2230-377	1,8 S 0.3	F405/J0766 (105,29)
*AM 2230-273 [2230-2730]	22 30 57 −27 30.5	3.337 0.309	24.63 −59.48	257.8 21.8	CLOSE PAIR (#2: E) MCG-5-53-15 = 468-G-13	23 D 0.3	F468/J0659 (251,313)
AM 2231-371	22 31 00 −37 18.2	3.458 0.309	5.30 −59.41	249.2 16.3	CLOSE I/A DBL.	2 S 0.2	F345/J0844 (186,322)
AM 2231-522 [2231-5221]	22 31 08 −52 20.4	3.727 0.309	339.27 −54.24	236.9 7.3	PAIR of E's (#1)	23 S 2.4	F190/J1765 (195,322)
AM 2231-522 [2231-5220]	22 31 22 −52 21.9	3.726 0.309	339.20 −54.25	236.9 7.3	PAIR of E's (#2)	23 S 2.4	F190/J1765 (193,321)
*AM 2231-375	22 31 31 −37 59.3	3.466 0.309	3.94 −59.40	248.7 15.8	RICH CLUSTER of GALAXIES	19 D 4.5	F345/J0844 (181,285)
AM 2231-374	22 31 42 −37 42.4	3.461 0.310	4.47 −59.48	248.9 16.0	CHAIN of I/A GALAXIES B 2231-377?	2,17 D 1.0	F345/J0844 (179,300)
AM 2232-373	22 32 38 −37 38.1	3.456 0.310	4.55 −59.68	249.1 15.8	CLOSE PAIR of SPIRALS 345-G-32/33	23 S 2.7	F406/J0756 (340,33)
*AM 2232-510	22 32 39 −51 08.9	3.690 0.310	340.84 −55.04	238.0 7.9	SPIRAL w. ONE BRIGHT ARM 238-G-18	10 D 0.7	F238/J0961 (157,118)
AM 2232-394	22 32 43 −39 49.2	3.487 0.310	0.33 −59.26	247.3 14.6	DWARF (RESOLVED)	20 D 0.6	F345/J0844 (168,187)
AM 2232-470	22 32 55 −47 04.8	3.606 0.310	347.29 −56.93	241.3 10.3	DWARF	20 D 0.4	F289/J1593 (98,66)
*AM 2233-261	22 33 01 −26 18.5	3.318 0.310	27.12 −59.75	259.2 22.0	SPIRAL w. 4 APP. COMP'NS. NGC 7314 = ARP 14 = 533-G-53	8 S 11.2	F533/J4469 (46,108)
AM 2233-384	22 33 08 −38 40.2	3.469 0.310	2.50 −59.59	248.3 15.2	COMPACT	13 D 0.2	F406/J0756 (332, −22)
AM 2233-345	22 33 12 −34 58.8	3.419 0.310	9.77 −60.14	251.5 17.3	CHAIN of 6	17 S 11.2	F405/J0766 (75,177)
*AM 2233-613	22 33 20 −61 39.5	3.987 0.310	326.81 −49.00	229.7 1.4	E + INTEGRAL-SIGN SPIRAL 147-IG-3 + APP. COMP'N.	8,9,23 S 0.8	F146/J1873 (31,84)
AM 2233-243	22 33 39 −24 36.3	3.298 0.311	30.48 −59.55	260.8 22.8	CLUSTER of GALAXIES MCG-4-53-20 = 534-G-2	19 D 5.4	F533/J4469 (37,200)
AM 2233-594 [2233-5946]	22 33 45 −59 48.6	3.917 0.311	328.94 −50.23	231.1 2.5	I/A DBL. (#1: E)	2 D 0.4	F147/J1579 (273,185)
AM 2233-594 [2233-5948]	22 33 57 −59 46.3	3.914 0.311	328.96 −50.27	231.2 2.5	I/A DBL. (#2: SPIRAL)	2 D 0.5	F147/J1579 (272,187)
AM 2234-245	22 34 19 −24 57.2	3.300 0.311	29.87 −59.78	260.6 22.5	CLOSE PAIR MCG-4-53-24 = 534-G-5	23 S 0.8	F533/J4469 (29,181)
AM 2234-553	22 34 25 −55 34.1	3.784 0.311	334.27 −52.88	234.6 5.0	PAIR of I/A SPIRALS 190-G-17 = B 2234-555	2 S 0.5	F190/J1765 (168,149)
AM 2234-265	22 34 41 −26 54.2	3.320 0.311	26.06 −60.21	258.8 21.4	CLOSE PAIR	23 S 0.7	F468/J0659 (207,346)
AM 2234-371 [2234-3712]	22 34 50 −37 12.1	3.441 0.311	5.26 −60.18	249.7 15.7	I/A DBL.(#1)	2 D 7.4	F345/J0844 (145,327)

Identification	R.A. (1950.0) Declination	Precession R.A./Dec.	Gal. Long. Gal. Lat.	S.G.L. S.G.B.	Description Cross-Identification	Coding Size (Min.)	Field/Plate (x, y)
AM 2234-371	22 34 57	3.441	5.23	249.7	I/A DBL.(#2)	2	F345/J0844
[2234-3712]	−37 12.8	0.311	−60.20	15.7		D 7.4	(144,327)
AM 2235-363	22 35 47	3.430	6.33	250.3	NUCLEUS w. LATERAL COMPACTS	8,13	F405/J0766
	−36 38.0	0.312	−60.46	15.9		D 0.3	(49,87)
AM 2236-362	22 36 23	3.425	6.71	250.6	GROUP of 4 in RICH FIELD	18	F406/J0756
	−36 25.7	0.312	−60.61	15.9		D 1.3	(301,99)
AM 2236-665	22 36 27	4.194	320.77	225.6	I/A DBL. (2 E's in ENVELOPE)	2,24	F109/J0954
	−66 59.2	0.312	−45.57	−2.1	109-G-10	S 1.1	(216,69)
AM 2236-260	22 36 28	3.306	27.79	259.8	SPIRAL w. I/A COMP'N.	1	F534/J0651
	−26 06.2	0.312	−60.47	21.5	MCG-4-53-28 = 534-G-12	S 0.7	(267,118)
AM 2236-331	22 36 35	3.385	13.14	253.3	COMPACT	13	F405/J0766
	−33 16.3	0.312	−60.96	17.6		D 0.3	(35,268)
AM 2236-421	22 36 35	3.506	355.24	245.6	I/A DBL.(CHAIN of 3)	2	F345/J0844
	−42 18.8	0.312	−59.28	12.5	345-G-36 = B 2236-423	D 1.5	(131,53)
AM 2236-364	22 36 44	3.426	6.21	250.4	COMPACT I/A DBL. w. MATERIAL	2,13,15	F405/J0766
	−36 40.0	0.312	−60.64	15.7		D 0.9	(39,85)
AM 2236-664	22 36 55	4.175	320.98	225.9	SPIRAL w. APP. COMP'NS.	8	F109/J0954
	−66 44.0	0.312	−45.78	−2.0	NGC 7329 = 109-G-12	D 4.9	(214,83)
AM 2237-375	22 37 34	3.440	3.57	249.4	M51-TYPE (?)	9	F345/J0844
	−37 57.0	0.313	−60.58	14.9		D 0.9	(117,287)
AM 2237-481	22 37 41	3.598	344.79	240.8	I/A TRIPLE	3	F238/J0961
	−48 12.1	0.313	−57.16	9.0	238-IG-20	D 0.5	(111,275)
AM 2237-310	22 37 57	3.355	17.70	255.5	DWARF	20	F468/J0659
	−31 03.9	0.313	−61.28	18.6	468-G-20	D 1.1	(168,122)
AM 2237-610	22 37 57	3.918	326.99	230.5	I/A DBL. (#1)	2	F147/J1579
[2238-6101]	−61 00.9	0.313	−49.84	1.3		D 0.3	(242,121)
AM 2238-272	22 38 08	3.316	25.18	258.7	I/A DBL.	2	F534/J0651
	−27 27.0	0.313	−61.05	20.4		S 1.1	(246,45)
AM 2238-621	22 38 12	3.962	325.50	229.5	COMPACT	13	F109/J0954
	−62 18.0	0.313	−48.99	0.5	Fa 195	D 0.1	(213,321)
AM 2238-653	22 38 12	4.100	322.01	226.9	I/A DBL. (SPIRALS)	2	F109/J0954
	−65 34.3	0.313	−46.71	−1.4		S 1.0	(209,146)
AM 2238-610	22 38 14	3.915	326.95	230.5	I/A DBL. (#2)	2	F147/J1579
[2237-6100]	−61 01.2	0.313	−49.86	1.3	IC 5238 = 147-G-4	D 0.3	(240,121)
*AM 2238-575	22 38 26	3.814	330.72	233.1	SPIRAL + APP. COMP'N.	8	F147/J1579
	−57 52.1	0.313	−51.93	3.1	147-G-5	S 1.3	(245,290)
AM 2238-351	22 38 32	3.401	9.04	251.9	ONE-ARMED SPIRAL	10	F406/J0756
	−35 13.5	0.313	−61.20	16.2		D 0.5	(280,164)
AM 2238-650	22 38 33	4.071	322.54	227.3	CLOSE PAIR	23	F109/J0954
	−65 01.3	0.313	−47.12	−1.1		S 2.0	(208,175)
AM 2238-242	22 38 43	3.284	31.29	261.6	I/A DBL w. TAILGV180,4;ES,3	2	F534/J0651
	−24 27.1	0.313	−60.64	21.9	534-IG-16	S 1.0	(240,206)
AM 2238-323	22 38 44	3.369	14.53	254.2	COMPACT	13	F405/J0766
	−32 35.1	0.313	−61.43	17.6		D 0.2	(9,304)
*AM 2238-575	22 38 50	3.813	330.55	233.0	I/A TRIPLE	3	F190/J1765
	−57 57.5	0.313	−51.91	3.0		D 0.2	(138,20)
AM 2238-541	22 38 52	3.716	335.45	236.0	PAIR (incl. RING)	6,23	F190/J1765
	−54 15.4	0.313	−54.16	5.2		S 0.9	(133,219)
AM 2238 462	22 38 53	3.559	347.71	242.5	COMPACT + FAINT OUTER DISK + ARM	13,15	F290/J2406
	−46 21.6	0.313	−58.16	9.9	290-G-1	D 0.4	(298,102)
AM 2238-450	22 38 57	3.536	350.01	243.6	SB(r) w. APP. COMP'NS.	8	F289/J1593
	−45 01.9	0.313	−58.72	10.6	IC 5240 = 290-G-2 = R 84	D 4.5	(37,174)
AM 2238-425	22 38 59	3.502	353.95	245.4	CLUSTER (I/A PAIR DOMINANT)	2,19	F345/J0844
	−42 51.4	0.313	−59.52	11.9	290-IG-3	D 9.0	(108,23)
AM 2239-320	22 39 07	3.362	15.58	254.7	DISTURBED (HI.S.BR.DWARF?)	16	F468/J0659
	−32 04.7	0.313	−61.53	17.8	468-G-22	D 0.5	(155,68)
AM 2239-303	22 39 10	3.346	18.61	256.0	VERY CLOSE PAIR	23	F468/J0659
	−30 38.1	0.313	−61.53	18.6		S 0.3	(154,145)
AM 2239-371	22 39 20	3.424	4.79	250.2	COMPACT	13	F345/J0844
	−37 16.6	0.314	−61.05	14.9		D 0.2	(97,322)
*AM 2239-301	22 39 32	3.342	19.27	256.3	SPIRAL w. CLOSE PAIRS off MI.AX.	8,23,23	F468/J0659
	−30 19.3	0.314	−61.60	18.7	NGC 7361 = IC 5237 = 468-G-23	D 13.4	(150,162)
AM 2239-575	22 39 34	3.805	330.52	233.1	I/A DBL.	2	F190/J1765
	−57 54.4	0.314	−52.02	3.0		S 0.1	(132,23)
AM 2239-372	22 39 43	3.424	4.42	250.1	E w. EXTENDED MATERIAL	15	F345/J0844
	−37 26.7	0.314	−61.09	14.8	406-G-4	D 1.1	(93,313)
AM 2239-760	22 39 43	4.916	312.54	218.5	I/A DBL.	2	F049/J3647
	−76 01.9	0.314	−38.87	−7.7	49-IG-2	S 0.3	(264,119)
AM 2239-285	22 39 44	3.325	22.38	257.7	DISTURBED SINGLE GALAXY	16	F468/J0659
	−28 50.4	0.314	−61.55	19.4		D 0.6	(147,242)
AM 2239-431	22 39 44	3.503	353.27	245.2	GALAXY w. FAINT OUTER MATERIAL	16	F289/J1593
	−43 10.5	0.314	−59.54	11.6	290-G-4 = MC 2239-431	D 0.7	(25,273)

Identification	R.A. (1950.0) Declination	Precession R.A./Dec.	Gal. Long. Gal. Lat.	S.G.L. S.G.B.	Description Cross-Identification	Coding Size (Min.)	Field/Plate (x, y)
AM 2239-331	22 39 48 −33 15.1	3.373 0.314	13.11 −61.63	253.7 17.1	COMPACT	13 D 0.2	F406/J0756 (267,271)
*AM 2240-400	22 40 24 −40 08.0	3.456 0.314	358.99 −60.61	247.9 13.2	SPIRAL + COMP'NS.(I/A ?) Ag-75 = 345-G-46 = UKS 2240-401	8 D 3.4	F345/J0844 (90,169)
AM 2240-370	22 40 41 −37 07.8	3.416 0.314	4.99 −61.34	250.5 14.8	SPIRAL w. I/A COMP'N. 406-G-6	1 D 0.9	F345/J0844 (83,330)
AM 2240-325	22 40 46 −32 57.7	3.367 0.314	13.70 −61.85	254.1 17.1	GALAXY w. I/A COMP'NS.	1 S 2.0	F406/J0756 (257,286)
AM 2240-892 [2247-8924]	22 40 49 −89 23.2	0.754 0.314	303.37 −27.91	207.3 −15.3	I/A DBL. + SP.w. PEC.ARMS (#1) NGC 2573a = 1-IG-8	2,10 S 1.2	F001/J3648 (190,210)
*AM 2240-304	22 40 51 −30 44.6	3.342 0.314	18.40 −61.89	256.1 18.2	ISOLATED RING GALAXY	6 D 0.4	F468/J0659 (135,139)
AM 2240-364	22 40 51 −36 41.7	3.410 0.314	5.88 −61.45	250.9 15.0	I/A DBL. w. PLUME 406-IG-7	2 S 0.2	F406/J0756 (253,86)
*AM 2240-285	22 40 59 −28 51.3	3.322 0.314	22.40 −61.82	257.8 19.2	GALAXY w. FILAMENTS from END 468-G-24	15,22 D 0.6	F468/J0659 (132,241)
AM 2241-330	22 41 23 −33 00.4	3.365 0.315	13.59 −61.97	254.1 16.9	GALAXY w. 3 I/A COMP'NS. 406-G-8	1 S 2.0	F468/J0659 (130,18)
AM 2242-575	22 42 00 −57 55.4	3.784 0.315	330.16 −52.26	233.3 2.7	SPIRAL w. I/A COMP'N.	1 S 0.3	F190/J1765 (115,21)
AM 2242-630	22 42 00 −63 05.5	3.952 0.315	324.17 −48.76	229.1 −0.3	CLUSTER of FAINT GALS. (in ARC) 109-IG-17 = DN 2242-631	19 D 9.0	F109/J0954 (189,279)
AM 2242-342	22 42 02 −34 28.6	3.379 0.315	10.44 −61.99	252.9 16.0	I/A DBL. w. TAILS 406-G-10	2,15 S 0.4	F406/J0756 (242,205)
AM 2242-595	22 42 36 −59 51.8	3.835 0.315	327.73 −51.04	231.8 1.5	I/A DBL. (SMALL)	2 S 0.2	F147/J1579 (213,184)
AM 2242-333	22 42 59 −33 30.9	3.365 0.315	12.45 −62.27	253.9 16.4	CLOSE TRIPLE in GROUP of 5	18,24 D 1.1	F406/J0756 (232,257)
*AM 2243-525	22 43 15 −52 58.7	3.655 0.315	336.60 −55.43	237.4 5.5	ELONG.(DBL?)cD + HI.S.BR. I/A GAL B 2243-529 in DN 2243-530	1,2,13,19 D 11.2	F239/J3659 (314,15)
AM 2243-650	22 43 17 −65 09.4	4.021 0.315	321.87 −47.37	227.5 −1.6	I/A DBL. (incl.THIN GALAXY) IC 5246 = 109-G-19	2 D 5.6	F109/J0954 (181,168)
AM 2243-523	22 43 22 −52 30.1	3.644 0.316	337.27 −55.72	237.8 5.7	SMALL CONTACT PAIR	2 S 0.1	F190/J1765 (94,312)
*AM 2243-341	22 43 41 −34 18.5	3.371 0.316	10.73 −62.35	253.3 15.8	GROUP of 6 PKS 2243-345?	18 D 4.0	F406/J0756 (223,214)
*AM 2244-651	22 44 01 −65 19.7	4.020 0.316	321.62 −47.30	227.4 −1.8	I/A DBL. + PLUMES (CENTRE of 6) IC 5250 = 109-IG-22 = Fa 197 [B]	2,15,18 S 0.4	F109/J0954 (177,159)
AM 2244-390	22 44 04 −39 01.6	3.425 0.316	0.81 −61.58	249.2 13.2	GALAXY w. TAIL (COUNTER TAIL?)	15 D 0.9	F345/J0844 (50,227)
*AM 2244-583	22 44 29 −58 32.7	3.780 0.316	329.05 −52.09	233.0 2.1	LOOSE CHAIN of 12 147-G-7	17 D 17.9	F147/J1579 (201,255)
AM 2244-621	22 44 38 −62 12.1	3.891 0.316	324.81 −49.60	230.0 0.0	PROBABLE DEFECT	22 D 0.9	F109/J0954 (173,327)
AM 2244-520	22 44 42 −52 06.1	3.627 0.316	337.65 −56.11	238.3 5.8	DISTURBED SINGLE SPIRAL 238-IG-27	10 D 0.9	F239/J3659 (305,63)
*AM 2244-614	22 44 44 −61 40.2	3.872 0.316	325.39 −49.98	230.4 0.3	SB(r) + I/A SP. = "WENDY'S SYSTEM" Fa 134/5 = 147-IG-8 = B 2244-616	2 S 1.1	F147/J1579 (198,87)
AM 2244-390	22 44 45 −39 03.9	3.423 0.316	0.66 −61.70	249.3 13.1	DWARF (SOME RESOLUTION)	20 D 0.4	F346/J0829 (315,224)
AM 2244-702	22 44 58 −70 27.6	4.284 0.316	316.73 −43.48	223.3 −4.8	M51-TYPE (DIFFUSE COUNTER ARM) 76-IG-27 = B 2244-704	9 S 0.9	F076/J1792 (119,150)
*AM 2245-665	22 45 09 −66 59.4	4.082 0.316	319.87 −46.14	226.1 −2.8	PEC. SPIRAL w. TAIL and JET 109-IG-23	7,12 D 0.5	F109/J0954 (171,70)
AM 2245-583	22 45 10 −58 35.9	3.775 0.316	328.88 −52.13	233.0 2.0	I/A TRIPLE in CHAIN = DN 2245-585 Fa 136/7/8/9 = 147-IG-9	3 D 1.3	F147/J1579 (196,252)
AM 2245-342	22 45 26 −34 26.3	3.366 0.316	10.36 −62.69	253.4 15.4	I/A DBL. w. BRIDGE	2 S 0.6	F406/J0756 (204,207)
AM 2245-620	22 45 28 −62 03.4	3.878 0.316	324.87 −49.77	230.2 0.0	CLOSE DBL. (E + SPIRAL)	23 S 0.5	F109/J0954 (168,335)
*AM 2245-401	22 45 30 −40 13.2	3.434 0.317	358.25 −61.52	248.3 12.3	SP. + I/A COMP'N.(MULTIPLE NUCLEUS)	1,22 D 0.4	F345/J0844 (38,162)
AM 2245-584	22 45 31 −58 40.1	3.774 0.317	328.75 −52.11	233.0 1.9	SB(r) w. 3 ARMS Fa-138/9?	11,17 D 0.5	F147/J1579 (194,248)
AM 2245-842	22 45 40 −84 23.1	7.403 0.317	306.40 −32.17	211.6 −12.6	SP. w. STELLAR NUCLEUS (CHAIN:5)	17 D 5.4	F011/J2562 (89,199)
AM 2245-620	22 45 48 −62 02.8	3.874 0.317	324.83 −49.81	230.2 0.0	I/A DBL.	2 S 0.6	F147/J1579 (191,67)
AM 2246-330	22 46 07 −33 06.8	3.349 0.317	13.23 −62.96	254.6 16.0	COMPACT	13 D 0.2	F406/J0756 (196,279)
AM 2246-581	22 46 12 −58 12.1	3.755 0.317	329.22 −52.49	233.4 2.1	RING in GROUP of 4 GALAXIES Fa-143/4 = R 86	6,18 D 6.3	F147/J1579 (189,274)

Identification	R.A. (1950.0) Declination	Precession R.A./Dec.	Gal. Long. Gal. Lat.	S.G.L. S.G.B.	Description Cross-Identification	Coding Size (Min.)	Field/Plate (x, y)
AM 2246-295	22 46 17 −29 58.2	3.317 0.317	20.15 −63.04	257.4 17.6	COMPACT	13 D 0.2	F468/J0659 (71,180)
AM 2246-374	22 46 35 −37 44.3	3.399 0.317	3.22 −62.37	250.6 13.5	I/A DBL. in CHAIN of 11 346-G-3 = PKS 2246-377	2,17 S 0.4	F406/J0756 (191,30)
*AM 2246-490	22 46 43 −49 07.0	3.558 0.317	341.93 −58.01	240.9 7.2	SPIRAL + I/A COMP'N.+ TAIL 239-IG-2 = Fa 358 = B 2246-491	1,15 D 2.2	F239/J3659 (296,224)
AM 2247-892 [2240-8923]	22 47 00 −89 24.1	0.718 0.317	303.34 −27.90	207.2 −15.3	I/A DBL. + SP.w. PEC.ARMS (#2) NGC 2573b = 1-IG-9	2,10 S 1.2	F001/J3648 (189,210)
AM 2247-283	22 47 07 −28 39.9	3.301 0.317	23.05 −63.15	258.7 18.1	M51-TYPE	9 S 0.5	F468/J0659 (60,250)
AM 2247-430	22 47 23 −43 03.3	3.462 0.317	352.46 −60.88	246.1 10.5	EDGE-ON w. PECULIAR DUST LANE 290-G-17	14 D 0.6	F290/J2406 (222,281)
AM 2247-350	22 47 34 −35 08.2	3.365 0.317	8.72 −63.04	253.0 14.7	ONE-ARMED SPIRAL + COMP'NS.	8,10 D 0.4	F406/J0756 (180,170)
*AM 2247-515	22 47 34 −51 53.2	3.602 0.317	337.51 −56.60	238.7 5.5	COMPACT w. TAIL	13,15 D 0.5	F239/J3659 (282,76)
*AM 2247-440	22 47 35 −44 05.4	3.475 0.317	350.49 −60.51	245.3 9.9	RING + COMP'N. + TAIL 290-IG-18	6,8,15 S 0.8	F290/J2406 (219,226)
*AM 2247-515	22 47 37 −51 54.9	3.603 0.317	337.46 −56.59	238.7 5.5	COMPACT w. TAIL	13,15 D 0.5	F191/J1824 (328,338)
AM 2248-323	22 48 04 −32 39.0	3.337 0.318	14.19 −63.40	255.2 15.9	I/A DBL.	2 S 1.3	F406/J0756 (174,304)
AM 2248-271	22 48 07 −27 15.4	3.285 0.318	26.21 −63.22	260.1 18.6	COMPACT	13 D 0.3	F468/J0659 (46,325)
AM 2248-332	22 48 10 −33 20.3	3.344 0.318	12.66 −63.37	254.6 15.5	RING GALAXY	6 D 0.3	F406/J0756 (174,267)
AM 2248-324	22 48 49 −32 49.2	3.336 0.318	13.79 −63.54	255.2 15.7	DISK w. DISTURBED MATR'L (RING?)	6,22 D 0.4	F406/J0756 (166,294)
*AM 2249-403	22 49 45 −40 35.6	3.418 0.318	356.96 −62.17	248.5 11.4	CHAIN of 6 incl. 346-G-6	17 D 15.7	F346/J0829 (261,143)
*AM 2249-291	22 49 54 −29 19.4	3.299 0.318	21.67 −63.80	258.4 17.3	SPIRAL w. PECULIAR ARMS MCG-5-54-3 = 469-G-4	10 D 0.9	F469/J3508 (297,215)
AM 2250-605	22 50 01 −60 58.0	3.797 0.319	325.46 −50.93	231.4 0.2	SP. w. PEC. ARMS + DIAM. COMP'NS Fa 145 = 147-G-13	10 D 1.6	F147/J1579 (164,125)
AM 2250-603	22 50 16 −60 32.2	3.782 0.319	325.90 −51.25	231.8 0.4	LOOSE CHAIN of GALAXY 148-G-12/13	17 D 9.0	F147/J1579 (162,148)
AM 2250-294 [2250-2950]	22 50 35 −29 49.4	3.301 0.319	20.55 −63.97	258.1 16.9	PAIR (#1: E)	23 S 1.4	F469/J3508 (289,188)
AM 2250-533	22 50 37 −53 37.6	3.614 0.319	334.51 −55.92	237.5 4.2	SMALL I/A DBL. in GROUP of 5 191-IG-2	2,18 S 0.4	F190/J1765 (39,249)
AM 2250-295 [2250-2949]	22 50 39 −29 50.9	3.301 0.319	20.50 −63.98	258.0 16.9	PAIR (#2: SPIRAL)	23 S 1.4	F469/J3508 (288,187)
AM 2250-633	22 50 48 −63 30.5	3.870 0.319	322.65 −49.14	229.3 −1.3	VERY FAINT GALAXY CLUSTER	19 D 6.7	F109/J0954 (136,256)
*AM 2251-341	22 51 15 −34 19.6	3.343 0.319	10.29 −63.90	254.1 14.5	SB(r) w. LINEAR ARM + COMP'NS 406-G-17	8,10 D 1.1	F406/J0756 (140,213)
AM 2251-372	22 51 26 −37 21.2	3.373 0.319	3.56 −63.39	251.5 12.9	SP. w. I/A COMP'N. near NUCLEUS 406-G-18	1 D 1.1	F346/J0829 (247,318)
AM 2251-453	22 51 27 −45 37.6	3.475 0.319	347.08 −60.44	244.3 8.5	KNOTTY SPIRAL w. APP.COMP'N. NGC 7400 = 290-G-22	8 S 1.7	F290/J2406 (182,144)
AM 2251-360	22 51 33 −36 07.0	3.360 0.319	6.27 −63.67	252.6 13.5	I/A DBL.	2 S 0.3	F406/J0756 (137,117)
AM 2251-330	22 51 35 −33 05.1	3.329 0.319	13.09 −64.10	255.2 15.1	I/A TRIPLE	3 D 0.7	F406/J0756 (135,280)
AM 2251-500	22 51 50 −50 04.1	3.541 0.319	339.57 −58.19	240.6 6.0	COMPACT	13 D 0.3	F239/J3659 (250,175)
AM 2251-705	22 51 56 −70 54.9	4.202 0.319	315.73 −43.48	223.2 −5.5	CLOSE PAIR (#1) 76-G-30/31 = Fa 359	23 D 1.0	F076/J1792 (90,124)
AM 2252-362	22 52 08 −36 28.1	3.361 0.319	5.43 −63.72	252.3 13.2	COMPACT	13 D 0.1	F406/J0756 (131,98)
AM 2252-705	22 52 27 −70 50.4	4.190 0.320	315.74 −43.56	223.3 −5.5	CLOSE PAIR (#2) 76-G-31	23 D 0.8	F076/J1792 (87,128)
AM 2252-340	22 52 33 −34 09.4	3.336 0.320	10.60 −64.19	254.4 14.3	I/A DBL. + BRIDGE in GROUP of 3 IC 5262 = 406-IG-20	2,24 D 1.8	F406/J0756 (125,222)
AM 2252-425 [2254-4340]	22 52 57 −42 55.2	3.431 0.320	351.88 −61.87	246.8 9.7	PAIR (#1: SPLIT-ARMED SPIRAL) NGC 7412 = 290-G-24	8,10,23 D 3.1	F290/J2406 (167,289)
AM 2253-353	22 53 54 −35 36.6	3.345 0.320	7.20 −64.24	253.3 13.4	I/A DBL. w. LOOPS	2,15 S 0.2	F406/J0756 (111,144)
*AM 2254-373	22 54 07 −37 37.0	3.365 0.320	2.70 −63.84	251.5 12.3	SP. w. OUTER REGION DISPLACED NGC 7421 = 346-G-17	12,16 D 2.7	F346/J0829 (219,304)
*AM 2254-480	22 54 20 −48 03.0	3.493 0.320	342.39 −59.65	242.5 6.7	SPIRAL w. HI.S.BR. COMP'N. 239-G-4 = Fa 614	8,13 S 1.5	F239/J3659 (230,283)

Identification	R.A. (1950.0) Declination	Precession R.A./Dec.	Gal. Long. Gal. Lat.	S.G.L. S.G.B.	Description Cross-Identification	Coding Size (Min.)	Field/Plate (x, y)
AM 2254-434 [2252-4255]	22 54 22 −43 40.4	3.433 0.320	350.20 −61.79	246.3 9.1	PAIR (#2: SPIRAL w. COMP'NS) IC 5267 = 290-G-29	8,23 D 3.6	F290/J2406 (154,248)
*AM 2254-364	22 54 24 −36 43.9	3.354 0.320	4.63 −64.11	252.3 12.7	E (CENTRE of DN 2254-367 [PKS]) IC 1459 = IC 5265 = 406-G-30	12,17,20 D 3.6	F406/J0756 (107,84)
AM 2254-652	22 54 39 −65 29.3	3.897 0.321	320.22 −47.90	227.9 −2.8	I/A DBL.(SMALL)	2 S 0.1	F109/J0954 (118,149)
AM 2254-342	22 54 44 −34 21.6	3.330 0.321	9.99 −64.61	254.4 13.8	GALAXY w. JETS	7 D 0.5	F406/J0756 (101,211)
AM 2254-524	22 54 57 −52 46.3	3.565 0.321	335.00 −56.97	238.6 4.1	I/A DBL. (2 E's + FUZZ)	2 S 0.4	F239/J3659 (220,31)
AM 2255-450	22 55 07 −45 04.4	3.447 0.321	347.46 −61.27	245.2 8.2	DISTURBED SPIRAL 290-IG-30	12 D 0.6	F290/J2406 (147,173)
*AM 2255-652	22 55 08 −65 23.9	3.888 0.321	320.25 −48.01	228.0 −2.8	IRR. (CENTRE of GROUP of 8) 109-G-39 in NGC 7417 Group	16,18 D 15.7	F109/J0954 (115,154)
*AM 2255-323	22 55 42 −32 30.5	3.309 0.321	14.27 −65.01	256.2 14.6	SP. w. (3?)PEC. ARMS + APP.COMP'N MCG-5-54-6 = 406-G-35	8,10,11 D 0.9	F406/J0756 (88,310)
*AM 2255-475	22 55 43 −47 56.9	3.483 0.321	342.31 −59.90	242.8 6.6	PEC. SP. (FILAMENT to STAR ?) 239-G-5	12 D 0.9	F290/J2406 (143,19)
AM 2255-322	22 55 44 −32 24.9	3.308 0.321	14.48 −65.02	256.3 14.6	3/4-ARMED SPIRAL + APP. COMP'N.	8,11 D 0.9	F469/J3508 (228,50)
*AM 2256-304	22 56 14 −30 45.8	3.291 0.321	18.39 −65.20	257.8 15.4	M51-TYPE + 2 APP. COMP'NS. MCG-5-54-7 = 469-G-7	8,9 S 0.8	F469/J3508 (223,138)
*AM 2256-254	22 56 19 −25 47.8	3.250 0.321	30.13 −64.81	262.4 17.7	M51-TYPE w. CIRCULAR ARMS MCG-4-54-4 = 535-G-1	9,10 D 0.7	F535/J0747 (296,135)
AM 2256-614	22 56 25 −61 49.6	3.756 0.321	323.64 −50.82	231.1 −1.0	SPIRAL w. PEC. ARM + COMP'N.	8,10 S 0.9	F109/J0954 (98,345)
AM 2256-584	22 56 28 −58 42.7	3.675 0.321	327.08 −53.11	233.7 0.7	I/A TRIPLE 147-IG-16	3 D 1.1	F147/J1579 (118,245)
*AM 2256-480	22 56 33 −48 09.7	3.481 0.321	341.80 −59.90	242.6 6.4	M51-TYPE 239-IG-6	9 S 0.4	F239/J3659 (210,278)
AM 2256-675	22 56 43 −67 56.4	3.972 0.321	317.77 −46.10	225.9 −4.3	LINEAR MARK (DEFECT?)	22 D 0.9	F109/J0954 (113,17)
AM 2256-524	22 56 55 −52 46.5	3.551 0.321	334.63 −57.19	238.8 3.9	CHAIN of 5 191-G?-4	17 D 2.5	F191/J1824 (250,295)
AM 2256-675	22 56 55 −67 57.4	3.970 0.321	317.73 −46.10	225.9 −4.3	COMPACT near LINEAR OBJECT	13 D 0.2	F109/J0954 (112,16)
AM 2256-221	22 56 57 −22 17.6	3.222 0.321	38.12 −64.13	265.7 19.2	SMALL I/A QUARTET	4 D 0.5	F535/J0747 (292,324)
AM 2257-563	22 57 02 −56 34.5	3.622 0.321	329.55 −54.67	235.6 1.8	RICH ELONGATED CLUSTER	19 D 20.2	F191/J1824 (243,92)
AM 2257-401	22 57 26 −40 11.0	3.377 0.322	356.72 −63.69	249.6 10.4	DWARF 346-G-23	20 D 0.7	F346/J0829 (183,167)
AM 2257-651	22 57 27 −65 11.8	3.852 0.322	320.16 −48.31	228.3 −2.9	I/A DBL.	2 S 0.6	F109/J0954 (101,164)
AM 2257-620	22 57 31 −62 07.4	3.753 0.322	323.18 −50.68	230.9 −1.2	CHAIN of 5	17 D 8.7	F109/J0954 (92,329)
AM 2257-372	22 57 35 −37 28.4	3.348 0.322	2.61 −64.54	252.0 11.8	LO.S.BR. DWARF 406-G-40	20 D 0.7	F406/J0756 (74,43)
AM 2257-661	22 57 47 −66 17.5	3.889 0.322	319.11 −47.47	227.4 −3.5	PECULIAR SPIRAL	12 D 0.9	F109/J0954 (102,105)
AM 2257-635	22 57 50 −63 55.3	3.804 0.322	321.33 −49.33	229.4 −2.2	DISTURBED GALAXY B 2257-639	16 S 1.0	F109/J0954 (95,232)
AM 2258-274	22 58 01 −27 45.3	3.261 0.322	25.64 −65.46	260.8 16.5	CHAIN of 4	17 D 7.2	F535/J0747 (274,30)
AM 2258-353	22 58 02 −35 38.5	3.329 0.322	6.75 −65.05	253.7 12.6	PECULIAR SPIRAL w. APP. COMP'NS. IC 5269c = 406-G-41	8,12 D 2.2	F406/J0756 (66,141)
AM 2258-614	22 58 09 −61 48.0	3.737 0.322	323.43 −50.97	231.2 −1.1	CHAIN of 5 (E at CENTRE = Fa 146) DN 2258-618 = PKS 2258-619?	17 D 15.7	F147/J1579 (112,79)
AM 2258-441 [2258-4412]	22 58 17 −44 13.8	3.419 0.322	348.47 −62.17	246.2 8.2	SMALL I/A DBL. (E + SP.?)(#1)	2 S 0.2	F290/J2406 (116,218)
AM 2258-441 [2258-4413]	22 58 21 −44 12.8	3.418 0.322	348.49 −62.18	246.2 8.2	SMALL I/A DBL. (E + SP.?)(#2)	2 S 0.2	F290/J2406 (116,219)
*AM 2258-595	22 58 27 −59 54.3	3.685 0.322	325.42 −52.41	232.8 −0.1	I/A DBL. (VERY DISTURBED) 147-IG-19 = B 2258-598	2 S 0.4	F147/J1579 (106,181)
AM 2258-291	22 58 29 −29 17.8	3.272 0.322	21.94 −65.66	259.4 15.6	CHAIN of 4	17 D 1.6	F469/J3508 (197,217)
AM 2258-291	22 58 40 −29 18.4	3.271 0.322	21.92 −65.71	259.4 15.6	COMPACT	13 D 0.2	F469/J3508 (194,217)
AM 2258-412	22 58 52 −41 23.5	3.383 0.322	353.97 −63.49	248.7 9.6	COMPACT	13 D 0.3	F346/J0829 (168,102)
AM 2259-221	22 59 32 −22 17.6	3.216 0.322	38.53 −64.70	266.0 18.7	RICH CLUSTER of FAINT GALAXIES (DBL. E in CENTRE)	19 D 5.4	F535/J0747 (260,324)

Identification	R.A. (1950.0) Declination	Precession R.A./Dec.	Gal. Long. Gal. Lat.	S.G.L. S.G.B.	Description Cross-Identification	Coding Size (Min.)	Field/Plate (x, y)
AM 2259-252	22 59 32	3.239	31.18	263.0	COMPACT	13	F535/J0747
	−25 28.8	0.322	−65.47	17.2		D 0.2	(258,153)
AM 2259-692	22 59 44	4.001	316.15	224.8	GALAXY w. LO.S.BR. MATERIAL	15	F077/J0910
	−69 28.8	0.323	−45.02	−5.3	IC 5279? = 77-IG-1	D 0.9	(292,198)
AM 2259-243	22 59 51	3.232	33.31	263.9	SMALL M51-TYPE	9	F535/J0747
	−24 35.5	0.323	−65.37	17.6		D 0.4	(254,201)
AM 2300-383	23 00 04	3.349	359.82	251.3	SPIRAL w. I/A COMP'N.	1	F346/J0829
	−38 34.7	0.323	−64.70	10.8		S 0.6	(156,253)
*AM 2300-465	23 00 07	3.443	343.15	244.0	CHAIN of 4 (I/A DBL. at CENTER)	2,17	F239/J3659
	−46 58.5	0.323	−61.05	6.5	R 93 = DN 2300-447	D 11.4	(179,342)
AM 2300-511	23 00 12	3.502	336.32	240.4	I/A DBL.	2	F239/J3659
	−51 12.0	0.323	−58.58	4.3	239-IG-8	S 0.4	(177,115)
AM 2300-261	23 00 20	3.242	29.58	262.5	PAIR	23	F535/J0747
	−26 11.0	0.323	−65.77	16.7		S 1.0	(247,115)
AM 2300-672	23 00 32	3.900	317.74	226.5	CLOSE TRIPLE of COMPACTS	13,24	F109/J0954
	−67 29.0	0.323	−46.69	−4.3	DN 2300-670?	D 2.0	(92,40)
AM 2300-461	23 00 36	3.431	344.21	244.6	DWARF	20	F290/J2406
	−46 18.9	0.323	−61.48	6.8	290-G-39	D 1.3	(97,106)
*AM 2301-263	23 01 03	3.243	28.59	262.2	PAIR (THIN ARMS CIRCLING NUCLEUS)	10,23	F535/J0747
	−26 36.8	0.323	−65.99	16.4		S 2.7	(239,92)
AM 2301-290	23 01 06	3.261	22.69	260.0	DWARF	20	F469/J3508
	−29 00.8	0.323	−66.23	15.3	469-G-8	D 13.4	(166,232)
AM 2301-472	23 01 12	3.442	342.19	243.7	GALAXY + DISTURBED COMP'N.	8	F239/J3659
	−47 24.7	0.323	−60.96	6.1	290-G-42	S 4.1	(169,318)
AM 2301-634	23 01 12	3.759	321.10	229.8	SP. w. APP. COMP'NS. either SIDE	8	F109/J0954
	−63 42.1	0.323	−49.74	−2.4	109-G-34	D 1.1	(75,243)
AM 2301-441	23 01 27	3.402	347.77	246.4	I/A DBL.	2	F290/J2406
	−44 17.2	0.323	−62.64	7.7	290-IG-43	S 0.4	(86,214)
AM 2301-843	23 01 35	6.673	305.85	211.6	CLOSE PAIR (#1: E)	23	F011/J2562
[2303-8438]	−84 39.3	0.323	−32.13	−13.1		S 2.6	(75,179)
AM 2301-500	23 01 43	3.474	337.84	241.5	GROUP of 5	18	F239/J3659
	−50 01.0	0.323	−59.50	4.7	DN 2302-500	D 2.5	(164,179)
*AM 2302-502	23 02 22	3.474	337.15	241.2	PEC. SP. w. ABS'N. + 4 APP.COMP'NS	8,14	F239/J3659
	−50 22.8	0.323	−59.36	4.4	239-G-9	S 7.2	(159,159)
AM 2302-300	23 02 28	3.265	20.06	259.1	DWARF (DEFECT?)	20	F469/J3508
	−30 04.5	0.324	−66.55	14.5		D 0.4	(150,175)
AM 2302-224	23 02 43	3.212	37.92	265.9	PAIR of I/A SPIRALS	2	F535/J0747
	−22 46.8	0.324	−65.55	17.8		S 0.4	(220,298)
AM 2302-322	23 02 43	3.283	14.19	257.0	RING in CLOSE PAIR	6,23	F406/J0756
	−32 24.8	0.324	−66.49	13.3		D 0.4	(8,313)
AM 2302-515	23 02 51	3.494	334.71	239.9	M51-TYPE:	9	F239/J3659
	−51 57.1	0.324	−58.41	3.5	239-IG-10	S 0.5	(155,75)
*AM 2303-305	23 03 05	3.269	18.03	258.5	GAL. w. LOOPS + 6 SMALL COMP'NS.	8,15	F469/J3508
	−30 52.9	0.324	−66.66	14.0	MCG-5-54-13 = 469-G-11	D 1.6	(144,132)
AM 2303-474	23 03 17	3.433	341.17	243.6	CLOSE TRIPLE	24	F239/J3659
	−47 46.1	0.324	−61.05	5.6	239-IG-11	D 0.8	(150,299)
AM 2303-843	23 03 23	6.554	305.81	211.6	CLOSE PAIR (#2: SPIRAL)	23	F011/J2562
[2301-8439]	−84 38.4	0.324	−32.16	−13.1		S 2.6	(72,179)
AM 2303-321	23 03 49	3.277	14.63	257.3	I/A DBL (STAR to [W.])	2	F407/J0737
	−32 12.9	0.324	−66.74	13.2		S 0.4	(268,327)
*AM 2303-443	23 03 57	3.391	346.85	246.4	GALAXY w. JETS	7	F291/J2533
	−44 31.2	0.324	−62.91	7.2	290-IG-50 in R 98? = DN 2303-447	D 1.0	(333,201)
AM 2304-431	23 04 08	3.375	349.42	247.6	GROUP of 3 w. APP. COMP'NS.	2,8,24	F291/J2533
	−43 10.4	0.324	−63.60	7.8	MCG-7-47-17/8 = 290-IG-51 = B2304-431	S 1.6	(334,274)
AM 2304-610	23 04 15	3.654	323.33	232.3	ELONGATED RICH, DISTANT CLUSTER	19	F148/J1760
	−61 01.7	0.324	−52.03	−1.3	(3 ALIGNED CONDENSATIONS)	D 17.9	(320,118)
AM 2304-431	23 04 24	3.374	349.38	247.7	2 E's in COMMON ENVELOPE (+7)	2	F346/J0829
	−43 10.1	0.324	−63.64	7.8	DN 2304-431	S 0.6	(114,6)
AM 2304-410	23 04 26	3.353	353.48	249.4	ARM ENDS in ARC (as in NGC3310)	15	F346/J0829
	−41 09.9	0.324	−64.54	8.8	346-G-31	D 1.1	(112,113)
AM 2304-273	23 04 47	3.240	26.35	261.6	SPIRAL w. FAINT OUTER ARMS + COMP'N	8,10	F535/J0747
	−27 36.7	0.324	−66.93	15.2		S 1.2	(194,39)
*AM 2304-490	23 04 53	3.439	338.78	242.6	MULTI-ARMED SP. w. I/A COMP'NS.	1,10	F239/J3659
	−49 01.2	0.324	−60.52	4.7	239-IG-12	S 0.9	(136,232)
AM 2305-282	23 05 32	3.243	24.23	260.9	DWARF (POSSIBLE DEFECT)	20	F469/J3508
	−28 27.3	0.325	−67.16	14.7		D 0.4	(113,262)
AM 2305-684	23 05 54	3.875	316.12	225.7	EXTENDED CLOUD of GALAXIES	19	F077/J0910
	−68 42.9	0.325	−45.96	−5.4			(266,242)
AM 2306-400	23 06 00	3.335	355.54	250.5	DWARF	20	F346/J0829
	−40 04.7	0.325	−65.25	9.0	346-G-33	D 1.1	(95,171)
*AM 2306-465	23 06 04	3.406	342.01	244.5	M51-TYPE	9	F239/J3659
	−46 56.7	0.325	−61.91	5.6		D 0.4	(124,343)

Identification	R.A. (1950.0) Declination	Precession R.A./Dec.	Gal. Long. Gal. Lat.	S.G.L. S.G.B.	Description Cross-Identification	Coding Size (Min.)	Field/Plate (x, y)
AM 2306-220	23 06 05 −22 02.2	3.199 0.325	40.22 −66.07	267.0 17.4	PAIR (incl. DISTURBED + MATERIAL)	16,23 S 1.7	F535/J0747 (178,338)
*AM 2306-721	23 06 20 −72 17.0	4.045 0.325	313.37 −42.96	222.6 −7.2	SPIRAL w. I/A HI.S.BR. COMP'N. 77-IG-4	1,2,13 S 1.6	F049/J3647 (178,324)
AM 2306-632	23 06 21 −63 23.4	3.692 0.325	320.69 −50.33	230.4 −2.8	GAL. w. TAILS 110-IG-1	15 D 0.5	F109/J0954 (43,257)
AM 2306-621	23 06 22 −62 16.5	3.662 0.325	321.76 −51.22	231.3 −2.2	GAL. w. APP. COMP'NS. + PEC. ARM 147-G-22	8 D 1.0	F110/J1811 (314,320)
AM 2306-222	23 06 45 −22 26.4	3.200 0.325	39.38 −66.34	266.7 17.1	SMALL CLUSTER of FAINT GALAXIES	19 D 1.8	F535/J0747 (170,317)
*AM 2306-485	23 06 58 −48 51.0	3.424 0.325	338.63 −60.89	242.9 4.5	I/A SEXTET 239-IG-13	5,22 D 2.0	F239/J3659 (118,241)
AM 2307-664	23 07 18 −66 45.2	3.782 0.325	317.57 −47.66	227.5 −4.5	I/A DBL. (CLOSE) 110-IG-3	2 S 0.4	F109/J0954 (53,77)
AM 2307-254	23 07 24 −25 44.9	3.220 0.325	31.30 −67.26	263.6 15.5	GROUP of 3 (ONE DISTURBED)	16,24 D 1.7	F535/J0747 (162,139)
AM 2307-861	23 07 30 −86 14.0	0.128 0.325	304.88 −30.78	210.2 −13.9	PAIR of CONTACT E GALAXIES	2 S 0.1	F002/J2563 (225,108)
AM 2307-331	23 07 58 −33 17.2	3.270 0.325	11.63 −67.48	256.8 12.0	SPIRAL w. I/A COMP'N.	1 S 0.4	F407/J0737 (221,270)
AM 2308-383	23 08 07 −38 34.7	3.312 0.325	358.56 −66.18	252.1 9.4	PECULIAR GALAXY (KNOTTY) Ag-77 = 346-G-35	12,16 D 0.7	F346/J0829 (71,251)
AM 2308-634	23 08 24 −63 45.5	3.678 0.326	320.06 −50.17	230.2 −3.1	CHAIN of 4 (3 I/A) 110-IG-4 = Fa 204/5/6	2,17 D 1.4	F109/J0954 (32,236)
AM 2308-682	23 08 32 −68 22.0	3.823 0.326	316.11 −46.39	226.2 −5.4	GALAXY w. FAINT LOOP IC 5288 = 77-G-6	12,15 D 2.0	F077/J0910 (254,262)
*AM 2308-324	23 08 34 −32 43.5	3.264 0.326	13.05 −67.68	257.3 12.1	RING GALAXY (NUCLEUS in RIM) IC 5289 = MCG-5-54-20 = 407-IG-8	6 D 0.7	F407/J0737 (214,300)
AM 2308-271	23 08 39 −27 11.2	3.225 0.326	27.68 −67.74	262.4 14.6	I/A DBL.	2 S 0.4	F535/J0747 (148,62)
AM 2309-311	23 09 19 −31 11.0	3.250 0.326	17.07 −67.99	258.8 12.7	I/A DBL.	2 S 0.1	F470/J0777 (333,112)
AM 2309-284	23 09 26 −28 48.7	3.234 0.326	23.41 −68.04	261.0 13.7	E + 5 COMP'NS. NGC 7507 = 469-G-19	8 S 4.5	F469/J3508 (68,242)
AM 2309-235	23 09 33 −23 52.8	3.202 0.326	36.30 −67.35	265.6 15.9	PAIR (incl. RING w. ABS'N.?)	6,14,23 S 3.0	F535/J0747 (136,239)
AM 2309-440	23 09 34 −44 03.8	3.355 0.326	346.56 −64.01	247.3 6.5	DWARF NGC 7496a = Se 159/4 = 291-G-7	20 D 1.6	F290/J2406 (7,223)
AM 2309-614	23 09 40 −61 44.3	3.614 0.326	321.79 −51.88	232.0 −2.3	GROUP of 4 Fa 207/8	18 D 11.2	F147/J1579 (39,78)
*AM 2310-372	23 10 00 −37 29.2	3.295 0.326	0.84 −66.90	253.2 9.6	GAL. + 2 APP.COMP'NS + I/A COMP'N. 407-G-9	1,8,13 S 2.2	F347/J0839 (326,310)
AM 2310-222	23 10 10 −22 26.6	3.192 0.326	39.98 −67.10	267.0 16.4	LARGE CLUSTER of FAINT GALAXIES	19 D 4.5	F535/J0747 (127,316)
AM 2310-275	23 10 12 −27 52.4	3.225 0.326	25.95 −68.15	261.9 14.0	PAIR incl. 469-G-20	23 S 1.9	F535/J0747 (129,25)
AM 2310-283	23 10 19 −28 39.2	3.230 0.326	23.86 −68.22	261.2 13.6	BACKGROUND CLUSTER of GALAXIES 469-G-21	19 D 3.6	F469/J3508 (57,250)
AM 2310-283	23 10 32 −28 37.8	3.229 0.326	23.93 −68.27	261.3 13.6	SPIRAL w. DWARF COMP'N. NGC 7513 = 469-G-22	8 S 4.0	F469/J3508 (55,251)
AM 2310-404	23 10 39 −40 44.3	3.319 0.326	353.21 −65.78	250.4 8.0	I/A TRIPLE	3 D 0.7	F347/J0839 (313,135)
*AM 2310-685	23 10 42 −68 53.7	3.812 0.326	315.46 −46.05	225.8 −5.9	PEC. GAL. + SP. + I/A DBL. to SIDE	12,2 D 3.4	F077/J0910 (242,234)
AM 2310-615	23 10 44 −61 55.5	3.607 0.326	321.45 −51.81	231.9 −2.5	I/A DBL.	2 S 1.0	F147/J1579 (33,67)
AM 2310-283	23 10 54 −28 39.0	3.228 0.326	23.88 −68.35	261.3 13.5	DWARF COMP'N.	20 S 4.0	F469/J3508 (50,250)
AM 2310-503 [2311-5037]	23 10 56 −50 37.0	3.419 0.326	335.04 −60.24	241.7 3.1	I/A SPIRALS (#1) 239-IG-16	2 S 0.6	F240/J0962 (341,141)
AM 2311-503 [2310-5037]	23 11 01 −50 37.0	3.418 0.326	335.02 −60.25	241.7 3.1	I/A SPIRALS (#2) 239-IG-16	2 S 0.6	F240/J0962 (341,141)
AM 2311-424	23 11 02 −42 45.0	3.335 0.326	348.87 −64.91	248.6 6.9	COMPACT in a CLUSTER Se 159/3 = DN 2311-429	13 D 0.1	F347/J0839 (305,27)
*AM 2311-425	23 11 06 −42 52.7	3.336 0.326	348.59 −64.85	248.5 6.9	RING in CHAIN w. other RINGS	6,17 D 0.4	F347/J0839 (304,20)
AM 2311-472	23 11 32 −47 28.2	3.379 0.327	339.93 −62.34	244.5 4.6	M51-TYPE (STAR?)	9 D 0.4	F239/J3659 (75,314)
AM 2311-654 [2312-6538]	23 11 51 −65 40.7	3.689 0.327	317.90 −48.81	228.7 −4.4	CLOSE PAIR (#1) 110-G-5	23 S 2.2	F110/J1811 (269,140)
AM 2311-561	23 11 59 −56 14.8	3.489 0.327	327.31 −56.33	236.9 0.2	CLOSE PAIR	23 S 0.4	F191/J1824 (132,110)

Identification	R.A. (1950.0) Declination	Precession R.A./Dec.	Gal. Long. Gal. Lat.	S.G.L. S.G.B.	Description Cross-Identification	Coding Size (Min.)	Field/Plate (x, y)
AM 2312-435	23 12 01 −43 52.4	3.340 0.327	346.40 −64.48	247.7 6.2	PEC. ARM SP. + COMPACT + COMP'NS. [A82]	8,10,15 D 5.4	F291/J2533 (256,239)
AM 2312-653 [2311-6540]	23 12 02 −65 38.6	3.686 0.327	317.91 −48.85	228.7 −4.4	CLOSE PAIR (#2)	23 S 2.2	F110/J1811 (268,142)
AM 2312-380	23 12 10 −38 07.8	3.290 0.327	358.92 −67.08	252.8 8.9	HI.S.BR. GALAXY w. DUST LANE 347-G-3	14 D 0.6	F346/J0829 (28,274)
AM 2312-535	23 12 14 −53 55.7	3.453 0.327	330.14 −58.07	238.9 1.3	GALAXY w. TAIL	15 D 0.5	F191/J1824 (127,234)
AM 2312-494	23 12 22 −49 40.2	3.398 0.327	336.17 −61.04	242.7 3.4	LO.S.BR. SP. + LO.S.BR. COMP'NS. 239-G-17 + COMPACT	8,13,20 D 1.3	F240/J0962 (332,192)
AM 2312-384	23 12 48 −38 48.6	3.293 0.327	357.18 −66.95	252.3 8.5	PECULIAR SPIRAL NGC 7545 = Ag-78 = 347-G-4	12 D 1.0	F346/J0829 (22,237)
*AM 2312-591	23 12 51 −59 19.6	3.533 0.327	323.71 −54.03	234.3 −1.4	SINGLE GALAXY w. TWO TAILS 148-IG-2 = B 2312-593	15 D 0.8	F148/J1760 (269,213)
AM 2312-380	23 12 52 −38 00.1	3.286 0.327	359.10 −67.26	253.0 8.9	LARGE RICH CLUSTER	19 D 22.4	F347/J0839 (294,283)
AM 2312-511 [2312-5118]	23 12 56 −51 19.0	3.413 0.327	333.58 −60.00	241.3 2.5	PAIR (#2: E) 239-IG-18	23 S 1.0	F240/J0962 (322,104)
AM 2312-511 [2312-5119]	23 12 57 −51 18.0	3.413 0.327	333.60 −60.01	241.3 2.5	PAIR (#1: SPIRAL) 239-IG-18	23 S 0.5	F240/J0962 (322,105)
*AM 2313-261	23 13 06 −26 10.7	3.206 0.327	30.72 −68.60	263.8 14.2	TWO HI.S.BR. KNOTS in ENVELOPE 535-IG-12	6 D 0.4	F535/J0747 (94,115)
AM 2313-504	23 13 40 −50 41.7	3.401 0.327	334.33 −60.51	241.9 2.7	DISTURBED SINGLE 239-G-19	16 D 0.6	F240/J0962 (318,138)
*AM 2313-222	23 13 47 −22 25.6	3.183 0.327	40.68 −67.89	267.4 15.6	SPIRAL w.DIAMETRIC COMP'NS. NGC 7573 = 604-G-8	8 D 1.9	F535/J0747 (82,317)
AM 2314-225	23 14 03 −22 57.1	3.186 0.327	39.39 −68.10	266.9 15.4	I/A TRIPLE incl. ONE-ARMED SPIRAL	3,10 D 1.6	F535/J0747 (80,288)
AM 2315-460	23 15 04 −46 04.8	3.343 0.328	341.54 −63.67	246.0 4.7	SMALL I/A DBL.	2 S 0.1	F291/J2533 (225,121)
AM 2315-561	23 15 22 −56 15.4	3.460 0.328	326.65 −56.63	237.1 −0.2	CHAIN of 5 (CENTRED on E)	17 D 6.3	F191/J1824 (107,108)
AM 2315-690	23 15 22 −69 05.3	3.749 0.328	314.80 −46.10	225.8 −6.3	PECULIAR GALAXY (DWARF ?) 77-IG-10 = B 2315-690	20 D 0.6	F077/J0910 (219,225)
AM 2315-335	23 15 33 −33 59.2	3.247 0.328	9.03 −68.91	256.9 10.2	CHAIN of 4 GALAXIES	17 D 5.2	F407/J0737 (136,232)
AM 2315-495	23 15 33 −49 52.8	3.379 0.328	335.13 −61.29	242.7 2.8	SMALL RING	6 D 0.3	F240/J0962 (304,183)
*AM 2315-423	23 15 39 −42 38.9	3.310 0.328	348.05 −65.69	249.1 6.2	GROUP of 4 (2 I/A) in DN 2315-431 NGC 7582 = 291-G-16 = R 109 [PKS + A82]	2,18 D 40.3	F347/J0839 (259,34)
*AM 2316-272	23 16 04 −27 24.3	3.205 0.328	27.55 −69.40	263.0 13.1	DISTURBED SPIRAL w. 2 COMP'NS. MCG-5-55-1 = 535-IG-14 (JET?)	7,8 S 2.2	F535/J0747 (59,49)
*AM 2316-665	23 16 06 −66 56.9	3.670 0.328	316.33 −47.97	227.7 −5.4	DISTURBED SP. + 3 APP. COMP'NS. 110-G-8 [A82]	1,7,8,12 D 0.5	F110/J1811 (242,73)
*AM 2316-225 [2316-2255]	23 16 23 −22 59.4	3.180 0.328	39.71 −68.63	267.1 14.9	PAIR (#1: SPIRAL) MCG-4-55-2 = 535-G-15	23 S 4.0	F535/J0747 (51,286)
AM 2316-535 [2317-5356]	23 16 28 −53 57.5	3.420 0.328	329.23 −58.47	239.2 0.7	WIDE PAIR (#1: SPIRAL)	23 D 0.3	F191/J1824 (94,231)
AM 2316-520	23 16 30 −52 02.1	3.396 0.328	331.78 −59.88	240.9 1.7	RING w. I/A COMP'N	1,6 D 0.5	F240/J0962 (290,68)
AM 2316-393	23 16 34 −39 33.0	3.281 0.328	354.68 −67.31	252.0 7.5	DISTURBED SINGLE GALAXY MCG 2316-395	16 D 0.9	F347/J0839 (254,201)
*AM 2316-225 [2316-2259]	23 16 37 −22 55.7	3.179 0.328	39.91 −68.66	267.2 14.8	PAIR (#2: IRR. w. I/A DIAMETRICS) MCG-4-55-4 = 535-G-17	1,23 S 4.0	F535/J0747 (48,289)
AM 2316-421	23 16 37 −42 15.2	3.301 0.328	348.65 −66.05	249.6 6.3	CLOSE GROUP of 9 GALAXIES R 110 = PKS 2316-423?	18 D 1.1	F347/J0839 (250,56)
AM 2316-315	23 16 51 −31 56.8	3.229 0.328	14.57 −69.52	258.9 10.9	I/A DBL. (ISOLATED in FIELD)	2 D 0.6	F470/J0777 (246,73)
AM 2316-294	23 16 54 −29 48.5	3.216 0.328	20.69 −69.67	260.8 11.8	GROUP of 7	18 D 3.8	F470/J0777 (247,187)
AM 2316-395	23 16 58 −39 53.8	3.281 0.328	353.79 −67.23	251.7 7.3	SMALL I/A DBL.	2 S 0.2	F347/J0839 (249,182)
AM 2317-555	23 17 02 −55 59.7	3.442 0.328	326.63 −56.98	237.5 −0.3	SMALL GALAXY w. JET	7 D 0.7	F191/J1824 (94,122)
AM 2317-224	23 17 03 −22 44.2	3.177 0.328	40.50 −68.70	267.4 14.8	CLUSTER of FAINT GALAXIES	19 D 2.7	F535/J0747 (42,299)
*AM 2317-672	23 17 06 −67 25.3	3.671 0.328	315.84 −47.61	227.4 −5.7	RING GALAXY w. CONDENSATION in RIM + COMP'N.	6,8 D 0.6	F077/J0910 (214,314)
AM 2317-535 [2316-5357]	23 17 07 −53 56.7	3.414 0.328	329.11 −58.55	239.3 0.7	WIDE PAIR (#2: E) PKS 2316-538?	23 D 0.2	F191/J1824 (88,232)
AM 2317-564 [2317-5645]	23 17 09 −56 43.4	3.451 0.328	325.77 −56.43	236.8 −0.7	I/A DBL. (#1)	2 D 0.4	F191/J1824 (94,83)

Identification	R.A. (1950.0) Declination	Precession R.A./Dec.	Gal. Long. Gal. Lat.	S.G.L. S.G.B.	Description Cross-Identification	Coding Size (Min.)	Field/Plate (x, y)
*AM 2317-672	23 17 10 −67 25.5	3.670 0.328	315.83 −47.61	227.4 −5.7	RING w. DBL. NUCLEUS + 2 COMP'NS.	6,8 S 0.5	F110/J1811 (235,48)
AM 2317-564 [2317-5643]	23 17 18 −56 45.3	3.451 0.328	325.71 −56.41	236.8 −0.7	I/A DBL. (#2) 191-G-8	2 D 0.6	F191/J1824 (93,81)
AM 2317-554	23 17 25 −55 43.2	3.435 0.328	326.88 −57.23	237.7 −0.2	CLOSE PAIR (incl. COMPACT)	13,23 S 0.4	F191/J1824 (90,137)
AM 2317-420	23 17 50 −42 4.4	3.294 0.328	348.74 −66.33	249.8 6.1	I/A DBL. w. BRIDGE in GROUP of 6 347-IG-7 = Se 159/2 = DN 2317-423	2,18 D 1.3	F347/J0839 (238,66)
AM 2317-692	23 17 58 −69 29.3	3.724 0.328	314.22 −45.87	225.6 −6.7	I/A DBL. w. TAILS 77-IG-14 = B 2317-694	2,15 S 0.3	F077/J0910 (206,204)
AM 2318-461	23 18 01 −46 11.7	3.327 0.328	340.63 −64.00	246.2 4.2	SMALL I/A DBL.	2 S 0.2	F291/J2533 (198,115)
AM 2318-365	23 18 04 −36 56.1	3.256 0.328	0.81 −68.59	254.5 8.4	DISTURBED GAL.w. FNT.OUTER PLUMES	15,16 D 0.4	F407/J0737 (111,73)
AM 2318-420	23 18 04 −42 00.5	3.292 0.328	348.82 −66.40	249.9 6.1	LO.S.BR. DWARF (RESOLUTION ?) Ka-57 = Se 159/1 = 347-G-8	8,20 D 3.1	F347/J0839 (236,69)
AM 2318-225	23 18 15 −22 54.9	3.175 0.328	40.25 −69.02	267.4 14.5	I/A DBL.	2 S 0.5	F535/J0747 (28,289)
AM 2318-461	23 18 15 −46 13.2	3.326 0.328	340.53 −64.02	246.2 4.1	HI.S.BR. INTEGRAL + CLOSE COMP'N	2,13 S 0.2	F291/J2533 (196,114)
AM 2318-695	23 18 33 −69 56.6	3.731 0.329	313.84 −45.50	225.2 −7.0	CLUSTER of GALAXIES	19 D 6.7	F077/J0910 (203,179)
AM 2318-622	23 18 44 −62 23.6	3.530 0.329	319.77 −51.96	231.9 −3.5	EDGE-ON w. APP. COMP'NS. NGC 7622 = 148-G-8	8 S 2.7	F110/J1811 (237,318)
AM 2318-481 [2319-4811]	23 18 53 −48 17.3	3.340 0.329	336.85 −62.76	244.4 3.1	TRIPLE (#1: S0)	24 D 0.8	F240/J0962 (279,269)
AM 2319-274	23 19 00 −27 49.2	3.198 0.329	26.51 −70.08	262.9 12.3	PAIR of I/A DBL.	2 S 0.3	F535/J0747 (25,26)
*AM 2319-481 [2318-4817]	23 19 05 −48 12.7	3.338 0.329	336.92 −62.84	244.5 3.1	TRIPLE (#2: PLUMED GALAXY) 240-G-1 = B 2319-482	7,15 D 0.9	F240/J0962 (277,274)
AM 2319-534	23 19 06 −53 45.8	3.397 0.329	328.92 −58.87	239.6 0.5	LINE of 3	24 D 3.6	F191/J1824 (72,241)
*AM 2319-481 [2319-4811]	23 19 11 −48 11.6	3.338 0.329	336.93 −62.86	244.5 3.1	TRIPLE (#3: EDGE-ON)	24 D 0.5	F240/J0962 (276,275)
*AM 2319-424	23 19 15 −42 45.3	3.291 0.329	346.96 −66.19	249.3 5.6	S0 w. DBL. I/A COMP'NS. [A82] NGC7632 = Ag-79 = IC 5313 = 291-G-21	2,8 S 7.6	F347/J0839 (224,29)
AM 2319-325	23 19 18 −32 54.2	3.226 0.329	11.63 −69.89	258.2 10.0	COMPACT	13 D 0.2	F470/J0777 (218,21)
AM 2319-234	23 19 23 −23 46.8	3.177 0.329	38.10 −69.50	266.7 13.9	E + I/A COMP'N. in CHAIN of 4 MCG-4-55-6 = 536-G-2	1,17 S 0.5	F535/J0747 (15,243)
AM 2319-333	23 19 23 −33 38.8	3.230 0.329	9.50 −69.76	257.6 9.7	SPIRAL w. COMP'N. 407-G-16	8 S 0.6	F407/J0737 (93,249)
AM 2319-425	23 19 26 −42 52.2	3.291 0.329	346.68 −66.16	249.3 5.5	I/A DBL 291-G-22/23 = B 2319-428	2 S 0.6	F347/J0839 (222,23)
AM 2319-595	23 19 34 −59 58.5	3.479 0.329	321.90 −54.02	234.1 −2.5	SMALL M51-TYPE	9 S 0.3	F148/J1760 (222,180)
AM 2319-640	23 19 34 −64 08.5	3.557 0.329	318.13 −50.54	230.4 −4.4	CLOSE PAIR of SPIRALS	23 S 0.5	F110/J1811 (229,225)
AM 2320-524	23 20 11 −52 45.7	3.377 0.329	329.98 −59.73	240.5 0.8	EDGE-ON SPIRAL + I/A COMP'N. 191-G-9	1 S 1.0	F239/J3659 (15,26)
AM 2320-314	23 20 22 −31 41.6	3.215 0.329	15.05 −70.28	259.4 10.4	I/A DBL.	2 S 0.1	F470/J0777 (206,87)
AM 2320-733	23 20 53 −73 33.9	3.842 0.329	311.26 −42.38	222.0 −8.8	CLOSE TRIPLE	24 D 0.7	F028/J1935 (327,244)
AM 2321-670	23 21 04 −67 08.0	3.609 0.329	315.57 −48.06	227.8 −5.9	PEC. GAL. + CONDENS. on FILAMENT	7 D 0.4	F077/J0910 (194,330)
AM 2321-504	23 21 12 −50 47.9	3.349 0.329	332.45 −61.28	242.3 1.6	COMPACT 240-G-2	13 D 0.3	F240/J0962 (254,136)
AM 2321-683	23 21 19 −68 38.4	3.647 0.329	314.45 −46.76	226.5 −6.6	GALAXY w. I/A COMP'N. 77-G-16	1 D 0.6	F077/J0910 (191,249)
AM 2321-344	23 21 40 −34 43.4	3.227 0.329	6.14 −69.96	256.8 8.8	I/A DBL. w. JET 407-IG-17 = PKS 2321-347	2,7 S 0.3	F407/J0737 (69,191)
AM 2322-533 [2322-5335]	23 22 25 −53 33.7	3.368 0.330	328.45 −59.35	240.0 0.2	PAIR(#1: COMPACT)	13,23 D 0.2	F192/J1839 (312,252)
AM 2322-533 [2322-5333]	23 22 28 −53 35.1	3.368 0.330	328.41 −59.33	240.0 0.1	PAIR(#2)	23 D 0.2	F192/J1839 (311,251)
AM 2322-580	23 22 30 −58 03.8	3.422 0.330	323.29 −55.79	236.0 −1.9	GROUP of 3 DISTURBED SP. Se 160/1 NGC 7650/2 = 148-IG-10/11 = Fa 149[B]	24 S 9.9	F148/J1760 (204,283)
*AM 2322-821	23 22 42 −82 11.3	4.650 0.330	306.48 −34.61	214.1 −12.5	ONE-ARMED SPIRAL + I/A COMP'N. NGC 7637 = 12-G-1 [A82]	1,8,10,12,13 S 2.2	F027/J2485 (84,51)
*AM 2322-671	23 22 44 −67 19.9	3.591 0.330	315.21 −47.96	227.7 −6.1	RING GALAXY w. I/A COMP'N. 110-IG-10	1,6 S 0.2	F077/J0910 (185,320)

Identification	R.A. (1950.0) Declination	Precession R.A./Dec.	Gal. Long. Gal. Lat.	S.G.L. S.G.B.	Description Cross-Identification	Coding Size (Min.)	Field/Plate (x, y)
AM 2323-702	23 23 05 −70 28.7	3.677 0.330	313.00 −45.21	224.9 −7.6	I/A DBL.	2 S 0.2	F050/J0955 (341,139)
AM 2323-382	23 23 22 −38 23.5	3.241 0.330	355.97 −69.00	253.6 6.9	I/A DBL. (HI.S.BR.) 347-IG-14	2 S 0.1	F347/J0839 (183,264)
AM 2323-544	23 23 28 −54 47.1	3.373 0.330	326.72 −58.49	239.0 −0.5	CLOSE PAIR of SPIRALS 192-G-1	23 S 0.9	F191/J1824 (41,184)
AM 2323-533	23 23 32 −53 34.5	3.360 0.330	328.18 −59.44	240.0 0.0	RICH CLUSTER	19 D 6.7	F191/J1824 (36,249)
AM 2323-392	23 23 42 −39 29.8	3.247 0.330	353.16 −68.57	252.7 6.3	I/A DBL. w. CONNECTING LOOP NGC 7658b = 347-G-16	2 S 1.1	F347/J0839 (180,204)
AM 2323-512	23 23 45 −51 24.7	3.337 0.330	330.98 −61.11	242.0 1.0	SMALL RING w. OFF-SET NUCLEUS	6 D 0.3	F240/J0962 (231,103)
*AM 2323-323	23 23 46 −32 39.8	3.208 0.330	11.84 −70.85	258.9 9.3	DWARF (RESOLVED) UKS 2323-326 MCG-5-55-12 = 407-G-18	20 D 1.1	F408/J0767 (314,301)
AM 2324-574 [2324-5748]	23 24 23 −57 48.5	3.401 0.330	323.19 −56.15	236.3 −2.0	I/A DBL. (#1)	2 S 0.5	F148/J1760 (191,297)
AM 2324-653	23 24 26 −65 32.9	3.527 0.330	316.32 −49.60	229.4 −5.5	SPIRAL + 2 APP. COMP'NS. 110-G-11	8 D 1.6	F110/J1811 (199,150)
AM 2324-574 [2324-5748]	23 24 28 −57 48.2	3.401 0.330	323.18 −56.16	236.3 −2.0	I/A DBL. (#2)	2 S 0.5	F148/J1760 (190,297)
*AM 2324-602	23 24 39 −60 26.3	3.435 0.330	320.57 −53.98	234.0 −3.2	I/A DBL. w. TAILS or JET B 2324-604	2,7,15 S 1.2	F148/J1760 (188,156)
AM 2324-690	23 24 51 −69 09.4	3.609 0.330	313.68 −46.45	226.1 −7.1	COMPACT	13 D 0.2	F050/J0955 (344,210)
AM 2324-335	23 24 59 −33 59.2	3.210 0.330	7.74 −70.81	257.8 8.5	DBL. E in COMMON ENVELOPE 408-IG-2	2 S 0.2	F408/J0767 (299,230)
AM 2325-672	23 25 04 −67 26.1	3.561 0.330	314.85 −47.98	227.7 −6.4	V.LO.S.BR. SPIRAL (STRUCTURED)	12,20 D 0.5	F110/J1811 (195,49)
*AM 2325-473	23 25 05 −47 39.4	3.295 0.330	336.31 −63.95	245.4 2.4	SP. w. PECULIAR ARM + I/A COMP'N. 240-G-3 = B 2325-476	1,10 S 1.4	F240/J0962 (224,305)
AM 2325-455	23 25 06 −45 58.3	3.283 0.330	339.20 −65.09	246.9 3.2	DISTURBED SINGLE 291-G-25	16 D 0.7	F291/J2533 (132,127)
*AM 2325-621 [2325-6215]	23 25 09 −62 11.5	3.457 0.330	318.92 −52.53	232.4 −4.1	WIDE PAIR of PECULIARS (#1)	23 S 4.2	F110/J1811 (197,330)
AM 2325-541	23 25 17 −54 15.1	3.353 0.330	326.96 −59.08	239.6 −0.5	SMALL GALAXY w. LOOP + TAIL	15 D 0.4	F192/J1839 (287,217)
AM 2325-691	23 25 18 −69 19.5	3.607 0.330	313.52 −46.32	226.0 −7.2	COMPACT	13 D 0.1	F050/J0955 (340,202)
AM 2325-680	23 25 21 −68 05.8	3.573 0.330	314.35 −47.41	227.1 −6.7	TIGHT GROUP (6 EARLY-TYPE GAL.) NGC 7655 + IC 5320/2/3/4 = 77-G-19	18 D 11.2	F077/J0910 (171,279)
*AM 2325-621 [2325-6211]	23 25 26 −62 15.8	3.455 0.330	318.82 −52.49	232.4 −4.1	WIDE PAIR of PECULIARS (#2) 148-G-15 (FILAMENTS)	15,23 S 4.2	F110/J1811 (195,326)
AM 2325-832	23 25 31 −83 22.4	4.796 0.330	305.83 −33.55	213.0 −13.1	GROUP of 3	24 D 4.3	F002/J2563 (233,263)
AM 2325-691	23 25 33 −69 12.1	3.600 0.330	313.57 −46.44	226.1 −7.2	FAINT GALAXY CLUSTER	19 D 11.2	F077/J0910 (170,219)
AM 2325-413	23 25 59 −41 36.6	3.248 0.330	347.67 −67.86	250.9 5.0	SPIRAL + APP. COMP'NS. IC 5325 = 347-G-18	8 S 3.6	F347/J0839 (157,91)
AM 2326-535	23 26 08 −53 54.0	3.343 0.330	327.19 −59.43	239.9 −0.5	SMALL CLUSTER of V.V.FAINT GALS.	19 D 4.0	F192/J1839 (281,236)
AM 2326-522 [2326-5230]	23 26 09 −52 29.1	3.329 0.330	328.96 −60.53	241.2 0.1	PAIR in GROUP of FOUR (#1) 192-G-3	18,23 D 0.3	F192/J1839 (285,312)
AM 2326-523 [2326-5229]	23 26 14 −52 30.9	3.329 0.330	328.91 −60.52	241.2 0.1	PAIR in GROUP of FOUR (#2) 192-G-3	18,23 D 0.3	F192/J1839 (284,310)
AM 2326-595	23 26 14 −59 59.6	3.413 0.330	320.70 −54.47	234.5 −3.2	E6 w. FILAMENT and CONDENSATION NGC 7676 = 148-G-16	7,13 S 1.9	F148/J1760 (178,180)
AM 2326-392	23 26 15 −39 29.7	3.235 0.330	352.52 −69.00	252.9 5.9	PAIR of COMPACTS	13 S 0.1	F347/J0839 (154,204)
AM 2326-542	23 26 18 −54 23.5	3.346 0.330	326.56 −59.06	239.5 −0.7	I/A DBL.	2 S 0.2	F192/J1839 (279,209)
AM 2326-661 [2326-6621]	23 26 25 −66 19.6	3.518 0.330	315.47 −49.02	228.8 −6.0	PAIR (#1: SB(r)) Fa 617	23 S 2.7	F110/J1811 (188,108)
AM 2326-393	23 26 36 −39 30.3	3.233 0.330	352.41 −69.05	252.9 5.8	I/A DBL. COMPACTS (CHAIN of 12)	2,13,17 D 0.4	F347/J0839 (150,204)
AM 2326-790 [2329-7907]	23 26 36 −79 01.2	4.073 0.330	307.85 −37.58	217.1 −11.4	CLOSE GROUP of 3 (#1)	24 S 3.4	F027/J2485 (36,215)
AM 2326-662 [2326-6619]	23 26 45 −66 21.8	3.514 0.330	315.39 −49.00	228.8 −6.1	PAIR (#2: DISTURBED E)	16,23 S 2.7	F110/J1811 (186,106)
*AM 2327-422	23 27 00 −42 26.5	3.248 0.331	345.58 −67.55	250.3 4.5	M51-TYPE	9 S 0.4	F291/J2533 (110,316)
AM 2327-312	23 27 26 −31 24.3	3.189 0.331	15.34 −71.81	260.4 9.1	SP. w. I/A COMP'NS. (on ARM?) MCG-5-55-17 = 470-G-14	1 D 1.5	F470/J0777 (125,102)

Identification	R.A. (1950.0) Declination	Precession R.A./Dec.	Gal. Long. Gal. Lat.	S.G.L. S.G.B.	Description Cross-Identification	Coding Size (Min.)	Field/Plate (x, y)
AM 2327-471	23 27 36	3.277	336.32	246.0	SPIRAL w. I/A COMP'NS.	1	F291/J2533
	−47 14.5	0.331	−64.55	2.3	B 2327-472	S 0.4	(110,58)
AM 2327-351	23 27 43	3.205	3.66	256.9	CLUSTER (20:) MEDIUM BR. GALAXIES	19	F408/J0767
	−35 12.5	0.331	−71.01	7.5	incl. 408-G-4	D 17.9	(267,165)
AM 2327-574	23 27 57	3.367	322.61	236.6	COMPACT	13	F191/J1824
	−57 40.9	0.331	−56.52	−2.4		D 0.1	(19,26)
AM 2328-790	23 28 51	4.007	307.72	217.2	CLOSE GROUP of 3 (#2)	24	F027/J2485
[2326-7902]	−79 02.1	0.331	−37.61	−11.5	PKS 2329-793?	S 3.4	(31,212)
AM 2329-380	23 29 05	3.214	355.29	254.4	CLOSE PAIR (#1: E + SP. w. JETS)	7,12,23	F347/J0839
[2329-3806]	−38 06.6	0.331	−70.14	6.0		D 0.6	(123,278)
AM 2329-380	23 29 09	3.214	355.21	254.3	CLOSE PAIR (#2: E + SP. w. JETS)	7,12,23	F347/J0839
[2329-3808]	−38 08.1	0.331	−70.14	6.0		D 1.1	(122,277)
*AM 2329-410	23 29 13	3.229	348.04	251.7	GALAXY w. JETS	7	F347/J0839
	−41 02.7	0.331	−68.68	4.7		D 0.9	(124,121)
AM 2329-414	23 29 34	3.231	346.28	251.1	I/A TRIPLE in LINE	3	F347/J0839
	−41 46.9	0.331	−68.32	4.3	PKS 2329-415?	D 2.7	(121,81)
AM 2329-280	23 29 46	3.167	26.23	263.7	CLOSE PAIR of SPIRALS	23	F536/J0845
	−28 02.8	0.331	−72.47	10.0	MCG-5-55-24 = 470-G-19	S 5.4	(158,16)
AM 2329-790	23 29 49	3.986	307.62	217.1	CLOSE GROUP of 3 (#3)	24	F027/J2485
[2328-7902]	−79 07.8	0.331	−37.54	−11.6	PKS 2329-793?	S 3.4	(29,207)
AM 2329-295	23 29 53	3.173	20.28	262.0	COMPACT	13	F470/J0777
	−29 50.1	0.331	−72.49	9.3		D 0.1	(96,185)
AM 2330-295	23 30 11	3.172	20.22	262.0	COMPACT	13	F470/J0777
	−29 51.1	0.331	−72.55	9.2		D 0.1	(92,185)
AM 2330-363	23 30 11	3.201	359.38	255.9	SPIRAL w. I/A COMP'N.(GROUP of 5)	1,18	F408/J0767
	−36 30.5	0.331	−71.01	6.5		D 1.3	(239,96)
AM 2330-375	23 30 11	3.208	355.45	254.6	CHAIN of 6	17	F408/J0767
	−37 56.5	0.331	−70.40	5.8		D 11.2	(238,19)
*AM 2330-451	23 30 33	3.246	338.89	248.0	E (ABS'N. EDGE) w. CLOSE COMP'N.	8	F291/J2533
	−45 17.6	0.331	−66.25	2.6	IC 5328/28a = 291-G-28/29 [MC + A82]	S 0.8	(80,162)
AM 2330-650	23 30 39	3.441	315.73	230.0	GALAXY w. TAIL (RINGING?)	15	F110/J1811
	−65 09.0	0.331	−50.26	−5.9		D 0.6	(164,171)
*AM 2330-385	23 30 47	3.210	352.74	253.8	CLOSE PAIR (#1)	23	F347/J0839
[2330-3852]	−38 55.1	0.331	−70.05	5.3		D 0.4	(106,235)
*AM 2330-385	23 30 49	3.210	352.84	253.8	CLOSE PAIR (#2: LARGE JET)	7,23	F347/J0839
[2330-3855]	−38 52.5	0.331	−70.07	5.3	347-G-22	D 0.6	(106,237)
AM 2331-624	23 31 04	3.400	317.48	232.2	2 E's in COMMON ENVELOPE	2	F110/J1811
	−62 45.9	0.331	−52.38	−4.9		S 0.2	(160,299)
*AM 2331-705	23 31 30	3.552	311.84	224.8	DISTORTED RING GALAXY	6,16	F050/J0955
	−70 54.7	0.331	−45.12	−8.4	77-G-23 = Fa 618	D 1.1	(301,121)
AM 2331-461	23 31 39	3.245	336.82	247.2	DWARF w. CENTRAL CONDENSATION	20	F291/J2533
	−46 16.4	0.331	−65.72	2.0	291-G-31	D 0.9	(71,109)
AM 2331-343	23 31 57	3.185	4.88	257.9	SPIRAL w. COMP'N. on ROOT of ARM	8,10	F408/J0767
	−34 30.7	0.332	−72.05	7.0		D 1.3	(221,203)
*AM 2332-361	23 32 00	3.193	359.50	256.3	COMPACT w. LOOP (near BR. SPIRAL)	13,15	F408/J0767
	−36 18.6	0.332	−71.42	6.2	408-G?-10	D 0.7	(220,107)
AM 2332-654	23 32 03	3.432	315.15	229.6	PAIR (#1: HI.S.BR. GAL.w. PLUMES)	15,23	F110/J1811
[2332-6552]	−65 40.5	0.332	−49.86	−6.3	IC 5333 = 110-G-12	S 8.1	(157,143)
AM 2332-665	23 32 09	3.454	314.26	228.4	SPIRAL w. DIAMETRIC COMP'NS.	8	F110/J1811
	−66 57.4	0.332	−48.72	−6.8	110-G-13	D 4.9	(158,74)
AM 2332-440	23 32 11	3.229	340.85	249.3	CLOSE PAIR	23	F291/J2533
	−44 00.6	0.332	−67.32	2.9		S 1.1	(62,230)
AM 2332-623	23 32 15	3.383	317.49	232.5	2 E's in COMMON ENVELOPE	2,13	F110/J1811
	−62 30.0	0.332	−52.68	−4.9		S 0.1	(153,313)
AM 2332-404	23 32 17	3.212	347.88	252.3	CLOSE GROUP of 3 (CHAIN of 10)	17,24	F347/J0839
	−40 43.3	0.332	−69.34	4.3		D 2.0	(93,138)
AM 2332-655	23 32 21	3.432	314.98	229.4	PAIR (#2: LO.S.BR. SPIRAL)	15,20,23	F110/J1811
[2532-6540]	−65 52.0	0.332	−49.70	−6.4	NGC 7697 = 110-G-14	S 8.1	(156,133)
AM 2332-542	23 32 31	3.296	325.12	239.9	I/A DBL. in GROUP of 4	2,18	F192/J1839
	−54 24.5	0.332	−59.58	−1.6		S 0.4	(230,210)
AM 2332-291	23 32 38	3.162	22.12	262.8	GALAXY w. I/A COMP'N.	1	F471/J0840
	−29 16.0	0.332	−73.11	9.0	PKS 2332-293	S 0.2	(333,214)
AM 2332-431	23 32 41	3.222	342.38	250.1	DISTURBED SINGLE	16	F291/J2533
	−43 10.2	0.332	−67.93	3.2	291-G-34	D 0.6	(55,275)
*AM 2332-674	23 32 57	3.456	313.68	227.8	DISTURBED GAL w. OUTER MATERIAL	14,16	F110/J1811
	−67 40.5	0.332	−48.11	−7.2	IC 5335 = 77-G-25 (PEC.DUST LANE)	D 0.6	(154,35)
AM 2333-742	23 33 11	3.630	309.76	221.6	CLOSE GROUP of 4 (incl. I/A)	2,18	F028/J1935
	−74 22.1	0.332	−42.01	−9.9	49-IG-12 = DN 2333-744	D 0.9	(276,207)
AM 2333-324	23 33 28	3.172	10.11	259.6	CHAIN of 4 + COMPACTS	13,17	F408/J0767
	−32 47.0	0.332	−72.82	7.4		D 9.0	(204,296)
AM 2333-315	23 33 35	3.169	13.11	260.5	E + 6 CLOSE COMP'NS.	8	F408/J0767
	−31 52.7	0.332	−73.03	7.7	470-G-20 = PKS 2333-318	D 1.2	(203,345)

Identification	R.A. (1950.0) Declination	Precession R.A./Dec.	Gal. Long. Gal. Lat.	S.G.L. S.G.B.	Description Cross-Identification	Coding Size (Min.)	Field/Plate (x, y)
AM 2333-844	23 33 35 −84 49.9	4.771 0.332	304.97 −32.26	211.7 −13.8	COMPACT	13 D 0.2	F011/J2562 (43,154)
AM 2333-324	23 33 36 −32 47.5	3.172 0.332	10.06 −72.85	259.6 7.4	CHAIN of 5 CENTERED on E PKS 233-327	17 D 9.9	F471/J0840 (317,25)
AM 2333-381	23 33 36 −38 12.9	3.194 0.332	353.77 −70.87	254.6 5.1	GALAXY + APP. COMP'NS. NGC 7713 = 347-G-28	8 S 2.7	F347/J0839 (75,272)
AM 2333-315	23 33 38 −31 52.7	3.168 0.332	13.11 −73.04	260.5 7.7	SMALL NEST of 6 E GALAXIES 470-G-20	18 D 4.5	F471/J0840 (318,74)
*AM 2333-491	23 33 52 −49 12.2	3.249 0.332	331.49 −63.83	244.7 0.5	COMPACT w. JET (I/A DBL.?) 240-IG-9	2,7,13 S 0.3	F240/J0962 (145,222)
*AM 2333-630	23 33 54 −63 05.6	3.372 0.332	316.76 −52.25	232.1 −5.4	GALAXY w. JET w. OFF-SET NUCLEUS 110-G-16 + PLUME	7,15 D 0.8	F110/J1811 (143,281)
AM 2334-271	23 34 00 −27 15.9	3.151 0.332	29.08 −73.36	264.8 9.5	PECULIAR SPIRAL w. APP. COMP'NS. MCG-5-55-27 = 536-G-10	8,12 D 0.9	F471/J0840 (320,322)
AM 2334-782 [2335-7821]	23 34 05 −78 22.7	3.806 0.332	307.73 −38.31	217.9 −11.5	TRIPLE (#1: SPIRAL?) 12-G-3	24 S 4.3	F012/J2486 (250,262)
*AM 2334-292	23 34 29 −29 22.5	3.157 0.332	21.69 −73.51	262.9 8.6	RING w. OFF-CENTRE NUCLEUS	6 D 0.4	F470/J0777 (42,209)
AM 2334-375	23 34 29 −37 59.5	3.189 0.332	354.11 −71.13	254.9 5.1	BRIGHT-RIMMED SPIRAL NGC 7713a = 347-G-30	12 D 2.0	F348/J3586 (344,283)
AM 2334-614	23 34 34 −61 42.7	3.348 0.332	317.73 −53.50	233.3 −4.9	CLOSE QUARTET	18 D 0.9	F111/J0921 (338,78)
AM 2334-435	23 34 49 −43 54.2	3.214 0.332	340.24 −67.75	249.6 2.5	DWARF 292-G-2	20 D 0.7	F291/J2533 (36,235)
*AM 2335-474 [2335-4759]	23 35 02 −47 46.8	3.233 0.332	333.32 −65.03	246.0 0.9	PAIR(#1:S0 w.TAIL + LO.S.BR. JET) 240-G-10 = MCG 2335-477 in R116 [B]	1,24 S 1.7	F240/J0962 (134,299)
AM 2335-475 [2335-4746]	23 35 07 −47 59.8	3.234 0.332	332.95 −64.88	245.9 0.8	PAIR (#2: EDGE-ON SPIRAL) [A82] 240-G-11 in R116 = DN 2335-471	1,24 S 1.7	F240/J0962 (133,287)
AM 2335-383 [2335-3833]	23 35 12 −38 32.9	3.188 0.332	352.41 −70.98	254.5 4.7	I/A TRIPLE (#1) 347-G-32	3 D 0.6	F347/J0839 (59,253)
AM 2335-383 [2335-3832]	23 35 13 −38 31.4	3.188 0.332	352.47 −70.99	254.5 4.7	I/A TRIPLE (#2)	3 D 0.5	F347/J0839 (59,255)
AM 2335-781 [2334-7822]	23 35 14 −78 17.7	3.768 0.332	307.70 −38.41	218.0 −11.5	TRIPLE(#2: COMPACT)	13,24 S 4.3	F012/J2486 (247,267)
AM 2335-782 [2335-7817]	23 35 25 −78 21.9	3.767 0.332	307.65 −38.35	217.9 −11.5	TRIPLE(#3: E)	24 S 4.3	F012/J2486 (246,263)
AM 2335-254	23 35 26 −25 40.0	3.142 0.332	214.78 −73.47	273.8 −22.4	CLOSE TRIPLE 536-G-13	24 D 2.2	F536/J0845 (89,143)
AM 2335-383 [2335-3831]	23 35 26 −38 33.9	3.187 0.332	352.29 −71.01	254.5 4.7	I/A TRIPLE (#3)	3 D 0.5	F347/J0839 (57,252)
AM 2335-513	23 35 42 −51 31.2	3.251 0.332	327.80 −62.20	242.7 −0.8	SMALL PAIR of CONTACT E's	2 S 0.1	F240/J0962 (131,98)
AM 2335-662	23 35 49 −66 23.9	3.395 0.332	314.13 −49.38	229.1 −6.9	PAIR of E's	23 S 0.9	F110/J1811 (137,104)
AM 2335-671	23 35 51 −67 13.8	3.408 0.332	313.59 −48.63	228.3 −7.2	DWARF (NUCLEATED)	20 D 0.2	F050/J0955 (302,320)
AM 2335-643 [2336-6438]	23 35 57 −64 39.5	3.368 0.332	315.29 −50.95	230.7 −6.2	I/A DBL. SPIRALS (#1)	2 S 1.8	F110/J1811 (134,197)
AM 2336-564 [2336-5641]	23 36 02 −56 45.5	3.286 0.332	321.83 −57.89	237.9 −3.0	CLOSE PAIR (#1) 192-IG-11 = Fa 619 = B 2336-567	8,23 D 0.4	F192/J1839 (201,85)
AM 2336-643 [2335-6439]	23 36 14 −64 38.1	3.365 0.332	315.27 −50.99	230.7 −6.2	I/A DBL. SPIRALS (#2) 110-G-18	2 S 1.8	F110/J1811 (132,198)
AM 2336-580	23 36 20 −58 05.4	3.294 0.332	320.51 −56.77	236.7 −3.6	I/A TRIPLE	3 D 1.2	F148/J1760 (106,281)
AM 2336-255	23 36 24 −25 56.9	3.140 0.332	33.93 −73.73	266.2 9.5	SP. w. 4 APP.COMP'NS.(I/A TRIPLE) MCG-4-55-16 = 536-G-14	3,8 S 2.7	F537/J3384 (343,126)
AM 2336-564 [2336-5645]	23 36 24 −56 41.8	3.282 0.332	321.81 −57.97	238.0 −3.0	CLOSE PAIR (#2) 192-G-12 = Fa 620	8,23 D 0.7	F192/J1839 (198,88)
AM 2336-315	23 36 32 −31 51.1	3.158 0.332	12.80 −73.65	260.7 7.2	LOOSE CHAIN of 4 (SP. + COMP'N)	8,17 D 2.0	F471/J0840 (285,77)
AM 2336-521	23 36 48 −52 19.6	3.248 0.332	326.50 −61.65	242.0 −1.3	GALAXY + COMPACT + 4 APP. COMP'NS 240-G-14	8,13,23 S 4.3	F192/J1839 (198,323)
AM 2337-420	23 37 12 −42 09.1	3.193 0.332	343.09 −69.22	251.3 2.9	GROUP of 4	18 D 2.7	F347/J0839 (46,59)
AM 2337-384	23 37 14 −38 42.7	3.179 0.332	351.35 −71.23	254.5 4.3	E w. FAINT JETS 348-G-6	7 D 0.9	F347/J0839 (38,244)
*AM 2337-230	23 37 15 −23 02.1	3.129 0.332	44.04 −73.22	269.0 10.4	I/A DBL. (COMP'N. I/A ?)	1,2 S 2.1	F537/J3384 (337,283)
AM 2337-222	23 37 20 −22 22.0	3.127 0.332	46.23 −73.01	269.7 10.7	DWARF 605-G-24	20 D 0.6	F537/J3384 (337,319)
AM 2337-624	23 37 23 −62 43.2	3.328 0.332	316.46 −52.76	232.6 −5.6	I/A DBL. + PECULIAR SPIRAL COMP'N.	2,8 S 1.0	F110/J1811 (121,301)

Identification	R.A. (1950.0) Declination	Precession R.A./Dec.	Gal. Long. Gal. Lat.	S.G.L. S.G.B.	Description Cross-Identification	Coding Size (Min.)	Field/Plate (x, y)
AM 2337-624 [2337-6243]	23 37 27 −62 41.6	3.327 0.332	316.47 −52.79	232.6 −5.6	WIDE I/A PAIR (#1)	2 S 3.1	F148/J1760 (108,34)
AM 2337-405	23 37 41 −40 58.7	3.186 0.332	345.57 −70.02	252.4 3.3	I/A DBL.	2 S 1.5	F347/J0839 (38,122)
AM 2337-444	23 37 44 −44 47.4	3.202 0.332	337.60 −67.52	249.0 1.7	SB + 7 APP. + I/A COMP'N. Se 162/3 = 292-G-9 = B 2337-447	1,8 S 4.5	F292/J3733 (277,188)
AM 2337-612	23 37 54 −61 21.3	3.308 0.332	317.42 −54.02	233.9 −5.1	E + 3 APP. COMP'NS.	8 S 1.3	F111/J0921 (318,100)
AM 2337-624 [2337-6241]	23 37 56 −62 43.3	3.322 0.332	316.37 −52.79	232.6 −5.6	WIDE I/A PAIR (#2)	2 S 3.1	F148/J1760 (105,32)
AM 2338-421	23 38 19 −42 11.5	3.187 0.333	342.63 −69.35	251.4 2.7	3 HI.S.BR. CONDENSATIONS w. TAIL	15,22 D 0.4	F347/J0839 (35,56)
AM 2338-450	23 38 23 −45 05.9	3.199 0.333	336.82 −67.38	248.7 1.5	SPIRAL w. JET 292-G-11 = B 2338-450	7 D 1.3	F292/J3733 (270,172)
AM 2338-334	23 38 32 −33 44.3	3.156 0.333	5.94 −73.59	259.2 6.0	V.SMALL GROUP of 6 E's 408-IG-15	18 D 0.6	F408/J0767 (148,245)
AM 2338-570	23 38 37 −57 01.3	3.265 0.333	321.01 −57.84	237.9 −3.4	LUMPY LINEAR GALAXY 192-IG-14	16 D 0.4	F111/J0921 (332,331)
AM 2338-582	23 38 43 −58 25.1	3.275 0.333	319.71 −56.64	236.6 −4.0	DWARF (4 BR. KNOTS in a LINE) 148-IG-21 = Se 160/2 = PKS 2338-58?	20 S 0.5	F148/J1760 (90,263)
AM 2338-312	23 38 48 −31 20.0	3.148 0.333	14.32 −74.22	261.4 6.9	CLOSE PAIR incl. BULL'S EYE 471-G-3	6,23 S 1.3	F471/J0840 (259,105)
AM 2339-395	23 39 23 −39 55.1	3.173 0.333	347.54 −70.91	253.5 3.4	CLOSE PAIR + COMPACT in LINE	13,23,24 S 1.6	F348/J3586 (288,181)
AM 2339-424	23 39 24 −42 44.8	3.184 0.333	341.07 −69.14	251.0 2.3	I/A DBL. 292-IG-12	2 S 0.3	F292/J3733 (266,298)
AM 2339-674	23 39 33 −67 40.9	3.363 0.333	312.82 −48.36	228.1 −7.7	ONE-ARMED SPIRAL 77-G-27	10 D 0.7	F110/J1811 (121,34)
AM 2339-570	23 39 38 −57 03.9	3.256 0.333	320.74 −57.87	237.9 −3.6	GALAXY w. APP. COMP'N. + PECULIAR 192-G-15 = B 2339-570 ARM	8,10 D 1.1	F111/J0921 (324,330)
AM 2339-675	23 39 38 −67 57.3	3.366 0.333	312.65 −48.12	227.8 −7.9	CLOSE PAIR (E + SPIRAL)	23 S 4.9	F110/J1811 (121,19)
AM 2339-583	23 39 39 −58 38.2	3.267 0.333	319.32 −56.50	236.4 −4.2	DOUBLE M51-TYPES ? (SMALL)	9 S 0.4	F111/J0921 (318,245)
*AM 2339-661 [2339-6613]	23 39 46 −66 14.0	3.340 0.333	313.68 −49.70	229.4 −7.2	I/A SPIRALS (#1: SB(r)) NGC 7733 = 110-G-22 = B 2339-662	2 D 1.3	F078/J0928 (286,107)
AM 2339-404	23 39 47 −40 43.8	3.174 0.333	345.43 −70.48	252.8 3.0	COMPACT (OVAL)	13,16 D 0.3	F348/J3586 (283,138)
AM 2339-271	23 39 48 −27 13.0	3.133 0.333	29.64 −74.64	265.3 8.3	I/A DBL.	2 S 1.1	F471/J0840 (250,326)
*AM 2339-661 [2339-6614]	23 39 55 −66 13.4	3.338 0.333	313.67 −49.71	229.4 −7.2	I/A SP.(#2: ONE-ARMED KNOTTY SB) NGC 7734 = 110-G-23 = B 2339-662	2,10,12 D 1.2	F078/J0928 (285,107)
AM 2340-630	23 40 03 −63 09.8	3.302 0.333	315.69 −52.50	232.3 −6.0	DWARF w. OFF-SET LUMP 110-G-24	20 D 0.6	F110/J1811 (106,276)
AM 2340-425	23 40 16 −42 52.8	3.180 0.333	340.50 −69.17	250.9 2.1	SMALL I/A DBL.	2 S 0.2	F292/J3733 (257,292)
AM 2340-263	23 40 24 −26 36.3	3.130 0.333	32.00 −74.71	265.9 8.4	DISTURBED SINGLE 537-G-1	16 D 0.7	F537/J3384 (294,92)
AM 2340-431	23 40 29 −43 11.0	3.180 0.333	339.80 −68.99	250.6 1.9	I/A DBL. 292-IG-15	2 S 0.4	F292/J3733 (254,275)
AM 2340-335	23 40 42 −33 58.4	3.149 0.333	4.63 −73.93	259.1 5.5	I/A TRIPLET	3 D 0.7	F408/J0767 (124,232)
AM 2340-310	23 40 44 −31 09.7	3.141 0.333	14.68 −74.66	261.7 6.6	CLOSE PAIR	23 S 0.5	F471/J0840 (237,115)
AM 2340-592	23 40 58 −59 20.3	3.260 0.333	318.46 −55.96	235.8 −4.6	SPIRAL. w. CONDENSATION in DISK	12 D 0.7	F111/J0921 (306,208)
*AM 2341-321	23 41 08 −32 14.1	3.142 0.333	10.66 −74.53	260.8 6.1	EDGE-ON + PEC. SP. + 3 I/A COMP'NS MCG-5-56-1 = 471-G-6 [A82]	3,8 D 4.5	F471/J0840 (232,57)
AM 2341-252	23 41 12 −25 25.6	3.125 0.333	216.51 −74.70	273.1 −21.3	SMALL I/A DBL. w. TAILS	2,9,15 S 0.4	F536/J0845 (19,155)
AM 2341-363	23 41 13 −36 33.5	3.154 0.333	356.15 −72.99	256.8 4.4	I/A DBL. (COMPACTS in AREA)	2,13 D 0.4	F408/J0767 (120,93)
AM 2341-563	23 41 21 −56 34.1	3.237 0.333	320.82 −58.42	238.4 −3.6	COMPACT	13 D 0.2	F192/J1839 (162,95)
AM 2341-275	23 41 41 −27 55.7	3.130 0.333	27.01 −75.10	264.8 7.7	SMALL SPIRAL w. I/A COMP'N.	1 S 1.1	F471/J0840 (228,288)
AM 2341-652 [2341-6518]	23 41 47 −65 22.7	3.304 0.333	313.93 −50.56	230.3 −7.1	PAIR w. COMP'NS (#1: SPIRAL)	8,23 S 4.1	F110/J1811 (102,157)
AM 2341-651 [2341-6522]	23 41 57 −65 18.4	3.301 0.333	313.95 −50.63	230.4 −7.0	PAIR w. COMP'NS (#2: E)	8,23 S 4.1	F110/J1811 (101,161)
AM 2342-240	23 42 03 −24 05.3	3.120 0.333	41.60 −74.58	268.4 9.0	I/A DBL. MCG-4-56-1/2 = 537-IG-3	2 S 0.4	F537/J3384 (277,227)

Identification	R.A. (1950.0) Declination	Precession R.A./Dec.	Gal. Long. Gal. Lat.	S.G.L. S.G.B.	Description Cross-Identification	Coding Size (Min.)	Field/Plate (x, y)
AM 2342-680	23 42 03 −68 03.8	3.332 0.333	312.27 −48.10	227.8 −8.1	SPIRAL + COMP'N.?	8 D 0.4	F077/J0910 (88,278)
AM 2342-373	23 42 09 −37 33.9	3.153 0.333	352.87 −72.66	255.9 3.8	SPIRAL w. THICK JET from NUCLEUS	7,8 D 0.5	F348/J3586 (263,309)
AM 2342-241	23 42 12 −24 14.8	3.120 0.333	41.06 −74.66	268.3 8.9	I/A DBL. in GROUP of 3	2,24 S 0.4	F537/J3384 (275,219)
AM 2342-431	23 42 21 −43 11.3	3.169 0.333	339.13 −69.23	250.8 1.6	PEC. S0 + APP. COMP'NS (DWARF) NGC 7744 = IC 5348 = 292-G-17	8,20 D 1.3	F292/J3733 (236,276)
AM 2342-271	23 42 55 −27 18.2	3.124 0.333	29.51 −75.34	265.5 7.6	I/A DBL.	2 S 0.9	F471/J0840 (213,322)
AM 2342-270	23 42 59 −27 00.4	3.123 0.333	30.68 −75.33	265.8 7.7	I/A DBL.	2 S 0.9	F471/J0840 (212,338)
*AM 2343-633	23 43 02 −63 35.1	3.272 0.333	314.91 −52.25	232.0 −6.5	M51 TYPE 110-IG-25 = B 2342-635	9 S 0.8	F078/J0928 (277,250)
AM 2343-633	23 43 07 −63 30.6	3.270 0.333	314.94 −52.33	232.1 −6.5	SMALL I/A DBL.	2 S 1.4	F110/J1811 (89,256)
AM 2343-753	23 43 21 −75 33.1	3.449 0.333	308.37 −41.15	220.8 −10.9	SPIRAL w. PECULIAR ARMS + COMP'N. 28-G-2	8,10 S 1.6	F028/J1935 (235,148)
*AM 2343-370	23 43 35 −37 02.6	3.145 0.333	353.91 −73.17	256.5 3.8	CHAIN of 5 + MATERIAL 408-G-28	17 D 5.6	F408/J0767 (95,66)
AM 2343-770	23 43 41 −77 03.8	3.486 0.333	307.69 −39.72	219.3 −11.5	GALAXY w. CONDENSATIONS 28-G-3	16 D 0.9	F012/J2486 (229,335)
*AM 2343-703	23 43 49 −70 39.6	3.341 0.333	310.65 −45.75	225.4 −9.2	RING GAL. + I/A + 5 APP. COMP'NS 50-G-3	1,6,8,13 S 0.5	F050/J0955 (248,139)
*AM 2343-575	23 43 58 −57 59.7	3.222 0.333	318.96 −57.34	237.2 −4.5	SPIRAL w. 3 DIAMETRIC COMP'NS.	2,8 S 1.1	F148/J1760 (52,284)
AM 2344-525	23 44 01 −52 57.9	3.196 0.333	323.83 −61.72	241.9 −2.5	DISTURBED GALAXY	16 D 0.3	F192/J1839 (139,288)
AM 2344-503	23 44 08 −50 39.3	3.186 0.333	326.53 −63.67	244.0 −1.6	SMALL GALAXY w. PLUMES and JET to COMP'N.	1,7,15 S 1.0	F193/J3714 (314,140)
AM 2344-350	23 44 15 −35 04.0	3.137 0.333	359.93 −74.20	258.4 4.4	I/A DBL	2 D 0.2	F408/J0767 (86,172)
AM 2344-633	23 44 27 −63 38.1	3.255 0.333	314.63 −52.28	232.0 −6.7	BOX NUCLEUS w. DIFFUSE MATERIAL 110-G-27 + JET	7,22 D 0.9	F110/J1811 (81,249)
AM 2344-471	23 44 32 −47 14.9	3.170 0.333	331.16 −66.47	247.2 −0.4	I/A DBL. in CLUSTER R 121 = 2344-473	2,19 S 0.1	F292/J3733 (209,58)
AM 2344-595	23 44 47 −59 58.0	3.226 0.333	317.19 −55.62	235.5 −5.3	SMALL I/A DBL.	2 S 0.3	F148/J1760 (53,177)
AM 2344-350	23 44 48 −35 00.5	3.135 0.333	359.95 −74.33	258.5 4.3	SPIRAL w. I/A COMP'N. 408-IG-30	1 S 0.2	F408/J0767 (79,175)
AM 2344-252	23 44 49 −25 28.5	3.115 0.333	36.94 −75.52	267.4 7.9	I/A DBL. E w. ENVELOPE 537-G-10	2 S 0.1	F537/J3384 (242,153)
*AM 2344-282	23 44 59 −28 24.2	3.120 0.333	25.16 −75.83	264.7 6.8	2 ELONG. cD GALS. in DN 2345-284 IC 5353 = 471-G-17 in KLEM 44	19 D 13.4	F471/J0840 (188,263)
AM 2345-572	23 45 10 −57 20.7	3.208 0.333	319.25 −57.99	237.9 −4.4	IRR. w. BR. KNOTS 149-G-1 (I/A DBL. off MI.AX.)	2,163 D 3.4	F148/J1760 (41,318)
*AM 2345-304	23 45 15 −30 47.9	3.124 0.333	15.43 −75.68	262.4 5.9	SPIRAL w. COMP'NS (ONE on ARM) NGC 7755 = 471-G-20 [A82]	1,8 S 1.8	F471/J0840 (185,134)
AM 2345-625	23 45 17 −62 58.1	3.241 0.333	314.94 −52.92	232.7 −6.5	GALAXY w. FAINT OUTER ARMS	10 D 0.7	F110/J1811 (74,285)
AM 2345-665	23 45 41 −66 53.3	3.268 0.333	312.45 −49.32	229.0 −8.0	ONE-ARMED SPIRAL	10 D 0.5	F050/J0955 (252,342)
AM 2345-480	23 45 52 −48 09.2	3.165 0.333	329.36 −65.88	246.4 −0.9	DWARF 193-G-1	20 D 0.9	F193/J3714 (306,275)
AM 2345-295	23 45 57 −29 59.9	3.120 0.333	18.57 −75.94	263.2 6.0	COMPACT	13 D 0.2	F471/J0840 (177,177)
*AM 2346-731	23 46 04 −73 13.7	3.342 0.333	309.18 −43.40	223.0 −10.3	V.POOR, FAINT STAR CLUSTER AM-3 = 28-SC-4	22 D 0.4	F049/J3647 (25,262)
AM 2346-303	23 46 41 −30 31.7	3.119 0.333	16.31 −76.03	262.8 5.7	DISTURBED w. PECULIAR ABS'N. MCG-5-56-17 = 471-G-22	14,16 D 1.1	F471/J0840 (169,149)
AM 2346-380	23 46 47 −38 03.1	3.133 0.333	349.79 −73.16	255.8 2.8	EDGE-ON DWARF (RESOLUTION:) 348-G-9 = UKS 2346-380	20 D 2.2	F348/J3586 (214,284)
AM 2346-270	23 46 59 −27 01.8	3.112 0.333	30.92 −76.22	266.1 6.9	CLOSE PAIR (ONE COMPACT)	13,23 S 0.8	F471/J0840 (165,337)
*AM 2347-292	23 47 08 −29 23.1	3.115 0.333	21.01 −76.26	263.9 6.0	CHAIN of 4 (incl. I/A DBL.)	2,17 D 2.9	F471/J0840 (163,210)
AM 2347-365	23 47 10 −36 51.5	3.129 0.333	353.17 −73.88	257.0 3.2	COMPACT	13 D 0.2	F408/J0767 (56,75)
AM 2347-385	23 47 12 −38 50.2	3.133 0.333	347.46 −72.77	255.1 2.4	SPIRAL w. JET from COMP'N. 293-G-2 = B2347-388 (in GRP of 7)	7,8,18 D 1.1	F348/J3586 (209,242)
AM 2347-633	23 47 16 −63 36.5	3.223 0.333	314.20 −52.42	232.2 −6.9	LARGE LUMPY COMPACT	13,16 D 0.4	F110/J1811 (65,250)

Identification	R.A. (1950.0) Declination	Precession R.A./Dec.	Gal. Long. Gal. Lat.	S.G.L. S.G.B.	Description Cross-Identification	Coding Size (Min.)	Field/Plate (x, y)
AM 2347-354	23 47 21 −35 45.4	3.126 0.333	356.60 −74.46	258.0 3.6	D + 8 COMP'NS. across DIAMETER 408-G-37 = MC 2347-357	8 D 6.7	F408/J0767 (53,134)
AM 2347-723	23 47 40 −72 38.3	3.303 0.334	309.30 −43.99	223.6 −10.2	DWARF ? 28-G-5	20 D 0.4	F077/J0910 (82,31)
*AM 2347-614	23 47 41 −61 48.4	3.207 0.334	315.29 −54.10	233.9 −6.3	GALAXY w. STRONG JET or PLUME 111-IG-3 = Fa 209/10(I/A TRIPLE?)	3,7,15 D 0.3	F148/J1760 (41,77)
AM 2347-575	23 47 43 −57 56.3	3.187 0.334	318.18 −57.61	237.5 −4.9	COMPACT	13 D 0.1	F192/J1839 (117,21)
AM 2347-594	23 47 44 −59 47.3	3.196 0.334	316.73 −55.94	235.8 −5.6	SMALL GALAXY w. BROAD JET 111-IG-4	7 D 0.8	F111/J0921 (259,187)
AM 2347-684	23 47 48 −68 42.4	3.255 0.334	311.18 −47.70	227.4 −8.8	FAINT GALAXY CLUSTER ("V"-SHAPED)	19 D 4.5	F077/J0910 (62,241)
AM 2347-320	23 47 51 −32 07.6	3.117 0.334	9.65 −75.93	261.4 4.8	CLOSE TRIPLE	24 D 2.2	F471/J0840 (156,63)
AM 2347-314	23 47 58 −31 45.5	3.116 0.334	11.08 −76.05	261.8 5.0	COMPACT	13 D 0.2	F409/J2693 (317,83)
AM 2348-284	23 48 03 −28 42.7	3.111 0.334	23.81 −76.50	264.6 6.1	CHAIN of 4	17 D 3.0	F409/J2693 (320,247)
AM 2348-410	23 48 17 −41 00.8	3.132 0.334	341.58 −71.53	253.2 1.4	DISTURBED SINGLE GALAXY NGC 7764 = 298-G-4	16 D 1.8	F348/J3586 (196,125)
AM 2348-405	23 48 22 −40 59.9	3.132 0.334	341.57 −71.55	253.2 1.4	DISTURBED SINGLE (KNOTTY) NGC 7764 = 293-G-4	16 D 1.1	F293/J3578 (297,123)
AM 2348-313	23 48 34 −31 37.0	3.114 0.334	11.53 −76.20	261.9 4.9	QUARTET	18 D 4.3	F409/J2693 (310,91)
AM 2348-281	23 48 55 −28 18.0	3.108 0.334	25.57 −76.70	265.1 6.0	2 PAIR + PEC. SP. (GROUP of 11)	10,18,23 D 13.4	F409/J2693 (310,269)
AM 2348-344	23 48 56 −34 43.5	3.118 0.334	359.56 −75.22	259.1 3.7	E + COMP'NS. (at CLUSTER CENTRE) 349-G-1	8 D 4.0	F408/J0767 (33,189)
*AM 2349-635	23 49 00 −63 51.0	3.204 0.334	313.74 −52.27	232.0 −7.2	SPIRAL w. HI.S.BR. ARM	10 D 0.4	F110/J1811 (55,236)
AM 2349-392	23 49 06 −39 27.2	3.125 0.334	345.06 −72.68	254.7 1.9	GALAXY w. TAILS	15 D 0.4	F348/J3586 (189,209)
AM 2349-260 [2349-2608]	23 49 08 −26 07.8	3.104 0.334	34.96 −76.59	267.1 6.8	CLOSE PAIR of DWARFS (#1) MCG-4-56-4 = 472-G-1	20,23 S 4.0	F472/J3428 (311,118)
AM 2349-305	23 49 19 −30 52.7	3.110 0.334	14.44 −76.52	262.7 5.0	SPIRAL + 2 APP. COMP'NS. 471-G-28	8 S 1.3	F409/J2693 (302,131)
AM 2349-632	23 49 20 −63 24.8	3.197 0.334	313.95 −52.69	232.4 −7.1	COMPACT	13 D 0.4	F078/J0928 (240,261)
AM 2349-391	23 49 21 −39 17.5	3.124 0.334	345.38 −72.81	254.9 1.9	I/A DBL. 293-IG-6 = B 2349-392	2 S 0.5	F293/J3578 (290,215)
*AM 2349-525	23 49 26 −52 51.4	3.154 0.334	322.45 −62.23	242.3 −3.2	IRR. w. KNOTS and FAINT JET 149-G-3 = UKS 2349-528	7,16 D 0.8	F240/J0962 (21,21)
AM 2349-260 [2349-2607]	23 49 28 −26 08.3	3.103 0.334	34.98 −76.66	267.1 6.7	CLOSE PAIR of DWARFS (#2) MCG-4-56-5 = 472-G-2	20,23 S 4.0	F472/J3428 (306,117)
AM 2349-625 [2349-6250]	23 49 31 −62 51.2	3.192 0.334	314.28 −53.22	233.0 −6.9	GROUP of 3 (#1: EDGE-ON) 110-G-29 = 78-G-8	24 D 1.5	F078/J0928 (240,291)
*AM 2349-294	23 49 41 −29 49.2	3.107 0.334	18.91 −76.77	263.7 5.3	CHAIN of 9 (MOSTLY EDGE-ON) incl. MCG-5-56-27 = 471-G-30	17 D 24.2	F409/J2693 (299,188)
AM 2349-303 [2349-3027]	23 49 47 −30 30.5	3.108 0.334	15.93 −76.69	263.1 5.1	CLOSE PAIR (#1) MCG-5-56-28 = 471-G-32	23 D 0.3	F471/J0840 (133,150)
AM 2349-624 [2349-6251]	23 49 48 −62 48.9	3.189 0.334	314.25 −53.26	233.0 −6.9	GROUP of 3 (#2: SPIRAL)	24 D 0.4	F078/J0928 (238,293)
AM 2349-625 [2349-6248]	23 49 48 −62 50.7	3.189 0.334	314.23 −53.24	233.0 −6.9	GROUP of 3 (#3: SPIRAL) 78-G-9	24 D 0.7	F078/J0928 (238,292)
AM 2349-624	23 49 49 −62 47.4	3.188 0.334	314.26 −53.29	233.0 −6.9	I/A DBL. in GROUP of 3 near 78-G-9	2,24 D 2.9	F110/J1811 (45,292)
AM 2349-302 [2349-3030]	23 49 57 −30 27.6	3.107 0.334	16.11 −76.73	263.1 5.0	CLOSE PAIR (#2) MCG-5-56-31 = 471-G-34	23 D 0.5	F471/J0840 (131,152)
*AM 2350-605	23 50 06 −60 55.0	3.177 0.334	315.45 −55.03	234.8 −6.3	3-ARMED SPIRAL 111-G-6	11 D 1.4	F111/J0921 (241,127)
AM 2350-320	23 50 38 −32 00.1	3.107 0.334	9.47 −76.53	261.7 4.3	LINE of 3 (SP. + E + SP.) 471-G-36	24 D 8.5	F409/J2693 (286,71)
AM 2350-800	23 50 43 −80 05.6	3.383 0.334	306.10 −36.94	216.5 −12.8	CHAIN of 4 12-G-9?	17 D 4.5	F012/J2486 (201,174)
*AM 2350-302	23 50 44 −30 25.9	3.105 0.334	16.11 −76.90	263.2 4.9	ONE-ARMED SPIRAL + DBL. COMP'N. 471-IG-37	8,10,24 D 1.6	F409/J2693 (286,155)
AM 2350-404	23 50 44 −40 42.6	3.119 0.334	341.27 −72.07	253.7 1.1	GAL. w. DIAMETRIC JET off MI.AX.	7 D 0.4	F293/J3578 (273,140)
*AM 2350-410	23 50 47 −41 05.2	3.120 0.334	340.37 −71.82	253.3 1.0	I/A TRIPLE: Ag-83= MCG-7-1-1 [Lu] NGC 7764a = 293-IG-8 = B 2350-410	3 D 1.4	F348/J3586 (171,121)
AM 2350-315	23 50 48 −31 58.4	3.106 0.334	9.54 −76.57	261.8 4.3	GROUP of 3 (ALIGNED) 471-G-38	24 D 9.0	F471/J0840 (122,71)

Identification	R.A. (1950.0) Declination	Precession R.A./Dec.	Gal. Long. Gal. Lat.	S.G.L. S.G.B.	Description Cross-Identification	Coding Size (Min.)	Field/Plate (x, y)
AM 2350-514	23 50 48 −51 42.9	3.141 0.334	323.28 −63.33	243.4 −3.0	SMALL I/A DBL. in RING	2,6 S 0.1	F193/J3714 (256,86)
AM 2350-250	23 50 58 −25 07.7	3.098 0.334	39.57 −76.80	268.2 6.8	CHAIN of 4	17 D 0.4	F472/J3428 (289,172)
AM 2350-250	23 50 58 −25 09.2	3.098 0.334	39.47 −76.81	268.2 6.7	LINE of 3	24 D 1.3	F472/J3428 (289,171)
AM 2351-350	23 51 05 −35 09.5	3.110 0.334	357.24 −75.41	258.8 3.1	COMPACT	13 D 0.2	F408/J0767 (11,165)
AM 2351-542	23 51 10 −54 28.8	3.145 0.334	320.36 −60.92	240.9 −4.1	CLOSE PAIR of SPIRALS in TRIPLE	23,24 S 0.7	F149/J1739 (248,206)
AM 2351-401	23 51 12 −40 18.3	3.116 0.334	342.04 −72.42	254.1 1.2	E + CLOSE COMP'N. 293-IG-10	8 S 0.3	F348/J3586 (167,163)
AM 2351-251	23 51 33 −25 10.4	3.096 0.334	39.51 −76.94	268.2 6.6	COMPACT	13 D 0.4	F472/J3428 (282,170)
*AM 2351-301	23 51 38 −30 17.1	3.101 0.334	16.61 −77.12	263.4 4.8	V.FNT. COMPACT GROUP (STARS?)	18,22 D 1.3	F409/J2693 (276,163)
AM 2352-621	23 52 00 −62 15.0	3.161 0.334	314.21 −53.88	233.6 −7.0	CLOSE DOUBLE (COMPACTS)	13,13,23 S 0.2	F111/J0921 (226,56)
AM 2352-275	23 52 01 −27 50.7	3.098 0.334	27.62 −77.38	265.7 5.6	I/A DBL. (in LINE of 3)	2,24 S 0.8	F471/J0840 (105,292)
AM 2352-275	23 52 11 −27 57.6	3.097 0.334	27.10 −77.42	265.7 5.5	15 MEDIUM-BRIGHT GALAXIES	19 D 9.0	F409/J2693 (272,288)
AM 2352-280	23 52 11 −28 07.9	3.097 0.334	26.31 −77.42	265.5 5.4	DBL. M51-TYPE (SMALL)	9 S 0.3	F471/J0840 (103,277)
AM 2352-315	23 52 33 −31 51.5	3.100 0.334	9.57 −76.96	262.0 4.0	GALAXY w. APP. COMP'N. MCG-5-1-3 = 471-G-41	8 S 1.6	F471/J0840 (102,77)
AM 2352-323	23 52 41 −32 37.3	3.100 0.334	6.34 −76.75	261.3 3.7	I/A DBL. (SMALL, COMPACT)	2,13 S 0.2	F349/J0738 (258,305)
AM 2352-565	23 52 42 −56 51.3	3.138 0.334	317.91 −58.87	238.7 −5.1	I/A DBL. 149-IG-5	2 S 0.4	F111/J0921 (229,345)
AM 2352-273	23 52 49 −27 32.3	3.095 0.334	29.06 −77.54	266.1 5.5	CHAIN of 4 CENTRED on E	17 D 1.8	F471/J0840 (95,309)
AM 2352-524	23 52 57 −52 48.8	3.127 0.334	321.49 −62.52	242.5 −3.7	COMPACT	13 D 0.1	F192/J1839 (67,294)
AM 2353-400 [2353-4004]	23 53 27 −40 02.3	3.105 0.334	341.71 −72.91	254.5 0.9	PAIR + COMP'NS. (#1: E)	8,23 S 2.3	F293/J3578 (247,176)
AM 2353-411	23 53 27 −41 10.4	3.106 0.334	339.03 −72.11	253.4 0.5	ONE-ARMED SP. + I/A DBL. COMP'N. 293-IG-14	2,10 D 0.8	F348/J3586 (144,117)
AM 2353-412	23 53 28 −41 27.3	3.107 0.334	338.39 −71.91	253.2 0.3	GROUP of 3	24 D 2.9	F293/J3578 (245,100)
AM 2353-400 [2353-4002]	23 53 33 −40 04.8	3.105 0.334	341.56 −72.90	254.4 0.8	PAIR + COMP'NS. (#2: SPIRAL)	8,23 S 2.3	F293/J3578 (246,174)
*AM 2353-605	23 53 33 −60 57.3	3.141 0.334	314.75 −55.15	234.9 −6.7	CHAIN of 5 (I/A SPIRALS at END) 111-G-9	2,17 D 11.2	F111/J0921 (218,126)
AM 2353-274	23 53 34 −27 45.0	3.093 0.334	28.10 −77.72	266.0 5.3	3 in a LINE MCG-5-1-5 = 471-G-44	24 D 6.5	F537/J3384 (137,32)
*AM 2353-815	23 53 34 −81 51.0	3.335 0.334	305.38 −35.28	214.8 −13.4	Sd/DWARF (CONDENSATION) 12-G-10 = UKS 2353-818	20 D 3.6	F012/J2486 (191,80)
AM 2353-434	23 53 40 −43 42.6	3.108 0.334	333.73 −70.24	251.1 −0.5	SINGLE IRR. 241-G-6	16 D 1.1	F292/J3733 (125,249)
AM 2353-411	23 53 44 −41 11.3	3.105 0.334	338.86 −72.14	253.4 0.4	I/A DBL. 293-IG-16	2 S 0.2	F293/J3578 (242,115)
AM 2353-411	23 53 44 −41 16.0	3.105 0.334	338.69 −72.08	253.3 0.4	GROUP of 5	18 D 1.6	F293/J3578 (242,110)
AM 2353-572	23 53 46 −57 20.1	3.130 0.334	317.27 −58.50	238.3 −5.4	GALAXY w. JET (I/A COMPACT)	7,13 D 0.8	F149/J1739 (224,53)
*AM 2353-291	23 53 49 −29 18.3	3.093 0.334	20.78 −77.72	264.5 4.7	I/A DBL. (E + IRR.) POSS.RING IC 5364 = MCG-5-1-7/8 = 471-IG-47	2,6 S 0.4	F409/J2693 (251,216)
*AM 2353-392	23 53 53 −39 26.6	3.102 0.334	343.01 −73.38	255.1 1.0	CHAIN of 6 incl. I/A TRIPLE 293-IG-17 in DN 2353-392	3,17 D 6.7	F293/J3578 (243,208)
AM 2354-375	23 54 09 −37 50.1	3.099 0.334	347.32 −74.46	256.6 1.5	PAIR 293-IG-18 in DN 2354-379	23 S 0.4	F293/J3578 (242,295)
AM 2354-382	23 54 19 −38 29.7	3.099 0.334	345.36 −74.07	256.0 1.3	SMALL I/A DBL. 293-IG-19	2 S 0.3	F293/J3578 (239,259)
AM 2354-385	23 54 20 −38 56.7	3.100 0.334	344.12 −73.78	255.6 1.1	GALAXY w. DIAMETRIC JETS 293-G-20	7 D 0.9	F293/J3578 (239,235)
AM 2354-413	23 54 25 −41 33.1	3.102 0.334	337.76 −71.96	253.1 0.1	GROUP of 3	24 D 0.9	F293/J3578 (235,95)
AM 2354-645	23 54 38 −64 57.5	3.140 0.334	312.18 −51.45	231.2 −8.2	CLOSE PAIR + APP. COMP'N. 78-G-10	8,23 D 1.3	F110/J1811 (28,173)
AM 2354-411	23 54 43 −41 10.4	3.100 0.334	338.46 −72.28	253.5 0.2	I/A DBL. (COMPACT + SPIRAL) 293-IG-21 = B 2354-411	2,13 S 0.6	F293/J3578 (232,116)

Identification	R.A. (1950.0) Declination	Precession R.A./Dec.	Gal. Long. Gal. Lat.	S.G.L. S.G.B.	Description Cross-Identification	Coding Size (Min.)	Field/Plate (x, y)
AM 2354-304	23 54 54 −30 43.4	3.091 0.334	14.01 −77.73	263.3 4.0	I/A DBL. in GROUP of 3	2,24 S 0.4	F471/J0840 (74,137)
*AM 2354-221	23 54 59 −22 16.0	3.085 0.334	52.73 −76.73	271.2 6.9	I/A DBL. or GALAXY w. I/A COMP'N. MCG-4-1-7/8 = VV332a/b = 538-G-8/9	2,8 S 1.1	F472/J3428 (241,326)
AM 2355-313	23 55 09 −31 31.3	3.090 0.334	10.33 −77.58	262.5 3.6	PAIR of SPIRALS	23 S 0.8	F409/J2693 (235,97)
AM 2355-325	23 55 16 −32 52.0	3.091 0.334	4.47 −77.17	261.3 3.1	BR. GAL.(MANY KNOTS) + 11 COMP'NS NGC 7793 = 349-G-12	8 D 11.2	F409/J2693 (233,25)
*AM 2355-383	23 55 16 −38 39.9	3.095 0.334	344.44 −74.10	255.9 1.0	CHAIN of 4 incl. DISTURBED SINGLE DN 2355-384	16,17 D 6.7	F293/J3578 (229,250)
AM 2355-414	23 55 28 −41 48.4	3.097 0.334	336.74 −71.90	253.0 −0.1	GALAXY w. JET + DBL. COMPACT MCG-7-1-2 = 293-G-22 = B 2355-418	7,8,13 S 4.5	F292/J3733 (107,351)
AM 2355-274	23 55 31 −27 43.3	3.087 0.334	28.29 −78.15	266.1 4.9	I/A DBL. w. TAIL	2,15 S 0.3	F471/J0840 (63,298)
*AM 2356-244 [2356-2453]	23 56 02 −24 49.2	3.084 0.334	42.21 −77.84	268.9 5.8	PAIR (#1: DWARF w. RESOLUTION)	20,23 D 0.5	F537/J3384 (106,189)
AM 2356-225	23 56 11 −22 51.0	3.082 0.334	50.88 −77.22	270.8 6.4	COMPACT	13 D 0.2	F472/J3428 (226,295)
AM 2356-245 [2356-2449]	23 56 18 −24 53.6	3.083 0.334	41.95 −77.92	268.9 5.7	PAIR (#2: ONE-ARMED SPIRAL)	10,20,23 D 0.6	F537/J3384 (103,185)
AM 2356-571	23 56 19 −57 10.0	3.106 0.334	316.79 −58.79	238.6 −5.7	M51-TYPE (SMALL) 149-IG-6	9 D 0.3	F111/J0921 (202,329)
AM 2356-554	23 56 25 −55 44.2	3.103 0.334	317.88 −60.11	240.0 −5.2	LARGE E w. 8 COMP'NS. NGC 7796 = Se 164/2 = 149-G-7	8 S 6.5	F192/J1839 (47,136)
AM 2356-320 [2356-3200]	23 56 27 −32 04.5	3.086 0.334	7.48 −77.68	262.1 3.2	WIDE PAIR (#1: I/A DBL.)	2,23 D 0.2	F471/J0840 (58,64)
AM 2356-610	23 56 28 −61 00.2	3.110 0.334	314.14 −55.24	235.0 −7.0	ELONGATED FAINT CLUSTER PKS 2356-61	19 D 21.5	F111/J0921 (199,123)
AM 2356-690	23 56 42 −69 07.0	3.124 0.334	309.87 −47.57	227.3 −9.7	I/A DBL. (CONNECTED)	2 S 0.3	F050/J0955 (192,224)
AM 2356-320 [2356-3204]	23 56 45 −32 00.5	3.085 0.334	7.68 −77.76	262.2 3.1	WIDE PAIR (#2: DISTURBED) 471-IG-54	16,23 D 0.3	F471/J0840 (54,68)
AM 2356-714	23 56 50 −71 46.3	3.129 0.334	308.73 −45.03	224.7 −10.6	VERY SMALL CLUSTER (15 GALS.)	19 D 1.6	F077/J0910 (40,73)
*AM 2356-244	23 56 53 −24 42.3	3.081 0.334	42.98 −78.00	269.1 5.6	I/A SPIRALS MCG-4-1-12 = 472-G-13	1,2 S 1.7	F472/J3428 (217,195)
AM 2357-702	23 57 30 −70 29.3	3.114 0.334	309.18 −46.28	226.0 −10.2	COMPACT	13 D 0.1	F050/J0955 (188,151)
AM 2357-773	23 57 32 −77 37.0	3.138 0.334	306.54 −39.42	219.0 −12.4	I/A DBL. 12-IG-11	2 S 0.2	F028/J1935 (186,39)
AM 2357-252	23 57 39 −25 27.9	3.079 0.334	39.59 −78.35	268.4 5.2	2 E's in COMMON ENVELOPE	2 S 0.2	F537/J3384 (86,154)
AM 2357-270	23 57 40 −27 00.6	3.080 0.334	31.94 −78.58	267.0 4.7	I/A DBL.	2 S 0.5	F409/J2693 (207,340)
AM 2357-810	23 57 44 −81 04.3	3.157 0.334	305.44 −36.07	215.6 −13.3	BRIGHT-RIMMED SPIRAL 12-G-12	10 D 1.3	F012/J2486 (184,122)
AM 2357-242	23 57 51 −24 21.0	3.079 0.334	44.93 −78.11	269.5 5.6	SMALL I/A TRIPLE	3,13 D 1.0	F537/J3384 (83,214)
*AM 2357-404	23 57 56 −40 45.9	3.083 0.334	337.89 −72.99	254.1 −0.2	SPIRAL w. I/A DBL. COMP'N. 293-G-27	2,8 D 1.8	F348/J3586 (98,138)
AM 2357-502	23 57 57 −50 25.7	3.087 0.334	322.44 −64.99	245.0 −3.6	PAIR in COMMON ENVELOPE 192-IG-8	2 S 0.4	F193/J3714 (197,156)
AM 2357-682	23 57 59 −68 25.2	3.103 0.334	310.01 −48.27	228.0 −9.6	M51-TYPE	2,9 S 0.2	F050/J0955 (186,262)
AM 2358-821 [2358-8213]	23 58 02 −82 14.9	3.157 0.334	305.08 −34.93	214.5 −13.7	CLOSE PAIR (#1: S0)	23 S 2.0	F002/J2563 (183,327)
AM 2358-535	23 58 15 −53 57.3	3.087 0.334	318.89 −61.84	241.7 −4.9	PAIR of COMPACTS	13,13,23 S 3.1	F192/J1839 (27,231)
AM 2358-433	23 58 29 −43 36.7	3.081 0.334	331.84 −70.84	251.5 −1.3	E + COMP'NS. (LOOSE CHAIN of 5) 241-G-10 = R 124 = DN 2358-442	2,8,17 D 13.4	F292/J3733 (78,254)
AM 2358-522	23 58 34 −52 29.3	3.084 0.334	320.14 −63.19	243.1 −4.4	LO.S.BR. PAIR 193-G-10 = B 2358-524	23 S 1.0	F192/J1839 (20,309)
AM 2358-821 [2358-8214]	23 58 39 −82 13.2	3.131 0.334	305.06 −34.97	214.5 −13.7	CLOSE PAIR (#2: E:)	23 S 2.0	F002/J2563 (182,329)
AM 2358-390	23 58 48 −39 02.3	3.079 0.334	341.72 −74.35	255.8 0.3	E + 6 COMP'NS.	8 S 1.8	F293/J3578 (192,231)
AM 2359-381	23 59 02 −38 18.8	3.077 0.334	343.59 −74.88	256.5 0.5	GALAXY w. TAIL and JET	7,15 D 0.5	F293/J3578 (190,270)
AM 2359-771	23 59 05 −77 13.3	3.096 0.334	306.57 −39.82	219.4 −12.3	GALAXY w. I/A COMP'N.	1 S 0.7	F049/J3647 (22,42)
*AM 2359-275	23 59 21 −27 54.2	3.075 0.334	27.43 −78.98	266.3 4.0	SPIRAL w. APP. COMP'NS. [W.] MCG-5-1-23 = 409-G-3	8 S 1.3	F471/J0840 (19,287)

Identification	R.A. (1950.0) Declination	Precession R.A./Dec.	Gal. Long. Gal. Lat.	S.G.L. S.G.B.	Description Cross-Identification	Coding Size (Min.)	Field/Plate (x, y)
*AM 2359-441	23 59 36 −44 15.0	3.075 0.334	330.23 −70.44	250.9 −1.7	E + 6 COMP'NS.(at CLUSTER CENTRE) R 124	8,19 D 5.4	F292/J3733 (68,219)

APPENDIX

Total number of galaxies inspected per plate centre

Field Number	Galaxies Counted	Plate Centre R.A. (1950)	Plate Centre Declination
1	64	16 00	–90
2	183	00 00	–85
3	116	02 24	–85
4	88	04 48	–85
5	114	07 12	–85
6	49	09 36	–85
7	77	12 00	–85
8	96	14 24	–85
9	69	16 48	–85
10	100	19 12	–85
11	219	21 36	–85
12	76	00 00	–80
13	276	01 30	–80
14	185	03 00	–80
15	102	04 30	–80
16	77	06 00	–80
17	54	07 30	–80
18	56	09 00	–80
19	54	10 30	–80
20	22	12 00	–80
21	42	13 30	–80
22	51	15 00	–80
23	135	16 30	–80
24	172	18 00	–80
25	89	19 30	–80
26	86	21 00	–80
27	162	22 30	–80
28	103	00 00	–75
29	196	01 06	–75
30	111	02 12	–75
31	227	03 18	–75
32	218	04 24	–75
33	205	05 30	–75
34	87	06 36	–75
35	71	07 42	–75
36	81	08 48	–75
37	73	09 54	–75
38	50	11 00	–75
39	48	12 06	–75
40	34	13 12	–75
41	42	14 18	–75
42	39	15 24	–75
43	70	16 30	–75
44	109	17 36	–75
45	105	18 42	–75
46	107	19 48	–75
47	241	20 54	–75
48	200	22 00	–75
49	165	23 06	–75
50	547	00 00	–70
51	155	00 52	–70
52	406	01 44	–70
53	276	02 36	–70
54	266	03 28	–70
55	227	04 20	–70
56	46	05 12	–70
57	129	06 04	–70
58	340	06 56	–70
59	131	07 48	–70
60	291	08 40	–70
61	152	09 32	–70
62	60	10 24	–70
63	67	11 16	–70
64	10	12 08	–70
65	18	13 00	–70
66	15	13 52	–70
67	26	14 44	–70
68	66	15 36	–70
69	89	16 28	–70
70	60	17 20	–70
71	87	18 12	–70
72	112	19 04	–70
73	256	19 36	–70
74	204	20 48	–70
75	756	21 40	–70
76	178	22 32	–70
77	207	23 24	–70
78	173	00 00	–65
79	174	00 44	–65
80	230	01 28	–65
81	157	02 12	–65
82	217	02 56	–65
83	212	03 40	–65
84	523	04 24	–65
85	534	05 08	–65
86	433	05 52	–65
87	826	06 36	–65
88	203	07 20	–65
89	74	08 04	–65
90	87	08 48	–65
91	69	09 32	–65
92	33	10 16	–65
93	4	11 00	–65
94	0	11 44	–65
95	10	12 28	–65
96	0	13 12	–65
97	3	13 56	–65
98	6	14 40	–65
99	31	15 24	–65
100	53	16 08	–65
101	94	16 52	–65
102	103	17 36	–65
103	152	18 20	–65
104	224	19 04	–65
105	129	19 48	–65
106	256	20 32	–65
107	400	21 16	–65
108	555	22 00	–65
109	424	22 44	–65
110	285	23 28	–65
111	523	00 00	–60
112	182	00 38	–60
113	188	01 16	–60
114	170	01 54	–60
115	275	02 32	–60
116	249	03 10	–60
117	279	03 48	–60
118	175	04 26	–60
119	256	05 04	–60
120	180	05 42	–60
121	174	06 20	–60
122	91	06 58	–60
123	106	07 36	–60
124	54	08 14	–60
125	30	08 52	–60
126	57	09 30	–60
127	10	10 08	–60
128	0	10 46	–60
129	2	11 24	–60
130	1	12 02	–60
131	3	12 40	–60
132	16	13 18	–60
133	5	13 56	–60
134	3	14 34	–60
135	4	15 12	–60
136	112	15 50	–60
137	153	16 28	–60
138	80	17 06	–60
139	173	17 44	–60
140	147	18 22	–60
141	180	19 00	–60
142	244	19 38	–60
143	273	20 16	–60
144	355	20 54	–60
145	766	21 32	–60
146	416	22 10	–60
147	536	22 48	–60
148	185	23 26	–60
149	155	00 00	–55
150	443	00 33	–55
151	184	01 06	–55
152	168	01 39	–55
153	222	02 12	–55
154	206	02 45	–55
155	183	03 18	–55
156	366	03 51	–55
157	253	04 24	–55
158	232	04 57	–55
159	141	05 30	–55
160	135	06 03	–55
161	193	06 36	–55
162	95	07 09	–55
163	117	07 42	–55
164	40	08 15	–55
165	15	08 48	–55
166	0	09 21	–55
167	5	09 54	–55
168	5	10 27	–55
169	5	11 00	–55
170	27	11 33	–55
171	60	12 06	–55
172	26	12 39	–55
173	38	13 12	–55
174	28	13 45	–55
175	24	14 18	–55
176	6	14 51	–55
177	3	15 24	–55
178	2	15 57	–55
179	15	16 30	–55
180	15	17 03	–55
181	28	17 36	–55
182	98	18 09	–55
183	151	18 42	–55
184	267	19 15	–55
185	272	19 42	–55
186	387	20 21	–55
187	349	20 54	–55
188	191	21 27	–55
189	328	22 00	–55
190	274	22 33	–55
191	204	23 06	–55
192	215	23 39	–55
193	281	00 00	–50
194	274	00 30	–50
195	771	01 00	–50
196	175	01 30	–50
197	523	02 00	–50
198	392	02 30	–50
199	244	03 00	–50
200	186	03 30	–50
201	447	04 00	–50
202	666	04 30	–50
203	331	05 00	–50
204	314	05 30	–50
205	236	06 00	–50
206	161	06 30	–50
207	165	07 00	–50
208	122	07 30	–50
209	46	08 00	–50
210	6	08 30	–50

Field Number	Galaxies Counted	Plate Centre R.A. (1950)	Plate Centre Declination
211	1	09 00	–50
212	7	09 30	–50
213	26	10 00	–50
214	91	10 30	–50
215	97	11 00	–50
216	95	11 30	–50
217	94	12 00	–50
218	87	12 30	–50
219	100	13 00	–50
220	74	13 30	–50
221	94	14 00	–50
222	42	14 30	–50
223	31	15 00	–50
224	6	15 30	–50
225	6	16 00	–50
226	4	16 30	–50
227	2	17 00	–50
228	10	17 30	–50
229	35	18 00	–50
230	93	18 30	–50
231	115	19 00	–50
232	108	19 30	–50
233	272	20 00	–50
234	445	20 30	–50
235	307	21 00	–50
236	384	21 30	–50
237	197	22 00	–50
238	200	22 30	–50
239	308	23 00	–50
240	175	23 30	–50
241	210	00 00	–45
242	488	00 28	–45
243	263	00 56	–45
244	333	01 24	–45
245	253	01 52	–45
246	178	02 20	–45
247	93	02 48	–45
248	194	03 16	–45
249	295	03 44	–45
250	355	04 12	–45
251	704	04 40	–45
252	239	05 08	–45
253	287	05 36	–45
254	300	06 04	–45
255	160	06 32	–45
256	50	07 00	–45
257	45	07 28	–45
258	14	07 56	–45
259	4	08 24	–45
260	0	08 52	–45
261	10	09 20	–45
262	50	09 48	–45
263	91	10 16	–45
264	171	10 44	–45
265	413	11 12	–45
266	132	11 40	–45
267	196	12 08	–45
268	155	12 36	–45
269	233	13 04	–45
270	194	13 32	–45
271	71	14 00	–45
272	96	14 28	–45
273	50	14 56	–45
274	51	15 24	–45
275	12	15 52	–45
276	0	16 20	–45
277	0	16 48	–45
278	1	17 16	–45
279	4	17 44	–45
280	58	18 12	–45
281	92	18 40	–45
282	140	19 08	–45
283	69	19 36	–45
284	216	20 04	–45
285	254	20 32	–45
286	449	21 00	–45
287	242	21 28	–45
288	230	21 56	–45
289	258	22 24	–45
290	198	22 52	–45
291	264	23 20	–45
292	341	23 48	–45
293	288	00 00	–40
294	269	00 26	–40
295	285	00 52	–40
296	290	01 18	–40
297	197	01 44	–40
298	254	02 10	–40
299	157	02 36	–40
300	380	03 02	–40
301	406	03 28	–40
302	246	03 54	–40
303	241	04 20	–40
304	321	04 46	–40
305	181	05 12	–40
306	253	05 38	–40
307	313	06 04	–40
308	162	06 30	–40
309	138	06 56	–40
310	55	07 22	–40
311	38	07 48	–40
312	4	08 14	–40
313	6	08 40	–40
314	25	09 06	–40
315	57	09 32	–40
316	333	09 58	–40
317	198	10 24	–40
318	168	10 50	–40
319	113	11 16	–40
320	120	11 42	–40
321	123	12 08	–40
322	359	12 34	–40
323	260	13 00	–40
324	249	13 26	–40
325	190	13 52	–40
326	136	14 18	–40
327	417	14 44	–40
328	137	15 10	–40
329	41	15 36	–40
330	29	16 02	–40
331	3	16 28	–40
332	1	16 54	–40
333	2	17 20	–40
334	1	17 46	–40
335	41	18 12	–40
336	78	18 38	–40
337	117	19 04	–40
338	77	19 30	–40
339	193	19 56	–40
340	12	20 22	–40
341	295	20 48	–40
342	349	21 14	–40
343	213	21 40	–40
344	208	22 06	–40
345	617	22 32	–40
346	376	22 58	–40
347	386	23 24	–40
348	211	23 50	–40
349	266	00 00	–35
350	253	00 24	–35
351	229	00 48	–35
352	309	01 12	–35
353	297	01 36	–35
354	311	02 00	–35
355	283	02 24	–35
356	148	02 48	–35
357	244	03 12	–35
358	381	03 36	–35
359	223	04 00	–35
360	186	04 24	–35
361	241	04 48	–35
362	165	05 12	–35
363	171	05 36	–35
364	276	06 00	–35
365	397	06 24	–35
366	152	06 48	–35
367	113	07 12	–35
368	39	07 36	–35
369	5	08 00	–35
370	19	08 24	–35
371	29	08 48	–35
372	53	09 12	–35
373	121	09 36	–35
374	174	10 00	–35
375	340	10 24	–35
376	121	10 48	–35
377	175	11 12	–35
378	204	11 36	–35
379	201	12 00	–35
380	177	12 24	–35
381	149	12 48	–35
382	399	13 12	–35
383	325	13 36	–35
384	258	14 00	–35
385	227	14 24	–35
386	98	14 48	–35
387	119	15 12	–35
388	74	15 36	–35
389	97	16 00	–35
390	7	16 24	–35
391	7	16 48	–35
392	1	17 12	–35
393	1	17 36	–35
394	1	18 00	–35
395	19	18 24	–35
396	25	18 48	–35
397	57	19 12	–35
398	127	19 36	–35
399	42	20 00	–35
400	155	20 24	–35
401	222	20 48	–35
402	126	21 12	–35
403	170	21 36	–35
404	248	22 00	–35
405	232	22 24	–35
406	489	22 48	–35
407	108	23 12	–35
408	394	23 36	–35
409	353	00 00	–30
410	421	00 23	–30
411	237	00 46	–30
412	212	01 09	–30
413	427	01 32	–30
414	309	01 55	–30
415	248	02 18	–30
416	208	02 41	–30
417	183	03 04	–30
418	243	03 27	–30
419	309	03 50	–30
420	238	04 13	–30
421	176	04 36	–30
422	112	04 59	–30
423	255	05 22	–30
424	242	05 45	–30
425	151	06 08	–30
426	215	06 31	–30
427	108	06 54	–30
428	66	07 17	–30
429	97	07 40	–30
430	25	08 03	–30
431	57	08 26	–30
432	58	08 49	–30
433	90	09 12	–30
434	149	09 35	–30
435	214	09 58	–30

Field Number	Galaxies Counted	Plate Centre R.A. (1950)	Declination
436	358	10 21	–30
437	372	10 44	–30
438	117	11 07	–30
439	267	11 30	–30
440	234	11 53	–30
441	153	12 16	–30
442	152	12 39	–30
443	300	13 02	–30
444	519	13 25	–30
445	426	13 48	–30
446	370	14 11	–30
447	150	14 34	–30
448	72	14 57	–30
449	78	15 20	–30
450	126	15 43	–30
451	85	16 06	–30
452	7	16 29	–30
453	3	16 52	–30
454	2	17 15	–30
455	5	17 38	–30
456	0	18 01	–30
457	1	18 24	–30
458	26	18 47	–30
459	61	19 10	–30
460	92	19 33	–30
461	117	19 56	–30
462	256	20 19	–30
463	201	20 42	–30
464	193	21 05	–30
465	124	21 28	–30
466	149	21 51	–30
467	340	22 14	–30
468	325	22 37	–30
469	153	23 00	–30

Field Number	Galaxies Counted	Plate Centre R.A. (1950)	Declination
470	212	23 23	–30
471	471	23 46	–30
472	291	00 00	–25
473	287	00 22	–25
474	449	00 44	–25
475	313	01 06	–25
476	153	01 28	–25
477	215	01 50	–25
478	292	02 12	–25
479	234	02 34	–25
480	210	02 56	–25
481	167	03 18	–25
482	280	03 40	–25
483	130	04 02	–25
484	151	04 24	–25
485	219	04 46	–25
486	336	05 08	–25
487	317	05 30	–25
488	393	05 52	–25
489	127	06 14	–25
490	122	06 36	–25
491	60	06 58	–25
492	22	07 20	–25
493	14	07 44	–25
494	59	08 04	–25
495	81	08 26	–25
496	104	08 48	–25
497	144	09 10	–25
498	145	09 32	–25
499	121	09 54	–25
500	169	10 16	–25
501	288	10 38	–25
502	142	11 00	–25
503	184	11 22	–25

Field Number	Galaxies Counted	Plate Centre R.A. (1950)	Declination
504	103	11 44	–25
505	147	12 06	–25
506	182	12 28	–25
507	349	12 50	–25
508	374	13 12	–25
509	741	13 34	–25
510	471	13 56	–25
511	201	14 18	–25
512	160	14 40	–25
513	131	15 02	–25
514	92	15 24	–25
515	59	15 46	–25
516	56	16 08	–25
517	18	16 30	–25
518	21	16 52	–25
519	4	17 14	–25
520	2	17 36	–25
521	0	17 58	–25
522	0	18 20	–25
523	4	18 42	–25
524	78	19 04	–25
525	100	19 26	–25
526	148	19 48	–25
527	76	20 10	–25
528	191	20 32	–25
529	512	20 54	–25
530	212	21 16	–25
531	316	21 38	–25
532	261	22 00	–25
533	319	22 22	–25
534	250	22 44	–25
535	223	23 06	–25
536	140	23 28	–25
537	254	23 50	–25